International Conference on Tribology (ROTRIB'19)

IOP Conference Series: Materials Science and Engineering
Volume 724

Cluj-Napoca, Romania
19 - 21 September 2019

ISBN: 978-1-7138-2583-8
ISSN: 1757-8981

Printed from e-media with permission by:

Curran Associates, Inc.
57 Morehouse Lane
Red Hook, NY 12571

Some format issues inherent in the e-media version may also appear in this print version.

Institute of Physics
Dirac House, Temple Back
Bristol BS1 6BE UK

Phone: 44 1 17 929 7481
Fax: 44 1 17 920 0979

techtracking@iop.org

Additional copies of this publication are available from:

Curran Associates, Inc.
57 Morehouse Lane
Red Hook, NY 12571 USA
Phone: 845-758-0400
Fax: 845-758-2633
Email: curran@proceedings.com
Web: www.proceedings.com

TABLE OF CONTENTS

TRIBOLOGY OF MACHINE ELEMENTS

FRICTION AND WEAR

CONTACT MECHANICS

CONDITION MONITORING, TRIBOLOGY IN MANUFACTURING PROCESSES

LUBRICATION

LUBRICANTS, EMERGING TECHNOLOGIES AND INDUSTRIAL TRIBOLOGY

MICRO AND NANOTRIBOLOGY

BIOTRIBOLOGY AND TRIBOLOGY IN MEDICINE

SURFACE METROLOGY, SURFACE ENGINEERING AND COATINGS

#	Author	Email	Country
1	Iulian Patirnac	ipatirnac@gmail.com	Romania
2	Mihai Tiberiu Lates	latesmt@unitbv.ro	Romania
3	Cornel Catalin Gavrila	cgavrila@unitbv.ro	Romania
4	Gabriela Cristina Ionescu	ionescu_g_r@yahoo.com	Romania
5	Radu Velicu	rvelicu@unitbv.ro	Romania
6	Ionut Geonea	igeonea@yahoo.com	Romania
7	Sorin Cananau	s_cananau@yahoo.com	Romania
8	Stelian Alaci	alaci@fim.usv.ro	Romania
9	Dumitru N. Olaru	dolaru@mail.tuiasi.ro	Romania
10	Mihail Bogdan -Roth	rothinimagic@yahoo.com	Romania
11	Valentin Gatu	Valentin.Gatu@gmail.com	Romania
12	Aleksandar Vencl	avencl@mas.bg.ac.rs	Serbia
13	Musca Ilie	iliem@fim.usv.ro	Romania
14	Mara Kandeva	kandevam@gmail.com	Bulgaria
15	Todor Penyashki	tpeniashki@abv.bg	Bulgaria
16	Iulia Graur	iulia.graur@ugal.ro	Romania
17	Corina Birleanu	Corina.Barleanu@omt.utcluj.ro	Romania
18	Gencaga Purcek	gpurcek@gmail.com	Turkey
19	Virgil Iliuta	virgil.iliuta@ugal.ro	Romania
20	Ionut Cristian Romanu	ionutromanucristian@usm.ro	Romania
21	Carmen Bujoreanu	carmen.bujoreanu@gmail.com	Romania
22	Ionut-Tudor Maries	ionut.maries@omt.utcluj.ro	Romania
23	Spiridon Cretu	spiridon.cretu@tuiasi.ro	Romania
24	Alexandru Valentin Radulescu	varrav2000@yahoo.com	Romania
25	Marcelin Benchea	marcelin_ben@yahoo.com	Romania
26	Sergiu Spinu	sergiu.spinu@fim.usv.ro	Romania
27	Sergiu Spinu	sergiu.spinu@fim.usv.ro	Romania
28	Laura Babici	lbabici@yahoo.com	Romania
29	Cristea Aurora Felicia	cristea_fa@yahoo.de	Romania
30	Jamal Ahmed	jamalahmedkhan20@gmail.com	Pakistan
30	Stelian Cazan	stelian.cazan09@gmail.com	Romania
31	Gabriel Popescu	gpopescu65@yahoo.com	Romania
32	Claudiu Ovidiu Popa	claudepopa@yahoo.com	Romania
33	Adrian Catalin Drumeanu	drumeanu@upg-ploiesti.ro	Romania
34	Adrian Catalin Drumeanu	drumeanu@upg-ploiesti.ro	Romania
35	Irina Radulescu	irena_sandu@yahoo.com	Romania
36	Horea George Crișan	horea.crisan@omt.utcluj.ro	Romania
37	Adrian Lospa	adrian.lospa@gmail.com	Romania
38	Xingnan Zhang	xingnan.zhang@sussex.ac.uk	United Kingdom
39	Xingnan Zhang	xingnan.zhang@sussex.ac.uk	United Kingdom
40	Andrei Popescu	andrey151289@gmail.com	Romania
41	Traian Cicone	traian.cicone@upb.ro	Romania
42	Petrică Turtoi	petrica.turtoi@upb.ro	Romania
43	Andrei-Iancu Stoices	stoicescu_andrei_iancu@yahoo.com	Romania
44	Ovidiu Buiga	Ovidiu.Buiga@omt.utcluj.ro	Romania

45	Constantin Georgescu	constantin.georgescu@ugal.ro	Romania
46	Constantin Georgescu	constantin.georgescu@ugal.ro	Romania
47	Andreia Molea	andreia.molea@auto.utcluj.ro	Romania
48	Thami Zeghloul	thami.zeghloul@univ-poitiers.fr	France
49	Catalin Pirvu	pirvu.catalin@incas.ro	Romania
50	Thami Zeghloul	thami.zeghloul@univ-poitiers.fr	France
51	Aurelio Somà	AURELIO.SOMA@POLITO.IT	Italy
52	Marius Pustan	Marius.Pustan@omt.utcluj.ro	Romania
53	Rodica Voicu	rodica.voicu@imt.ro	Romania
54	Lucel Sirghi	lsirghi@uaic.ro	Romania
55	Simona Noveanu	simona.noveanu@mdm.utcluj.ro	Romania
56	Mihaela Buciumeanu	mihaela.buciumeanu@ugal.ro	Romania
57	Liliana-Laura Badita	badita_l@yahoo.com	Romania
58	Andrei Calin	acalin05@gmail.com	Romania
59	Dumitru N Olaru	dolaru@mail.tuiasi.ro	Romania
60	Violeta Merie	violeta.merie@stm.utcluj.ro	Romania
61	Viorel Paleu	vpaleu@tuiasi.ro	Romania
62	Florina Șerdean	florina.rusu@omt.utcluj.ro	Romania
63	Corneliu Munteanu	cornelmun@gmail.com	Rwanda
64	Cornelia Cirlan Paleu	cpaleu@yahoo.com	Romania

Affiliation
University Petroleum-Gas of Ploiesti
Transilvania University of Brasov
Transilvania University of Brasov
Petroleum-Gas University of Ploiesti
Transilvania University of Brasov
University of Craiova
Politehnica University of Bucharest
Stefan cel Mare University Suceava
Technical University "Gheorghe Asachi" Iasi
Petroleum -Gas University of Ploiesti
University Politehnica of Bucharest
University of Belgrade, Faculty of Mechanical Engineering
University Stefan cel Mare Suceava
Technical University Sofia
ISSAPP "N.Pushkarov"
Dunarea de Jos
Technical University of Cluj-Napoca
Karadeniz Technical university
"Dunarea de Jos" University of Galati
University of Suceava
Technical University Iasi
Technical University of Cluj-Napoca
TUIASI
University POLITEHNICA Bucharest
"Gheorghe Asachi" Technical University of Iaşi
Stefan cel Mare University of Suceava
Stefan cel Mare University of Suceava
Polytechnic University of Bucharest
Technical University of Cluj-Napoca
College of Electrical and Mechanical Engineering(NUST)
Technical University "Gh. Asachi" Iasi
Technical University Gheorghe Asachi of Iasi
Technical University of Cluj-Napoca
Universitatea Petrol-Gaze din Ploiesti
Universitatea Petrol-Gaze din Ploiesti
University POLITEHNICA Bucharest
Technical University of Cluj-Napoca
Petroleum and Gas University of Ploiesti
University of Sussex
University of Sussex
"Gheorghe Asachi" Technical University of Iasi-Romania, Faculty of Mechanical Engineering
University POLITEHNICA of Bucharest
University POTEHNICA of Bucharest, Spl. Independenței 313, Bucharest, 060042, Romania
University Politehnica of Bucharest
Technical University of Cluj-Napoca

IOP Conf. Series: Materials Science and Engineering **724** (2020) 011001 doi:10.1088/1757-899X/724/1/011001

Dunarea de Jos University of Galati
"Dunarea de Jos" University
Technical University of Cluj-Napoca, Automotive Engineering and Transports Department
Institut Pprime
INCAS - National Institute for Aerospace Research "Elie Carafoli", Bucharest
Institut Pprime
Department of Mechanical and Aerospace Engineering
Technical University of Cluj-Napoca
National Institute for R&D in Microtechnologies - IMT Bucharest
Alexandru Ioan Cuza University
Technical University of Cluj-Napoca
Dunarea de jos University of Galati
National Institute of Research and Development in Mechatronics and Measurement Technique
Politehnica University Bucharest
Technical University "Gheorghe Asachi" Iasi
Technical Univesity of Cluj-Napoca
Gheorghe Asachi Technical University of Iasi
Technical University of Cluj-Napoca
Gheorghe Asachi Technical University of Iaşi
Gheorghe Asachi Technical University of Iasi

#	Author	Email	Country
2	Cristea Aurora Felicia	cristea_fa@yahoo.de	Romania
3	Paul Olaru	polaru70@yahoo.com	Romania
4	Iulian Patirnac	ipatirnac@gmail.com	Romania
4	Razvan George Ripeanu	rrapenau@upg-ploiesti.ro	Romania
4	Eugen Laudacescu	elaudacescu@yahoo.com	Romania
5	Thami Zeghloul	thami.zeghloul@univ-poitiers.fr	France
6	Musca Ilie	iliem@fim.usv.ro	Romania
7	Marcelin Benchea	marcelin_ben@yahoo.com	Romania
8	Mihai Tiberiu Lates	latesmt@unitbv.ro	Romania
9	Cornel Catalin Gavrila	cgavrila@unitbv.ro	Romania
10	Adrian Lospa	adrian.lospa@gmail.com	Romania
11	Xingnan Zhang	xingnan.zhang@sussex.ac.uk	United Kingdom
11	Romeo Glovnea	R.P.Glovnea@sussex.ac.uk	United Kingdom
12	Xingnan Zhang	xingnan.zhang@sussex.ac.uk	United Kingdom
12	Romeo Glovnea	R.P.Glovnea@sussex.ac.uk	United Kingdom
13	Carmen Bujoreanu	carmen.bujoreanu@gmail.com	Romania
15	Mihaela Buciumeanu	mihaela.buciumeanu@ugal.ro	Romania
16	Liliana-Laura Badita	badita_l@yahoo.com	Romania
19	Gabriela Cristina Ionescu	ionescu_g_r@yahoo.com	Romania
21	Ovidiu Buiga	Ovidiu.Buiga@omt.utcluj.ro	Romania
22	Iulia Graur	iulia.graur@ugal.ro	Romania
23	Radu Velicu	rvelicu@unitbv.ro	Romania
24	Sergiu Spinu	sergiu.spinu@fim.usv.ro	Romania
25	Laura Babici	lbabici@yahoo.com	Romania
25	Andrei Tudor	andreitudor1947@gmail.com	Romania
25	Jordi Romeu Garbi	jordi.romeu@upc.edu	Spain
25	Marilena Stoica	stoica.marilena23@gmail.com	Romania
26	Virgil Iliuta	virgil.iliuta@ugal.ro	Romania
27	Ionut Cristian Romanu	ionutromanucristian@usm.ro	Romania
29	Andrei Calin	acalin05@gmail.com	Romania
30	Jamal Ahmed	jamalahmedkhan20@gmail.com	Pakistan
30	Raja Amer Azim	hamza.hussain95@gmail.com	Pakistan
30	Waleed Hussain	w.hussain9253@gmail.com	Pakistan
33	Claudiu Ovidiu Popa	claudepopa@yahoo.com	Romania
37	Sergiu Spinu	sergiu.spinu@fim.usv.ro	Romania
38	Stelian Cazan	stelian.cazan09@gmail.com	Romania
38	Claudiu Dănilă	klausosho@yahoo.com	Romania
38	Spiridon Crețu	sp_cretu@yahoo.com	Romania
40	Adrian Catalin Drumeanu	drumeanu@upg-ploiesti.ro	Romania
41	Adrian Catalin Drumeanu	drumeanu@upg-ploiesti.ro	Romania
42	Aurelio Somà	AURELIO.SOMA@POLITO.IT	Italy
43	Spiridon Cretu	spiridon.cretu@tuiasi.ro	Romania
44	Constantin Georgescu	constantin.georgescu@ugal.ro	Romania
44	Lorena Deleanu	lorena.deleanu@ugal.ro	Romania
45	Mara Kandeva	kandevam@gmail.com	Bulgaria
47	Ionut Geonea	igeonea@yahoo.com	Romania
48	Viorel Paleu	vpaleu@tuiasi.ro	Romania
49	Catalin Pirvu	pirvu.catalin@incas.ro	Romania
49	Lorena Deleanu	lorena.deleanu@ugal.ro	Romania
50	Constantin Georgescu	constantin.georgescu@ugal.ro	Romania
50	Lorena Deleanu	lorena.deleanu@ugal.ro	Romania
51	Violeta Merie	violeta.merie@stm.utcluj.ro	Romania
52	Florina Șerdean	florina.rusu@omt.utcluj.ro	Romania
53	Andrei Popescu	andrey151289@gmail.com	Romania

53	Dumitru Olaru	dolaru@tuiasi.ro	Romania
54	Dumitru N Olaru	dolaru@mail.tuiasi.ro	Romania
55	Irina Radulescu	irena_sandu@yahoo.com	Romania
56	Rodica Voicu	rodica.voicu@imt.ro	Romania
57	Sorin Cananau	s_cananau@yahoo.com	Romania
58	Gencaga Purcek	gpurcek@gmail.com	Turkey
59	Andreia Molea	andreia.molea@auto.utcluj.ro	Romania
60	Alexandru Valentin Radulescu	varrav2000@yahoo.com	Romania
61	Stefan Iliev Dishliev	stefan_dishliev@abv.bg	Bulgaria
62	Horea George Crișan	horea.crisan@omt.utcluj.ro	Romania
62	Oana Adriana Crișan	oana.crisan@staff.utcluj.ro	Romania
64	Gabriel Popescu	gpopescu65@yahoo.com	Romania
65	Aleksandar Vencl	avencl@mas.bg.ac.rs	Serbia
66	Thami Zeghloul	thami.zeghloul@univ-poitiers.fr	France
67	Stelian Alaci	alaci@fim.usv.ro	Romania
68	Todor Penyashki	tpeniashki@abv.bg	Bulgaria
68	Dimitar Radev	ddradev@gmail.com	Bulgaria
68	Mara Kandeva	kandevam@gmail.com	Bulgaria
68	Georgi Kostadinov	gdkostadinov@gmail.com	Bulgaria
69	Corneliu Munteanu	cornelmun@gmail.com	Rwanda
70	Marius Pustan	Marius.Pustan@omt.utcluj.ro	Romania
71	Corina Birleanu	Corina.Barleanu@omt.utcluj.ro	Romania
72	Simona Noveanu	simona.noveanu@mdm.utcluj.ro	Romania
73	Traian Cicone	traian.cicone@upb.ro	Romania
73	Alice Alina Marinescu	alicemarinescu96@gmail.com	Romania
73	Ștefan Sorohan	stefan.sorohan@upb.ro	Romania
74	Petrică Turtoi	petrica.turtoi@upb.ro	Romania
74	Traian Cicone	traian.cicone@upb.ro	Romania
76	Ionut-Tudor Maries	ionut.maries@omt.utcluj.ro	Romania
77	Dumitru N. Olaru	dolaru@mail.tuiasi.ro	Romania
80	Mihail Bogdan -Roth	rothinimagic@yahoo.com	Romania
80	Mirela Romanet	mirelamagic25@yahoo.com	Romania
80	Razvan George Ripeanu	rrapeanu@upg-ploiesti.ro	Romania
81	Cornelia Cirlan Paleu	cpaleu@yahoo.com	Romania
84	Lucel Sirghi	lsirghi@uaic.ro	Romania
87	Andrei-Iancu Stoicescu	stoicescu_andrei_iancu@yahoo.co m	Romania
87	Adrian Predescu	adrian_omtr@yahoo.com	Romania
87	Valentin Gatu	Valentin.Gatu@gmail.com	Romania

Affiliation
Technical University of Cluj-Napoca
SCO Expert Pool & DGM
University Petroleum-Gas of Ploiesti
University Petroleum-Gas of Ploiesti
University Petroleum-Gas of Ploiesti
Institut Pprime
University Stefan cel Mare Suceava
"Gheorghe Asachi" Technical University of Iaşi
Transilvania University of Brasov
Transilvania University of Brasov
Petroleum and Gas University of Ploiesti
University of Sussex
University of Sussex
University of Sussex
University of Sussex
Technical University Iasi
Dunarea de jos University of Galati
National Institute of Research and Development in Mechatronics and Measurement Technique
Petroleum-Gas University of Ploiesti
Technical University of Cluj-Napoca
Dunarea de Jos
Transilvania University of Brasov
Stefan cel Mare University of Suceava
Polytechnic University of Bucharest
Polytechnic University of Bucharest
Universitat Politècnica de Catalunya
Polytechnic University of Bucharest
"Dunarea de Jos" University of Galati
University of Suceava
Politehnica University Bucharest
College of Electrical and Mechanical Engineering(NUST)
College of Electrical and Mechanical Engineering(NUST)
Institute of Space technology, Pakistan
Technical University of Cluj-Napoca
Stefan cel Mare University of Suceava
Technical University "Gh. Asachi" Iasi
Technical University "Gh. Asachi" Iasi
Technical University "Gh. Asachi" Iasi
Universitatea Petrol-Gaze din Ploiesti
Universitatea Petrol-Gaze din Ploiesti
Department of Mechanical and Aerospace Engineering
TUIASI
Dunarea de Jos University of Galati
Dunarea de Jos University of Galati
Technical University Sofia
University of Craiova
Gheorghe Asachi Technical University of Iasi
INCAS - National Institute for Aerospace Research "Elie Carafoli", Bucharest
"Dunarea de Jos" University
"Dunarea de Jos" University
"Dunarea de Jos" University
Technical Univesity of Cluj-Napoca
Technical University of Cluj-Napoca
"Gheorghe Asachi" Technical University of Iasi-Romania, Faculty of Mechanical Engineering

IOP Conf. Series: Materials Science and Engineering **724** (2020) 011001 doi:10.1088/1757-899X/724/1/011001

"Gheorghe Asachi" Technical University of Iasi-Romania, Faculty of Mechanical Engineering
Technical University "Gheorghe Asachi" Iasi
University POLITEHNICA Bucharest
National Institute for R&D in Microtechnologies - IMT Bucharest
Politehnica University of Bucharest, Machine Elements and Tribology Dept. 313 Spl.Independentei, Bucharest, Romania
Karadeniz Technical university
Technical University of Cluj-Napoca, Automotive Engineering and Transports Department
University POLITEHNICA Bucharest
University of Food Technologies, Plovdiv
Technical University of Cluj-Napoca
Technical University of Cluj-Napoca
Technical University Gheorghe Asachi of Iasi
University of Belgrade, Faculty of Mechanical Engineering
Institut Pprime
Stefan cel Mare University Suceava
ISSAPP "N.Pushkarov"
BAS
Technical University, Sofia
ISSAPP "N.Pushkarov"
Gheorghe Asachi Technical University of Iaşi
Technical University of Cluj-Napoca
Technical University of Cluj-Napoca
Technical University of Cluj-Napoca
University POLITEHNICA of Bucharest
University POLITEHNICA of Bucharest
University POLITEHNICA of Bucharest
University POTEHNICA of Bucharest, Spl. Independenței 313, Bucharest, 060042, Romania
University POTEHNICA of Bucharest, Spl. Independenței 313, Bucharest, 060042, Romania
Technical University of Cluj-Napoca
Technical University "Gheorghe Asachi" Iasi
Petroleum -Gas University of Ploiesti
Petroleum- Gas University of Ploiesti
Petroleum -Gas University of ploiesti
Gheorghe Asachi Technical University of Iasi
Alexandru Ioan Cuza University
University Politehnica of Bucharest
University Politehnica of Bucharest
University Politehnica of Bucharest

#	Author	Email	Country
2	Cristea Aurora Felicia	cristea_fa@yahoo.de	Romania
4	Iulian Patirnac	ipatirnac@gmail.com	Romania
4	Razvan George Ripeanu	rrapenau@upg-ploiesti.ro	Romania
4	Eugen Laudacescu	elaudacescu@yahoo.com	Romania
5	Imed Eddine Achouri	imed.eddine.achouri@univ-poitiers.fr	France
5	Thami Zeghloul	thami.zeghloul@univ-poitiers.fr	France
5	Karim Medles	kmedles@gmail.com	France
5	Gontran Richard	g.richard@citf-group.com	France
5	Lucien Dascalescu	lucien.dascalescu@univ-poitiers.fr	France
6	Musca Ilie	iliem@fim.usv.ro	Romania
6	Fomin Michaela	fomin_michaela@yahoo.com	Romania
7	Marcelin Benchea	marcelin_ben@yahoo.com	Romania
7	Spiridon Cretu	spcretu@yahoo.com	Romania
8	Cornel Catalin Gavrila	cgavrila@unitbv.ro	Romania
8	Mihai Tiberiu Lates	latesmt@unitbv.ro	Romania
9	Mihai Tiberiu Lates	latesmt@unitbv.ro	Romania
9	Cornel Catalin Gavrila	cgavrila@unitbv.ro	Romania
10	Adrian Lospa	adrian.lospa@gmail.com	Romania
10	Razvan George Ripeanu	rrapeanu@upg-ploiesti.ro	Romania
10	Alin Dinita	adinita@upg-ploiesti.ro	Romania
11	Xingnan Zhang	xingnan.zhang@sussex.ac.uk	United Kingdom
11	Romeo Glovnea	R.P.Glovnea@sussex.ac.uk	United Kingdom
12	Xingnan Zhang	xingnan.zhang@sussex.ac.uk	United Kingdom
12	Romeo Glovnea	R.P.Glovnea@sussex.ac.uk	United Kingdom
13	Carmen Bujoreanu	carmen.bujoreanu@gmail.com	Romania
13	Razvan Constantin Iordache	i.razvan11@yahoo.com	Romania
13	Narcis Daniel Petrea	petrea_narcis09@yahoo.com	Romania
15	Bogdan Mircea Achimescu	bogdanachimescu@yahoo.com	Romania
15	Sorin Ciortan	sorin.ciortan@ugal.ro	Romania
15	Mihaela Buciumeanu	mihaela.buciumeanu@ugal.ro	Romania
16	Lucian Capitanu	lucian.capitanu@yahoo.com	Romania
16	Liliana-Laura Badita	badita_l@yahoo.com	Romania
16	Virgil Florescu	florescuvirgil@yahoo.com	Romania
16	Constantin Tiganesteanu	tctin2002@yahoo.com	Romania
16	Liviu Florin Isvoranu	florinisvo@yahoo.com	Romania
19	Adrian Hagianu	adi_hagianu@yahoo.com	Romania
19	Ion Nae	inae59@gmail.com	Romania
19	Gabriela Cristina Ionescu	ionescu_g_r@yahoo.com	Romania
19	Razvan George Rîpeanu	rrapeanu@upg-ploiesti.ro	Romania
21	Simion Haragâș	sharagas@yahoo.com	Romania
21	Dumitru Pop	pdcluj@gmail.com	Romania
21	Felicia Cristea	cristea_fa@yahoo.de	Romania
21	Ovidiu Buiga	Ovidiu.Buiga@omt.utcluj.ro	Romania
22	Iulian Păduraru	paduraru_iulian@yahoo.com	Romania
22	Gabriel Săracu		Romania
22	Mihail Bogdan		Romania
22	Iulia Graur	iulia.graur@ugal.ro	Romania
22	Cristian Muntenita		Romania
22	Adrian Circiumaru		Romania
23	Radu Velicu	rvelicu@unitbv.ro	Romania
24	Sergiu Spinu	sergiu.spinu@fim.usv.ro	Romania
24	Delia Cerlinca	delia@usm.ro	Romania
25	Laura Babici	lbabici@yahoo.com	Romania
25	Andrei Tudor	andreitudor1947@gmail.com	Romania
25	Jordi Romeu Garbi	jordi.romeu@upc.edu	Spain
25	Marilena Stoica	stoica.marilena23@gmail.com	Romania
26	Virgil Iliuta	virgil.iliuta@ugal.ro	Romania

26	Minodora Ripa	minodora.ripa@ugal.ro	Romania
26	Juliana Javorova	julianata@abv.bg	Bulgaria
26	Gabriel Andrei	gabriel.andrei@ugal.ro	Romania
27	Ionut Cristian Romanu	ionutromanucristian@usm.ro	Romania
29	Andrei Calin	acalin05@gmail.com	Romania
29	Andrei Tudor	andreitudor1947@gmail.com	Romania
29	Marilena Stoica	stoica.marilena23@gmail.com	Romania
29	Nicolae Stoica	nicu_alex_stoica@yahoo.com	Romania
29	Kussay Subhi	qusay_abunoor@yahoo.com	Iraq
30	Jamal Ahmed	jamalahmedkhan20@gmail.com	Pakistan
30	Raja Amer Azim	hamza.hussain95@gmail.com	Pakistan
30	Waleed Hussain	w.hussain9253@gmail.com	Pakistan
33	Claudiu Ovidiu Popa	claudepopa@yahoo.com	Romania
37	Sergiu Spinu	sergiu.spinu@fim.usv.ro	Romania
37	Delia Cerlinca	delia@usm.ro	Romania
38	Stelian Cazan	stelian.cazan09@gmail.com	Romania
38	Claudiu Dănilă	klausosho@yahoo.com	Romania
38	Spiridon Creţu	sp_cretu@yahoo.com	Romania
40	Adrian Catalin Drumeanu	drumeanu@upg-ploiesti.ro	Romania
41	Adrian Catalin Drumeanu	drumeanu@upg-ploiesti.ro	Romania
42	Aurelio Somà	AURELIO.SOMA@POLITO.IT	Italy
43	Spiridon Cretu	spiridon.cretu@tuiasi.ro	Romania
44	Traian Florian Ionescu	Traian.Ionescu@ugal.ro	Romania
44	Dionis Guglea	guglea.dionis@gmail.com	Romania
44	Constantin Georgescu	constantin.georgescu@ugal.ro	Romania
44	Dumitru Dima	dimad@ugal.ro	Romania
44	Lorena Deleanu	lorena.deleanu@ugal.ro	Romania
45	Mara Kandeva	kandevam@gmail.com	Bulgaria
45	Todor Penyashki	peniashki@abv.bg	Bulgaria
45	Georgi Kostadinov	gdkostadinov@gmail.com	Bulgaria
45	Zhecho Kalichin	kalitchin@gmail.com	Bulgaria
47	Ionut Geonea	igeonea@yahoo.com	Romania
47	Nicolae Dumitru	nicolae_dtru@yahoo.com	Romania
47	Cristian Copilusi	cristache03@yahoo.co.uk	Romania
47	Alexandru Margine	fam_margine@yahoo.com	Romania
48	Shubrajit Bhaumik	shubrajit.b@ktr.srmuniv.ac.in	India
48	Viorel Paleu	vpaleu@tuiasi.ro	Romania
48	Dhrubajyoti Chowdhury	dhrubajyoti_debabrata@srmuniv.edu.in	India
48	Cornelia Cîrlan Paleu	cpaleu@yahoo.com	Romania
48	Shubhabrata Datta	shu.datt@gmail.com	India
49	George Ghiocel Ojoc	george.ojoc@gmail.com	Romania
49	Catalin Pirvu	pirvu.catalin@incas.ro	Romania
49	Simona Sandu	smsbadea@yahoo.com	Romania
49	Lorena Deleanu	lorena.deleanu@ugal.ro	Romania
50	Dionis Guglea	guglea.dionis@gmail.com	Romania
50	Traian Florian Ionescu	traian.ionescu@ugal.ro	Romania
50	Dumitru Dima	dimadumitru@yahoo.com	Romania
50	Constantin Georgescu	constantin.georgescu@ugal.ro	Romania
50	Lorena Deleanu	lorena.deleanu@ugal.ro	Romania
51	Violeta Merie	violeta.merie@stm.utcluj.ro	Romania
51	Gavril Negrea	gavril.negrea@ispm.utcluj.ro	Romania
51	Corina Bîrleanu	corina.barleanu@omt.utcluj.ro	Romania
51	Horea Crișan	horea.crisan@omt.utcluj.ro	Romania
51	Marius Pustan	Marius.Pustan@omt.utcluj.ro	Romania

52	Florina Șerdean	florina.rusu@omt.utcluj.ro	Romania
52	Violeta Merie	violeta.merie@stm.utcluj.ro	Romania
52	Gavril Negrea	gavril.negrea@ispm.utcluj.ro	Romania
52	Corina Bîrleanu	corina.barleanu@omt.utcluj.ro	Romania
52	Ionut Măries	Ionut.Maries@omt.utcluj.ro	Romania
53	Andrei Popescu	andrey151289@gmail.com	Romania
53	Dumitru Olaru	dolaru@tuiasi.ro	Romania
54	Cezara Mariuca Oprisan	mariuca.oprisan@yahoo.com	Romania
54	Bogdan Chiriac	bogdanchiriac24@yahoo.com	Romania
54	Vlad Carlescu	vlad.carlescu@tuiasi.ro	Romania
54	Dumitru N Olaru	dolaru@mail.tuiasi.ro	Romania
55	Irina Radulescu	irena_sandu@yahoo.com	Romania
55	Alexandru Valentin Radulescu	varrav2000@yahoo.com	Romania
56	Rodica Voicu	rodica.voicu@imt.ro	Romania
56	Marius Pustan	Marius.Pustan@omt.utcluj.ro	Romania
56	Corina Birleanu	Corina.Barleanu@omt.utcluj.ro	Romania
56	Raluca Gavrila	raluca.gavrila@imt.ro	Romania
56	Cosmin Romanitan	cosmin.romanitan@imt.ro	Romania
56	Oana Tutunaru	oana.tutunaru@imt.ro	Romania
56	Alina Cismaru	alina.cismaru@imt.ro	Romania
57	Sorin Gabroveanu	sorin.gabroveanu@gmail.com	Romania
57	Sorin Cananau	s_cananau@yahoo.com	Romania
57	Radu-Florin Mirica	radu_mirica@yahoo.com	Romania
57	Alexandru Ilies	ilies_alex86@yahoo.com	Romania
58	Harun Yanar	harunyanarr@gmail.com	Turkey
58	Gencaga Purcek	gpurcek@gmail.com	Turkey
58	H. Hüseyin Ayar	metisafe@gmail.com	Turkey
59	Dan Moldovanu	dan.moldovanu@auto.utcluj.ro	Romania
59	Andreia Molea	andreia.molea@auto.utcluj.ro	Romania
59	Istvan Barabas	istvan.barabas@auto.utcluj.ro	Romania
60	Alexandru Valentin Radulescu	varrav2000@yahoo.com	Romania
60	Irina Radulescu	irena_sandu@yahoo.com	Romania
61	Velko Rupetsov	velko_r@abv.bg	Bulgaria
61	Stefan Iliev Dishliev	stefan_dishliev@abv.bg	Bulgaria
61	Georgi Mishev	gmishevpl@yahoo.com	Bulgaria
62	Horea George Crișan	horea.crisan@omt.utcluj.ro	Romania
62	Oana Adriana Crișan	oana.crisan@staff.utcluj.ro	Romania
62	Marius Sorin Pustan	marius.pustan@omt.utcluj.ro	Romania
62	Corina Julieta Bîrleanu	corina.barleanu@omt.utcluj.ro	Romania
62	Violeta Valentina Merie	vio1919@yahoo.com	Romania
64	Gabriel Popescu	gpopescu65@yahoo.com	Romania
65	Aleksandar Vencl	avencl@mas.bg.ac.rs	Serbia
65	Nikolaos Vaxevanidis	nvaxev@gmail.com	Greece
65	Mara Kandeva	kandevam@gmail.com	Bulgaria
66	Omar Tilmatine	omar.tilmatine@univ-poitiers.fr	France
66	Thami Zeghloul	thami.zeghloul@univ-poitiers.fr	France
66	Aurelian Fatu	aurelian.fatu@univ-poitiers.fr	France

66	Lucien Dascalescu	lucien.dascalescu@univ-poitiers.fr	France
67	Stelian Alaci	alaci@fim.usv.ro	Romania
67	Carmen Bujoreanu	cbujorea@tuiasi.ro	Romania
67	Narcis Daniel Petrea	petrea_narcis09@yahoo.com	Romania
67	Florina Carmen Ciornei	florina.ciornei@usm.ro	Romania
67	Luminita Irimescu	luminita.irimescu@usm.ro	Romania
68	Todor Penyashki	tpeniashki@abv.bg	Bulgaria
68	Dimitar Radev	ddradev@gmail.com	Bulgaria
68	Mara Kandeva	kandevam@gmail.com	Bulgaria
68	Georgi Kostadinov	gdkostadinov@gmail.com	Bulgaria
69	Cornelia Cîrlan Paleu	cpaleu@yahoo.com	Romania
69	Bogdan Istrate	bogdan_istrate1@yahoo.com	Romania
69	Viorel Paleu	vpaleu@tuiasi.ro	Romania
69	Corneliu Munteanu	cornelmun@gmail.com	Rwanda
70	Marius Pustan	Marius.Pustan@omt.utcluj.ro	Romania
70	Corina Birleanu	Corina.Barleanu@omt.utcluj.ro	Romania
70	Violeta Merie	Violeta.MERIE@stm.utcluj.ro	Romania
70	Liviu Zarbo	liviu.zarbo@itim-cj.ro	Romania
70	Sorina Garabagiu	sorina.garabagiu@itim-cj.ro	Romania
70	Daniel Marconi	daniel.marconi@itim-cj.ro	Romania
71	Corina Birleanu	Corina.Barleanu@omt.utcluj.ro	Romania
71	Marius Pustan	Marius.Pustan@omt.utcluj.ro	Romania
71	Cosmin Cosma	cosmin.cosma@tcm.utcluj.ro	Romania
71	Violeta Merie	violeta.merie@stm.utcluj.ro	Romania
71	Ovidiu Dranda	ovidiu.dranda@volvo.com	Romania
72	Simona Noveanu	simona.noveanu@mdm.utcluj.ro	Romania
72	Calin Rusu	calin.rusu@mdm.utcluj.ro	Romania
72	Camelia Rancea	camelia.rancea@staff.utcluj.ro	Romania
72	Daniel Lates	daniellates@yahoo.fr	Romania
73	Traian Cicone	traian.cicone@upb.ro	Romania
73	Alice Alina Marinescu	alicemarinescu96@gmail.com	Romania
73	Ștefan Sorohan	stefan.sorohan@upb.ro	Romania
74	Georgian Lupu	georgianlupu89@gmail.com	Romania
74	Petrică Turtoi	petrica.turtoi@upb.ro	Romania
74	Traian Cicone	traian.cicone@upb.ro	Romania
76	Ionut-Tudor Maries	ionut.maries@omt.utcluj.ro	Romania
76	Vilau Cristian	cristian.vilau@tcm.utcluj.ro	Romania
76	Pustan Marius Sorin	marius.pustan@omt.utcluj.ro	Romania
76	Dudescu Cristian	mircea.dudescu@rezi.utcluj.ro	Romania
76	Crisan Horea George	horea.crisan@auto.utcluj.ro	Romania
77	Gelu Ianus	gianus2002@yahoo.com	Romania
77	Denis Cojocaru	darckreign2004@gmail.com	Romania
77	Cezara Mariuca Oprisan	mariuca.oprisan@yahoo.com	Romania
77	Viorel Paleu	vpaleu@yahoo.com	Romania
77	Dumitru N. Olaru	dolaru@mail.tuiasi.ro	Romania
80	Mihail Bogdan -Roth	rothinimagic@yahoo.com	Romania
80	Mirela Romanet	mirelamagic25@yahoo.com	Romania
80	Razvan George Ripeanu	rrapeanu@upg-ploiesti.ro	Romania
81	Viorel Paleu	vpaleu@yahoo.com	Romania
81	Cornelia Cirlan Paleu	cpaleu@yahoo.com	Romania
81	Bogdan Istrate	bogdan_istrate1@yahoo.com	Romania
81	Shubrajit Bhaumik	shubrajit.b@ktr.srmuniv.ac.in	India

81	Corneliu Munteanu	cornelmun@gmail.com	Romania
84	Sabina Teodoroff-Onesim	teodoroff_onesim_sabina@yahoo.com	Romania
84	Alexandra Besleaga	alexandra_besleaga@yahoo.com	Romania
84	Lucel Sirghi	lsirghi@uaic.ro	Romania

IOP Conf. Series: Materials Science and Engineering **724** (2020) 011001 doi:10.1088/1757-899X/724/1/011001

Affiliation
Technical University of Cluj-Napoca
University Petroleum-Gas of Ploiesti
University Petroleum-Gas of Ploiesti
University Petroleum-Gas of Ploiesti
Institut Pprime & CITF Company
Institut Pprime
Institut Pprime
CITF Company
Institut Pprime
University Stefan cel Mare Suceava
Technical College, Radauti
"Gheorghe Asachi" Technical University of Iaşi
"Gheorghe Asachi" Technical University of Iaşi
Transilvania University of Brasov
Transilvania University of Brasov
Transilvania University of Brasov
Transilvania University of Brasov
Petroleum and Gas University of Ploiesti
Petroleum and Gas University of Ploiesti
Petroleum and Gas University of Ploiesti
University of Sussex
University of Sussex
University of Sussex
University of Sussex
Technical University Iasi
Technical University Iasi
Technical University Iasi
Dunarea de jos University of Galati
Dunarea de jos University of Galati
Dunarea de jos University of Galati
Institute of Solid Mechanics of the Romanian Academy
National Institute of Research and Development in Mechatronics and Measurement Technique
University of Civil Engineering
Institute of Solid Mechanics of the Romanian Academy
Institute of Solid Mechanics of the Romanian Academy
Petroleum-Gas University of Ploiesti
Petroleum-Gas University of Ploiesti
Petroleum-Gas University of Ploiesti
Petroleum-Gas University of Ploiesti
Technical University of Cluj-Napoca
Technical University of Cluj-Napoca
Technical University of Cluj-Napoca
Technical University of Cluj-Napoca
Dunarea de Jos
Dunarea de Jos
Dunarea de Jos
Dunarea de Jos
Dunarea de Jos
Dunarea de Jos
Transilvania University of Brasov
Stefan cel Mare University of Suceava
Stefan cel Mare University of Suceava
Polytechnic University of Bucharest
Polytechnic University of Bucharest
Universitat Politècnica de Catalunya
Polytechnic University of Bucharest
"Dunarea de Jos" University of Galati

"Dunarea de Jos" University of Galati
University of Chemical Technology and Metallurgy, Sofia
"Dunarea de Jos" University of Galati
University of Suceava
Politehnica University Bucharest
Politehnica University Bucharest
Politehnica University Bucharest
Politehnica University Bucharest
Al-Furat Al-Awsat Technical University
College of Electrical and Mechanical Engineering(NUST)
College of Electrical and Mechanical Engineering(NUST)
Institute of Space technology, Pakistan
Technical University of Cluj-Napoca
Stefan cel Mare University of Suceava
Stefan cel Mare University of Suceava
Technical University "Gh. Asachi" Iasi
Technical University "Gh. Asachi" Iasi
Technical University "Gh. Asachi" Iasi
Universitatea Petrol-Gaze din Ploiesti
Universitatea Petrol-Gaze din Ploiesti
Department of Mechanical and Aerospace Engineering
TUIASI
Dunarea de Jos University of Galati
Dunarea de Jos University of Galati
Dunarea de Jos University of Galati
Dunarea de Jos University of Galati
Dunarea de Jos University of Galati
Technical University Sofia
Institute of Soil Science Agrotechnologies and Plant Protection N. Pushkarov
Institute of Soil Science Agrotechnologies and Plant Protection N. Pushkarov
SciBuCom 2 Ltd
University of Craiova
University of Craiova
University of Craiova
University of Craiova
SRM Institute of Science and Technology
Gheorghe Asachi Technical University of Iasi
SRM Institute of Science and Technology
Gheorghe Asachi Technical University of Iasi
SRM Institute of Science and Technology
"Dunarea de Jos" University
INCAS - National Institute for Aerospace Research "Elie Carafoli", Bucharest
Scientific Research Center for CBRN Defense and Ecology, Bucharest
"Dunarea de Jos" University
"Dunarea de Jos" University
"Dunarea de Jos" University
"Dunarea de Jos" University
"Dunarea de Jos" University
"Dunarea de Jos" University
Technical Univesity of Cluj-Napoca
Technical Univesity of Cluj-Napoca
Technical Univesity of Cluj-Napoca
Technical Univesity of Cluj-Napoca
Technical University of Cluj-Napoca

IOP Conf. Series: Materials Science and Engineering **724** (2020) 011001 doi:10.1088/1757-899X/724/1/011001

Technical University of Cluj-Napoca
Technical University of Cluj-Napoca
Technical University of Cluj-Napoca
Technical University of Cluj-Napoca
Technical University of Cluj-Napoca
"Gheorghe Asachi" Technical University of Iasi-Romania, Faculty of Mechanical Engineering
"Gheorghe Asachi" Technical University of Iasi-Romania, Faculty of Mechanical Engineering
Technical University "Gheorghe Asachi" Iasi
Technical University "Gheorghe Asachi" Iasi
Technical University "Gheorghe Asachi" Iasi
Technical University "Gheorghe Asachi" Iasi
University POLITEHNICA Bucharest
University POLITEHNICA Bucharest
National Institute for R&D in Microtechnologies - IMT Bucharest
Technical University of Cluj-Napoca
Technical University of Cluj-Napoca
IMT Bucharest
IMT Bucharest
IMT Bucharest
IMT Bucharest
National Research and Development for Gas Turbines SC COMOTI SA, 220 D Iuliu Maniu Bd., 061126, Bucharest, Romania
Politehnica University of Bucharest, Machine Elements and Tribology Dept. 313 Spl.Independentei, Bucharest, Romania
Politehnica University of Bucharest, Machine Elements and Tribology Dept. 313 Spl.Independentei, Bucharest, Romania
National Research and Development for Gas Turbines SC COMOTI SA, 220 D Iuliu Maniu Bd., 061126, Bucharest, Romania
Karadeniz Technical university
Karadeniz Technical university
Metisafe, Cleanroom and Biosafety Systems
Technical University of Cluj-Napoca, Automotive Engineering and Transports Department
Technical University of Cluj-Napoca, Automotive Engineering and Transports Department
Technical University of Cluj-Napoca, Automotive Engineering and Transports Department
University POLITEHNICA Bucharest
University POLITEHNICA Bucharest
Plovdiv University "Paisii Hilendarski"
University of Food Technologies, Plovdiv
Plovdiv University "Paisii Hilendarski"
Technical University of Cluj-Napoca
Technical University of Cluj-Napoca
Technical University of Cluj-Napoca
Technical University of Cluj-Napoca
Technical University of Cluj-Napoca
Technical University Gheorghe Asachi of Iasi
University of Belgrade, Faculty of Mechanical Engineering
School of Pedagogical and Technological Education, 14121 Athens, Greece
Faculty of Industrial Technology, Technical University of Sofia
Institut Pprime
Institut Pprime
Institut Pprime

Institut Pprime
Stefan cel Mare University Suceava
„Gheorghe Asachi" Technical University, Iasi, Romania
„Gheorghe Asachi" Technical University, Iasi, Romania
Stefan cel Mare University Suceava
Stefan cel Mare University Suceava
ISSAPP "N.Pushkarov"
BAS
Technical University, Sofia
ISSAPP "N.Pushkarov"
Gheorghe Asachi Technical University of Iaşi
Gheorghe Asachi Technical University of Iaşi
Gheorghe Asachi Technical University of Iaşi
Gheorghe Asachi Technical University of Iaşi
Technical University of Cluj-Napoca
Technical university of Cluj-Napoca
Technical University of Cluj-Napoca
National Institute for Research and Development of Isotopic and Molecular Technologies
National Institute for Research and Development of Isotopic and Molecular Technologies
National Institute for Research and Development of Isotopic and Molecular Technologies
Technical University of Cluj-Napoca
Technical University of Cluj-Napoca
Technical University of Cluj-Napoca
Technical University of Cluj-Napoca
Technical University of Cluj-Napoca
Technical University of Cluj-Napoca
Technical University of Cluj-Napoca
Technical University of Cluj-Napoca
University of Medicine, Pharmacy, Science and Technology of Târgu Mureş
University POLITEHNICA of Bucharest
University POLITEHNICA of Bucharest
University POLITEHNICA of Bucharest
Military Equipment and Technologies Research Agency, Aeroportului 16, Clinceni, Ilfov, 077025, Romania
University POTEHNICA of Bucharest, Spl. Independenței 313, Bucharest, 060042, Romania
University POTEHNICA of Bucharest, Spl. Independenței 313, Bucharest, 060042, Romania
Technical University of Cluj-Napoca
Technical University of Cluj-Napoca
Technical University of Cluj-Napoca
Technical University of Cluj-Napoca
Technical University of Cluj-Napoca
Technical University "Gheorghe Asachi" Iasi
Technical University "Gheorghe Asachi" Iasi
Technical University "Gheorghe Asachi" Iasi
Technical University "Gheorghe Asachi" Iasi
Technical University "Gheorghe Asachi" Iasi
Petroleum -Gas University of Ploiesti
Petroleum- Gas University of Ploiesti
Petroleum -Gas University of ploiesti
Gheorghe Asachi Technical University of Iasi
Gheorghe Asachi Technical University of Iasi
Gheorghe Asachi Technical University of Iasi
SRM Institute of Science and Technology

Gheorghe Asachi Technical University of Iasi
Alexandru Ioan Cuza University
Alexandru Ioan Cuza University
Alexandru Ioan Cuza University

#	Author	Email	Country
2	Cristea Aurora Felicia	cristea_fa@yahoo.de	Romania
3	Paul Olaru	polaru70@yahoo.com	Romania
4	Iulian Patirnac	ipatirnac@gmail.com	Romania
4	Razvan George Ripeanu	rrapenau@upg-ploiesti.ro	Romania
4	Eugen Laudacescu	elaudacescu@yahoo.com	Romania
5	Imed Eddine Achouri	imed.eddine.achouri@univ-poitiers.fr	France
5	Thami Zeghloul	thami.zeghloul@univ-poitiers.fr	France
5	Karim Medles	kmedles@gmail.com	France
5	Gontran Richard	g.richard@citf-group.com	France
5	Lucien Dascalescu	lucien.dascalescu@univ-poitiers.fr	France
6	Musca Ilie	iliem@fim.usv.ro	Romania
6	Fomin Michaela	fomin_michaela@yahoo.com	Romania
7	Marcelin Benchea	marcelin_ben@yahoo.com	Romania
7	Spiridon Cretu	spcretu@yahoo.com	Romania
8	Cornel Catalin Gavrila	cgavrila@unitbv.ro	Romania
8	Mihai Tiberiu Lates	latesmt@unitbv.ro	Romania
9	Mihai Tiberiu Lates	latesmt@unitbv.ro	Romania
9	Cornel Catalin Gavrila	cgavrila@unitbv.ro	Romania
10	Adrian Lospa	adrian.lospa@gmail.com	Romania
10	Razvan George Ripeanu	rrapeanu@upg-ploiesti.ro	Romania
10	Alin Dinita	adinita@upg-ploiesti.ro	Romania
11	Xingnan Zhang	xingnan.zhang@sussex.ac.uk	United Kingdom
11	Romeo Glovnea	R.P.Glovnea@sussex.ac.uk	United Kingdom
12	Xingnan Zhang	xingnan.zhang@sussex.ac.uk	United Kingdom
12	Romeo Glovnea	R.P.Glovnea@sussex.ac.uk	United Kingdom
13	Carmen Bujoreanu	carmen.bujoreanu@gmail.com	Romania
13	Razvan Constantin Iordache	i.razvan11@yahoo.com	Romania
13	Narcis Daniel Petrea	petrea_narcis09@yahoo.com	Romania
15	Bogdan Mircea Achimescu	bogdanachimescu@yahoo.com	Romania
15	Sorin Ciortan	sorin.ciortan@ugal.ro	Romania
15	Mihaela Buciumeanu	mihaela.buciumeanu@ugal.ro	Romania
16	Lucian Capitanu	lucian.capitanu@yahoo.com	Romania
16	Liliana-Laura Badita	badita_l@yahoo.com	Romania
16	Virgil Florescu	florescuvirgil@yahoo.com	Romania
16	Constantin Tiganesteanu	tctin2002@yahoo.com	Romania
16	Liviu Florin Isvoranu	florinisvo@yahoo.com	Romania
19	Adrian Hagianu	adi_hagianu@yahoo.com	Romania
19	Ion Nae	inae59@gmail.com	Romania
19	Gabriela Cristina Ionescu	ionescu_g_r@yahoo.com	Romania
19	Razvan George Rîpeanu	rrapeanu@upg-ploiesti.ro	Romania
21	Simion Haragâș	sharagas@yahoo.com	Romania
21	Dumitru Pop	pdcluj@gmail.com	Romania
21	Felicia Cristea	cristea_fa@yahoo.de	Romania
21	Ovidiu Buiga	Ovidiu.Buiga@omt.utcluj.ro	Romania
22	Iulian Păduraru	paduraru_iulian@yahoo.com	Romania
22	Gabriel Săracu		Romania
22	Mihail Bogdan		Romania
22	Iulia Graur	iulia.graur@ugal.ro	Romania
22	Cristian Muntenita		Romania
22	Adrian Circiumaru		Romania
23	Radu Velicu	rvelicu@unitbv.ro	Romania
24	Sergiu Spinu	sergiu.spinu@fim.usv.ro	Romania
24	Delia Cerlinca	delia@usm.ro	Romania
25	Laura Babici	lbabici@yahoo.com	Romania
25	Andrei Tudor	andreitudor1947@gmail.com	Romania
25	Jordi Romeu Garbi	jordi.romeu@upc.edu	Spain
25	Marilena Stoica	stoica.marilena23@gmail.com	Romania

26	Virgil Iliuta	virgil.iliuta@ugal.ro	Romania
26	Minodora Ripa	minodora.ripa@ugal.ro	Romania
26	Juliana Javorova	julianata@abv.bg	Bulgaria
26	Gabriel Andrei	gabriel.andrei@ugal.ro	Romania
27	Ionut Cristian Romanu	ionutromanucristian@usm.ro	Romania
29	Andrei Calin	acalin05@gmail.com	Romania
29	Andrei Tudor	andreitudor1947@gmail.com	Romania
29	Marilena Stoica	stoica.marilena23@gmail.com	Romania
29	Nicolae Stoica	nicu_alex_stoica@yahoo.com	Romania
29	Kussay Subhi	qusay_abunoor@yahoo.com	Iraq
30	Jamal Ahmed	jamalahmedkhan20@gmail.com	Pakistan
30	Raja Amer Azim	hamza.hussain95@gmail.com	Pakistan
30	Waleed Hussain	w.hussain9253@gmail.com	Pakistan
33	Claudiu Ovidiu Popa	claudepopa@yahoo.com	Romania
37	Sergiu Spinu	sergiu.spinu@fim.usv.ro	Romania
37	Delia Cerlinca	delia@usm.ro	Romania
38	Stelian Cazan	stelian.cazan09@gmail.com	Romania
38	Claudiu Dănilă	klausosho@yahoo.com	Romania
38	Spiridon Crețu	sp_cretu@yahoo.com	Romania
40	Adrian Catalin Drumeanu	drumeanu@upg-ploiesti.ro	Romania
41	Adrian Catalin Drumeanu	drumeanu@upg-ploiesti.ro	Romania
42	Aurelio Somà	AURELIO.SOMA@POLITO.IT	Italy
43	Spiridon Cretu	spiridon.cretu@tuiasi.ro	Romania
44	Traian Florian Ionescu	Traian.Ionescu@ugal.ro	Romania
44	Dionis Guglea	guglea.dionis@gmail.com	Romania
44	Constantin Georgescu	constantin.georgescu@ugal.ro	Romania
44	Dumitru Dima	dimad@ugal.ro	Romania
44	Lorena Deleanu	lorena.deleanu@ugal.ro	Romania
45	Mara Kandeva	kandevam@gmail.com	Bulgaria
45	Todor Penyashki	peniashki@abv.bg	Bulgaria
45	Georgi Kostadinov	gdkostadinov@gmail.com	Bulgaria
45	Zhecho Kalichin	kalitchin@gmail.com	Bulgaria
47	Ionut Geonea	igeonea@yahoo.com	Romania
47	Nicolae Dumitru	nicolae_dtru@yahoo.com	Romania
47	Cristian Copilusi	cristache03@yahoo.co.uk	Romania
47	Alexandru Margine	fam_margine@yahoo.com	Romania
48	Shubrajit Bhaumik	shubrajit.b@ktr.srmuniv.ac.in	India
48	Viorel Paleu	vpaleu@tuiasi.ro	Romania
48	Dhrubajyoti Chowdhury	dhrubajyoti_debabrata@srmuniv.edu.in	India
48	Cornelia Cîrlan Paleu	cpaleu@yahoo.com	Romania
48	Shubhabrata Datta	shu.datt@gmail.com	India
49	George Ghiocel Ojoc	george.ojoc@gmail.com	Romania
49	Catalin Pirvu	pirvu.catalin@incas.ro	Romania
49	Simona Sandu	smsbadea@yahoo.com	Romania
49	Lorena Deleanu	lorena.deleanu@ugal.ro	Romania
50	Dionis Guglea	guglea.dionis@gmail.com	Romania
50	Traian Florian Ionescu	traian.ionescu@ugal.ro	Romania
50	Dumitru Dima	dimadumitru@yahoo.com	Romania
50	Constantin Georgescu	constantin.georgescu@ugal.ro	Romania
50	Lorena Deleanu	lorena.deleanu@ugal.ro	Romania
51	Violeta Merie	violeta.merie@stm.utcluj.ro	Romania
51	Gavril Negrea	gavril.negrea@ispm.utcluj.ro	Romania
51	Corina Bîrleanu	corina.barleanu@omt.utcluj.ro	Romania
51	Horea Crișan	horea.crisan@omt.utcluj.ro	Romania

IOP Conf. Series: Materials Science and Engineering **724** (2020) 011001 doi:10.1088/1757-899X/724/1/011001

51	Marius Pustan	Marius.Pustan@omt.utcluj.ro	Romania
52	Florina Șerdean	florina.rusu@omt.utcluj.ro	Romania
52	Violeta Merie	violeta.merie@stm.utcluj.ro	Romania
52	Gavril Negrea	gavril.negrea@ispm.utcluj.ro	Romania
52	Corina Bîrleanu	corina.barleanu@omt.utcluj.ro	Romania
52	Ionut Mărieș	Ionut.Maries@omt.utcluj.ro	Romania
53	Andrei Popescu	andrey151289@gmail.com	Romania
53	Dumitru Olaru	dolaru@tuiasi.ro	Romania
54	Cezara Mariuca Oprisan	mariuca.oprisan@yahoo.com	Romania
54	Bogdan Chiriac	bogdanchiriac24@yahoo.com	Romania
54	Vlad Carlescu	vlad.carlescu@tuiasi.ro	Romania
54	Dumitru N Olaru	dolaru@mail.tuiasi.ro	Romania
55	Irina Radulescu	irena_sandu@yahoo.com	Romania
55	Alexandru Valentin Radulescu	varrav2000@yahoo.com	Romania
56	Rodica Voicu	rodica.voicu@imt.ro	Romania
56	Marius Pustan	Marius.Pustan@omt.utcluj.ro	Romania
56	Corina Birleanu	Corina.Barleanu@omt.utcluj.ro	Romania
56	Raluca Gavrila	raluca.gavrila@imt.ro	Romania
56	Cosmin Romanitan	cosmin.romanitan@imt.ro	Romania
56	Oana Tutunaru	oana.tutunaru@imt.ro	Romania
56	Alina Cismaru	alina.cismaru@imt.ro	Romania
57	Sorin Gabroveanu	sorin.gabroveanu@gmail.com	Romania
57	Sorin Cananau	s_cananau@yahoo.com	Romania
57	Radu-Florin Mirica	radu_mirica@yahoo.com	Romania
57	Alexandru Ilies	ilies_alex86@yahoo.com	Romania
58	Harun Yanar	harunyanarr@gmail.com	Turkey
58	Gencaga Purcek	gpurcek@gmail.com	Turkey
58	H. Hüseyin Ayar	metisafe@gmail.com	Turkey
59	Dan Moldovanu	dan.moldovanu@auto.utcluj.ro	Romania
59	Andreia Molea	andreia.molea@auto.utcluj.ro	Romania
59	Istvan Barabas	istvan.barabas@auto.utcluj.ro	Romania
60	Alexandru Valentin Radulescu	varrav2000@yahoo.com	Romania
60	Irina Radulescu	irena_sandu@yahoo.com	Romania
61	Velko Rupetsov	velko_r@abv.bg	Bulgaria
61	Stefan Iliev Dishliev	stefan_dishliev@abv.bg	Bulgaria
61	Georgi Mishev	gmishevpl@yahoo.com	Bulgaria
62	Horea George Crișan	horea.crisan@omt.utcluj.ro	Romania
62	Oana Adriana Crișan	oana.crisan@staff.utcluj.ro	Romania
62	Marius Sorin Pustan	marius.pustan@omt.utcluj.ro	Romania
62	Corina Julieta Bîrleanu	corina.barleanu@omt.utcluj.ro	Romania
62	Violeta Valentina Merie	vio1919@yahoo.com	Romania
64	Gabriel Popescu	gpopescu65@yahoo.com	Romania
65	Aleksandar Vencl	avencl@mas.bg.ac.rs	Serbia
65	Nikolaos Vaxevanidis	nvaxev@gmail.com	Greece
65	Mara Kandeva	kandevam@gmail.com	Bulgaria
66	Omar Tilmatine	omar.tilmatine@univ-poitiers.fr	France
66	Thami Zeghloul	thami.zeghloul@univ-poitiers.fr	France

66	Aurelian Fatu	aurelian.fatu@univ-poitiers.fr	France
66	Lucien Dascalescu	lucien.dascalescu@univ-poitiers.fr	France
67	Stelian Alaci	alaci@fim.usv.ro	Romania
67	Carmen Bujoreanu	cbujorea@tuiasi.ro	Romania
67	Narcis Daniel Petrea	petrea_narcis09@yahoo.com	Romania
67	Florina Carmen Ciornei	florina.ciornei@usm.ro	Romania
67	Luminita Irimescu	luminita.irimescu@usm.ro	Romania
68	Todor Penyashki	tpeniashki@abv.bg	Bulgaria
68	Dimitar Radev	ddradev@gmail.com	Bulgaria
68	Mara Kandeva	kandevam@gmail.com	Bulgaria
68	Georgi Kostadinov	gdkostadinov@gmail.com	Bulgaria
69	Cornelia Cîrlan Paleu	cpaleu@yahoo.com	Romania
69	Bogdan Istrate	bogdan_istrate1@yahoo.com	Romania
69	Viorel Paleu	vpaleu@tuiasi.ro	Romania
69	Corneliu Munteanu	cornelmun@gmail.com	Rwanda
70	Marius Pustan	Marius.Pustan@omt.utcluj.ro	Romania
70	Corina Birleanu	Corina.Barleanu@omt.utcluj.ro	Romania
70	Violeta Merie	Violeta.MERIE@stm.utcluj.ro	Romania
70	Liviu Zarbo	liviu.zarbo@itim-cj.ro	Romania
70	Sorina Garabagiu	sorina.garabagiu@itim-cj.ro	Romania
70	Daniel Marconi	daniel.marconi@itim-cj.ro	Romania
71	Corina Birleanu	Corina.Barleanu@omt.utcluj.ro	Romania
71	Marius Pustan	Marius.Pustan@omt.utcluj.ro	Romania
71	Cosmin Cosma	cosmin.cosma@tcm.utcluj.ro	Romania
71	Violeta Merie	violeta.merie@stm.utcluj.ro	Romania
71	Ovidiu Dranda	ovidiu.dranda@volvo.com	Romania
72	Simona Noveanu	simona.noveanu@mdm.utcluj.ro	Romania
72	Calin Rusu	calin.rusu@mdm.utcluj.ro	Romania
72	Camelia Rancea	camelia.rancea@staff.utcluj.ro	Romania
72	Daniel Lates	daniellates@yahoo.fr	Romania
73	Traian Cicone	traian.cicone@upb.ro	Romania
73	Alice Alina Marinescu	alicemarinescu96@gmail.com	Romania
73	Ștefan Sorohan	stefan.sorohan@upb.ro	Romania
74	Georgian Lupu	georgianlupu89@gmail.com	Romania
74	Petrică Turtoi	petrica.turtoi@upb.ro	Romania
74	Traian Cicone	traian.cicone@upb.ro	Romania
76	Ionut-Tudor Maries	ionut.maries@omt.utcluj.ro	Romania
76	Vilau Cristian	cristian.vilau@tcm.utcluj.ro	Romania
76	Pustan Marius Sorin	marius.pustan@omt.utcluj.ro	Romania
76	Dudescu Cristian	mircea.dudescu@rezi.utcluj.ro	Romania
76	Crisan Horea George	horea.crisan@auto.utcluj.ro	Romania
77	Gelu Ianus	gianus2002@yahoo.com	Romania
77	Denis Cojocaru	darckreign2004@gmail.com	Romania
77	Cezara Mariuca Oprisan	mariuca.oprisan@yahoo.com	Romania
77	Viorel Paleu	vpaleu@yahoo.com	Romania
77	Dumitru N. Olaru	dolaru@mail.tuiasi.ro	Romania
80	Mihail Bogdan -Roth	rothinimagic@yahoo.com	Romania
80	Mirela Romanet	mirelamagic25@yahoo.com	Romania
80	Razvan George Ripeanu	rrapeanu@upg-ploiesti.ro	Romania
81	Viorel Paleu	vpaleu@yahoo.com	Romania
81	Cornelia Cirlan Paleu	cpaleu@yahoo.com	Romania
81	Bogdan Istrate	bogdan_istrate1@yahoo.com	Romania

81	Shubrajit Bhaumik	shubrajit.b@ktr.srmuniv.ac.in	India
81	Corneliu Munteanu	cornelmun@gmail.com	Romania
84	Sabina Teodoroff-Onesim	teodoroff_onesim_sabina@yahoo.com	Romania
84	Alexandra Besleaga	alexandra_besleaga@yahoo.com	Romania
84	Lucel Sirghi	lsirghi@uaic.ro	Romania
87	Andrei-Iancu Stoicescu	stoicescu_andrei_iancu@yahoo.com	Romania
87	Adrian Predescu	adrian_omtr@yahoo.com	Romania
87	Valentin Gatu	Valentin.Gatu@gmail.com	Romania

Affiliation
Technical University of Cluj-Napoca
SCO Expert Pool & DGM
University Petroleum-Gas of Ploiesti
University Petroleum-Gas of Ploiesti
University Petroleum-Gas of Ploiesti
Institut Pprime & CITF Company
Institut Pprime
Institut Pprime
CITF Company
Institut Pprime
University Stefan cel Mare Suceava
Technical College, Radauti
"Gheorghe Asachi" Technical University of Iaşi
"Gheorghe Asachi" Technical University of Iaşi
Transilvania University of Brasov
Transilvania University of Brasov
Transilvania University of Brasov
Transilvania University of Brasov
Petroleum and Gas University of Ploiesti
Petroleum and Gas University of Ploiesti
Petroleum and Gas University of Ploiesti
University of Sussex
University of Sussex
University of Sussex
University of Sussex
Technical University Iasi
Technical University Iasi
Technical University Iasi
Dunarea de jos University of Galati
Dunarea de jos University of Galati
Dunarea de jos University of Galati
Institute of Solid Mechanics of the Romanian Academy
National Institute of Research and Development in Mechatronics and Measurement Technique
University of Civil Engineering
Institute of Solid Mechanics of the Romanian Academy
Institute of Solid Mechanics of the Romanian Academy
Petroleum-Gas University of Ploiesti
Petroleum-Gas University of Ploiesti
Petroleum-Gas University of Ploiesti
Petroleum-Gas University of Ploiesti
Technical University of Cluj-Napoca
Technical University of Cluj-Napoca
Technical University of Cluj-Napoca
Technical University of Cluj-Napoca
Dunarea de Jos
Dunarea de Jos
Dunarea de Jos
Dunarea de Jos
Dunarea de Jos
Dunarea de Jos
Transilvania University of Brasov
Stefan cel Mare University of Suceava
Stefan cel Mare University of Suceava
Polytechnic University of Bucharest
Polytechnic University of Bucharest
Universitat Politècnica de Catalunya
Polytechnic University of Bucharest

IOP Conf. Series: Materials Science and Engineering **724** (2020) 011001 doi:10.1088/1757-899X/724/1/011001

"Dunarea de Jos" University of Galati
"Dunarea de Jos" University of Galati
University of Chemical Technology and Metallurgy, Sofia
"Dunarea de Jos" University of Galati
University of Suceava
Politehnica University Bucharest
Politehnica University Bucharest
Politehnica University Bucharest
Politehnica University Bucharest
Al-Furat Al-Awsat Technical University
College of Electrical and Mechanical Engineering(NUST)
College of Electrical and Mechanical Engineering(NUST)
Institute of Space technology, Pakistan
Technical University of Cluj-Napoca
Stefan cel Mare University of Suceava
Stefan cel Mare University of Suceava
Technical University "Gh. Asachi" Iasi
Technical University "Gh. Asachi" Iasi
Technical University "Gh. Asachi" Iasi
Universitatea Petrol-Gaze din Ploiesti
Universitatea Petrol-Gaze din Ploiesti
Department of Mechanical and Aerospace Engineering
TUIASI
Dunarea de Jos University of Galati
Dunarea de Jos University of Galati
Dunarea de Jos University of Galati
Dunarea de Jos University of Galati
Dunarea de Jos University of Galati
Technical University Sofia
Institute of Soil Science Agrotechnologies and Plant Protection N. Pushkarov
Institute of Soil Science Agrotechnologies and Plant Protection N. Pushkarov
SciBuCom 2 Ltd
University of Craiova
University of Craiova
University of Craiova
University of Craiova
SRM Institute of Science and Technology
Gheorghe Asachi Technical University of Iasi
SRM Institute of Science and Technology
Gheorghe Asachi Technical University of Iasi
SRM Institute of Science and Technology
"Dunarea de Jos" University
INCAS - National Institute for Aerospace Research "Elie Carafoli", Bucharest
Scientific Research Center for CBRN Defense and Ecology, Bucharest
"Dunarea de Jos" University
"Dunarea de Jos" University
"Dunarea de Jos" University
"Dunarea de Jos" University
"Dunarea de Jos" University
"Dunarea de Jos" University
Technical Univesity of Cluj-Napoca
Technical Univesity of Cluj-Napoca
Technical Univesity of Cluj-Napoca
Technical Univesity of Cluj-Napoca

Technical University of Cluj-Napoca
Technical University of Cluj-Napoca
Technical University of Cluj-Napoca
Technical University of Cluj-Napoca
Technical University of Cluj-Napoca
Technical University of Cluj-Napoca
"Gheorghe Asachi" Technical University of Iasi-Romania, Faculty of Mechanical Engineering
"Gheorghe Asachi" Technical University of Iasi-Romania, Faculty of Mechanical Engineering
Technical University "Gheorghe Asachi" Iasi
Technical University "Gheorghe Asachi" Iasi
Technical University "Gheorghe Asachi" Iasi
Technical University "Gheorghe Asachi" Iasi
University POLITEHNICA Bucharest
University POLITEHNICA Bucharest
National Institute for R&D in Microtechnologies - IMT Bucharest
Technical University of Cluj-Napoca
Technical University of Cluj-Napoca
IMT Bucharest
IMT Bucharest
IMT Bucharest
IMT Bucharest
National Research and Development for Gas Turbines SC COMOTI SA, 220 D Iuliu Maniu Bd., 061126, Bucharest, Romania
Politehnica University of Bucharest, Machine Elements and Tribology Dept. 313 Spl.Independentei, Bucharest, Romania
Politehnica University of Bucharest, Machine Elements and Tribology Dept. 313 Spl.Independentei, Bucharest, Romania
National Research and Development for Gas Turbines SC COMOTI SA, 220 D Iuliu Maniu Bd., 061126, Bucharest, Romania
Karadeniz Technical university
Karadeniz Technical university
Metisafe, Cleanroom and Biosafety Systems
Technical University of Cluj-Napoca, Automotive Engineering and Transports Department
Technical University of Cluj-Napoca, Automotive Engineering and Transports Department
Technical University of Cluj-Napoca, Automotive Engineering and Transports Department
University POLITEHNICA Bucharest
University POLITEHNICA Bucharest
Plovdiv University "Paisii Hilendarski"
University of Food Technologies, Plovdiv
Plovdiv University "Paisii Hilendarski"
Technical University of Cluj-Napoca
Technical University of Cluj-Napoca
Technical University of Cluj-Napoca
Technical University of Cluj-Napoca
Technical University of Cluj-Napoca
Technical University Gheorghe Asachi of Iasi
University of Belgrade, Faculty of Mechanical Engineering
School of Pedagogical and Technological Education, 14121 Athens, Greece
Faculty of Industrial Technology, Technical University of Sofia
Institut Pprime
Institut Pprime

Institut Pprime
Institut Pprime
Stefan cel Mare University Suceava
„Gheorghe Asachi" Technical University, Iasi, Romania
„Gheorghe Asachi" Technical University, Iasi, Romania
Stefan cel Mare University Suceava
Stefan cel Mare University Suceava
ISSAPP "N.Pushkarov"
BAS
Technical University, Sofia
ISSAPP "N.Pushkarov"
Gheorghe Asachi Technical University of Iaşi
Gheorghe Asachi Technical University of Iaşi
Gheorghe Asachi Technical University of Iaşi
Gheorghe Asachi Technical University of Iaşi
Technical University of Cluj-Napoca
Technical university of Cluj-Napoca
Technical University of Cluj-Napoca
National Institute for Research and Development of Isotopic and Molecular Technologies
National Institute for Research and Development of Isotopic and Molecular Technologies
National Institute for Research and Development of Isotopic and Molecular Technologies
Technical University of Cluj-Napoca
Technical University of Cluj-Napoca
Technical University of Cluj-Napoca
Technical University of Cluj-Napoca
Technical University of Cluj-Napoca
Technical University of Cluj-Napoca
Technical University of Cluj-Napoca
Technical University of Cluj-Napoca
University of Medicine, Pharmacy, Science and Technology of Târgu Mureş
University POLITEHNICA of Bucharest
University POLITEHNICA of Bucharest
University POLITEHNICA of Bucharest
Military Equipment and Technologies Research Agency, Aeroportului 16, Clinceni, Ilfov, 077025, Romania
University POTEHNICA of Bucharest, Spl. Independenței 313, Bucharest, 060042, Romania
University POTEHNICA of Bucharest, Spl. Independenței 313, Bucharest, 060042, Romania
Technical University of Cluj-Napoca
Technical University of Cluj-Napoca
Technical University of Cluj-Napoca
Technical University of Cluj-Napoca
Technical University of Cluj-Napoca
Technical University "Gheorghe Asachi" Iasi
Technical University "Gheorghe Asachi" Iasi
Technical University "Gheorghe Asachi" Iasi
Technical University "Gheorghe Asachi" Iasi
Technical University "Gheorghe Asachi" Iasi
Petroleum -Gas University of Ploiesti
Petroleum- Gas University of Ploiesti
Petroleum -Gas University of ploiesti
Gheorghe Asachi Technical University of Iasi
Gheorghe Asachi Technical University of Iasi
Gheorghe Asachi Technical University of Iasi

SRM Institute of Science and Technology
Gheorghe Asachi Technical University of Iasi
Alexandru Ioan Cuza University
Alexandru Ioan Cuza University
Alexandru Ioan Cuza University
University Politehnica of Bucharest
University Politehnica of Bucharest
University Politehnica of Bucharest

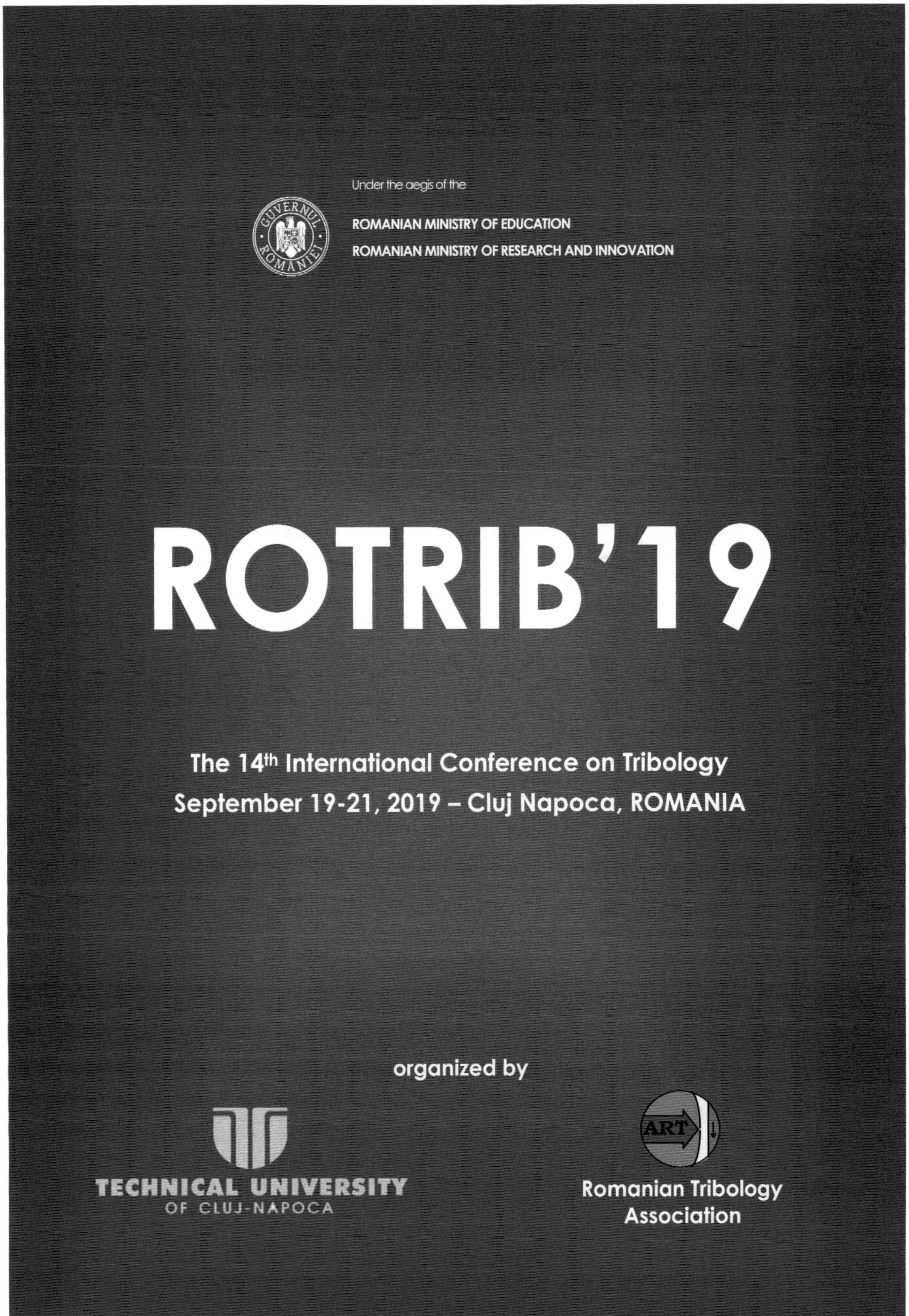
Under the aegis of the
GUVERNUL ROMÂNIEI
ROMANIAN MINISTRY OF EDUCATION
ROMANIAN MINISTRY OF RESEARCH AND INNOVATION
ROTRIB'19
The 14th International Conference on Tribology
September 19-21, 2019 – Cluj Napoca, ROMANIA
organized by
TECHNICAL UNIVERSITY
OF CLUJ-NAPOCA
ART
Romanian Tribology
Association

IOP Conf. Series: Materials Science and Engineering **724** (2020) 011002 doi:10.1088/1757-899X/724/1/011002

FOREWORD ROTRIB'19

ROTRIB'19 is the 14th Conference organized in Romania in a series of international scientific events on the subject of TRIBOLOGY. The general objective of this conference was to bring contributions to the fundamental applications and experiments in the field of Tribology and Nanotribology. The conference provided an interdisciplinary forum for discussions and networking as well as the aim to initiate collaborations among specialists regarding materials in Tribology and their application in product design and reliability at macro/ micro and nano-scale. The exchange of ideas, theories, developments, innovations, improved research methods and future trends was a key scope of this conference. This goal was achieved by bringing together specialists from the industry and the academic environment as well as leading and young researchers from various scientific fields.

Papers presented at ROTRIB 2019 are included in the following topics:

1. Friction and Wear
2. Contact Mechanics
3. Biotribology, tribology in medicine
4. Micro and nanotribology
5. Lubricants
6. Lubrication
7. Tribochemistry and tribocorrosion
8. Tribology of machine elements
9. Tribology in manufacturing processes
10. Surface metrology, surface engineering and coatings
11. Emerging technologies and industrial tribology
12. Condition monitoring

These topics were discussed in the different technical sessions. The official language of the ROTRIB'19 Conference was English. The conference received an excellent number of submissions from which, after a thorough peer reviewing process, 64 scientific contributions were accepted for presentation. The authors of the accepted papers were invited to present their work at the Conference, and so, we were pleased to welcome 87 participants from different countries.

The Conference program was enriched significantly by plenary presentations of renowned experts in their respective domains: Distinguished Professor Dr. Aurelio Somà (Italy) and Distinguished Professor Dr. Spiridon Cretu (Romania). The ROTRIB'19 Conference was organized by the Technical University of Cluj-Napoca and the Romanian Tribology Association, under the aegis of the Romanian Ministry for Research and Innovation. The Conference was held in Cluj-Napoca, ROMANIA at the Technical University of Cluj-Napoca.

The Conference was financially supported by the Technical University of Cluj-Napoca, the Romanian Tribology Association, and sponsors which participated at ROTRIB'19 with company profile presentations or product displays. We would like to express our gratitude to the authors for submitting their work. We would like to thank all those who have contributed, in one way or another, to the success of the Conference. We are grateful to the Scientific Committee members, as well as to the external reviewers for their volunteer hard work of reading and discussing the submissions. We would also like to express our thanks to the members of the Organizing Committee for their support.

It was a great pleasure for us to organize this Conference and we hope that the ROTRIB'19 Conference will stimulate future developments and innovations in the field of tribology.

We hope all participants had a successful conference and a pleasant time in Cluj-Napoca.

Organizing Committee
October 2019, Cluj-Napoca, Romania

ROTRIB'**19**

Committees

Chair of Conference:

Prof. Marius PUSTAN

Honorary Committee:

Prof. Napoleon ANTONESCU
Prof. Spiridon CRETU
Prof. Mihai GAFITANU
Prof. Mircea PASCOVICI
Prof. Liviu PALAGHIAN
Prof. Dumitru POP

Scientific Committee:

Assoc. Prof. Emilia ASSENOVA	**BG**
Prof. Arturo AYON	**US**
Prof. Bou-Said BENYEBKA	**FR**
Prof. Corina BIRLEANU	**RO**
Prof. Mihaela BUCIUMEANU	**RO**
Prof. Traian CICONE	**RO**
Prof. Lorena DELEANU	**RO**
Assoc Prof. Adrian DRUMEANU	**RO**
Prof. Valeriu DULGHERU	**MD**
Prof. Nuria ESPALLARGAS	**NO**
Prof. Aurelian FATU	**FR**
Prof. Michel FILLON	**FR**
Dr. Romeo GLOVNEA	**UK**
Prof. Simion HARAGAS	**RO**
Prof. Hakan KALELI	**TR**
Prof. Mara KANDEVA	**BG**
Prof. M. Baki KARAMIS	**TR**
Prof. Jan MEHNER	**DE**
Prof. Athanassios MIHAILIDIS	**GR**
Prof. Slobodan MITROVIC	**RS**
Dr. Raluca MULLER	**RO**
Prof. Ilie MUSCA	**RO**
Prof. Nikolai MYSHKIN	**BY**
Prof. Iuliu NEGREAN	**RO**
Prof. Daniel NELIAS	**FR**
Prof. Ludovic NOELS	**BE**
Prof. Dumitru OLARU	**RO**
Prof. Razvan George RIPEANU	**RO**
Prof. Alexandru RADULESCU	**RO**
Prof. Filipe Samuel SILVA	**PT**
Prof. Aurelio SOMÀ	**IT**
Prof. Marian SZCZEREK	**PL**
Prof. Andrei TUDOR	**RO**
Prof. Aleksandar VENCL	**RS**

Organizing Committee

Chair:

Prof. Corina BIRLEANU

Members:

Prof. Traian CICONE
Lecturer Stefan CRACIUN
Assist. Horea CRISAN
Lecturer Felicia CRISTEA
Prof. Simion HARAGAS
Drd. Ionut MARIES
Lecturer Violeta MERIE
Prof. Marius PUSTAN
Prof. Minodora RIPA
Lecturer Florina SERDEAN

Reviewers

Stelian Alaci, Marius Alexandrescu, Gabriel Andrei, Sorin Besoiu , Corina Bîrleanu , Mihaela Buciumeanu , Ovidiu Buiga, Traian Cicone, Adrian Circiumaru, Cornel Ciupan, Radu Cotetiu, Stefan Ionut Craciun, Spiridon Cretu, Lorena Deleanu, Adrian Catalin Drumeanu, Cristian Dudescu, Romeo Glovnea, Simion Haragas, Crisan Horea George , Ilie Musca, Luminita Irimescu, Juliana Javorova, Violeta Valentina Merie, Ludovic Noels, Simona Noveanu, Dumitru Olaru, Liviu Palaghian, Viorel Paleu, Dumitru Ovidiu Pop , Claudiu Popa, Alexandru Valentin Radulescu, Minodora Ripa, Razvan George Ripeanu, Florina Maria Serdean, Iacob Liviu Scurtu, Chippa Shriniwas, Aurelio Soma', Marian Szczerek, Nacer Tala Ighil, Radu Velicu, Aleksandar Vencl, Sorin Cristian Vladescu.

Sponsors

IOP Conf. Series: Materials Science and Engineering **724** (2020) 011003 doi:10.1088/1757-899X/724/1/011003

Peer review statement

All papers published in this volume of *IOP Conference Series: Materials Science and Engineering* have been peer reviewed through processes administered by the proceedings Editors. Reviews were conducted by expert referees to the professional and scientific standards expected of a proceedings journal published by IOP Publishing.

IOP Conf. Series: Materials Science and Engineering **724** (2020) 012001 doi:10.1088/1757-899X/724/1/012001

Abrasive flow modelling through active parts water jet machine using CFD simulation

I Pătîrnac[1], R G Ripeanu[1], E Laudacescu[1]

[1] University Petroleum-Gas of Ploiesti, Mechanical Engineering Department, 39 Bucharest Blvd., 100680 Ploiesti, Romania

E-mail: ipatirnac@gmail.com

Abstract. This study has the main objective the flow modelling using CFD through active parts waterjet cutting machine, model YCWJ-380-1520. More precisely, the study of flow through the cutting head and estimate the erosion wear caused by abrasion of these parts. The active parts of cutting head are the body of head, high pressure nozzle with jewel and mixing tube made from wolfram carbide. On all these components has previously done measurement in order to create the geometrical model. The most wearing subjected parts water-jet machine are the water nozzle and mixing tube where the erosion wear is accelerated by the high velocity. In the first part of this study is presented the water flow model without regarding the abrasive inlet geometry, using Ansys FLOTRAN CFD. Has been considered three dimensions of water nozzle along with three dimensions of mixing tube in order to evaluate velocity variation at mixing tube outlet and how geometry influences outlet velocity. The second part of this paper is made using CFD software Ansys Fluent regarding a tridimensional geometry and a multiphasic flow. In the framework of flow model is established the abrasive particles velocity, difference of pressure inside tube, contour of streamlines, shear stress on the tube wall in order to accomplish the erosion rate during the process flow. It can be noticed that there are two main streamlines: first at the outlet of water nozzle and second from the inlet of abrasive material. Particles velocity is higher at the contact with water stream and lower at the end of cutting head because the difference of diameters between jewel and mixing tube. The maximum erosion rate is recorded where the shear stress are high values, that means to the interior wall of water nozzle and inlet of mixing tube. Pressure becomes high to the jewel of water nozzle and decreases in value toward at outlet of cutting head.

1. Introduction

The AWJ process is an erosive manufacturing method which causes important damage at active parts of waterjet machines. The aim of this paper is to make an analytical and experimental analysis of water flow throughout active parts of AWJ machine WUXI YCWJ-380-1520 and to estimate the erosion wear on these elements. The flowing model is made with finite element method CFD using Ansys FLUENT. The water jet machine is presented in figure 1 [4], and the cutting head along with the components is shown in figure 2. The geometrical model is accomplished by real measurements on the cutting head elements like: high pressure water nozzle, mixing tube and mixing chamber, considering these elements being the most intensive subjected to wear. The abrasive material used by the AWJ machine is garnet and the dimension of the particle is considered to be between 0.16mm and 0.2mm.

Figure 1. Photo of AWJ Machine WUXI YCWJ-380-X-1520 [4].

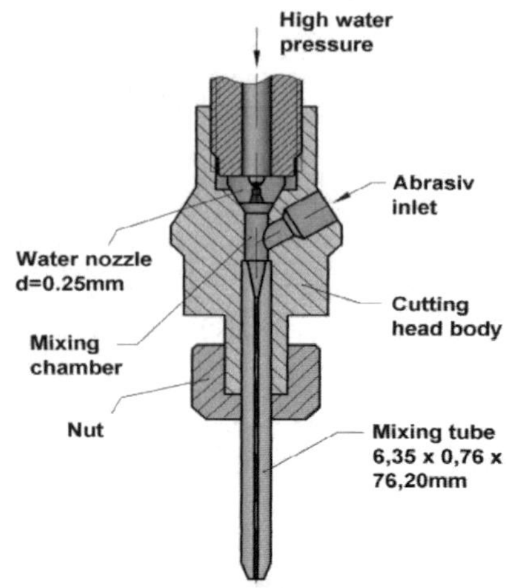

Figure 2. Schematic of Cutting head and the active parts.

In the first part of this study it is made the flowing model using Ansys FLUID 141, only for the liquid part of cutting head without regarding the abrasive inlet orifice. It is made the assumption that the flow throughout the water nozzle is laminar having a parabolic distribution of velocities at the end of the jewel. The first part of the study has the main purpose velocity analysis at the mixing tube outlet, regarding three dimensions of water nozzle along with three other diameters of mixing tubes.

In the second part of this paper it is made a tridimensional CFD analysis using Ansys Fluent upon active parts of cutting machine, regarding the pressure on the water nozzle being 280 MPa and two fluid circuits: one for high velocity water and other for mixing of air and abrasive particles, in order to estimate the erosion wear upon these elements [3].

2. Pure waterjet CFD model

2.1. Theoretical and experimental data

The flow modelling through cutting head of AWJ Machine it is made using Ansys FLOTRAN-CFD and element type FLUID 141. For the FLOTRAN CFD elements, the velocities are obtained from the conservation of momentum principle, and the pressure is obtained from the conservation of mass principle. A segregated sequential solver algorithm is used; that is, the matrix system derived from the finite element discretization of the governing equation for each degree of freedom is solved separately. The flow problem is nonlinear and the governing equations are coupled together. The sequential solution of all the governing equations, combined with the update of any temperature or pressure-dependent properties, constitutes a global iteration [6]. In this model has been considered only the water jet flow without regarding the geometry of abrasive tube. For each model has been adopted three dimensions for water nozzle and tree dimensions for mixing tube, according to manufacturer documents [7], as is shown in table 2.

Table 1. Water nozzle and mixing tube dimensions.

Water nozzle diameter, *mm*	Mixing tube diameter, *mm*
0.20	0.76
0.25	0.90
0.30	1.02

The water flow generated by WJ machine has been obtained through experimental attempts measuring the water level difference which fills the machine bath into a specified time period. Knowing the surface of machine bath and measuring the water level difference, can be calculated the water volume using equation below.

$$V = A_b \cdot \Delta h = 3.8824 \cdot 12 \cdot 10^{-3} = 4.658 \cdot 10^{-2}\ \mathrm{m}^3 \qquad (1)$$

where: A_b is the surface of machine bath, A_b = 3.8824 m^2; Δh – water level difference, Δh = 12 mm.

Knowing the water volume which fills the machine bath in a specified period of time, it can be calculated the water flow using relation (2)

$$Q = V/t = 4.658 \cdot 10^{-2}/1525 = 3.055 \cdot 10^{-5}\ \mathrm{m}^3/\mathrm{s} \qquad (2)$$

where: t is the measured time period to fill the machine bath with Δh, t = 25' 25'' = 1525secundes.

Making the assumption that the water flow is the same for all water nozzle dimensions and knowing the flowing aria for all of them, can be calculated the maximum velocity using equation (3) for each water nozzle diameter [2, 8].

$$v_{max,D} = 2 \cdot v_m = 2 \cdot Q/A_D \qquad (3)$$

where: $v_{max,D}$ is the maximum velocity value for each water nozzle diameter; Q is the water flow; A_D is the surface aria of water nozzle for each diameter; v_m is the average value of velocity.

Regarding a parabolic distribution of outlet velocity on water nozzle and the flowing is made in opposite direction of Y axis, the physical distribution law of velocity along the symmetry axis of geometrical model is given by a parabolic equation [1]. The equation (4) is used to write loads as input velocity in Ansys CFD model.

$$v_y(x) = v_{max,D}(R^{-2} \cdot x^2 - 1) \qquad (4)$$

where: $v_{max,D}$ is the maximum velocity value for each water nozzle diameter; R is the radius of water jewel.

2.2. Experimental results

The experimental results for outlet mixing tube velocity given by Ansys FLOTRAN CFD are depicted in diagrams shown below. The flow analysis it is made considering an incompressible fluid (water) and the reference temperature is 20^{0}C. The velocity vectors generated in this analysis are shown in figure 3.

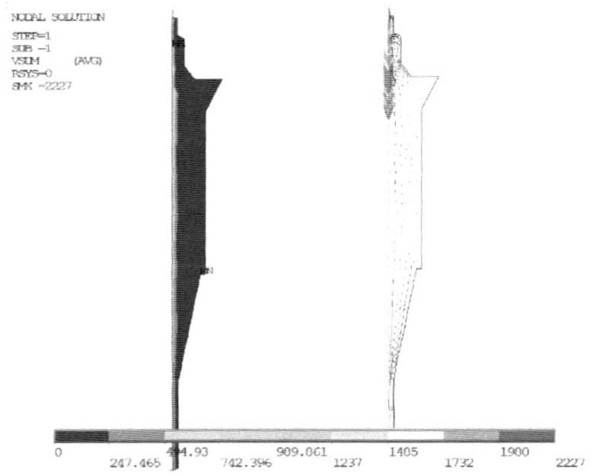

Figure 3. Velocity contour and vectors of flowing model.

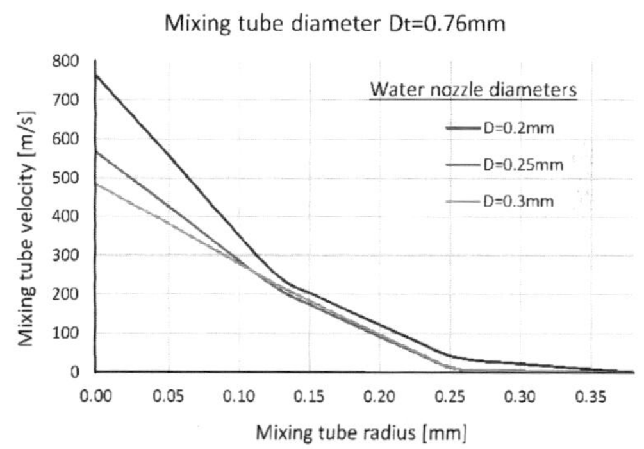

Figure 4. Velocity diagrams for mixing tube diameter 0.76 mm.

On figure 4, can be observed that in the case of 0.76mm in diameter of the mixing tube, the biggest value of outlet velocity can be obtained in the middle of waterjet (symmetry axis of jet) using 0.2 mm diameter of water nozzle and lowest value for the 0.3 mm. Also, can be noticed a slowing down of velocity from jet axis toward the extremity, once with the approach nearby mixing tube wall. The reduction rate becomes maxim using 0.2 mm water nozzle and minimal using 0.3mm of water nozzle jewel. At values of mixing tube radius around 0.11 mm, the outlet velocities obtained for 0.25 mm and 0.3 mm diameters of water nozzle become equal and over that value both water nozzle have approximately the same behaviour.

In the case of 0.9mm mixing tube diameter, as shown in figure 5, the biggest value of the jet axis velocity it is obtained using 0.25 mm water nozzle and the lowest value using 0.3 mm. The outlet velocities of mixing tube are reduced from the jet axis toward the mixing tube wall. On the diagram, it is observed that over 0.07mm of mixing tube radius, the obtained velocity using 0.2mm water nozzle becomes higher rather than using 0.25mm and 0.3mm. Around this very same value of radius, the outlet velocities become equal using both 0.2mm and 0.25mm water nozzle. Can be noticed that, in this case, using a 0.3mm water nozzle can be obtained lower values of outlet velocities in comparison with two others dimension.

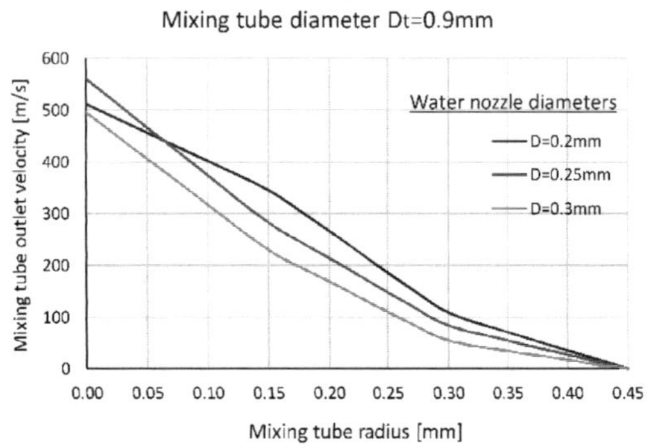

Figure 5. Velocity diagrams for mixing tube diameter 0.9mm.

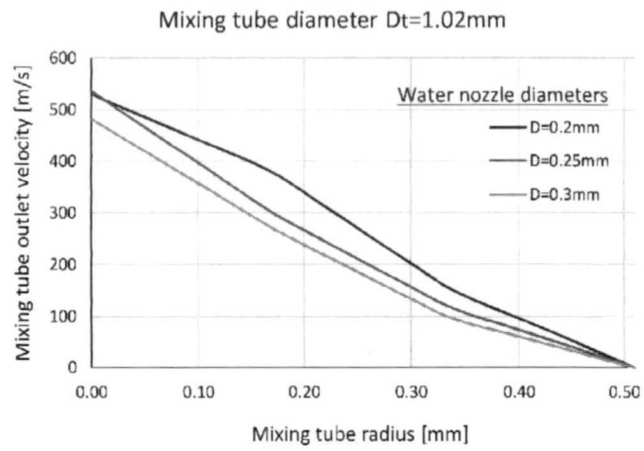

Figure 6. Velocity diagrams for mixing tube diameter 1.02mm.

In the case of 1.02mm mixing tube diameter the highest outlet velocity value in jet axis it is obtained using 0.25mm water jet nozzle (figure 6) and the lowest using 0.3mm. It can be noticed that to 0.03mm towards axis jet on radius, the maximum velocity value can be obtained using 0.2mm water nozzle and the reduction rate is lower in comparison with two other diameters of nozzle. Such in the previous case, lower values of outlet velocities can be obtained using 0.3mm water nozzle rather than using the two other dimensions.

Figures 4,5,6 point out that the maximum outlet velocities values are obtained using 0.76mm mixing tube diameter (about 761.52 m/s) and 0.2mm water nozzle, 565.86 m/s using 0.25mm water nozzle and 483.76 m/s using 0.3mm water nozzle diameter. At 0.9mm mixing tube diameter the highest velocity is 560.26 m/s obtained with 0.25mm water nozzle and in case of 1.02mm mixing tube diameter, the highest velocity value (536.22 m/s) is obtained using 0.25mm water nozzle.

3. Erosion CFD model of active parts

The erosive flow through active elements AWJ Machine WUXI YCWJ-380-1520 has been made using Ansys Workbench – Fluid flow (Fluent) and 3D model of geometry [5]. The materials considered in the framework of simulation is solid-aluminum for body elements and anthracite particles like abrasive material (these materials are provided by Workbench Library). The analysis has beem made considering an isothermal model and the reference temperature is ambient (20^0C). The erosive CFD model is accomplished regarding them. The real material of mixing tube is tungsten carbide (WC), abrasive material used is garnet, ruby jewel for water nozzle and the other elements are made from stainless-steel. There are two circuits of flow: first circuit is pure water at 280MPa and the second is mixed air and abrasion particles. It is considered that particles are spherical with 180 μm diameter and total flow rate is $5 \cdot 10^{-7}$ kg/m. The fraction of air on abrasive inlet is 0.85.

In figure 7 can be observed the streamlines created by the high pressure of waterjet along the cutting head. It is noticed that the maximum fluid velocity is about 751.2 m/s and appears throughout the waterjet ruby jewel of 0.25mm diameter, this velocity is created by the jewel implicitly and the high pressure of water (280MPa). The velocity value throughout the mixing tube is about 187.8 m/s, this decreasing appears because the flowing section increases from the 0.25mm to 0.76mm diameter of mixing tube. It can be noticed a vortex created by the turbulences which appear in the region of abrasive material inlet. This vortex is created because the water strives to go outside through the higher diameter of abrasive inlet. Finally, some of the waterjet streamlines are directed through the mixing tube orifice and others are striking the wall of mixing chamber.

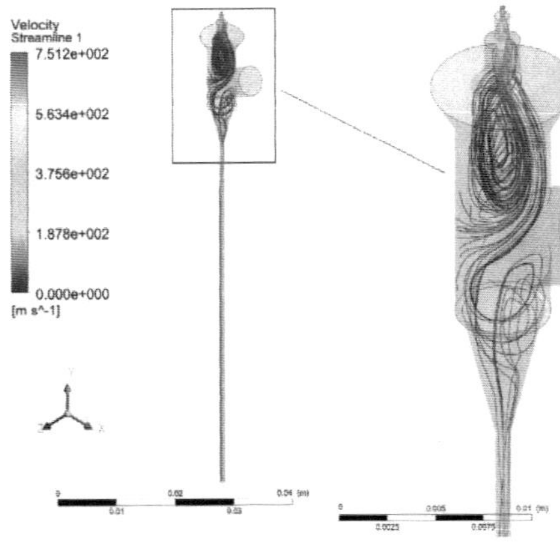

Figure 7. Waterjet streamline provided by 280MPa.

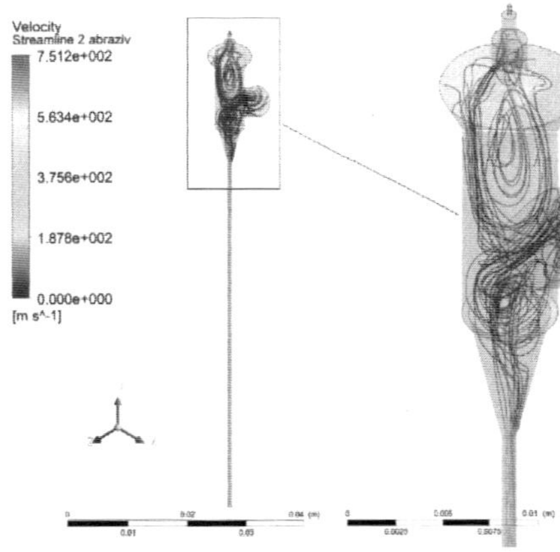

Figure 8. Abrasive particles track generated during 0.337secunds.

There is another fluid circuit created by air which carry away abrasive particles, as is shown in figure 8. The abrasive particles are carried out and trained by the high velocity waterjet up to the value of 187.8 m/s throughout the mixing tube. Also, can be observed some turbulences, mainly on the cone region of mixing tube and to the vortex created by the high velocity waterjet.

The erosion wear produced by the abrasive particles in the interval of 0.337 seconds with their specific features presented previously, is estimated to be around $2.34 \cdot 10^{-5}$ kg/m^2, as is depicted in figure 9 and can be noticed that there are two regions with maximum value of DPM erosion. The first zone is located on the mixing chamber surface and the mixing tube, right on the entrance of abrasive particles into cutting head and the maximum value of erosion wear is estimated to be around $2.11 \cdot 10^{-5}$ kg/m^2. The second zone is located on the surface of water nozzle, having a maximum value of erosion wear about $2.34 \cdot 10^{-5}$ kg/m^2 and it is created by the turbulences of waterjet. In practice, it is proved that these two zones are most intensively subjected to erosion wear.

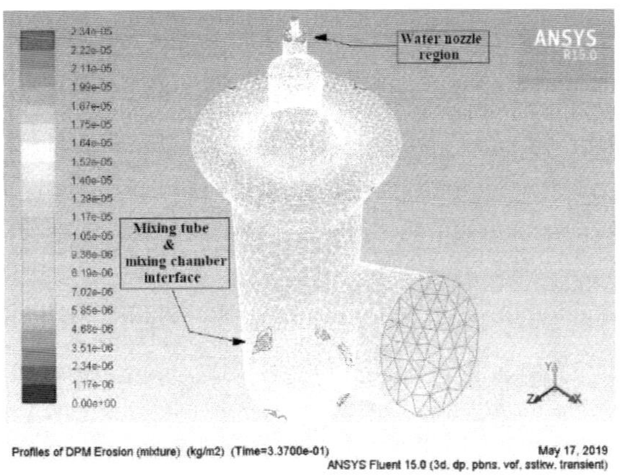

Figure 9. Main regions of maximum DPM erosion on CFD simulation.

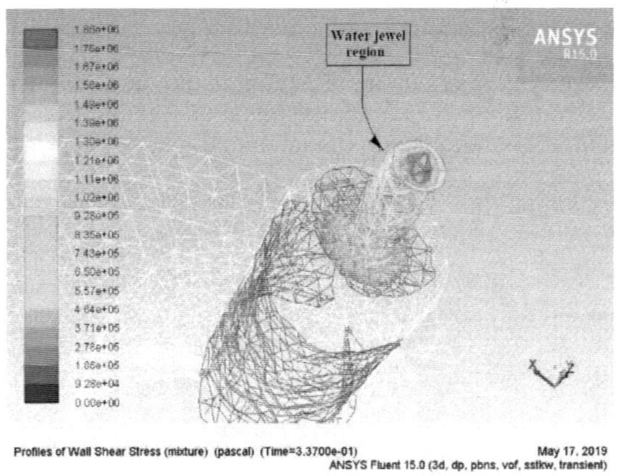

Figure 10. Main regions of shear stress.

In figure 10 presents variation of wall shear stress along the active parts of cutting head. It can be observed that the maximum value is recorded to be 1.86 MPa and it is located on the wall of water nozzle jewel. That is the reason why this part is made from hard materials like sapphire, ruby and diamond.

4. Conclusions

Regarding a laminar flow along water nozzle and a parabolic distribution law of velocity as input data of Ansys boundary conditions, it is estimated that using 0.76mm mixing tube along with 0.2mm water nozzle drives to the highest value of outlet velocity placed in jet axis of mixing tube.

According to real investigation and calculate Reynolds number it is observed a turbulent flow along water nozzle, because the real value of Re is estimated to be around 155000. Subsequent studies will be done regarding a turbulent flow upon water nozzle and a logarithmic distribution law of velocities along ruby nozzle.

Using a 280 MPa water pressure upon 0.25mm diameter water jewel orifice can be obtained a maximum fluid velocity of 751.2 m/s on water nozzle and 187.8 m/s on mixing tube. These values have been obtained assuming an isothermal flow analysis.

According to initial boundary conditions of active parts and properties and material of abrasive particles, it is estimated that the maximum value of erosion wear of active elements is about $2.34 \cdot 10^{-5}$ kg/m^2 made in interval of 0.337secunds.

The maximum value of wall shear stress is calculated to be 1.86 MPa and located on water jewel wall. Knowing these wall shear stress can be estimated the erosion rate using analytical and empirical equations given by various researchers.

For more accurate investigation of erosion behavior upon active parts of AWJ Machine WUXI YCWJ-380-1520 studies will be continued in order to take in consideration the real material of water nozzle and mixing tube. Also, will be made experimental investigation to establish influence of real shape of abrasive particles and considering a running program period longer than 0.337seconds.

References

[1] Oroveanu T, Stan A, Talle V, *Transportul petrolului* 1985 (Bucuresti, Romania, Ed. Tehnica) pp 32-32

[2] Kamarudin N H, Prasada R A K, Azhari A, *CFD Based Erosion Modelling of Abrasive Waterjet Nozzle using Discrete Phase Method*, IOP Conf. Ser.: Mater. Sci. Eng. 114 012016 https://doi.org/10.1088/1757-899X/114/1/012016,

[3] Narayanan C, Caviezel D, Lakehal D, *Optimization of abrasive waterjet nozzle design for precision and reduced wear using compressible multiphase CFD modelling* , http://ascomp.ch/wp-content/uploads/BHR_Waterjet2016_Narayanan_Final.pdf (accessed: may, 2019)

[4] https://chinawaterjet.com/ycwaterjet/products/full-set-yc-water-jet-cutting-machine/ (accessed: may, 2019)

[5] https://www.sharcnet.ca/Software/Ansys/15.0.7/en-us/help/ai_rn_0410.pdf

[6] http://research.me.udel.edu/~lwang/teaching/MEx81/ansys56manual.pdf

[7] https://www.ceratizit.com/en/products/wear-protection/list/detail/product-detail/water-jet-nozzles/ (accessed: may, 2019)

[8] Ionescu O.N. Ionescu G. C., Minescu M. and Nae, I. 2013 Research on life prolongation for temperature sensors placed in highly erosive environment *Journal of the Balkan Tribological Association* **19** (2) 214 -220

Friction Phenomenon Study For the Chain Link – Pin Joint

C C Gavrila[1] and M T Lates[1]

[1]Product Design, Mechatronics and Environment Department, "Transilvania" University of Brasov, Brasov, Romania

E-mail: latesmt@unitbv.ro

Abstract. The friction inside the automobile mechanical transmissions leads to a large influence of the classic fuel consumption and also to the environment impact. Actually, the chain drive is one of the most important transmission inside a car and has few friction sources inside, as the properly chain friction between its components (links, pins), the friction between the chain links and the sprockets and also the friction between the chain links and the tensioning guide active part. The paper main objective is to study the friction phenomenon in the chain link – pin contact for different loads. The tests are achieved for different sets of motion rotational speeds and normal loads; so according to these, there are determined the dynamic friction coefficient (during the motion) and the static friction coefficient (at the stroke ends). The results led to the static and dynamic friction coefficient variation, with the motion rotational speed and the normal loading force. In the final part, the conclusions are presented.

1. Introduction

The chain drive (with silent chains or bush chains inside) is one of the most important transmission inside a car and has, as disadvantages, few friction sources inside, as the properly chain friction between its components (links, pins), the friction between the chain links and the sprockets and also the friction between the chain links and the tensioning guide active part. The aim of the paper is to study the friction phenomenon in the chain link – pin contact for different considered loads.

Generally, the friction in a bush chain drive transmission can be split in the friction inside the chain links (between the links and pins or between the links and the bushes), the friction between the chain links and the sprockets and between the chain links and the tensioning guides [1, 2, 3].

2. Experimental section

The tribometers are used to study the friction between two surfaces in contact [3, 4], which belong to real pieces adjusted to be a mechanical structure. The tribometer used for tests is equipped with a force sensor which allows the forces measurements along the vertical and also, a horizontal direction, in the interval of 0.1 to 1000 N, with the resolution of 50 μN. The force sensor is placed on two sliders, for vertical motions, with a stroke up to 75 mm and a speed between 0.01 and 10 m/s. The lateral position resolution is R_r =2 μm. The oscillatory motion module, allow the alternant rotational oscillating motion. To complete the testing and acquisition data system, the tribometer is connected to an acquisition plate, to a computer, as is presented in Figure 1, a.

For testing various pieces, the tribometer has its own devices, to fit the parts together, but, there are situations when the pieces to be tested are in particular shape, so is not possible to be fitted in the apparatus device without some cutting, trimming or polishing operations. Usually, for small pieces, as bolt or chain links, this could lead to the part destruction [5]. To avoid this, it was used an adequate special device to keep small pin and chain links in relative motion, for the friction coefficient study in the chain link - pin joint. The device needs to fit in the tribometer adequate oscillatory motion module, as is presented in Figure 1, b.

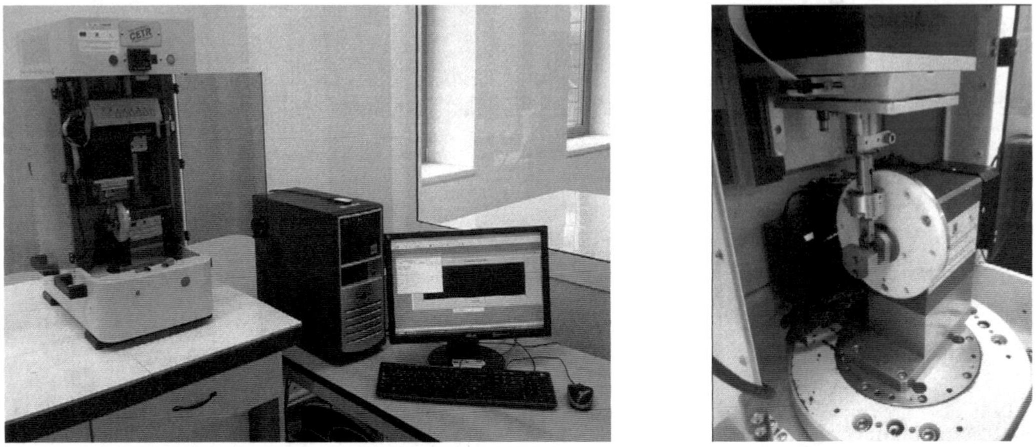

a. b.

Figure 1. The testing and data acquisition system.

a. b. c. d.

Figure 2. The chain link in the adequate device.

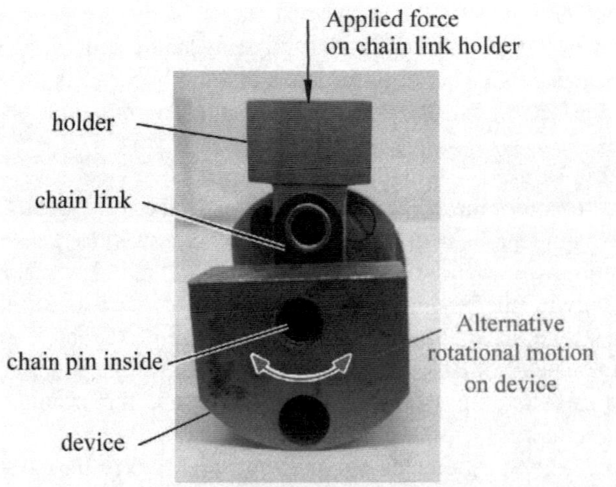

Figure 3. The load and the motion on device.

The pin is included by pressing in the special device and has to fit to the chain link, as in real functioning conditions of the chain joint [6, 7, 8]; then, the chain link is attached thru a special holder to the tribometer, as in presented in Figure 2. The detailed device ensemble with the applied force and motion is presented in Figure 3.

3. The tests and results

The tests are performed by using a chain link and a chain pin, fitted in the described device, without lubrication. The rotational motion is an alternative, in the interval of ±15 degrees. There are chosen three sets of angular speeds as: 2.5 rpm, 5 rpm, 10 rpm, and each equivalent to 15°/s, 30°/s and 60°/s respectively. The load normal force is variable between 10 to 50 N. As results, there are determined the dynamic friction coefficient, during the motion, and, respectively, the static friction coefficient, at the stroke ends.

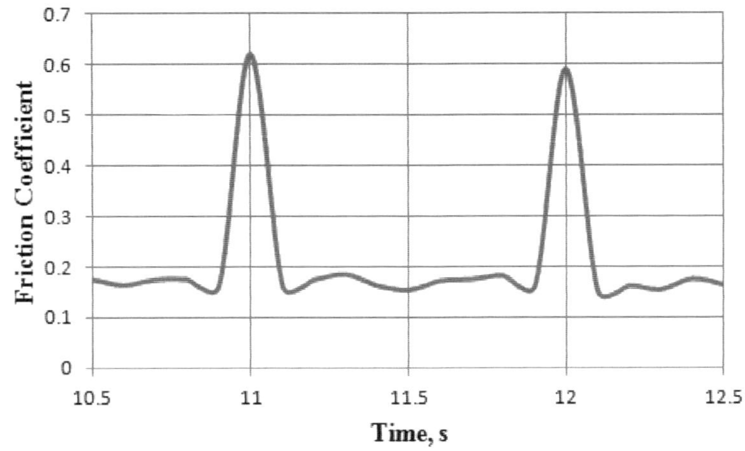

Figure 4. The variation of the friction coefficient during the motion.

In Figure 4 is presented the variation of the friction coefficient during the motion. Along the corresponding motion portion of the curve, the dynamic friction coefficient can be observed. At the ends of the angular stroke, it can be observed the increased values of the static friction coefficient. The situation corresponds to the real chaining on the chain wheel, at the beginning of the relative rotation between the chain links and pins respectively.

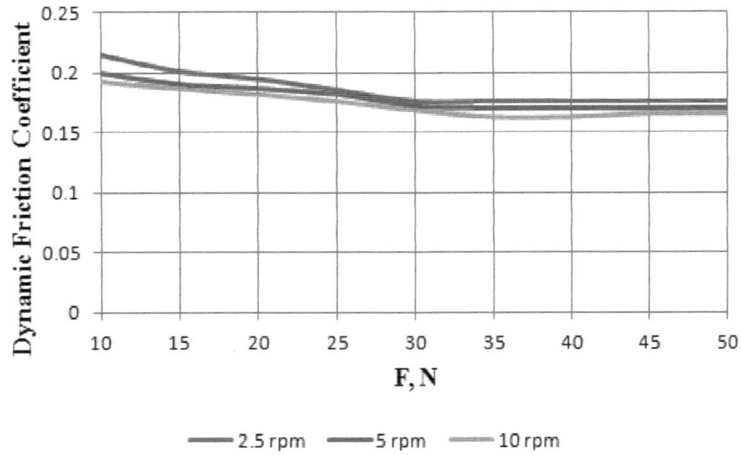

Figure 5. The dynamic friction coefficient.

In Figure 5, the dynamic friction coefficient is presented, for different load force from 10 to 50 N and, respectively different rotation values. For decreased load force, the friction dynamic coefficient value is close to 0.2. For all the rotation speeds, at the load force increasing, the dynamic friction coefficient value is slightly decreasing around 0.17.

In Figure 6, the static friction coefficient is presented, for the same loads and rotation values. For different applied load force, the friction dynamic coefficient value is around 0.6.

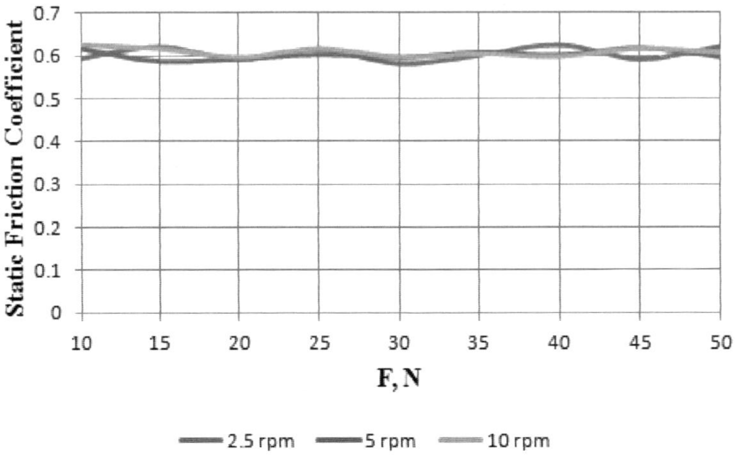

Figure 6. The static friction coefficient.

4. Conclusions

From the variation of the friction coefficient during the motion, it can be observed the increased values of the static friction coefficient at the ends of the angular stroke, which corresponds to the beginning of the links chaining on the chain wheel.

For studied case, the static friction coefficient value, in the mentioned testing inputs, is around 0.6 and the dynamic friction coefficient value is close to 0.2. For all the rotation speeds, at the load force increasing, the dynamic friction coefficient value is slightly decreasing around 0.17, which can be explained by the tested chain link - pin adjustment behaviour under the load.

Following the previously results, it can be concluded that the existing misalignments and clearances are influencing the chain link – pin joint friction condition. The chain adequate use and lubrication can improve the chain link – pin joint behaviour under load, in order to reduce the chain wear.

References

[1] Pedersen S L 2005 Model of contact between rollers and sprockets in chain-drive systems *Archive of Applied Mechanics* **74** pp 489-508

[2] War X and Dwyer-Joyce R S Model experiments on automotive chain drive systems *Elsevier Science* **39** pp 851-861

[3] Velicu R Lates MT Papuc R 2012 Friction test ball on flat during running-in period on UMT tribometer *Annals of the Oradea University 2012, Fascicle of Management and Technological Engineering* XI pp 2.142 - 2.147

[4] *** 2009 UMT Multi-Specimen Test System. Hardware Installation & Application Manual. Center for Tribology, Dell Ave, Campbell, USA

[5] Gavrilă C C and Velicu R 2016 Virtual Modeling, Detail Design and FEM Analysis for a Testing Device *Current Solutions in Mechanical Engineering* Trans Tech Publications pp. 3 - 6.

[6] Papuc R Velicu R Gavrilă C C 2014 Geometrical Study Of Guide - Chain Contact, For General

Chain Transmission *Annals of the Oradea University, Fascicle of Management and Technological Engineering* XIII pp 163 –166.

[7] Lateş M T and Gavrilă C C 2019 Influence of the silent chain link geometry on the link / chain guide contact *IMT Oradea Proceedings of the Annual Session of Scientific Papers* XVIII pp 188 - 192

[8] Lateş M T 2012 Bush chains design process *Annals of the Oradea University, Fascicle of Management and Technological Engineering* XXI pp 2.51 - 2.55

IOP Conf. Series: Materials Science and Engineering **724** (2020) 012003 doi:10.1088/1757-899X/724/1/012003

Comparative study of the friction phenomenon in the contact between the guide and the bush and silent chain links

M T Lates[1] and C C Gavrila[1]

[1]Product Design, Mechatronics and Environment Department, "Transilvania" University of Brasov, Brasov, Romania

E-mail: cgavrila@unitbv.ro

Abstract. The optimisation process of the chain drives includes the development of new chain links in order to reduce the noises, the vibrations and the non-linearity of the motions. One solution is the development of the silent chains in order to replace the bush chains. According to the literature, the silent chains have a better dynamic behaviour instead of bush chains but there are only few studies regarding their frictional behaviour. As part of the global friction in a chain drive transmission, the paper presents an experimental comparative study of the friction between a PA46 polyamide material (used to manufacture the active part of the tensioning guide) and the bush chain and silent chains respectively. The tests are performed on a tribometer by considering different rotational speeds, local pressures and temperatures. The results are presented as variations of the dynamic and static friction coefficients depending on the test parameters. Finally there are offered conclusions regarding the practical applications of the results.

1. Introduction

Chain drive transmissions are widely used in the case of mechanical systems which should transmit high power with application in: automotives, machine tools, naval engineering etc [1, 2]. The main disadvantages of the chain drive transmissions are referring on: the non-linearity of the transmitted motion (due to the gearing and due to their construction, there are shocks and are developed noises during their functioning), the necessity to use the tensioning guides, the wear influence on the chain's stiffness; the friction which influence the transmission's efficiency [1, 2].

In order to improve their dynamic behaviour there are developed new types of chains with different geometries of their links. For instance, the silent chains have a better dynamic behaviour than the bush chains due to the reason that the toothed links have a better gearing process with the sprockets, instead of the links from the bush chains [1, 2]. According to these, in the case of the silent chains, there are smaller shocks, smaller noises during their functioning period than in the case of the bush chains.

The frictional loses are characterising both the silent chains and the bush chains. Generally, the friction in a chain drive transmission can be split in the friction inside the chain links (between the links and pins or between the links and the bushes), the friction between the chain links and the sprockets [3] and between the chain links and the tensioning guides [4].

The literature [5, 6] provides measuring methods and test results in order to study the friction of the chain drive transmissions; one of the basic references used to develop measuring methods to determine the friction between the chain links and the polyamide-made guides is represented by [7]. Polyamides are widely used in the construction of the active part of the tensioning guides due to their good frictional behaviour and internal structure stability with the variation of the temperature [8, 9, 10, 11].

The aim of the present paper is to make a comparative study of the friction between a PA46 polyamide type material used in the design of the active part of a tensioning guide and the bush chain and silent chain links, respectively.

2. The tests

The measuring and acquisition system used for the tests is presented in figure 1. The tribometer is equipped with a force sensor which allows measurements of forces about the vertical direction and about a horizontal direction in the interval of 0.1 ... 1000 N with the resolution of 50 mN; the friction coefficient is determined by calculating the ratio between the measured horizontal force and the normal force [12]. The sensor force is mounted on two sliders which allow vertical motions with a stroke equal with 150 mm and with a speed between 0.001 ... 10 m/s and lateral motions with a stroke equal with 150 mm and with a speed between 0.001 ... 10 m/s.

On the force sensor is mounted a suspension in order to damp the oscillations. On the base plate of the force sensor is mounted a holder which allows the mounting of different pieces as ball, pins, plates. The vertical stroke is equal with 75 mm and it can be reached with a speed situated in an interval of 0.01 ... 10 m/s [12].

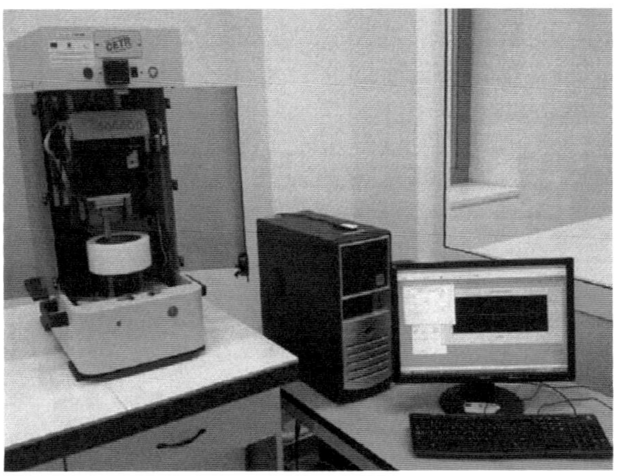

Figure 1. The test rig.

In the tribometer is mounted a rotary unit which allows rotational motions with a rotational speed in the interval of 0.001 ... 5000 rpm and a resolution equal with 1 μm [12]. The rotary unit is equipped with and oil bath connected to a heater which allows a heating up to 150 °C [12].

The tests are performed in two situations: a bush chain link in contact with a PA46 polyamide disk and a bush silent link in contact with a PA46 polyamide disk – figure 2. The polyamide disk is made from a PA46 material which is used in the manufacturing of the active part of the tensioning guides used in chain drive transmissions from the car distribution transmission. The chain links have the same pitch – 8 mm.

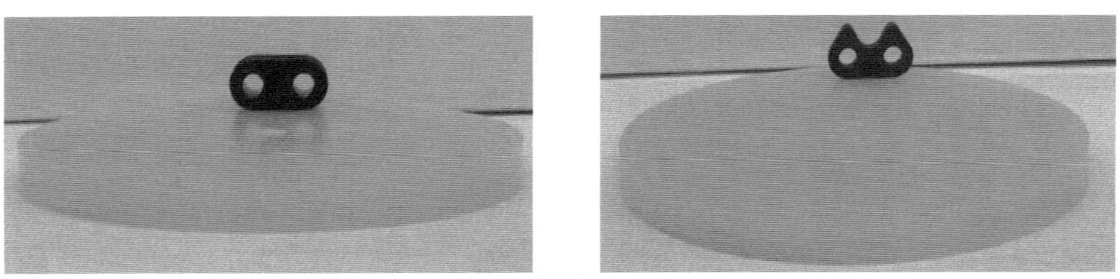

Figure 2. The chain links in contact with the disk.

The chain links are mounted inside a holder – figure 3. The links are mounted inside the solder with a clip which allows small relative motions – as in real functioning conditions of the chain drive transmission

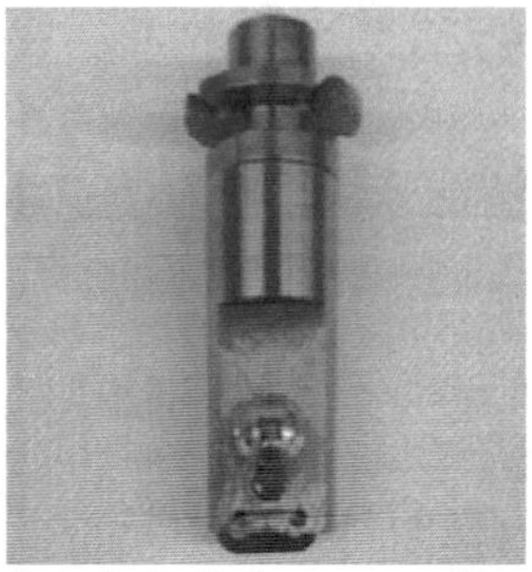

Figure 3. The chain link mounted in the holder.

The disk is mounted inside an oil bath of the rotary unit of the tribometer and the holder with the chain link is mounted on the upper slider as it can be observed in figure 4. The contact between the chain link and the disk is accomplished by moving the sliders about the vertical and the lateral directions.

The tests are performed by using two disks, made by the same materials, with the same geometry, used – one for the bush chain link and the other for the silent chain link tests. At the beginning a running-in procedure is tested during 2 hours for a pressure of 5 MPa, a rotational speed of 500 rpm about two directions in every 10 minutes and an environmental temperature of 22 ^{o}C. The tests continue with a set of rotary speeds of 50, 100, 500, 1000, 2000 rpm about two directions at every 5 minutes, at local pressures between 5 ... 10 MPa and at the temperatures of the oil bath equals with 90 ^{o}C and 120 ^{o}C.

Figure 4. The holder mounted on the tribometer.

3. Results

The results present the evolution of the wear and of the friction coefficient during the running-in process and the variation of the friction coefficient during the tests, depending on the local pressure, the rotational speeds and the temperature of the oil bath.

Figure 5 presents the evolution of the wear during the running-in process. The wear evolution is slightly the same for the tested chain links; the value of the wear is stabilised after approximately 25 minute at a value around 0.12 mm.

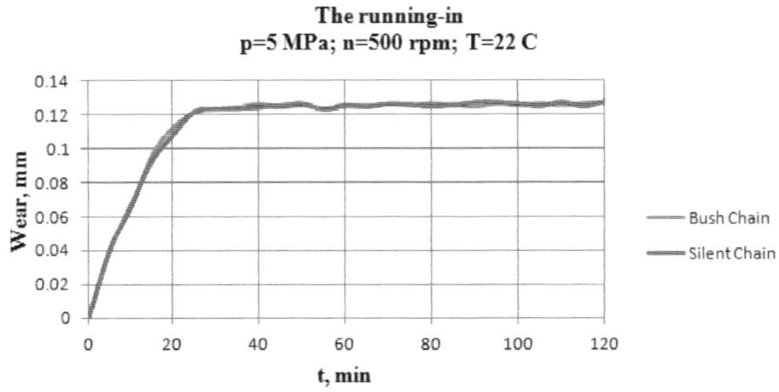

Figure 5. The evolution of the wear during the running-in process.

During the running-in period and during the tests, due to the rotations about two directions, there are identified and measured two types of friction coefficients: the dynamic friction coefficient (during the motion) and the static friction coefficient (when the motion changes the direction) – figure 6. Due to inertial reasons, when the motion slows down in order to change the direction, it is a small decrease of the friction coefficients value; from the same reason, the value of the static friction coefficient in the case of the silent chain link is smaller than the value of the static friction coefficient in the case of the bush chain link. The values of the dynamic friction coefficients are mainly the same.

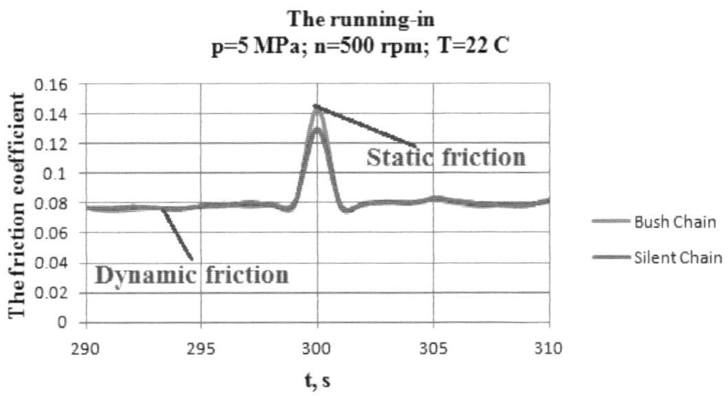

Figure 6. The friction coefficient during the change of the motion direction.

The evolution of the dynamic and of the static friction coefficient during the running-in process is presented in figure 7 and figure 8, respectively. The dynamic friction coefficient has the same variation for both chain links: it is stabilised after 15 minutes at a value around 0.08. The static friction coefficient has smaller values for the silent chain link thank the bush chain links; an explanation of this is based on the different inertias of the links, due to their shapes. The value of the static friction coefficient is stabilised after 60 minutes at around 0.135 for the bush chain and at around 0.125 for the silent chain.

Figure 9 and figure 10 present the values of the dynamic friction coefficient and the static friction coefficient, respectively, for a local pressure of 5 MPa; the values of the dynamic friction coefficients are almost identically for the bush chain and for the silent chain. The dynamic friction coefficient decreases with the increasing of the rotational speed and increases with the increasing of the temperature.

According to figure 10, the value of the static friction coefficients decreases with the increasing of the rotational speed and increases with the increasing of the temperature. The static friction coefficient has smaller values for the silent chain link than the static friction coefficient for the bush chain link; the difference between these two parameters decreases with the increasing of the temperature.

IOP Conf. Series: Materials Science and Engineering **724** (2020) 012003 doi:10.1088/1757-899X/724/1/012003

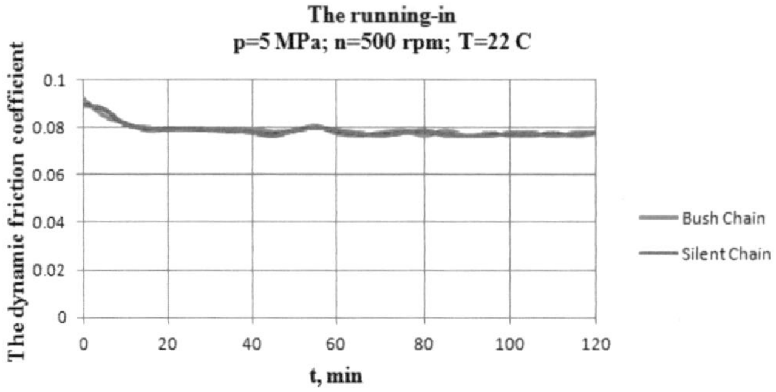

Figure 7. The dynamic friction coefficient.

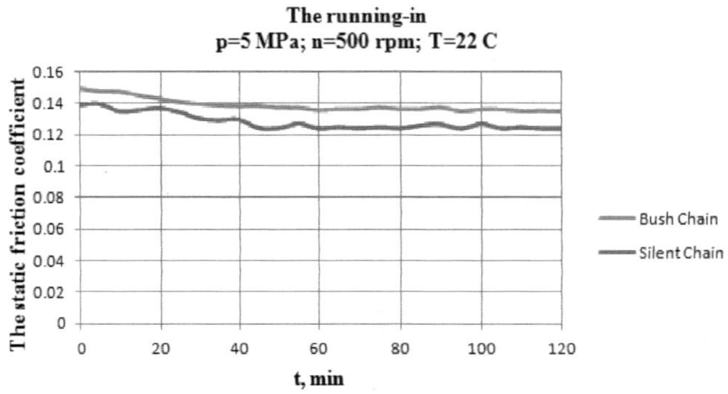

Figure 8. The static friction coefficient.

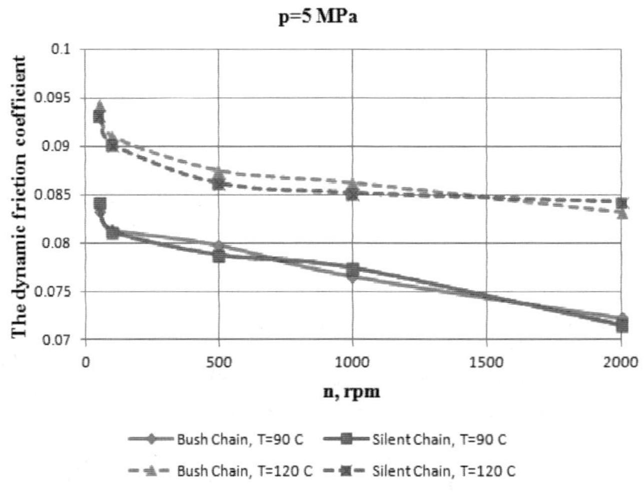

Figure 9. The dynamic friction coefficient.

4. Conclusions

As a general conclusion the values of all the measured friction coefficients decrease with the increasing of the local pressure and of the rotational speed; the values of the friction coefficients increase with the increasing of the temperature.

Regarding the comparative study of the friction coefficients for the bush chain link and the friction coefficients for the silent chain links, it can be concluded that the values of the dynamic friction coefficients are similar. In the case of the static friction coefficients, their values are smaller for the silent chain link than the values for the bush chain link. This can be explained by taking into account to the different inertias of the chain links. Practically, these conclusions shows that, in the case of start/stop situations of the car's engine, it is a smaller frictional lose if the silent chains are used in the distribution transmission of the car.

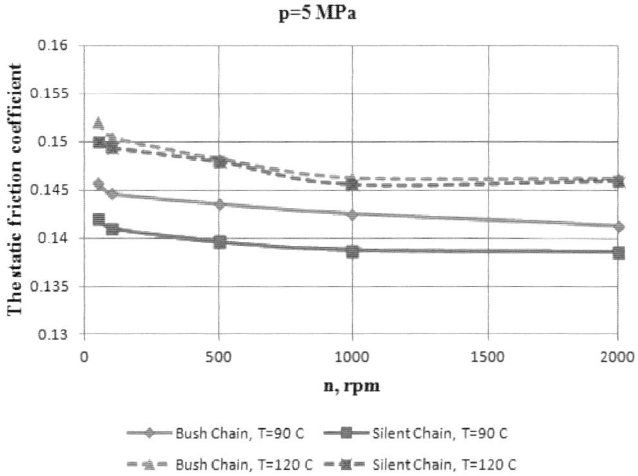

Figure 10. The static friction coefficient.

5. References

[1] Heisler H 2005 Advanced vehicle technology *The bath press*

[2] Vlasrnik J 2009 Chain drive computional model as virtual engine module *Ph.D. Thesis. Brno University of Technology* pp 7-87

[3] Pedersen S L 2005 Model of contact between rollers and sprockets in chain-drive *systems Arch. Appl. Mech.* **74** pp 489-508

[4] War X and Dwyer-Joyce R S Model experiments on automotive chain drive systems *Els. Sci.* **39** pp 851-861

[5] Fink T and Bodestein H 2011 Friction reduction potentials in chain drives *MTZ Worldwide Edition* **72** pp 46-51

[6] Hyakutake T, Inagaki M, Matduda M, Hakamada N and Teramaki Y 2001 Measuring of friction in timing chain *Soc. Autom. Eng. Jap.* **22** pp 343-347

[7] J. van Ruiten van J, Proost R and Meuwissen M 2012 How the choise of the polyamide type in timing chains tensioning systems affects the CO_2 emission and fuel economy of internal combustion engines *Presentation at VDI Vertiltrieb un Zylinderkopf*

[8] Buccella M, Dorigato A, Pasqualini E, Caldara M and Fambri L 2012 Thermo-mechanical properties of Polyamide 6 chemically modified by chain extension with Polyamide/Polycarbonate blend *J. Polym. Res.* **19** pp 9935-9944

[9] Gordon D H and Kukureka S N 2009 The wear and friction of polyamide 46 and polyamide 46/aramid-fibre composites in sliding–rolling contact *Wear* pp 669-678

[10] Chiu F-C and Kao G-F 2012 Polyamide 46/multi-walled carbon nanotube nanocomposites with enhanced thermal, electrical and mechanical properties *J. Comp.* **43** pp 208-218

[11] Hsiao S, Wang H, Chou J, Guo W and Tsai T 2012 Synthesis and characterisation of novel organosoluble and thermally stable polyamides bearing triptycene in their backbones *J. Polym. Res.* **19** pp 9902-9911

[12] *** 2009 UMT Multi-Specimen Test System. Hardware Installation & Application Manual. Center for Tribology, Dell Ave, Campbell, USA

IOP Conf. Series: Materials Science and Engineering **724** (2020) 012004 doi:10.1088/1757-899X/724/1/012004

Research on mechanical and geometrical characteristics of materials used for flexible tubing production

A Hagianu[1], I Nae[1], G C Ionescu[1,2] and R G Ripeanu[1]

[1] Department of Mechanical Engineering, Petroleum-Gas University of Ploiesti, 39 Bucharest Blvd., 100680 Ploiesti, Romania.
[2] correspondence author

E-mail: adi_hagianu@yahoo.com, inae59@gmail.com, ionescu_g_r@yahoo.com, rrapeanu@upg-ploiesti.ro.

Abstract. Coiled tubing used in operational exploitation of oil and gas wells are made of special steels which have to resists to some special requirements such as: tensile strength, internal pressure resistance, collapsing resistance, fatigue strength, corrosion resistance and action of weakening factors. It is interesting that various producers are using different compositions of their steels obtaining quite different mechanical characteristics of the flexible tubing delivered on the market. This paper is presenting the experimental programme for mechanical tests conducted on materials used for production of flexible tubing and the results obtained during these tests. In addition to the materials testing there were studied geometrical characteristics of the flexible tubing such as wall thickness, ovality, cilindricity and their impact on operational life of the flexible tubing.

1. Introduction

Coiled tubing equipment has become a necessity in the petroleum and oil industry, being required in many important applications such as horizontal or arborescent drilling, unclogging old wells or in recovery of broken tools.

Whatever the type of applications it is used for, the coiled tubing is significantly reducing the operational costs due to the reduction of assembly time as well as possibility of introduction of the coiled tube into the well bore without dismantling eruption head. Areas were the flexible tubing is stressed (bending, stretching, compression, traction, inside outside pressure) are presented in figure 1 as follows: introduction step (1-2-3), extraction step (4-5-6) [1].

Whenever the coiled tube is uncoiled from the drum, and it is passing through the guidance head, there are complex bending/unbending processes which could have a great impact on material structure. In addition, the exploitation of the coiled tube, according to the working cycles presented on figure 1, generates complex stress in the material and could lead to permanent deformations, fatigue and material deteriorations. Another source of stress is the vibration induced by the pump which generates the pressure of the working fluid, however, before proceeding to laboratory tests it should be measured the pattern of these vibrations in the field using specialized equipment [2 - 4].

In order to ensure the required state of health of the coiled tubing all the stress induced in the material should not be greater than the one which may generate permanent deformations (to maintain the material into its elastic limits). Any physical deformation of the coiled tubing could lead to irremediable damage both to the equipment from the bore well and to the injector.

In table 1 are presented in a centralised manner all the parameters which should be monitored, in order to establish the state of health of coiled tubing.

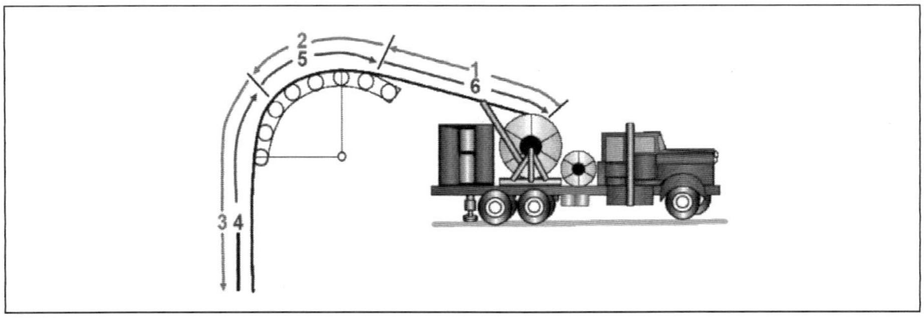

Figure 1. Coiled tube areas which are under a cyclic, repeated stress [1].
Well bore introduction cycle (1-2-3):
1. compression affected area, 2. bending and compression affected area; 3. area affected by traction;
Well bore extraction cycle (4-5-6):
4. traction affected area, 5. bending and compression affected area, 6. compression affected area.

Table 1. Parameters to be monitored in order to establish coiled tubing's state of health

Parameter/characteristics	Characteristic parameter
Functional technical characteristics of the equipment	• Traction force; • Speed of coiled tube introduction/extraction • Pressure on injector clamps; • Pressure of working fluid (pressure from inside of coiled tubing); • Pressure from outside (on the external walls of coiled tubing). • Vibration introduced by the working fluid pump (100-200Hz)
Mechanical characteristics of material from which coiled tubing is made off	• Traction breaking resistance (R_m) • Flow limit (Rp0,2) • Elongation after braking (percent) (A)
Geometric characteristics of coiled tubing	• External nominal diameter, D (mm) • Nominal thickness of coiled tubing wall, t (mm) • Ovality (O) • Coaxiality (C)
Coiled tubing resistance to corrosion	• Corrosion potential (Ecorr) • Corrosion current densities (Icorr) • Corrosion speed (Crr).

2. Exploitation generated stress in coiled tubing

Coiled tubing is affected by various and complex stress both during the operational phase (figure 2) and stationary phase (figure 3), all these accumulating and conducting to a reduction of its operational life.

The stress effects generated by exploitation could lead to:

- Ovalization of coiled tubing and this will decrease the surface of contact between the coiled tube and the clamps of the injection head. In this situation there could be situations when the

IOP Conf. Series: Materials Science and Engineering **724** (2020) 012004 doi:10.1088/1757-899X/724/1/012004

injection head will not be capable to support the weight of the tubing, and this will get inside of the well bore driven by its weight, without any control.

- Reduction of coiled tube wall thickness and consequently decreasing the resistance to the external or internal pressure. The reduction of wall thickness is the result of coiled tube elongation (overloads in exploitation) and/or erosion due to the various corrosive agents encountered in the drilling process (acidization, salty water, hydrogen sulphide H_2S etc).

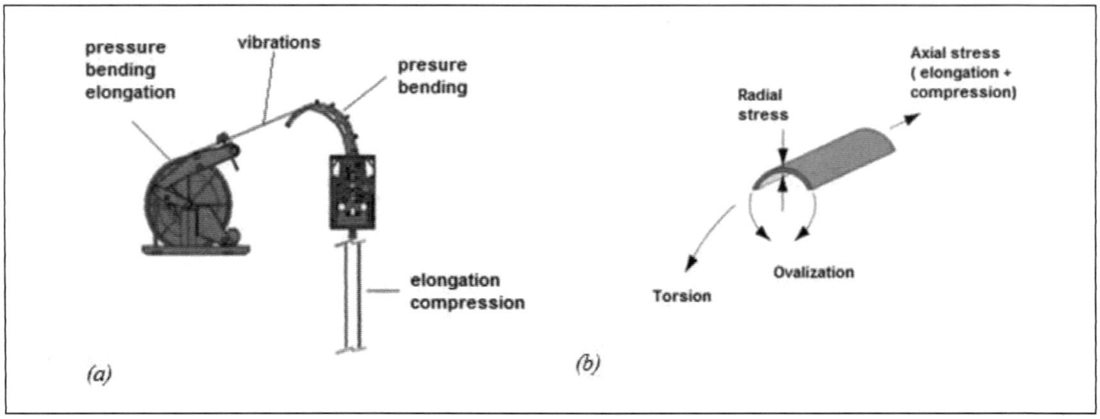

Figure 2. Stress affecting coiled tubing, coiled tubing introduction/extraction from/to well bore (a), stress represented in section of coiled tubing (b) [5]

Taking into account the issues presented a reliable system capable to provide reliable information concerning the state of health of the coiled tubing should monitor and record data regarding: nominal diameter wall thickness and ovalization of the coiled tubing.

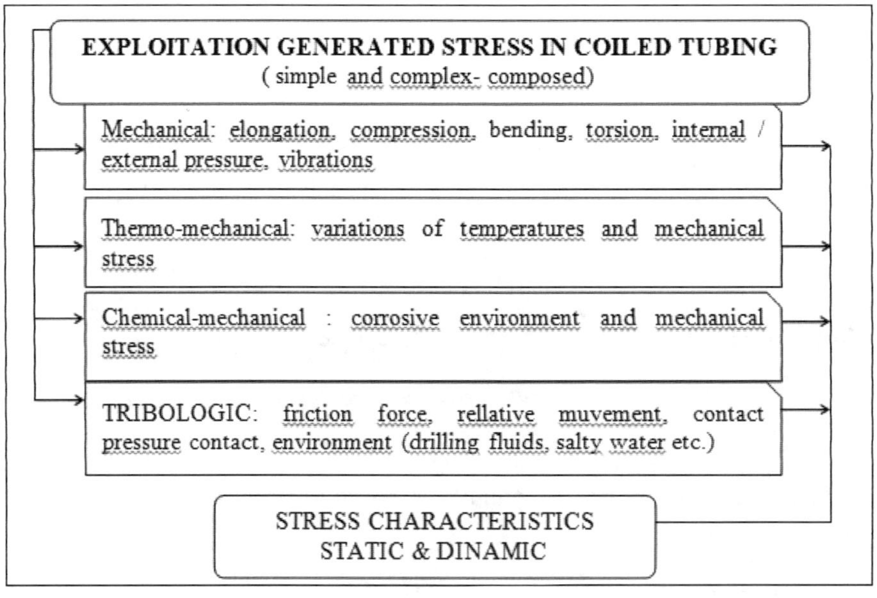

Figure 3. Exploitation generated stress in coiled tubing

3. Experimental research conducted

The experimental work conducted had the goal to establish the necessary state of health for the coiled tubing in order to ensure the required safety both for people and well during the operation of insertion/extraction. The coiled tubing is inserted/extracted in/from the well bore by using a device known as injector. The injector should provide the required force to introduce, extract and sustain the coiled tubing as presented in figure 4.

The injector assembly is designed to fulfil three main functions:

- provides the required force to introduce the coiled tubing inside the well bore acting against the internal pressure of the well and friction forces.
- control the speed of coiled tubing introduction when the weight of the tubing is pooling down the coil.
- sustain the weight of the column and speed up the extraction of the tubing during the extraction phase.

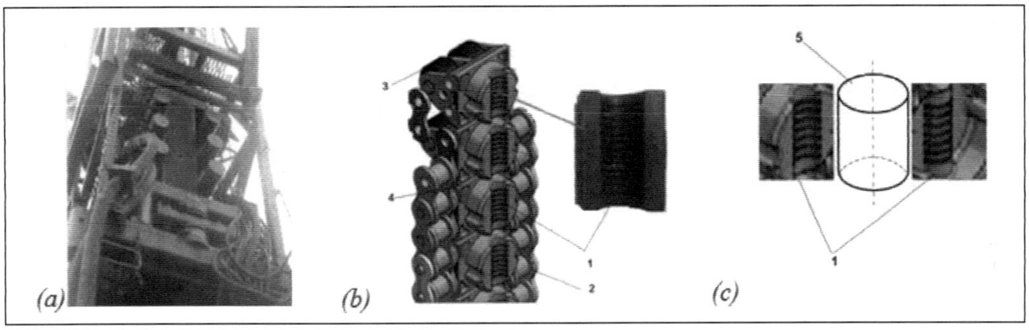

Figure 4. Coiled tubing injector assembly, injector assembly (a), fixing system with clamps (b) [6], clamps contact with the coiled tubing (c): 1. Clamps; 2. Chain; 3. Liaison roller; 4. Safety pin; 5. Coiled tubing.

Based on 4 pairs of clamps, hydraulically actuated, the injector generates the contact pressure (on the external side of coiled tubing), in order to pull and ensure guidance for it. The main working parameters for the injector assembly are: force pressing on clamp, geometrical dimensions of clamps, speed of chain, and the flow of working fluid, all of them being established in well working plan.

The state of health of equipment is established by the capability of injector to sustain the maximum weight of coiled tubing introduced into the bore well. This capability is based on the system of forces developed at the contact between the external surface of coiled tubing and internal surface of injector's clamps as presented in figure 5.

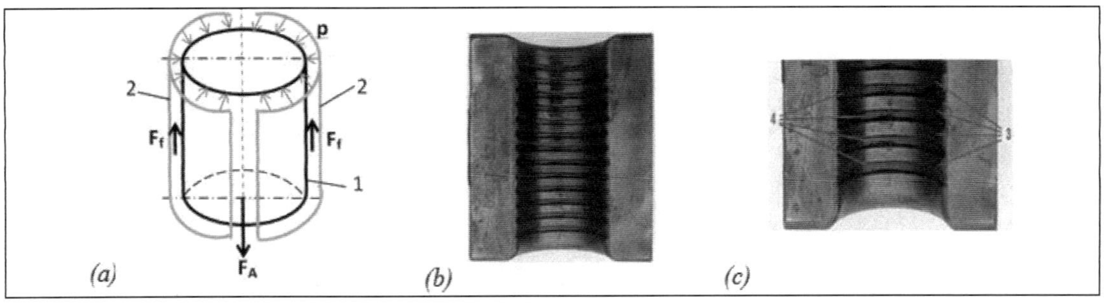

Figure 5. Injector assembly, Forces distribution schematics (a),clamp internal profile (b), geometry of clamp surface of contact with coiled tubing (c): F_f - slip friction forces, F_A- force generated by the weight of the column, p - pressure, 1. Coiled tubing, 2. Clamps, 3. Clamps usable surface of contact with coiled tubing, 4. Channel.

Injector functioning is based on slip friction forces (F_f) which occurs between the clamp and the external surface of the coiled tubing, as a result of perpendicular forces on the surface of contact is resulting the pressure p as presented in figure 5. During the extraction on the coiled tubing is auctioning a force (F_A) generated by the weight of the column:

$$F_A = m \times g \quad (1)$$

were: *m*, the mass of the column of flexible tubing inside of the bore well and *g*, the gravitational acceleration ($g = 9.81\ ms^{-2}$).

A proper functioning of the equipment requires to be fulfilled the condition:

$$F_A < F_f \quad (2)$$

According to [7] slip friction force is calculated with the relationship:

$$F_f = \mu \times A_r \times \tau_f \quad (3)$$

where: μ is the coefficient of slip friction at the contact of clamp with the coiled tubing; A_r is the area of contact between clamp and coiled tubing; τ_f – shear resistance of coiled tube material.

The value of slip friction force depends on slip friction coefficient μ, which could be affected by several factors such as clamp geometry, greasing material, clamp or coiled tube oxidation, the type of material used for clamps etc. The recommended values for slip friction coefficient μ are μ = (0.07 ...0.16) [1].

According to [7] shear resistance τ_f of the coiled tubing material is calculated by using formula:

$$\tau_f = (0.2 \ldots\ldots 0.3) \times R_{p0.2} \quad (4)$$

where: $R_{p0.2}$, the flow limit of the material..

Based on the previous presented issues it was decided to focus the research on the establishment on impact of stress affecting the geometrical characteristics of coiled tubing. According to the manufacturing specifications the coiled tube is made off A 606 steel, and based on the ASTM A standard the mechanical characteristics of this material are presented in table 2. The experiments were conducted on coiled tubing samples as presented in figure 6.

Table 2. Physical characteristics of A606 steel [8]

Material	Mechanical characteristics (min)		
	Traction resistance R_m, MPa (ksi)	Elasticity limit $R_{p\,0.2}$, MPa (ksi)	Elongation, %
A606	450 (65)	310 (45)	22

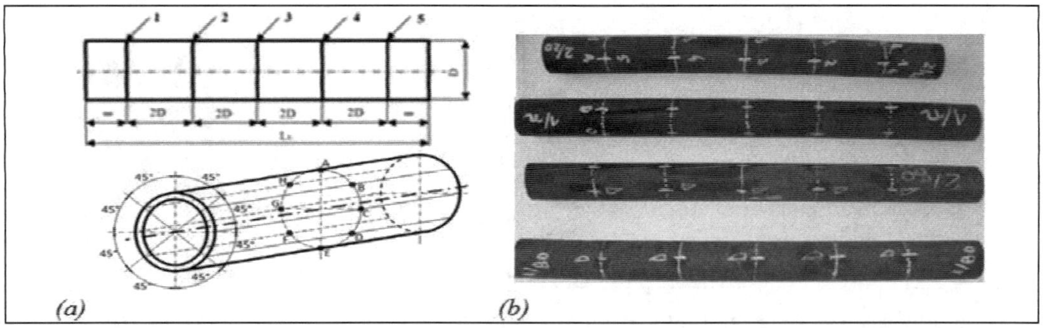

Figure 6. Geometrical characteristics of samples, sample design (a), Four different samples (b): *D* – External diameter, L_1 – minimum length of tested sample, *1, 2, 3, 4, 5* – Five different zones for each being measured external diameter and wall thickness.

IOP Conf. Series: Materials Science and Engineering **724** (2020) 012004 doi:10.1088/1757-899X/724/1/012004

The equipment used for tests is INSTRON 600 LX. The tests results are presented in table 3. The values obtained during the conducted tests (table 3) are certifying that the material is fulfilling the provision of ASTM A 606 Standard as presented in table 2. According to the types of samples measured (figure 6) and exploitation conditions (table 3) there were conducted measurements of geometrical characteristics of sample P3 from coiled tubing as presented in table 4.

Table 3. Experimental results

Sample code[a]	Measured characteristics[b]
P1 – sample taken from coiled tubes with 0 hours of functioning	Elasticity limits $R_{p\,0.2}$ = 509 (MPa) Traction resistance R_m= 645 (MPa) Elongation = 22.78 (%)
P2 – sample taken from coiled tubing used for 30 operational cycles	Elasticity limits $R_{p\,0.2}$ = 502 (MPa) Traction resistance R_m= 636 (MPa) Elongation = 22 (%)
P3 – sample taken from coiled tubing used for 80 operational cycles	Elasticity limits $R_{p\,0.2}$ = 435 (MPa) Traction resistance R_m= 587 (MPa) Elongation = 22 (%)

[a] Geometrical characteristics of samples are: external nominal diameter, D = 38.1 (mm), nominal wall thickness, t =2.76 (mm), sample length L = 400 (mm).
[b] Presented values are calculated as the average of tree measurements conducted in similar conditions.

Variations of coiled tubing external diameter correlated to each measured areas ("1", "2", "3", "4", "5") and each marked position (A-E, B-F, C-G, D-H) is presented in figure 7.

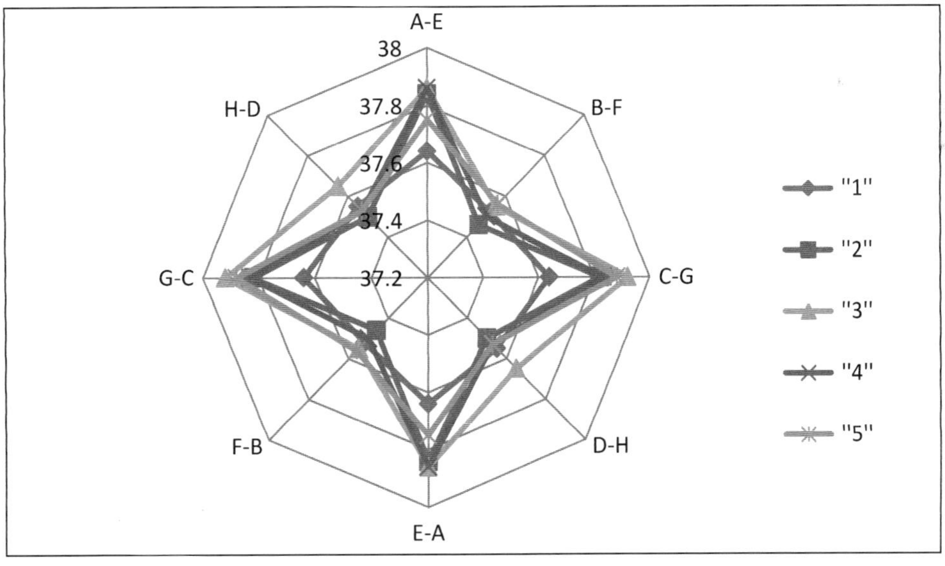

Figure 7. Variations of coiled tubing external diameter, D for sample P3.

Table 4. Measurements of geometrical characteristics conducted on Sample P3.

Sample P3 / (D = 1 1/2 in, 38.1 mm), 80 cycles of operation						
Measured values of wall thickness, t (mm)						
Position	"1"	"2"	"3"	"4"	"5"	Average
"A"	2.820	2.850	2.840	2.980	2.800	2.858
"C"	2.700	2.900	2.950	2.840	2.750	2.828
"E"	2.920	2.980	2.970	2.980	2.970	2.964
"G"	2.820	2.890	2.940	2.850	2.900	2.880
Average	2.815	2.905	2.925	2.9125	2.855	2.882
Eccentricity (%)	$E = [(t_{max} - t_{min}) / t_{med}] \times 100$					9.859
Ovality (%)	$O = [(D_{max} - D_{min}) / D_{med}] \times 100$					1.220
External diameter measured values, D (mm)						
Position	"1"	"2"	"3"	"4"	"5"	
A-E	37.640	37.840	37.860	37.860	37.750	
B-F	37.520	37.460	37.550	37.500	37.560	
C-G	37.640	37.820	37.920	37.830	37.880	
D-H	37.550	37.500	37.650	37.530	37.530	
Average	37.588	37.660	37.745	37.680	37.680	
Wall thickness, t (mm)			External diameter, D (mm)			
t_{max}		2.980	D_{max}		37.920	
t_{min}		2.700	D_{min}		37.460	
Average, t_{med} (mm)		2.840	Average, D_{med} (mm)		37.690	

Geometrical characteristics are used in the process of state of health estimation of coiled tubing due to their impact on real surface of contact with the injector clamps A_r calculated with ecuation (3), as well as wall thickness (t) and ovality (O).

4. Case study

The continuous monitoring of mechanical and geometrical characteristics of coiled tubing it is a key element in establishing the state of health of this equipment and the data collected during all these measurements could be used with a high degree of confidence in the prediction of operational life of the coiled tubing.

This case study has the goal of establishing the requirements for establishing the operational status of coiled tubing (2). The input data of experiments are presented in table 5.

In order to verify the state of health of the equipment (injector capability to sustain the coiled tubing weight) as presented by (2), in table 6 there are presented calculation elements and in table 7 surface of contact area. The loading mass is considered as for a coil of tubing with total length of 3300 m, external diameter 38.1 mm (1.5 inch), wall thickness 2.84 mm and specific weight, 2.41 kg/m.

Table 5. Input data

No. crt.	Characteristic	Calculation formula	Measured value	Observations
1	Average, external nominal diameter of coiled tubing	D_{med}	D_{med} = 37.69 mm	According table 3
2	Average value of coiled tube wall thickness	t_{med}	t_{med} = 2.84mm	According table 3
3	Area of contact between clamps and coiled tubing	A_r	$A_r = 56176 \times 10^{-6}$ m^2	There are used 9 pairs of clamps see figure 5, *c*
4	Slip friction force coefficient at the contact clamp-coiled tubing	$\mu = (0.07 ..0.16)$	$\mu = 0.1$	Contact surfaces are greased according to exploiting instructions.
5	The flow limit of material	$R_{p0.2}$	$R_{p0.2}$= 435 MPa	There it was considered minimum value experimental measured (Table 2)
6	Shear resistance of material	$\tau_f = (0.2 \ldots 0.3) \times R_{p0.2}$	$\tau_f = 0.2 \times R_{p0.2} = 0.2 \times 435 = 87$ MPa	

Table 6. Calculation elements

No. crt.	Wall thickness, g (mm)	Internal diameter of clamp, D (mm)	Load mass, m (kg)	Axial force $F_A = m \times g$ (N)	Slip friction force $F_f = \mu \times A_r \times \tau_f$ (N)
1.	2.59	32.91	14234	139635.54	
2.	2.76	32.56	15133	148454.73	
3.	2.84	32.10	16276	159667.56	
4.	3.17	31.75	17157	168310.17	
5.	3.40	31.29	18269	179218.89	488731.2
6.	3.68	30.73	19613	192403.53	
7.	3.96	30.17	20929	205313.49	
8.	4.45	29.21	23144	227042.64	
9.	4.77	28.55	24619	241512.39	

Table 7. Surface of contact area

Wall thickness, *g* (mm)	2.59	2.76	2.84	3.17
Load mass, *m* (kg)	14234	15133	16276	17157
Axial force, F_A	139635.54	148454.73	159667.56	168310.17
Surface of contact, $Ar \times 10^{-6}$ (m^2)	0.1×Ar	0.2×*Ar*	0.3×*Ar*	0.4×*Ar*
Slip friction force, F_f (N)	48873	97746	146619	195492

5. Conclusion and future work

According to the results presented in table 6 for t = 2.84 mm and a column of coiled tubing with the weight of 16276kg is resulting a force $F_A = 159667.56$ N < $F_f = 488731.2$ N. This results in a loading of about 33% and a reserve of 64%. The reserve was specially designed for special situation when the coiled tubing is stacked in the bore well or it is used for recovering stacked tools inside of the bore well and there it is a need for a higher axial force F_A.

One very important factor in the equipment well-functioning as presented in formula (2) is the surface of contact between the clamp and the coiled tubing (Ar). Great impacts of this surface have the clamp geometry as well as the ovality of the coiled tubing. Thus it is very important to monitor both deformation of clamp and the ovality of the coiled tubing. Considering this, a brief calculation will lead to an acceptable maximum deformation which still provides 40% contact between the coiled tubing and clamps ($0.4 \times Ar$).

Another important factor which main lead to malfunctioning of the device could be some local transversal deformations/constrictions, generating also reductions of the surface of contact between the clamp and the coiled tubing (Ar)

The results obtained in this article as well as the one presented in [9] are to be used in an intelligent system for coiled tube state of health monitoring.

6. References

[1] *** https://wildwell.com/wp-content/uploads/Coiled-Tubing.pdf
[2] Drumeanu A C and Rîpeanu R G 2010 Metallic element design of the tribo-thermal stressed dry friction couples, *J. of the Balkan Tribological Association*,**16** (3) p.362-372
[3] Ionescu G et al 2013 Wireless AIN sensor for condition based monitoring of industrial equipment, *CAS 2013 vol.I (International Semiconductor Conf.)*, p. 55-58
[4] Ionescu O N et al 2013 Research on life prolongation for temperature sensors placed in highly erosive environment, *J. of the he Balkan Tribological Association*,**19** (2) p.214 -220
[5] *** https://www.youtube.com/watch?v=kTD2N83aN3o
[6] *** https://www.reasontek.com/oilfield-equipment/coiled-tubing-injector-chain-assembl
[7] Gafițeanu M et al 1981 *Organe de mașini*, **I**, ed.Tehnică (București)
[8] *** ASTM A 606 Standard Specification for Steel, Sheet and Strip, High-Strength, Low-Alloy, Hot-Rolled and Cold-Rolled, with Improved Atmospheric Corrosion Resistance
[9] Hagianu A et al 2019 Research and studies on flexible tubing wearing, *Proc.16th International Conference on Tribology (*Kragujevac, Serbia) p. 321-330

IOP Conf. Series: Materials Science and Engineering **724** (2020) 012005 doi:10.1088/1757-899X/724/1/012005

Oil temperature influence on friction losses in silent chains

R Velicu

Transilvania University of Brasov, Romania, Dept. of Product Design Mechatronics and Environment

Abstract. Evaluation of friction in chain drives is the first step in the process of reducing the friction. This paper presents the experiments on the analyses of temperature influence on friction in silent chains. A rig dedicated for friction measurements on parallel axes transmissions is used. Friction is measured as friction torque with possibility to control and measure the input rotational speed, chain tensioning, pressure and temperature of the oil for chain lubrication, pressure and temperature of the oil for bearings lubrication. Temperature depending viscosity of chain lubrication oil is measured separately. The functioning conditions of the bearings are maintained constant during the testing and only the temperature of chain oil lubrication is increased and then decreased in the range of functioning temperature of a timing chain. Oil temperature influence on friction from silent chains joints gives an image on the type of lubrication. The reduction of silent chain friction by reducing the functioning temperature may be a very simple solution and this paper shows how much power may be saved. Friction torque is measured for three distinct rotational speeds in the same tensioning parameters. Conclusions on the influence of rotational speed on friction losses with temperature are finally drawn.

1. Introduction

The importance of reducing friction in automobiles transmission is as high as possible in the context of the regulations imposed by the 20/20/20 goals from the Directive for Energy Efficiency of the European Union (2012/27/UE), imposing targets for year 2020, like: 20% reduction of CO_2 emissions compared to 1990 levels and 20% reduction of final energy consumptions.

In the same directions have been moved the goals of automotive industry for year 2020. Targets in fuel consumptions and also in the equivalent CO_2 emissions have been drawn. American standards CAFE (Corporate Average Fuel Economy) impose, for 2020, a one third reduction of fuel consumption, relative to year 2008. European Commission set a goal for 2021 of average CO_2 emissions of 95 grams CO_2 / km for the whole fleet produced in 2020.

Bush chains are used in the timing system of combustion engines, where tensioning and speed reach some of the extreme values of the field of use. Reducing friction in timing chains is an important goal since, according to [1, 2], the part of the fuel energy dedicated to overcome friction in the engine system is 7–18% of total losses. This is why, it is of high importance to evaluate influences on chain friction in order to take the right measures for reducing it.

Results on measuring losses in the timing chain and the loss in the guides of an engine, using equipment consisting in a full engine are presented in [2]. The rig allows a good replica of the real engine conditions. The contribution of chain friction and guide friction on the global timing system friction are only presented for one value of oil temperature.

Friction in chain drives friction is subject of several papers referring to friction coefficients [3], friction contribution [4] or theoretical approaches [5].

Research on friction in chain drives, mainly used for the timing system of combustion engines, has been developed with the finance of Schaeffler Group, at the Tribology Centre of Transylvania University of Brasov. For a basic chain drive with identical sprockets, the procedure for measuring chain friction is presented in [6]. The chain rig used for these experiments is presented in figure 1. Friction losses are measured as global friction torque depending on speed, tensioning and oil temperature, based on the diagram from figure 1, a.

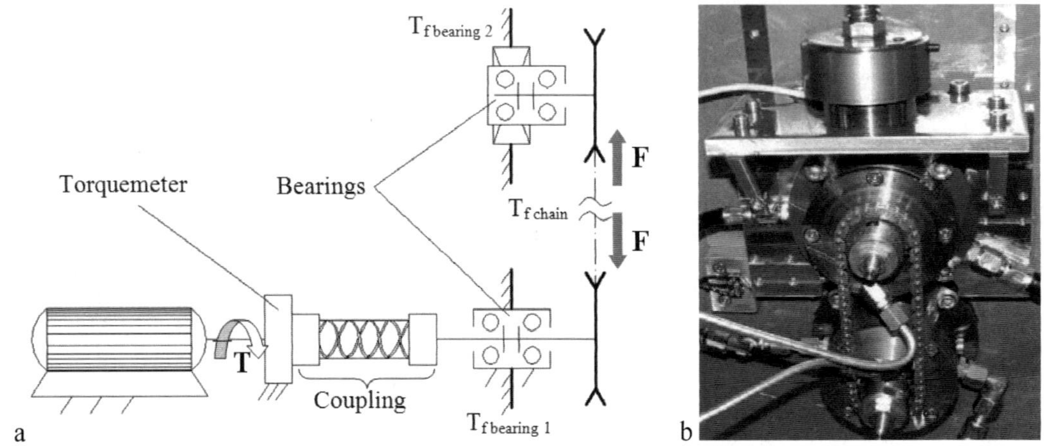

Figure 1. Chain friction rig: a – functioning diagram, b – front picture of the chain drive [6].

The aim of the paper is to experimentally determine the influence of oil temperature on the friction in the chain, without considering bearing friction. Influence of speed is also investigated.

2. Experimental on temperature influence on bush chain friction

The tests have been developed on the chain rig described in [6], based on the procedures presented in the same reference.

The subject of the test is a silent chain with 8 mm pitch, 64 links, with identical 23 teeth sprockets. Tensioning is created by enlarging the centre distance. For all the tests, the tensioning is automatic, ensuring a constant tensioning force F of 1 kN. The transmission is a vertical basic transmission with transmission ratio equal to one.

Lubrication of the chain is made with two jets of oil oriented to the points of chain entering the two sprockets (see figure 1, b). Lubrication of the bearings is made with a lower pressure flow of oil inside the bearing boxes. The rig has two separated circuits for lubrication of chain and bearings. Temperature and pressure in each circuit are measured with a precision of ± 0.1 °C. The temperature of oil in each circuit is controlled with a maximum deviation of ± 2 °C. The oil used for tests is fresh 5W30 Castrol Edge synthetic oil. Figure 2 presents the measured dynamic viscosity η of the oil depending on temperature. The same oil is used for chain and bearing lubrication.

Global friction torque (including friction resistance in chain and bearings) is measured for constant speed and tensioning, automatically keeping constant the temperature of the bearing oil and only changing the temperature of the oil for chain lubrication. Since the conditions of working of the bearings are kept constant, the change in friction is only coming from temperature influence on chain friction.

With a software controlled interface, the averaged values of the measurements (time, friction torque, chain oil temperature and pressure, bearing oil temperature and pressure, tensioning, rotational speed) are saved at each 5 seconds. The database is accessed as an .xls file.

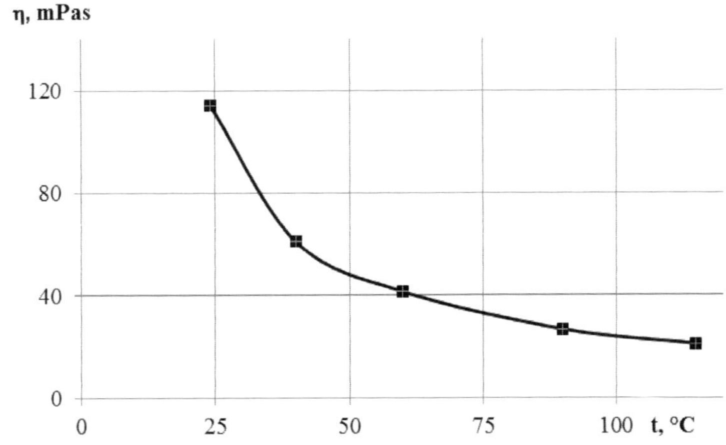

Figure 2. 5W30 Castrol Edge oil dynamic viscosity vs. temperature.

The tested chain has been previously subject of running in for 50 hours, at constant tensioning of 1 kN, 2000 rpm rotational speed and 50 °C chain oil temperature.

Tests have been developed for constant bearing oil temperature of 60 °C, constant tensioning of 1 kN. The chain oil is heated from 70 °C to 100 °C and then cooled back to 70 °C. The tests have been repeated 3 times separately for three values of rotational speed, 1000, 2000 and 4000 rpm.

Figure 3 presents time depending chain and bearings oil temperature for constant 2000 rpm rotational speed.

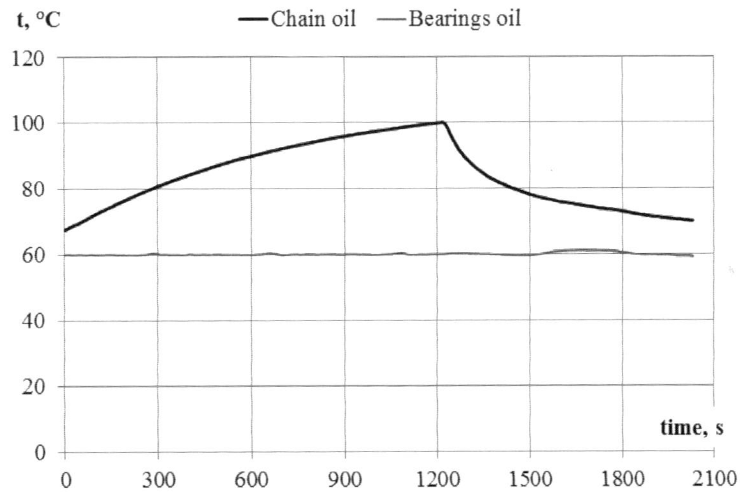

Figure 3. Time depending chain and bearings oil temperature during a test.

Figure 4 presents, for the same test, time depending chain and bearings oil pressure.

The analyses of the results presented in figures 3 and 4 allow us to draw few conclusions:

- Chain oil temperature is increasing slower and faster decreasing, especially at the beginning of cooling;
- Chain oil pressure decreases with increasing temperature and increases back during cooling; we tested this influence on friction and it is not important.

Figure 5 presents the measured global friction torque depending on chain oil temperature, for the two stages of increasing temperature and decreasing temperature.

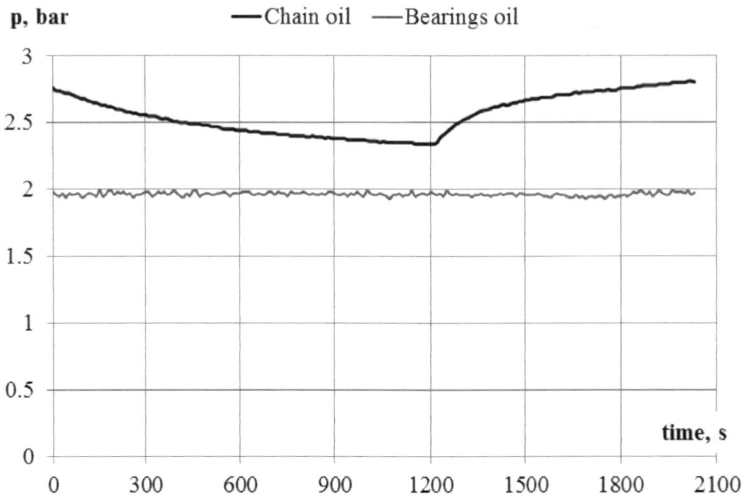

Figure 4. Time depending chain and bearings oil pressure during a test.

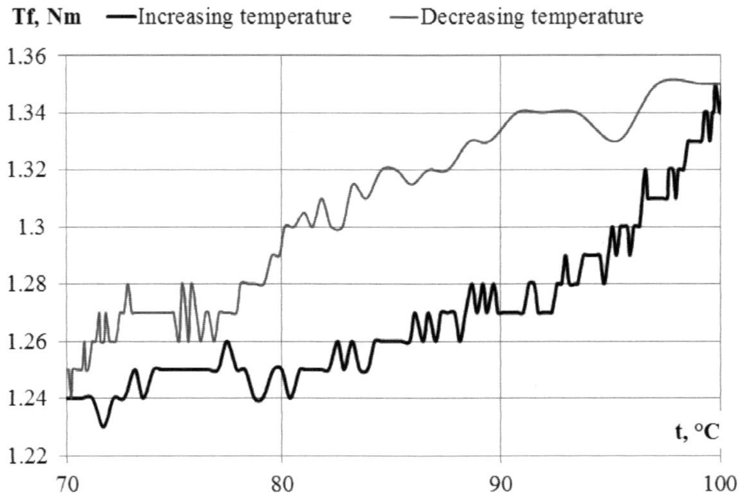

Figure 5. Global friction depending on chain oil temperature.

The change of global friction comes from the change in chain friction since the bearing friction conditions are about the same. This means that chain friction increases with increasing oil temperature (decreasing oil viscosity). Friction in the cylindrical joints of the chains is the main friction that must be considered [7]. According to Stribeck's curve, this variation shows the presence of boundary or mixed friction in the joints of the chain. A decrease of chain oil temperature from 100 °C to 70 °C determines a decrease of chain friction of about 9 % of the global friction in this basic transmission.

The values of the global friction torque are bigger during decreasing temperature than during increasing temperature. The difference could be explained by the difference between the temperature of the oil and the temperature of the parts involved in friction (links, sprockets, even bearings casings). During increasing temperature, temperature of the parts is lower than the measured oil temperature, so at the contact in joints, conditions of lower temperature create lower friction. For the cooling process, temperature of the parts is bigger than the measured oil temperature, so, at the contact in joints, conditions of higher temperature create higher friction.

The difference between measured torque at 100 °C and measured torque at 70 °C, ΔTf 100-70, represents chain friction reduction with reducing chain oil temperature. Figure 6, a presents ΔTf 100-70 resulted from tests at different rotational speed. Figure 6, b presents the calculated gain in power obtained by reducing chain temperature from 100 to 70 °C, ΔP 100-70. Influence of speed on torque reduction is not very important, but at level of power reduction it becomes really important.

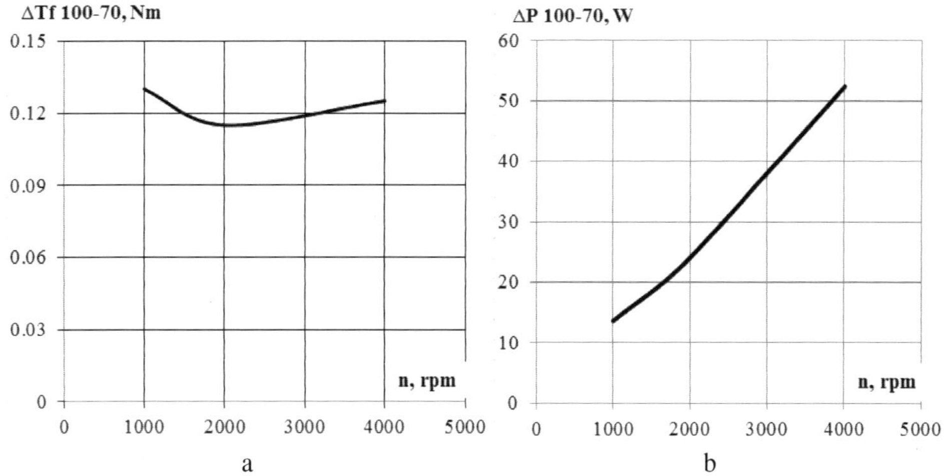

Figure 6. Torque (a) and power (b) gain reducing chain temperature from 100 to 70 °C vs. rotational speed.

3. Conclusion

This paper shows that friction in silent chains increases with increasing oil temperature.

The reduction of silent chain friction by reducing the functioning temperature of the chain may be a very simple direction in improving the efficiency of combustion engines and this paper gives information on how much power may be saved with this solution.

4. References

[1] Holmberg K, Andersson P and Erdemir A 2012 Global energy consumption due to friction in passenger cars *Tribol. Int.* **47**

[2] Hyakutake T, Inagaki M, Matsuda M, Hakamada N and Teramachi Y 2001Measurement of friction in timing chain *JSAE Review* **22:5**

[3] Papuc R, Gavrila C and Lates M T 2017 Tribological tests of steel on polyamide 66, polyamide 46 type surface contacts *MATEC Web Conf.* **126** 02003

[4] Papuc R, Gavrila C and Lates M T 2017 The study of chain link/guide contact in relative alternative translational motion *MATEC Web Conf.* **126** 01003

[5] Lates M T, Velicu R G and Papuc R 2014 Multiscale modeling of chain-guide contact by using tests and FEM *WCCM 2014*

[6] Velicu R and Lates M 2015 On the measurement procedure for testing friction in bearing mountings *AUO-FMTE* **XXIV (XIV)**

[7] Niemann G 1975 *Machinenelemente Band II* (Berlin: Springer Verlag)

IOP Conf. Series: Materials Science and Engineering **724** (2020) 012006 doi:10.1088/1757-899X/724/1/012006

Experimental Determination of the Loading Capacity of the Elastic Bracelet Assembly

I Geonea[1], N Dumitru[1], C Copilusi[1] and A Margine[1]

[1]University of Craiova, Faculty of Mechanics, e-mail: igeonea@yahoo.com

Abstract. In this paper we aim to experimentally determine the bearing capacity (the momentum) of the elastic bracelet assemblies. The moment of torsion transmitted by the assembly depends on the clamping force of the assembly and on the friction coefficient between the bushing mounted on the shaft and the hub of the assembly. An experimental stand is used which comprises a hub shaft assembly with a segmented elastic bracelet. The load bearing capacity of the elastic bracelet is determined depending on the strength of the screws. Experimental determination is made on three types of materials. For this purpose, on the shaft is assembled a bush made of different materials: aluminum, bronze, steel. In order to determine the capable assembly moment, strain gauge transducers are used to measure the elastic strains. The recording of the deformations is done by means of an efficient data acquisition system. The assembly characteristic is plotted, namely the dependence between the torque transmitted according to the tightening of the bolts, for three couples of materials.

1. Introduction

The aspect of experimental tests of assemblies is the subject of numerous published articles. Thus, experimental researches are presented regarding the experimental determination of the bearing capacity of the cone clamping shaft assemblies [1], the experimental determination of the efficiency of the cylindrical gears [2], the experimental determination of the friction torque in the threaded assemblies [3] or the experimental determination of the friction torque in the bearings [4].

In this paper we aim to perform the experimental determination of the bearing capacity of the hub shaft assembly with elastic brace according to the clamping force of the clamping screws. The elastic bracelet, also called bracket or clamp, is part of the shaft-hub assemblies by tightening. The operation of these assemblies is based on the frictional forces occurring between the elements, resulting from the action of normal forces on the contact surfaces (figure 1 and figure 2). The normal force required to transmit the torque is achieved with the clamping screws. The magnitude of the friction forces depends on the value of the friction coefficient μ and it is influenced by a number of factors: the pair of materials, the macro and micro geometry of the surfaces in contact, the existence and nature of the lubricant, the existence of oxides, the existence of vibrations, etc.

Elastic brace assembly is used to lock circular shafts of cranks, bearings, transmission wheels, etc. when the load capacity is reduced or parts require frequent adjustment. These joints allow mounting and adjustment of the conjugated parts in any angular or axial position along the length of a constant diameter shaft section. The bracelet connections can be with cover (figure 2) or with the unilaterally cut hub (figure 1) when is called clamps. The contact surface of the joint parts may be smooth or roughed.

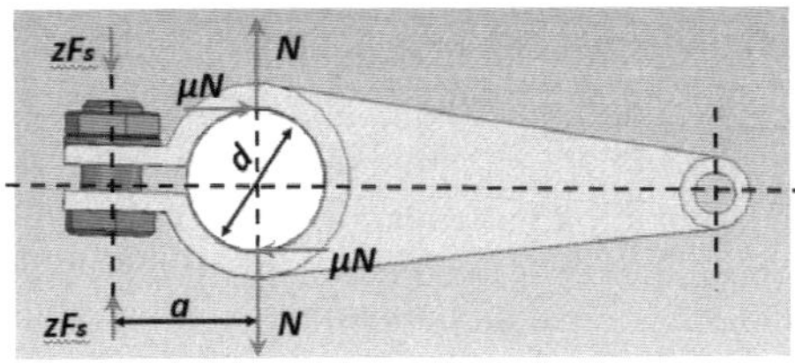

Figure 1. Elastic bracket from one part.

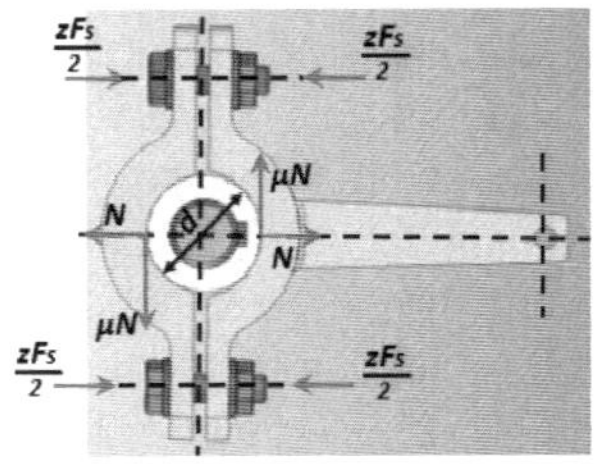

Figure 2. Elastic bracket from two parts.

The bearing capacity of the joint (the capable moment of transmission) in both constructive variants can be expressed in terms of the tightening torque in the bolts and the displacement law that is accepted for the contact stresses between the joint surfaces of the shaft and the hub.

Assuming that the contact stresses are evenly distributed over the contact surface, the momentum of the assembly is:

- for elastic bracelet with cover [5]:

$$T = \frac{z \cdot \mu \cdot \pi \cdot d \cdot F_S}{2 \cdot k} \text{ [N·mm]} \quad (1)$$

- for the elastic bracelet sectioned [5]:

$$T = \frac{z \cdot \mu \cdot (2 \cdot a + d) \cdot \pi \cdot F_S}{2 \cdot k} \text{ [N·mm]} \quad (2)$$

where: T - the momentum of the assembly, [N·mm]; z-number of elastic bracelet clamping screws; μ- the coefficient of friction between the shaft and the elastic brace; d- the diameter of the shaft on which the bracelet is mounted, in [mm]; k-safety coefficient of the assembly. Typically, k = 1,2 ... 1,5 and in shock operation k = 2 ... 4. F_s - the tightening force of the bolts, in [N] (see figure 2); a=145 mm, is the distance from the axis of the clamping screws to the assembly axis, in [mm];

2. Experimental setup

An experimental test bench is made to determine the load capacity of the assembly. The actuation is by turning the crank 1 which drives the worm and worm gear 2-3 mounted in the casing 6 that is attached to the base 7. The worm wheel 3 is provided with the nut 4 acting on the force bolt 5. This cannot be rotated thanks to the stop 8, and will perform a translation motion. This movement is transmitted via pin 9 to the transducer 10 which, by means of the bolt 11, acts on the arm of the elastic bracket 12. The brace cover 20 is fastened to the body 12 by means of two screws 13. The bolts are tightened by means of the nuts 14 via the transducers 15 and the special washers 16.

The shaft on which the assembly is fastened is constituted by the bush 18 (there are four identical bushings of various materials) which are mounted by means of the parallel wedge 17 on the fixed shaft 19.

The clamping forces of the bolts 13 are measured by means of the force transducers formed by the bush 15 and the washers 16. On the outer surface of the sleeves 15 there are applied two strain gauge transducers (TERs) disposed in a half bridge configuration and connected to an acquisition system.

With the aid of the transducer 10, on which two other strain gauge (arranged in a half Wheatstone bridge) are applied, and the force acting on the arm of the brace is measured. This force acting at a distance of 154.5 mm from the center of the assembly creates an active moment, opposed to the moment of friction forces that develop on the bracelet-tree contact surfaces.

To determine the possible moment of assembly, is proceed as follows:

a) one of the bushings 18 is mounted on the fixed shaft 19;

b) installing the elastic brace without tightening the nuts 14 until the clearance between the parts is cancelled;

c) the transducer 10 is mounted with bolts 9 and 11;

d) zero balancing the electronic equipment;

e) tighten the bolts 13 taking care that the forces developing in them are equal (since the calibration diagrams of the two transducers are practically identical, it will be observed that the indications of the electronic equipment are the same); the indicated values are entered in the result table;

f) rotate evenly the crank 1 so that the screw 5 moves to the left (figure 3), thus tensioning the transducer 10. When the moment created by the force in the transducer equals the moment of the friction forces in the assembly, the slip occurs and the indication of the electronic device no longer grows. That value will be recorded as being appropriate to the capable moment;

g) Unload the transducer 10 by turning the crank in the reverse direction and unscrew the bolts 13. Change the tightening force of the screws and resume the experience.

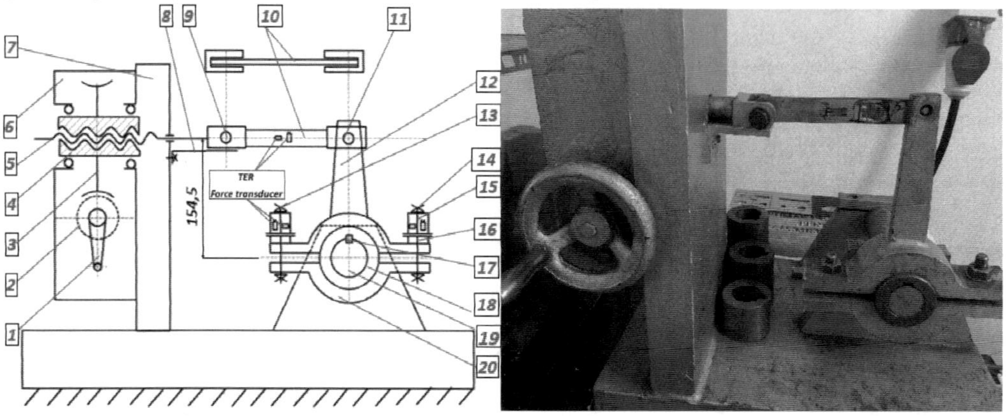

Figure 3. The experimental setup.

The load on bolts 13 will be limited to maximum 1900 N, that means $\varepsilon_1^i = \varepsilon_2^i = \max \ 400 \mu m / m$.

The experimental moment is calculated with the equation (3) [6]:

$$T_e = 154{,}5 \cdot F_3 \ [\mathrm{N \cdot mm}] \tag{3}$$

The equation (1) determines the theoretically capable torque depending on the tightening forces in the screws.

For each pair of materials, the graph of the torque variation is plotted on Excel, depending on the tightening force of the bolts (experimental and theoretical).

The figure 4 shows the experimental recording of force F_3, which acts in the end of the brace, corresponding to the clamping force F_s of 900 N. The recording is made for the steel and bronze used as pair of materials of the assembly. Note that the force increases to a certain value until slippage occurs. The value of the measured force multiplied by the distance a, gives us the momentum of the assembly.

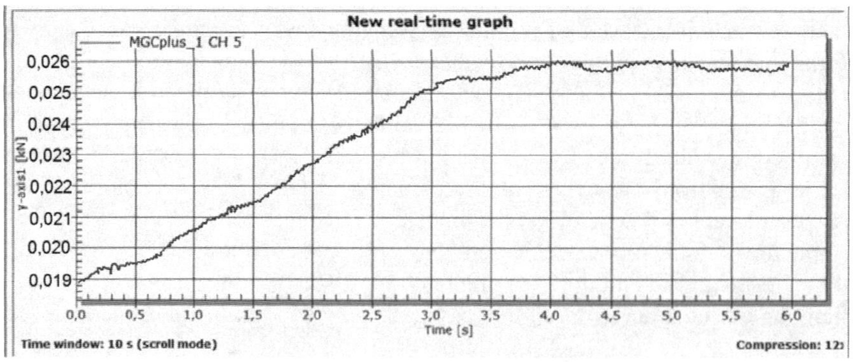

Figure 4. The experimental results.

The graph in figure 4 is obtained for the clamping force of 230 N. In order to determine the dependence of figure 5 is performed progressive tightening of the bolts with force F_s, for which the assembly capable torque is measured. It is noted that the torque transmitted by the assembly increases as the clamping force of the screws increases. Dependence is obtained for the couple of steel and bronze materials.

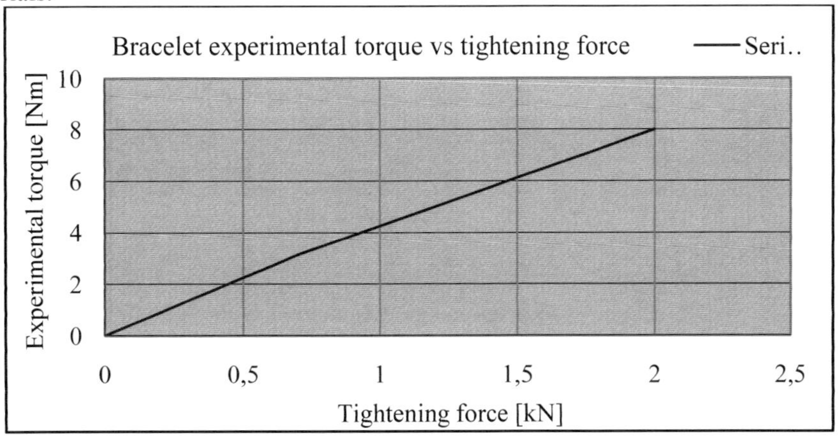

Figure 5. The experimental results.

3. Bracelet structural analysis

To perform the structural analysis of the elastic bracket assembly, we performed the simulation of the experimental stand in SolidWorks. The design of the experimental bench is shown in figure 6.

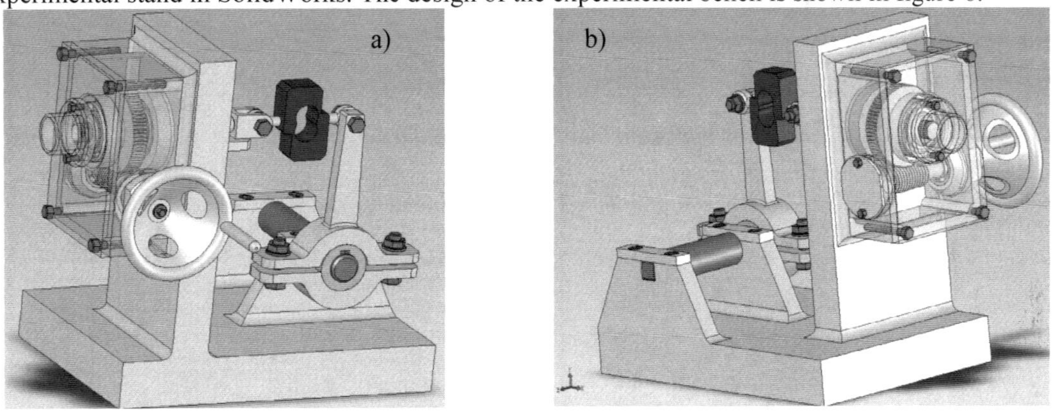

Figure 6. The experimental bench in SolidWorks.

The simplified model of the experimental bench is imported into the ANSYS finite element analysis program. Elements of component elements are defined, contact between elements is defined by the bonded type. The detailing of the contact between the components is shown in figure 7.

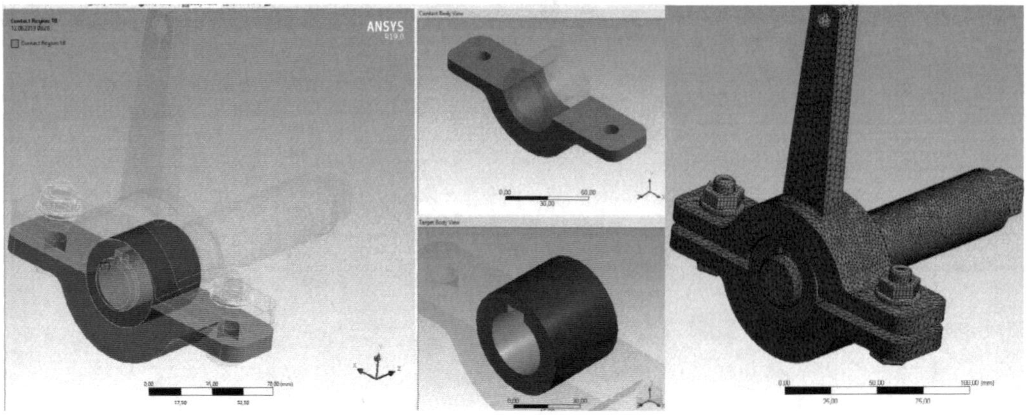

Figure 7. The contact details and mesh in ANSYS.

To perform the structural verification of the elastic bracket, a force of 2000 N is defined in the free end of the brace. As a fixed support is defined the assembly shaft.

The results of the finite element analysis, namely the distribution of the equivalent stresses and the distribution of the elastic strain, are shown in figure 8 and figure 9. The maximum elastic strain of the bracelet assembly is 0.00108 and the total deformation is 0.522 mm. The maximum amount of equivalent stress is 212 MPa.

Figure 8. The equivalent elastic strain distribution in ANSYS.

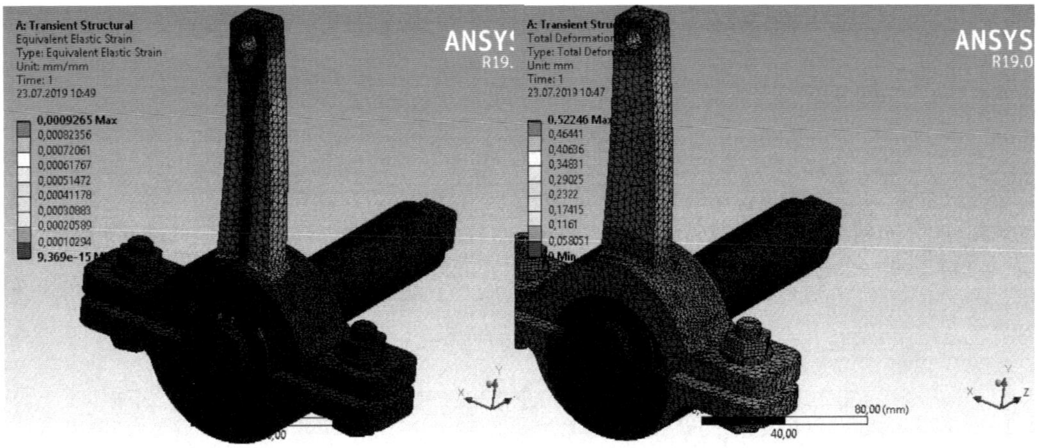

Figure 9. The equivalent von Mises stress and total deformation in ANSYS.

4. Conclusion

Elastic bracelets are used in the field of agricultural machinery. The carrying torque capacity of the assembly depends on the tightening force of the screws and by the material pair between the shaft and the hub. For the couple of steel-bronze materials, the load bearing capacity of the assembly is shown. Dependence is a linear one. Also stress analysis of the assembly is performed in ANSYS.

5. References

[1] Geonea I, Dumitru N, Copilusi C, Otat O and Tutunea D 2018 Theoretical and experimental analysis for the load capacity of the cone tightening assemblies *Proc. in Automotive Eng.* **1** 437-445

[2] Dumitru N, Dragut E, Craciunoiu N and Geonea I 2018 Experimental bench for spur gears efficiency measurement *Mech. and Mach. Sci.* **57** pp 477-485

[3] Geonea I, Dumitru N, Margine A and Copilusi C 2018 Experimental investigations concerning friction from threaded assemblies *App. Mech. and Mat.* **880** pp 33-38

[4] Geonea I, Dumitru N and Dumitru I 2017 Experimental and theoretical study of friction torque from radial ball bearings *IOP Conf. Ser.: Mater. Sci. Eng.* **252** 012048

[5] Mott R L and Tang J 1999 *Machine Elements in Mechanical Design* (London: Prentice-Hall) pp 17-150

[6] Catrina G, Margine A, Ungureanu C and Dumitru N 2009 *Machine Parts* vol 3 (Craiova: Universitaria) p 122

IOP Conf. Series: Materials Science and Engineering **724** (2020) 012007 doi:10.1088/1757-899X/724/1/012007

On thermal calculation for helical gear transmission system

I S Gabroveanu[1], S Cananau[2,3,], R F Mirica[2] and A A Ilies[1]

[1] National Research and Development for Gas Turbines SC COMOTI SA, 220 D Iuliu Maniu Bd., 061126, Bucharest, Romania
[2] Politehnica University of Bucharest, Machine Elements and Tribology Department, 313 Spl.Independentei, 060042, Bucharest, Romania

Abstract. The purpose of the study in the paper is to evaluate the thermal comportment of the transmission systems with helical gears. Also, the influences of heat on characteristics of gear transmission system, based on the meshing thermal stiffness of gears are analyzed. An experimental testing stand of the system will be used to obtain data. The respond of the transmission under different working conditions is also analyzed. The study concerns the comparison of theoretical results obtained using calculation methods (from literature) and the experimental results, in order to obtain an evaluation of the friction in the meshing gears. The heat (temperature) affects the dynamic characteristics of the system by the meshing stiffness of the gear. The experimental results are used to verify the correctness of the theoretical model.

1. Introduction

In the last years, the design of modern gear transmissions is concerning to improve energy-dense property of the new product. As consequence the development of new mechanical systems with gears require much detailed knowledge and experiments. One of interesting area of the gears transmissions research is to evaluate all kind of energy losses in the all detailed components in function in the system for a better running performance. Energy losses in a gearbox are converted into heat which must be dissipated. Thus, this heat is the main responsible for the temperature increase while the system is running. In this paper we are going to investigate, within a thermal model who evaluates each kind of power loss, the thermal calculation for gears.

Many authors were concern of this area of research: Durand de Gevigney et alt. are concern to study thermal *modelling* of a back-to-back gearbox test machine with an experimental FZG test rig [1], Hlebanja and Kulovec focusses on the differences in thermal load of cylindrical, spur, plastic involute and S-gears [2], Kanatnikov et alt. presents a model for predicting the thermal processes arising during shaping gears with internal non-involute teeth [3]. The variations of the temperature in the function time are also a subject of research interest. Luo and Fei focus on some common mechanical parts in precision technology and propose mathematical models for hollow piece, gear and cube. The experimental results also make it clear that these models are more logical than traditional models [4]. Other authors are involved in the study of comportment of the lubricants in these mechanical systems. Badrinarayanan et alt. said that fluid transmission losses and inherent friction, heat energy is generated in the system which rises the fluid temperature. This heat has to be dissipated by proper means to

[3] Author to whom any correspondence should be addressed

maintain the system temperature within safe limits. In their paper the authors attempt to study the thermal aspects of hydraulic system, heat generation and dissipation to estimate temperature rise profile and steady state temperature [5]. Other authors are interested to establish the dynamic behaviour of the gear pairs under different operating conditions, as Radu et alt. [6]

2. Thermal balance and heat dissipation

By hypothesis the gearbox at a *thermal balance*, heat evacuated by conduction, convection and radiation through the all mechanical elements of the transmission is a result of a power loss generated inside the system. So, the balance will be expressed as:

$$Q_{diss} = P_{loss} \tag{1}$$

Each part of this equation has the following explanation:

2.1. The heat dissipation

2.1.1. The heat dissipation by radiation. The heat dissipation Q_{diss} is a sum of evacuated heat by conduction, convection and radiation for all elements of the system that in time are supposed to be in the process of dissipation

$$Q_{diss} = \sum_{i=1}^{n} Q_{i,rad} + \sum_{j=1}^{m} Q_{j,conv} + \sum_{k=1}^{p} Q_{k,cond} \tag{2}$$

According to other works, example [1], the internal heat transfer is not taken into account for the stationary gearboxes. The dissipation by radiation is due to the heat flow from the surfaces of the gearboxes housing.

2.1.2. The heat dissipation by convection. The convection is due to the contact between the oil and the mechanical parts interaction, at various temperatures

$$Q_{conv} = c_{conv} \sum_{j=1}^{m} A_{j,conv} (T_{wall} - T_{oil}) \tag{3}$$

where c_{conv} is the oil convection coefficient, A_j is the area of a surface j involved in heat dissipation by convection, T_{wall} is gearbox housing temperature, T_{oil} is the fluid (oil) temperature.

2.1.3. The heat dissipation due to conduction is not taken into account in this paper.
Concerning the second part of the Eq.(1) we will take into account: *2.2.1.* teeth friction loss, 2.2.2. shaft-bearing power loss, *2.2.3.* shaft seal power loss, *2.2.4.* churning losses.

2.2. The power losses

2.2.1. Teeth friction losses. In the functioning conditions the pinion (driving) and the wheel (driven) teeth are in meshing. In this paper we assume the calculus for the teeth friction losses in conditions of mixed film lubrication, which means that the normal load for the teeth contact is not fully supported by EHD lubricant film. In the real case of functioning there are different stages of meshing with possible various film thickness, due to variation of the contact geometry, pressure and load value along the meshing line. For general purpose we consider the loss power in mesh as is considered in [1]:

$$P_{mesh} = \mu_f \cdot P_{in} \cdot H_v \tag{4}$$

where H_v is the geometry parameter:

$$H_v = \frac{\pi(u+1)}{(z_1 \cdot u \cdot \cos\beta_b)(1-\varepsilon)} \tag{5}$$

and P_{in} is the input power, μ_f friction coefficient.

Regarding the friction coefficient μ_f there is a lot of discussions and researches done to agree with an expression to fit a sustainable value of this parameter. We take for reference the expression of Höhn at al. [7] and developed by Castro and Seabra [8]. In this paper we adopt the following expression:

$$\mu_f = 0{,}171\left(\frac{W_L}{R'_{eq(C)} \cdot V_R}\right)^{0,2} m^{-0,05}\left(\frac{R_{aEq}}{d_1}\right) X_L X_C \tag{6}$$

where W_L is specific normal load, V_R is the rolling speed, $R'_{eq(C)}$ is the equivalent radius of curvature of the contact points of teeth (C), m is dynamic oil viscosity at oil temperature, d_1 is the pinion pitch diameter, R_{aEq} is the equivalent arithmetic average of the roughness of surfaces in contact, X_L is lubricant correction factor (function of the nature of additives) and X_C is a correction factor for gears with coated surfaces. (in our case, X_C is equal to unit).

2.2.2. Rolling bearing power loss. Taking into account the considerations due to Niemann et. all. [9] we will consider only the following causes for the friction in shaft-bearings system:

- rolling at a contact area, having different radiuses with elastic deformations and material hysteresis,
- sliding between guiding surfaces and rolling elements,
- lubricant internal friction, drag losses, churning, splashing,
- air resistance by high speed,
- rolling resistance caused by various contaminations in contact area.

In [9] a simplified calculation method is obtained. The frictional moment can be estimated with sufficient accuracy using a constant coefficient of friction, μ_{bearing}, the mean diameter of rolling bearing, d_m, and the equivalent dynamic bearing load, F_{eq}: According to SKF [10], the recommend following formula to estimate the frictional torque, T_{fr}, is:

$$T_{fr} = 0{,}5 \cdot \mu_{bearing} \cdot d_m \cdot F_{eq} \tag{7}$$

where $\mu_{bearing}$ = 0.002 (for spherical roller bearings that are used in experimental rig) is an approximated constant coefficient of friction for the bearing, F_{eq} is the equivalent dynamic bearing load. SKF [10] recommend the same formula but using $\mu_{bearing} = 0.0018$.

The power loss is estimated using the angular velocity of shaft, ω:

$$P_{fr} = T_{fr} \cdot \omega \tag{8}$$

2.2.3. Shaft seal power losses. In many studies researchers evaluate the seal power loss with less influence compared to other losses in the period of functioning. Seal losses are the power losses due to friction between the shaft and its seal. We must pay attention to consider, in our study, both gearboxes with possible heat dissipation. We take into account in this paper the expression from Höhn et al. [7]:

$$P_{seal} = 7{,}69 \cdot 10^{-6} \cdot d_{sh} \cdot n \tag{9}$$

where d_{sh} is the diameter of the seal.

2.2.4. Churning losses. Churning losses is also a subject of various opinions from many researches. This is a consequence of the experimental set up of the test gear, for each research considered. For example, there are different designs for the case of the gears. By other hand, the expression of this calculation is in direct relation with the geometry of housing, the design and also the dimensions. In this paper we had as reference the work of Hai Xu and Kahraman [11], Changenet and Velex [12].

The expression we will use is obtained from [12]:

$$P_{churn} = \frac{1}{4}(d_{w2} \cdot \omega)^3 \cdot \rho \cdot A \cdot C_m \qquad (10)$$

where ω is the angular velocity, ρ is the oil density, A is the lateral and teeth immersed surface of contact between gear and lubricant, C_m is the dimensionless drag torque. In our case, for the calculus, we will consider that the gear is immersed in lubricant with a dimension of three times the gear module.

3. Validation of presented method

With a view to validate the previously presented method, some experiments have been effectuated. The used test rig is a testing machine with closed power circuit, ensuring the application of different torques, in the range 0 - 200 Nm and number of revolutions in the range of 1000 - 3000 rpm. The test rig was instrumented with: optical precision encoder, ROC 425 type, Heidenhain production, Germany, two pieces; torque converter without contact T10FS Hottinger Baldwin Messtechnik, precision class 0.05%; lines for measuring the oil temperature in the lubrication baths of the two gearboxes, test and return; AC electric motor speed measurement. In figure 1 is shown the test rig. The geometry of tested gears and other characteristics used in the calculation method are given in table 1.

Figure 1. Test rig.

Table 1. Geometry and other characteristics of the test rig.

Characteristic	Notation	Data	Characteristic	Notation	Data
Centre distance [mm]	a_w	125	Pinion hardness	HB	270-290
Pinion teeth number	z_1	15	Wheel hardness	HB	290-310
Wheel teeth number	z_2	46	Flank roughness [μm]	Ra	0.4
Helix angle [°]	β	10	Type of the spherical roller bearings	-	21310 E
Profile shift coefficient at pinion	x_1	0.427			
Profile shift coefficient at wheel	x_2	-0.138	Distance between bearings [mm]	-	200
Face width [mm]	b	72.37	Oil type	-	H46EP
Transverse contact ratio	ε_α	1.4532	Oil volume [l]	V_{oil}	6
Overlap ratio	ε_β	0.871	Oil density [kg/m³]	ρ	877
Gear accuracy grade	-	5	Mean of thermal capacity of oil [kJ/kg K]	c_p	2.121
Gear material (case hardened)	-	41MoCr11			

At pinion torque values of 50 Nm, 100 Nm and 150 Nm are determined the heat quantity accumulated in oil sump at number of revolutions of 1000 rpm and 2000 rpm. The temperature increase is registered. The convectively dissipated heat through housing of test rig was neglected in this phase.

Using the experimental thermal capacity of oil

$$c_p = 1.826 + 4.213 \cdot t_{med} / 1000 \tag{11}$$

the temperature variation, Δt and test duration (time), the power loss gets into oil sump is calculated:

$$P = Q / time = \rho \cdot V_{oil} \cdot c_p \cdot \Delta t / time \tag{12}$$

The experimental data have been centralized in the table 2. In the figures 2 and 3 are shown the calculated power losses using the presented method.

Table 2. Experimental power loss [kW] / at mean temperature [° C].

Torque [Nm]	50 Nm	100 Nm	150 Nm
Number of revolutions [rpm] / time [min]			
1000 / 10	0.019 / 38.85	0.053 / 25.25	0.053 / 32.75
1000 / 30	0.04 / 27.5	0.045 / 35.7	0.064 / 29.95
2000 / 10	0.089 / 27.675	0.091 / 29.45	0.096 / 37.05
2000 / 30	0.067 / 36.75	0.08 / 46.35	0.102 / 44.3

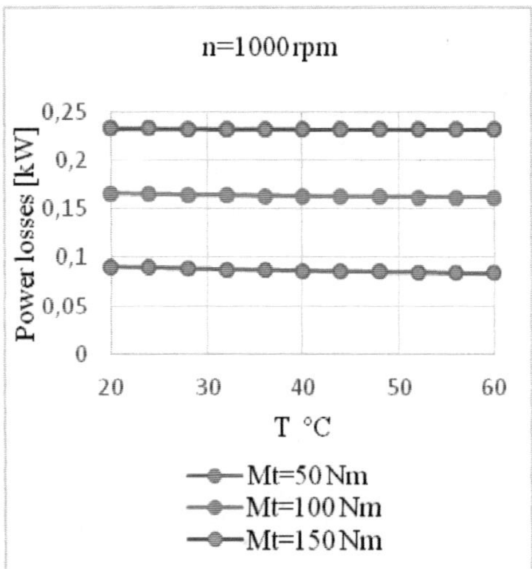

Figure 2. Calculated power loss at 1000 rpm.

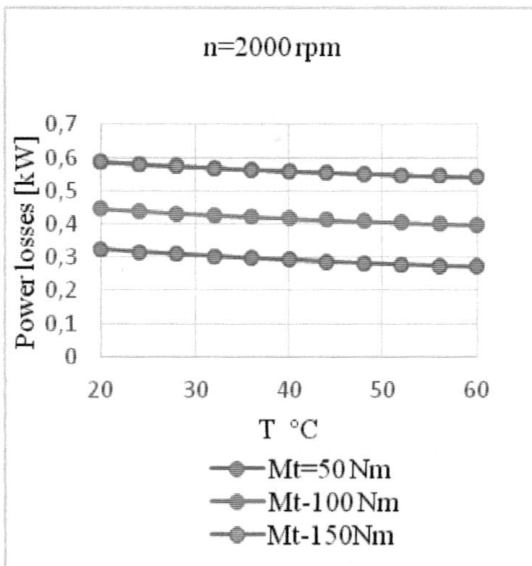

Figure 3. Calculated power loss at 2000 rpm.

Comparing the experimental results from table 2 with the calculated ones from figures 2 and 3, it can be observed that the experimental values are considerably lower than the calculated ones. The possible explanation is the use of simplified calculation methods (that must give sure results) and the influence of convective heat transfer from test rig housing. It is to observe, that at same torque and number of revolutions the experimental power loss (table 2) is lower than the calculated one (figures 2,3) at higher mean temperature. The explanation is the difference of temperature between test rig and environment which is greater and, as consequence, also the convection is greater.

4. Conclusions

1. The thermal calculation of gearboxes was analyzed.
2. A test rig for experimental study of thermal behavior of gearboxes was realized.
3. Preliminary results of experiments are compared with calculation method. The calculated results are greater than the experimental ones. An explanation is the convective dissipation, which was neglected. Therefore the experimental assessment of convective heat dissipation of test rig is necessary.
4. An improvement of calculation especially of power losses in gears is needed.

5. References

[1] Durand de Gevigney J, Changenet C, Ville F and Velex P, 2012, Thermal modeling of a back-to-back gearbox test machine: Application to the FZG test rig. *Proceedings of the Institution of mechanical Engineers, Part J: Journal of Engineering Tribology*, vol **226** issue: 6, pp 501-515, https://doi.org/10.1177/1350650111433243

[2] Hlebanja G and Kulovec S, 2018, Thermal properties of S-gears in comparison with involute gears, *IOP Conf. Series: Materials Science and Engineering* 393 012067 doi:10.1088/1757-899X/393/1/012067

[3] Kanatnikov N V, Kanatnikova P A, Vlasov V V and Pashmentova A S, 2018, Modeling of thermal processes arising during shaping gears with internal non-involute teeth, *IOP Conf. Series: Materials Science and Engineering* 327 022056 doi:10.1088/1757-899X/327/2/022056

[4] Luo Z and Fei Y T, 2006, Research on Non-Similarity about Thermal Deformation Error of Mechanical Parts in Highaccuracy Measurement, *J. Phys.: Conf. Ser.* **48** **41**, doi:10.1088/1742-6596/48/1/008 *International Symposium on Instrumentation Science and Technology, Institute of Physics Publishing Journal of Physics: Conference Series* **48** (2006) pp 41–45

[5] Badrinarayanan S, Bhinder Kuldeep Singh, Ramesh Kumar V and Anupprasad, 2018, Thermal analysis of hydraulic system of landing gear, IOP Conf. Series: Materials Science and Engineering 402 012100 doi:10.1088/1757-899X/402/1/012100

[6] Radu M C, Andrei L and Andrei G, 2018, A perspective on gear meshing quality based on transmission error analysis, *IOP Conf. Series: Materials Science and Engineering* 444 052011 doi:10.1088/1757-899X/444/5/052011

[7] Höhn B-R, Michaelis K, and Vollmer T, 1996, *Thermal rating of Gear Drives: Balance Between Power Loss and Heat Dissipation*, AGMA Technical paper

[8] Castro, M J D and Seabra, J, 2007, Coefficient of friction in mixed film lubrication: Gears versus Twin-discs, Proceedings of the Institution of Mechanical Engineers, *Part J: Journal of Engineering Tribology*, vol **221**, No. 3, pp 399-411

[9] Niemann, G, Winter, H, 1985, *Maschinenelemente*, Band II, Springer-Verlag, Berlin

[10] *** SKF rolling bearings catalogue, 2013, ed. SKF Group

[11] Xu, H., Kahraman, A, 2005, A Frictional Efficiency Loss Model for Helical Gears, DETC2005-85243, Proceedings of IDETC/CIE 2005, *ASME 2005 International Design Engineering Technical Conferences and Computers and Information in Engineering Conference*, Long Beach, CA

[12] Changenet C, Velex P, 2007, A model for the Prediction of Churning Losses in Geared Transmissions-Preliminary Results, *Journal of Mechanical Design,* **129(1),** doi:10.1115/1. 2403727.

IOP Conf. Series: Materials Science and Engineering **724** (2020) 012008 doi:10.1088/1757-899X/724/1/012008

Proposed parameter for the characterization of friction in cylindrical gears teeth contact

S Alaci[1], C Bujoreanu[2], N D Petrea[2], F C Ciornei[1], L Irimescu[1]

[1]Department of Mechanics and Technologies, „Stefan cel Mare" University, Suceava, Romania

[2]Department of Mechanical Engineering, Mecahatronics and Robotics, „Gheorghe Asachi" Technical University, Iasi, Romania

Abstract. Gear mechanisms are widely used in technical applications, mainly for transmitting rotational motion between two axes with constant transmission ratio via a higher pair. Significant friction energy loss in the pair between teeth highlights the importance of estimating the mechanical work lost by friction within this contact. In a previous study, the coefficient of rolling friction between a cylindrical body and a flat face body was found based on the decrease of oscillations amplitude of a cycloidal pendulum. The basic idea of the present paper follows the observation that in the case of a gear mechanism, the relative motion between the wheels can be taken as relative motion of their axodes. In this regard, a pendulum obtained from a mobile cylindrical gear with attached rod and a fixed corresponding rack is considered. The law of motion of the cycloidal pendulum is applied for the proposed pendulum and an "equivalent rolling friction coefficient" which characterises the friction between the teeth of the pinion and the rack is found. The differential equation modelling the motion of the pinion-rack pendulum is difficult to obtain, which is explained by the analysis of the involute gear field of a general cylindrical gear mechanism.

1. Theoretical considerations

The gearing phenomenon occurring between two geared wheels is quite complex [1-2] Several aspects should be considered for the proper running of a gear mechanism, such as continuity of action, avoiding primary interference, secondary interference, radial clearance limitation etc. In figure 1, the gearing field of an involute gearing [3] is presented. The gearing process takes place in the region between the addendum circles of the wheels, c_{a1} and c_{a2}.

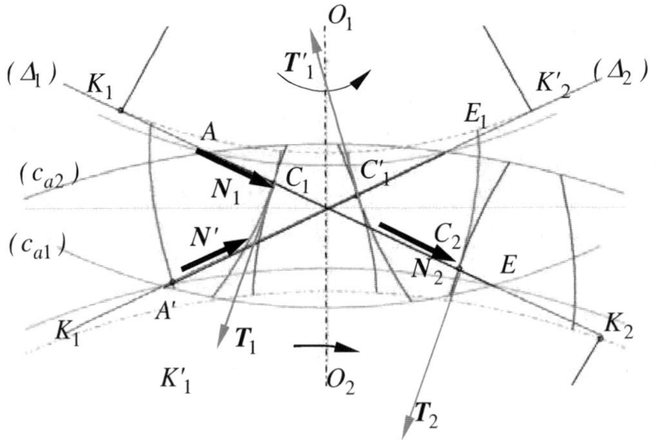

Figure 1. The gearing field of an involute gearing.

The contact points between wheels move along the lines of action K_1K_2 and $K'_1K'_2$, which are common tangents to the base circles. In the case when the pinion is denoted 1, the points where the force between the wheels is transmitted are C_1 and C_2, the gearing starts in point A and ends in point E. The normal lines in the contact points are directed along the straight line K_1K_2 while the friction forces $\boldsymbol{T}_1$ and $\boldsymbol{T}_2$ are perpendicular to the line of action. In the instance when wheel 2 becomes the driving one, with the rotation direction as in figure 1 the point where the force is transmitted is C'_1 from $K'_1K'_2$, the gearing starts in the point A' and ends in the point E', the normal force $\boldsymbol{N}'_1$ is now along the straight line K'_1K_2 and the friction force $\boldsymbol{T}'_1$ is normal to it. Regardless of which wheel is the driving one, the gearing is double point contact at the ends of the segment of action, while in the vicinity of the gearing pole, it is single point contact. During the double point contact gearing, the mechanical system becomes statically undetermined, since from the equations of equilibrium only the sum between the magnitudes of the normal forces can be found. This results in the difficulty in estimation of the friction forces between the flanks.

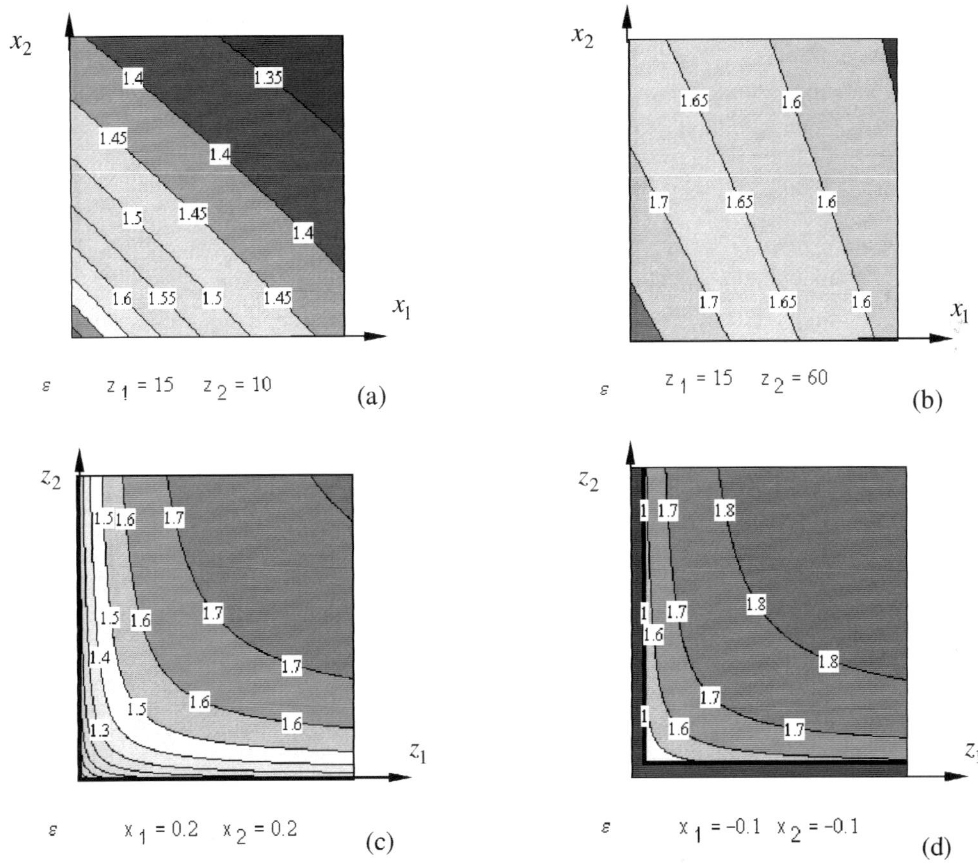

Figure 2. The effect of profile shifts (a), (b) and of number of teeth (c), (d) upon the contact ratio.

The ratio between the periods of uni-point and double-point contact gearing is characterised by ε, the contact ratio [4]. The higher the contact ratio, the greater the percentage of double point contact gearing. In figure 2, the effects of the variance of number of teeth and of the profile shifts upon the contact ratio ε are presented. Regardless of the two wheels' parameters values, the double point contact gearing is always present; therefore, the friction between the flanks of the teeth is always difficult to characterise.

2. Principle of the method. Theoretical model

In engineering applications, an estimation of the global effect of friction forces is seldom required, instead the instantaneous values of the friction torsor is preferred. For this purpose, we start from the remark that during the gearing process, the relative motion of the wheels is completely characterized by the motion between the axodes of the relative motion. In [5], based on the hypothesis of proportionality between the normal force and the rolling friction torque, the coefficient of rolling friction was found using a cycloidal pendulum made by a spherical end rolling on a plane surface.

For the purpose of this paper, for the pinion-rack gearing, the axodes of the motion are the pitch cylinder and the rolling plane of the rack. The cycloidal pendulum from figure 3 is used in characterizing the friction between the flanks of the pinion and the rack. A circle of r radius, the pitch radius of the wheel, rolls with no slip over a straight line Δ_0. G is the centre of mass of the pendulum, M is the mass of the pendulum and d is the distance with respect to the centre of the circle. The central moment of inertia of the pendulum is J_G. The normal reaction N, the friction force $\boldsymbol{T}$ and the rolling friction torque M_r act at the contact point between the circle and the straight line:

$$M_r = s_r N \tag{1}$$

where s_r is the coefficient of rolling friction. Under the assumption of pure rolling in the contact point, the position of the pendulum is given by the angle φ between the straight line through the centre of the circle and the centre of mass and the vertical direction:

$$\ddot{\varphi} = -\frac{(r\,sin\varphi/d + s_r\cdot cos\varphi\, sgn\dot{\varphi}/d)\dot{\varphi}^2 + (s_r\, sgn\dot{\varphi}/d + sin\varphi)g/d}{1 + J_g/Md^2 + s_r\cdot sin\varphi\, sgn\dot{\varphi}/d - 2r\,cos\varphi/d + r^2/d^2}. \tag{2}$$

The equation (2) is nonlinear and must be integrated numerically. It is necessary to stipulate the initial amplitude of the pendulum, φ_0 and the angular velocity ω_0. For $s_r = 0$ and $r \to 0$ the above equation is simplified as:

$$\ddot{\varphi} = -\frac{g\,sin\varphi/d}{1 + J_G/Md^2} = -\frac{Mgd\,sin\varphi}{J_G + Md^2} = -\frac{Mgd\,sin\varphi}{J_O}. \tag{3}$$

Equation (3) is the equation of motion of the physical pendulum, oscillating about an axis passing through O which validates equation (2). After integration, the magnitude of the normal force N and of the friction force T from the contact point C are found applying the relations:

$$N = \ddot{\varphi} Md\,sin\varphi + \dot{\varphi}^2 Md\,cos\varphi + Mg \tag{4}$$

$$T = \ddot{\varphi} M(r - d\,cos\varphi) + \dot{\varphi}^2 Md\,sin\varphi \tag{5}$$

3. Experimental set-up. Description and functioning

In figures 3 and 4, the experimental device is shown; it consists of a rack 1 of module $m = 2$ (mm), fixed horizontally on the working table by two rods of square cross-section. A mobile pinion with $z = 50$ is gearing the fixed rack. During the pinion-rack gearing, the pitch circle of the pinion rolls without sliding over the pitch line of the rack, therefore the points of the wheel describe a family of cycloids.

Two aluminum rods 2 in which equidistant holes were made (the arm) are fixed on the pinion 3 and on the cylindrical part (the bob) from the lower end of the pendulum. When the pinion and the rack teeth are brought into contact as a gear mechanism (the pivot), a cycloidal pendulum which oscillates under its own weight is obtained. As shown above, the position of the pendulum is entirely given by the angle made by the straight line passing through the centers of the pinion and of the bob with the vertical direction. In order to find the tilt angle of the pendulum, a small laser pointer 4 is attached, tangent to the pitch circle of the pinion and normal to the arm. The displacement of the spot emitted by the laser pointer can be estimated on a vertical ruler 5, fixed at a distance D from the axis of the vertical pendulum.

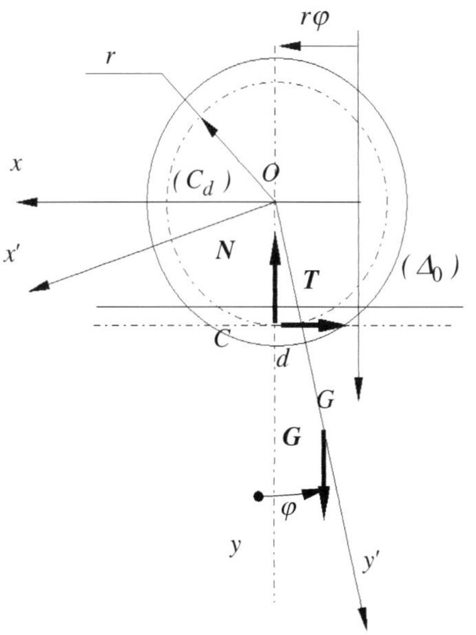

Figure 3. The principle scheme of the device.

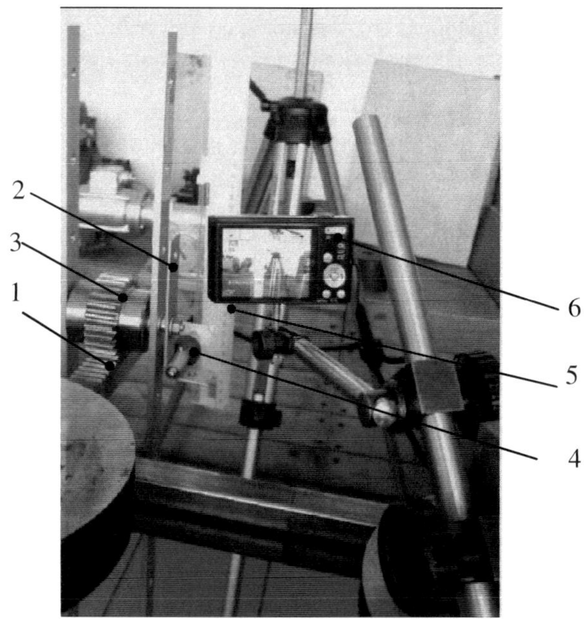

Figure 4. The actual device.

The motion of the mobile spot along the ruler is filmed using a camera 6. The moments when the spot attains the extreme positions can be precisely identified by analyzing the frames of the film. Thus is obtained the law of variation of the amplitude of the spot oscillation.

$$z_k = z_k(t_k) \tag{6}$$

By neglecting the effect of the horizontal displacement of the pinion during the oscillation, the angular amplitude of the pendulum can be expressed as:

$$\theta_k = atan\frac{z_k - z0}{D} \tag{7}$$

where $z0$ is the coordinate of the laser spot for the immobile pendulum. The angular amplitude of the pendulum versus time described by relation (7), obtained experimentally, is presented in figure 5. *The linear decrease of the amplitude validates that the attenuation of the motion is due only to the dry friction* as shown in [6-7]. When other responsible factors occur, the shape of the plot is curved.

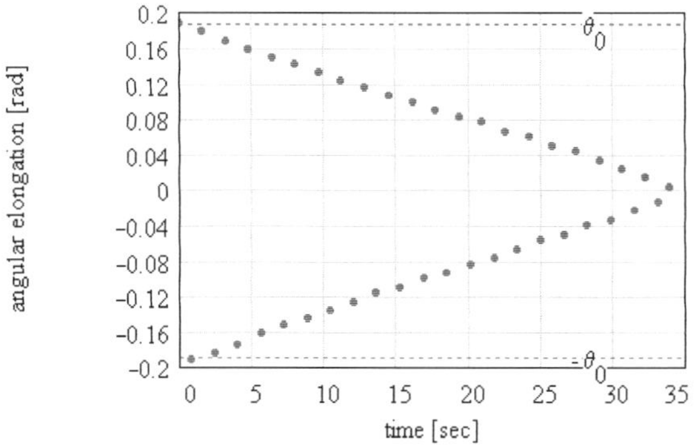

Figure 5. The experimental variation of the angular amplitude of the pendulum.

4. Finding the characteristic parameter of the friction between the flanks

The parameter that characterizes the friction of the pivot of the pendulum, named "equivalent rolling friction coefficient" since provides a global illustration of the friction phenomenon in the teeth contact and denoted by s, is obtained following the methodology presented in [5]. In brief, the equation of motion of the pendulum is numerically integrated with the following imposed initial conditions:

$$\varphi\big|_{t=0} = \theta_0;\ \omega\big|_{t=0} = 0 \tag{8}$$

where θ_0 is the maximum value of experimental amplitude. Subsequently, the value of equivalent rolling friction coefficient s is found by trials from the condition that the theoretical solution passes through the experimental points. In order to integrate the equation of motion (2), the mass of the pendulum M, the position of the center of mass with respect to the center of the pinion d and the moment of inertia J_G with respect to an axis parallel to the axis of oscillation passing through the center of gravity G are required. The mass M is found through weighting and the value obtained is $M = 2.837kg$. The position of the center of mass and the moment of inertia J_G were obtained (as in [5], [8]) and $d = 0.480m$ was obtained. The central moment of inertia was found based on the period of oscillation of the physical pendulum. From the oscillations of the pendulum about the shaft results the period of oscillation T_0:

$$T_0 = \frac{nr.frames \cdot times\_of\_frame}{nr.ocilations} = \frac{1563frames \cdot (1/30)sec}{30oscilations} = 1.737\,sec \tag{9}$$

The physical pendulum's expression of period leads to the value of the central moment of inertia J_G:

$$J_G = \left(\frac{g\zeta T_0^2}{4\pi^2} - \zeta^2\right)M = 0.302\,kg \cdot m^2 \tag{10}$$

The equation of motion (2) was integrated using the above data, for several values of the friction parameter (equivalent rolling friction coefficient) s; after some trials (following the methodology from [9-10]) the value $s = 0.00105$ proved an excellent interpolation of the experimental data, as shown in figure 6. Full concordance is observed between the period of oscillation of the experimental pendulum and the period of the theoretical model. Another remark that should be highlighted concerns the sensibility of the interpolation curves versus the parameter of friction s. In order to support the affirmation, the plots from figure 7 are presented, where the value of the s parameter was modified by 5% in both directions. As expected, the normal force differs slightly from the weight Mg of the pendulum and it can be considered constant.

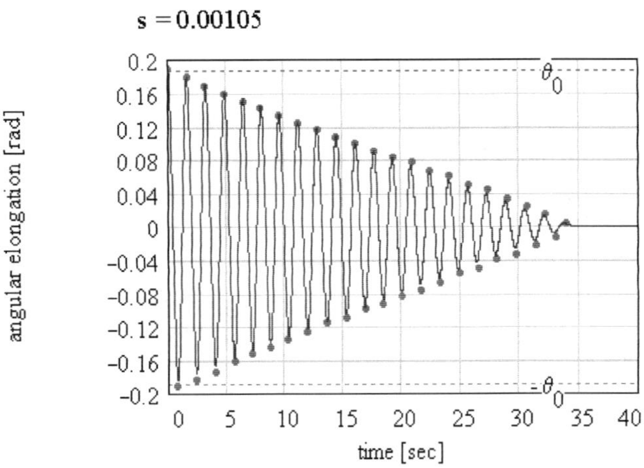

Figure 6. The interpolation of experimental data with the solution of the equation of motion, and establishing the parameter of friction.

Concerning the friction force, its amplitude decreases linearly, and in the points of maximum elongation, finite discontinuities exist, characteristic to the transition from dynamic to static friction.

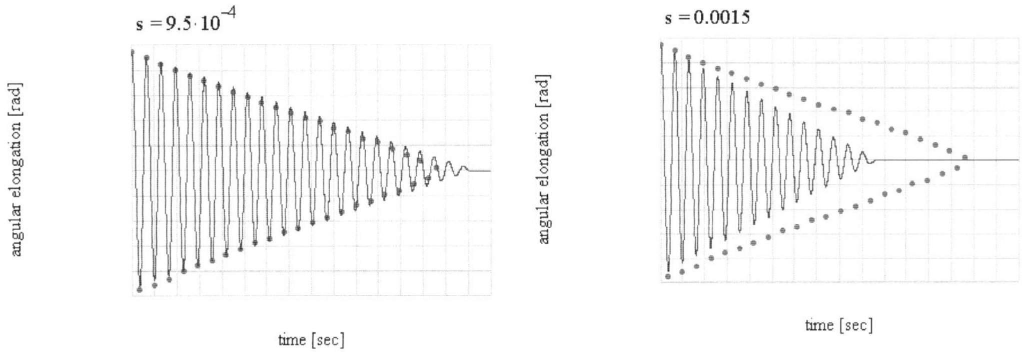

Figure 7. The effect of the coefficient s upon the solution of the theoretical model.

5. Conclusions

The paper presents a method for the estimation of the friction between the flanks of cylindrical gears with the concrete example of application on spur gears. An experimental device was designed and constructed, based on the method of finding the coefficient of rolling friction between a sphere and a plane, and on the remark that the motion between a pinion and a rack can be replaced by the motion of rolling without sliding between the pitch circle and the rack's line of action. The device consists of a pendulum for which the pinion- rack contact is the pivot. For this pendulum, the linear decrease of angular amplitude is obtained experimentally, attenuation for which only the dry friction between the flanks of the gear is responsible. The experimental data are interpolated with the signal generated by the integration of the cycloidal pendulum's equation of motion, and the "coefficient of rolling friction" correlated to the actual pendulum is found. This global parameter allows for the quantitative comparison of frictions occurring in different pairs of gears. This parameter can be employed for the global characterization of friction between the pinion-rack flanks. The effects of the gear parameters - like the module, the rack shift coefficient, and the inertial characteristics of the pendulum upon the "equivalent rolling friction coefficient" are intended for future work.

6. References

[1] Rider M J 2015 *Design and Analysis of Mechanisms: A Planar Approach* 1st Ed Wiley 316 p
[2] Sclater N 2011 *Mechanisms and Mechanical Devices Sourcebook* 5th Ed McGraw-Hill Education p560
[3] Handra Luca V and Stoica I A 1982 *Introduction in the theory of mechanism* (in Romanian *Introducere in teoria mecanismelor*) Vol. I Ed Dacia Cluj
[4] Maros D 1958 *Kinematics of gear mechanisms* (in Romanian *Cinematica Rotilor Dintate)* Ed Tehnica Bucuresti p327
[5] Ciornei M-C, Alaci S, Ciornei F-C and Romanu I-C 2017 *IOP-MSE* **227** 012027
[6] Torzo G and Pearnzoni P 2009 The real pendulum: theory, simulation, experiment *Lat. Am. J. Phys. Educ.* **3**(2) pp 221-228
[7] Zonetti L F C, Camargo S S, Sartori J, de Sousa D F and Nunes L A O 1999 A demonstration of dry and viscous damping of an oscillating pendulum *Eur. J. Phys.* **20** pp 85–88
[8] Hibbeler R C 2015 *Engineering Mechanics: Dynamics* 14th Ed Pearson p784
[9] Alaci S, Ciornei F-C, Bujoreanu C,Ciornei M-C and Acsinte I-L 2018 doi: 10.1088/1757-899X/444/2/022015 *IOP-MSE* **444** 022015
[10] Alaci S, Ciornei F-C, Ciogole A and Ciornei M-C 2017 doi:10.1088/1757-899X/200/1/012005 *IOP-MSE* **200** 012005

Acknowledgement
This work was supported by a grant of the Romanian Ministry of Research and Innovation, CCCDI – UEFISCDI, project number PN-III-P1-1.2-PCCDI-2017-0404 / 31PCCDI/2018, within PNCDI III.

Power loss in grease lubricated ball bearings

G Ianuş[1], D Cojocaru[1], M C Oprişan[1], V Paleu[1] and D N Olaru[1*]
[1] Gheorghe Asachi Technical University of Iaşi, Mechanical Engineering Faculty, Mechanical Engineering, Mechatronics and Robotics Department, 63 D. Mangeron Blvd., 700050, Iaşi, RO

Abstract. The authors theoretically investigated and then validated the power loss generated both by sliding and rolling in a modified thrust ball bearing with 3 balls operating at low axial load and grease lubrication. Based on ball forces and moment equilibrium, it has been determined the tangential forces on contact ellipses. The hydrodynamic rolling friction forces have been determined by using Biboulet & Houpert's methodology considering base oil viscosity of the grease. Total power loss has been obtained as a sum of two components: power loss generated by sliding and power loss generated by rolling. Also, total friction torque has been theoretically determined and, as an alternative method, the total power loss has been calculated as a product between total friction torque and angular speed. Finally, based on previously experimental results, the power loss has been validated as the product between experimentally total friction torque and angular speed. A good correlation between the power losses determined by all three methods for 100, 200, 300 and 400 rpm have been obtained.

1. Introduction

The lubrication mechanism of the greases is a complex one and depends on grease penetration point, film formation, rheological behaviour, grease life [1]. The power loss dissipated in a greased thrust ball bearing is the result of some friction sources as: rolling and sliding in balls - race contacts, sliding between balls and cage, drag loss of balls and cage in grease. In practical applications the power loss dissipated in a ball bearing can be estimated as a product between friction torque and angular speed. To determine the friction torque the SKF methodology can be used if the applied load is equal to or larger than the recommended minimum load of the bearing [2]. This methodology generally includes ball bearing geometry, rotational speed, applied loads and base oil viscosity of the grease.

Cousseau et al. [3,4] determined by experiments the friction torque in a 51107 thrust ball bearing axially loaded with 7000 N and rotational speed ranging between 500 rpm and 2000 rpm. Various types of greases have been used by authors. Considering the SKF methodology, the authors obtained different values for the friction coefficient μ_{sl} as function of the grease type and rotational speed.

For very low axial loads, Bălan et al. [5] determined the friction torque in a modified thrust ball bearing with 3 balls and no cage. The authors evaluated the influence of the lubricant on the friction torque using the Biboulet & Houpert model for hydrodynamic rolling forces in ball and races contacts [6] and experimentally determined the friction torque by spin-down method to validate the theoretical model.

Ianuş et al. [7] determined both theoretical and experimental the friction torque in a modified thrust ball bearing lubricated with two types of greases. Considering the Biboulet & Houpert model for hydrodynamic rolling forces, the authors obtained good correspondence between theoretical and experimental friction torque values.

Popescu et al. [8] proposed a theoretical model to evaluate the ball motion in a modified 7205B angular contact ball bearing based on minimisation of the power loss in ball and race contacts. It considered a constant friction torque in ball-race contacts and obtained total power loss in a ball-race contact as a sum of sliding and rolling power losses.

In the present paper the authors propose an extension of the methodology used in [8] considering supplementary the influence of the hydrodynamic forces generated as result of grease lubrication and ball inertial effects. Also, to determine only the power loss in ball-race contacts, the authors applied the proposed methodology on a 51205 modified thrust ball bearing having only 3 balls and no cage. As an alternative method, the total power loss has been calculated as a product between total friction torque and angular speed. Finally, based on previously experimental results obtained in [7], the power loss has been validated as the product between experimentally total friction torque and angular speed.

2. Analytical model

2.1 Sliding speeds on contact ellipses

For a thrust ball bearing axially loaded with upper race rotating and fixed lower race, the angular and orbital speeds of the balls ω_b and ω_c respectively, can be determined with the following simplified relations [9]

$$\omega_b = \frac{\pi \cdot n_u}{60} \cdot \frac{d_m}{d_b} \tag{1}$$

$$\omega_c = \frac{\pi \cdot n_u}{60} \tag{2}$$

where n_u is rotational speed of the upper race, d_m is mean diameter of the ball bearing and d_b is ball diameter.

On a slice from the lower and upper ball-race contact ellipses, denoted with l and u indexes, respectively the sliding speeds (in the rolling direction) v_{sl} and v_{su} are determined by the following relations [9]:

$$v_{sl,u}(x, n_u, \omega_c, \omega_b) = v_{l,u}(x, n_u, \omega_c) - v_{bl,u}(x, \omega_b) \tag{3}$$

where x is the distance from the centre of contact ellipse to a slice, as it is presented in Fig. 1 [8].

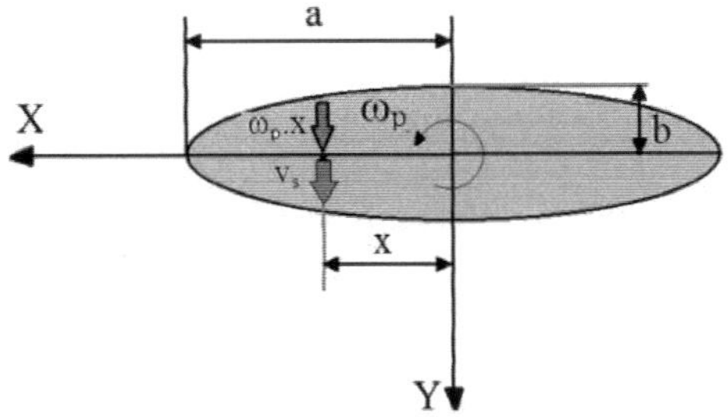

Figure 1. Sliding speeds on contact ellipse [8].

The tangential speeds v_l and v_u in the rolling direction can be expressed using Harris's equations adapted for a thrust ball bearing [9]:

$$v_l(x, n_u, \omega_c) = \left[\frac{d_m}{2} + x\right] \cdot \omega_c \quad (4)$$

$$v_u(x, n_u, \omega_c) = \left[\frac{d_m}{2} + x\right] \cdot (\pi \cdot \frac{n_u}{30} - \omega_c) \quad (5)$$

The tangential speeds of a point from the ball $v_{bl,u}$ are determined by Harris's equations:

$$v_{bl,u}(x, \omega_b) = \omega_b \cdot B_{l,u}(x) \quad (6)$$

The distance from the centre of the ball to the point on the contact ellipses, $B_{l,u}(x)$ are determined by Harris's relations [9]:

$$B_{l,u}(x) = \left\{\left(Ra_{l,u}{}^2 - x^2\right)^{0.5} - \left(Ra_{l,u}{}^2 - a_{l,u}{}^2\right)^{0.5} + \left[(0.5 \cdot d_b)^2 - a_{l,u}{}^2\right]^{0.5}\right\} \quad (7)$$

where $Ra_{l,u}$ are determined with relations [9]:

$$Ra_{l,u} = d_b \cdot \frac{2 \cdot f_{l,u}}{(2 \cdot f_{l,u} + 1)} \quad (8)$$

The semi major and semi minor contact ellipse axis $a_{l,u}$ and $b_{l,u}$ are determined according to Houpert's relations [10].

$$a_{l,u} \approx 1.3085 \cdot R_{Yl,u} \cdot k_{l,u}{}^{0.4091} \cdot \left(\frac{Q}{E^* \cdot R_{Yl,u}{}^2}\right)^{\frac{1}{3}} \quad (9)$$

$$b_{l,u} \approx 1.1502 \cdot R_{Yl,u} \cdot k_{l,u}{}^{-0.1876} \cdot \left(\frac{Q}{E^* \cdot R_{Yl,u}{}^2}\right)^{\frac{1}{3}} \quad (10)$$

where E^* is the equivalent Young's modulus for the materials in contact, Q is the normal load, $R_{Yl,u}$ are the equivalent radii in the rolling direction determined by relation [10]:

$$R_{Yl} = R_{Yu} = \frac{d_b}{2} \quad (11)$$

The radius ratio $k_{l,u}$ are expressed by the relations [10]:

$$k_l = \frac{2 \cdot f_l}{(2 \cdot f_l - 1))} \qquad k_u = \frac{2 \cdot f_u}{(2 \cdot f_u - 1))} \quad (12)$$

where f_l and f_u are the conformities of the two races [9].

In a thrust ball bearing the ball is supplementary rotated with an angular pivoting speeds ω_p expressed by the following equation [9]:

$$\omega_p = \omega_c \quad (13)$$

The total sliding speeds in the rolling directions $v_{tl,u}$ are determined with the following equations:

$$v_{tl,u}(x, n_u, \omega_b, \omega_c) = v_{sl,u}(x, n_u, \omega_b, \omega_c) + \omega_p \cdot x \quad (14)$$

2.2. Friction coefficient on contact ellipses

The presence of the grease in the ball-race contacts develops an EHD film with the minimum thickness determined by Hamrock & Dowson's equations [10,11]:

$$h_{min\,l,u} = 3.63 \cdot R_{yl,u} \cdot U_{l,u}{}^{0.68} \cdot G^{0.49} \cdot W_{l,u}{}^{-0.073} \cdot \left(1 - 0.61 \cdot e^{-0.68 \cdot k_{l,u}}\right) \quad (15)$$

where ***U, G*** and ***W*** are the dimensionless speed parameter, material parameter and load parameter respectively, determined by the classical relations:

$$U_{l,u} = \frac{\eta_0 \cdot v_{l,u}}{E^* \cdot R_{yl,u}}; \quad G = \alpha_p \cdot E^*; \quad W_{l,u} = \frac{Q}{E^* \cdot R_{l,u}^2} \quad (16)$$

In Eqs. (16) η_0 is the base oil dynamic viscosity at the operating temperature, α_p is the piesoviscosity exponent, E^* is equivalent elastic modulus, Q is normal load and $v_{l,u}$ are the average entrainment speed in ball-race contacts defined by equation:

IOP Conf. Series: Materials Science and Engineering **724** (2020) 012009 doi:10.1088/1757-899X/724/1/012009

$$v_{l,u} \approx \frac{\pi \cdot n_u \cdot d_m}{120} \tag{17}$$

In a mixed lubrication regime, the friction coefficient is dependent of the lubricant parameters $\lambda_{l,u}$ and it can be estimated by the following equations [12]:

$$\mu_{sl,u} = \mu_e \cdot 0.82 \cdot \lambda_{l,u}{}^{0.28} + \mu_a \cdot \left(1 - 0.82 \cdot \lambda_{l,u}{}^{0.28}\right) \tag{18}$$

where μ_e and μ_a are the friction coefficient in full lubrication and in limit lubrication conditions, respectively.

The lubrication parameters $\lambda_{l,u}$ are determined by the following relations:

$$\lambda_{l,u} = \frac{h_{min\,l,u}}{\sqrt{R_{ql,u}{}^2 + R_{qb}{}^2}} \tag{19}$$

where $R_{ql,u}$ are the RMS roughness on the two race surfaces and R_{qb} is the RMS roughness on the ball surfaces.

2.3. Forces and moments on contact ellipses

When upper race is rotating and lower race is stopped, the forces and moments acting on a ball in a thrust ball bearing without cage are presented in figure 2 [12].

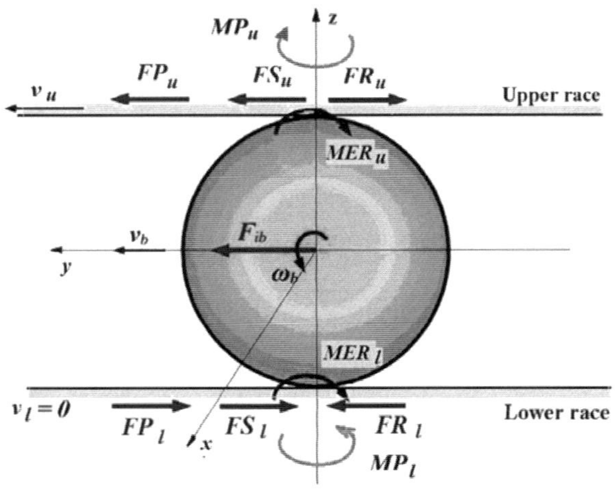

Figure 2. Forces and moments acting on a ball [12].

The presence of the lubricant in ball-race contacts leads to both hydrodynamic rolling forces $FR_{l,u}$ and pressure forces $FP_{l,u}$.

The hydrodynamic rolling forces are determined for the two contacts by Biboulet & Houpert's transition equations considering both IVR and EHL lubrication regimes in ball race contacts [5],[6]:

$$FR_{l,u} = \left[1/(1 + M_{l,u}/6.6)\right] \cdot FR_{IVR,l,u} + \left[(M_{l,u}/6.6)/(1 + M_{l,u}/6.6)\right] \cdot FR_{EHL,l,u} \tag{20}$$

where the IVR and EHL components of FR are determined by Biboulet's relations [5]:

$$FR_{IVR,l,u} = 2.9766 \cdot E^* \cdot R_{yl,u}^2 \cdot k_{l,u}{}^{0.3316} \cdot W_{l,u}{}^{1/3} \cdot U_{l,u}{}^{2/3} \tag{21}$$

$$FR_{EHL,l,u} = 7.5826 \cdot E^* \cdot R_{yl,u}^2 \cdot k_{l,u}{}^{0.4055} \cdot W_{l,u}{}^{1/3} \cdot U_{l,u}{}^{3/4} \tag{22}$$

The transition parameter $M_{l,u}$ from the IVR to EHL lubrication conditions can be determined with equations [6]:

$$M_{l,u} = 0.5549 \cdot k_{l,u}^{-0.6029} \cdot W_{l,u} \cdot U_{l,u}^{-0.75} \tag{23}$$

When the race radius is infinite in the rolling direction (as in a thrust ball bearing), the pressure forces are the followings $FP_{l,u} = 2\ FR_{l,u}$.

The inertial force acting on the ball centre F_{ib} is given by equation:

$$F_{ib} = -\frac{m_b \cdot d_m}{2} \cdot \frac{d\omega_c}{dt} \tag{24}$$

where m_b is the ball mass.

$MER_{l,u}$ are the rolling resistant moments due to elastic hysteresis losses in compression during the rolling process, and can be determined by equations [10]:

$$MER_{l,d} = 7.4810^{-7} \cdot \left(\frac{d_b}{2}\right)^{0.33} .Q^{1.33}.\left\{1 - 3.519\ 10^{-3}(k_{l,u} - 1)^{0.8063}\right\} \tag{25}$$

The traction forces $FS_{l,u}$ are determined as a result of the equilibrium of the forces and moments acting on a ball and results:

$$FS_{l,u} = \frac{MER_l + MER_u}{d_b} \pm \frac{Fib}{2} + FR_{u,l} \tag{26}$$

where the negative sign is for upper ball- race contact.

The total tangential force developed between a ball and the upper race Ft_u is the sum of sliding, rolling and pressure forces, according to following relation:

$$Ft_u = \frac{MER_l + MER_u}{d_b} + (FR_u + FR_l) - \frac{Fib}{2} \tag{27}$$

The pivoting friction moments $MP_{l,u}$,normal to the center of the contact ellipse, are determined with the following relations [10]:

$$MP_{l,u} = \frac{3}{8} \cdot \mu_{sl,u} \cdot Q \cdot a_{l,u} \tag{28}$$

The friction torque generated at upper race for a ball is obtained by relation:

$$dTz_u = Ft_u \cdot \frac{d_m}{2} + MP_u \tag{29}$$

For a modified thrust ball bearing with only three balls, the total friction torque acting on upper race can be written:

$$Tz_u = \frac{3 \cdot (MER_l + MER_u) \cdot d_m}{2 \cdot d_b} + \frac{3 \cdot (FR_u + FR_l) \cdot d_m}{2} + 3 \cdot MP_u - \frac{3 \cdot Fib \cdot d_m}{4} \tag{30}$$

Four components of the total friction torque have been included in the Eq. (30) as following: the components of the rolling friction torque generated by hysteresis, by hydrodynamic effect, by pivoting effects, and by inertial effect, respectively.

2.4. Evaluation of the total power loss

A complex method to estimate the power loss consisting in summing the power loss generated by sliding effect and power loss generated by rolling effect has been proposed by the authors.

The power losses generated by sliding speeds $v_{sl,u}$ are determined using the following relations [8]:

$$P_{S,l,u} = \frac{1}{a_{l,u}} \int_{-a_{l,u}}^{a_{l,u}} \left| v_{tl,u}(x, n_o, \omega_b, \omega_c) \right| \cdot dF_{s,l,u}(x) \cdot dx \tag{31}$$

For the elementary forces $dF_{sl,u}(x)$ acting in the rolling direction on contact ellipses, have been adapted the Houpert's equation [10] as follows:

$$dF_{s,l,u}(x) = \frac{3}{4} \cdot \mu_{sl,u} \cdot \frac{Q}{a_{l,u}} \cdot \sqrt{1 - \left(\frac{x}{a_{l,u}}\right)^2} \tag{32}$$

The total power loss due to sliding in the rolling direction for a single ball is considered the sum of the power loss on lower and upper race contact:

$$P_S = P_{S,l} + P_{S,u} \qquad (33)$$

The total power loss generated by rolling for a single ball includes the power loss due to the moments $MER_{i,u}$ and the power loss due to the inertial effect. It can be expressed with the following relation:

$$P_R = (MER_l + MER_u) \cdot \omega_b + (FR_l + FR_u) \cdot d_b \cdot \omega_b - Fbi \cdot \frac{d_m}{2} \cdot \omega_c \qquad (34)$$

The total power loss for the modified thrust ball bearing with three balls is determined by the following equation:

$$P_t = 3 \cdot (P_S + P_R) \qquad (35)$$

In order to validate the results obtained with Eq. (35) two methods have been proposed:

A first method consist in estimating the total power loss in the modified thrust ball bearing according to the total friction torque determined with Eq. (30), using following relation:

$$P_{t,I} = Tz_u \cdot \frac{\pi \cdot n_u}{30} \qquad (36)$$

The second method consist in estimating the total power loss according to experimental friction torque using the following relation:

$$P_{t,II} = Tz_{u,exp} \cdot \frac{\pi \cdot n_u}{30} \qquad (37)$$

where $Tz_{u,exp}$ is the measured total friction torque [7].

3. Results and discussion

The simulation of the total power loss has been realized for a modified 51205 thrust ball bearing having only 3 balls, without cage and operating in deceleration process according to spin-down method [5, 7]. The axial load Fa, was 4.26 N and normal load for each ball Q, was 1.42N. The grease type used was MOL Liton 00 and the viscosity of the base oil at 40°C of 40 mm²/s has been considered for simulation. The piesoviscosity exponent α_p was considered of $2.5 \cdot 10^{-8} Pa^{-1}$. The following values for geometrical parameters of ball bearing have been considered: d_b = 7.938 mm (5/16"), $f_l = f_u$ = 0.53, d_m = 36 mm, the average roughness for races $R_{ql,u}$ = 0.06 μm and the average roughness of the balls was R_{qb} = 0.03 μm. Rotational speeds n_u between 100 and 400 rpm have been considered for simulation. Also, the laboratory temperature of 23°C -24°C has been considered for base oil viscosity. In Eq. 18 the following values for the coefficients μ_a and μ_e have been adopted: μ_a=0.11 and μ_e=0.05 [10].

For the identical geometry, load and tangential speed on lower and upper raceways, the ball-race contact ellipses have the same dimensions ($a_l = a_u$). Also the moments MER, MP and the hydrodynamic forces FR have the same values for both ball-race contacts.

The experimental methodology and the values for the total friction torque determined with the grease by using spin down procedure were presented in detail by Ianuş et al. [7]. Based on these experiments, it were determined the values for upper race decelerations $d\omega_c/dt$ and the experimental total friction torques $T_{zu,exp}$ corresponding to 100, 200, 300 and 400 rpm rotational speed of upper race.

The most important parameters obtained by simulation are presented in the table 1.

A very good correlation between the power loss has been obtained for four different rotational speeds: n_u = 100, 200, 300 and 400 rpm using any of the following three approaches:

(a) The sum of sliding, rolling and inertial power loss - P_t;

(b) The product between theoretical total friction torque and angular speed- $P_{t,I}$;

(c) The product between experimentally measured total friction torque and angular speed- $P_{t,II}$.

Table 1. The simulation results obtained for upper race rotational speed of 100, 200, 300 and 400 rpm.

	Rotational speed, n_u (rpm)			
Parameters	100	200	300	400
ω_b [rad/s]	23.74	47.49	71.24	94.98
ω_c [rad/s]	5.23	10.47	15.70	20.94
$h_{minl,u}$ [μm]	0.11	0.17	0.23	0.27
$\lambda_{I,u}$	1.42	2.26	2.96	3.6
$\mu_{I,u}$	0.06	0.05	0.05	0.05
$FR_{I,u}$ [N]	0.59×10^{-2}	1.1×10^{-2}	1.5×10^{-2}	1.9×10^{-2}
$d\omega_c/dt$ [rad/s^2]	3.68	7.3	10.4	13.5
$Fib_{I,u}$ [N]	1.36×10^{-4}	2.68×10^{-4}	3.86×10^{-4}	4.9×10^{-4}
Ft_u [N]	1.17×10^{-2}	2.19×10^{-2}	2.98×10^{-2}	3.78×10^{-2}
$MP_{i,u}$ [Nm]	3.26×10^{-6}	3.25×10^{-6}	3.25×10^{-6}	3.25×10^{-6}
$MER_{i,u}$ [Nm]	1.75×10^{-7}	1.75×10^{-7}	1.75×10^{-7}	1.75×10^{-7}
Tz_u [Nm]	0.64×10^{-3}	1.19×10^{-3}	1.62×10^{-3}	2.06×10^{-3}
P_S [W]	5.03×10^{-4}	4.17×10^{-4}	4.19×10^{-4}	4.22×10^{-4}
P_R [W]	2.67×10^{-2}	4.98×10^{-2}	6.79×10^{-2}	8.6×10^{-2}
P_t [W]	2.72×10^{-2}	5.03×10^{-2}	6.83×10^{-2}	8.64×10^{-2}
$P_{t,\,I}$ [W]	2.69×10^{-2}	4.99×10^{-2}	6.79×10^{-2}	8.58×10^{-2}
$P_{t,\,II}$ [W]	2.68×10^{-2}	4.94×10^{-2}	6.7×10^{-2}	8.5×10^{-2}

As seen from table 1, the hydrodynamic rolling friction forces *FR* have major contribution on total power loss for all four rotational speeds. Also, was demonstrated that the use of base oil viscosity of the grease in the theoretical Biboulet&Houpert model for hydrodynamic rolling friction forces given by Eqs: (20) – (23), can conduct to correct values for total power loss. For given operation conditions (very low load on ball- race contacts and low values of friction coefficient as result of the film thickness) the total power loss generated by sliding friction (sliding in rolling direction and pivoting sliding on contact ellipses) does not exceed 2% from the total power loss generated by rolling friction. Based on these results, to reduce the total power loss in a very low loaded ball bearing, is recommended to use grease with a very low base oil viscosity.

4. Conclusions

To evaluate the power loss of grease lubricated thrust ball bearings with no cage, the authors developed a simulation program consisting in the establishment of sliding, rolling and inertial power loss. Using an analytical method, the kinematics, sliding and rolling forces, rolling and pivoting moments, total friction torque and total power loss in a modified 51205 thrust ball bearing were determined. The modified 51205 rolling bearing had only 3 balls and no cage, being lubricated by MOL Liton 00 grease.

The simulations were realized for an axial load of 4.26N and a variation of the rotational speed from 100 to 400 rpm. The most important conclusions are:

- The total power loss obtained by the proposed analytical model is in good correlation with the total power loss obtained using two other methods:
  - the product between analytical friction torque and angular speed;
  - the product between the experimentally measured friction torque and the angular speed.
- It was demonstrated that using the base oil viscosity of the grease in the theoretical model for hydrodynamic rolling friction forces computation, can draw to correct values of computed total power loss.
- For imposed conditions, the hydrodynamic rolling friction forces *FR* have major contribution on total power loss. The total power loss generated by sliding friction does not exceeds 2% from the total power loss generated by rolling friction.
- To reduce the total power loss in a ball bearing very low loaded, is recommended to use grease with a very low base oil viscosity.

5. Acknowledgments

This paper was supported by Mechanical Engineering Doctoral School of "Gheorghe Asachi" Technical University, Iaşi.

6. References

[1] Lught P M 2009, A Review on Grease Lubrication in Rolling Bearings, *Tribol. Trans.* **52**: 470-480
[2] SKF General Catalogue, 2008
[3] Cousseau T, Graça B, Campos A and Seabra, J 2011 Friction torque in grease lubricated thrust ball bearings, *Tribol. Int.* **44**
[4] Cousseau T, Graca B M, Campos A V and Seabra J 2012, Influence of grease rheology on thrust ball bearings friction torque, *Tribol. Int.* **46**, 106-113
[5] Bălan M R, Stamate V C, Houpert L and Olaru D N 2014, The influence of the lubricant viscosity on the rolling friction torque, *Tribol. Int.* **72**, 1–12
[6] Biboulet N and Houpert L 2010, Hydrodynamic force and moment in pure rolling lubricated contacts: part II, point contacts, *Proceedings of the Institution of Mechanical Engineers, Part J: Journal of Engineering Tribology* **224** 777-787
[7] Ianuş G, Dumitraşcu A C, Cârlescu V and Olaru D N 2016, Friction torque in thrust ball bearings grease lubricated, *IOP Conf. Series: Materials Science and Engineering* **147**
[8] Popescu, A, Nazare, M I, Carlescu, V and Olaru, D N 2017, An energetic method to determine existence of the raceway control in an angular contact ball bearing operating at low axial load, *Int. Conf. COMEC*, Braşov, **1** 34-39
[9] Harris T 1991 Rolling Bearing Analysis, *John Wiley & Sons*
[10] Houpert L 1999, Numerical and Analytical Calculations in Ball Bearings, *European Space Agency*, **ESA-S 438 283**
[11] Houpert L 2002, Ball Bearing and Tapered Roller Bearing Torque: Analytical, Numerical and Experimental Results, *STLE Trib. Trans.* **45(3)** 345-353
[12] Balan M R, Houpert L, Tufescu A and Olaru D 2015 Rolling Friction Torque in Ball- Race Contacts Operating in Mixed Lubrication Conditions, *LUBRICANTS* **3(2)** 222-243

Eccentric device for varying the gear ratio

M B - Roth[1], M Romanet[2] and R G Ripeanu[3]

[1] University Petroleum-Gas of Ploiesti, Mechanical Engineering Department, 39 Bucharest Blvd., 100680 Ploiesti, Romania
[2] University Petroleum-Gas of Ploiesti, Mechanical Engineering Department, 39 Bucharest Blvd., 100680 Ploiesti, Romania
3 University Petroleum-Gas of Ploiesti, Mechanical Engineering Department, 39 Bucharest Blvd., 100680 Ploiesti, Romania

E-mail:rothinimagic@yahoo.com

Abstract. In the paper is presented the working principle of the device as well as the constructive-functional scheme for its realization. The use of such a device allows variable gear ratios to be obtained in a very wide range. It has the advantage of a compact construction relative to other mechanisms with varying transmission ratios, such as the crank mechanism. The device contains a worm gear that allows for the eccentricity of a toothed wheel in relation to the cone wheel. The worm gear mechanism also ensures that it is self-locking while it is in operation without the need for other insurance items. The device can be used in the control of actuators in the field of machining or packaging in an automated process.

1. General considerations regarding wheel drive with eccentric wheels

Practice demonstrates that there are many situations where a variable gearing ratio is required for ordering assembly or machining processes. Variable gear ratios can be the basis for rotational motion transformation in linear or oscillating displacements with variable speeds,[8].

The realization of these variable or variable gear ratios can also be implemented with camshafts, non-circular wheels (figure 1 and figure 2), but also with eccentric circular wheels. The use of non-circular wheels involves special problems both in design and especially in manufacturing,[1].

Figure 1. Logarithmic non-circular gears.

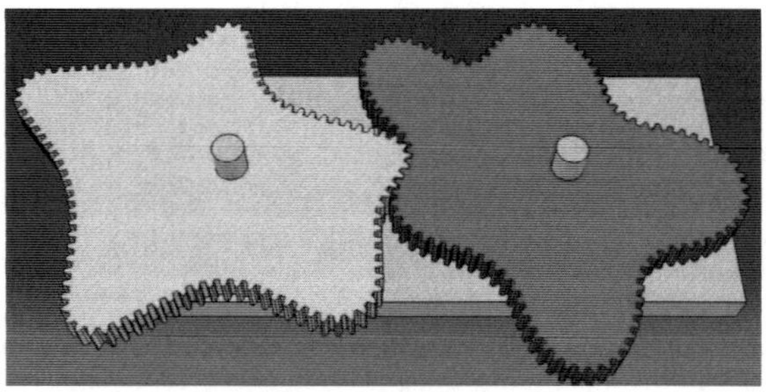

Figure 2. Complex non-circular roots.

The concave non-circular wheels can only be made by copying, a technological process that gives low precision.

The eccentric circular wheels can be run by rolling and centric but also eccentric. In the use of eccentric wheels we studied: circular rings with identical eccentrics, with constant axis spacing (figure 3); circular, eccentric, both eccentric or one non-eccentric; the center of rotation of the eccentric wheel moving in a fixed direction (figure 4); circular, non-eccentric and the other eccentric; the center of rotation of the eccentric wheel moving on a trajectory in a circular arc (figure 5),[3].

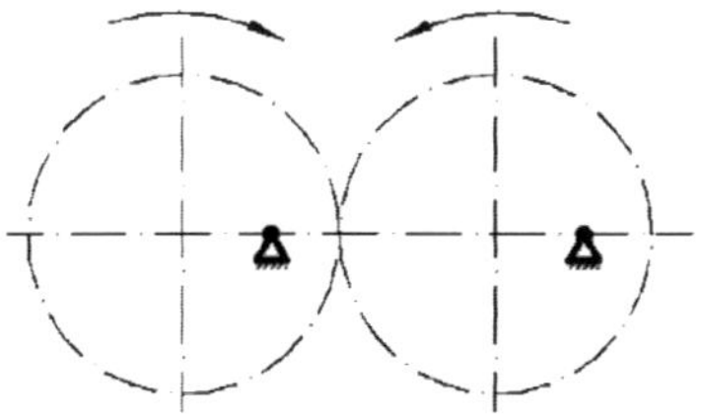

Figure 3. Non-eccentric wheel and eccentric wheel with fixed center of rotation.

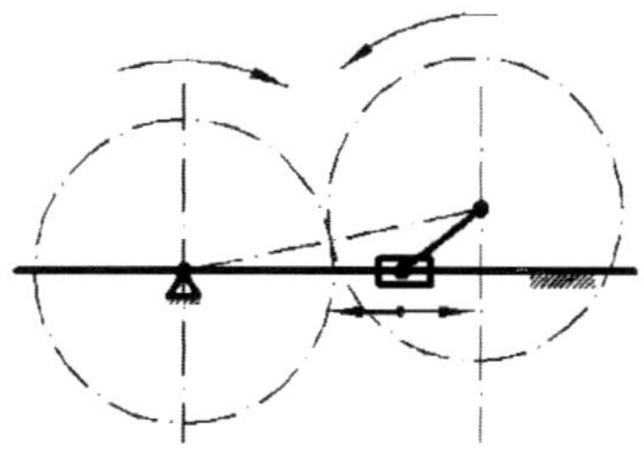

Figure 4. Identical identical eccentric circles with constant axis spacing.

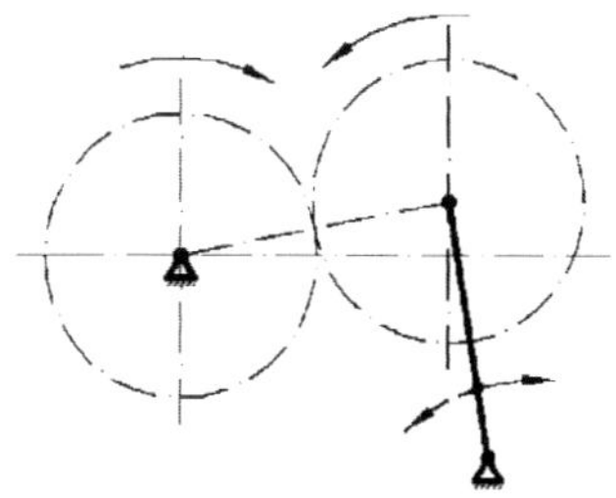

Figure 5. Non-eccentric wheel and eccentric wheel with oscillating center.

2. Determining the mathematical expression of the gear ratio

In order to establish thematic expression of the ratio of gearing, it is considered the case of the engagement between two toothed wheels, one centered and the other one eccentric.

It is also necessary, as a condition of operation, to keep the center of rotation of the eccentric wheel on a fixed line as presented in figure 6,[7].

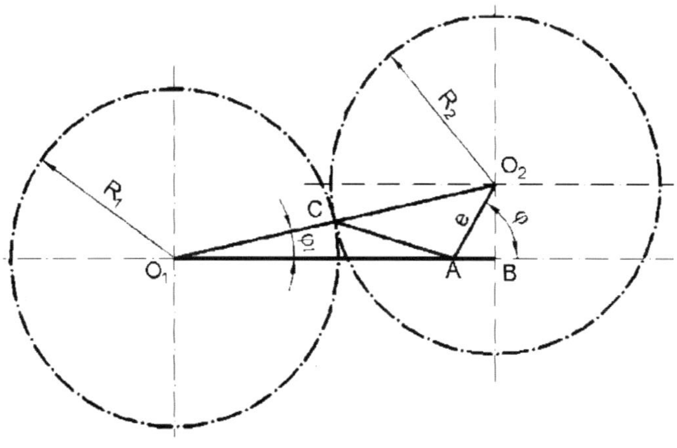

Figure 6. Calculation scheme of the gear ratio for the variant in figure 1.

- $O_1A\ O_2$ center wheel as steering wheel with angle;
- O_1 cent wheel as a driven wheel, normal center wheel;
- R_2 - is the radius of the driving wheel;
- R_1 - driven wheel radius;
- the guiding point A moves;
- e represents the wheel eccentricity

It has been admitted that the variable center of wheel 2 is the leading wheel for mathematical approach because the angle can go through the entire trigonometric circle compared to the angle that covers a range.

It can also be mentioned that the two wheels will be permanently in contact at point C located on the line of the O_1O_2 centers and the purpose of establishing the mathematical relationship of the gear ratio has as an intermediate element the determination of the length of the AC segment that will enter the gearing relation as follows,[6]:

Is considered ΔO_1O_2A in which it is noted $\angle O_2O_1A = \varphi_1$ and it results $\angle O_1O_2A=\varphi-\varphi_1$ angle φ_1 can be calculated from ΔO_1O_2B where $O_2B = e \cdot \sin\varphi$ It is obtained, [5]:

$\sin\varphi_1 = \dfrac{e \cdot \sin\varphi}{R_1 + R_2}$ with the particular solution $\varphi_1 = \arcsin\dfrac{e \cdot \sin\varphi}{R_1 + R_2}$ The sinus theorem is applied ΔO_1O_2A

$$\frac{e}{\sin\varphi_1} = \frac{O_1A}{\sin(\varphi - \varphi_1)} \tag{1}$$

$$O_1A = \frac{e \cdot \sin(\varphi - \varphi_1)}{\sin\varphi_1} \text{ getting} \tag{2}$$

$$AC = \sqrt{R_1^2 + O_1A^2 - 2 \cdot R_1 \cdot O_1A \cdot \cos\varphi_1} \tag{3}$$

Into the ΔO_1CA we apply the generalized Pythagoras theorem for calculating the AC segment that will enter the calculation ratio of the gear ratio:

For Mathcad representation of the gear ratio as a function of all the sizes used should be expressed as simple functions or composed of this angle. Thus, the expressions are used, [4]:

$$O_1A(\varphi) = \frac{e \times sin\,(\varphi - \varphi_1)}{sin\varphi_1(\varphi)} \tag{4}$$

$$\varphi\varphi_1(\varphi) = arcsin\frac{e \times sin\varphi}{R_1 + R_2} \tag{5}$$

The angle is the angle of the O_1O_2 center line with the direction of rotation of the center of rotation A. To determine the value range where the variation is located, the derivation of its expression will be made according to:

$$\varphi_1'(\varphi) = \left[\arcsin\frac{e \cdot \sin\varphi}{R_1 + R_2}\right]' = \frac{e}{R_1 + R_2} \cdot \frac{\cos\varphi}{\sqrt{1 - \left(\dfrac{e \cdot \sin\varphi}{R_1 + R_2}\right)^2}} \tag{6}$$

$$AC(\varphi) = \sqrt{R_1^2 + O_1A(\varphi)^2 - 2 \cdot R_1 \cdot O_1A(\varphi) \cdot \cos\varphi_1(\varphi)} \tag{7}$$

For: $e = 25$, $R_1 = R_2 = 50mm$, $\varphi \in [0^0, 360^0]$, $\varphi_1'(\varphi) = 0$ follows: $\varphi_{1\max} = 14^0 30'$; $\varphi_{1\min} = -14^0 30'$

The gear ratio function becomes: $u_{12}(\varphi) = \dfrac{\omega_1}{\omega_2} = \dfrac{AC(\varphi)}{R_1}$; $u_{21}(\varphi) = \dfrac{\omega_2}{\omega_1} = \dfrac{R_1}{AC(\varphi)}$

where, $u_{1,2}$ is the ratio of the angular speed; $\omega_{1,2}$ the angular speed of the wheels.

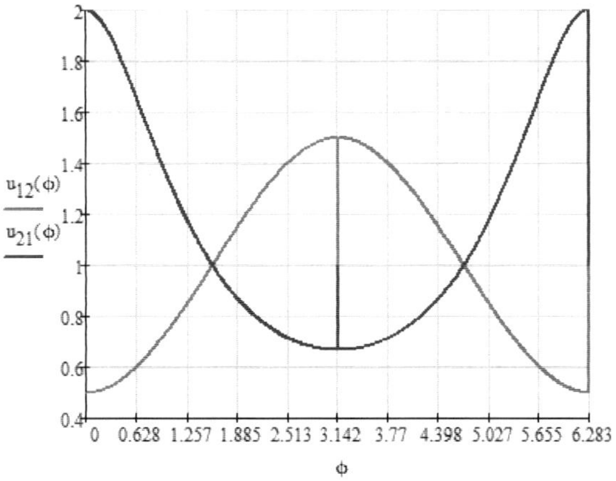

Figure 7. Engagement reports between the two wheels.

The graphs in figure 7 were obtained in the case of driven wheel radius, $R_1 = 50$ mm, the radius of the driving wheel, $R_2 = 50$ mm and wheel eccentricity $e = 25$ mm.

Curves have critical points for $\varphi = \pi$ or $\varphi = 2\cdot\pi$ points from which the mechanism exits through the action of the motor or as the effect of inertia. The mechanism achieved in practice demonstrates this, in the case of $u_{12}> 1$ and as the speed multiplier on the interval where $u_{12} <1$. Analog can be concluded for the other expression of the gear ratio. It can also be seen that the mechanism works with the reduction of the angular velocity, on the interval in which the variation of the gear ratio for different eccentricity values (figure 6) and for different values of the radius of the circular wheel without eccentricity (figure 7) is shown below.

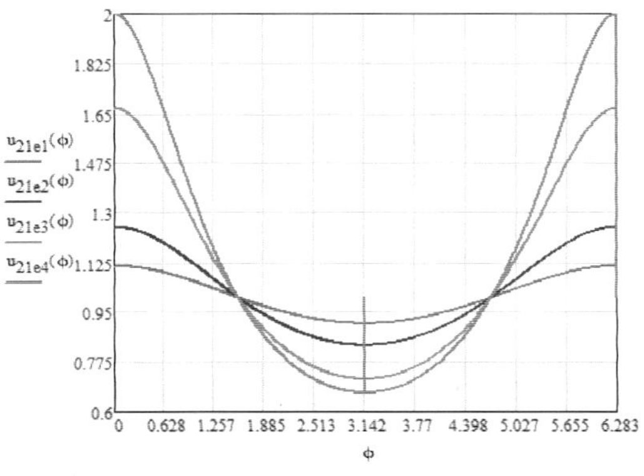

Figure 8. Variation of gear ratio according to eccentricity.

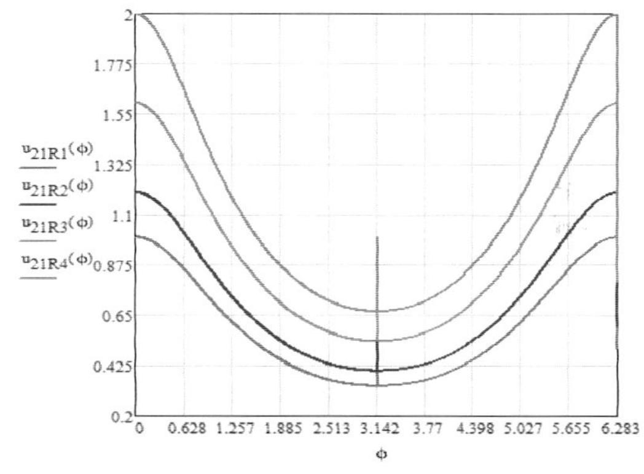

Figure 9. Variation of gear ratio according to centric wheel radius variation.

In figure 8 it is the variation of the drive ratio between two circular wheels with eccentric:

$$R_1 = 50\,\text{mm}; \quad R_2 = 50\text{mm}; \ e \in \{5; 10; 20; 25\}$$

In figure 9 is the variation of the drive ratio between two circular wheels with eccentric having

$$R_2 = 50\text{mm}; \quad e = 25\text{mm}; \ R_1 \in \{5; 10; 20; 25\}$$

The paper presents the working principle of a device as well as its functional scheme. The device allows a variable gear ratio. It consists of two circular gears with eccentric gear, with the possibility to change the eccentricity and the distance between the axes, given the variation of the transmission ratio.

IOP Conf. Series: Materials Science and Engineering **724** (2020) 012010 doi:10.1088/1757-899X/724/1/012010

The construction contains a worn gear allowing a fine variation of the eccentricities and implicitly of the transmission ratio. The device can be used in the control drive devices in the process of processing and packaging in an automated process, [9].

3. Construction and operation of the eccentric device

For the practical embodiment of the device, the embodiment of figure 3 it is adopted in which the wheel 2 performs a pivotal movement around a fixed point 1 the center of rotation of the wheel moving after a trajectory in the arch arc,[11].

Wheel 3 it is an eccentric wheel. Its eccentricity is achieved by means of the worm gear consisting of the auger 4 and the worm gear 5. By rotation of the screw fixed on the wheel 3 and the eccentric piece 6, the eccentricity of the wheel is changed. Changing the eccentricity can be done in a continuous range of values.

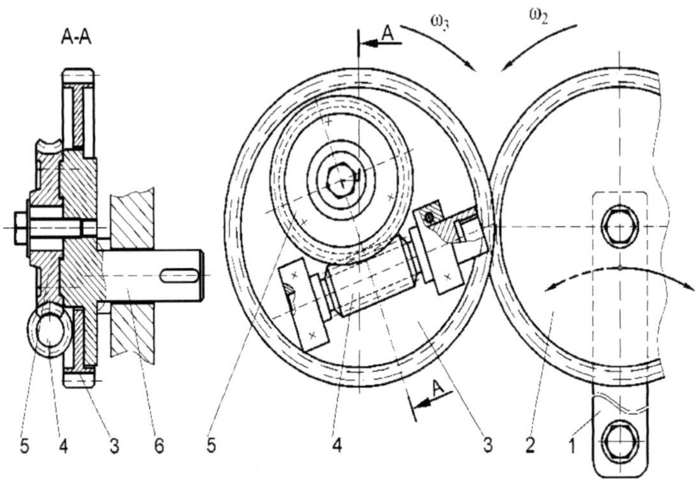

Figure 10. The eccentric wheel drive.

Devices designed to obtain a variable gear ratio can also be used to multiply torque; the reduction of the angular velocity is achieved with torque amplification, [10].

It is well known that the moments are established with the relation:

$$M_t \times \eta \frac{P}{n} \qquad (8)$$

where: M_t represents the moment of torsion transmitted by the gear $[N \cdot m]$

P = driving power [kW] and n = speed of rotation [rpm]

η = efficiency (is the friction losses according to the type of gear and the material working conditions. In the case of the device presented in the paper we considered the value of η = 0.98).

It can be seen that the speed reduction with $u_{1,2} = 2$ leads to the twisting torque multiplying twice. In this assumption, the device can be used both for cinematic purposes and for dynamic purposes, [9].

4. Conclusions

The use of eccentric cylindrical wheels is a solution for making variable gear ratios used in various areas of machine building or automation of technological processes.

The eccentric cylindrical wheels have the advantage that they can be machined by rolling to produce high productivity and precision. The cost of manufacturing them is reduced compared to that of non-circular wheels or different mechanisms that can perform similar functions.

The physically created device is an original idea of the authors and its practical realization confirms the theoretical support.

The different wear of the tooth surface is due to the variation of speeds, moments, contact forces and bending as well as to the variation of the distance between axes leading to the different wear of the teeth on the same wheel.

With the built model we aim to accelerate some types of degradation such as pinching, exfoliation, grip, cracking and breaking of teeth with and without lubricant.

References

[1] Litvin F L, Gonzales-Perez I, Fuentes-Azanar A and Hayasaka K 2009 *Noncircular Gears Design and Generation* (NY, USA, Cambrige University Press) pp 10-103
[2] Litvin F L, Faydor L and Fuentes A 2004 *Gear Geometry and Applied Theory* (NY, USA, Cambrige University Press) pp 26-84
[3] David B, 2001 Function generation utilizing an eight-link mechanism andoptimized non-circular gear element with application to automotive steering *Proc. Inst. Mech. Eng. C* **215** (7) 847-587
[4] Litvin F L, 1956 *Noncircular Gears: Design, Theory of Gearing, and Manufacture* (Moscow, Russia, Gas Tech Isdat) pp 56-93
[5] Litvin F L, Gonzales-Perez I, Fuentes-Azanar A and Hayasaka K 2009 Tandem design of mechnisms for function generation and output speed variation *Comput. Method. Appl. M.* **198** (5-8) 860-876
[6] Florea I 2014 *Machine parts* (Ploiesti, Romania, Petroleum-Gas Publ. House) pp 115-205
[7] Florea I, Rizea N and Popa I 2010 Graphics of the involute profile for gear drawing *Petroleum Gas Bull. Tech. Ser.* **62** (4B) 7-12
[8] Niculescu M 2014 *Researches regarding the design and generation of non-circular gears with applicability in the metallurgical industry, doctoral thesis summary,* (Galati, Romania) pp 63-205
[9] Tong S and Yang C H 1998 *J. Mech. Des.* **120** (June) 337-343
[10] Litvin F L and Ketov C F 1949 Planetary reducer *Certificate of Invention* Russia
[11] Bogdan-Roth M, Romanet M, Ripeanu R G 2019 Device for machining non- circular gears *Proc. on Eng. Sci. 16th Int. Conf. on Tribology* (Kragujevac, Serbia May 15-17) pp 449-453

IOP Conf. Series: Materials Science and Engineering **724** (2020) 012011 doi:10.1088/1757-899X/724/1/012011

Preliminary theoretical results of limited slip couplings

V Gatu[1] , A Predescu[1] and A I Stoicescu[1]

[1] Department of machine elements and tribology, "Politehnica" University of Bucharest, 313 Splaiul Independentei, Bucharest, Romania

"Politehnica" University of Bucharest, 313 Splaiul Independentei, Bucharest, Romania

Abstract Limited slip couplings are a solution for the traction of an all-wheel drive vehicle. The limited slip coupling can be mounted between the front and rear differentials for torque distribution, but also on a front or rear axle to distribute torque between two wheels
Theoretical analysis is performed with a software plug-in with computed fluid dynamics. One of the most important features of a four wheel drive system is the control of the torque transmitted to the wheels. In order to get as much power as possible to the ground, the torque has to be divided among the four wheels to minimize the wheel spin. The limited slip coupling deals with this problem in a new, and, from many points a view, better way. The coupling can be compared to a hydraulic piston pump, which is driven by the slip between input and output shafts. The oil flow from the pump results in an oil pressure that is controlled by a throttle valve. The oil pressure compresses a wet multi-plate clutch that derivers torque to the output shaft. The torque on the output shaft is proportional to the oil pressure. Without the throttle valve the coupling would transfer the torque whenever there is differential rotational speed between the two shafts. One of the objectives of this paper is to present the importance of using CAD and CFD programs and to simulate the behavior of the coupling introducing a momentum, an energy and the material used for the disks, and to obtain results for heat flux and temperature.

1. Introduction

When driving under demanding conditions such as snow, ice and mud, a regular two wheel driven car will fall short. To increase safety and performance many car manufacturers offer a four wheel driven alternative. However, there are a vast variety of four-wheel drive setups. Some systems are full-time AWD (AWD- All Wheel Drive), which means that all wheels receive torque from the engine at all time. The amount of torque distributed between the front and rear axis can have a fixed ratio, typically 50/50 or could be variable. These kinds of full time systems are common among the Asian car manufacturers, such as Subaru and Toyota [1].

As an alternative to the full time systems there are systems called part time systems. Cars using a part time system can either be front or back wheel driven under normal driving conditions. To distribute torque to the other axis some kind of torque transferring device is needed (Figure 1). There are a variety of different solutions achieving it. To transfer torque between the right and left wheels, a differential able to split the available torque is needed.

One of the objectives of this paper is to present the importance of using CAD and CFD programs and to simulate the behavior of the coupling introducing a momentum, an energy and the material used for the disks, and to obtain results for heat flux and temperature.

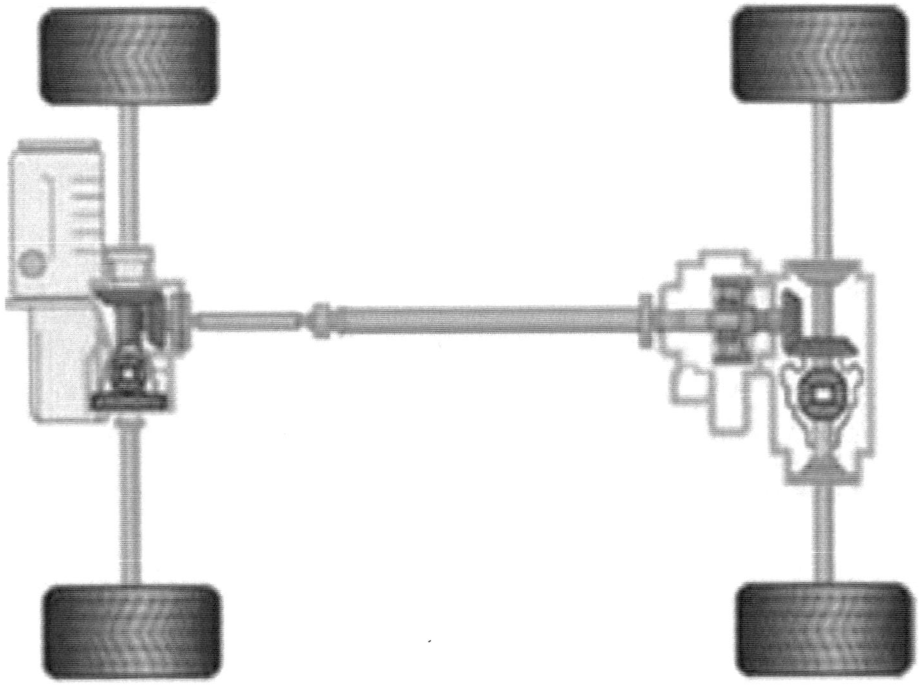

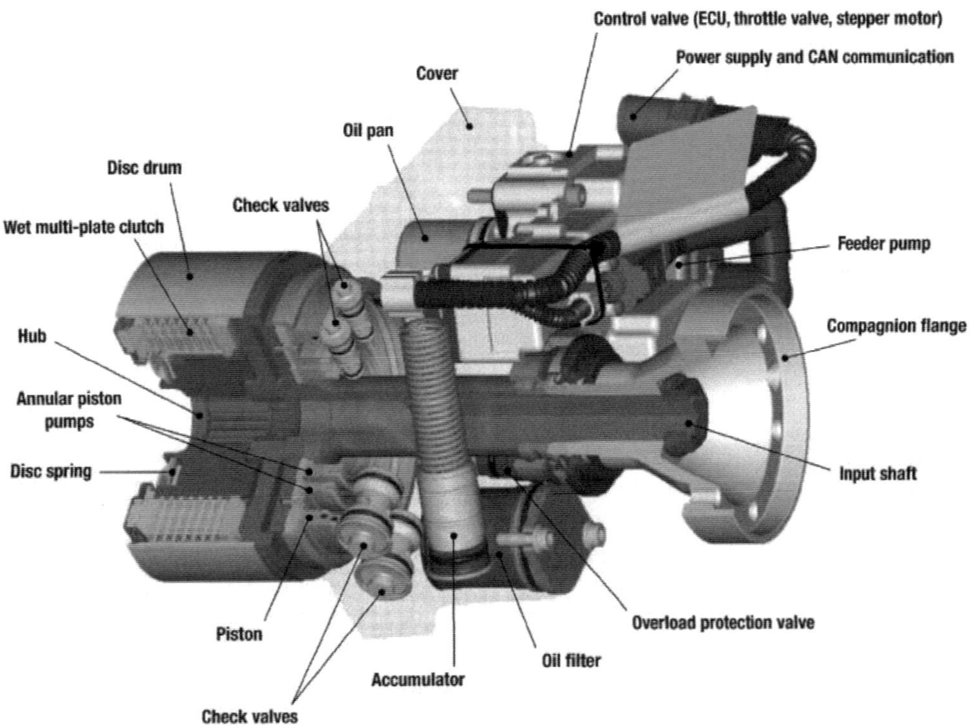

Figure 1. Illustration of the drive line. Highlighted areas are the rear and forward differentials and the AWD coupling [1].

2. Theoretical considerations

A viscous coupling is defined by torque transmission with the help of fluid friction. The viscous coupling is defined by shear stress of a permanent fluid film between two surfaces in relative motion. The shear stress is dependent on the velocity of the film fluid, by the dynamic viscosity [3] - Newton equation, eq. (1).

$$\tau = \eta \frac{\delta u}{\delta y} \quad (1)$$

Where τ – shear stres in the lubricant oil; η- dynamic viscosity; u- velocity; y-film fluid thickness;

For the simulation, the model is based on 3D model, computed in CAD program, using the specified dimensions. Also, the velocity is predefined for $\Delta\omega$= 100 rad/s, (ω1-ω2).

The fluid friction torque is calculated regarding the minimum radius and maximum radius of the interior and exterior dimension of the coupling (Ri= 38 mm and Ro= 53.7 mm) [3].

$$M = z\pi\eta(\omega_1 - \omega_2)\frac{Ro^4 - Ri^4}{2h} \quad (2)$$

Where: M- torque friction, z-number of disks, η-dynamic viscosity, ω-angular speed, Ri- inner radius, Ro-outer radius, h- film thickness;

The value calculated for torque is M= 90.7 Nm ;

The simulation for energy, momentum and force are done in CFD (computed fluid dynamics) . The film fluid is a hydraulic oil type, with the film dimension h=0.05 mm, with the fluid viscosity η=0.12 Pas. The disk is based on steel , with the thickness of 1 mm, the number of the disks are z=11 and the convective coefficient estimated (Vauto≥60 km/h) [3] is $\alpha_1 = 50\frac{W}{m^2}K$.

Because of the geometry of the coupling, it is more convenient to use a cylindrical system of coordinates (rΘz). Each axis has a corresponding velocity, namely uΘ, ur and uz. We will presume that the velocity distribution in the oil films is linear, eq. (3) [4].

$$u_{\theta}(r) = \omega r \frac{z}{h} \quad (3)$$

The temperature distribution within the oil film is governed by the transient energy equation [4], eq.(4) .

$$\rho c \frac{\delta T}{\delta t} = \lambda\left(\frac{\delta^2 T}{\delta r^2} + \frac{\delta^2 T}{\delta z^2}\right) + \eta(T)\left(\frac{\delta u_\theta}{\delta z}\right)^2 \quad (4)$$

Where: ρ- density of the fluid; c- convective term; T- temperature of the fluid; r- radius of the coupling; u_θ- velocity; z- axis coordinate.

The equation is used for calculate the values of temperature field in the coupling presented in Figure 5.

3. Numerical simulation

For a numerical simulation a computed fluid dynamic is used (Fig 2.). Using a schematic figure (Fig.4) of the viscous coupling, a 2D model can be obtained using a CAD program (Catia V5 and Gambit-FluentInc.). Pre-processing require a 2D model (Fig 3.), using Catia V5 and GambitFluent Inc., regarding the dimensions. The boundary conditions are presented in Fig2, where the disks that are engaged by the central shaft with (Ro=53,7 mm) are considered moving walls and the disks that are engaged by the exterior shaft (Ri=38 mm) are considered stationary walls. The movement of the disks, are analyzed and presented in Fig5. The input torque is defined by the exterior (Ro) and the output torque is defined by the interior shaft (Ri) in Fig.2. In figure 3 is presented the assemble of the disks mounted on the shafts.

The difference between the temperature variation for the axial direction of the coupling (Figure 5), is maximum in the film fluid, because the heat flux is maximum in liquid state and the fluid (hydraulic oil) , has the physic property of thermal transfer. Where the difference is increasing linear but slowly, is because the heat flux is minimum in solid state (steel disks).

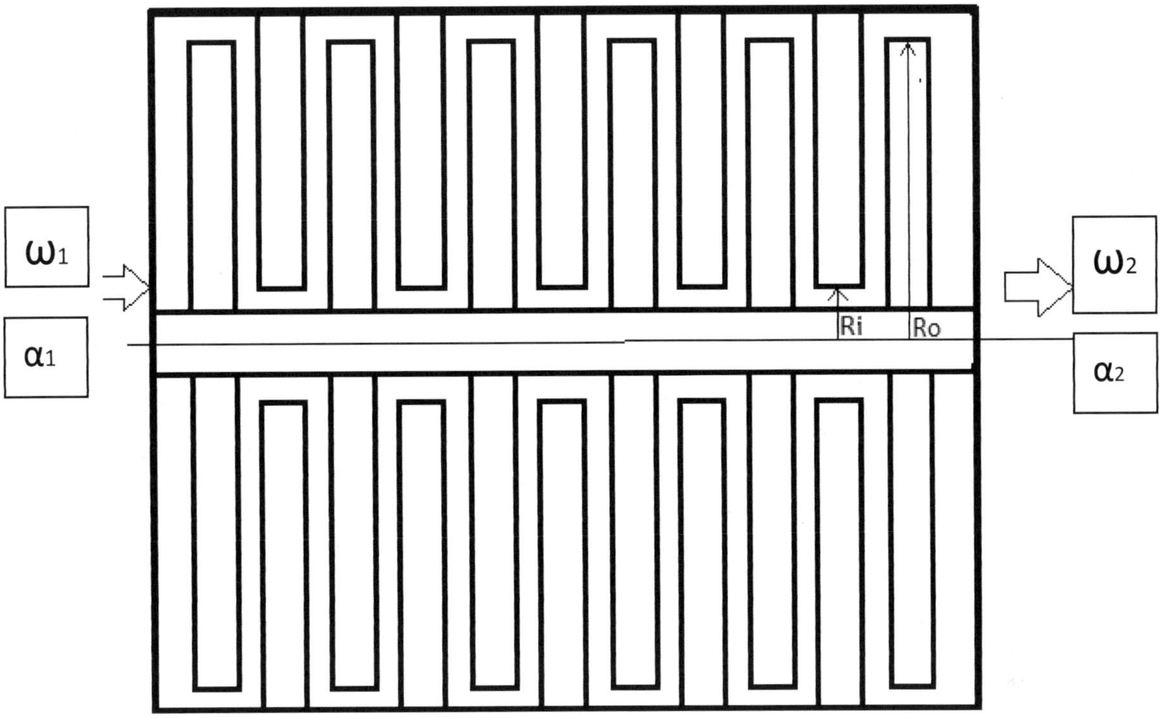

Figure 2. The model used for introducing the CFD software.

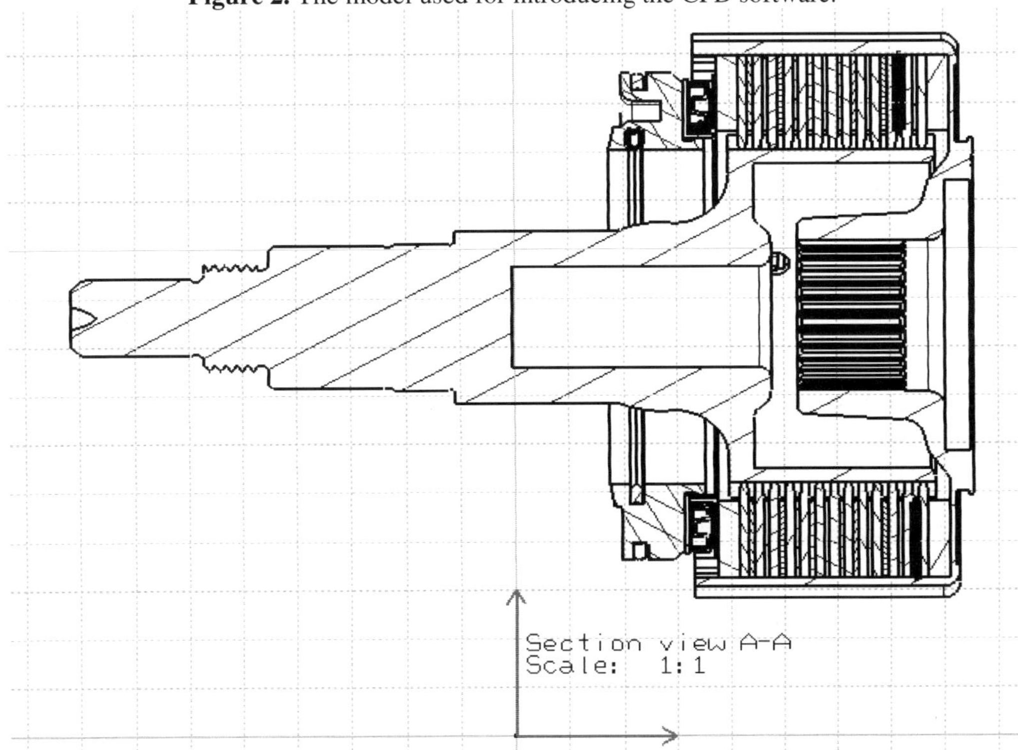

Figure 3. The 2D model for CAD.

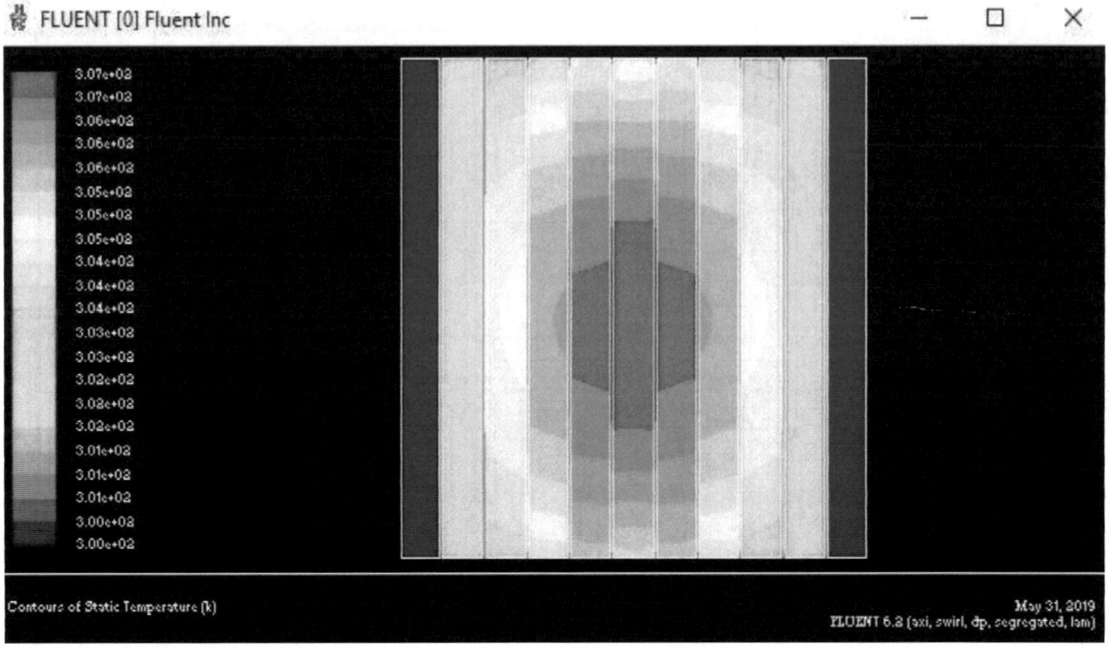

Figure 4. CFD simulation of the internal temperature distribution in the coupling.

The temperature difference is about 7 degrees Kelvin and the maximum value is obtained in the core of the coupling, because of the symmetry of the heat flux between the disks. We could improve the heat flux by increasing the convective coefficient α but this is limited by the vehicle structure.

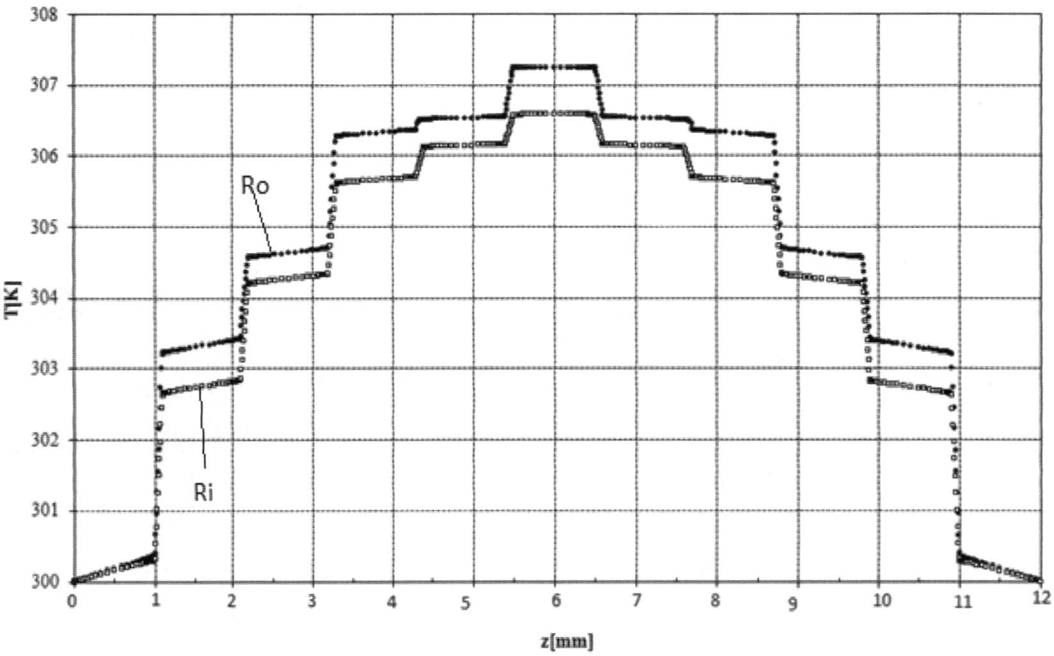

Figure 5. Temperature variation for the axial direction of the coupling.

Figure 5 presents the temperature difference between the disks, the first line is the temperature for the interior disk (Ri) and the second line is the temperature for the exterior disk (Ro).

4. Conclusions

In the automotive industry, especially in the mechanical departments, one of the issues regarding the innovations in this field, is the economically side. Real simulations and tests are very expensive, and it takes a lot of time preparing the models. The materials used, for the mechanical parts, are based on metals, improved by alloy them with other materials. A good way for testify and check all the dimensions and forces , used in mechanical engineering, is using a 3D model designed with a computed aided program, and a simulations with a computed aided engineering program.
Using CAD and CFD, we could obtained the results that we could improve with a real simulation, but we have a big step regarding the new prototype that we are studying.

As a mechanical part, the viscous coupling have a big problem in heat flux, because the core of the mechanical part is loaded with a lot of energy. This energy must be transferred to the exterior , for controlling the over heating situation, when a vehicle has issues in controlling hard terrain for example.

The temperature distribution for the axial direction of the coupling, after a short simulation, is rising very rapidly in the film fluid, as figure 5 is presented. The lower temperatures are simulated for the disks, because the solid state of a material, is heating slowly then the fluid state of a material.

5. References

[1] David B 2011, *Deriving and implementing a model of fifth generation Haldex AWD actuator* p 53
[2] Bulent Y 2002, *Viscous Heating Effects in Viscoelastic Flow between Rotating Parallel Disk Turkish J. Eng. Env. Sci.* p 511
[3] Adrian P 2011 , *Cuplaje cu frecare viscoasa* p 313
[4] Hamrock B.J., Schmid S.R., Jacobson Bo.O. 2004 , *Fundamentals of fluid film lubrication Second Edition* p 699

A bibliometric analysis of scientific research on tribology of composites in Southeastern Europe

Aleksandar Vencl[1,2], Nikolaos M Vaxevanidis[3] and Mara Kandeva[2,4]

[1] University of Belgrade, Faculty of Mechanical Engineering, Kraljice Marije 16, 11120 Belgrade 35, Serbia
[2] South Ural State University, Lenin prospekt 76, 454080 Chelyabinsk, Russia
[3] School of Pedagogical and Technological Education, 14121 Athens, Greece
[4] Faculty of Industrial Technology, Technical University of Sofia, 8 Kliment Ohridski Blvd, 1000 Sofia, Bulgaria

E-mail: avencl@mas.bg.ac.rs

Abstract. In recent years there is a small but increasing number of articles presenting and/or analyzing the scientific production from discrete geographical regions on a certain scientific topic. These articles applied the so-called bibliometric methods in order to evaluate the contribution of different countries in a scientific research field. In the present work, the research output of all countries in Southeastern Europe on the scientific field of composites tribology is presented by using bibliometric indicators such as the total number of publications and citations, the average number of citations per publication, and the *h*-index. Analysis spans the last ten years and the required scientific data in order to calculate the bibliometric indicators were retrieved using the Scopus® scientific database.

1. Introduction

One of the definitions of bibliometrics say that it is the quantitative study of physically published units, or of bibliographic units, or of alternatives of either [1]. Bibliometric analysis uses the statistical methods to analyse books, articles, and other publications published in certain time period, and to measure the impact of the analysed subjects (research field/topic, particular paper/researcher, set of researchers, journal, institution/department, geographical region/country, etc.), through the defined bibliometric indicators. It is also a useful tool to obtain information on the state-of-the-art of research in particular topic, which helps the researchers to identify their research directions.

Composite material is a mixture of two or more materials or phases of the same material, insoluble in one another, possessing properties which are superior to any of the component materials. The use of different types of metal/ceramic/polymer matrix composites (MMCs/CMCs/PMCs) is constantly growing since they possess better physical and/or mechanical and/or tribological properties compared to the matrix materials. Better tribological properties are of great importance in all machines where parts are in contact and relative motions (tribological systems). The fact is that tribological properties define possible application of material far more than their mechanical or physical properties, since they are in better correlation with behaviour in practice [2].

This work presents the research output of all countries in Southeastern Europe (SEE) on the scientific field of composites tribology, in last ten years, by analysing bibliometric indicators such as the total number of publications and citations, the average number of citations per publication, and the h-index. Similarly to our previous studies [3,4] the countries of SEE are defined geographically, as follows: Albania, Bosnia and Herzegovina, Bulgaria, Croatia, Greece, Montenegro, North Macedonia, Romania, Serbia, Slovenia and Turkey.

2. Methodology

Selecting the appropriate bibliometric indicators is very important for the reliable evaluation and presentation of the results in bibliometric analysis. It is also better to use a range of performance indicators than just one or two [5]. There are many different bibliometric indicators [6], but basically, there are three types: quantity indicators, which measure the productivity of a particular researcher or research group; performance indicators, which measure the quality of a publication, researcher or research group; and structural indicators, which measure connections between publications, authors and areas of research [7].

Indicators used in this study correspond to four research pillars: productivity (P: total number of publications on country level, in the given time-period); impact (C: total number of citations, excluding self-citations, on country level, in the given time-period); efficiency (C_{av}: average number of citations per publication, excluding self-citations, on country level, in the given time-period); and hybrid, i.e. productivity + impact (*h*-index, excluding self-citations, on country level, in the given time-period).

Productivity indicator, i.e. total number of publications on country level (P) is usually considered as a indicator for scientific size of a country, which of course is not the same as scientific strength of a country [8]. Total number of citations on country level (C), received by these P publications in the examined period, is another standard indicator which is considered as a indicator for scientific impact of a country. Average number of citations per publication on country level (C_{av}) is calculated as the ratio of total number of publications (P) and citations (C) in the given time-period, i.e. $C_{av} = C/P$. This indicator is used as an of efficiency indicator or as an impact index [8]. The *h*-index [9] is considered as a single indicator which combine productivity and impact indicators. The *h*-index on country level is calculated considering P and C of a country as the outcome of a single researcher. Characterization of certain number of publications (P) or citations per publication (C_{av}) as low or high, was performed with the field-based normalization method [8].

The required scientific data for each country, for the last ten years, necessary to calculate the bibliometric indicators were retrieved using the Scopus® scientific database. During search, document type was limited, i.e. only following types was counted: Article; Conference Paper; Book; Book Chapter; Review, while types such as: Erratum; Editorial; Conference Review; Retracted; etc. were not counted. For the co-authored publications the "whole counting" method was applied, i.e. every country gets full credit for internationally co-authored publications, and every co-author gets full credit for co-authored papers. There were no special corrections of the results except from some obvious errors, as well as more detail check-up of the most cited publications in the field of composites tribology (Table 3). The used keywords, as well as the hierarchical order in which they should be inserted in the searching machine in order to get the search results were as follows:

(TITLE (tribo***) OR KEY (tribo***) OR TITLE (friction) OR KEY (friction) OR TITLE (wear) OR KEY (wear) OR TITLE (lubrication) OR KEY (lubrication) AND TITLE (***composite*) OR KEY (***composite*) OR TITLE (*mc*) OR KEY (*mc*) AND AFFILCOUNTRY (each of SEE country)) AND PUBYEAR > 2008 AND PUBYEAR < 2019 AND (LIMIT-TO (PUBSTAGE, "final"))

where term: TITLE refer to the classification "Article title", KEY refer to the classification "Keywords", AFFILCOUNTRY refer to the classification "Affiliation country", PUBYEAR refer to the limit "Published year", and PUBSTAGE refer to the classification "Publication stage (final or article in press)".

The asterisk/s at the end or the beginning of some terms is used to overcome the underestimation resulting in by the use of different but equivalent terms, i.e. "tribo<u>logy</u>" or "tribo<u>logical</u>" or "tribo<u>corrosion</u>" or "tribo<u>system</u>", etc.; "composite" or "composite<u>s</u>" or "<u>nano</u>composite", etc.; "<u>M</u>MC" or "<u>M</u>MC<u>s</u>" or "<u>C</u>MC" or "<u>P</u>MC<u>s</u>", etc. The keywords selection used in study is actually the result of an optimization, based on large number of trials. All data presented in this work were collected on 20th June 2019 and involved scientific work and citations published until the end of 2018.

3. Results and discussion

3.1. Trends in publications

Results of the basic search for all countries in SEE on the scientific field of composites tribology is presented in Table 1. They are presented through the selected bibliometric indicators for the last decade (2009 – 2018). For comparison purposes, search for all countries in the World was also performed. First thing that can be noticed is that for Albania, Bosnia and Herzegovina, and Montenegro searches return zero results, whilst for North Macedonia output is one. For this reason, these countries are omitted from further analysis.

Table 1. Selected bibliometric indicators for the last decade.

Country	Number of publications (P)	Number of citations (C)	Number of citations per publication (C_{av})	*h*-index
Albania	0	0	–	0
Bosnia and Herzegovina	0	0	–	0
Bulgaria	42	223	5.31	6
Croatia	14	23	1.64	2
Greece	83	857	10.33	15
Montenegro	0	0	–	0
North Macedonia	1	0	0	0
Romania	183	505	2.76	12
Serbia	37	301	8.14	10
Slovenia	35	295	8.43	9
Turkey	508	3596	7.08	31
Word	18151	n/a	n/a	n/a

n/a – not available

By comparing the number of publications (P) in Table 1, it is evident that SEE countries account only 5 % of the literature on tribology of composites published worldwide. Turkey has the largest publication output of the SEE countries, i.e., 56.3 % of total publications, followed by Romania (20.3 %) and Greece (9.2 %). These three countries contribute more than 85 % of the total SEE production on the topic. The variation of publications with year for SEE countries is not discussed in details, due to space restrictions. In general, a non-monotonic increase was identified with the annual total of publications increasing from 50 (2009) to 111 (2018).

Total number of citations (C) which is obtained by the P publications is usually used as a indicator of the cited research impact, while the number of citations per publication (C_{av}) is considered as an index of the scientific research efficiency. From Table 1 it is clear that Turkey holds the highest C which is expectable since more than half of all SEE countries publications come from Turkey. Similarly to the number of publications distribution, Turkey (62.0 %), Greece (14.8 %) and Romania (8.7 %) contribute more than 85 % of the total SEE production on the topic. On the other hand, order of SEE countries is completely different when considering the number of citations per publication (C_{av}). Greece holds the highest index (C_{av} = 10.3) followed by Slovenia (C_{av} = 8.4), Serbia (C_{av} = 8.1) and Turkey (C_{av} = 7.1). Only these four countries are above the average value of 6.2 calculated for the whole region. It is also interesting that Romania have very low C_{av} value of 2.8, the second lowest of all analysed SEE countries, although it has relatively high production of the publications.

The *h*-index [9] was originally intended to combine in a single, hybrid, indicator a metric for quantity and impact of a researcher's scientific output, i.e. it was originally to evaluate individual researchers. Nevertheless, it may also be applied to research teams, faculties and faculty/university departments, journals, countries, etc. According to the obtained results (Table 1), the order of SEE countries according to the *h*-index is similar as the order according to the total number of citations (C), i.e. Turkey holds the highest *h*-index of 31 followed in descending order by Greece (15) and Romania (12).

Since there are no data available to reliably compare publication outputs to inputs in terms of number of researchers, publication output is usually compared to the size of the population in the different countries – although differences in population do not necessarily reflect differences in research efforts [10]. Much of the difference among SEE countries in publication output shown in Table 1 can be attributed to differences in country size, number and quality of researchers and investments in research, as well as to social-economic indicators.

3.2. Authorship characteristics

As it can be noticed from Table 1, the analysis is based on research output of approximately 900 publications (articles; conference papers, books and books chapters) indexed in Scopus® during the period 2009 – 2018. All these publications include at least one author listing an affiliation in one of the SEE countries. During search it was observed that the majority of publications are written by the authors employed at the universities and public research institutions associated with higher education organizations. Private research centers and industry seem to have very small contribution and participation. Moreover, as it is expected for a multi-disciplinary scientific topic such as tribology of composites, the vast majority of published articles have been co-authored. The most productive authors (with a threshold of $P \geq 5$) in the field of composites tribology, from each SEE country and worldwide (for compassion), are presented in Table 2. It should be noted that the P, C and *h*-index values for authors in Table 2 refer only to publications concerning "tribology of composites", so the overall output of an author might be much greater.

By comparing the number of publications of different authors (P) in Table 2, it is evident that the authors from SEE countries are much less productive that the most productive authors worldwide. The most productive authors in the field of composites tribology publish approximately 9 publications per year, which is quite impressive if they write all the publications. On the other hand, the difference between worldwide authors and SEE authors are not so obvious when considering the total number of citations per publication (C). The diversity of institutions of the most productive SEE authors is questionable, since in some cases all of the authors are from the same institution, e.g. Sakarya University in Turkey, or in other cases, there are not enough authors that pass the threshold ($P \geq 5$), e.g. Bulgaria and Croatia. It is also interesting that some authors have more publications than citations. This could be either due to the fact that the publications are published in the last year (2018) of the analysed period, and is not recognised as relevant literature yet, or that the publications do not have the potential to impact other researches.

The most cited publications in the field of composites tribology, from each SEE country and worldwide (for compassion), are presented in Table 3. Three publications at maximum from each country were included, with a limit that the publication received at least 10 citations ($C \geq 10$). It is very hard to compare different publications according to the citations they obtain, since they were not the same period of time available for citations, i.e. some of them was published in 2009 (or even in 2008 as on-line first) and some at the end of 2008. More appropriate criteria would be citation per year, but this data was not available for this analysis. As expected review papers usually get more citations. Another interesting aspect that was investigated is the "journal preference" [3,4], but due to space restrictions it was omitted from the final version of this paper.

3.3. International scientific collaborations

Nowadays there is an increasing international scientific collaboration in many disciplines [8]. Table 4 shows the international collaboration within the SEE counties and with other countries. For each SEE country the three countries with the highest number of common papers are presented. First thing that can be noticed is that the percentage of international publications is for all countries moderate, and mostly with non SEE countries. This can be observed when total number of publications obtained as a sum of each country publications (903) is compared with total number of publications without overlapping (889), since for the co-authored publications the "whole counting" method was applied, i.e. every country gets full credit for internationally co-authored publications. This means that 14 or less publications have co-authors from 2 or more SEE countries, i.e. less than 2 %. During search, it was also noticed that almost all (881) publications are written in English.

Table 2. The most productive authors in the field of composites tribology for the last decade.

Country (P/C)	Author	P*	C*	Affiliation
Bulgaria (42/223)	M. Kandeva	11	10	Technical University of Sofia, Faculty of Industrial Technology, Sofia
Greece (83/857)	C.A. Charitidis	10	228	National Technical University of Athens, School of Chemical Engineering, Athens
	G.D. Papadimitriou	8	89	National Technical University of Athens, School of Mining and Metallurgical Engineering, Athens
	E.P. Koumoulos	8	64	National Technical University of Athens, School of Chemical Engineering, Athens
	A.G. Lekatou	8	62	University of Ioannina, School of Engineering, Ioannina
	A.E. Karantzalis	8	61	University of Ioannina, School of Engineering, Ioannina
Romania (183/505)	M. Buciumeanu	16	93	"Dunărea de Jos" University of Galaţi, Faculty of Engineering, Galaţi
	G. Andrei	13	5	"Dunărea de Jos" University of Galaţi, Faculty of Engineering, Galaţi
	R. Căliman	11	1	Vasile Alecsandri University of Bacău, Faculty of Engineering, Bacău
Serbia (37/301)	B. Stojanović	15	94	University of Kragujevac, Faculty of Engineering, Kragujevac
	M. Babić	14	192	University of Kragujevac, Faculty of Engineering, Kragujevac
	A. Vencl	11	123	University of Belgrade, Faculty of Mechanical Engineering, Belgrade
Slovenia (35/295)	M. Kalin	7	130	University of Ljubljana, Faculty of Mechanical Engineering, Ljubljana
	B. Podgornik	6	16	Institute of Metals and Technology Ljubljana, Ljubljana
	P. Panjan	5	49	Jožef Stefan Institute, Ljubljana
Turkey (508/3596)	H. Akbulut	26	405	Sakarya University, Faculty of Engineering, Adapazarı
	A. Mimaroğlu	22	98	Sakarya University, Faculty of Engineering, Adapazarı
	M. Uysal	18	224	Sakarya University, Faculty of Engineering, Adapazarı
	H. Ünal	18	91	Sakarya University, Faculty of Technology, Adapazarı
World (18151)	X. Shi	89	465	Wuhan University of Technology, School of Mechanical and Electronic Engineering, Wuhan, China
	W.M. Liu	87	779	Chinese Academy of Sciences, Lanzhou Institute of Chemical Physics, Lanzhou, China
	A. Patnaik	83	675	Malaviya National Institute of Technology Jaipur, Department of Mechanical Engineering, Jaipur, India

* P and C values for authors refer only to papers concerning "tribology of composites", so the overall output of an author might be much greater

Table 3. The most cited publications in the field of composites tribology for the last decade.

Country (P)	Authors and publication	C
Bulgaria (223)	Kelly P.J., Li H., Whitehead K.A., Verran J., Arnell R.D., Iordanova I., *A study of the antimicrobial and tribological properties of TiN/Ag nanocomposite coatings*, Surface and Coatings Technology, 204, 6-7, 2009, 1137-1140	83
	Kelly P.J., Li H., Benson P.S., Whitehead K.A., Verran J., Arnell R.D., Iordanova I., *Comparison of the tribological and antimicrobial properties of CrN/Ag, ZrN/Ag, TiN/Ag, and TiN/Cu nanocomposite coatings*, Surface and Coatings Technology, 205, 5, 2010, 1606-1610	83
	Xu D., Karger-Kocsis J., Apostolov A.A., *Hybrids from HNBR and in situ polymerizable cyclic butylene terephthalate (CBT): Structure and rolling wear properties*, European Polymer Journal, 45, 4, 2009, 1270-1281	14
Greece (857)	Charitidis C.A., *Nanomechanical and nanotribological properties of carbon-based thin films: A review*, International Journal of Refractory Metals and Hard Materials, 28, 1, 2010, 51-70	164
	Patnaik A., Satapathy A., Chand N., Barkoula N.M., Biswas S., *Solid particle erosion wear characteristics of fiber and particulate filled polymer composites: A review*, Wear, 268, 1-2, 2010, 249-263	85
	Lekatou A., Karantzalis A.E., Evangelou A., Gousia V., Kaptay G., Gácsi Z., Baumli P., Simon A., *Aluminium reinforced by WC and TiC nanoparticles (ex-situ) and aluminide particles (in-situ): Microstructure, wear and corrosion behaviour*, Materials & Design, 65, 2015, 1121-1135	45
Romania (505)	Benea L., Wenger F., Ponthiaux P., Celis J.P., *Tribocorrosion behaviour of Ni-SiC nano-structured composite coatings obtained by electrodeposition*, Wear, 266, 3-4, 2009, 398-405	41
	Benea L., Başaa S.-B., Dănăilă E., Caron N., Raquet O., Ponthiauxd P., Celis J.-P., *Fretting and wear behaviors of Ni/nano-WC composite coatings in dry and wet conditions*, Materials & Design, 65, 2015, 550-558	39
	Benea L., *Electrodeposition and tribocorrosion behaviour of ZrO_2-Ni composite coatings*, Journal of Applied Electrochemistry, 39, 10, 2009, 1671-1681	36
Serbia (301)	Vencl A., Bobic I., Arostegui S., Bobic B., Marinković A., Babić M., *Structural, mechanical and tribological properties of A356 aluminium alloy reinforced with Al_2O_3, SiC and SiC + graphite particles*, Journal of Alloys and Compounds, 506, 2, 2010, 631-639	68
	Mitrović S., Babić M., Stojanović B., Miloradović N., Pantić M., Džunić D., *Tribological potential of hybrid composites based on zinc and aluminium alloys reinforced with SiC and graphite particles*, Tribology in Industry, 34, 4, 2012, 177-185	27
	Babic M., Slobodan M., Džunic D., Jeremic B., Ilija B., *Tribological behavior of composites based on ZA-27 alloy reinforced with graphite particles*, Tribology Letters, 37, 2, 2010, 401-410	21
Slovenia (295)	Kalin M., Zalaznik M., Novak S., *Wear and friction behaviour of poly-ether-ether-ketone (PEEK) filled with graphene, WS_2 and CNT nanoparticles*, Wear, 332-333, 2015, 855-862	48
	Basu B., Kalin M., *Tribology of Ceramics and Composites: A Materials Science Perspective*, John Wiley & Sons, Hoboken, 2011	48
	Çalışkan H., Kurbanoğlu C., Panjan P., Čekada M., Kramar D., *Wear behavior and cutting performance of nanostructured hard coatings on cemented carbide cutting tools in hard milling*, Tribology International, 62, 2013, 215-222	37

Table 3. *Continued.*

Turkey (3596)	Gül H., Kılıç F., Aslan S., Alp A., Akbulut H., *Characteristics of electro-co-deposited Ni-Al_2O_3 nano-particle reinforced metal matrix composite (MMC) coatings*, Wear, 267, 5-8, 2009, 976-990	120
	Kurt A., Uygur I., Cete E., *Surface modification of aluminium by friction stir processing*, Journal of Materials Processing Technology, 211, 3, 2011, 313-317	72
	Gül H., Kılıç F., Uysal M., Aslan S., Alp A., Akbulut H., *Effect of particle concentration on the structure and tribological properties of submicron particle SiC reinforced Ni metal matrix composite (MMC) coatings produced by electrodeposition*, Applied Surface Science, 258, 10, 2012, 4260-4267	67
World (n/a)	Geetha M., Singh A.K., Asokamani R., Gogia A.K., *Ti based biomaterials, the ultimate choice for orthopaedic implants – A review*, Progress in Materials Science, 54, 3, 2009, 397-425	2081
	Ferracane J.L., Resin composite – State of the art, Dental Materials, 27, 1, 2011, 29-38	617
	Liu D., Tang Y., Cong W.L., *A review of mechanical drilling for composite laminates*, Composite Structures, 94, 4, 2012, 1265-1279	292

Table 4. Collaborations of SEE countries with other countries in the field of composites tribology.

Country (P)	Collaboration country* (P)		
Bulgaria (42)	United Kingdom (5)	Poland (4)	CZE (3) FRA (3) DEU (3) USA (3)
Croatia (14)	Slovenia (4)	CHN (1) CZE (1) FRA (1) USA (1)	
Greece (83)	Hungary (6)	United Kingdom (6)	DEU (5) ITA (5) USA (5)
Romania (183)	Portugal (17)	Germany (9)	BRA (8)
Serbia (37)	France (2)	Slovakia (2)	CHE (2)
Slovenia (35)	Croatia (5)	Turkey (3)	IND (2) ESP (2)
Turkey (508)	United States (20)	Iran (11)	GBR (10) DEU (10)

* name of the countries in the last column was abbreviated by the three letters UN codes

The share of internationally collaborative publications in the overall scientific output of a country can be considered as first indication of the extent to which researchers in a country cooperate with colleagues outside their own country. This indicator can be seen as a measure of the international orientation of a country. On the other hand, the share of non-internationally collaborative publications, is an indicator of the size of a country's own scientific basis and of its "scientific independence" [8].

4. Conclusions

Bibliometric indicators were used to describe the scientific activity in the field of composites tribology of in the 11 Southeastern Europe countries during the last decade. Based on the authors' knowledge, this is the first article that analyse the quantity and quality of scientific research on this topic from this region. The research output is presented through bibliometric indicators such as the total number of publications and citations, the average number of citations per publication, and the *h*-index.

The total number of publications found in Scopus® between 2009 and 2018 is approximately 900, which is only 5 % of the literature on tribology of composites published worldwide. Turkey has the largest publication output of the SEE countries, i.e., 56.3 %, followed by Romania (20.3 %) and Greece (9.2 %). On the other hand, order of SEE countries is completely different when considering the scientific research efficiency i.e. the average number of citations per publication. Greece holds the highest number of citations per publication (10.3) followed by Slovenia (8.4), Serbia (8.1) and Turkey (7.1). Romania have very low number of citations per publication value (2.8), the second lowest of all analysed SEE countries, although it has relatively high production of the publications.

The majority of publications are written in English, and by the authors employed at the universities and public research institutions associated with higher education organizations. The diversity of institutions of the most productive SEE authors is questionable, since in some cases all of the authors are from the same institution. In addition, some authors have more publications than citations, which can be due to several reasons among which is that the publications do not have the potential to impact other researches. The percentage of international publications is for all SEE countries moderate, and mostly with non SEE countries (more than 98 %).

Acknowledgments
Aleksandar Vencl acknowledges the project TR 35021, supported by the Republic of Serbia, Ministry of Education, Science and Technological Development. Nikolaos M. Vaxevanidis acknowledges the funding program "Strengthening research of ASPETE faculty members", supported by the "Special Account for Research of ASPETE", Greece. Mara Kandeva acknowledges the project ДН 07/28-15.12.2016, funded by the National Science Fund at the Ministry of Education and Science, Bulgaria. Collaboration through the CEEPUS network CIII-BG-0703 is also acknowledged.

References
[1] Broadus R N 1987 Toward a definition of "Bibliometrics", *Scientometrics* **12** 373–9
[2] Vencl A 2012 Tribology of the Al-Si alloy based MMCs and their application in automotive industry, in: Magagnin L (Ed.) *Engineered Metal Matrix Composites: Forming Methods, Material Properties and Industrial Applications* (New York; Nova Science Publishers) pp 127–66
[3] Vaxevanidis N M, Vencl A and Psyllaki P 2015, Research on tribology in Southeastern Europe: A bibliometric study, *FME Transactions* **43** 259–68
[4] Vaxevanidis N M, Vencl A, Assenova E, Kandeva M and Psyllaki P 2019 Scientific literature on thermal spray coatings from Southeastern Europe: A ten years bibliometric analysis, *FME Transactions* **47** 649–57
[5] Martin B R 1996 The use of multiple indicators in the assessment of basic research, *Scientometrics* **36** 343–62
[6] Todeschini R and Baccini A 2016 *Handbook of Bibliometric Indicators: Quantitative Tools for Studying and Evaluating Research* (Weinheim; Wiley-VCH)
[7] Durieux V and Gevenois P A 2010 Bibliometric indicators: Quality measurements of sientific publication, *Radiology* **255** 342–51
[8] van Raan A F J, van Leeuwen T N and Calero-Medina C 2009 Performance of European science: research networks and profiles of EU countries in a global perspective, in: Delanghe H, Muldur U and Soete L (Eds.) *European Science and Technology Policy: Towards Integration or Fragmentation?* (Cheltenham: Edward Elgar Publishing) pp 256–90
[9] Hirsch J E 2005 An index to quantify an individual's scientific research output, *Proc. Natl. Acad. Sci. U.S.A.* **102** 16569–72
[10] dos Santos Rubem A P, de Moura A L and Soares de Mello J.C.C.B 2015 Comparative analysis of some individual bibliometric indices when applied to groups of researchers, *Scientometrics* **102** 1019–35

Numerical considerations regarding the occurrence of plastic shearing with implications in scuffing

M Fomin[1] **and I Musca**[2]

[1]Technical College, Radauti, Romania
[2]Mechanical Engineering Faculty, Stefan cel Mare University, Suceava, Romania

E-mail: iliem@usm.ro

Abstract. Scuffing occurrence is also correlate with rheological behaviour of the lubricant. Using a viscous-plastic rheological lubricant model and an idealized roughness, as a symmetrical planar pad, a theoretical model of the pad-plane contact behaviour was developed. Based on this theoretical lubricating model, a numerical study was conducted in order to reveal the possibility of plastic shearing occurrence, in a pair formed by a symmetric pad, with dimensions comparable to the real roughness of the specimens previously used in previous experiments and a plane for values of operating parameters comparable to the real ones. The effect of the relative speed in contact, the effect of the minimum film thickness and the effect of the tilt angle of the idealized asperities flanks was analyzed. The numerical results, obtained based on the analytical model developed, correspond as order of magnitude and tendency of the parameters influence with the experimental scuffing results obtained by the authors and those presented in the literature and highlight the possibility of the occurrence of plastic shearing in the vicinity of the peak of the asperity and therefore the possibility of initiating the scuffing phenomena into contact.

1. Introduction

Mechanical equipment operates on the basis of a transfer of energy from the leading elements to the controlled elements. This transfer involves the presence of controlled moving parts, subjected to links in the mechanical couplings pairs that support them, and the relative movement of the couplings is accompanied by the natural phenomena of friction and wear.

Over time, researchers have been constantly concerned with the reduction of these effects, which lead to a decrease in the efficiency of the machines but also to their breakdown.

Between the types of the damages that occur in mechanical pairs, the catastrophic damage is the most dangerous, because it leads to unexpected stop working of the machinery, with serious consequences on economic implications. The catastrophic failure occurs suddenly and the recovery is impossible.

Of the catastrophic damage, which can suddenly stop working the mechanical pairs, the most dangerous is the tribological damage known as scuffing, characterized by the avalanche of adhesion between the micro asperities of the surfaces. Main subject of research many of authors, as: Block, [1], Ludema, [2], Jacobson, [3], Castro, [4], Dyson, [5], Bujoreanu, [6], Diaconescu, [7,8,9] Balan, [10], Fomin, [11] and others, scuffing remains a catastrophic deterioration whose causes are not fully elucidated and understood.

Of the parameters accepted as having a decisive influence on the occurrence of scuffing, the sliding velocity into the contact, the load and the size of the surface roughness are distinguished. Less studied from the point of initiation of the phenomenon of scuffing (but having a particular phenomenological significance) is the rheological manifestation of the lubricant, to which authors such as Diaconescu, Huang and Wen, [12], Balan, [10], Nelias, [13], Fomin , [11], pay a special attention, showing that the lubricant transition from viscous to plastic behavior, there are effects that may favor the occurrence of scuffing, which is especially noted to reduce the lift effect of the lubricant film.

The present paper proposes, in this context, to carry out a numerical two dimensional simulation of the lubricant behavior in a micro-contact, formed between a micro-asperity and a plane, in view of reaching the plastic shear stress and thus eventual occurrence of the plastic behavior.

2. Mathematical and geometrical model

To study the viscous plastic behavior of the lubricant in a micro-contact, we consider the geometric model formed by an idealized asperity, in the form of double symmetrical pads, as used by Fomin, [11], figure 1, and the analytical model of equations:

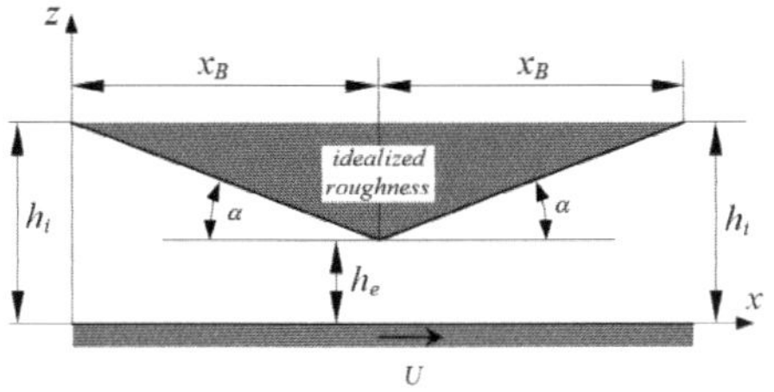

Figure 1. Geometrical model of the idealized roughness.

-The film's thickness equation

$$h(x)=\begin{cases} h_i+k_1x & x\in[0,\ x_B] \\ h_e-k_2x_B+k_2x & x\in[x_B,\ 2x_B] \end{cases} \tag{1}$$

where $k_1 = - k_2$ are the pads tilts (symmetrical pads).

- Lubricant behaviour,

$$\begin{cases} \tau=\eta\left|\dfrac{\partial u}{\partial z}\right| & \left|\dfrac{\partial u}{\partial z}\right|<\left|\dfrac{\partial u}{\partial z}\right|_{cr} \\ \tau=\tau_l & \left|\dfrac{\partial u}{\partial z}\right|\geq\left|\dfrac{\partial u}{\partial z}\right|_{cr} \end{cases} \tag{2}$$

where τ_l is the plastic limit shear stress and the critical shear $\left|\dfrac{\partial u}{\partial z}\right|_{cr}$ are an intrinsic characteristic of the lubricant

-Flow speed equations, in Newtonian flow:

$$u(x,z)=\begin{cases} \dfrac{1}{2\eta}\dfrac{dp_1}{dx}(z^2-zh_1)+U\left(1-\dfrac{z}{h_1}\right) & 0\leq x\leq x_B \\ \dfrac{1}{2\eta}\dfrac{dp_2}{dx}(z^2-zh_2)+U\left(1-\dfrac{z}{h_2}\right) & x_B\leq x\leq 2x_B \end{cases} \tag{3}$$

where η is the lubricant viscosity, p pressure and U relative speed.

By introducing the velocities into the continuity equation, the expressions of pressure along of the contact can be computed,

$$p(x)=\begin{cases}\dfrac{6\eta U}{k_1}\left(-\dfrac{1}{h_1(x)}+\dfrac{h_{m1}}{2h_1^2(x)}\right)+C_1 & 0\le x\le x_B\\ \dfrac{6\eta U}{k_2}\left(-\dfrac{1}{h_2(x)}+\dfrac{h_{m2}}{2h_2^2(x)}\right)+C_2 & x_B\le x\le 2x_B\end{cases} \tag{4}$$

where: $C_1=\dfrac{6\eta U}{k_1}\left(\dfrac{1}{h_i}-\dfrac{h_{m1}}{2h_i^2}\right)$, $C_2=\dfrac{6\eta U}{k_2}\left(\dfrac{1}{h_i}-\dfrac{h_{m2}}{2h_i^2}\right)$ and $h_m=\dfrac{2h_ih_e}{h_i+h_e}$.

-Tangential stresses distribution,

$$\tau(x,z)=\begin{cases}\dfrac{dp_1}{dx}\left(z-\dfrac{h_1}{2}\right)-\dfrac{\eta U}{h_1} & 0\le x\le x_B\\ \dfrac{dp_2}{dx}\left(z-\dfrac{h_2}{2}\right)-\dfrac{\eta U}{h_2} & x_B\le x\le 2x_B\end{cases} \tag{5}$$

-The limit shear stress,

$$\tau_{\lim}=\begin{cases}\tau_{l0}+\beta p & 0\le x\le x_B\\ \tau_{l0}+\beta p & x_B\le x\le 2x_B\end{cases} \tag{6}$$

where the pressure is taken according to the geometric position and β is a coefficient.

The transition from the Newtonian to the plastic regime is highlighted by the graphical comparison of the calculated viscous shear stresses with the plastic shear stresses, in agreement with Balan, [10], and Fomin, [11]. If somewhere, along the pads, the computed numerically viscous shear stress exceeds the calculated plastic shear stress value, which is physically unfeasible, it is considered that along these domains of the model the lubricant has a plastic behavior.

3. Numerical results

The presented analytical model was implemented in a numerical computation software. The numerical values of the dimensions of the asperity were adopted based on the evaluation of the specimen micro-topography of the experiments presented by Fomin, [11].

Table 1. The numerical values of the dimensions of the idealized asperity.

Specimen	1	2
Finished	x_B=2.5 10^{-5} (m), h_{rug}=1,89 10^{-6} (m)	x_B=6.8 10^{-5} (m), h_{rug}=1,54 10^{-6} (m)
Superfinished	x_B=3.8 10^{-5} (m), h_{rug}=1,77 10^{-6} (m)	x_B=3.5 10^{-5} (m), h_{rug}=1,37 10^{-6} (m)

For these values the calculated values of the tilt parameter are in the range from 0.02 to 0.07.

The considered values of the kinematic, load and lubricant parameters are similar with the values used in experiment, [15], corresponding to the scuffing occurence: viscosity, $\eta=0.3$ (Pas), plastic limit shear stress, $\tau_l=0.5\cdot10^6$ (Pa), $\beta=0.15$, normal load 100 (N) and 18 (ms^{-1}) relative speed. Minimum film thickness, h_{min}, calculated using Hamrock&Dowson formulas, [14], was evaluated as

$h_{min} = 7-8 \cdot 10^{-7}$ (m).

Three categories of influences were analyzed respectively: sliding effect, film thickness effect and pad tilt effect.

3.1. The effect of the relative sliding speed into the contact

By maintaining the considered geometry and the mentioned parameters unchanged, the developed software was run for a minimum film thickness of 0.5 (μm) and a relative contact velocity U = 10-190 (ms^{-1}), with $-k_1 = k_2 = 0.04$ and $x_B = 0.025$ (mm).

From the obtained results, presented in figure 2, 3 and 4 it can be concluded that by increasing the relative speed, the viscous shearing stress increases progressively and the shear limit stresses decreases with speed. At the beginning of the diverging zone of the pad, from a certain value of the relative speed, there is a domain where the computed viscous stress value near the pad surface exceeds the limit of the plastic shear stress, so there occur a plastic shear zone into the contact, figure 2. At higher speed values, as it can be seen in figure 3, also the computed viscous stress value near the plane surface exceeds the limit shear stress. An intuitive evolution of the plastic limit shear stress and the viscous stress value near the pad are presented in figure 4, that confirm the previous conclusions.

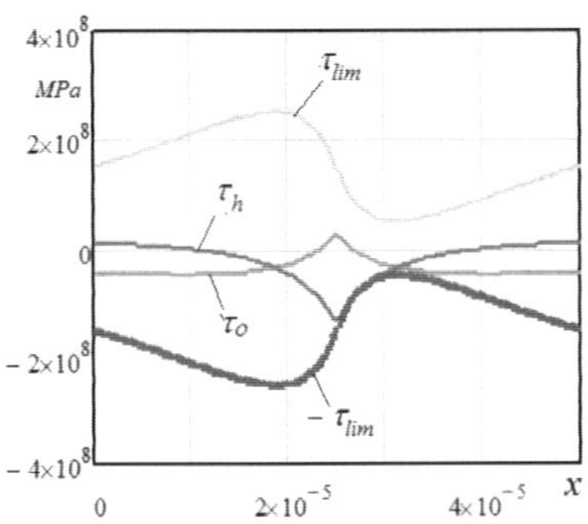

Figure 2. Speed effect at U=90 (ms^{-1})

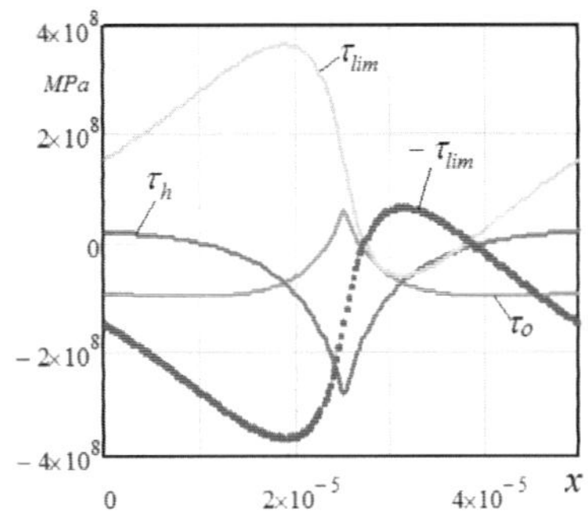

Figure 3. Speed effect at U=190 (ms^{-1})

3.2. Effect of the minimum film thickness

By maintaining the geometry mentioned above considered as undeformable and the characteristics of the lubricant and the speed U = 12 (ms^{-1}) unchanged, the software was run to study the effect of the variation of the minimum film thickness of the lubricant for film thicknesses h_{min} = 0.1 (μm) and h_{min} = 0.2 (μm).

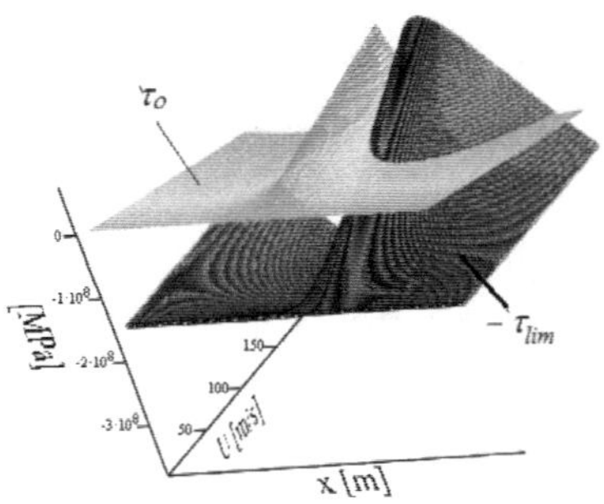

Figure 4. Speed effect - 3D View.

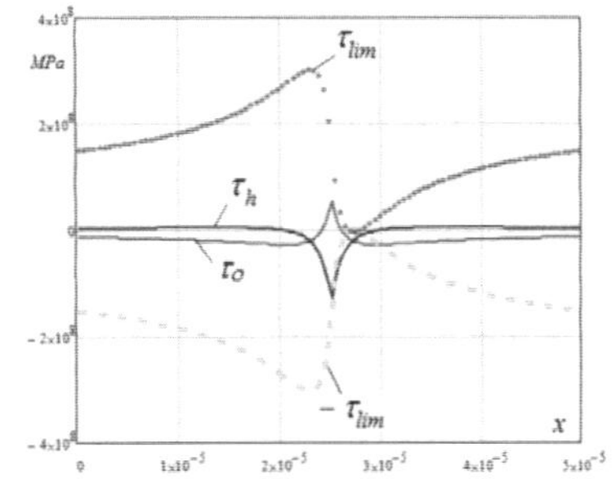

Figure 5. Effect of the minimum film thickness for U=12 (ms^{-1}) and h_{min}=0.1 (μm).

The graphical representation of the viscous and plastic shear stresses shows that the diminishing of the minimum thickness of the film causes the increase of the limit shear stress gradient, but also the increase in the values of this stress, which increase on the first pad but decrease on the second one, thus the possibility of plastic shearing occurring in the first part of the diverging pad, when the thickness of film is decreased. At lower values of minimum film thickness the plastic shearing condition can occur.

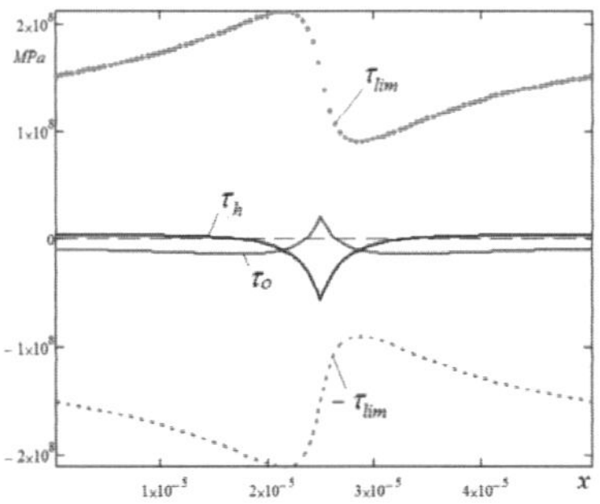

Figure 6. Effect of the minimum film thickness for U=12 (ms^{-1}) and h_{min}=0.2 (μm).

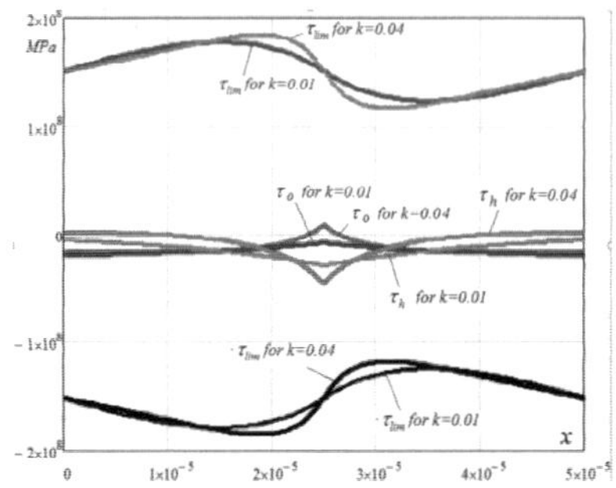

Figure 7. The effect of the pads tilt.

3.3. The effect of the pads tilt

Under the same conditions of constant preservation of the other parameters, only the angle of inclination of the pads from 0.01 to 0.04, for U = 30 (ms^{-1}), h_{min} = 0.5 (μm) and x_B = 0.025 (mm) was varied. The results show that the variation in the angle of inclination of the lateral slopes of the micro-asperity has a much lower effect compared to the variation of the minimum film thickness and the relative sliding velocity. Changing the angle of inclination of micro-asperity flanks in the range of analyzes could not significantly alter the rheological behavior of the lubricant and thus could not favor the initiation of scuffing.

4. Validation and conclusions

The numerical results were obtained for values of the parameters comparable to the values at which experimental initiation of scuffing was revealed, [15], the results being consistent with other results from the literature mentioned by Nelias and Bujoreanu.

It has been found that the numerical results agree as order of magnitude with the experimental results. However, appreciating the degree of simplification adopted, namely, to consider the roughness as having the simple shape of symmetrical pads with plane flanks and infinitely long to deal with the two-dimensional problem, and also the neglect of the presence of a rounding region at the tip of the asperity, the authors appreciate the numerical results obtained as satisfactory.

Another aspect, in support of the validation of this model, is that of matching the phenomenological trends encountered in the experiments, of scuffing initiating faster to larger roughnesses (i.e., to smaller film thickness) at higher speeds (hence higher shear speed) are also found in the numerical study performed.

5. References

[1] Block H 1937 Theoretical Study of Temperature Rise at Surfaces of Actual Contact Under Oiliness Lubricating Conditions *Proc. ImechE, Gen. Disc. Lubr.* **2** 222-35

[2] Ludema and Kenneth C 1984 A Review of Scuffing and Running-in of Lubricated Surfaces, With Asperities and Oxides in Perspective *Wear* **100.1** 315-31

[3] Jacobson B O and Vinet P 1986 A Model for the Influence of Pressure on The Bulk Modulus and the Influence of Temperature on the Solidification Pressure for Liquid Lubricants *J. Tribol.* **109** 709-1
[4] Castro J and Seabra J 2008 Global and Local Analysis of Gear Scuffing Tests Using a Mixed Film Lubrication Model *Tribol. Int.* **41** 244–55
[5] Dyson A 1975 Scuffing-A Review Part I, II *Tribol. Int.* 77-87, 117-22
[6] Bujoreanu C 2002 Studies and Researches on Scuffing in Rolling Contact Applied on Ball Bearings (in Romanian) *Ph. D. Thesis* (Iasy) 235
[7] Diaconescu E 1996 Lubricant Plastic Shearing in Plane Pad – Part I (in Romanian) *VAREHD 8* (Suceava) 70-8
[8] Diaconescu E 1996 Lubricant Plastic Shearing in Plane Pad – Part II *VAREHD 8* (Suceava) 80-6
[9] Diaconescu E 1996 About Isothermal Shearing Region in Hydrodynamic Plane Pad (in Romanian) *VAREHD 8* (Suceava) 78-80
[10] Balan M 2003 Limit Shear Stress Effect in Lubrication (in Romanian) *Ph. D. Thesis* (Suceava) 155
[11] Fomin M 2014 Speed Distribution Analysis in a Lubricating Fluid Near an Ideal Rougosity *TEHNOMUS Journ.* **1** 44-9
[12] Huang P and Wen S Z 1994 A New Model of Visco-Plastic Fluid Lubrication for Sliding Problems *Acta Tribol.* **2** 23-30
[13] Nelias D 1997 Étude expérimentale et théorique du microgrippage dans le contacts élastohydrodynamiques *ELSEVIER Rev. Gén. de Therm.* **36** 26-39
[14] Hamrock B J and Dowson D 1977 Isothermal Elastohydrodynamic Lubrication of Point Contacts: Part III—Fully Flooded Results *J. Tribol.* **99.2** 264-75
[15] Fomin M 2014 Lubricant plastic shearing effect implications in scuffing occurence (in Romanian) *Ph. D. Thesis* (Suceava) 185

Abrasive and Erosion Wear of Composite NiCrSiB Coatings Inflicted with Subsonic Flammable Structure through a Cold Process

M Kandeva[1,2,5], T Penyashki[3], G Kostadinov[3] and Zh Kalichin[4]

[1]Faculty of Industrial Technology, Technical University of Sofia, Bulgaria
[2]South Ural State University, 76 Prospekt Lenina, Chelyabinsk, Russia
[3]Institute of Soil Science Agrotechnologies and Plant Protection "N. Pushkarov", Sofia, Bulgaria
[4] SciBuCom 2 Ltd., P.O.Box 249, 1113 Sofia, Bulgaria
[5]Corresponding author: kandevam@gmail.com

Abstract. Ni-Cr-B-Si alloys are commonly used in the thermal spraying of steels. Despite their widespread use, their role in protecting metal surfaces from wear is not yet fully clarified. In order to obtain data and understanding of the wear and wear resistance of these coatings, the present work is studied the characteristics of the abrasive and erosive wear of new Ni-Cr-B-Si composite coatings applied by means of a sonic flame jet when heating the substrate to 180 °C (cold process). The powdered compositions used contain particles with an average size of 40 μm and a different percentage of the chemical elements. The abrasion wear is investigated using the thumb-disc pattern for dry surface friction with rigidly attached abrasive particles. Erosion wear is investigated by the interaction of air-jet coatings carrying abrasion particles. All samples in abrasion and erosion are tested under the same contact conditions. The mechanism, speed, and intensity of wear, and wear resistance of coatings at different normal load and slip speeds were obtained.

1. Introduction

The protection of the environment from pollution and the exhaustion of natural resources and the saving of energy are top-priority tasks of the scientific research and engineering practice during the last decade. These activities have direct connection with the tribology as a contemporary interdisciplinary science and technology for such contact processes as friction, wear, lubrication in technical systems. The decrease in the wear degree leads to reduction of the expenses for materials and consumables and in this way it decreases the extraction of raw materials from the natural environment, which results in changes in the equilibrium in ecological systems [1-3].

The abrasive and erosive wearing processes are among the most often met kinds of wear in various branches of the industry, especially in ore-mining and processing, foodstuffs producing industry, agricultural technics, energy production, road-construction technics, transportation, aircraft building, space technics and others. More than 50% of all the observed failures, connected with the wearing of industrial details, are due to abrasive wear. The expenses on a world-wide scale, caused by abrasive wear, amount to 1 – 4% of the gross domestic product of the industrially developed countries in the world [4-5].

There are several main tendencies to be outlined in the tribology, aimed at solving the problems, connected with wear [6]:

- The elaboration of new construction and lubrication materials having enhanced tribological characteristics;
- The designing of new production technologies, which ensure qualitatively new tribological properties of the conventional materials;
- The invention of new tribological coatings and modification of surface layers;
- Upgrading of the already existing constructions of the tribosystems and efficient lubricating and filtering systems;
- The discovering of new effects of self-organization in tribological systems, which lead to lowering of the entropy, expressed in minimization of friction and wear, as well as realization of such conditions of no wearing off process.

During the last 15 years there appeared many promising research results in the field of highly resistant to wear composite coatings and surface layers, containing micro- and nano-modifiers of different materials, obtained by different methods [7-10]. It is of special interest to study composite coatings, deposited by the method of thermal spraying by flame stream. The subsonic flame spraying can be realized by two different methods – a cold process and a hot process, which differ in the temperature of heating of the substrate. In the case of hot process the surface is heated in advance to 600-800°C, while the temperature in the cold process is from 50°C up to 200°C depending on the nature and on the quantity of the powder particles and the substrate.

The aim of the present research work was to investigate the characteristics of wearing of composite Ni-Cr-B-Si coatings in both regimes of interaction: dry friction along the surface by solid fixed abrasive particles and interaction with the air stream, carrying the abrasive particles.

2. Materials and technology

Four types of coatings have been studied, obtained by cold thermal spraying upon the substrate of steel having chemical composition: C – 0.15%; S - 0.025%; Mn – 0.8%; P – 0.011%; Si – 0.21%; Cr – 0.3%; Ni – 0.3% and hardness 210.5 HV2. Composite mixtures on the basis of nickel have been used having different percentage of Si, B and Cr in their composition (table 1).

Table 1. Chemical composition, hardness, thickness and roughness of the studied coatings

Sample	Coating designation	Chemical composition (wt. %)	Hardness	Thickness (μm)
1	FS-CP-1 (1355-20)	**Cr**:16; **Si**:4; **B**:3.4; **Fe**:2.7; **C**:0.6; **Mo**:3; **Cu**:3; **Ni** Balance	57 HRC	450
2	FS-CP-2 (80M60)	**Cr**:14; **Si**:4.2; **B**:2,9; **Fe**:4,6; **C**:0.6; **Mo**:2,5; **Cu**:2,4; **Ni** Balance	58 HRC	450
3	FS-CP-3 (72M40)	**Cr**:9,9; **Si**:3.1; **B**:1.7; **Fe**:3.2; **C**:0.35; **Mo**:3; **Cu**:3; **Ni** Balance	57 HRC	450
4	FS-CP-4 (HN40)	**Cr**: 20%; **Ni**: 80%	54 HRC	450
5	Substrate Steel	**Cr**:0.3; **Si**:0.21; **S**:0.025; **Mn**:0.8; **P**:0.01; **C**:0.21; **Ni**:0.3; Fe Balance	215.5 HV	-

The powder particles have average size of 45 ± 2.5 μm. Before the deposition of the coating the surface of the work piece has been treated in three stages: cleaning, blasting and mechanical processing. The cleaning was done using a solvent, which removes mechanical components, moisture, organic molecules, adsorbed on the surface. The blasting was carried out using abrasive material „Grit" in accordance with the requirements of ISO 11126.

The coatings were deposited using the system MICROJET+Hybrid, whereupon a combustion mixture of acetylene and oxygen was applied at a ratio 1:1. The velocity of the flame stream was 100 m/s, while the preliminary heating up of the substrate was done at a temperature 200°C. The coatings have equal thickness of 450 μm, which was measured by means of the device Pocket Leptoskop 2021 Fe. After polishing and finishing all the surfaces of the coatings have equal ruggedness Ra=0,430÷0,435 μm, measured by profile-metering device „TESA Rugosurf 10-10G". The hardness of the coatings was measured by hardness-metering device "Bambino" on the basis of the scale of Rockwell (HRC).

3. Abrasive wear

3.1. Experimental procedures

The methodology consists in measurement of the mass wearing off degree of the coatings after a certain number of cycles of friction and calculation of the wear rate, the absolute and the relative wear resistance. The abrasive wear resistance is represented as the reciprocal value of the rate of the wearing off process, while the relative wear resistance is the ratio between the wear resistance of the tested specimen and the wear resistance of a sample, accepted as a standard under identical conditions of friction. The abrasive wear was studied by means of the device „Thumb-disk". The samples have cylindrical form of diameter 15 mm. The investigation was carried out under the following conditions: loading of 4.5 N; nominal pressure 2.0 N/cm^2, rate of sliding 0.155 m/s; abrasive surface - Corundum P 320, temperature of the environment 21°C.

3.2. Experimental results and discussion

The experimental results on the mass wear, the rate of wearing off and the wear resistance are represented in the figures 1, 2 3 and 4.

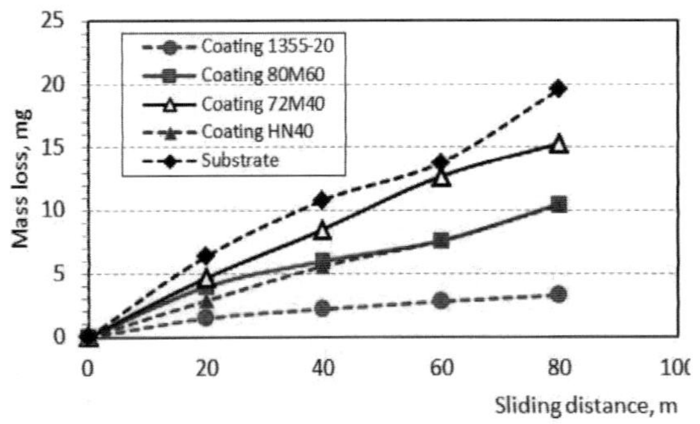

Figure 1. Mass loss vs. sliding distance for tested coatings and substrate.

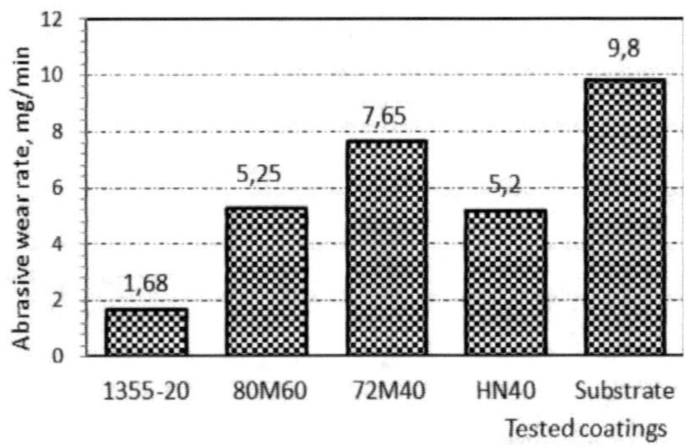

Figure 2. Abrasive wear rate for tested coatings and substrate in the friction path 80m.

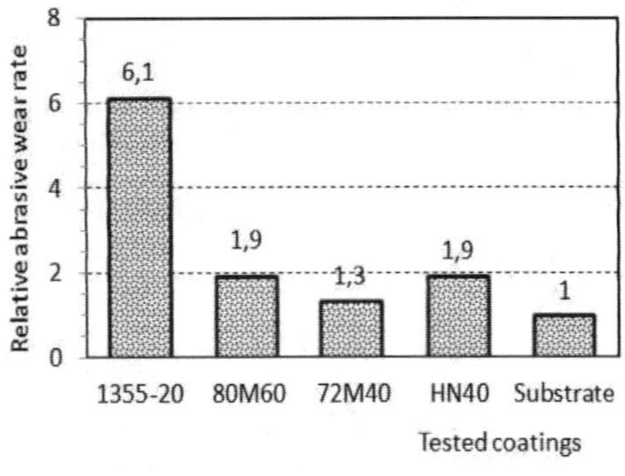

Figure 3. Diagram of relative wear rate to wear rate of substrate.

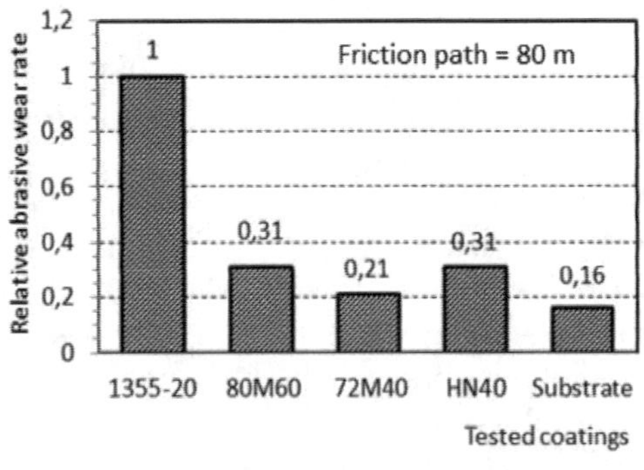

Figure 4. Diagram of relative wear rate to wear rate of coating with minimal wear rate (1355-20).

Based on the plotted kinetic curves of mass loss in Figure 1 one can see that the lowest wearing off degree is manifested by the coating 1355-20, which contains the highest amount of additional components: Cr-16%, B – 3.4%; Si – 4%. The greatest wear was manifested by the coating 72M40 having the lowest content of these elements (table 1). The wear rate value for the coating 72M40 is 4.55 greater than the wear rate of of the coating 1355-20 for one and the same length of the friction path of 80 m. The other two coatings 80M60 and HN40 have the same wear rate along the entire friction path length and values between those of the coatings 1355-20 and 72M40. The kinetic curves of mass wear rates of all the coatings have linear character, which is most distinctly expressed in the case of the coating 1355-20. The obtained results on the abrasive wear rate of the coatings cannot be explained on the basis of their hardness, there exists no correlation between the wear rate and the hardness (table 1).

4. Erosive wear

4.1. Experimental procedures

Erosive wear tests were carried out on jet nozzle type of erosion equipment (figure 5) in the ambient air, at room temperature [6]. This testing utilizes repeated gas-entrained solid particle impingement erosion, and involves a small nozzle delivering a stream of gas containing solid particles, which impacts the surface of a test specimen. Erosive wear tests were carried out on jet nozzle type erosion equipment in the ambient air, at room temperature. This testing utilizes repeated gas-entrained solid particle impingement erosion, and involves a small nozzle delivering a stream of gas containing solid particles, which impacts the surface of a test specimen. Solid particles are poured from the reservoir (1) by freefalling to the nozzle tube (2). Length of the nozzle is 200 mm, diameter is 8 mm, and exit diameter is 6 mm. Before the tests, solid particles material was sieved through a set of sieves and dried in an oven for removal of moisture from the particles. The air stream is provided by the compressed air at controlled pressure, purified from particles and moisture (3). Air stream also enters the nozzle tube (2), where the formation of two-phase (particle-air) working stream takes place. The test sample (4), in the rectangular shape (20×25 mm) and 3 mm thickness, is fixed in a holder (5) attached to the reversing mechanism (6). With reversing mechanism (6), two working parameters are controlled: (a) distance of the sample from the nozzle, and (b) impact angle of the particles. Parameters used in the erosive wear testing (solid particles material, maximum size of the particles, air stream pressure, particles flow, particles impact angle, distance between the sample and the nozzle and duration of the test) were the same for all tested coatings (table 1).

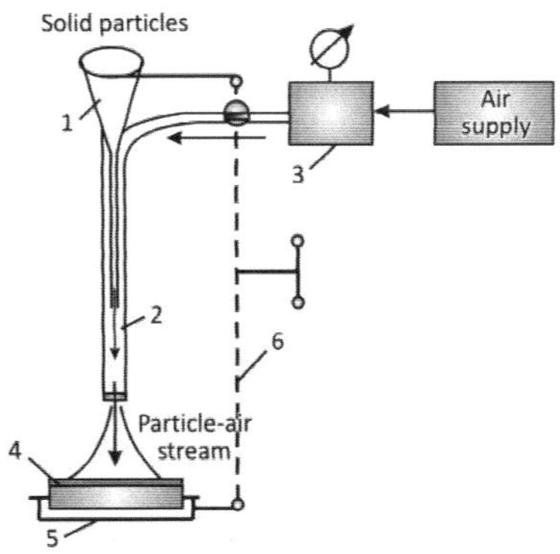

Figure 5. Schematic diagram of erosive wear testing.

The mass flow rate $\dot{m}_a$ of a given abrasive material with mass m_a in the device is determined by measuring the time t_a for the gravitational flowing out of the abrasive material - $\dot{m}_a = m_a/t_a$. The methodology for studying the erosion wear and wear resistance by means of the described device consists of measuring the mass of the sample before stream treatment and its mass after exposure to treatment with two-phase stream under the same fixed parameters: distance l angle of interaction α working pressure P, type and average size of the abrasive particles, mass flow rate of the abrasive material $\dot{m}_a$. The difference in the mass of the sample before and after the treatment represents the mass erosion wear – m_e. The rate of erosion wear is determined by the formula $\dot{m}_e = m_e/t_e$, where t_e is the time duration of the treatment of the coating. The intensity of the erosion is determined by the formula $i_e = \dot{m}_e/\dot{m}_a$. The erosion wear resistance I_e is defined as the reciprocal value of the intensity of erosion, i.e. $I_e = 1/i_e = \dot{m}_a/\dot{m}_e$. The relative erosion wear resistance $R_{i,j}$ is determined by the formula: $R_{i,j} = I_e^i/I_e^j$, where I_e^i is the erosion wear resistance of the tested sample, while I_e^j is the wear resistance of a sample, accepted as a standard for comparison, evaluated under the same conditions of wearing. The samples of the tested coatings and the substrate have the form of a parallelepiped of dimensions 25x30x5 mm. The parameters of the erosion are: solid particles material - black corundum (Al_2O_3); size of the particles – 600 μm; air stream pressure - 0.1 MPa; particles flow - 166.67 g min^{-1}; distance between the sample and the nozzle - 10 mm; duration of the test - 5 min; ambient temperature - 24°C.

4.2. Experimental results

Results have been obtained by the described methodology and testing device for erosive mass wear, for the rate, the intensity and the wear resistance during erosion at two values of the angle of interaction of the jet with the surface – 90° and 60° (figures 6, 7, 8, 9).

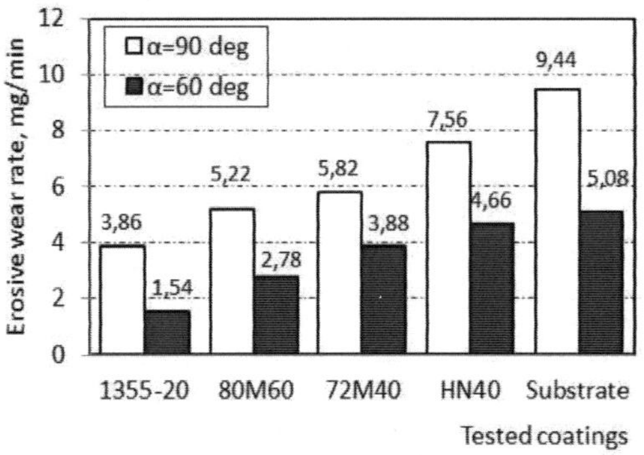

Figure 6. Influence of the angle of interaction of the jet with the surface of the coatings on the erosive wear rate.

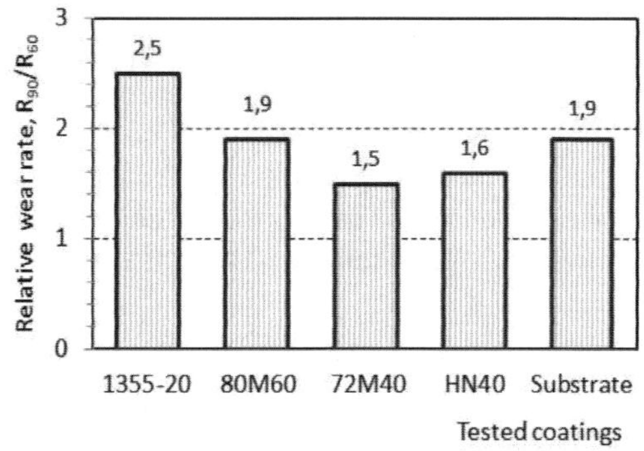

Figure 7. Relative erosive wear rate at two jet interactions with the surface area of the coatings - 90° and 60°.

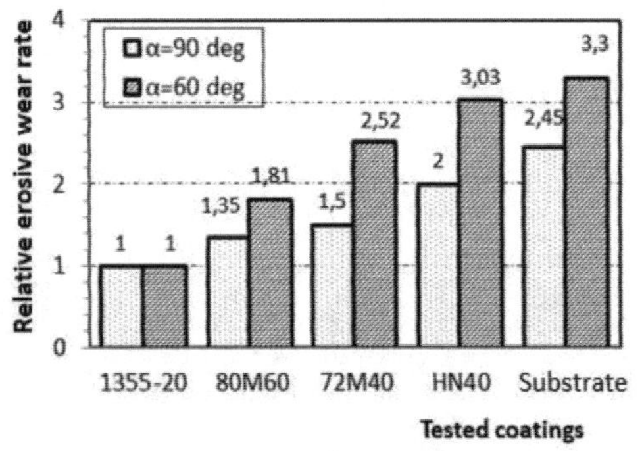

Figure 8. Relative erosive wear rate when compared to a coating with the lowest wear rate.

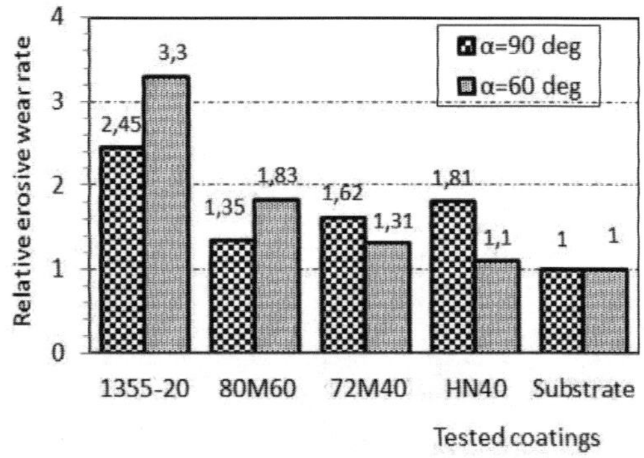

Figure 9. Relative erosive wear rate when compared to substrate coatings.

The analysis of the obtained results, shown in Figure 6 leads to the conclusion that the angle of interaction of the jet α with the surface exerts substantial influence on the rate of erosion. At $\alpha = 90^o$ the rate of erosion of all the coatings is larger than that at $\alpha = 60^o$. This is the result of the interaction of the surface with the field of distribution of rates of particles at the two angles. The lowest rate of wearing off at the two angle values is manifested by the coating 1355-20 having the highest content of Cr, B, Si. At angle $\alpha = 90^o$ its rate of erosion is 2.5 times higher than that at angle $\alpha = 60^o$. The highest wear rate is displayed by the coating HN40, which is 1.5 times higher at angle $\alpha = 90^o$ in comparison with that at $\alpha = 60^o$. The relative rates of wearing off with respect to the angle α for the coatings 72M40 and HN40 are almost identical – respectively 1.5 and 1.6. The coating 80M60 has the same relative rate of wear as that of the substrate – 1.9 (figure 7). The latter fact means that the coating 80M60 is not suitable for operation under conditions of erosion at angle $\alpha = 90^o$, that is - it does not fulfill its designed function.

4. Conclusions

Results have been obtained about the characteristics of the wearing off process for 4 kinds of coatings with nickel matrix having different contents of Cr, B and Si under two different regimes of contact interaction: dry friction along the surface with solid fixed attached abrasive particles (abrasion) and also in the case of interaction with air jet, carrying abrasive particles (erosion).

It has been ascertained that the lowest rate of abrasive and erosive wear is shown by the coating 1355-20, containing the highest concentration of the elements Cr-16%, B – 3.4%; Si – 4%. Its abrasive wear resistance is more than twice higher than its erosive wear resistance. It has been shown that the angle of interaction of air jet-abrasive stream exerts substantial influence on the rate of erosion wear, whereupon the wear rate is higher at an angle of 90 degrees.

The basic conclusion from this investigation is that the coating 1355-20 is appropriate for operation under conditions of dry abrasive friction and also under conditions of erosion at prevailing angles of interaction of 60°.

Acknowledgments

This study is connected with of the tasks under the Project DN 07/28-15.12.2016 "Research and creation of new wear-resistant coatings using composites and nanomaterials", funded by the National Science Fund at the Ministry of Education and Science, Bulgaria and under CEEPUS III Network: CIII BG 0703-"Modern Trends in Education and Research on Mechanical Systems – Bridging Reliability, Quality and Tribology".

5. References

[1] Kandeva M 2012 *The Contact Approach in Engineering Tribology* (Sofia: Publishing House TU-Sofia) p 504
[2] Bhushan B 2013 *Principles and Applications of Tribology* (New York: John Wiley & Sons) p 980
[3] Frolov KV 2014 *Modern Tribology Results and Prospects* (Moscow: LC p 580
[4] Kandeva M, Kamburov V, Zadorozhnaya E and Kalitchin Zh 2018 Abrasion Wear of Electroless Nickel Composite Coatings Modified with Boron Nitride Nanoparticles *JEPE of B.EN.A.* **19** (4) pp 1690-703
[5] Martin P M 2011 *Introduction to Surface Engineering and Functionally Engineered Materials.* (Hoboken: John Wiley & Sons) p 584
[6] Kandeva M, Karastoyanov D and Vencl A 2016 *Advanced Tribological Coatings for Heavy-Duty Applications: Case Studies* (Sofia: Marin Drinov Academic Publishing House) p 147
[7] Tashev P and Lazarova R 2013 Examination of layer overlaid with addition of TIN nanoparticles *MTC AJ* **11** (2) pp 7-12
[8] Wood R J K and Roy M 2013 *Tribology of thermal-sprayed coatings in Surface Engineering for Enhanced Performance against Wear* ed M Roy (Wien: Springer) pp 1-43
[9] Kamburov V, Dimitrova R and Nikolov K 2019 Electroless nickel coating of carbon microfibres and nanotubes intended for a reinforcing phase in metal / polymer matrix composites *J. Oxid. Commun.* **42** (1) pp 63–73
[10] Datcheva M, Dimitrova R, Iankov R, Sofronov Y and Kamburov V 2015 *Hardness characterization of laser cladding microsurfaced SiCp layers using instrumented indentation technique* Proc. Int. Conf. „70 years FIT" (Sozopol Bulgaria) pp 205-11

IOP Conf. Series: Materials Science and Engineering **724** (2020) 012015 doi:10.1088/1757-899X/724/1/012015

Structural and tribological properties of wear resistant coatings obtained by electrospark deposition

T Penyashki[1,4], D Radev[2], M Kandeva[3] and G Kostadinov[1]

[1]Institute of Soil Sciences, Agro-technologies and Plant Protection „N. Pushkarov",Sofia, Bulgaria
[2]Institute of General and Inorganic Chemistry, Bulgarian Academy of Sciences,
[3]Technical University – Sofia, Bulgaria
[4]Corresponding author: e-mail: tpeniashki@abv.bg

Abstract. In the present work, a new type of wear-resistant coatings obtained on steels by non-contact Electrical Spark Deposition (ESD), using a rotating electrode has been studied. Multicomponent electrodes for ESD containing WC-Co, Ni-Cr-B-Si semi-self-fluxing alloys and additives of superhard and refractory compounds of B_4C and TiB_2 have been obtained by the methods of mechanochemistry and powder metallurgy. The phase composition, microhardness and structural peculiarities of the coatings have been studied using methods of XRD, optical microscopy and SEM. The influence of the electrode composition and parameters of the deposition process on the structure, roughness, thickness, and abrasion wear of coatings have been studied. Tribological properties and wear resistance characteristics of coatings have also been examined. Coatings obtained using examined electrodes show higher friction and abrasion wear resistance than that obtained by conventionally used materials.

1. Introduction

The method of Electro Spark Deposition (ESD) is a progressive and dynamically developing direction of modern tribo technologies that meets modern requirements due to its unique set of advantages over the existing methods; high adhesion to the substrate, minor heating of the parts and lack of deformations, simplicity of the technology and equipment, low energy consumption, environmental safety. Selection of the composition of the electrode material is decisive for the complex of tribological and mechanical properties of coatings and for their service life. Traditional electrode materials for ESD are metals and their alloys - to increase the adhesion, the uniformity and the thickness of coatings [1,2,3], as well as carbide composites based mainly of tungsten [4,5] and titanium compounds [6,7] - to increase the hardness of the surface layers. Recently there has been a trend to used as electrodes composites based on non-metallic high melting super hard compounds such as B_4C, AlN, Al_2O_3 etc. [8,9,10] but its high brittleness limits their use.

In this connection, the aim of this work is to investigate and create suitable compositions for layering electrodes based on carbides, hard alloys, metals and additives from superhard and self-fluxing materials with which to obtain ESD coatings with improved wear resistance and triboefficiency.

The issue of improving adhesion of deposits and their wear resistance in the present work is solves by using of new composite materials with bonding components forming the unlimited solutions in the iron (Co, Ni, Cr). For maximally preserve the beneficial properties of the individual components of the coating material into the composition of the electrode material is introducing self-fluxing additives (B, Si, C,). The simultaneous introduction of the B_4C and the less fragile and high rigid components titanium diboride - TiB_2 and tungsten carbide - WC can reduces the brittles of B_4C, securing the necessary performance characteristics of the surface layer.

2. Materials and methods

The method of contactless Local Electro-Spark Deposition (LESD) [11] have been used to obtain layers on carbon steel with 0.45%C - (45) (ISO 683-1:2018) with hardness 160-180HB, and of 210Cr12 steel (ISO 4957:2018) heat treated to a hardness of HRC 59–61 and polished to a roughness Ra=0.63μm are used for substrate.

Depositing process has been performed on machine type "Elfa 541"[12] with single pulse energy up to 0.06 J. During the process, the substrate is moved with a controlled rate of axes X and Y. The high stability of the plasma spark discharges and the coating process is provided by an automatic regulation of the interelectrode distance.

During the experiments the following parameters of the regime for LESD have been used: current: 12.8÷32.0 A; voltage: 90 V; capacity: 0.22÷5.00 μF; impulse duration: 5, 8, 12, 20 μs; impulse fill factor: 0.1, 0.2 (impulse frequency - 5÷40 kHz); single impulse energy: 0.005÷0.06J; coating speed 0.5÷1.0 mm/s; number of passages of the electrode: 1÷4; rotation speed of the electrode 1500 rpm. Based on the literature analysis WC added Ni- and Co- based bonding electrode materials marked as NW and KW, respectively have been used (table 1).

Table 1. Chemical composition of the bonding metal mixtures

Designation/Element (%)	C	Si	Cr	Fe	B	Ni	Co	∑	WC
NW	0.6	2.9	12	3.9	3.6	balance	-	55	45
KW	1.5	1.5	23	0.5	1.5	30	42	45	55

The presence of self-fluxing additives (B, Si, C) and elements like Ni, Cr, Co forming unlimited solid solutions with Fe ensures obtaining of high-quality layers [8,10,13]. Different amounts of a hard alloy (WC-8%Co) and refractory and superhard B containing materials (TiB_2 and B_4C) have been added to the electrode materials designed as NW and KW to form composite materials with designations and compositions shown in table 2.

Table 2. Compositions and abbreviations of electrode materials

Abbrev./Comp.(%)	NW	KW	WC-8%Co	TiB_2	B_4C
NWW15T20	65	-	15	20	-
NWW15B10T20	55	-	15	20	10
NWW10T10B10	70	-	10	10	10
KWB10T20	-	70	-	20	10
KWB10T10	-	80	-	10	10

The electrodes have been produced by powder metallurgy methods including grinding, mixing, pressing and vacuum sintering at 1400 °C, and thereafter electrically discharge cutting with form and dimensions suitable for LESD. For comparison, a carbide electrodes with a composition of 2.8% (Ta, Nb) C + 9% CO + WC and indications WK9 [7] was used.

The surface roughness (Ra), and thickness (δ) of the resulting coatings were measured by using a Profilometer AR-132B and Leptoskop 2021 Fe (Germany). Density, uniformity, and morphology of the coatings have been monitored by a VT-300 digital microscope. The microstructure and microhardness of the coatings have been studied by optical microscopy metallographic microscope Neophot 22 (Germany) under indentation load of 5 and 10 g. The phase identification has been performed using X-ray diffractometer Bruker D8 Advance operating with Cu radiation tube. SEM methods (Bruker) have been used for structural investigations of coatings and detection the elemental distribution. The tribological properties and wear resistance of the coatings have been investigated by comparative tests of friction with tribotester type Thumb-disk under dry surface friction with hard-fixed abrasive particles. The fooling experimental conditions are used: Normal load – 5N, Nominal contact area - 2.25 cm^2 , Speed of rotation-212 min^{-1}, The friction path of the center of the contact site- 42.68 m, Abrasive surface - Corund P 320.

3. Results and discussion

3.1. Characterization of coatings

The obtained coatings have improved geometrical characteristics - higher density, uniformity, low porosity, acceptable repeatability, and lower roughness, which in most cases do not require further processing. Figure 1 shows the morphology of typical coatings from both cobalt and nickel based electrode materials. It can be seen that the surface of the coatings is mainly formed by a melt. The pictures show that the coatings have small in size structural components, with no visible cracks and pores. The morphology of coatings both cobalt and nickel-based electrodes onto the two steels is similar. Separated protuberances resulting from the fragile destruction of electrodes, whose quantity at the coatings from "NW" electrodes is bigger, can be seen. This indicates that the anode materials are eroded in both solid and liquid phases.

The minimal and maximal values of the thickness (δ), surface roughness (Ra), and microhardness (HV) of coatings, obtained from the used electrodes at different parameters of the LESD are shown in table 3. The higher amount of bonding metal in the compositions of electrodes leads to higher thickness of layers in both types of steel substrates. From the results obtained it is found that the specific values of the roughness and the thickness of coatings are different for the different electrodes but they are similar in value.The energy increase of pulses leads to increase of the roughness and thickness of the layers. The presence of B and Si slows down the formation of oxide films and has a favourable efect on the thickness of layers. The presence of B also reduces the erosion resistance of the electrode, thus increasing the transport of electrode material to the substrate surfaces [10].

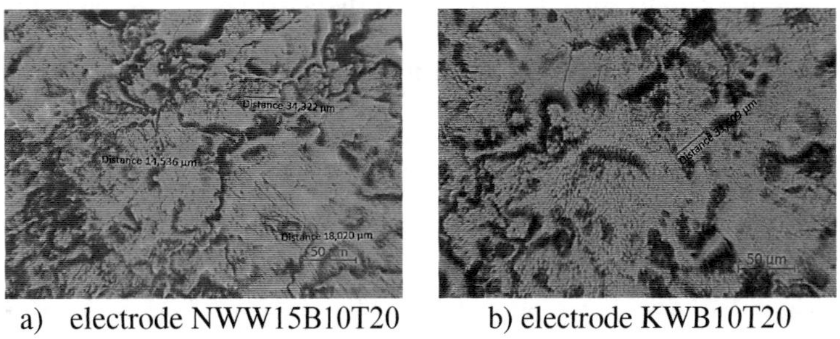

a) electrode NWW15B10T20 b) electrode KWB10T20

Figure 1. Microphotography of coating surface obtained by LESD on 45 steel with E=0.05J

Table 3. Properties of coatings obtained on 210Cr12 and 45 steel

Electrode	Coatings on 210Cr12 steel (E_i=0.005-0.05J)				Coatings on 45 steel (E_i=0.005-0.05J)			
	Ra (µm)	δ (µm)	HV (GPa)	Coeff. of hardening	Ra (µm)	δ (µm)	HV (GPa)	Coeff. of hardening
WK9	0.95÷2.5	4.5÷14	11÷14	1.2÷1.65	1.0÷ 2.6	4÷14	6÷11	2.6÷5
NWW15T20	1.3÷3.6	3.5÷20	12÷16	1.4÷1.75	1.3÷3.5	4÷20	7÷12	3÷5.4
NWW10B10T10	1.3÷3.5	3.5÷20	12÷16	1.4÷1.75	1.35÷3.7	4÷20	7÷14	3÷6.3
NWW15B10T20	1.3÷3.5	3.5÷18	12÷18	1.4÷1.75	1.35÷3.7	4÷16	7÷14	3÷6.3
KWB10 T10	1.1÷2.8	4÷18	11÷18	1.2÷1.85	1.0÷2.65	4÷18	6÷15	2.6÷5.4
KWB10T20	1.1÷3.2	4÷16	11÷17	1.2÷1.75	1.0÷2.65	4÷16	6÷15	2.6÷5.4
Substrate 210Cr12/ 45 steels	0.63	-	8.6	1.0	0.63	-	2.2	1.0

3.2. Structure and Microhardness of Coatings

Figure 2 represents the typical microstructure of some of the studied coatings. The coatings have a compact and uniform structure with hardness varing within the range 5÷18 GPa, depending on the electrode material and pulse energy of the LESD process. The initial hardness values of the hard compounds B_4C, TiB_2 and WC are 34, 33 and 25 GPa, respectively [13]. Due to the structural inhomogeneity of the coatings and the presence of a matrix of lower hardness and hard carbide phases, the microhardness of the surface layers widely varies. The maximal hardenss value of 18 GPa on 210Cr12 steel is achieved with KWB10T20 and NWW15B10T20 electrodes. The highest coefficient of increase, (K=3.8÷6 times) of the hardness, however, has been observed on 45steel, while the values of the same coefficient of tool 210Cr12 steel are 1.4÷2. Results indicate that the LESD method and the new electrodes are more effective for application to unhardened carbon steels. The observed results could be explained by the hindering influence of the alloying elements in the substrate composition.

3.3. Phase composition of coatings

Figure 3 shows the X-ray diffraction patterns of coatings deposited with NWW15B10T20 and NWW15T20 electrodes. The results show that the phase composition of layers is a mixture of disequilibrium heterogeneous finely grained martensitic-austenitic structures, ultra-disperse carbides and borides, solid solutions, intermetallic compounds and amorphous phases, whose part increases with increasing energy of deposition.

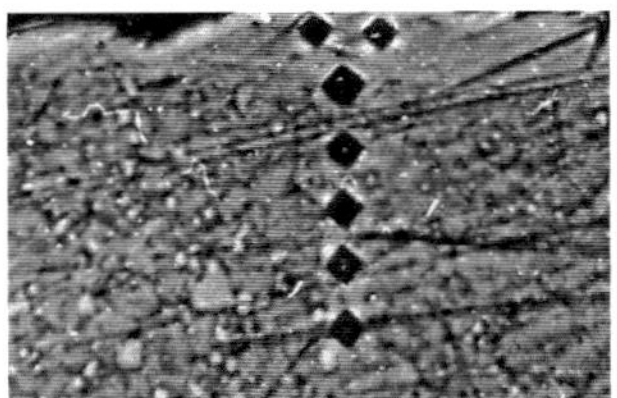

a) electrode KWB10T20/210Cr12 steel, E≈0.02J

в) electrode NWW15B10T20/ 45 steel, E≈0,05J

c) electrode KWB10T20: left - 210Cr12 steel, E≈ 0.05 J; right - 45 steel, E≈ 0.02 J

Figure 2. Cros sectional images of coatings from KW and NWW electrodes applied by LESD on 210Cr12 and 45 steels

The type and the number of the phases are determined by the anode and cathode composition and the energy of deposition. The following main phases arranged sequentially in the intensity of the characteristic lines have been registered in coatings deposited on steel 210Cr12: martensite (γ-Fe) and modified multi-component matrix of alloyed austenite (γ-Fe) - Cr-Ni-Fe-C, or Co- Cr-Ni-Fe-C when "KW" electrodes have been used. Modified carbides and borides: WC1-x, Cr_3B_4, TiB_2, TiB, B_4C, Fe_3W_3C and carbides of the basic material - (Cr Fe)$_7C_3$, Cr_3C_2, Fe_3C Fe_2C with a low intensity of peaks have also been registered. Traces of Co_3W_3C, CrNiW, WB, Fe_2B, WB as well as intermetallics and nitrides of Fe and Cr: - CrFe, $FeNi_3$ Cr_2N, Fe_2N, Ti_2Fe have been detected. On the samples of 45 steel martensite (γ-Fe) and multicomponent matrix of alloyed austenite (γ-Fe), modified carbides of type: WC1-x, and refractory borides and carbides: Cr_3B_4, TiB_2, TiB, B_4C, Cr_3C_2 have been registered. Traces of intermetallic compounds of Fe with W, Ti and Cr have also been observed. The short contact time of the particles with oxygen from the environment and the presence of self-flushing additives, create favorable conditions to prevent the coating from oxidation, which is of essential importance of the wear resistance.

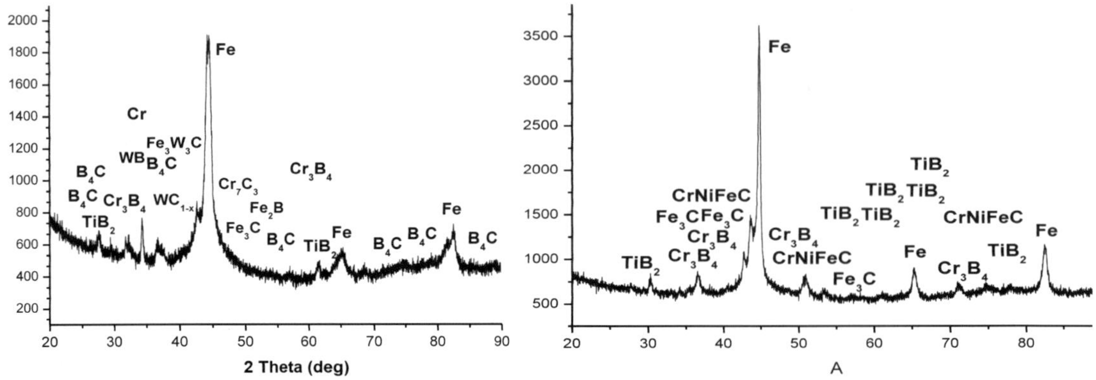

a) LESD coating from electrode NWW15B10T20 on 210Cr12 steel

b) LESD coating from electrode NWW15T20 on 45 steel

Figure 3. X-ray diffraction patterns of multi-component compositions coatings applied on 210Cr12 and 45 steels

New carbide and boride phases have been obtained by the reaction between the electrode bonding metals with boron and carbon. LESD process with electrodes NWW15B10T20, KW15B10T20 and KWB10T10 in addition to the upper shown phases leads to formation of high wear-resistant phase of the superhard compound B_4C. It indicates that B_4C has partially retained its original form during the transfer process, or which is also possible in the process of deposition secondary B_4C is synthesized. Volatilization of B leads to sytnesis of high wear resistant borides of Cr and W. It has been established that, although B_4C does not belong to the electrically conductive materials, the presence of titanium borides and tungsten carbides in the composition of electrodes make this material suitable for modifying the surface layers on steels by the method of LESD.

3.4. SEM analysis of coatings

Figure 4 shows a typical picture of the microrelief and the elemental distribution on 45 steel coated with NWW15B10T20 electrode.

El	AN	Series	unn. C (wt.%)	norm. C (wt.%)	Atom. C (at.%)	Error (1Sigma) (wt.%)
Fe	26	K-series	16.48	23.22	14.10	0.49
Ni	28	K-series	11.91	16.78	9.69	0.39
Ti	22	K-series	11.63	16.38	11.60	0.37
B	5	K-series	8.80	12.40	38.89	3.15
Cr	24	K-series	8.70	12.25	7.99	0.28
W	74	L-series	7.25	10.22	1.88	0.31
Si	14	K-series	3.27	4.60	5.55	0.18
C	6	K-series	2.50	3.52	9.92	0.85
Co	27	K-series	0.45	0.64	0.37	0.06
		Total:	71.00	100.00	100.00	

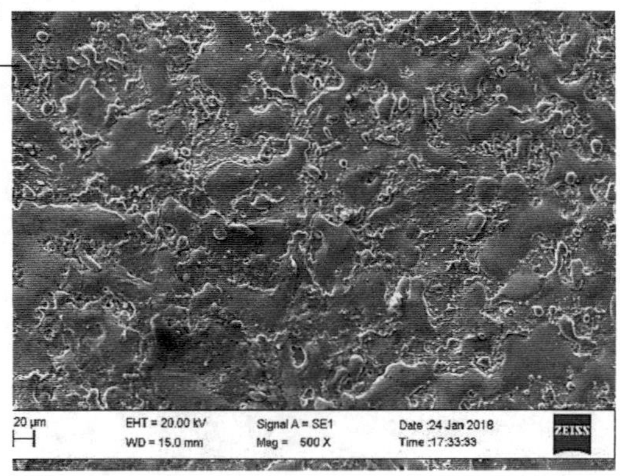

Figure 4. Microrelief and elemental composition of the coating of electrode NWW15B10T20 applied on 45 steel by vibration ESD in mode, Ee≈0.05J

The elemental analysis showed the presence of Fe, W, Ni, Co, Cr, Si, B, and C in the layer surface. Based on the data obtained, it can be concluded that the elements, which are not involved in the formation of carbides and borides form a solid solution with the iron. The microrelief of the obtained coatings is uneven and heterogeneous both in structure and in composition. The reverse diffusion of iron from substrate in direction of the coatings could be a reason for observed picture.

3.5. Wear of coatings

The results of the comparative studies of the influence of the electrode compositions and the LESD regimes on the tribo-technical properties of the coatings are summarized in table 4 and figures 5-7.

Table 4. Parameters of wear of test samples at 200 friction cycles

Pattern, Electrode / Designation, regime	Mass loss (mg)	Wear rate (gmin^{-1})	Intensity (mgm^{-1})	Wear resistance (mmg^{-1})
NWW15T20/45 steel, E≈0.03J	12.3	12.8	29 x 10^{-2}	3.45
NWW15T20/45 steel E≈0.05J	7.5	7.81	17.6 x 10^{-2}	5.68
NWW15B10T20/45 steel, E≈0.03 J	8.7	9.1	20.4 x 10^{-2}	4.9
NWW15B10T20 /45 steel, E≈0.05J	6.2	6.5	15.0 x 10^{-2}	6.87
KWB10T20 /45 steel, E≈0.03 J	5.8	6.0	13.6 x 10^{-2}	7.34
KWB10T20/45 steel, E ≈0.05 J	5.4	5.6	12.7 x 10^{-2}	7.87
KWB10T10/45 steel, E≈0.03 J	5.5	5.7	12.9 x 10^{-2}	7.75
KWB10T10 /45 steel, E≈0.05 J	5.6	5.8	13.1 x 10^{-2}	7.61
WK9/st.45 E≈0.05 J	11.7	12.2	27.4 x10^{-2}	3.64
NWW10B10T20 / 210Cr12 steel, E≈0.05 J	5.1	5.3	12 x 10^{-2}	8.35
NWW15T20/210Cr12 steel, E≈0.05J	6.2	6.5	15.0 x 10^{-2}	6.87
KWB10T20 /210Cr12 steel, I=22A, C=5μF, Ti=12 μs	4.6	4.8	10.8 x 10^{-2}	9.26
WK9/210Cr12 steel, E≈0.05 J	9.5	9.9	22.3 x 10^{-2}	4.48
Substrate 210Cr12 steel	17.8	18.5	41.7 x 10^{-2}	2.4
Substrate 45 steel	26.6	27.71	62.4x 10^{-2}	1.6

The coatings obtained reduced the wear intensity to a greater extent than these obtained using conventional tungsten hard alloys, slow down the development of wear over time, and can be used to increase the durability of friction steel surfaces as well as of parts subjected to abrasion wear. The level of wear of coated samples was 2-5 times lower than those of uncoated, and 2 times smaller than the deposited samples with WC-Co electrode. Comparing the influence of the electrode material it is found that at LESD with electrodes KWB10T20 and KWB10T10 the wear of the samples is up to 1.6 times lower than that of samples coated with the rest of electrodes shown in Table 4. Apparently, the combination of TiB_2, B_4C and WC in the composition of the Co-based electrodes allowed to use the full advantages of each of the individual component and to receive higher wear resistance of the layered surfaces compared to that obtained with WC - TiB_2 or with WC-Co electrodes. The fine-grained structure of the electrical-spark LESD coatings also contributes to the high wear resistance. Due to the lower roughness, better uniformity and the absence of thermal impact on the substrate the LESD with new electrodes is suitable for the initial layering of parts and tools with high demands on surface accuracy and quality.

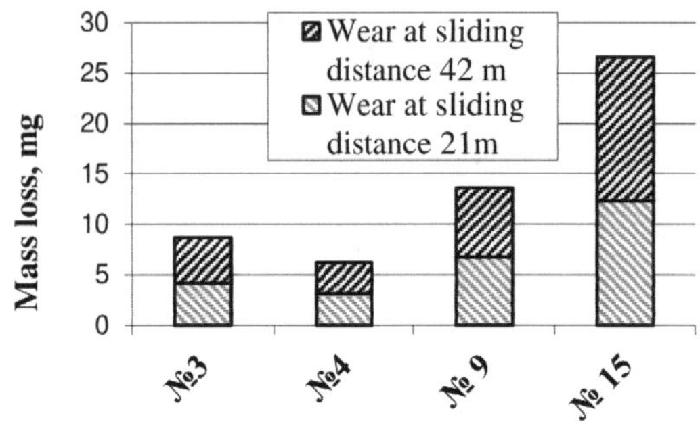

Figure 5. Mass loss as a function of sliding distance and puls energy for coatings from NWW15B10T20 electrode- table 4 on 45 steel

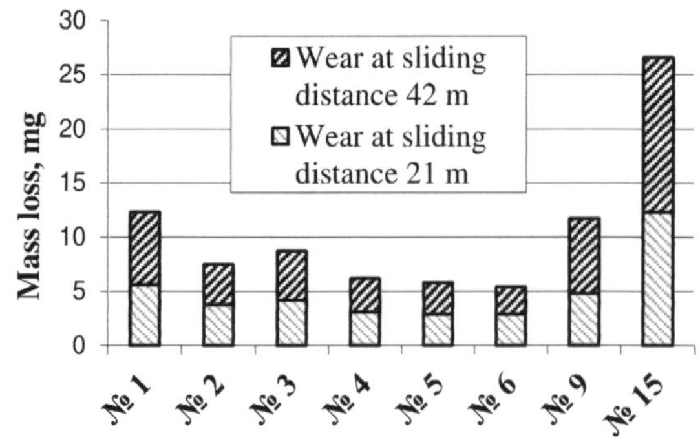

Figure 6. Mass loss as a function of sliding distance and puls energy for coatings from "NW"and "KW" electrodes - table 4 on 45 steel

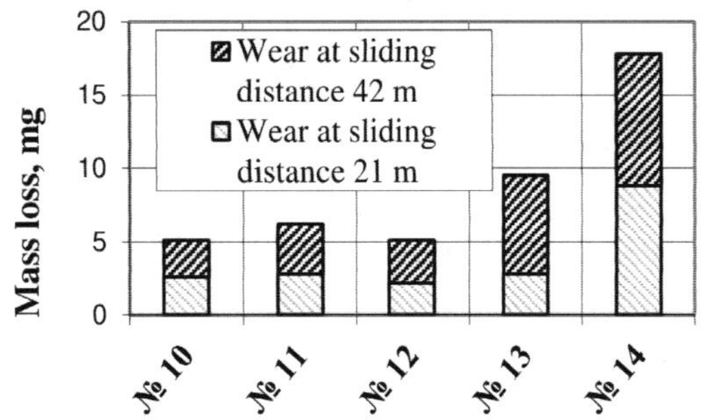

Figure 7. Mass loss as a function of sliding distance for coatings from "NW"and "KW" electrodes- table 4 on 210Cr12 steel

4. Conclusions

Multiphase electrodes based on mixtures of Ni-Cr-B-Si and Co-Ni-Cr-B-Si with additives of WC, TiB_2 and B_4C have been produced by the methods of powder metallurgy. Dense coatings with low roughness, thickness up to 20 μm and wear resistance over five times higher than that of non-coated surfaces have been obtained using LESD process.

The influence of the electrode composition and parameters of the LESD on the structure, roughness, thickness, and abrasion wear of coatings has been studied. The X-ray and SEM data show that the phase composition of coatings on steel 45 and alloyed tool steel 210Cr12 is enriched more significantly in carbide and boride phases than when conventional carbide electrodes have been used.

The optimal conditions and processing parameters to obtain the lowest wear of coated steels have been determined. The coatings obtained with KW10B20T20 electrode and the pulse energy of 0.05 J show the best wear properties.

The layers obtained may be efficiently used to strengthen rapidly wearing machine parts and tools of steels 45 and 210Cr12.

Acknowledgments

The present work is based on research funded by the Bulgarian National Science Fund of the Ministry of Education and Science under the project "Research and Development of New Wear - Resistant Coatings Using Compositional and Nano Materials".

5. References

[1] Verkhoturov A D and Nicolenko S V 2010 Development and creation of electrode materials for electric spark alloying *Streng. Tech. and Coat.* **2** pp 13-22

[2] Xiang Wei, Zhiguo Chen, Jue Zhong, and Yong Xiang 2016 Feasibility of preparing Mo_2FeB_2-based cermet coating by electrospark deposition on high speed steel *Surf. Coat. Technol.* **296** pp 58–64

[3] Xiao-Rong Wang, Zhao-Qin Wang, Tie-Song Lin and Peng He 2017 Mass transfer trends of AlCoCrFeNi high-entropy alloy coatings on TC11 substrate via electrospark–computer numerical controldeposition *J. Mater. Process. Technol.* **241** pp 93–102

[4] Ming, Wei Wang, Ren Pan, Wen Xin, Wen Qiang Chen, Xiu Jun Zhao, and Shu Li 2013 Effect of Electro-Spark Deposition Process Parameter on WC Coating Thickness of H13 Steel Surface *Adv. Mater. Res.* **690 - 93** pp 2112-15

[5] Burkov A A, Pyachin S A and Zaytsev A V 2012 Influence of Carbon Content of WC-Co Electrode Materials on the Wear Resistance of Electrospark Coatings *JSEMAT* **2** pp 65-70

[6] NIU Jing, ZHANG Li-wen, ZHANG Quan-ZHONG, WANG De-xin, CHEN Wen-hua, and LIU Dan, 2006 Microstructure of TiC Coating Deposited by Electric-spark Process on BT20 Titanium Alloy *Heat Treat. Met.* **31**(4) pp 59-61

[7] Penyashki T, Kostadinov G, Mortev I and Dimitrova E 2017 Investigation of properties and wear of WC, TiC and tin based multilayer coatings applied onto steels C45, 210Cr12 and HS6-5-2 deposited by non-contactelectrospark process *JBTA* **23** 2 pp 325–42

[8] Paustovskii A V, Tkachenko Yu G, Anisimov G M, et all 2006 Technology and materials for electroerosion hardening and restoration of machinery parts in: *Issues of Service Life and Operational Safety of Structures and Machines* [in Russian] (Ukrainy Kiev: Inst. Electrozvar. NAN) pp 553–58

[9] Paustovskii A V, Tkachenko Yu G, Alfintseva R A, Kirilenko S N and Yurchenko D Z 2013 Optimization of the composition, structure and properties of electrode materials and electrospark coatings during the hardening and restoration of metal surfaces *Surf. Engin. Appl. Electrochem.* **49**(1) pp 4–13

[10] Nikolenko V, Verkhoturov A D and Syui N A 2015 Generation and study of new electrode materials with self-fluxing additives to improve the efficiency of mechanical electrospark alloying *Surf. Engin. Appl. Electrochem.* **51** 1 pp 38-45

[11] Antonov B Device for local electric-spark layering of metals and alloys by means of rotating electrode *US Patent № 3832514* www.google.ch/patents/US3832514

[12] Antonov B, Panayotov St, Lyutakov O Apparatus for the spark deposition of metals, *US Patent 4226697* www.google.ch/patents/US4226697

[13] Radev D 2010 Pressureless Sintering of Boron Carbide-based Superhard Materials *Solid State Phenom.* **159** pp 145-48

Effects of organic compounds and nickel nitrate on friction coefficient of thermosetting

I Păduraru[1,2], G Săracu[1,2], M Bogdan[1,2], I Graur[1,2,3*], C Munteniță[1,2], A Cîrciumaru[1,2,3]

[1]"Dunărea de Jos" University of Galati, 47 Domneasca Str., Galati, Romania
[2] Research and Development Center for Thermoset Matrix Composites, 111 Domneasca Str, Corp K, Galati, Romania
[3] Diagnose and Measurement Group, 41 Rosiori Str, 800055, Galati, Romania

E-mail: iulia.graur@ugal.ro

Abstract. The principal aim of this study is to point out the effect of organic compounds mixed with nickel nitrate in determined weight ratios on friction coefficient of two distinct types of epoxy systems.The two used organic compounds are starch and agar. Ultrasound exposure is typically used to properly prepare the powder modified composite materials namely to obtain a uniform dispersion of nickel nitrate in to the epoxy. Friction coefficient behaviour of composite materials is essential for successful applications among various modern industries of these challenging materials. Friction coefficient of the modified epoxy matrix was accurately measured using pin-on-disc geometry with TRM 1000 tribometer from Wazau at the Research and Development Center for Thermoset Matrix Composites from Dunărea de Jos University of Galați.

1. Introduction

Epoxy resins are handled in a variety of applications due to their excellent properties such as thermal stability, mechanical response, electrical, corrosion resistance, and good dimensional stability [1]. It has been established in recent years that polymer based composites reinforced with a small percentages of strong fillers can significantly improve the mechanical, thermal and barrier properties of the pure polymer matrix [2].The use of organic or inorganic filler has become popular in polymeric systems [3]. In fact, it was verified experimentally by several research groups all over the world, that particles of metallic or inorganic type, prove the ability to reinforce effectively thermoplastic and also thermosetting polymer matrices [4]. Filling the polymer with various substances in different forms (powders, whiskers, and flakes) was, at the beginning, a way to reduce the amount of polymer used to form a structure but this method became a way to modify the basic properties of the polymer leading to improvements regarding thermal, electro-magnetic or tribological properties [5], [6]. An improvement in a property arises when a uniform dispersion of these fillers into the epoxy resin is

obtained. It has been shown, that a considerable improvement of the mechanical and tribological properties can already be achieved at very low filler volume content, somewhere in the range of 1–5 vol.% [7].
The dispersion of starch and agar together with nickel nitrate into epoxy resin is the biggest challenge. The commonly used methods to disperse organic or inorganic filler in a polymer matrix is by ultrasonication. Better mixing between these filler and the polymer matrix can provide higher strength and stiffness whereas poor mixing is seen to decrease those properties. There are a large number of industrial processes that employ sonication technique as an energy source for the generation of the fine emulsions and dispersions [8]. Several studies about the application of ultrasound in composite materials preparation found that ultrasound may assist in better filler dispersion [9].

Friction is a very common phenomenon in daily life and industry, which is governed by the processes occurring in the thin surfaces layers of bodies in moving contact [10].There are many polymers that have not yet studied in terms of tribology. Due to the importance of these properties which is equal to the mechanical properties, it is highly recommended for comprehensive study for such important materials [11].The most popular and also practically confirmed as the best tribological combination is steel-on-polymer frictional tribosystem [12] therefore the tests were done with this system. Most of the reported works mentioned that there is film transfer generated during the thermoset sliding against metal surfaces [13]. The transfer film formed on a non-polymer counterface is governed by the counterface material and roughness, and of course the sliding conditions [14].The transfer of material from polymer to metal surface initiates because of adhesion between the two materials which during sliding contributes to shear in the subsurface region of the contact [15]. Very wide possibilities to modify polymeric materials by fillers, and many other additives give very good perspectives to find polymeric composites that show excellent tribological properties as matched with non-polymer component [16].

The aim of this study is to point out the effect of organic compounds and nickel nitrate used as modifying agents, subjected to ultrasound exposure, for epoxy resins and to explore their role on friction coefficient of the composites materials. Friction coefficient of the composite materials was accurately measured using pin-on-disc geometry with TRM 1000 tribometer from Wazau.

2. Experimental section

To obtain these composite materials, organic agents such as starch with agar and nickel nitrate were used. The epoxy resins used are those of type HT and C with the hardener for each resin. the determination of the amount of material required was done according to Table 1.

Table 1 Amount of materials required to obtain composite materials.

Material	Resin	Hardener	Nickel nitrate	Starch	Agar	Ultrasonication
H0	100g	48g	5g	1g	0g	-
H1	100g	48g	5g	0g	1g	-
H2	100g	48g	5g	1g	0g	-
H3	100g	48g	5g	0g	1g	+
H4	100g	48g	5g	1g	0g	+
C0	100g	60g	5g	0g	1g	-
C1	100g	60g	5g	1g	0g	-
C2	100g	60g	5g	0g	1g	-
C3	100g	60g	5g	1g	0g	+
C4	100g	60g	5g	0g	1g	+

Ultrasonication was performed using an ultrasonic generator UP 100H from Hielscher (Fig. 1). The ultrasonic processor frequency is 30kHz. The diameter of the horn is 3mm.The time for ultrasonic treatment was for 2 hours.

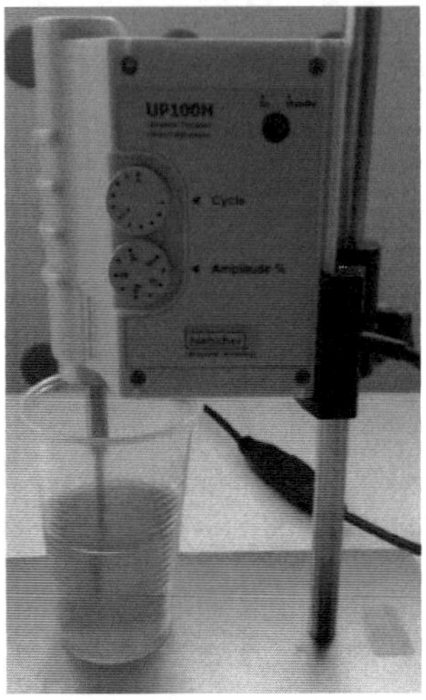

Figure 1. Ultrasonic Processor UP100H.

The friction coefficients behaviour were investigated using a TRM 1000 tribometer from Wazau (Fig.2). Friction coefficient studies of epoxy systems against steel disc are usually carried out at different load, sliding velocity, and distance conditions. R1, R2 and R3 represents the friction coeficient regimes used for composite materials tests. Tribological tests were done 23°C for 1000 m distance. The tests were performed at three different regimes with three loads and three sliding speeds in order to respect the same product load-speed (15N and 1m/s = **R1**; 20N and 0.75m/s = **R2**; 25N and 0.6 m/s = **R3**).

The cylindrical pin specimens of 8 mm diameter and 30 mm length were tested on pin-on-disk geometry, with the pin made of analyzed material and steel disk. The calculations of friction coefficient were carried out using Tribo Control Wazau software.

Figure 2. TRM 1000 Wazau tribometer.

3. Results and discussion

Evolution of friction coefficient during friction for composite materials based on epoxy resin HT are shown in Figures 4, 5 and 6. Each graph represents a regimen R1, R2 and R3 in which five curves are represented, corresponding to the five different materials: H0, H1, H2, H3 and H4.

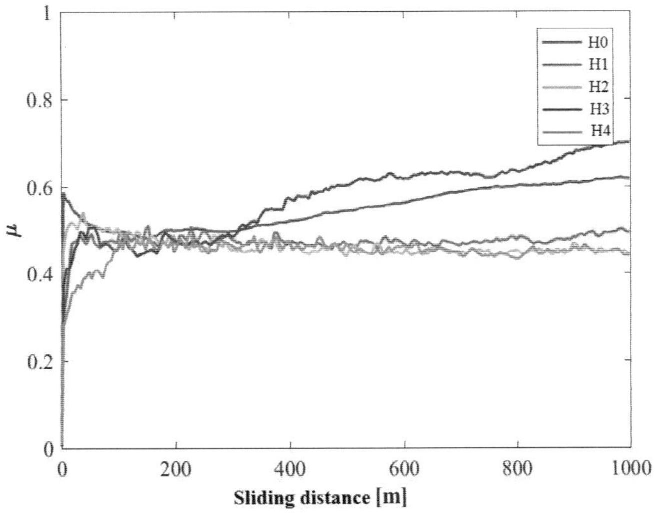

Figure 3. Friction coefficient of HT composites for R1 regime.

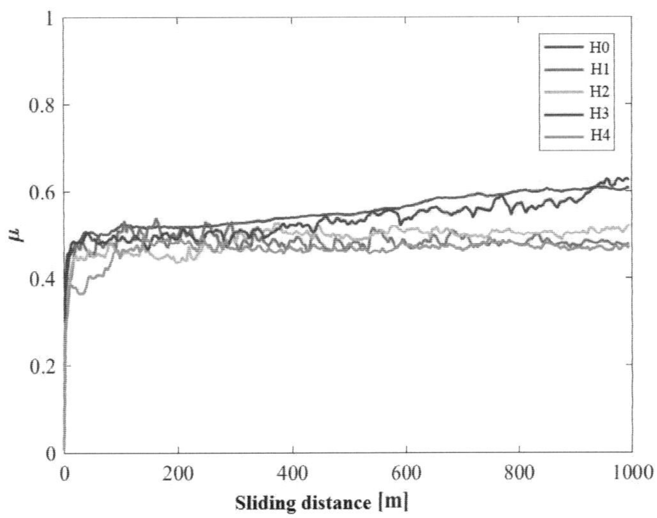

Figure 4. Friction coefficient of HT composites for R2 regime.

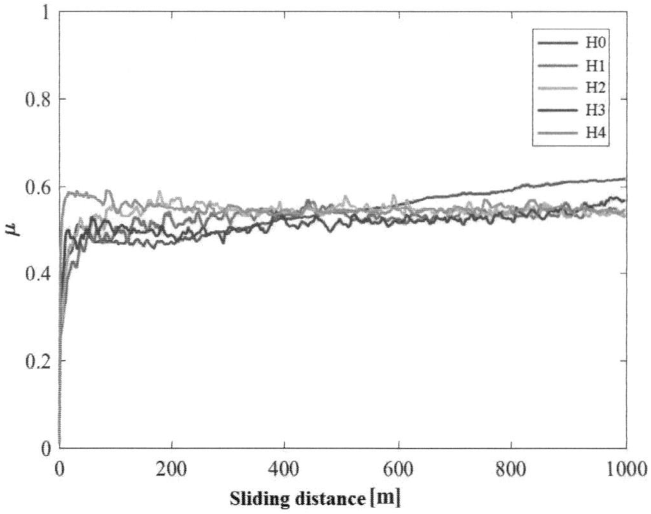

Figure 5. Friction coefficient of HT composites for R3 regime.

In Figures 3 and 4, H1, H2 and H4 materials exhibit lower friction coefficients. The materials represented in Figure 5 behave almost the same as the control material H0. The exception is the H3 material which in Figure 3 has higher coefficients of friction and in Figure 5 it behaves almost the same as the control material H0.

Evolution of friction coefficient during friction for composite materials based on epoxy resin C are shown in Figures 6, 7 and 8. Each graph represents one of the R1, R2 and R3 regimes in which five curves corresponding to the five different materials are represented: C0, C1, C2, C3 and C4.

The materials represented in Figures 6, 7 and 8 behave almost the same as the C0 control material except C3 and C4 materials exhibiting lower friction coefficients.

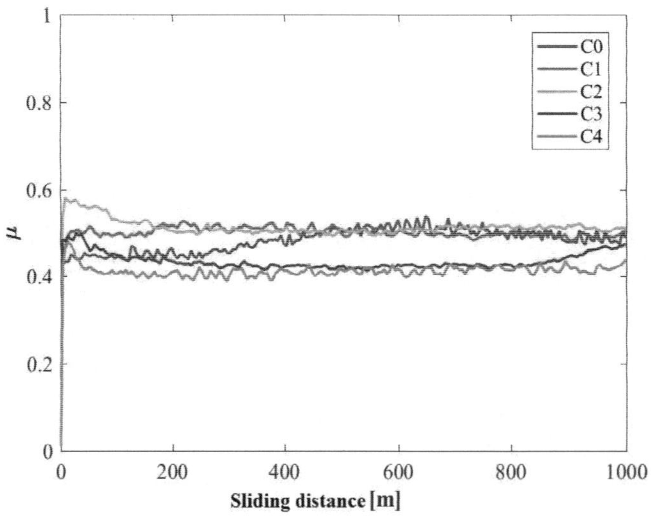

Figure 6. Friction coefficient of C composites for R1 regime.

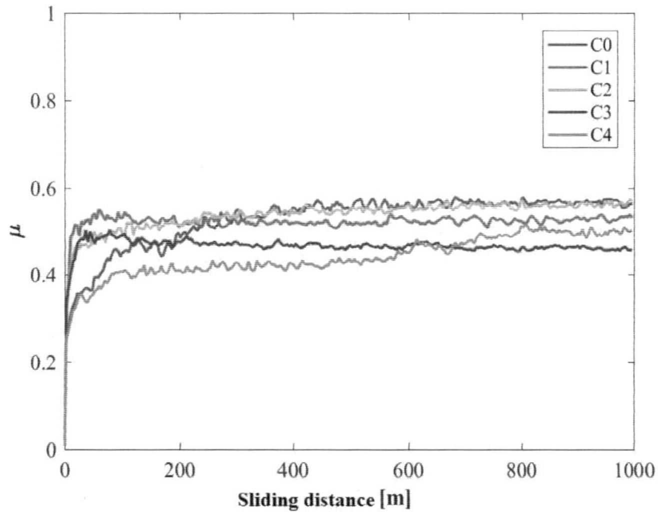

Figure 7. Friction coefficient of C composites for R2 regime.

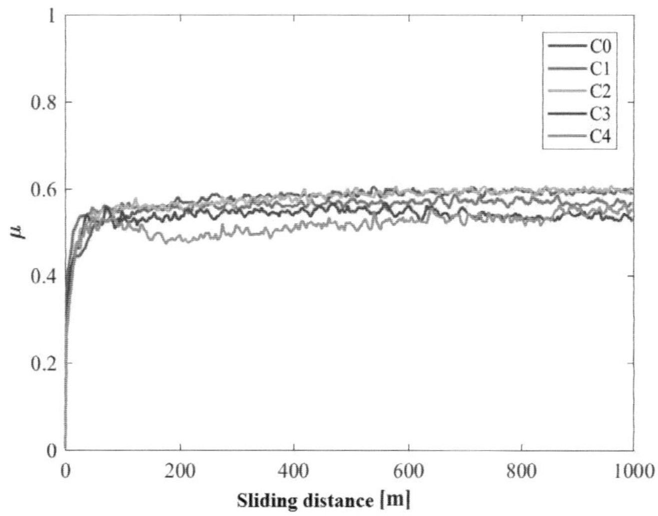

Figure 8. Friction coefficient of C composites for R3 regime.

To analyze the mass losses following the tribometric tests, the material was weighed before the test and after the test had been carried out. Analyzing mass losses, the results shown in Figures 9 and 10 were obtained.

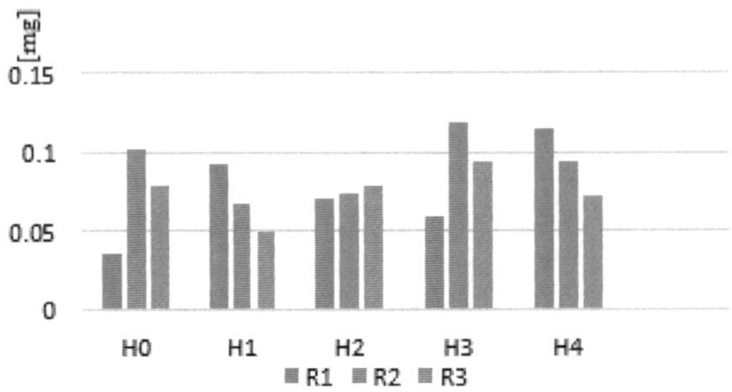

Figure 9. Loss of mass for HT composites.

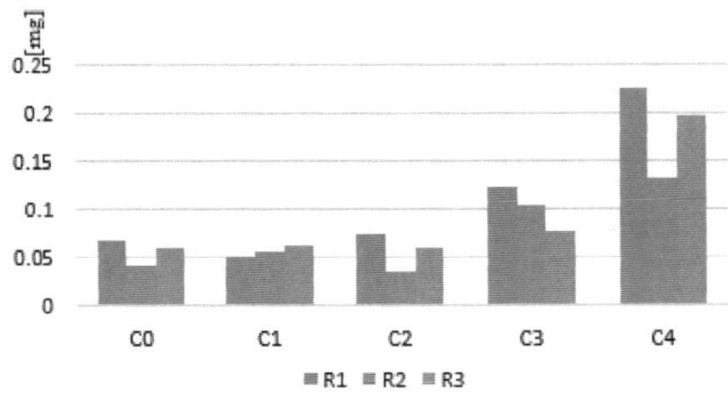

Figure 10. Loss of mass for C composites.

From Figures 9 and 10 it can be seen that ultrasound treated materials lead to higher mass losses in tribometric tests. Regarding the test regimes, one cannot establish a rule after which mass loss occurred.

4. Conclusions

In this paper we analyzed the evolution of the friction coefficient and the mass loss to identify the effect of the presence of additives (agar, starch and nickel nitrate) in the epoxy matrix and exposure to ultrasonic irradiation on materials of the same composition.
As a conclusion of the behavior of composite materials modified with additives (agar, starch and nickel nitrate) following tribological tests, we can state both the quantity and the type of the modifying agent have led to visible changes for some of thecomposite materials.

Acknowledgments

The authors would like to acknowledge the financial contribution of the Project 12 P01 024 21 (C11) /31.08.2012 (code SMIS 50414).

5. References

[1] Abdollah O, Saeedeh A and Mohammad A S 2010 Thermoset nanocomposites using hybrid nano TiO2–SiO2, *Mater Chem Phys* **122** 343–349
[2] Nathaniel C, Hassan M, Vijaya K R, Adnan A, Shaik J 2005 Fabrication and mechanical characterization of carbon/SiC-epoxy nanocomposites, *Compos Struct* **67** 115–124
[3] Farzana H, Mehdi H, Masami O and Russell E G 2006 Polymer-matrix Nanocomposites, Processing, Manufacturing, and Application: An Overview *J Compos Mater* **40** 1511-1575
[4] Singh RP, Zhang M and Chan D 2002 Toughening of a brittle thermosetting polymer: effects of reinforcement particle size and volume fraction *J Mater Sci* **37** 781–788
[5] Bălan I, Bosoancă R, Căpăţînă A, Graur I, Bria V and Ungureanu C 2017 A study regarding friction behaviour of lysine and isoleucine modified epoxy matrix. In IOP Conference Series:

IOP Conf. Series: Materials Science and Engineering **724** (2020) 012016 doi:10.1088/1757-899X/724/1/012016

Mater Sci Eng **174** 012023
[6] Bria V, Circiumaru A and Birsan IG 2011 Some Properties of Starch/Epoxy Composites, *MateR Plast*, **48** 189-194
[7] Wang Q, Shen W and Xue Q 1997 The friction and wear properties of nanometer SiO2 filled polyetheretherketone *Tribol Int* **30** 193–7
[8] Md E, Kabir MCS and Jeelani S 2007 Effect of ultrasound sonication in carbon nanofibers/polyurethane foam composite, *Mater Sci Eng A* **459** 111–116
[9] Lapshin S and Isayev AI 2006 Continuous Process for Melt Intercalation of PP–Clay Nanocomposites With Aid of Power Ultrasound, *J Vinyl Addit Techn* **12** 78-82
[10] Myshkin BK, Petrokovets MI and Kovalev AV 2005 Tribology of polymers: Adhesion, friction, wear, and mass-transfer, *Tribol Int* **38** 910–921
[11] Aldousiri B, Shalwan A and Chin CW 2013 A Review on Tribological Behaviour of Polymeric Composites and Future Reinforcements, *Adv Mater Sci Eng* **2013**
[12] Yamaguchi Y 1990 Tribology of Plastic Materials, *Tribology S* **16** Elsevier
[13] Jacko MG, Tsang PHS and Rhee SK 1989 Wear debris compaction and friction film formation of polymer composites, *Wear* **133** 23- 38
[14] Bahadu S 2000 The development of transfer layers and their role in polymer tribology, *Wear* **245** 92–99
[15] Tanaka K, Uchiyama Y and Toyooka S 1973 The mechanism of wear of polytetrafluoroethylene, *Wear*, **23(2)** 153-172
[16] Baltes L, Tierean M and Patachia S 2013 Investigation on the friction coefficient of the composite materials obtained from plastics wastes and cellulosic fibers, *J Optoelectron Adv M* **157(8)** 785 – 790

IOP Conf. Series: Materials Science and Engineering **724** (2020) 012017 doi:10.1088/1757-899X/724/1/012017

Tribological behaviour of sealing materials

C Birleanu[1], M Pustan[1], C Cosma[2], V Merie[3] and O Dranda[1]

[1]Mechanical Systems Engineering Department, Technical University of Cluj-Napoca, Cluj-Napoca, Romania
[2]Manufacturing Engineering Department, Technical University of Cluj-Napoca, Cluj-Napoca, Romania
[3]Materials Science and Engineering Department, Technical University of Cluj-Napoca, Cluj-Napoca, Romania

E-mail: Corina.Barleanu@omt.utcluj.ro

Abstract. The sealing elements are the most critical part of the mechanical sealing assembly and the main aspects of these are the tribological ones. The structural behaviour of the sealing components plays a crucial role in determining the sealing performance. Wide varieties of materials are used as seal face materials in practice. In this paper we focus on a couple of sealing materials CoCrWMo alloy / four different types of materials (copper-zinc alloy – brass, tin-copper alloy - bronze, graphite bronze and Teflon – PTFE). Dry and lubricant friction and ring on ring configuration are considered using a ring on ring tribometer. The rings from CoCrWMo alloy were additive manufactured using selective laser melting (SLM) process. Tribological tests were run at room temperature on different couples. These materials present different tribological behaviours in terms of sliding wear evaluation.

1. Introduction

Seals are used on a very large scale with direct influence on the reliability of mechanical systems (pumps, compressors, turbines, etc.). A dynamic seal is a mechanical device used to prevent liquid leakage when there is rotational movement between the sealing surfaces. In general, a mechanical seal works with an extremely thin film lubricant in the sealing space, the thickness thereof being typically in the range of 0.1-1.0 μm to reduce leakage and to limit friction and wear [1-3].

The best sealing materials have a low friction, high hardness differences, are not soluble in one another, good corrosion resistance, good bearing properties, good machining capacity and high thermal conductivity [4, 5].

Most dynamic seals, such as mechanical seals, fall into the category where friction surfaces are separated and lubricated with a thin film of lubricant [1]. Typical properties used for description of bulk material strength and deformation do not apply for thin films and micro devices. This size effect relationship is not yet well understood when structural dimensions decrease from millimeters to micro and nanometers [1].

A schematic presentation of a conventional mechanical face seal is presented in figure 1. Mechanical face seals mainly contain two rings adjusted on the rotating shaft. One part (face) is a rotating ring that is attached to the shaft and rotates with it and the other is stationary and is mounted in the housing. The rotating face can move along the shaft axis allowing relative movement and provide flexibility for small misalignment between the parts.

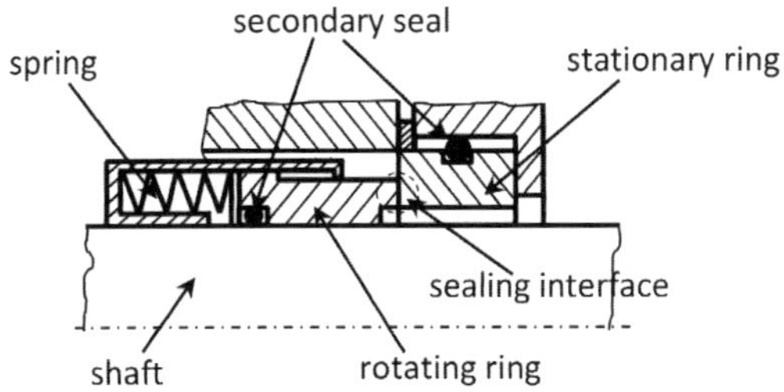

Figure 1. Schematic presentation of a mechanical face seal as a tribosystem.

We specified that the purpose of tribological studies in this paper is to establish a scientific approach to predict the friction state of mechanical seals based on higher material performance. CoCrWMo alloy widely used as a hard material is a potential candidate for but its tribomechanical properties need to be investigated in the aim of used in high performance mechanical seal working in severe conditions. CoCrWMo alloy offers good mechanical properties (hardness, Young modulus), resistance to corrosion and thermal conductivity that make it proper for tribological application in dry and lubricated sliding. Combined with matting face ring from a softer material, the sliding of CoCrWMo alloy can be sustained in severe working conditions [6].

In this paper the experimental investigations assesses the tribological behaviors of CoCrWMo alloy against four different sealing materials

2. Experimental procedures

2.1. Sample materials

The active semi-ring from the friction coupler made from CoCrWMo (commercially named Starbond CoS55) alloy was manufactured by selective laser melting (SLM) process. SLM is a complex thermo-physical process that largely depends on material, laser and process parameters (Figure 2). The main component is the solid Nd:YAG laser which emits continuous light with a wavelength of 1064 nm in infra-red spectrum. The system used was Realizer SLM 250 (Germany) and 50 μm the value of spot laser. The following SLM parameters could be set up on Realizer equipment: 20-200 W laser power, 20-100 μm layer thickness, 100-2000 mm/s scanning speed, 0.06-0.20mm hatch space and various scanning strategies (eq. X/Y, stripe hatch pattern, islands).

Two rings made of CoCrWMo alloy provided by Scheftner, Germany were manufactured by SLM (selective laser melting). This alloy is highly corrosion resistant and has a density of about 8.8 g/cm^3. The solidus-liquidus interval of it is between 1305-1400 °C. The mass percentages of each chemical element are presented in Table 1. The diameter of CoCr grains is between 10-55 μm and they are spherical.

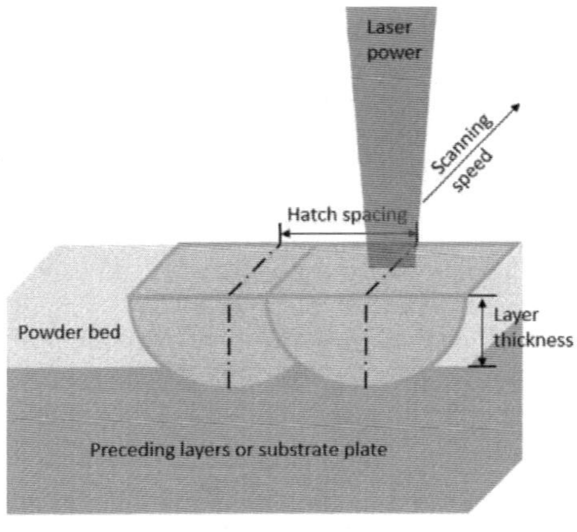

Figure 2. The main SLM process parameter [7].

Table 1. Chemical composition of the CoCrWMo powder.

Chemical element	Co	Cr	W	Mo	Si	Other elements (C, Fe, Mn, N)
Maximum weight percentage [%]	59	25	9.5	3.5	1	< 1

In order to avoid the unfavourable effects of residual stress, after SLM manufacturing the parts were exposed on a thermal treatment. This heat treatment was performed in an electric oven and air atmosphere. The parts were heated up to 860 °C with 7 °C/min. At this temperature the parts were hold for 1h and followed by cooling rate of 12°C/min. After the temperature decrease to 300 °C, we open the door of furnace for natural cooling. In the end, the parts were sandblasted with alumina. To archive the required accuracy, the rings were conventional processed (CNC turning).

Normally, after conventional manufacturing such as casting, the hardness of this CoCr alloy is between 407-601 HV1 [8]. Based on our previously research, the hardness of parts SLM processed is directly influenced by process parameters such as laser power, scanning speed or by the density of energy which is distributed on each powder slice [9]. Due to this know-how, its need to determine also the hardness obtained on rings manufactured with the following SLM conditions.

The dimensions for the primary ring from CoCrWMo are 62 mm for external diameter, D, and 4 mm thickness of hardness 540 HV1 (51,8 HRC). For the second ring (soft sample) diameter is 16 mm and 6 mm thickness and four different materials with a surface hardness presented in table 2. In the paper were defined as soft samples as follows: sample 1 made from copper-zinc alloy named brass (CuZn40PtSn); sample 2 from tin-copper alloy named bronze (BzA15), sample 3 from graphite bronze and sample 4 from teflon – PTFE.

The hardness values (table 2) are the average of 12-15 determinations per sample. The hardness was evaluated using a Microhardness Tester VMHT with the trace marks shown in figure 3.

Table 2. Rings hardness values.

Rings material	CoCrWMo primary ring	Material for soft samples Sample 1	Sample 2	Sample 3	Sample 4
Hardness HV0.5 (GPa)	5.4	1.906	1.634	0.460	0.033

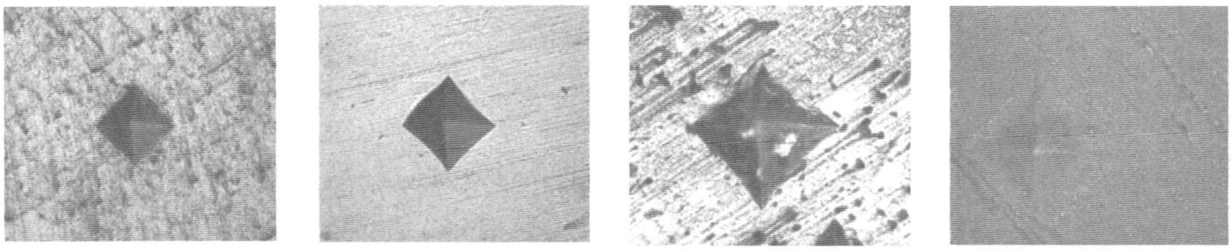

Sample 1 Sample 2 Sample 3 Sample 4

Figure 3. Indentation marks for the samples (soft samples).

2.2. Measurement of surface characteristics

Good surface roughness and flatness should be ensured to acquire low friction between seal faces. The surface of primary ring, SLM made from CoCrWMo alloy was polished to show an arithmetical mean height R_a less than 0.4 µm.

Four different sealing materials introduced above were selected as matting face rings. In the same way the surface were manufactured and R_a roughness for them is: sample 1 - 0.12 µm; sample 2 - 0.35 µm; sample 3 – 2.2 µm and sample 4 - 0.6 µm.

2.3. Tribometer and method

In this paper, we evaluated the sliding wear behaviours of a CoCrWMo alloy ring on four different materials rings in dry and lubricated environment using ring on ring tribometer in simulating the sliding wear process of material couples (figure 4). The using rings on ring tests were performed in dry and lubricated environment respectively to investigate the effect of lubrication on the wear behaviour. The lubricated test was performed in the T80W90 mechanical transmission oil. The wear track was examined using ZEISS Smartzoom 5 smart digital microscope.

Figure 4. Friction test set-up for sample 4.

Loading the friction coupler is done by a lever system by mounting the weights on the two arms. The load that sits on the plates is $F_1 = 10$ N, and second test with $F_1 = 30$ N.

The normal load that loads the friction coupler is:

$$F_{n1} = F_1 \cdot \frac{c}{d} \cdot \frac{a}{a+d} = 72.64\ N \text{ and for second test } F_{n1} = 217.91\ N \tag{1}$$

Where a, b, c, d represents the tribometric lever system. The total normal load will consider the weight of the arms and plates which are producing an additional charge $F_{n2} = 10\ N$.

$$F_n = F_{n1} + F_{n2} = 82.64\ N \text{ and respectively, } F_n = 227.91\ N. \tag{2}$$

The speed at the vertical shaft is $n = 980$ rpm and the peripheral speed is:

$$v = \frac{\pi \cdot D \cdot n}{60 \cdot 1000} = 3.18\ m \cdot s^{-1}. \tag{3}$$

The wear rate, *W*, was evaluated using the formula $W = V \cdot (F_n \cdot s)^{-1}$ where *V* is the worn volume, F_n is the normal load, *s* is the sliding distance. The wear volume *V* is determined with both ZEISS Smartzoom 5 smart digital microscope and through the difference between the sample masses before and after wear $(m_0 - m_1)$ [g] using an accurate balance type Kern AEJ with a measurement accuracy of 0.1 mg.

The specimens were washed with diluent and dried and then weighed before and after the test.

The surface topography of worn rings was determined using an optical microscope. All measurements were done at room temperature (RT). The test time was t = 60 minutes means 11448 m sliding distance per test.

3. Results and discussions

The wear track volumes and wear rates are listed in Table 3 and Table 4 and were calculated based on the formulas above. The density was determined for each sample based on mass measurements and exact volume determination. The density values are shown in Table 3 and 4.

Table 3. Result summary of dry wear tracks measured using different test parameters (F_1 = 10 N, operating time = 60 s, ambient temperature).

	Wear Track height (μm)	Wear Volume (mm³)	Wear Rate (mm³/N·m)
Sample 1 – ρ = 8.4515 g/cm³	288.00	13.8082	0.8757 x 10^{-3}
Sample 2 – ρ = 8.2123 g/cm³	202.30	25.0843	1.5908 x 10^{-3}
Sample 3 – ρ = 6.2206 g/cm³	96.60	2.25058	0.1427 x 10^{-3}
Sample 4 – ρ = 2.1310 g/cm³	373.00	44.2984	2.8094 x 10^{-3}

Table 4. Result summary of lubricated wear tracks measured using different test parameters (F_l = 10 N / F_l = 30 N, operating time = 3600 s, ambient temperature).

	Wear Track height (μm)	Wear Volume (mm^3)	Wear Rate (mm^3/N·m)
Sample 1 – ρ = 8.4515 g/cm³	49.10	0.07099	0.07503 x 10^{-6}
	84.20	0.13015	0.04988 x 10^{-6}
Sample 2 – ρ = 8.2123 g/cm³	48.70	0.02435	0.02574 x 10^{-6}
	54.00	inconclusive result	inconclusive result
Sample 3 – ρ = 6.2206 g/cm³	18.90	inconclusive result	inconclusive result
	25.70	inconclusive result	inconclusive result
Sample 4 – ρ = 2.1310 g/cm³	56.20	0.6912	0.73 x 10^{-6}
	163.50	7.0858	2.7158 x 10^{-6}

The soft samples after the dry wear test exhibits a bigger wear scar of volume. As expected, the most pronounced wear is at sample 4 (PTFE) in the amount of ~44.3 mm^3 and the least is worn sample 3 of bronze graphite in value of ~2.25 mm^3. In comparison, the wear test carried out in the mineral oil lubricant creates a substantially smaller wear track with a volume of only ~0.69 mm^3 at sample 4 (PTFE) loaded with 10N and ~7.1 mm^3 loaded with 30 N, respectively.

Figure 5 and 6 shows the optical images of the wear scars on the soft samples (the first and the third samples) after the dry and lubricated wear tests, respectively. Optical microscopy analyses were performed in order to study the surface of the samples. The tests were carried out on an Olympus GX 51 microscope. All the samples were investigated at a magnification of 50:1. We can affirm that severe wear takes place during the Ring-on-Ring test in the dry atmosphere, compared to mild parallel wear scars on the samples after the much longer lubricated wear test. The high heat (around 80-90 ℃) and intense vibration generated during the dry wear test promotes oxidation of the metallic debris and results in severe three-body abrasion. We can observe in the case of dry testing that they were affected by the sudden rise in temperature in the contact area between the sample and the rotating ring (sometimes the temperature reached 90-100 ℃).

Sample 1 **Sample 3**

Figure 5. Wear scars of the soft samples after dry wear tests under 10 N applied force and 60 s operating time at RT for the first and the third samples, magnification of 50:1.

In the lubricated test, however, the mineral oil ultimately reduces the friction and cools the contact face, as well as transports away abrasive debris created during wear, leading to significant reduction of wear rate. It was found that for the graphite bronze due to the structure of the material it absorbed oil during the lubricated test and thus the volume of wear material could not be expressed by the mass difference. Therefore, these results were not convinced by this method. Such a substantial difference in wear resistance measured in different environment shows the importance of proper simulating sliding wear in the realistic service condition.

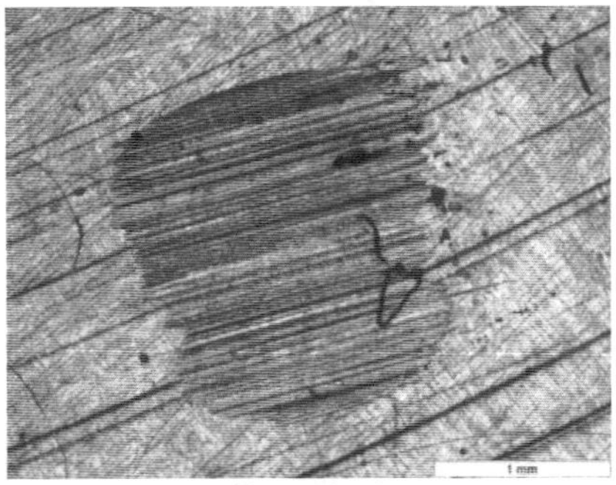

Sample 1

Sample 1

Sample 3

Sample 3

Figure 6. Wear scars of the soft samples after lubricated wear tests under 10 N and 30 N applied force and one-hour operating time and RT for the first and the third samples, magnification of 50:1.

CoCrWMo ring surface after lubricated test (30 N)

CoCrWMo ring surface after dry test (10 N)

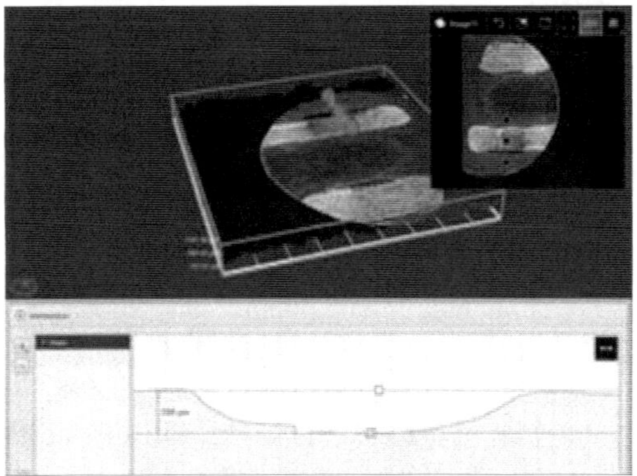

Samples 1 – surface images after dry test (10 N)

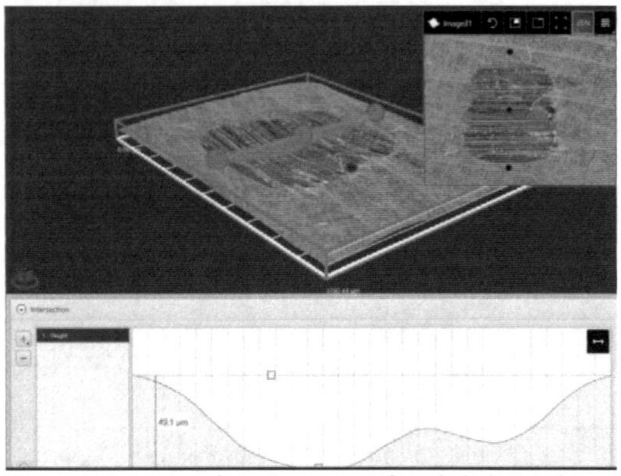

Samples 1 – surface images after lubricated test (10 N)

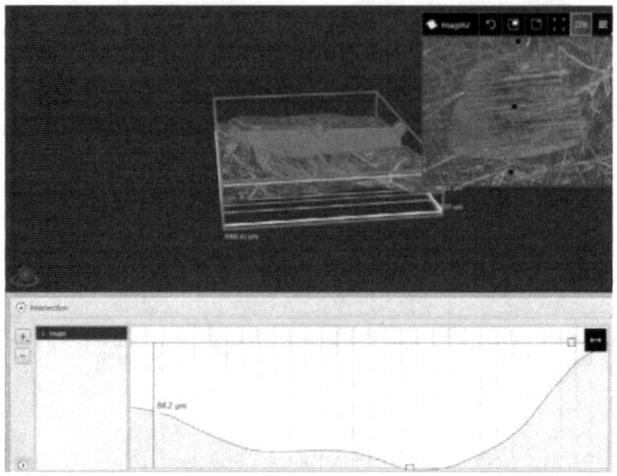

Samples 1 – surface images after lubricated test (30 N)

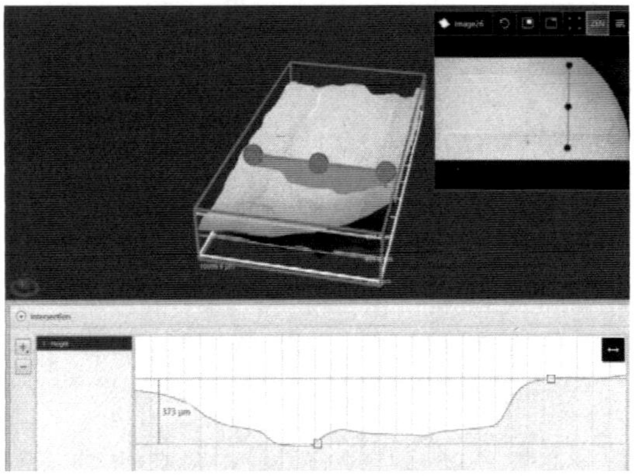

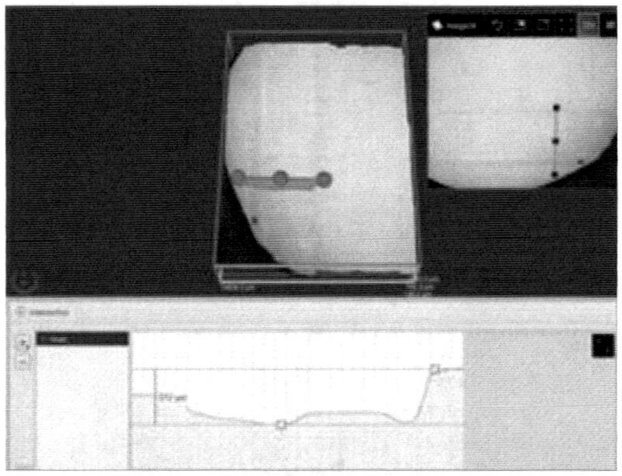

Samples 4 – surface images after dry test (10 N)

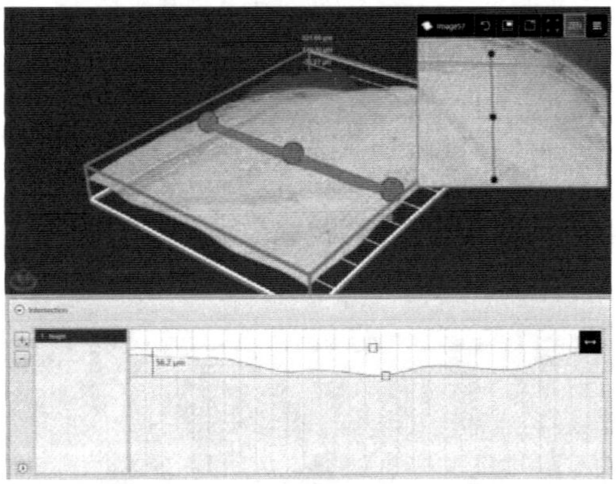

Samples 4 – surface images after lubricated test (10 N)

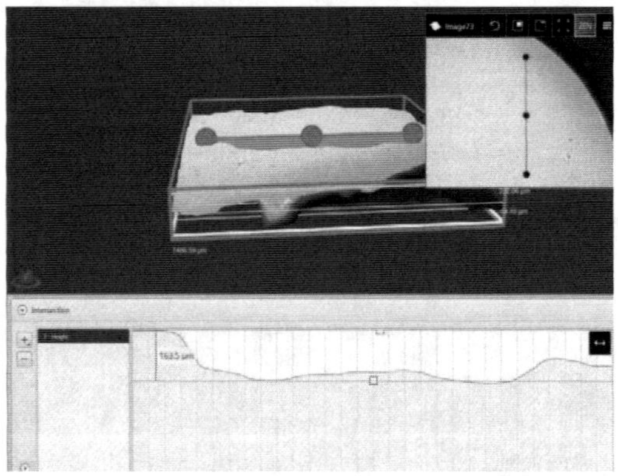

Samples 4 – surface images after lubricated test (30 N)

Figure 7. Wear scar profiles on the primary ring and soft samples for dry and lubricated wear tests.

The surface conditions of the CoCrWMo primarily ring and soft samples were examined by ZEISS Smartzoom 5 smart digital microscope. Figure 7 shows the surface morphology of the rings after the wear tests as an example. The four soft samples were analysed with this microscope under dry and lubricated conditions. In case of lubricant tests, two values for the force applied to the friction coupler were used 10 N and 30 N). All experimental results obtained on the wear track depth are presented in Tables 3 and 4. Significant roughening of the surface took place due to the three-body abrasion process during the dry wear test of 980 revolutions.

Severe wear quickly damages the contact surface without lubrication and leads to irreversible deterioration of the surface quality. Wear rate values for both primarily ring and soft samples are calculated from the volume of material lost during a specific friction run. The primarily ring scar depth is measured to calculate the ring scar volume, and soft samples scar volumes are calculated from samples weight loss. This simple method facilitates the determination and study of wear behaviour of almost every solid-state material combination, with varying time, contact pressure, velocity, temperature, humidity, lubrication, etc.

4. Conclusions

A higher wear loss is observed from the dry sliding test for CoCrWMo alloy/graphite bronze during a running-in period and accommodation. A third body is created mainly from the wear of the sealing materials ring. The impregnation of the graphite material plays a predominate role in the tribological performance of the matting face material. This third body acts like a solid lubricant reduces the stick-slip phenomenon inside the contact. A better resistance to the wear is achieved with these pairs of materials. The contact accommodations provide the best behaviour in terms of friction and then reduce also the mechanical vibrations and the noises.

In all cases, the surfaces of the friction pair are covered with oxides and adsorbed gases that fundamentally change the surface friction properties. The presence of the oxide layer is the result of a dynamic process of forming on one side and wear on the other. As it is different from static oxidation, oxidation of the friction surface is stimulated by both high temperature and surface activation and is the result of the friction process, more evident in dry friction.

When sealing rings connect each other, the tribological characteristics of the material combination determine the survival or failure of the seal. The best material used for sealing couplings have low friction, high hardness differences are not soluble with each other, have good corrosion resistance, good processing ability and high thermal conductivity.

In conclusion, based on the experimental experiments performed, we appreciate that in contact with the proposed CoCrWMo alloy material as the sealing friction semi coupling the mate material with the best tribological behaviour under both dry and lubricated conditions is graphite bronze. At lower loads it is not to be neglected either the CoCrWMo alloy material coupling with PTFE.

5. References

[1] Towsyfyan H 2017 Investigation of the Nonlinear Tribological Behaviour of Mechanical Seals for Online Condition Monitoring, Doctoral thesis, University of Huddersfield.

[2] Nau B 1997 *Proc. Inst. Mech. Eng. J: J. Eng. Tribol.* **211(3)** 165.

[3] Minet C, Brunetière N and Tournerie B 2012 *Proc. Inst. Mech. Eng. J: J. Eng. Tribol.* **226 (12)** p. 1109-1126

[4] Pustan M 2006 Contribuţii privind etanşările frontale cu impulsuri, Doctoral Thesis, Technical University of Cluj-Napoca

[5] Birleanu C and Sucala F 2008 *Tribol. Ind.* **30** 10.

[6] Zaidi H and Paulmier A 1991 *Surf. Sci.* **251** 778.

[7] Yap C Y, Chua C. K et all 2015 Review of selective laser melting: Materials and applications, Applied Physics Reviews **2**

[8] Young W C and Budynas R G 2002 *Roark's formulas for stress and strain* **7** (New York: McGraw-Hill).

[9] Cosma C, Balc N, Moldovan M, Morovic L, Gogola P and Borzan C 2017 *J. Optoelectron. Adv. Mater.* **19 (11-12)** 738.

Effect of steel fiber addition on the mechanical and tribological behavior of the composite brake pad materials

H Yanar[1], G Purcek[1,3] and H H Ayar[2]

[1] Department of Mechanical Engineering, Karadeniz Technical University, 61080, Trabzon, TURKEY
[2] Metisafe® Cleanroom and Biosafety Systems, 06980, Ankara, TURKEY

Abstract. Effect of steel fiber addition on the friction and wear characteristics and mechanical properties of non-asbestos composites brake pad material was investigated. For this purpose, the content of steel fiber was added in the range of 0 - 20 wt.% with changing the rock fiber of the matrix structure. The matrix has also phenolic binder resin, cashew nut shell liquid, aramid pulp, magnetite, rubber and graphite. Friction behavior of the samples prepared from the mixture were tested using a pin-on-disk type test configuration according to the NF F 11-292 French standard. It was observed that coefficient of friction (CoF) of composite samples increased with increasing steel fiber content. After 5wt.% steel fiber addition, the CoF of composite did not significantly change. Although the steel fiber addition caused a significant increase in the CoF of the composite, no significant change can be observed in the specific wear rate with steel fiber addition up to 15wt.%, above which it increased significantly. On the other hand, steel fiber addition to the composite matrix deteriorated the mechanical properties, but improved the fade resistance of the sample.

1. Introduction

Kinetic energy of the vehicle is converted to the heat energy at the interface of brake pad and disc/wheel during the braking application. The characteristics of the brake pad/shoes used as friction material in trains and automotive are directly affecting the braking performance of the vehicles. Therefore, it is expected that the brake pads should have an effective friction and wear performance at all stressing conditions such as high temperature and high speed. In last decades, these commercial products have been started to be produced from the composite materials where many ingredients are combined to achieve the desired requirements for the properties of the friction materials [1]. These ingredients used are collected in four groups: fibers, solid lubricants, friction modifier and space filler [2]. In recent years, the effects of types and amounts of these ingredients on the friction and wear performance of friction materials have been extensively studied by the researchers [3-6]. However, increasing demand to produce faster and more powerful vehicles has also dramatically changed the desired performance outputs from the brake friction materials. Therefore, these friction materials used need to be continuously improved.

Fibers are generally added to the brake pad matrix to increase its mechanical strength and thermal stability. Especially after asbestos fiber was banned, many different types of fiber have been added into the brake pad structure, and their tribological and mechanical effects have been comprehensively investigated [3, 7-10]. Among them, metal fibers seem to be more attractive due to their high thermal conductivity properties. Therefore, this research has attempted to examine the effect of steel fiber addition on the friction characteristic and mechanical properties of non-asbestos brake pad composite.

2. Experimental Procedure

All composite samples consist of eight constituents; binder (phenolic resin), filler (barite), fiber (steel wool, rockwool and kevlar), lubricant (graphite) and friction modifier (cashew nut shell liquid (CNSL)) and abrasive (magnetite). Steel fiber added between 0 wt.% - 20 wt.% was replaced by an equal amount of rock fiber into the matrix structure. All ingredients were mixed for 5 min in a high-speed mixer to ensure macroscopic homogenization of the mixture. Subsequently, ~10 g of the mixture was compressed under the pressure of 50 bar into a compression mold pre-heated to 90 °C. Temperature of the mold was then increased to 150 °C, and the samples were cured in the mold under a pressure of 50 bar for 10 min. Finally, post-curing was carried out by using a small size electrical furnace at 200 °C for 20 min to achieve curing all of the resin into the matrix.

Surface hardness of the test samples was measured by using a Rockwell hardness tester according to ASTM D785-65 standart with a 19.3 mm ball-indenter under the maximum load of 60 kg. This method is described as RSX. The compressive strength and elastic modulus (E) of the samples was determined using a tension/compression test machine (Instron 3382) with cylindrical test samples having the diameter of 20 mm and the thickness of 20 mm. Three samples of each composition were tested to estimate the repeatability of the results.

A specific pin-on-disc type test system designed according to the NF F 11-292 French standard was used to perform two types of friction and wear tests (Figure 1). The sample size for friction test was 20 mm in diameter and 15 mm in thickness. The counter disc was made of AISI 52100 steel with the surface roughness (Ra) of 0.05 mm and hardness of 55-58 HRC. The disc was rotated at the speed of 2800 rpm corresponding to the sliding speed of 26 m s^{-1} (95km. h^{-1}). The 80 kg normal load (corresponds to about 0.8 MPa) was applied to the samples by means of hydraulic cylinder for 20 s and then removed for 150 s for periodic braking test. This is called as one braking cycle and this was repeated 15 times for each brake pad sample. The second type of friction test (drag test) was performed to analyze the fade behavior of non-asbestos composite brake pad samples The same normal load was applied to the samples until the temperature of the disc surface reached to about 400 °C. During the test, coefficient of friction (CoF) and temperature of disc surface were recorded. The weight loss was determined using a microbalance with an accuracy of 0.1mg. The specific wear rate ($\boldsymbol{Q}$) of the samples was determined by using the Equation 1 given below. The worn surface of samples was also analyzed using a scanning electron microscope (SEM) in order to clarify the morphological features of the worn surfaces.

$$\boldsymbol{Q} = \mathrm{m}\,(\mathrm{g})\,.\;10^{6}.\,[\mathrm{X}\,.\;\mathrm{F}\,(N)\,.\,\mathrm{V}\,(m\,.\,s^{-1})\,.\;\mu\,.\;\mathrm{t}\,(\mathrm{s})\,.\;\rho\,(g.\,cm^{3})]^{-1} \qquad (1)$$

Where;

Q: Specific wear rate (cm^3. $(MJ)^{-1}$)

m: Weight Loss (g)

X: Number of braking period

F: Applied Normal Load (N)

V: Test speed (m. s^{-1})

μ: Average coefficient of friction

t: Braking time in one period (s)

ρ: Density of sample (g.cm^{-3})

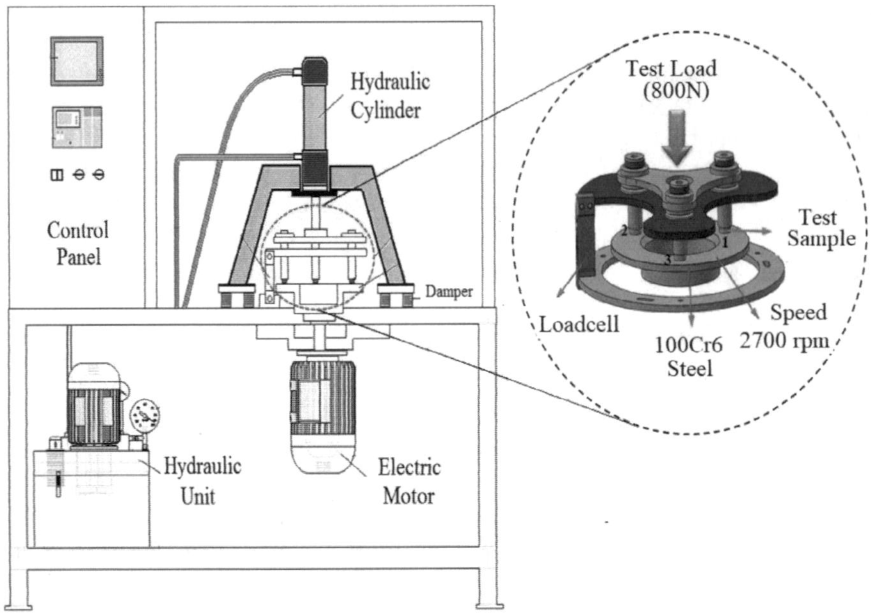

Figure 1. A schematic view of the pin-on-disc type friction and wear test rig used in the current study.

3. Results and Discussion

Table I shows the average CoF, specific wear rate, hardness, compressive strength and modulus of elasticity of all composite samples with or without steel fiber addition. Change in the amount of steel fiber affected the physical and mechanical properties of a friction material. As seen, the average CoF of composite samples increased from 0.20 (S1 base sample) to approximately 0.30 (S5 sample) with increasing the steel fiber content in the matrix structure. This is mainly because of the breakdown of the stable friction film (secondary plateaus) formed on the sliding surfaces of composite samples. Addition of 5wt.% steel fiber significantly increased the CoF of S1 sample, above which it did not cause a significant change in the CoF. The addition of steel fiber had no significant effect on the specific wear rate up to 15wt.%, above which it increased significantly. This unexpected increase is attributed to the instability of CoF. Addition steel fibre also affected the key mechanical properties of the samples. Steel fibre addition to the brake pad composite decreased generally the hardness and compressive strength of brake pad sample. However, 5% steel fibre addition into the matrix structure significantly increased the modules of elasticity of the base sample (S1). The E values were measured to be 770 MPa and 1100 MPa for the S1 and S2 samples, respectively. This could be due to the better binding capacity of steel fibre with the resin rather than rock wool fibres. The further increasing of steel fibre content in the matrix had a reducing effect on the E of composite brake pad samples. This is expected result since an increment in the amount of steel fibre in the composite's matrix is balanced only by decreasing the amount of rock wool having lower density than the steel fibre. It contributes to the formation of a more porous and void structure [11,12].

Table 1. Friction, wear and mechanical properties of composite samples including different amount of steel fiber

	Steel Fibre Content . (*wt %*)	*Average CoF (periodic braking)*	Specific Wear Rate (cm^3 (MJ)$^{-1}$)	Hardness (*HRX*)	Elastic Modulus (*MPa*)	*Compressive Strength (MPa)*
S1	0	0.205	0.026	60±0.7	770 ± 32.7	86±2.9
S2	5	0.263	0.023	61±1.2	1100 ± 64.1	83±5.1
S3	10	0.277	0.024	59±1.3	915 ± 35.0	80±2.6
S4	15	0.273	0.026	57±2.1	713 ± 47.2	76±5.3
S5	20	0.290	0.038	56±3.3	675 ± 41.8	72±4.4

The changes in the CoF of all samples with surface temperature of the disc during continuous braking tests are shown in Fig.1. The CoF of the base sample (S1) continuously decreased up to 120 °C and then remained constant to be 0.19 in-between 120 °C to 250 °C. After then, the CoF decreased again down to 0.14 level at about 400 °C. However, this change in the CoF with generated temperature was slightly different in the sample having steel fiber. It is important to note that the CoF of the samples having steel fiber did not drop below the 0.25 level until the disc surface reached to about 250 °C, above which the CoF started decreasing. Such a decrease in the CoF (Δμ) at high temperature is known as the fade. The fade values were obtained as 0.13, 0.12 0.10 and 0.07 for the S2, S3, S4 and S5 samples, respectively. These results show that the fade value decreases with increasing the steel fiber content in the matrix due to the increased thermal conductivity of the composite sample [12].

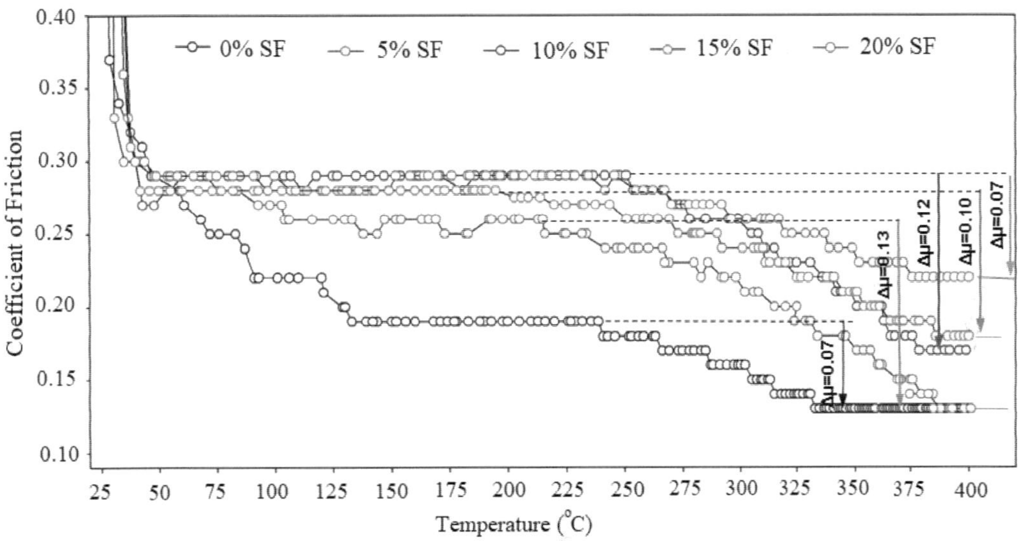

Figure 2. Variation of CoF with disc surface temperature during the continuous braking tests for the composite samples including different amount of steel fibre.

SEM images taken from the worn surfaces of brake pads after the fade (drag) tests are shown in Figure 3. In general, the worn surfaces of composite samples with and without steel fibre have different features. Worn surface of the S1 sample is almost fully covered with secondary plateaus (Figure 3 (b) and (c)). These secondary plateaus are formed by degradation of organic materials such as phenolic resin, CNSL and kevlar at the elevated braking temperature [11]. The secondary plateaus covering almost the entire surface of the S1 composite (Figure 3 (c)) are responsible for the lowest CoF and the highest degree of fade achieved in this sample (Figure 2). Distribution of primary and secondary plateaus on the worn surface varied according to the steel fibre content in the matrix structure. The areas covered by the primary plateaus (indicated by yellow arrows) on the worn surface of samples increased by increasing steel fibre content in the matrix. This resulted in an increase in the CoF of composite sample. On the other hand, pores and void (indicated by red arrow) were observed around the steel fiber on the worn surface of friction materials. This leads to poor bonding of steel fibers to the structure and reduced the mechanical properties and wear resistance of composite sample.

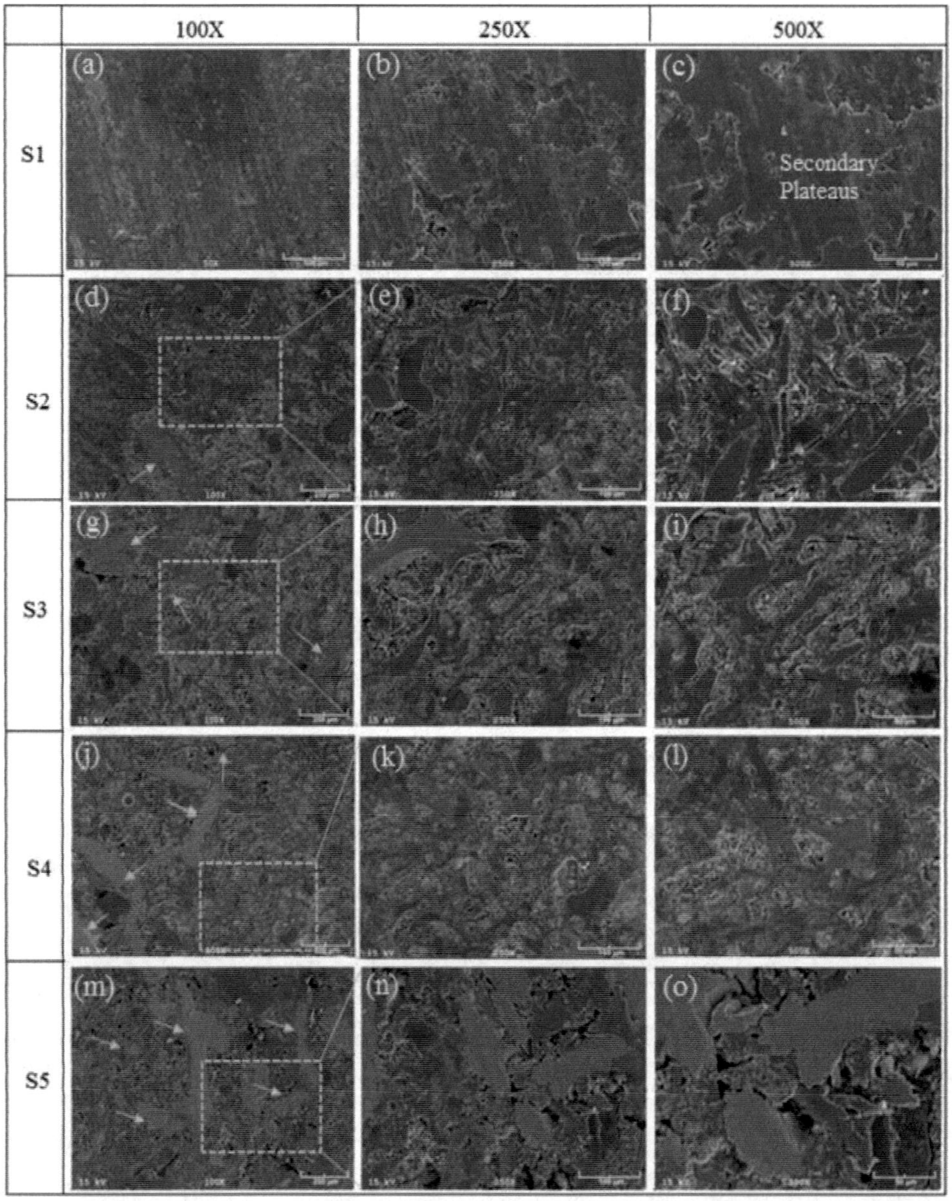

Figure 3. SEM images of worn surface of composite sample having different amount of steel fiber.

4. Conclusions

Steel fibre addition increases the CoF of non-asbestos composites including phenolic binder resin, cashew nut shell liquid, aramid pulp, magnetite, rubber and graphite. The CoF values increased from 0.20 for the sample having 0 wt.% steel fiber (S1) to about 0.30 for the sample having 20 wt.% steel fiber (S5). However, steel fiber addition does not considerably effect of the specific wear rate of the composite sample. Steel fiber addition to the brake pad composite decreases generally the hardness, modules of elasticity and compressive strength of sample. The fade resistance of the sample increases by increasing steel fiber content in the brake pad matrix structure.

5. References

[1] Cho M H, Kim S J, Kim D and Jang H 2005 *Wear* 258 1682-87.

[2] Lu Y 2006 *Compos. Sci. Technol.* 23 591-98.

[3] Jang H, Ko S, Kim S J, Bash R H and Fash J W 2004 *Wear* 256 406-14.

[4] Hong U S, Jung S L, Cho K H, Cho M H, Kim S J and Jang H 2009 *Wear* 266 739-44.

[5] Dureja N, Bijve J and Gurunath P V 2009 *J. Reinf. Plast. Comp.* 28 489-97.

[6] Cho M H, Ju J, Kim S J and Jang H 2006 *Wear* 260 855-60.

[7] Singh T, Patnaik A, Chauhan R and Rishiraj A 2015 *JKSUS* 29 183-90.

[8] Ho S C, Chern Lin J H and Ju C P 2005 *Wear* 258 861-69.

[9] Aranganathan N, Mahale V and Bijwe J 2016 *Wear* 354-355 69-77.

[10] Öztürk B, Arslan F and Öztürk B 2013 *Tribol T.* 56 536-45.

[11] Bijwe J and Kumar M 2007 *Wear* 263 1243-48.

[12] Kumar M and Bijwe J 2013 *Wear* 303 569-83.

Ball-on-flat reciprocating test to evaluate dry sliding wear behaviour of reinforced polymer composites

V Iliuta[1], M Ripa[1], J Javorova[2] and G Andrei[1]

[1] **Department of Mechanical Engineering, "Dunarea de Jos" University of Galati, Romania**
[2] **Department of Applied Mechanics, University of Chemical Technology and Metallurgy, Sofia, Bulgaria**

Abstract. This paper presents the results of the tribological investigations on polymeric composite materials reinforced with Cu, Zn, Sn, SiO_2 and MoS_2 microparticles, used for repairing metal parts made of brass and bronze. The tested material were: three composite materials (products of the company Diamant Metallplastic GmbH, Germany): Multimetall Messing (used for repair of brass parts), Multimetall Bronze (used for repair of bronze parts) and Moglice (for brass and bronze parts). The testing was conducted in dry conditions, in ball on flat configuration, by reciprocating method, on the CETR UMT-2 tribometer (Bruker Corporation) at normal loads 20, 30, 40 and 50N, over a sliding distance of 100m. The paper presents studies of the variations of the friction coefficient, linear wear, volumetric wear parameters, profilometric studies with SEM surface morphologies of the wear tracks. The researches allowed the classification of materials according to their tribological behaviour and the results recommend the tested reinforced composite materials as good candidates for repairing metal parts made of brass or bronzes.

1. Introduction

Due to their good mechanical properties (mechanical strength, stiffness, hardness, low density, etc.), polymer composites are successfully used in many industry sectors such as automotive industry, naval industry, aerospace, machine manufacturing and others [1, 2, 3]. The advantages of these materials recommend their use in the maintenance and repair activities of parts and equipment in various economic sectors.

For the tribological characterization of these materials, they are tested in the laboratory in a *ball-on-flat test configuration in reciprocating* sliding [4, 5, 6, 7]. In the cited papers, the authors studied the variation of the friction coefficient and the structure and topography of the used surfaces by modifying certain test parameters like loading forces, sliding distance and/or time, frequency.

The paper describes the results obtained through comparative analysis of three polymeric composites, as they were discussed previously [8, 9, 10, 11].

2. Experimental procedure

2.1. Materials characterization

The materials used for research purpose are provided by Diamant Metallplastic GmbH, Germany [12]: 1. Multimetall Messing, code SAC, used for repairing brass parts; 2. Multimetall Bronze, code SBC, used for repairing bronze parts; 3. Moglice, code SMC, composite material used to repair brass and bronze parts.

The composite materials SAC and SBC are epoxy matrix reinforced with Cu, Zn, Sn particles, and various allotropic forms of SiO_2 and belongs to Multimetall category [12].

The composite material SMC is an epoxy matrix reinforced with particles of cristobalite (αSiO_2) and molybdenum disulfide (MoS_2) [12].

These composites are recommended for the repair of surface defects such as cracks, holes executed incorrectly, pinching or squeezing, etc.

The counterpart was a steel ball (AISI 52100 GCr15, grade like SKF), Φ6, HV 946, surface roughness $R_a \approx 0.06$ μm.

The main mechanical characteristics of the composites materials [12] are shown in Table 1 and 2.

Table 1. The main mechanical characteristics of the composites materials

Material code	E-Modulus DIN 53457 [MPa]	Tensile Strength [MPa]	Compressiv e Strength [MPa]	Bending Strength [MPa]	Shear Strength [MPa]	Hardness [Shore D]
SAC	5800	63	155	95	35	87-89
SBC	5800	62	155	79.5	16.5	86
SMC	10400	-	120	66	-	88

Table 2. The main mechanical characteristics of the composites materials

Material code	Impact Strength [MPa]	Surface Pressure [MPa]	Specific Weight [g/cm^3]	Temperature Resistance permanent [°C]	Temperature Resistance temporary [°C]
SAC	4.8	-	1.6	-32 to +160	-
SBC	5.4	-	2.2	-32 to +160	+350
SMC		12.5	1.7	-20 to +60	-40 to +125

For all tested samples, the following average values of roughness (R_a) were determined: SAC - Ra=1.42 μm, SBC – Ra=1.81 μm and SMC – Ra=0.97 μm.

The composite is obtained by mixing two components. We obtain a putty that is applied on specially prepared metallic surfaces. After drying (4 ... 30 hours), the composite materials are machined. The test samples for composite materials were thus obtained. All samples for both composite materials had the final dimensions Φ76x7.

2.2. Experimental setup and procedures

The testing was conducted on the tribotester CETR UMT-2 (Bruker Corporation) at room temperature (20÷26°C) and relative humidity conditions of 50-60%. Dry sliding reciprocating tests were conducted in accordance with ASTM G133-05 standard. Testing parameters were: the stroke - 5mm, duration - 475 min, the sliding distance – 100 m and normal loading forces: 20, 30, 40 and 50N.

Before performing the tests, the samples were degreased with organic solvent and then dried with hot air at 50°C.

In order to study the intensity and type of the wear process, it is often used the profiometric analysis of the wear tracks. The acquisition of the digital profiles for all samples was done with a laser profilometer (μSCAN, ® NanoFocus) and they were processed with the software SPIP 6.2.6 (™ Image Metrology, Hørsholm

For each wear track, three cross sections were obtained as shown in Figure 1: on the center of the track (profile "m") and at the 2 mm distance on both side of the track (profile "i" and profile "e").

For the wear evaluation by profilometry, the area of cross-sections measured from the highest peaks adjacent to the track and the lower part of the profile was used.

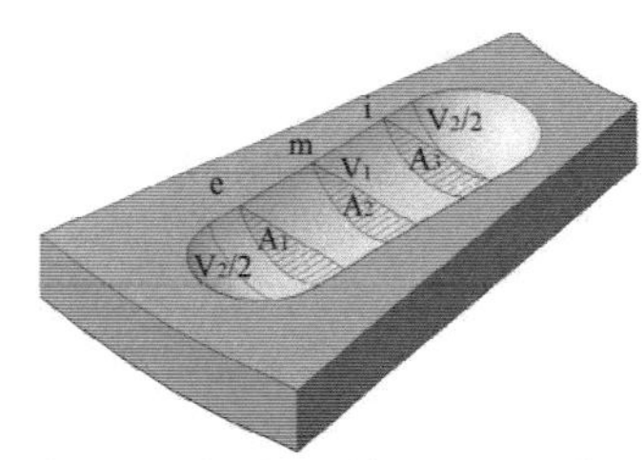

Figure 1. Graphical modeling of wear tracks [11]

3. Results and discussion

Figure 2 shows friction coefficient variation graphs for all studied materials and for the four loading forces according to the sliding distance. The values for COF were acquired at a frequency of 10 values per second.

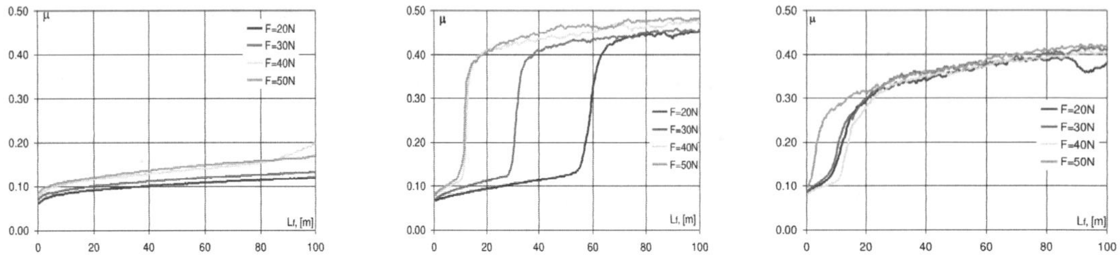

a) b) c)

Figure 2. Variation of the coefficient of friction for: a) composite (code SAC) [8]; b) composite (code SBC) [11]; c) Moglice (code SMC) [10]

Relative to composite materials the following is noticed:
- the friction coefficient increases with increasing load force (Figure 1a, 1c, 1e);
- there are three stages for all loading forces to composites SBC and SMC (Figure 1c, 1e): in first stage (from 10 m to 55m for composite code SBC and 0 to 10m for moglice) COF increases slowly; in the second stage COF increases very rapidly, and in the third stage a slow increase can be detected ;
- the length of the first stage (for SBC and SMC composite materials) decreases with the increase of the loading force;
- in step 3 of the friction coefficient for SBC and SMC composite materials shows instability;
- at SAC composite material, the friction coefficient features a steady and stable growth.

Figure 3 illustrates the graphs of the average friction coefficient for all samples and all loading forces. The friction coefficient was calculated with the values acquired over the last 25 m of the sliding distance.

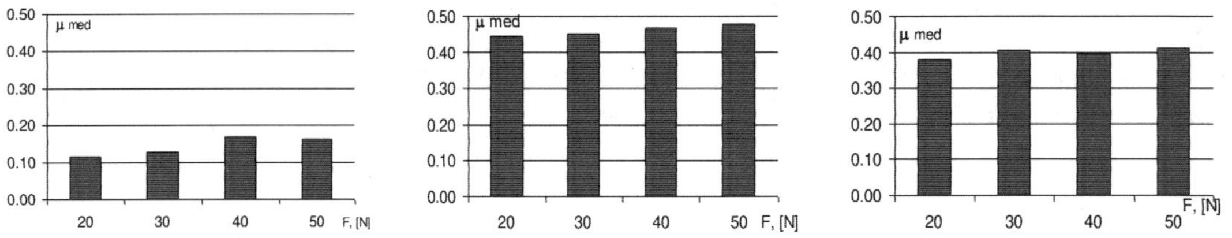

a) b) c)

Figure 3. Variation of the average coefficient of friction for:
a) composite (code SAC) [8]; b) composite (code SBC) [10]; c) Moglice (code SMC) [11]

It can be seen in Figure 3 that the smallest values of the average coefficient of friction are in the composite code SAC (Figure 3a); the other two composites have close values in terms of the average friction coefficient.

Figure 4 shows the variation graphs of the wear depth according to the distance for the four load forces. The following is noticed:
- low values for linear wear on SAC composite compared to other materials;
- small values for linear wear also occur in the SMC composite;
- different behaviour: on the brass composite, in the initial phase there is a marked increase, followed by a linear increase with a very low slope; in the other two composites, the linear wear has a small increase at first, followed by a significant increase; higher values are observed in the bronze composite.

Volumetric wear parameters considered for characterizing the intensity of wear are: the wear volume (V), the wear intensity (I_v) and the specific wear rate (k):
- the volume of wear (V) [11]:

$$V = A \cdot s \quad (1)$$

IOP Conf. Series: Materials Science and Engineering **724** (2020) 012019 doi:10.1088/1757-899X/724/1/012019

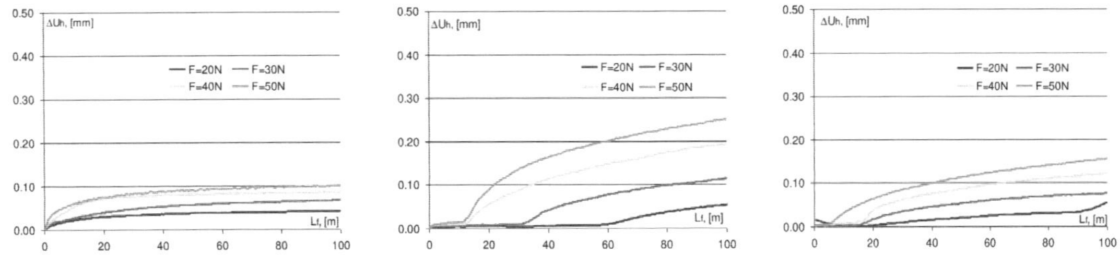

a) b) c)

Figure 4. Variation of linear wear depth (F=20, 30, 40 and 50N, L_f=100m):
a) composite (code SAC) [8]; b) composite (code SBC) [11]; c) Moglice (code SMC) [10]

- the volumetric wear intensity (I_v) [13, 14]:

$$I_v = \frac{V}{L_f} \tag{2}$$

- the specific rate of wear (k) [14]:

$$k = \frac{V}{F \cdot L_f} \tag{3}$$

where:

A - average cross-sectional area of wear track, mm^2;
s - length of the stroke, mm;
L_f – sliding distance of friction, m;
F – normal force load, N.

The average cross-sectional area of the wear tracks was calculated with:

$$A = \sum_{i=1}^{3} A_i \tag{4}$$

The Evolution of volumetric wear parameters is shown in Fig. 5. From this figure we can see:
- the composite material code SAC has the lowest values of volumetric wear parameters;
- all the volumetric parameters of the composite material for brass have a linear growth trend;
- the bronze composite has the highest values of the volumetric wear parameters;
- the composite moglice, compared to metallic materials in terms of volume wear parameters, seems to be a good material for refurbishing both brass and bronze parts.

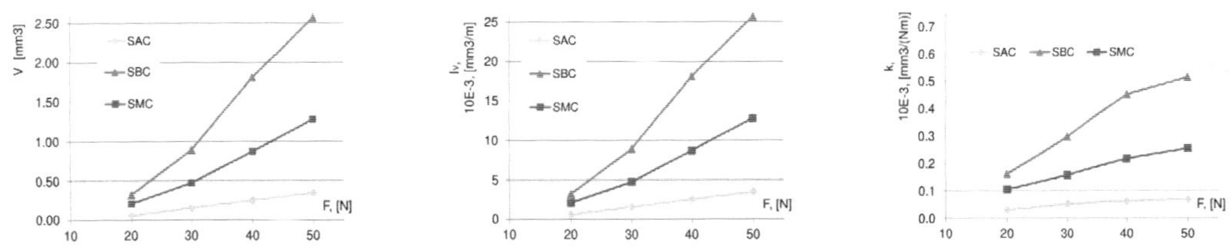

a) b) c)

Figure 5. Variation of volumetric wear parameters:
a) the wear volume (V); b) the wear intensity (I_v); c) the specific wear rate (k).

Figure 6 presents the most representative 2D profiles of the cross-section of the wear traces for each material according to the normal load force values.

It can be noticed :

- The rough appearance of the 2D profile shapes for brass composite material (Figure 6a) can be caused by the reinforcement particles unused during the test;

- The composite material moglice has 2D profiles (Figure 6c) as an order of magnitude of the width and depth of the wear parameter between the composite material cod SAC and the other materials. At the moglice composite, for the 20 N test force, the same rough profile of the profile as the SAC composite is observed.

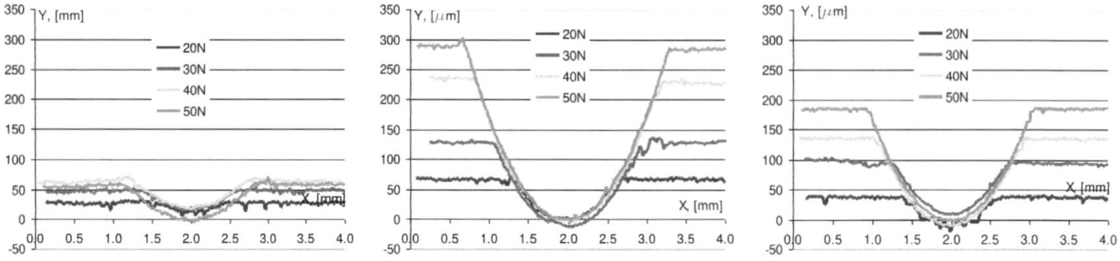

a) b) c)

Figure 6. Digital roughness profiles (2D) for the loading forces 20÷40N:
a) composite (code SAC) [10]; b) composite (code SBC) [13]; c) Moglice (code SMC) [12]

Figure 7 presents the 3D plots (scanning area: 0,5x0,5 mm) for the composite materials. Images were obtained with 3D profilometry module of the tribometer PRO-500 UMT-2.

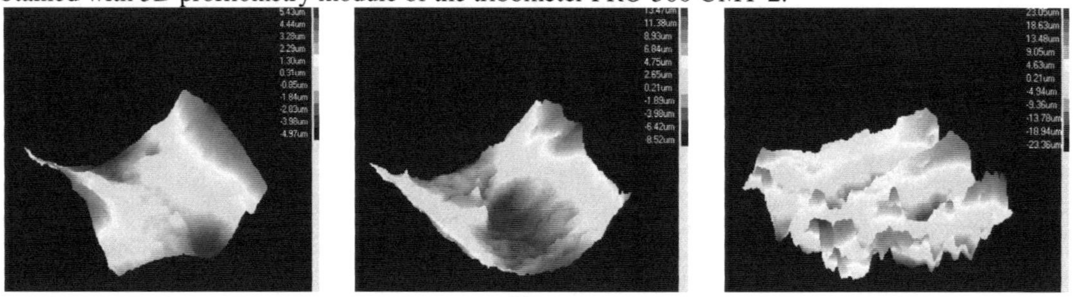

a) b) c)

Figure 7. Details of the wear track for loading force 30N, by the profilometric module of the tribotester CETR-UMT-2: a) composite (code SAC) [11]; b) composite (code SBC); c) Moglice (code SMC)

In Figure 7, it can be seen a gouges on the worn surface as an effect of adhesive wear when the microparticles are pulled away and removed from the surface to be rubbed. No abrasion wear is observed.

Evidence of wear phenomena results from the study of the wear traces with the Quanta 200 electronic microscope. The images are shown in Figure 8. The SEM images obtained at the 10000x magnification scale show cracks at the material grain boundary as well as bonding of the material onto the worn surface of the composites.

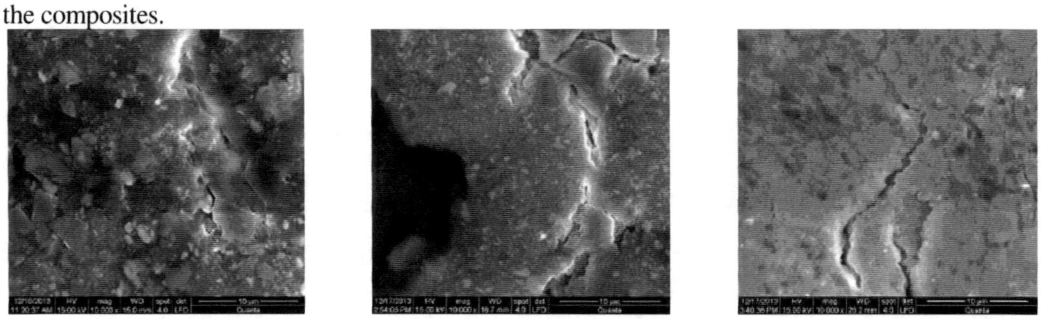

a) b) c)

Figure 8. SEM surface morphologies by the Quanta 200 electronic microscope (x10000):
a) composite (code SAC), F=40N; b) composite (code SBC), F=50N; c) Moglice (code SMC), F=20N

4. Conclusions

The purpose of this research was to make a comparison of the tribological behavior of three composite materials used in the repair of brass and bronze metal parts.

The results of testing composite materials with microparticles in dry friction reciprocating ball-on-flat conditions show the following:

1. The composite brass material has the lowest values for wear parameters such as: friction coefficient, linear wear depth and volumetric wear parameters (the cross sectional area, the volumetric wear, the volumetric wear intensity and the specific wear rate);
2. The moglice composite has good tribological characteristics and can be used successfully in reconditioning brass and bronze parts;
3. Adhesive wear is the dominant wear process for all three composite materials;

References

[1] Brydson J A 1999 *Plastics Materials*, Seventh Ed. (Oxford: Butterworth-Heinemann)
[2] Friedrich K 1993 *Advances in Composites Tribology* Composite Materials Series (Amsterdam: Elsevier)
[3] Ashby M F and Jones D R H 1998 *Engineering Materials 2: An Introduction to Microstructures, Processing and Design*, Second Ed. (Oxford: Butterworth-Heinemann)
[4] Li J and Xia Y C 2009 The Reinforcement Effect of Carbon Fiber on the Friction and Wear Properties of Carbon Fiber Reinforced PA6 Composites, *Fiber. Polym.* **10** 519-25
[5] Li J and Cheng X H 2007 Effect of rare earth solution on mechanical and tribological properties of carbon fiber reinforced thermoplastic polyimide composite *Tribol. Lett.* **25** 207-14
[6] Li J and Li X Z 2010 Evaluation of the Tribological Properties of Carbon Fiber Reinforced Poly (vinylidene fluoride) Composites, *J. Mater. Eng. Perform.* **19** 1025-30
[7] Nie W Z, Li J and Li X Z 2010 The addition of carbon fiber on the tribological properties of poly(vinylidene fluoride) composites, *Fiber. Polym.* **11** 559-64
[8] Iliuţă V, Preda A, Andrei G and Bîrsan I 2014 Wear Assessment of Polymer Composite Filled with Metal Particles Trough Ball-on-Flat Reciprocating Test *Mater. Plast.* **51** (4) 359-62
[9] Iliuţă V, Rîpă M, Andrei G, Preda A, Suciu C and Javorova J 2014 Profilometric Evaluation of the Worn Surfaces under Dry Reciprocating Wear Conditions of a Composite Material to Repair Brass Made Parts *Applied Mechanics and Materials* **657** 437-41
[10] Iliuţă V, Rîpă M, Preda A and Andrei G 2015 Friction and Wear Behavior of Moglice Polymer Composite through Dry Sliding Ball-on-Flat Reciprocating Test *Applied Mechanics and Materials* **808** 137-42
[11] Iliuţă V, Rîpă M, Javorova J and Andrei G 2017 A comparative study on tribological behavior between metal and polymeric composites used to repair bronze made parts in dry reciprocating sliding tests *IOP Conference Series-Materials Science and Engineering* **174**
[12] *** Diamant Metallplastic GmbH, prospects available at http://diamant-polymer.de/dia-downloads/02-TD-pm-GB.pdf (14.11.2013)
[13] Rîpă M and Deleanu L 2008 *Deteriorări în tribosisteme* (Galaţi: Ed. Zigotto)
[14] Stachowiak GW 2005 *Wear - Materials, Mechanisms and Practice* (Hoboken: John Wiley & Sons).

Preliminary study of friction in automotive ball joints

I Muscă, I C Românu and **A Gagea**
Stefan cel Mare University of Suceava-Romania, Department of Mechanics and Technologies, 13th University Street, 720229, Suceava, Romania

E-mail: ionutromanucristian@usm.ro

Abstract. The present paper offers a brief description of the testing equipment, experimental methodology and some results regarding the motion resistive effort in automotive ball joints. The local ball joints manufacturers have difficulties in controlling the clamping force of the ball joints in the production process. Due to this, an original experimental device was conceived and built in order to evaluate the magnitude and evolution of internal friction between the joint's surfaces. The experimental setup's main advantage is the capacity of controlling and maintaining a constant clamping force and that it ensures a constant motion of the ball stud. The force magnitude was recorded using resistive transducers connected to a data acquisition system. The numerical values were recorded and used to analyse the mobility of the tested ball joints.

1. Introduction

Modern equipment often uses ball joints to maintain the relative position between constitutive elements. One of the most important applications of ball joints is in the steering system of road vehicles. This equipment must determine the steering of front wheels in direct proportionally to the steering wheel's position and also ensure minimum clearances and as little as possible needed effort from the driver. Also, the steering system elements have to allow vertical movement of the wheels so that the suspension system of the car can be efficient. Due to these requirements, the clamping force of the ball stud must ensure a minimum clearance fit as well as a limited friction force. In order to highlight the friction force in this type of joints and to estimate their reliability, experimental investigations are required. The most common structure of a ball joint consists of: ball stud, cap, bearing, socket and gasket.

A proper function of automotive ball joints depends on the behavior of the ball stud-bearing contact. If the clamping force is too high, the friction between the ball stud and the bearing is also very high, the spinning motion will be cumbersome loading additionally different parts of the vehicles steering system. If the ball joints are affected by wear, then in addition to the rotating motion, a translation displacement appears, which is undesirable for a proper function of the vehicles steering system.

This paper appears as a consequence of discussions with local automotive ball joints manufacturers, which found large differences of the clamping force level applied on the ball stud of the joint as a consequence of the manufacturing process. One of the manufacturing solutions used in ball joints production process consists in applying the clamping force through a closure cap. The closure cap and the joint socket are assembled by plastic deformation of the socket upper edge. The clamping force between ball stud and the bearing depends on technological parameters as, the dimension of the socket's upper edge, the strain magnitude, manufacturing errors etc.

By analysing the technical literature regarding ball joints tests it is noted that many machines were developed for testing, [1-4]. Those machines are capable to simulate the most difficult situations similar to those found during operation. Given that, the issue raised in this paper is to find the cause of these great variations between the minimum and the maximum motion resistant force of the joint ball stud. A special setup was proposed for this purpose.

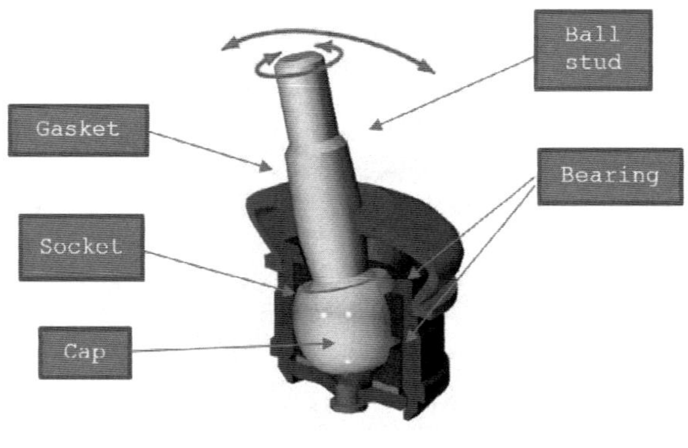

Figure 1. Ball joint structure, [5].

2. The experimental setup

An experimental setup was conceived and built. The experimental device must be capable to measure and to send further information regarding the clamping force level and the resistive motion force magnitude. The experimental setup has the following structure as is represented in Figure 2.

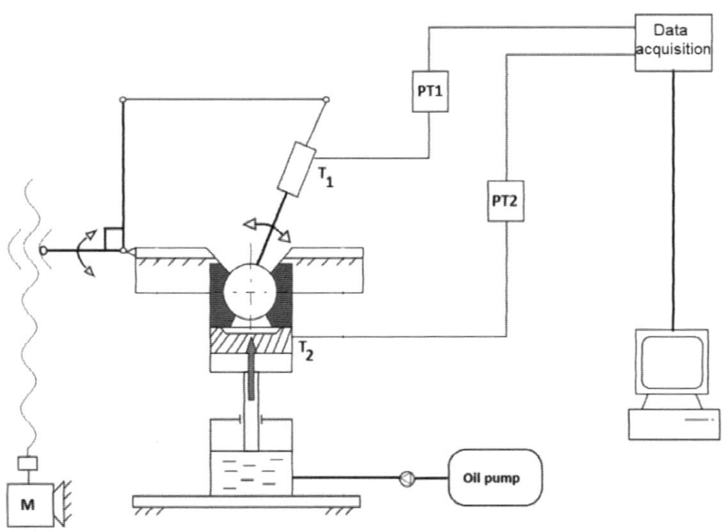

Figure 2. Schematic representation of the experimental setup, [6].

A DC motor is used to drive the experimental device at constant speed. The rotation motion generated by the motor is transformed into translation motion by means of a a screw - nut transmission. The translational motion of the nut is transmitted to a quadrilateral mechanism represented in Figure 2. The last element of the mechanism is connected firmly to the ball stud tail. The quadrilateral mechanism must provide a symmetrical alternating motion of the joint stud with constant speed.

IOP Conf. Series: Materials Science and Engineering **724** (2020) 012020 doi:10.1088/1757-899X/724/1/012020

The clamping force was generated using a hydraulic piston. The tested ball joints and the hydraulic piston were inserted in a rigid steel frame. In the upper region of the frame a hole was made to allow the pin to be connected with the transmission mechanism.

To determine the clamping force level and the resistive motion effort, two force transducers were used. One of them was inserted between the end of the hydraulic piston and the ball joint cap and the other was connected to the last element of the mechanism.

3. Testing methodology and results

Unpacked ball joints with 25 mm ball stud radius were used for tests. The ball stud was manufactured from C12E steel and the joint bearing was made from polyethylene. The experimental setup was connected to a DC power supply which provided constant DC voltage.

The force transducers were connected to a data acquisition system in order to record the numerical values. The recording rate was 50Hz. From the oil pump the clamping force was adjusted by reading the indication of the T1 transducer. After the clamping force adjustment, the electric motor was connected to the power supply. The numerical values recorded from the second transducer were recorded and processed.

The experimental values obtained by using those two force transducers were recorded and represented graphically. The following evolution of the motion effort of the ball joint corresponding to a 440N clamping force was obtained.

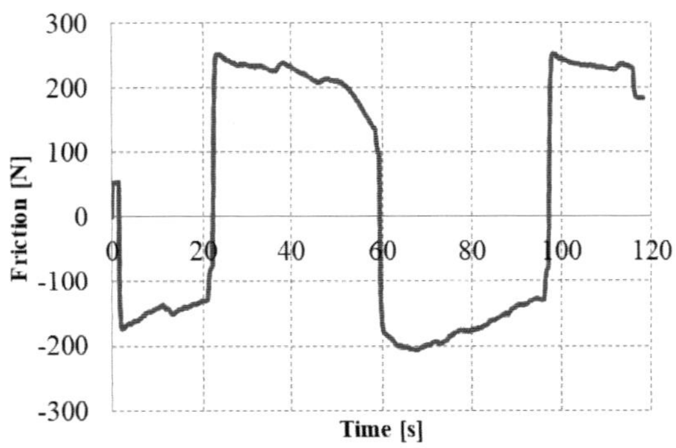

Figure 3. Ball joint friction evolution corresponding to 440N clamping force.

Several tests for different clamping forces (440N, 620N, 870N, 1080N) were conducted and the results were represented graphically on the same chart as shown in Figure 4.

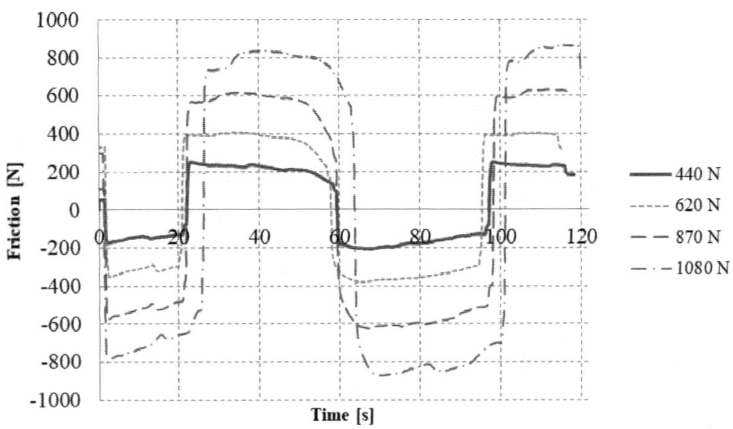

Figure 4. Ball joint friction evolution at different clamping levels.

To validate the reliability of the measurements, two tests were made at the same clamping loading level in different contact conditions. The evolution of the friction force corresponding to 650N clamping force with and without oil between contact surfaces was obtained and represented in the chart below. From the representation shown in Figure 5, it can be observed that the friction in the lubricated contact case is much smaller. The results show an evolution similar to those presented in the literature, [7].

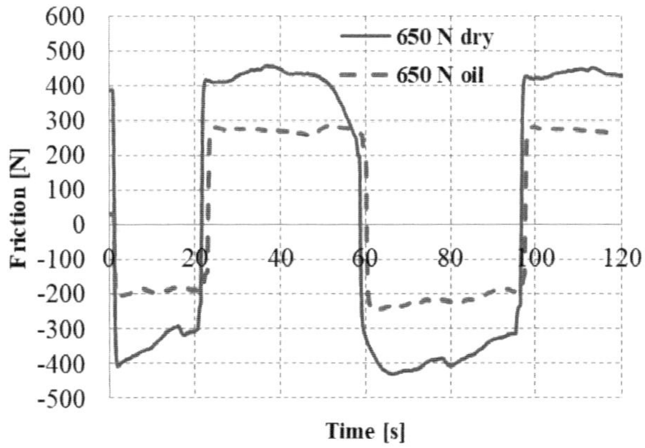

Figure 5. Ball joint friction evolution with and without oil between surfaces at 650N clamping force.

4. Conclusions

For the abovementioned experiments, an experimental equipment was conceived and built at the "Ştefan cel Mare" University of Suceava, in the Tribology Laboratory. The experimental setup permits to impose a planar oscillatory motion to a spherical joint, while applying a controlled clamping force by means of an original loading system. By aid of a special measuring device, the evolution of friction force in this type of ball joints was highlighted for different clamping force levels. Several tests were conducted with and without presence of lubricant in the ball joint.

Results obtained at low clamping force levels applied to the tested ball joints reveal that the static friction force under dry conditions has higher values than those obtained for kinetic friction. At high clamping force levels, the differences between the static and kinetic friction are increased.

Results obtained at 650N clamping force with and without oil between contact surfaces show a low magnitude of the friction force in the joint. This can be caused by the attenuation of the adhesion effects between surfaces due to presence of a thin film of oil and the thin oil film formed between contact surfaces. The magnitude of the friction decreases with approximately 50 % when the contact is lubricated.

Own results were compared to those obtained by Baroiu, [8]. Baroiu has made some tests on automotive ball joints and he determine the magnitude of the force necessary to tilt alternant a joint ball stud. His work shows a similar evolution of the friction force in the ball joint near the ends of the stroke.

5. References

[1] Sin B-S and Lee K-H 2014 Process design of a ball joint, considering caulking and pull-out strength *ScientificWorldJournal* doi.org/10.1155/2014/971679
[2] www.wpm-leipzig.de/index.php?id=51&L=1
[3] Raes S Devreese T De Pauw J and De Baets P. Design of a tribological ball joint tester, ojs.ugent.be/SCAD/article/download/1137/1164/
[4] www.servotestsystems.com/images/PDFs/ST_brochures/06_Ball_Joint_test.pdf
[5] www.tirereview.com/understanding-checking-and-replacing-ball-joints-for-vehicle-safety/
[6] Gagea A 2017 Stand pentru studiul frecării în pivoţi *Bachelor degree* University of Suceava
[7] Sage R M 1987 The mathematical modelling of ball-joints with friction *PhD Thesis* University of Leicester
[8] Baroiu N David M and Susac F 2014 Study concerning the ball joint functionality of a vehicle steering system *The Annals of "Dunărea de Jos" University of Galaţi Fasc.V* 1221- 4566

Wear's issues on high-pressure common rail pumps

R C Iordache, N D Petrea and C Bujoreanu
„Gheorghe Asachi" Technical University of Iasi, Department of Mechanical Engineering, Mechatronics and Robotics, Blvd. Mangeron, No. 43, Iasi, Romania

E-mail: carmen.bujoreanu@gmail.com

Abstract. High-pressure pumps are the most important components of modern fuel injection systems. A modern system is the common rail one, equipping vehicles with a diesel or a gasoline engine. Depending on the fuel type, the design of the pumps is completely different, and it is more complex for diesel engines that are more prone to wear. The components subjected to wear are those in contact and continuous motion like drive shaft with roller and shoe, plunger and hydraulic head assembly with valves. Wear is initiated as a result of a process and product deviations or inadequate exploitation of the high-pressure pump. The factors influencing the wear can be improper material hardness, improper coating, incomplete or wrong machining, wrong clearance or misalignment between components, inadequate surface finish, dimensions out of tolerances, wrong assembly process, poor lubrication, contamination or running at high loads and temperatures. Our paper aims to identify, describe and discuss the types of wear encountered at common rail high-pressure pumps, highlighting the effects on their functionality. The wear occurrence can be prevented by paying more attention to the design details, manufacturing processes and correct exploitation of the high-pressure pumps.

1. Introduction

The high-pressure common-rail system is a modern fuel injection system which equips most of the nowadays combustion engines. It is composed of a high-pressure pump, injectors, a high-pressure accumulator (rail) and the electronic control unit (ECU) which controls the whole system for a good performance of the engine and low emissions [1]. The pump is a complex mechanism that can deliver high-pressure fuel to the high-pressure accumulator. Pump components are subjected to high loads by compressing the fuel at high pressures. The most stressed subassemblies of the pump are the drivetrain components and the hydraulic head assembly.

The drivetrain components are the drive shaft, roller and shoe, which acts the plunger in order to compress the fuel in the compression chamber. These components are in permanent contact, and also they are lubricated with fuel. When the pump is operating, in some cases the parts may lose the fuel film and are subject to very high friction, resulting in high working temperatures. Obviously, this behaviour leads to a wear process.

The hydraulic head assembly with valves and plunger has the role in transforming the low pressure in high-pressure fuel. These valves have simple designs, and the outlet valve is composed of a spring, a ball with a spherical seat and the inlet valve is composed of a spring, a stem and a conical seat [2]. The valves are subjected to cyclic loads due to plunger's motion by compressing the fuel and are more prone to fatigue.

Figure 1 depicts the main possible types of wear that we have identified for each pump component.

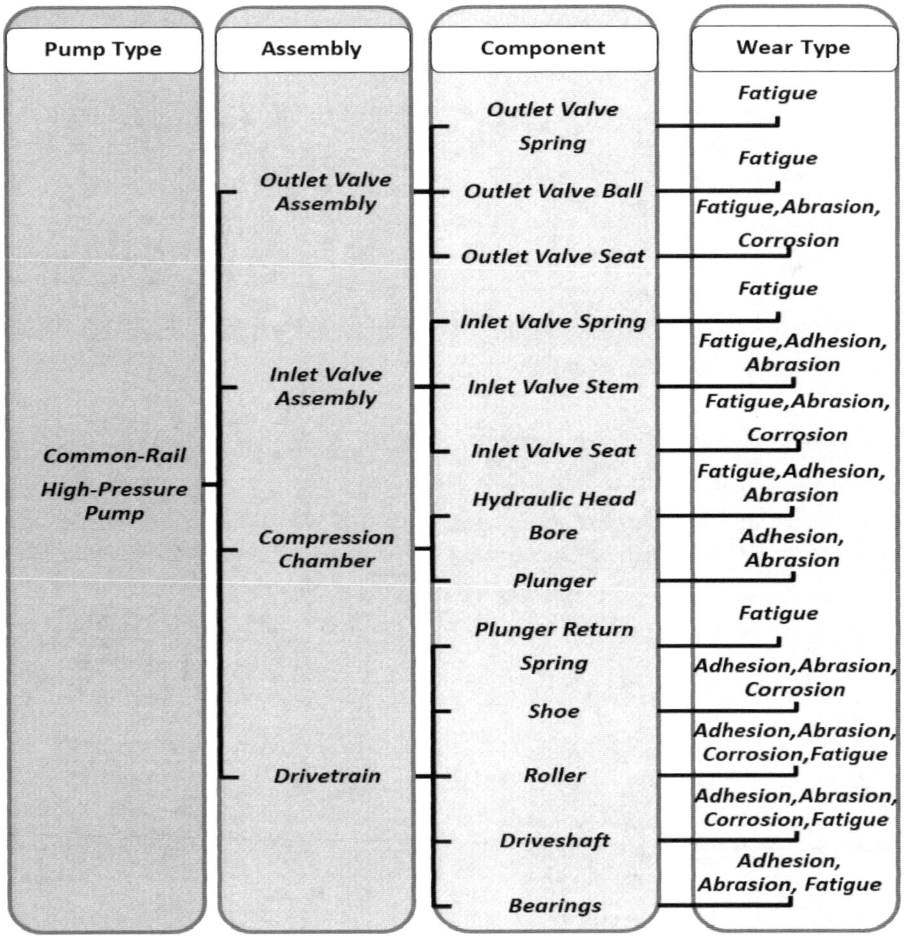

Figure 1. Wear types for common-rail high-pressure pump components.

2. Wear of the drivetrain components

Transmission of motion into the high-pressure pumps is achieved via the drive-shaft. It is driven directly by the camshaft or crankshaft indirectly via transmission belts, chains or gears. This rotating motion of the drive shaft is transformed into a translation one. The cam comes into direct contact with the roller and makes possible this movement transformation. The lubrication is carried out with the fuel flowing through the pump. The drive shaft is made of high-strength steel. The roller is a rolling element with a logarithmic profile made of steel [3]. The shoe is made of a material that has excellent compatibility with the roller material. In order to have high wear resistance, it is covered with DLC (Diamond Like Carbon). Considering that these elements are in permanent contact and the fluid used for lubrication is a fuel with limited lubrication properties, the appearance of wear is inevitable. Working conditions, surface irregularities, components clearance, improper dimensions, load distribution, vibration, or surface condition facilitate wear appearance. The main types of wear in this mechanism are adhesion, DLC removal, abrasion and corrosion. The working parameters to which the pump is directly subjected can influence the fuel properties and the contact between the components. According to a previous study [4], a high-pressure pump was subjected to an aggravated test, the Low Sommerfeld Number. This type of experiment was performed under critical conditions of speed, temperature and pressure. It has been shown that these parameters adapted to aggravate the test correlated with low clearance between components lead to the occurrence of scuffing (figure 2). During the run, because of the critical conditions, the surface of the roller adheres to the surface of the shoe, which leads in blocking the pump components.

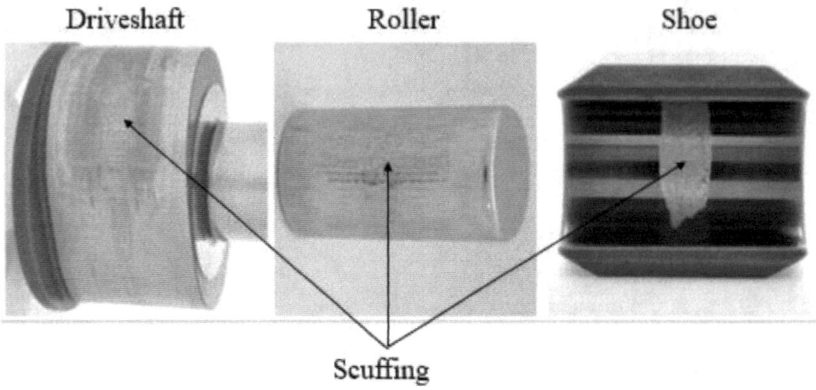

Figure 2. Roller, shoe and drive-shaft with scuffing after an aggressive test [4].

Mirror polishing (figure 3) is another important type of wear found in high pressure pumps and it appears on the drive shaft cam surface. This is not often seen because it is characterized as an effect due to the poor fuel properties. Over time, it leads to fatigue (figure 4).

Figure 3. Mirror polish.

Figure 4. Fatigue on drive-shaft.

The shoe internal diameter area is the most sensitive for removal of DLC. Depending on the roller rotation direction, the pressure on the shoe edges is uneven (figure 5). Thus, DLC removal may occur (figure 6).

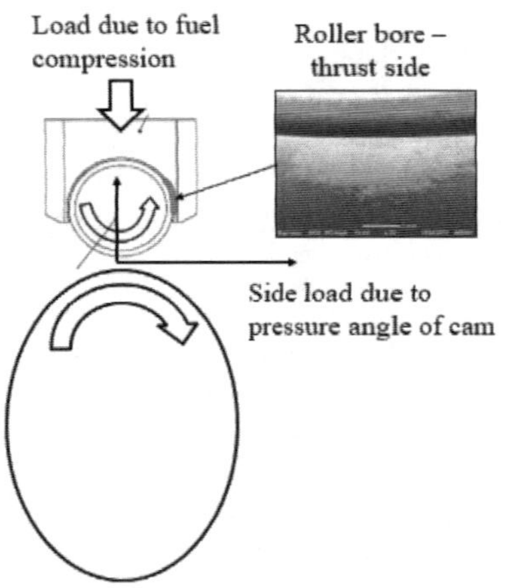

Figure 5. Shoe – load distribution.

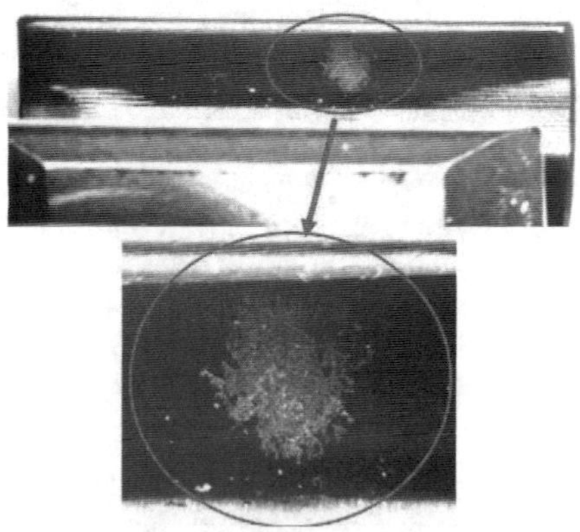

Figure 6. DLC Removal.

3. Wear of the hydraulic head components

The inlet valve with the outlet valve and the plunger assembled on the hydraulic head body, together form the hydraulic head assembly of the common-rail pump (figure 7). The hydraulic head is one of the most stressed parts of the pump, its role being to compress the fuel at pressures of over 2000 bar.

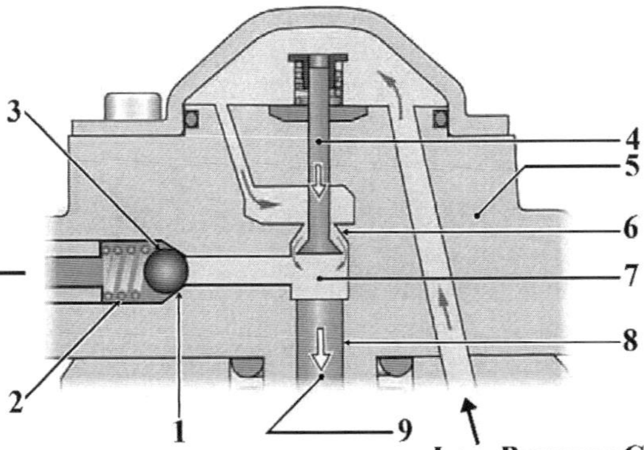

1 – Outlet Valve Seat; 2 – Outlet Valve Spring; 3 – Outlet Valve Ball; 4 – Inlet Valve Stem; 5 – Hydraulic Head Body; 6 – Inlet Valve Seat; 7 – Compression Chamber; 8 – Plunger Bore; 9 – Plunger

Figure 7. Hydraulic head assembly with valves and plunger [6].

3.1. Outlet valve

The outlet valve assembly is composed of a spring, a steel ball and a machined spherical seat, which is prone to impact wear and fatigue. The wear appears due to the opening and closing of the valve in which the ball hits the seat with high velocity and high force due to the high pressure from the high-pressure accumulator.

In figures 8 a, b (in a hydraulic head section) a crack on the outlet valve seat can be observed. During pump operation, this crack leads to high-pressure external leakage. By opening the fracture, the initiation point and the fracture pattern can be viewed in detail (figure 8 c). The crack initiation point is exactly on the machined spherical seat, in the worn area where the most considerable contact stress is due to the ball.

As a result of the analysis performed on the scanning electron microscope (SEM) in figure 8 d, an inclusion has been observed. It has a chemical composition different from that of the base steel which has an enormous influence on the mechanical properties of the material, mainly under high cyclic load conditions, resulting in a high dispersion of the fatigue life [7]. The main elements present in the inclusion are calcium (Ca), magnesium (Mg), aluminium (Al) and oxygen (O) and are suitable for a spinel inclusion, most likely formed during the steel manufacturing process.

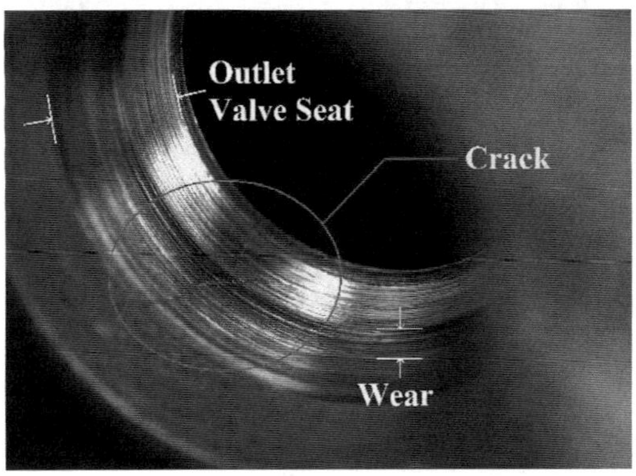

(a)

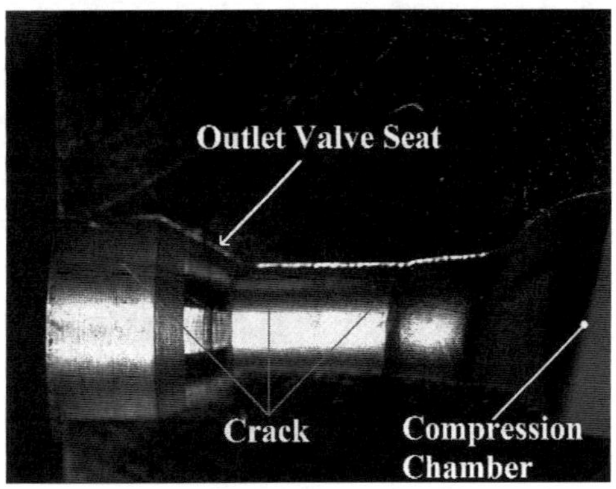

(b)

(c) (d)

Figure 8. Hydraulic head outlet valve failure due to material inclusion.

3.2. Inlet valve seat

The inlet valve assembly consists of a spring, a stem and a conical machined seat (figure 9 a) in which the components are more prone to abrasive wear, impact wear and fatigue. During operation of the pump, the stem moves up and down, making a translational movement. This movement can lead to abrasive wear due to a very small clearance with the guide, impact wear and fatigue due to the high-pressure cyclic pulsations in the compression chamber [8].

In figure 9 a, a crack on the inlet valve seat can be observed, which leads to external leakage during the pump compression stroke. The crack has been opened (one half in figure 9 b) and the fracture initiation point was identified on the inlet valve seat. At the scanning electron microscope, it was analyzed the fracture area in the region of the initiation point where a thin layer of material was observed (figure 9 c). This layer has a very fine fracture morphology indicating high hardness and high brittleness. Also, it has been found in the surrounding area signs of higher ductility due to over tempered martensite.

Half of the fracture was cut, polished and 2% Nital Etch Solution was used to reveal the microstructure (figure 9 d). A hardness check was performed, where the surface of the inlet valve, recognized on the image as a darker area, the material is much softer compared to other areas. The darker surface of the inlet valve seat is specific to the over tempered martensite which occurred during the machining process. The grinding process produces a very high temperature on the surface of the inlet valve seat in a short time that can change the metallurgical structure of the material and is called grinding burn [9, 10].

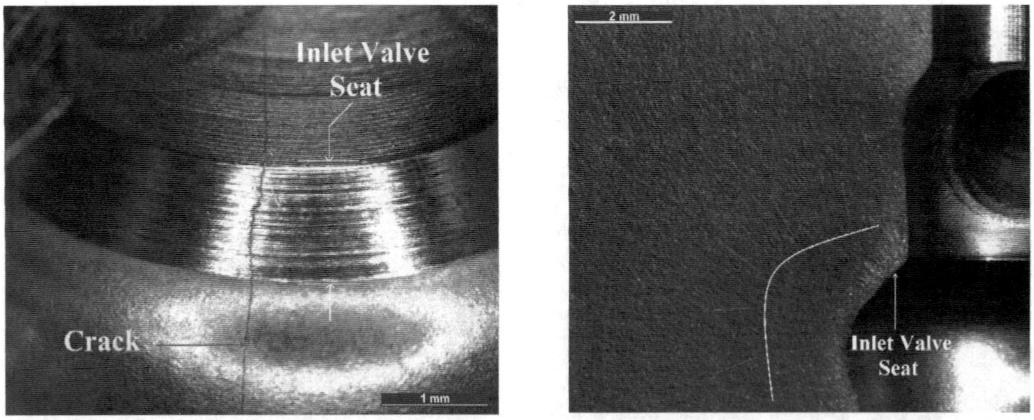

(a) (b)

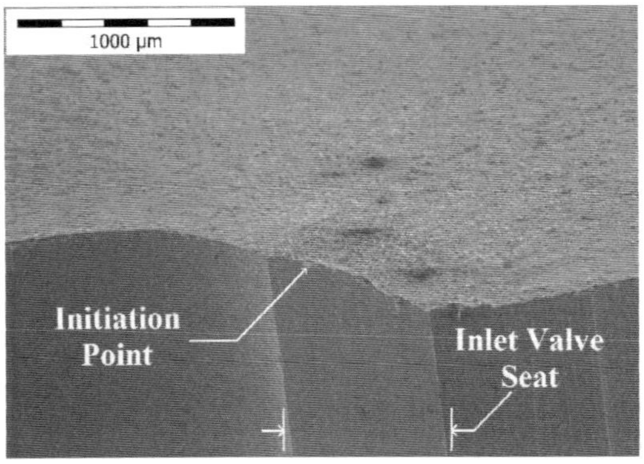

(c)

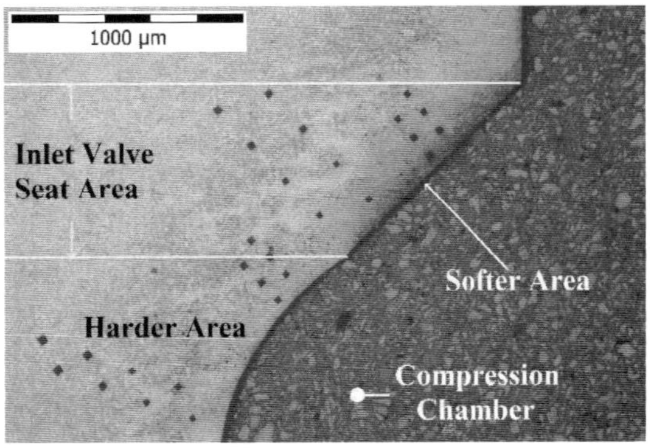

d)

Figure 9. Hydraulic head failure due to inlet valve machining burn.

4. Conclusions

The high-pressure common-rail injection system pump must have components with excellent wear resistance and cyclic loads because almost all components are mechanically driven and stressed by high-pressure fuel. Even so, after a certain period of operation, wear is inevitable. High pressures and load, incorrectly set of fuel injection parameters, heat transfer between components lead to overheating and geometry changes of common rail pumps. They are factors influencing decisively the pumps wear. The paper presents particular cases of wear and defects of the pump components, in which wear can take many forms for the drivetrain ones, such as the drive shaft, the roller and the shoe. For the hydraulic head components as the outlet valve and the inlet valve, wear is most often caused by several openings and closures where the high pressure acts with enormous force on the components. In some cases, as presented in this paper, the process of valve seat machining or material composition play an essential role in the life of the hydraulic head, and also, another role in terms of safety as these defects can lead in time to cracks and further to external leakage. Due to the tight clearance between hydraulic head bore and plunger, the components are susceptible to adhesion and abrasion wear. Therefore, the plunger is DLC coated in order to reduce friction during pump operation, but if the components are running in a system with poor lubrication fuel quality as low viscosity or contaminated with fine particles, the wear is accentuated, and the components will be out of use very quickly. The same issue can also occur for DLC coated shoe. To improve the wear resistance and hence the lifetime of the high-pressure pump, should be paid more attention to manufacturing processes of components or detection of the defects, but not least the assembly process of the pump, where there should be a controlled environment in terms of cleanliness. Our future work will focus on the theoretical study of some components (valves and roller-shoe) in order to quantitative estimate the wear rate.

5. References

[1] Teoh Y H, How H G, Masjuki H H, Nguyen H T, Kalam M A and Alabdulkarem A 2018 *Renew. Energ.* **136** 521-534
[2] Petrea N D, Bujoreanu C and Ripanu I E 2018 *IOP Conf. Ser.: Mater. Sci. Eng.* **444** 042020
[3] Fujiwara H and Kawase T 2007 *NTN Technical Review* **72** (75) 140-148
[4] Iordache C R, Bujoreanu C, Ciornei F C, Scuffing analysis of roller-shoe mechanism after an aggressive test (to be published in *IOP Conf. Ser.: Mater. Sci. Eng.*)
[5] Khan Z A, Saeed A, Gregory O and Ghafoor A 2016 *Tribology in Industry* **38** (2) 197-213
[6] Volkswagen AG 2010 *Self-Study Programme 465 - The 1.2L 3-cylinder TDI engine with common rail fuel injection system* (Wolfsburg, Germany)
[7] Krewerth D, Lippmann T, Weidner A and Biermann H 2016 *Int. J. Fatigue* **84** 40-52
[8] Yu H and Xu X 2019 *Eng. Fail. Anal.* **97** 145–160
[9] He B, Wei C, Ding S and Shi Z 2019 *Measurement* **134** 426–439
[10] Yao C, Wang T, Xiao W, Huang X and Ren J 2014 *J. Mater. Process. Tech.* **214** 2191–2199

Determining the tribological properties of different 3D printing filaments

I-T Maries[1], C Vilau[2], M S Pustan[1], C Dudescu[2], H G Crisan[1]

[1]Mechanical Systems Engineering Department, Faculty of Machine Building, Technical University of Cluj-Napoca, Cluj-Napoca, Romania
[2]Mechanical Engineering Department, Faculty of Automotive, Mechatronics and Mechanical Engineering, Technical University of Cluj-Napoca, Cluj-Napoca, Romania

E-mail: ionut.maries@omt.utcluj.ro

Abstract. The paper aims to determine the friction coefficient and wear behaviour of different 3D filaments in contact with a metallic disc. The method used for testing was the pin-on-disc method, where 3D half inch diameter pins with a half sphere end were printed out of the different filaments and rotated on the disc. The optimal time for measuring the friction coefficient was determined by observing when a sudden change in friction was recorded by the sensor as a result of material deposition on the disc. All filaments were tested for the same amount of time using the same normal loads. The friction behaviour was monitored and the mass wear was determined by weighting the spheres before and after each test. The research work of this paper was orientated to determine which 3D printing filament is most suitable for building different parts that come in contact with one another such as parts for creating an innovative 3D printed smart shoes sole.

1. Introduction

3D printing also known as rapid prototyping is widely used nowadays for quickly creating parts for mechanical systems [1]. The works of this paper will be focused on determining the friction coefficient and wear of different 3D printing materials. Although vast research was conducted in this area by Bellini et al [2] and Song et al [3], this paper will focus on 3D filaments produced and distributed by Zortrax, a Polish based company developing 3D printing solutions. Perepelkina et al [4] studied the friction coefficient and wear of PLA (a special type of thermoplastic derived from biomass) and ABS (the plastic that LEGO pieces are made of) materials with experimental parts being produced at 100% fill state, and concluded that it is possible to predict their tribological behaviour at this fill state. In accordance to their findings, this paper will analyse the tribological behaviour of 8 different polymers (different from the normal ABS and PLA) under a different fill state, printed on a Zortrax M200 Desktop 3D Printer. The research results will be used in determining which materials are the most suitable for building a new shoe sole. Mass printing custom fit 3D soles is considered one of the future directions in athletic shoe manufacturing and Adidas in collaboration with Carbon have already started research and production in this area (see figure 1).

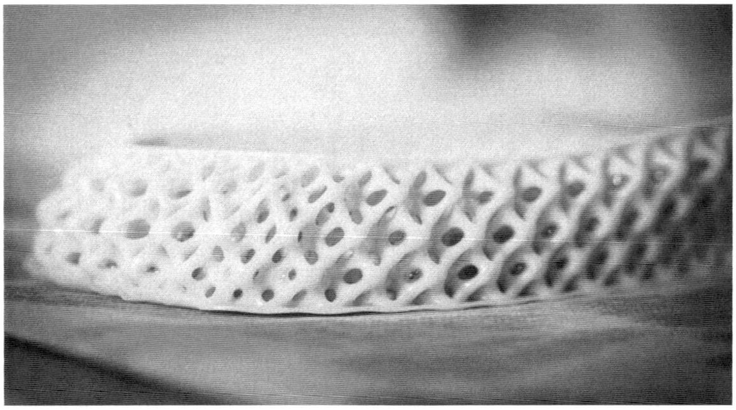

Figure 1. 3D printed shoe sole developed by Adidas and Carbon [5].

2. Sample description

For the purpose of this paper we analysed 8 different 3D printing materials that came in the form of filaments (see figure 2a) . The material filaments were the following: 1 (Z-ABS – optimal material for people who want to start 3D printing), 2 (Z-GLASS – material with a translucent structure), 3 (Z-HIPS – material designed for printing bigger parts), 4 (Z-PCABS – material with high strength and durability), 5 (Z-PETG – material resistant to acids, solvents, alkalis and salts), 7 (Z-ESD – material that provides electrostatic protection), 8 (Z-PLA PRO – material designed to reduce shrinkage in order to obtain fine details) and 9 (Z-ASA PRO – resistant to UV, humidity and heat, designed for outdoor conditions). For material number 6 (Z-ULTRAT), our sample filament broke before printing could be completed, so this material was excluded from the study. The detailed technical data sheets for all the materials can be found on the Zortrax website.

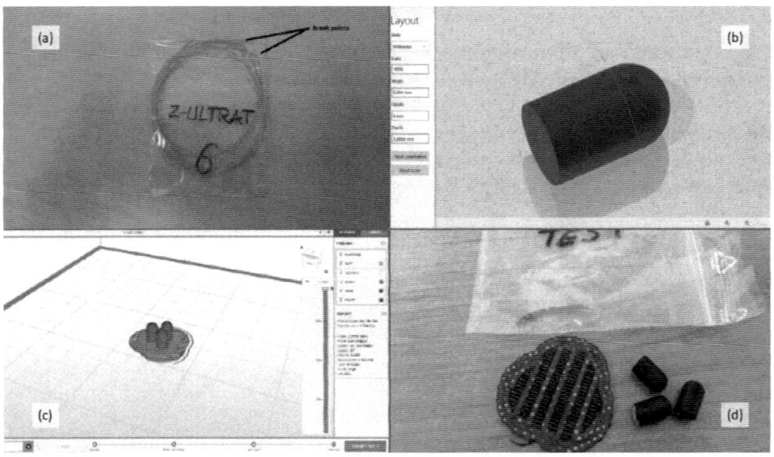

Figure 2. (a) Example of Z-ULTRAT filament **(b)** Pin modelled in SolidWorks and opened with Windows Print 3D (sideways orientation where width is actually the height) **(c)** Pre-production of 3D pins in Zortrax Z-Suite **(d)** Example of 3D printed test pins.

3D half inch diameter pins with a half sphere ends were modelled in SolidWorks, a 3D modelling software (see figure 2b), then added in the Zortrax Printing Software (Z-Suite) and arranged in series of

3 pins to be printed at once (see figure 2c). The pins from each material were then measured and sanded to proper dimensions in width to ensure they fit smoothly in the testing device (see figure 2d).

The technical dimensions and characteristics of the pins were the following: Height (10 mm), Diameter (6 mm), Half ball end (3 mm radius) (figure 2b). Distance from bottom to start of half ball end (7 mm). The pins were printed on a Zortrax M200 3D printed via FDM (Fused Deposit Modelling) where the filament was heated to a maximum of 290° C (depending on the materials used) and extruded through a 0.4 mm diameter nozzle. The nozzle would deposit the melted filament in a linear pattern (PATT. 0) with a layer thickness of 0.14 mm accounting for a 90% infill state for each printed object (figure 2c).

3. Experimental procedure

The scope of this paper was to determine the tribological properties of different 3D printing materials in order to draw conclusions on which of them would be more suitable in building an innovative 3D printed shoe sole. The friction coefficient as well as the wear behaviour were determined.

3.1 Determining the friction coefficient

The method used for testing the friction coefficient was the pin-on-disc method. With this method, 3 pins were placed in sockets on a disc at 100 mm from the centre (figure 3b), which in turn was placed on a metal disc, and then the disc was rotated. The speed of the rotating disc was 350 rmp (rotations per minute). Because small reside was left behind after each test, the disc was cleaned using 2 different sand papers with 1200 and 2000 granulation. The disc as well as the pins were then degreased using a degreasing solution and left to dry off leaving the contact surface smooth (figure 3a). The first load was the disc itself that accounted for a loading force of 10.17 N. After that, a first weight was added, and the total load force was 11.72 N. The second weight was added after that to increase the load force to 14.91 N, then the third to increase the load force to 17.13 N, and finally the forth for a total load force of 18.55 N. Figure 3b shows the slots where the pins were inserted, figure 6c shows the different weights, and figure 3d shows the full test assembly.

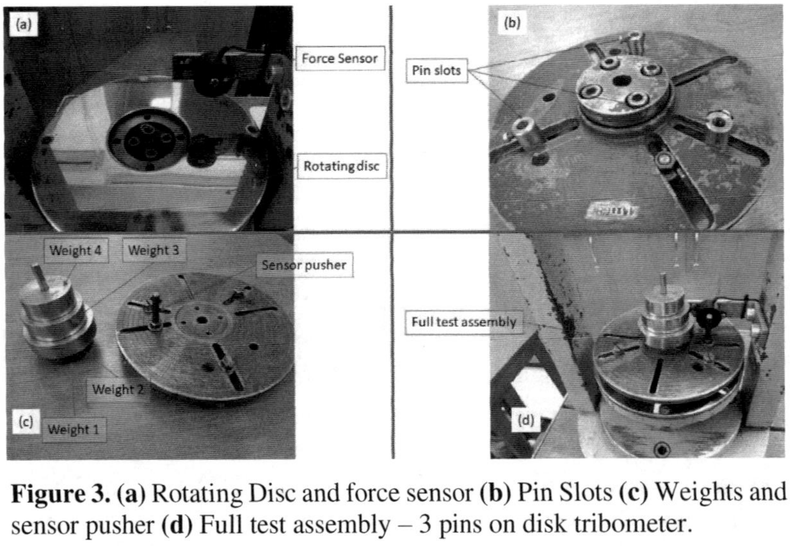

Figure 3. (a) Rotating Disc and force sensor **(b)** Pin Slots **(c)** Weights and sensor pusher **(d)** Full test assembly – 3 pins on disk tribometer.

3.2 Determining the wear behaviour

The wear behaviour was tested using the same test assembly as shown in figure 3d. The metal disc was rotated and the pins were placed on it with the maximum force load of 18.55 N for 5 minutes, then 10 minutes, 15 minutes and lastly 20 minutes. The pins were weighted before and after each run using with

IOP Conf. Series: Materials Science and Engineering **724** (2020) 012022 doi:10.1088/1757-899X/724/1/012022

an analytical KERN AEJ 200-4CM scale. As seen in figure 4, prominent residue was left after some runs, so the disc was cleaned with sand papers and solvent after each run. All in all, each pin sample was worn for 50 minutes in total. Photos of the end of the pins (the part that came in contact with the metal disc) were taken, using a Park System digital microscope (500 x zoom) before the start of the test and after each run. The pins were placed in the same position (in relation to the rotation angle) as to ensure micro parts don't break out so easily.

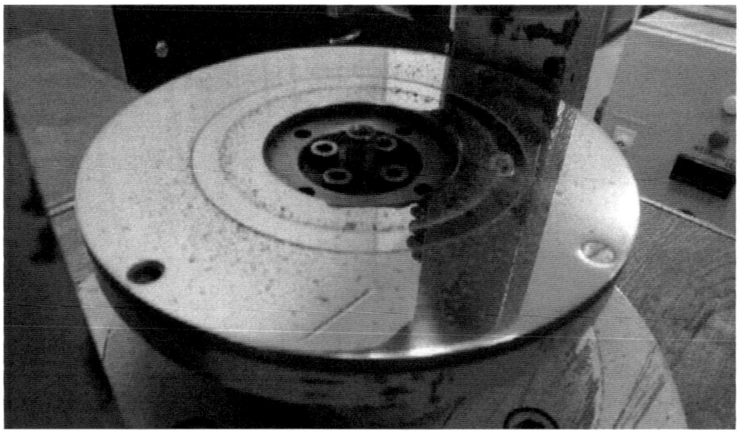

Figure 4. Residue left on disc after a wear test.

4. Results and Discussion

4.1 Friction coefficient results

Results for the friction coefficient were generated by the force sensor mounted on the test assembly. Figure 5 presents an example on how the results looked after each completed test. Table 1 shows the results for the friction coefficient constant (μ), where F_f is the friction force measured in Newtons (N) and F_n the normal loading force also measured in Newtons (N). In this relationship, $\mu = F_f/F_n$. The highlighted results were excluded from the calculation as they were too far off, while the last row (μ_a) in the table presents the average of all friction coefficients after all test runs.

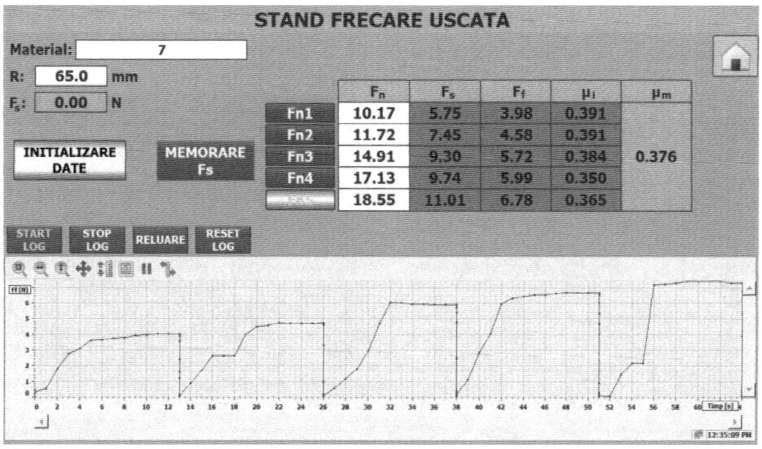

Figure 5. Example of the SIEMENS software used for the measurement.

Table 1. Results of friction coefficient test.

	1	**2**	**3**	**4**	**5**	**7**	**8**	**9**
F_{n1} (N)	10.17	10.17	10.17	10.17	10.17	10.17	10.17	10.17
F_{n2} (N)	11.72	11.72	11.72	11.72	11.72	11.72	11.72	11.72
F_{n3} (N)	14.91	14.91	14.91	14.91	14.91	14.91	14.91	14.91
F_{n4} (N)	17.13	17.13	17.13	17.13	17.13	17.13	17.13	17.13
F_{n5} (N)	18.55	18.55	18.55	18.55	18.55	18.55	18.55	18.55
F_{f1} (N)	2.23	4.38	0.09	2.74	0.72	3.98	3.15	4.97
F_{f2} (N)	2.79	4.54	5.71	3.88	0.31	4.58	1.93	3.96
F_{f3} (N)	3.26	5.89	7.49	5.09	2.54	5.72	4.32	7.04
F_{f4} (N)	5.07	6.39	8.76	8.4	5.73	5.99	5.16	8.02
F_{f5} (N)	5.68	7.77	9.07	0.59	0.37	6.78	0.03	8.67
μ_1	0.22	0.431	0.009	0.269	0.07	0.391	0.31	0.489
μ_2	0.238	0.387	0.488	0.331	0.026	0.391	0.165	0.338
μ_3	0.219	0.395	0.502	0.341	0.17	0.384	0.29	0.472
μ_4	0.296	0.373	0.511	0.49	0.335	0.35	0.301	0.468
μ_5	0.306	0.419	0.489	0.032	0.02	0.365	0.002	0.467
μ_a	**0.256**	**0.401**	**0.497**	**0.357**	**0.071**	**0.376**	**0.266**	**0.446**

4.2 Results of wear behaviour tests

The results shown in table 2 present the initial mass (M_i) of the pins and their mass, measured in grams (g), after 5 minutes of wear (M_1), 10 minutes of wear (M_2), 15 minutes of wear (M_3) and 20 minutes of wear (M_4). The graph in figure 6 presents the same wear behaviour in a visual manner, while figure 7 shows a time lapse of one of the pins to illustrate how the wear behaviour looks like.

Table 2. Results of wear behaviour tests.

	1	**2**	**3**	**4**	**5**	**7**	**8**	**9**
M_i (g)	0.7157	0.8753	0.6826	0.7426	0.9724	0.8000	1.0272	0.8277
5 minutes - M_1 (g)	0.7114	0.8749	0.6809	0.7426	0.9712	0.7992	1.0270	0.8268
10 minutes - M_2 (g)	0.7112	0.8746	0.6787	0.7419	0.9712	0.7958	1.0241	0.8154
15 minutes - M_3 (g)	0.7110	0.8746	0.6778	0.7416	0.9706	0.7927	1.0231	0.8042
20 minutes - M_4 (g)	0.7107	0.8746	0.6764	0.7405	0.9700	0.7860	1.0219	0.7702
Total wear (%)	**0.6986%**	**0.0799%**	**0.9082%**	**0.2827%**	**0.2468%**	**1.75%**	**0.5159%**	**6.9469%**

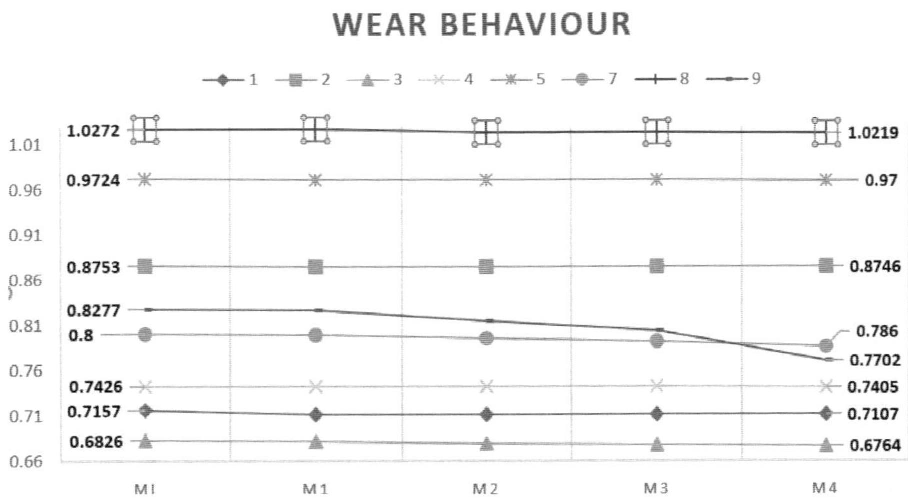

Figure 6. Graph of wear behaviour tests.

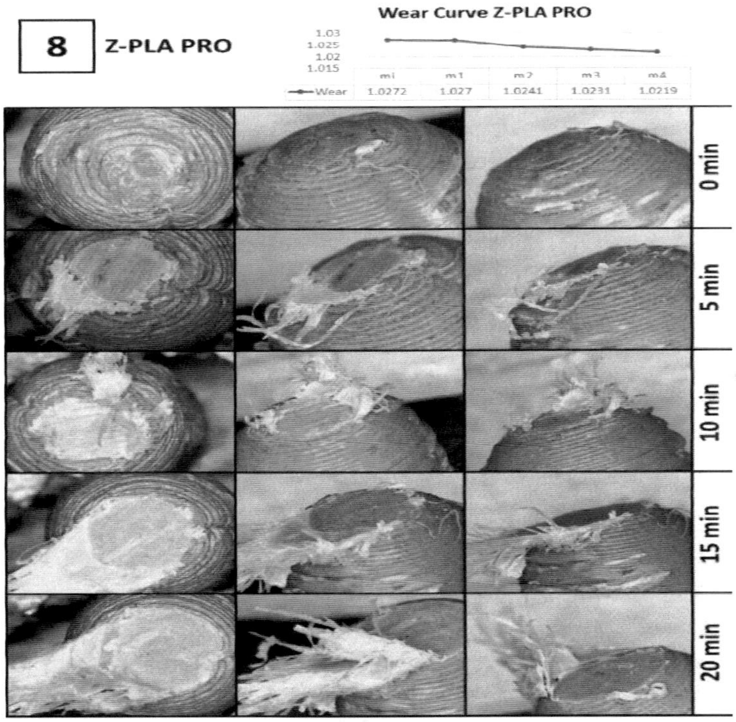

Figure 7. Time-lapse of wear behaviour on sample 8 and wear curve.

4.3 Discussions

The paper aims to determine the friction coefficient and wear behaviour of different 3D filaments in contact with a metallic disc. The method used for testing was the pin-on-disc method, where 3D half inch diameter pins with a half sphere end were printed out of the different filaments and placed on a rotating metal disc made out of C-45 steel.

The experiment was performed on 8 samples fabricated from different 3D printing materials, in room temperature conditions and ambient humidity.

The scope was to measure the friction coefficient and to evaluate the wear behaviour. The highest wear percentage (6.9469%) was determined in the case of sample 9 (Z-ASA PRO) followed by sample 7 (Z-ESD) and the lowest wear percentage (0.0799%) corresponds to the sample 2 (Z-GLASS), followed by sample 5 (Z-PETG). Figure 8 presents the comparative wear behaviour of the sample 9 (figure 8a) and sample 2 (figure 8b).

Moreover, in the case of some samples the material was removed during the wear tests while in other samples, although wear was clearly visible, the materials was only exfoliated without complete material removal (see figure 7).

The wear behaviour observed during the experiments was not in direct concordance with the friction coefficient of the different filaments. The highest friction coefficient was determined in the case of sample 3 (Z-HIPS) followed closely by sample 9 (Z-ASA PRO). The lowest friction coefficient was determined for sample 5 (Z-PETG) with a medium wear percentage of 0.2468% .

The wear behaviour observed during the experiment of sample 2 (Z-GLASS) is not in concordance with the measured friction coefficient, because a small wear percentage of 0.0799% was determined with relatively high friction coefficient of 0.401.

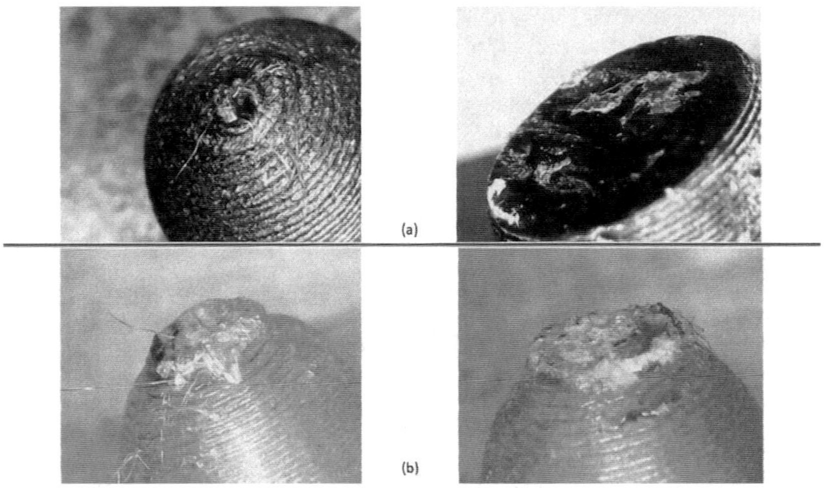

Figure 8. Wear behaviour in the case of **(a)** Z-ASA PRO and **(b)** in the case of Z-GLASS.

5. Conclusion

Product development based on different 3D printing materials is of growing interest worldwide. In order to produce reliable products, the materials' mechanical and tribological properties must be well known. This paper was orientated to investigate the tribological behaviours of same selected 3D printing materials. As a function of materials properties different friction and wear behaviours were determined. Depending on the applications, the product developers can choose which kind of materials correspond to the manufacturing needs.

In conclusion, if the interest is to develop soles with low wear behaviour and high friction, one recommendations is to use the Z-GLASS material, if the mechanical behaviour corresponds with the product requirements. If one needs a balance between friction and wear (like in the day-to-day shoes), then materials corresponding to the sample 1 (Z-ABS) and 8 (Z-PLA PRO) are recommended. For soles requiring low friction (for example sportive dancing) material nr 5 (Z-PETG) is recommended. If the wear is not considered a pressing issue, for good friction, material number 3 (Z-HIPS) is recommended.

6. References

[1] Pham D T and Gault R S 1998 A comparison of rapid prototyping technologies *Int. J. Mach. Tool. Manu.* **38** pp 1257-87

[2] Bellini A and Güçeri S 2003 Mechanical characterization of parts fabricated using fused deposition modelling *Rapid. Prototyp. J.* **9** pp 252-64

[3] Song Y, Li Y, Song W, Yee K, Lee K-Y and Tagarielli V L 2017 Measurements of the mechanical response of unidirectional 3D-printed PLA *Mater. Design.* **123** pp 154-64

[4] Perepelkina S, Kovalenko P, Pechenko R and Makhmudova K 2017 Investigation of Friction Coefficient of Various Polymers Used in Rapid Prototyping Technologies with Different Settings of 3D Printing *Tribology in Industry* **39** pp 519-26

[5] Carbon. (2017). The perfect fit: Carbon + adidas collaborate to upend athletic footwear - Carbon. [online] Available at: https://www.carbon3d.com/case-studies/adidas/ [Accessed 11 Apr. 2019]

IOP Conf. Series: Materials Science and Engineering **724** (2020) 012023 doi:10.1088/1757-899X/724/1/012023

Evaluation of stress risers severity in spur gears meshing

Spiridon S Crețu

Mechanical Engineering, Mechatronics and Robotics Department, "*Gheorghe Asachi*" Technical University of Iasi, 43 Dimitrie Mangeron Street, Iasi, Romania. E-mail: spiridon.cretu@tuiasi.ro

Abstract. The standards for gears design use the maximum value of the normal stress σ_{zz} , depicted σ_{H}, as critical stress for RCF of gears. This value σ_{H} can't sustain evidences regarding the stress risers influences. The paper uses the von Mises stress σ_{vM} as critical stress responsible for the initiation of RCF. For concentrated contacts analyses the drawback of FEM is the very long computing time and therefore the author used a semi-analytical method (SAM) based on the numerical formulation for the involved equations of linear half-space theory. The finite width of gears' teeth imposes quarter-space limitations and consequently the virtual pressures and correction coefficient were considered in the iterative process for pressures distribution. The pressure distribution was further used to evaluate the stress tensor components and 3D distributions of von Mises stresses. Examples are presented for spur gears with different values for end chamfers.

1.Introduction

Well designed gears, manufactured in good industrial practice and running in conditions to assure an adequate lubrication, are able to avoid common failures as fatigue and static fracture by root bending or various forms of wear, including scuffing, but eventually the rolling contact fatigue (RCF) determines gear's teeth failure [1, 2]. The different pressures risers accelerate the RCF that means shorter lives.

The maximum value σ_{H} of pressures distribution σ_{zz} attained under the pure normal loading figure 1, is used by standards for gears design [3]. This value known as Hertz pressure can't sustain evidences regarding the stress risers influences. Alternative fatigue criteria consider the maximum shear stress τ_{45}, or maximum von Mises stress σ_{vM} as critical stress responsible for the initiation of RCF [4-8]. The equation (1) of von Mises equivalent stress reflects the contribution of each tensor 's component :

$$\sigma_{\mathrm{vM}}(x,y,z) = \frac{1}{\sqrt{2}}\left[\left(\sigma_{xx}-\sigma_{yy}\right)^2+\left(\sigma_{yy}-\sigma_{zz}\right)^2+(\sigma_{zz}-\sigma_{xx})^2+6\left(\tau_{xy}^2+\tau_{yz}^2+\tau_{zx}^2\right)\right]^{1/2} \quad (1)$$

The paper uses the maximum von Mises stress σ_{vM} as critical stress responsible for the initiation of RCF, [9 - 13]. The drawback of FEM applied to concentrated contacts analyses is the very long computing time and therefore the author developed a fast and robust semi-analytical method (SAM).

2. Semi-analytical method

The developed SAM is based on Love's analytical formulation for Boussinesq equation of linear half-space theory and Hartnett approach [9] for numerical formulation:

a) geometric equation of the elastic contact:

$$g_{ij} = h_{ij} + w_{ij} - \delta_0 \quad (2)$$

IOP Conf. Series: Materials Science and Engineering **724** (2020) 012023 doi:10.1088/1757-899X/724/1/012023

where: g_{ij} is the gap between the normal loaded surfaces, h_{ij} is the separation between unloaded surfaces, w_{ij} is the elastic deformation of the two surfaces, measured along the normal load, and δ_0 is the rigid displacement of the contacting bodies;

b) equation of the normal surface displacement:

$$w_{ij} = \sum_{k=0}^{Nx-1} \sum_{j=0}^{Ny-1} (K_{i-k,j-l} * p_{kl}) \quad (3)$$

where: the influence coefficients function K_{ij} represents the value of the surface deformation created in the point *(i, j)* by the unit pressure acting in the elementary rectangle *(k, l)*, and $p_{k,l}$is the unknown value of pressure acting in the same elementary rectangle,

c) load balance equation:

$$\Delta x \Delta y \sum_{i=0}^{Nx-1} \sum_{j=0}^{Ny-1} p_{ij} = F \quad (4)$$

d) conditions regarding non-penetration, non-adhesion and elastic-perfect plastic behaviour are also included.

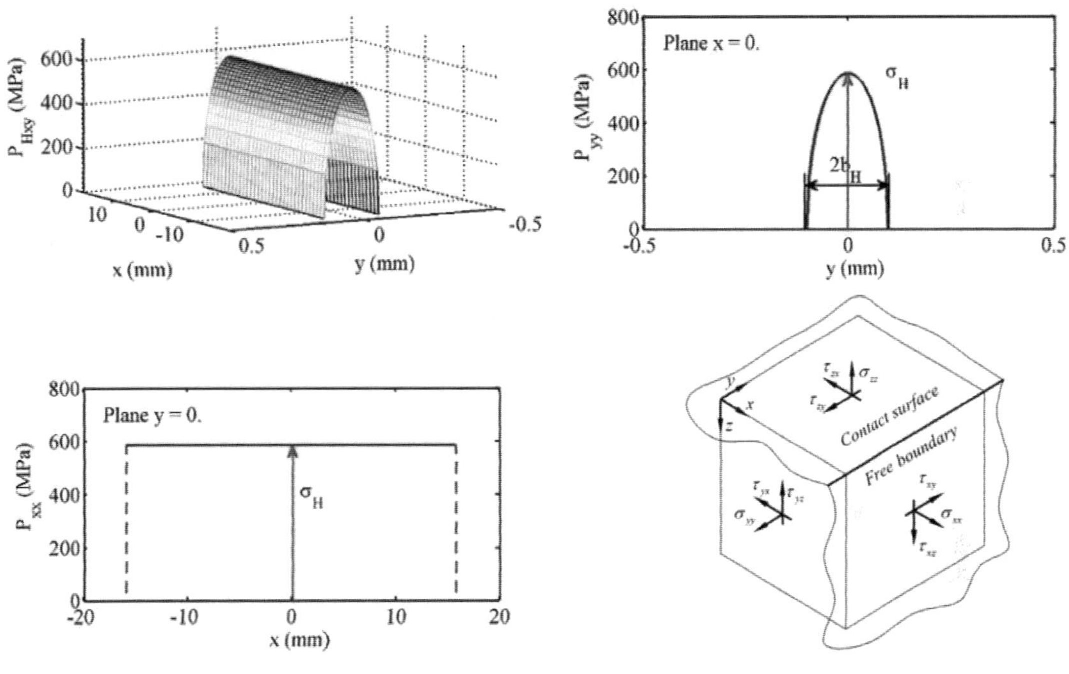

Figure 1. The pressures distributions in the linear half-space based on Hertz's hypothesis.

3. Solution and validation of SAM.

The system of algebraic equations was solved using conjugate gradients methods (GCM) including fast Fourier transform (FFT) to solve the convolution products, [10 - 13].

3.1 Point contacts.

The results obtained with SAM proved to be identical with those calculated using Hertz's formulae, when applicable or provided by FEM if the concentrated contact was a non-Hertz one, figure 2.

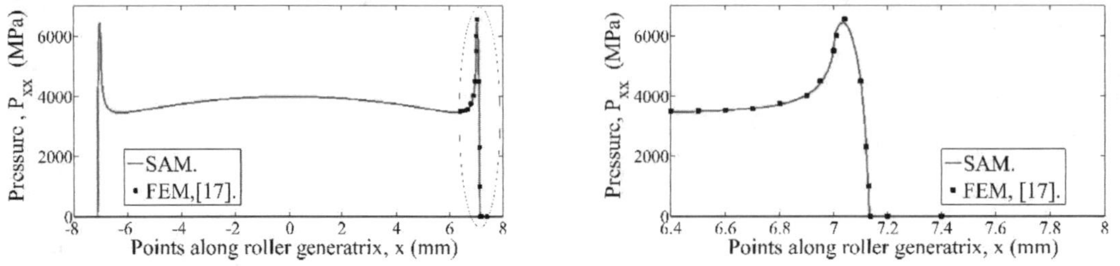

Figure 2. The pressure distributions developed in the normal contact between a crowned cylindrical roller with an end chamfer and a cylindrical inner ring [17].

3.2 Linear contacts and the effect of the finite length.

A real gear with two stress-free sides, has to be modelled as a quarter-space instead of a half-space, figure 1. To eliminate the two shear stresses and one normal stress in the width faces, it was proposed by Hettenyi [18] a two steps correction:

- firstly the application of mirror pressures, figure 3, able to eliminate the two shear stress, and,
- secondly an iterative overlapping half-space solutions till accomplish the boundary condition.

of the quarter-space. Theoretically, the number of iterations goes to infinite but practically the finite number of iterations depends on the accuracy needed. To speed up the entire computation process the virtual pressures and correction coefficient, figure 3 as recently were proposed by Guilbault and Najjari [18], have been considered in the iterative algorithm.

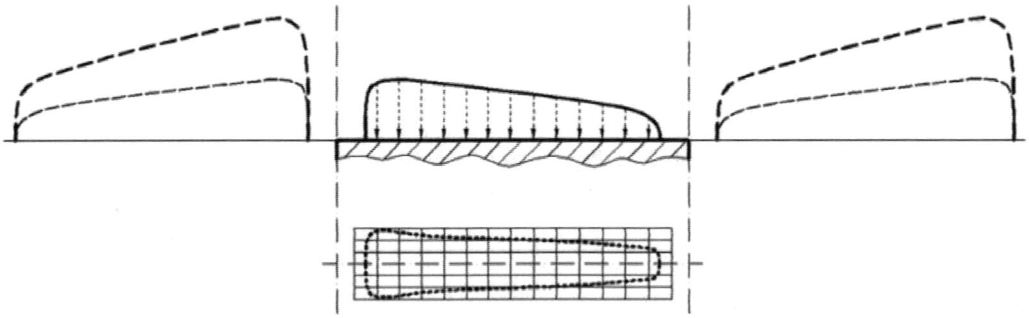

Figure 3. The virtual rectangular contact area and the mesh with coordinate system, [14].

4. The separation matrix

Spur involute gears are in line contact at every instant and a virtual rectangular contact area is built having the contact line as median line. figure 4. Using the involute proprieties the matrix of separations h_{ij} is possible to be expressed analytically for each mesh point (i, j) of the virtual rectangle [14 - 16].

5. Load repartition

The numerical analyses have been performed for a standard spur gars with the following data:

$\alpha_0 = 20^0$, $h^*_{ao} = 1$, $c^*_0 = 0.25$, $\rho^*_0 = 0.38$

$z_1 = 23$, $z_2 = 51$, $m_n = 4$ mm, $x_1 = 0$, $x_2 = 0$, $B = 32$ mm, $Rch = 0.3$ mm.

The nominal power parameters were: $T_2 = 15$ kW, $\varpi_2 = 53.77$ rad/s.

In this study the non-uniform model of load distribution proposed by Pedrero José et al. [19, 20] has been used. Results obtained with Hertz equations indicated the inner point B of single pair tooth contact as the point where the Hertz pressure σ_H attains the highest value.

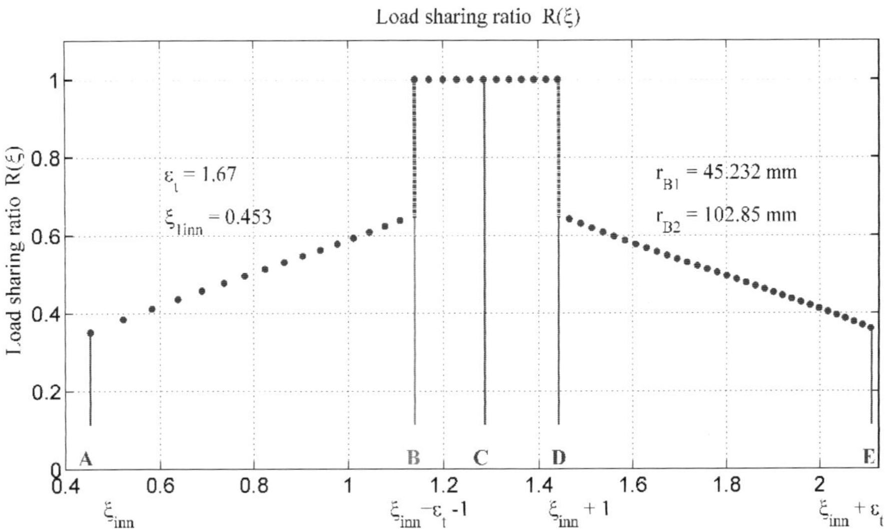

Figure 4. Load sharing ratio, [14].

6. Pressures distribution

The pressures distributions obtained for cases of half space and quarter space hypotheses are presented in figure 5 and figure 6, respectively.

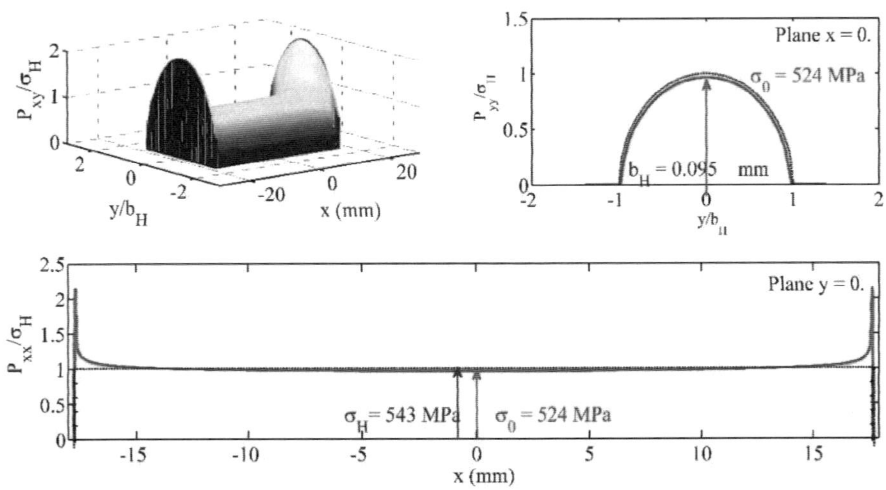

Figure 5. The pressures distribution provided by SAM method within half-space hypothesis.

7. Von Mises equivalent stress distribution

The pressures concentrations accelerate the rolling contact fatigue (RCF) and wear phenomena resulting shorter lives for gears. Popinceanu et al. [4] accomplished a comparative analysis of different components of stress tensor that have been hypothesized as the critical stress in RCF and concluded that von Mises equivalent stress, assures the best correlation with various theoretical and experimental findings. Zhu D. et. al [7], used von Mises stress to predict the pitting life in mixed lubricated line contacts. To evaluate the fatigue life of rolling contacts Morales-Espejel et. al [8] used von Mises equivalent stress in a general model for the surface-subsurface competing RCF mechanism.

8. Stress risers severity

For each infinitesimal volumetric element of significantly stressed volume the new approach takes into account the probability of survival *S* after a number *N* of contact cycles,[5, 6]:

$$\ln\frac{1}{S}\sim N^{e}\iiint_{0}^{V}\frac{(\sigma_{eq}-\sigma_{u})^{c}}{z'^{h}}dV \quad (5)$$

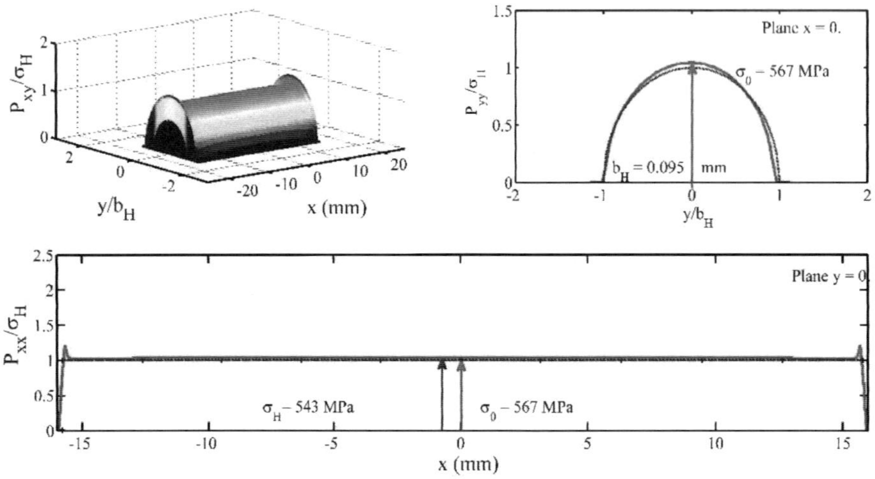

Figure 6. The pressures distribution provided by SAM method within quarter-space hypothesis.

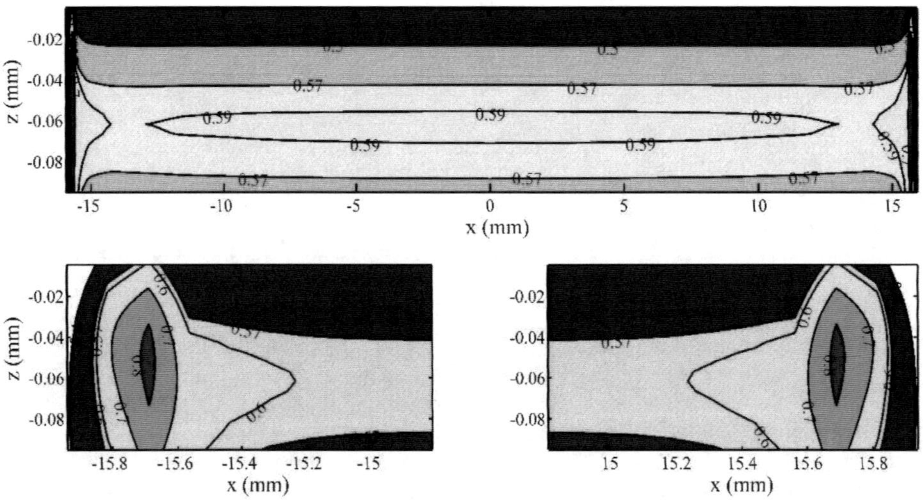

Figure 7. The depth distribution of von Mises stresses created by pressures depicted in figure 6.

where σ_{eq} is the equivalent stress value in the infinitesimal volumetric element, σ_u is the fatigue limit stress, ***c*** is the stress exponent, ***z'*** is the stress-weighted average depth, ***h*** is the depth exponent, ***e*** is the Weibull slope and ***V*** is the volume significantly stressed. For an imposed value of reliability the integration of last equation over the significant stressed volume lead to a correspondent number of loading cyles until rolling contact fatigue is evidenciated. The same operations done to the similar loading and geometrical conditions, but with no stress riser, provides the maximum number of loding cycles for the particular gear. The ratio of the two durabilities will represent the severity number.

9. Conclusions

The large industrial experience as well as a lot of experimental works pointed out the lack of Hertz stress to provide consistent explanations for important experimental outcomes. Theoretical and experimental studies proved that von Mises equivalent stress offers good explanations for the experimental and theoretical findings and provides a solid formulation to calculate the gears life as long as a value for reliability is chosen. The stress riser severity is proposed as the ratio between the optimum life obtained with no stress riser and the life obtained when the stress riser exists.

References

[1] Santus C, Beghini M. Bartilotta I and Facchini M, 2012, *Int. J. Fatigue,* **45**, 71-81.
[2] Nelias D, Dumont M.L, Champiot F, Vincent A, Girodin D, Fougeres R, and Flamand L,1999, *ASME Trans. J. Tribol.* ***128***, 240-251,1999.
[3] ISO 6336-1-5, 2006, *Calculation of load capacity of spur and helical gears*, Geneva, Switzerland.
[4] Popinceanu N, Diaconescu E and Creţu S, 1981, *WEAR,* **71**,265-82.
[5] Ioannides E and Harris A T, 1985, *ASME Trans.J.. Tribol.*,**107,**367-78.
[6] Ioannides E, Bergling G and Gabell A, 1999, *Acta Polytech. Scand.*, Mech. Eng. Series, **137,** Finnish Institute of Technology.
[7] Zhu D, Ren N and Wang Q J, 2009, *ASME Trans. J. Tribol.***131**, 4, 1-8.
[8] Morales-Espejel G E, Gabelli A and DeVries A J, 2015, *Tribol. Trans.,* **58**, 5, 894-96.
[9] Hartnett J M, 1979, *ASME Trans. J. Lub. Tech.*, 1979, **101,** 1, 106-109.
[10] Polonsky I A and Keer L, 2000, *ASME Trans.J. Tribo,l.* **122,** 1.
[11] Creţu S, Antalucă E and Creţu O, 2003, *Annals Univ. Galaţi* Romania **24**, 26-35.
[12] Creţu S, 2005, *Bull. Inst. Polit. Iaşi*, **LI, (LV)**, 1-2, 1-31.
[13] Creţu S, 2009, *Contactul concentrat elastic-plastic*, Ed. Politehnium, Iaşi, Romania,
[14] Crețu S, Pop N, Cazan S, 2017, MATEC Web of Conferences **112,** 07015, Ian E&E, May 25-26, Iaşi, Romania.
[15] Litvin Fand Fuentes A, 2004, *Gear geometry and applied theory*, Cambridge Univ. Press, 2
[16] Pop N, Tufescu A and Cretu S, 2014, *App. Mech. Mat.* **658**, 531-56.
[17] de Mul J M, Kalker J J and Fredriksson B, 1986, *ASME Trans. J. Lub. Tribol.*, **108**,140-48.
[18] Guibault R, 2011, *ASME Trans. J. Tribol.* **133**, 2.
[19] Pedrero I J, Pleguezuelos M and Artes M, 2010, Mech. Mach..Theory, **45**, 780-74.
[20] Pedrero I J, Pleguezuelos M, Munoz M, 2011, Mech. Mach. Theory, **46**, 425-47.

IOP Conf. Series: Materials Science and Engineering **724** (2020) 012024 doi:10.1088/1757-899X/724/1/012024

Theoretical Analysis of Spherical Bearings Lubricated with Greases

A V Radulescu[1], I Radulescu[1]

[1]University POLITEHNICA Bucharest, Romania

E-mail: alexandru.radulescu@upb.ro

Abstract. The paper presents theoretical researches concerning the spherical bearings lubricated with greases, in rheostatic conditions: high loads and low velocities. This type of bearing is very appropriate for modelling the human joints (hip, knee), spherical joints from automotive industry (pivots, connecting rods) or special joints from petroleum industry. Spherical bearings may be of the "fitted" type or the "clearance" type. The fitted type of nonflat bearing pad surface is lapped-in to fit the runner. When it is operating, the fitted pad will have a varying film thickness across the pad sill, between the pad and the runner. Similarly, in the clearance type the film shape will be non-uniform across the pad sill. The variation in film thickness across the bearing pad, for nonflat pads, influences the resulting pressure distribution in the bearing pad and, hence, its load carrying capacity. In this paper, the pressure distribution and the load capacity are determined for „clearance" type bearings, for two different cases: complete spherical bearings and partial spherical bearings. It has been found that the eccentricity ratio influences the angular position of the maximum pressure, which becomes nearer to the minimum angle of the bearing, once with the increasing of the eccentricity ratio.

NOMENCLATURE

$C_F = \bar{F}\psi$	Load coefficient	y	Cartesian coordinate
D	Spherical pad diameter	$\epsilon = 2e/J$	Eccentricity ratio
F	Load carrying capacity	η	Viscosity
$\bar{F} = \dfrac{F}{\pi D^2 \tau_0}$	Non-dimensional load carrying capacity	θ	Angular coordinate
J	Clearance	θ^*	Angle corresponding to the maximum pressure
d	Spherical ball diameter	θ_{max}	Maximum angle of the spherical bearing
e	Eccentricity	θ_{min}	Minimum angle of the spherical bearing
h	Film thickness	τ	Shear stress
p	Pressure	τ_0	Yield stress
$\bar{p} = p/p_{max}$	Non-dimensional pressure	$(\tau:\tau)/2$	Second invariant of the tangential stress tensor
u	Velocity	$\psi = J/D$	Relative clearance

IOP Conf. Series: Materials Science and Engineering **724** (2020) 012024 doi:10.1088/1757-899X/724/1/012024

1. Introduction

The classical hydrodynamic lubrication theory in its original form can be applied only if the lubricant behaves as a linearly viscous fluid (Newtonian). Applications involving non-Newtonian fluids require appropriate formulation [1]. A class of non-Newtonian fluid utilizes a so-called Bingham model in some tribological applications to describe the flow characteristics of some lubricants, such as grease or fine suspensions [2]. Two parameters, which characterize Bingham plastic models, are the yield stress and the viscosity. When the magnitude of the stress tensor is less than the yield stress, the behaviour of the fluid is akin to that of a rigid solid. When the yield stress is exceeded, an ideal Bingham plastic fluid commences to flow as a linear viscous (Newtonian) liquid [3].

The purpose of this paper is to study the spherical bearings lubricated with greases, in rheostatic conditions: high loads and low velocities. This type of bearing is very appropriate for modelling the human joints, such as the synovial joint (Figure 1) [4 - 7] or the spherical joints used in the motor vehicle industry, such as pivots and auxiliary connecting rods (Figure 2) [8 - 10].

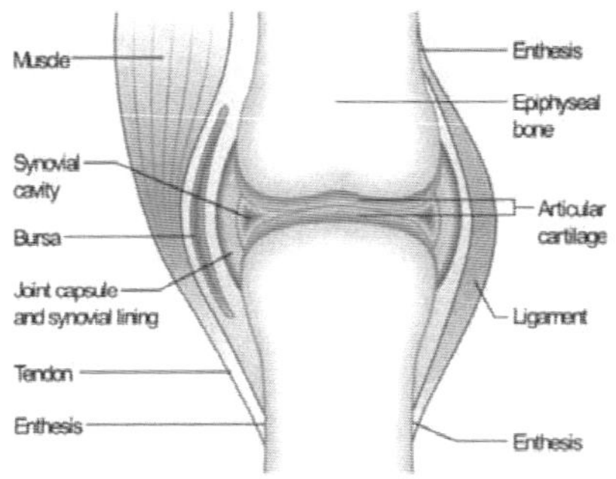

Figure 1. Human synovial joint [4]

Figure 2. Spherical joints from motor vehicle industry [8]

2. Theory

Special nonflat bearings pad can handle unusual load and motion requirements that flat thrust pads cannot. For instance, spherical-pad bearings are free to rotate about the three major axes without disrupting operation of the bearing. Spherical bearings may be of the "fitted" type or the "clearance" type [11 – 12]. The fitted type of nonflat bearing pad surface is lapped-in to fit the runner. When in operation, the fitted pad will have a varying film thickness across the pad sill between the pad and the runner. Similarly, in the clearance type the film shape will be nonuniform across the pad sill. An exception to this is when the combination of load and flow is such as to cause the bearing to operate in the "concentric" position. The variation in film thickness across the bearing pad for nonflat pads influences the resulting pressure distribution in the bearing pad and, hence, its load carrying capacity.

In this paper, only the clearance bearings are considered, for two different cases: complete spherical bearings and partial spherical bearings. The geometry of these bearings is presented in Figure 3 and Figure 4.

In these cases, the spherical pad radius is slightly larger than the spherical ball radius and the film thickness between the pad sill and the ball will vary with the angular coordinate such that:

$$h = \frac{J}{2} - e\cos\theta \tag{1}$$

The rheological model for the non-Newtonian fluid is ideal Bingham plastic model. The constitutive equation for the fluid is:

$$\tau = \tau_0 + \eta\frac{du}{dy} \tag{2}$$

For the general case of Bingham fluid, Oldroyd propose a three-dimensional rheostatic equation, based on the Mises plasticity criterion [13 - 14]:

$$\left|\frac{1}{2}(\tau:\tau)\right| \le \tau_0^2 \tag{3}$$

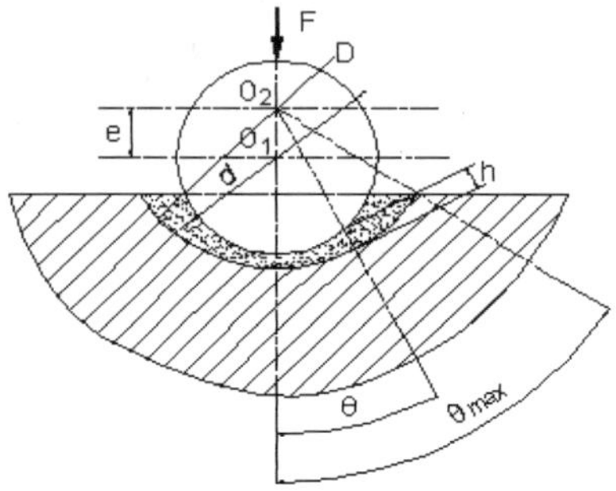

Figure 3. Complete spherical bearing

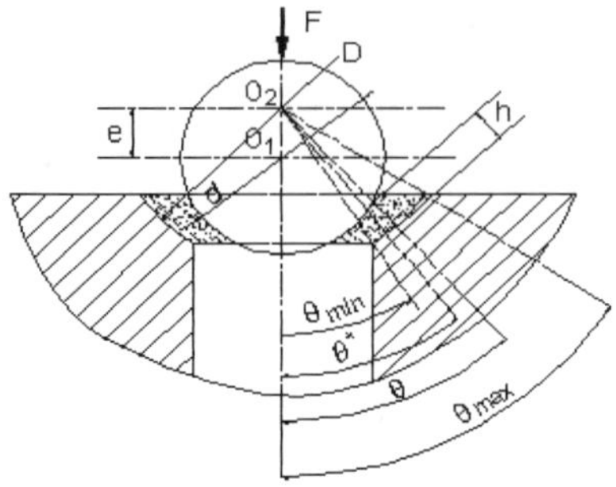

Figure 4. Partial spherical bearing

Due to the fact that the spherical bearings considered in Figures 3 and 4 are symmetrical, equation (3) becomes simpler and can be treated analytically [15]. Taking account that the lubricant flow is circumferential, the rheostatic equation for the spherical bearings will be:

$$\frac{dp}{d\theta} = -\frac{D\tau_0}{h} \tag{4}$$

The pressure distribution can be obtained integrating equation (4) and considering the film thickness given by the equation (1):

$$p(\theta) = \int \left(-\frac{D\tau_0}{\frac{J}{2} - e\cos\theta} \right) d\theta \tag{5}$$

If the surface of the spherical bearing is noted with Ω and $\Gamma = \bigcup_{i=1}^{n} \Gamma_i$ represent the curves adjacent to the surface Ω, the limit conditions for the pressure distribution (5) are:

$$p|_{(x,y)\in \bigcup_{i=1}^{n} \Gamma_i} = 0 \tag{6}$$

For this case, the load carrying capacity of the bearing is calculated by integrating the pressure distribution obtained from equation (5):

$$F = \int_{\Omega} p d\Omega \tag{7}$$

2.1 Complete spherical bearing

Considering the geometry of the complete spherical bearing (see Figure 3) and imposing the limit condition:

$$p = 0 \ \text{ for } \ \theta = \theta_{max} \qquad (8),$$

the pressure distribution results by integrating the equation (5):

$$p = \frac{4D\tau_0}{\sqrt{J^2 - 4e^2}} \left[arctg \frac{(J+2e)tg\frac{\theta_{max}}{2}}{\sqrt{J^2-4e^2}} - arctg \frac{(J+2e)tg\frac{\theta}{2}}{\sqrt{J^2-4e^2}} \right] \tag{9}$$

In order to obtain the load carrying capacity (eq. 7), the pressure distribution must be integrated:

$$F = \frac{\pi D^2}{2} \int_0^{\theta_{max}} p \sin\theta \cos\theta \, d\theta \tag{10}$$

Unfortunately, no analytical solution can be obtained and a numerical solution is imposed. Therefore, the following non-dimensional notations are necessary:

$$\varepsilon = \frac{2e}{J}; \ \ \psi = \frac{J}{D}; \ \ \bar{p} = \frac{p}{p_{max}}; \ \ \bar{F} = \frac{F}{\pi D^2 \tau_0}; \ \ C_F = \bar{F}\psi \tag{11}$$

2.2 Partial spherical bearing

The case of the partial spherical bearing (Figure 4) is treated similar to the case of the complete spherical bearings, excepting the limit conditions for the pressure distribution. These conditions must be imposed on two adjacent curves of the surface Ω, corresponding to the maximum and minimum angles of the spherical bearing:

$$p = 0 \ \text{ for } \ \theta = \theta_{max} \qquad \text{and} \qquad p = 0 \ \text{ for } \ \theta = \theta_{min} \tag{12}$$

Integrating the equation (5), the pressure distribution is obtained for two different zones:

$$p = \begin{cases} \frac{4D\tau_0}{\sqrt{J^2-4e^2}}\left[arctg\frac{(J+2e)tg\frac{\theta}{2}}{\sqrt{J^2-4e^2}} - arctg\frac{(J+2e)tg\frac{\theta_{min}}{2}}{\sqrt{J^2-4e^2}}\right] & \text{for } \theta \in [\theta_{min};\theta^*] \\ \frac{4D\tau_0}{\sqrt{J^2-4e^2}}\left[arctg\frac{(J+2e)tg\frac{\theta_{max}}{2}}{\sqrt{J^2-4e^2}} - arctg\frac{(J+2e)tg\frac{\theta}{2}}{\sqrt{J^2-4e^2}}\right] & \text{for } \theta \in (\theta^*;\theta_{max}] \end{cases} \quad (13),$$

where θ^*represents the angle corresponding to the maximum pressure. This value is obtained by solving numerically the following equation and using the non-dimensional notations (11):

$$arctg\frac{(1+\varepsilon)tg\frac{\theta^*}{2}}{\sqrt{1-\varepsilon^2}} = \frac{1}{2}\left[arctg\frac{(1+\varepsilon)tg\frac{\theta_{max}}{2}}{\sqrt{1-\varepsilon^2}} + arctg\frac{(1+\varepsilon)tg\frac{\theta_{min}}{2}}{\sqrt{1-\varepsilon^2}}\right] \quad (14)$$

Finally, the load carrying capacity is numerically obtained by integrating the pressure distribution (eq. 13) from minimum angle of the spherical bearing to the maximum angle:

$$F = \frac{\pi D^2}{2}\int_{\theta_{min}}^{\theta_{max}} p \sin\theta \cos\theta \, d\theta \quad (15)$$

3. Results

3.1 Complete spherical bearing

Figure 5 shows the spatial distribution of the non-dimensional pressure for four different maximum angles of the complete spherical bearing, in the case when the eccentricity ratio is $\varepsilon = 0.9$.

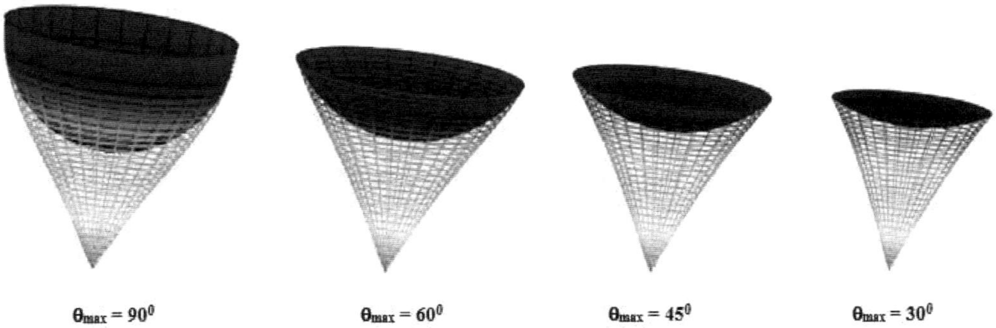

Figure 5. Spatial distribution of the non-dimensional pressure (ε = 0.9)

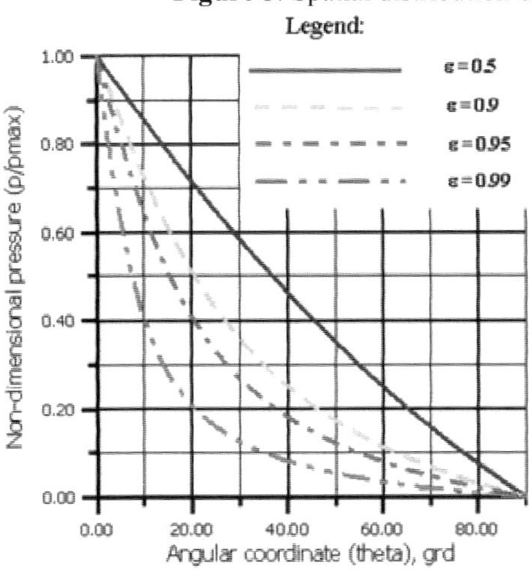

Figure 6. Pressure distribution (cross section θ_{max} = 90⁰)

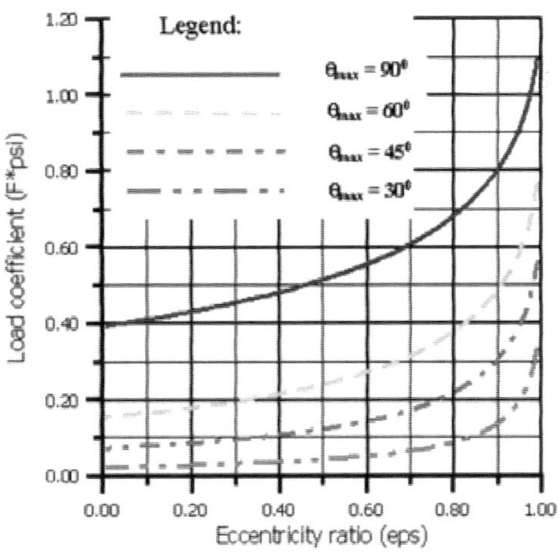

Figure 7. Variation of the load coefficient

Also, Figure 6 presents a cross section through the spatial distribution of the non-dimensional pressure, in order to study the influence of the eccentricity ratio on the pressure profile. Figure 7 presents the variation of the load coefficient versus eccentricity ratio for different values of the maximum angle of the complete spherical bearing.

3.2 Partial spherical bearing

Figure 8 shows the spatial distribution of the non-dimensional pressure for the eccentricity ratio $\varepsilon = 0.9$ and Figure 9 shows the same pressure distribution in the cross section for different values of minimum and maximum angles of the partial spherical bearing. Also, Figure 10 presents the position of the angle corresponding to the maximum pressure versus eccentricity ratio, for the same values of angles.

$\theta_{min} = 15^0$ **$\theta_{min} = 30^0$** **$\theta_{min} = 45^0$**

Figure 8. Spatial distribution of the non-dimensional pressure ($\varepsilon = 0.9$)

The variation of the load coefficient versus the eccentricity ratio for different values of minimum and maximum angle of the partial spherical bearing is presented in Figure 11.

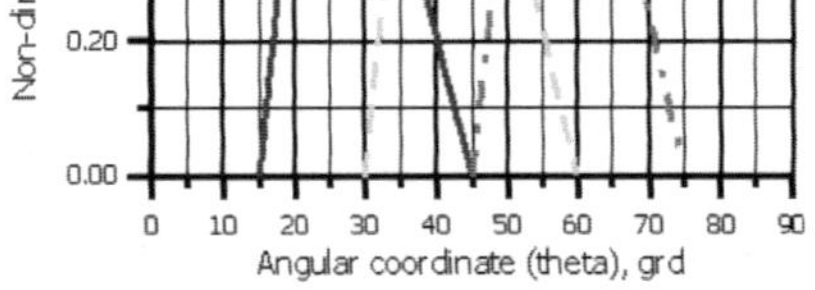

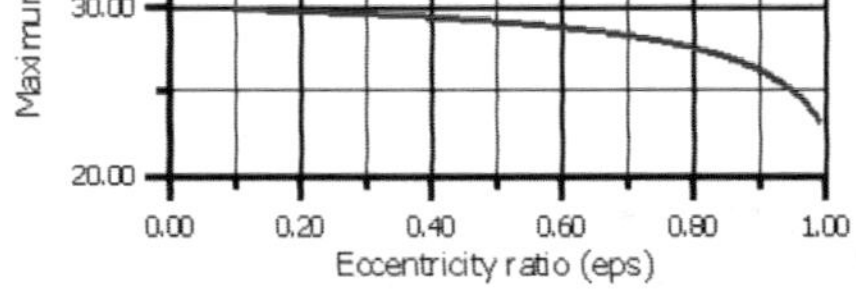

Legend:

Figure 9. Pressure distribution ——— $\theta_{min} = 15^0$ **Figure 10.** Maximum pressure angle

– – – – $\theta_{min} = 30^0$

– · – · – $\theta_{min} = 45^0$

4. Conclusions

1. The influence of the eccentricity ratio on the profile of the pressure distribution is very important for the complete spherical bearing.
2. The load carrying capacity for the complete spherical bearing increases with the increasing of the eccentricity ratio and the maximum angle of the spherical bearing.
3. In the case of the partial spherical bearing, the influence of the eccentricity ratio on the pressure distribution is practically negligible. It has been found that the eccentricity ratio influences

especially the angular position of the maximum pressure, which becomes nearer to the minimum angle of the bearing once with the increasing of the eccentricity ratio

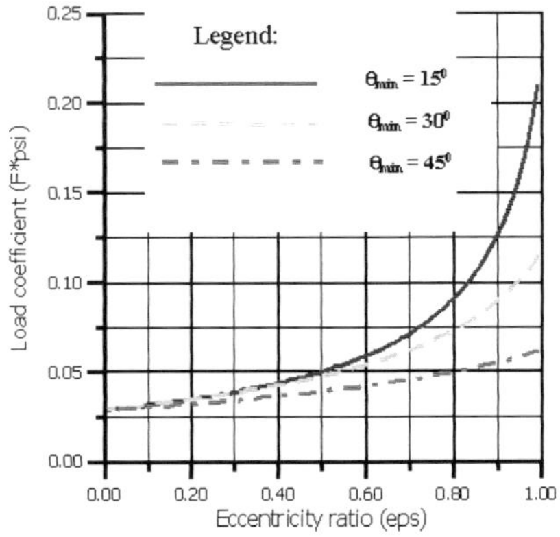

Figure 11. Variation of the load coefficient

4. For the partial spherical bearing, the load carrying capacity also increases with the increasing of the eccentricity ratio and with the decreasing of the minimum angle of the spherical bearing.

5. References

[1] C Balan 2000 *The rheology of lubricating grease* (Amsterdam: ELGI)
[2] T Mang and W Dresel 2007 *Lubricants and Lubrication* (Weinheim: Wiley-Vch)
[3] P M Lugt 2009 A review on grease lubrication in rolling bearings *Tribol. Trans.* **52** (4) pp 470-480
[4] *** https://www.pdhpe.net/the-body-in-motion/how-do-the-musculoskeletal-and-cardiorespiratory-systems-of-the-body-influence-and-respond-to-movement/skeletal-system/structure-function-synovial-joints/ (accessed on 11.04.2019)
[5] M Daniel, A Iglic and V Kralj-Iglic 2005 The shape of acetabular cartilage optimizes hip contact stress distribution *J. Anat.* **207** (1) pp 85-91
[6] M A Wimmer, G Sibylle, T Kaup, M Hänni, E Schneider, S Gogolewski and M Alini 2004 Tribology approach to the engineering and study of articular cartilage *Tissue Eng.* **10** (9-10) pp 1436-1445
[7] M Kothari, J F Booker and D L Bartel 1995 Analysis of artificial hip joints as spherical bearings *Proc. 21th Leeds-Lyon Symposium on Tribology (Leeds)* (Amsterdam: Elsevier) pp 93-98
[8] *** https://en.wikipedia.org/wiki/Ball_joint
[9] M M Khonsari and E R Booser 2017 *Applied tribology: bearing design and lubrication* (New York: John Wiley & Sons)
[10] M J Neale 2013 *Bearings: a tribology handbook* (Amsterdam: Elsevier)
[11] A Harnoy 2002 *Bearing design in machinery: engineering tribology and lubrication* (New York: CRC Press)
[12] B W Rowe 2012 *Hydrostatic, aerostatic and hybrid bearing design* (Amsterdam: Elsevier)
[13] J M Piau 1996 Flow of a yield stress fluid in a long domain. Application to flow on an inclined plane *J. Rheol.* **40** (4) pp 711-723
[14] R Mas and A Magnin 1996 Rheological and physical studies of lubricating greases before and after use in bearings *J. Tribol.* **118** (3) pp 681-686
[15] M D Pascovici and A V Radulescu 1984 Portanta statica a filmului de unsoare consistenta *Proc. Int. Conf. TRIBOTEHNICA '84 (Iasi)* vol 1 pp 295-302

IOP Conf. Series: Materials Science and Engineering **724** (2020) 012025 doi:10.1088/1757-899X/724/1/012025

Surface roughness influence on active surfaces geometry and modified rating life of rolling contacts

M Benchea[1] **and S Crețu**[1]

[1]Mechanical Engineering, Mechatronics and Robotics Department, "Gheorghe Asachi" Technical University of Iaşi, Iaşi, Romania

E-mail: marcelin.benchea@tuiasi.ro

Abstract. The active surfaces topography of bodies in rolling contact can be obtained by different manufacturing technology. The scope of this paper is to evidence the influence of the initial surface roughness on running-in process, working surface geometry and durability of rolling contacts. The paper presents a numerical study applied on a toroidal roller bearing where the modified rating lives have been evaluated with the methodologies mentioned by ISO 16281.

1. Introduction

The topography of the active surfaces in rolling contact depends on the surface manufacturing process technology. In the field of engineering the exact degree of roughness can be of considerable importance, affecting the function of a component or its cost [1,2,3].

The roughness of the surfaces that interacts has an important role on its tribological behaviour. The solutions of the models used in order to describe it [4,5] require the input of a three-dimensional rough profile. Since the experimental acquisition of such profiles does not allow a sufficiently large sample of profiles to be used, an essential requirement for any numerical algorithm is to generate arbitrarily rough surfaces with the same or similar proprieties of the real surfaces [6,7].

When two engineering surfaces are loaded for the first time changes in the topography of both surfaces generally occur. These changes appear between start-up and steady state and are associated with running-in process [8,9].

An analysis model was developed to model the nonlinear strain rate dependent deformation of materials stressed in elastic-plastic domain [10,11]. The model is developed in the frame of incremental theory of plasticity using the von Mises yield criterion and Prandtl-Reuss equations. Considering the isotropic and non-linear kinematic hardening laws of Lemaitre-Chaboche [12,13] the model accounts for the cyclic hardening phenomena.

The nano-asperities of contact surfaces generate high pressure peaks in pressure distributions that can severely diminish the modified rating life as is defined by [14].

2. Theory

2.1. Elastic-plastic model

The elastic-plastic model developed to evaluate the plastic deformation of the surfaces considers the material cyclic hardening characteristic that is described by a combined isotropic and nonlinear kinematic hardening laws [12,13].

The yield surface is defined by the von Mises yield criterion:

$$F = f(\sigma - \alpha) - \sigma_Y^0 = 0 \tag{1}$$

where $f(\sigma - \alpha)$ is the equivalent von Mises stress and σ_Y^0 is the yielding stress.

The non-linear kinematic hardening component describes the ratchetting effect by expressing the translation of the yield surface in the stress space through the back-stress α.

The isotropic hardening component of the model defines the yielding stress as a function of the equivalent plastic strain and is given by the relation:

$$\sigma_Y^0 = \sigma_{Y0} + Q_\infty \cdot \left(1 - e^{-b_\infty \cdot \bar{\varepsilon}^p}\right) \tag{2}$$

where Q_∞ is the limiting change in the yield surface size on the deviatoric plane and b_∞ describes how rapid the limit size is reached [15].

According to the energetic theory of fracture, the von Mises stress is an equivalent uniaxial traction stress [13]. For a 3D contact loading in each point of the stressed volume the equation of von Mises stress amplitude includes all 6 components of the stress tensor. For this consideration, when the yield limit is overcome, it can be used the pseudo-plastic Ramberg-Osgood's equation in the combined hardening laws, equation that gives the correspondence between the intensity of the stress tensor and the intensity of the strain tensor:

$$\bar{\varepsilon}_e = \frac{\bar{\sigma}_e}{E} + \left(\frac{\bar{\sigma}_e}{B}\right)^N \tag{3}$$

where B is the coefficient of plastic resistance and N is the hardening exponent [10,11].

The plastic deformation of the surfaces is given by the plastic strains accumulated in the sub-surface layer:

$$u^p(x,y) = \int_0^\infty \varepsilon^p(x,y,z) \cdot dz \tag{4}$$

2.2. Modified rating life

The classic computation of the rolling bearings basic rating life considers only the equivalent load on the bearing P, dynamic capacity load C and an exponent p depending of the rolling element:

$$L_{10} = \left[\frac{C}{P}\right]^p \tag{5}$$

where $p = 3$ for ball bearings and $p = 10/3$ for roller bearings [16].

To take into account the peaks of the pressure distributions generated by the presence of the nano-asperities on the surface, was developed a method to compute the basic rating life with methodology given by the standard ISO 16281 [14].

In the first instance, for the roller bearings, to evaluate the rating life is needed to determine the elastic deflection of each loaded roller j that depends on the radial load and internal radial clearance of the bearing:

$$\delta_j = \delta_r \cdot \cos\varphi_j - \frac{s}{2} \tag{6}$$

In the second instance, to consider de load on every loaded roller is used the lamina model as is described in [14] and the roller is divided into n_s identical laminae.

The load $q_{j,k}$ on lamina k of the roller j is:

$$q_{j,k} = c_s \cdot \delta_{j,k}^{10/9} \tag{7}$$

with the spring constant c_s:

$$c_s = \frac{35948 \cdot L_w^{8/9}}{n_s} \tag{8}$$

were L_w is the roller length and n_s is the laminae number [14].

In the third instance, the basic rating life L_{10r} of the roller bearings is calculated with:

$$L_{10r} = \left\{ \sum_{k=1}^{n_s} \left[\left(\frac{q_{kci}}{q_{kei}} \right)^{-9/2} + \left(\frac{q_{kce}}{q_{kee}} \right)^{-9/2} \right] \right\}^{-8/9} \tag{9}$$

where: q_{kci} - basic dynamic load rating of a bearing lamina of the inner ring,
q_{kce} - basic dynamic load rating of a bearing lamina of the outer ring,
q_{kei} - dynamic equivalent load on a lamina k of a rotating inner ring,
q_{kee} - dynamic equivalent load on a lamina k of a stationary outer ring.

According to standard [14], the dynamic equivalent loads on each lamina of the inner/outer ring q_{kei} and q_{kee} are computed as a function of the stress riser coefficients evaluated for each lamina of each roller. The stress risers are calculated considering the pressure distribution on the corresponding lamina.

In the last instance, the modified rating life L_{10mr} of the roller bearings is given by:

$$L_{10mr} = \left(\sum_{k=1}^{n_s} \left\{ \left[a_{ISO} \left(\frac{\eta_c \cdot P_u}{P_{ks}}, \kappa \right) \right]^{-9/8} \cdot \left[\left(\frac{q_{kci}}{q_{kei}} \right)^{-9/2} + \left(\frac{q_{kce}}{q_{kee}} \right)^{-9/2} \right] \right\} \right)^{-8/9} \tag{10}$$

where the life modification factor a_{ISO} is calculated for each bearing lamina k using its dynamic equivalent load P_{ks}:

$$P_{ks} = 0.323 \cdot Z \cdot n_s \cdot \cos\alpha \cdot \left[\frac{q_{kei}^{9/2} + \left(1.038 \cdot q_{kci}/q_{kce}\right)^{9/2} \cdot q_{kee}^{9/2}}{1 + \left(1.038 \cdot q_{kci}/q_{kce}\right)^{9/2}} \right]^{2/9} \tag{11}$$

In facts, the life modification factor a_{ISO} is a function that depends on the lubrication factor κ, oil contamination factor η_c and fatigue load limit P_u for the given application [16].

3. Results and Discussion

A toroidal roller bearing - CARB C2318, has been selected for the numerical case study. The main geometrical data are as follows: inner ring diameter $d_{ci} = 115\ mm$, outer ring diameter $d_{ce} = 165\ mm$, roller diameter $d_w = 25\ mm$, roller length $L_w = 54\ mm$, number of rollers $Z = 14$. The catalogue data [17]: dynamic capacity load $C = 610\ kN$ and fatigue load limit $P_u = 73.5\ kN$.

For numerical case study of the running-in process a medium radial loading $F_R = 0.185 \cdot C = 112.5\ kN$ and three magnitudes of the surface roughness parameters was used: a fine roughness with $R_{qw1} = 40\ nm$ for roller surface and $R_{qi1} = 70\ nm$ for inner ring raceway surface, similar to the surface roughness values measured with Taylor Hobson profilometer, a medium roughness with $R_{qw2} = 80\ nm$ and $R_{qi2} = 140\ nm$ and a grossly roughness with $R_{qw3} = 120\ nm$ and $R_{qi3} = 210\ nm$.

To consider the material behaviour, in the elastic-plastic model was used the following parameters of the roller bearing steel: the Young modulus $E = 209\ GPa$, the Poisson coefficient $\nu = 0.29$, the yielding stress $\sigma_{Y0} = 1650\ MPa$, the limiting change in the yield surface size on the deviatoric plane $Q_\infty = -100\ MPa$ and $b_\infty = 120$ which describes how rapid the limit size is reached [15], the coefficient of plastic resistance $B = 4320\ MPa$ and the hardening exponent $N = 12.6$ [10,11].

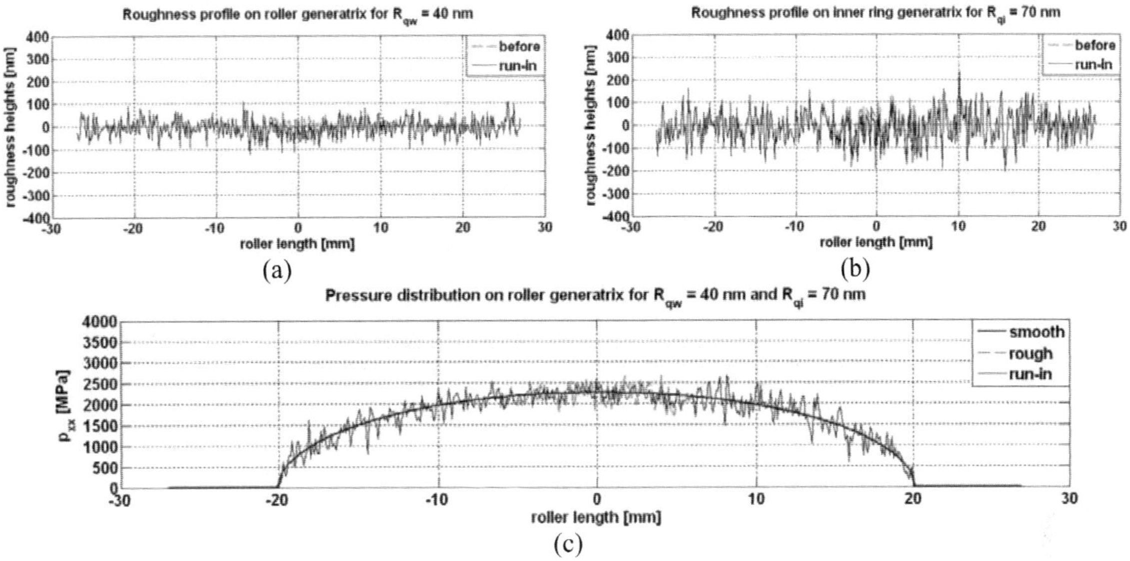

Figure 1. Numerical results for highest loaded roller-inner ring contact ($R_{qw1} = 40\ nm$, $R_{qi1} = 70\ nm$): (a) roughness profile on roller (before and after run-in), (b) roughness profile on inner ring (before and after run-in) and (c) pressure distribution.

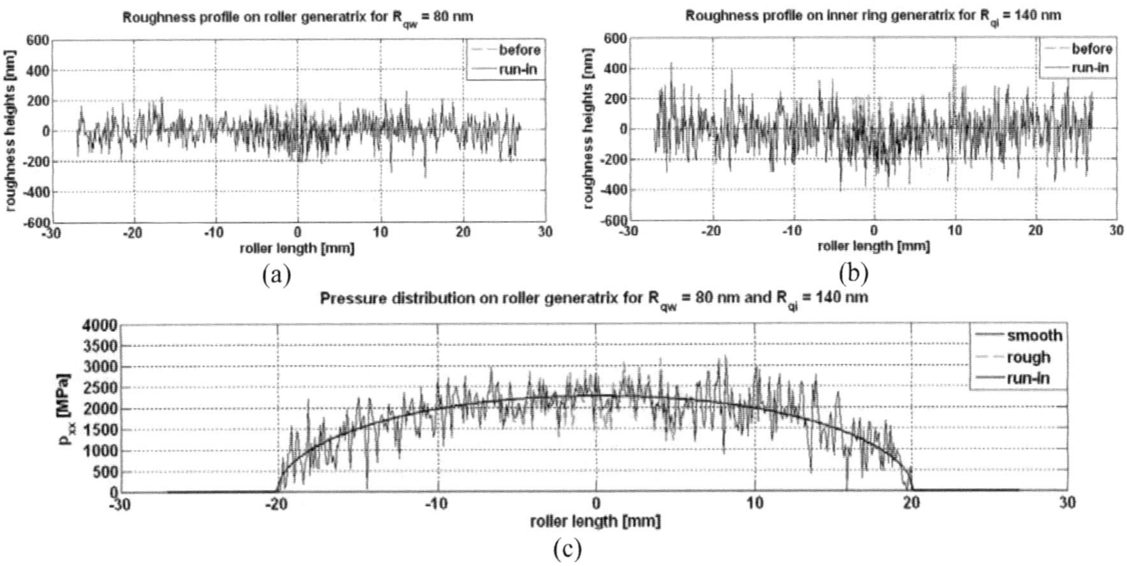

Figure 2. Numerical results for highest loaded roller-inner ring contact ($R_{qw1} = 80\ nm$, $R_{qi1} = 140\ nm$): (a) roughness profile on roller (before and after run-in), (b) roughness profile on inner ring (before and after run-in) and (c) pressure distribution.

In figures 1(a),(b), 2(a),(b) and 3(a),(b) are presented the 2D roller and inner ring roughness profiles before and after *300* running cycles. It can be observed that the running-in process is more pronounced for the inner ring and the roughness heights are plastic deformed in the middle contact area for *10 mm* in length for R_{qi1} = *70 nm*, for *20 mm* in length for R_{qi2} = *140 nm* and for *40 mm* in length for R_{qi3} = *210 nm*. The plastic deformations of the roughness peaks are higher as the roughness is coarser.

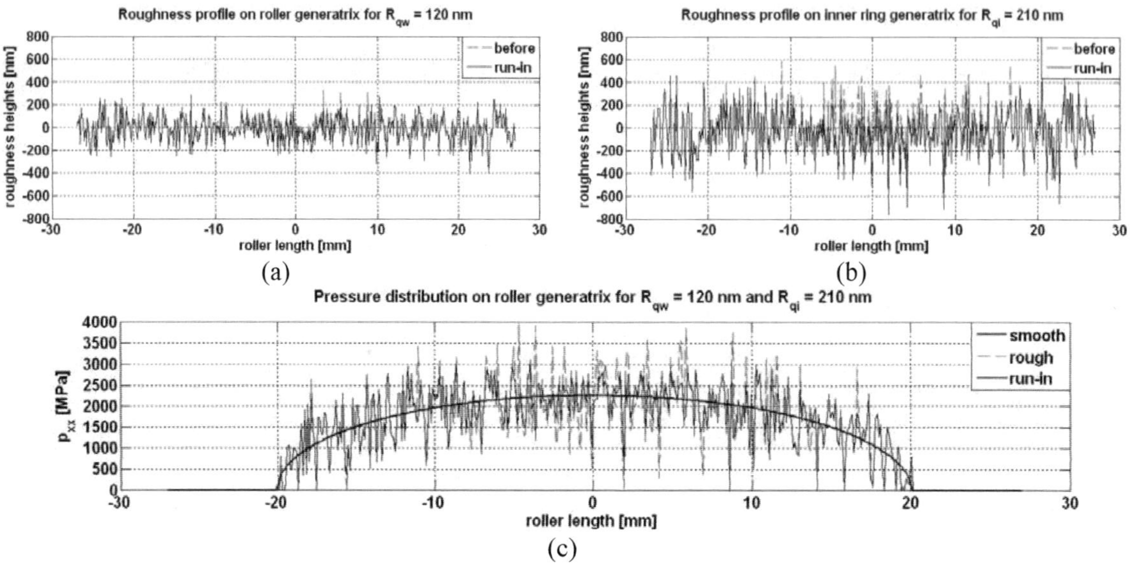

Figure 3. Numerical results for highest loaded roller-inner ring contact (R_{qw1} = *120 nm*, R_{qi1} = *210 nm*): (a) roughness profile on roller (before and after run-in), (b) roughness profile on inner ring (before and after run-in) and (c) pressure distribution.

The 2D pressure distributions on roller generatrix for the highest loaded roller-inner ring contact are presented in figures 1(c), 2(c) and 3(c). In comparison with a smooth contact surface it can be observed that the presence of the roughness on the contact surface generates peaks in pressure distributions. The pressure peaks are higher as the roughness is more grossly and have the maximum value of *2700 MPa* for the fine roughness, *3200 MPa* for the medium roughness and *4000 MPa* for the grossly roughness. In the middle of the contact length the maximum pressure for the smooth contact surface is *2250 MPa*.

It can be observed that after *300* running cycles the pressure peaks are slightly diminished in the middle of the contact length for the fine and medium roughness and are significantly diminished on the all contact length for the grossly roughness.

Table 1. Rating lives for light loading conditions ($F_R = 0.055 \cdot C = 33.5$ *kN*)

Roller roughness R_{qw} [*nm*]	Inner ring roughness R_{qi} [*nm*]	Rating life, L_{10}, *hours*, [16]	Modified rating life, L_{10mr}, *hours*, [14] Before	Run-in
smooth	smooth		*688750*	
40	*70*	*264700*	*218660*	*324140*
80	*140*		*28205*	*74993*
120	*210*		*2297*	*38602*

In table 1 are presented the modified lives L_{10mr} for light loading conditions ($F_R = 0.055 \cdot C = 33.5kN$). The values of the modified rating life are smaller for the rough surfaces than for the smooth surface. For the case of modified surfaces after *300* running cycles with medium radial loading, it can be observed an increase of the modified rating life, apparently a higher increase for the grossly roughness, approximately *16* times more than before run-in (from *2297 hours* to *38602 hours*), but a significantly increase for the fine roughness, approximately *105000 hours* more than before run-in (from *218660 hours* to *324140 hours*).

4. Conclusions

The topography of the active surfaces in rolling contact depends on the surface manufacturing process technology. Numerical results presented in this paper evidence the fact that the presence of nano-asperities on contact surfaces influence the running-in process and durability of rolling contacts.

It was used a numerical method to generate an arbitrarily defined three-dimensional rough surface, a elastic-plastic model to take into account the running-in process and the new methodology given by ISO 16281 to evaluate the modified rating life.

The presence of nano-asperities on contact surfaces generates peaks in pressure distributions, higher peaks for surfaces with grossly roughness. For surfaces with coarser roughness the running-in process is more pronounced than for surfaces with a finer roughness.

The modified rating life it is severely diminished for surfaces with coarser roughness (by two orders of magnitude), even after running-in process (by one order of magnitude) than for smooth surfaces or with a finer roughness.

5. References

[1] Robbe-Valloire F 2001 Statistical analysis of asperities on a rough surface *Wear* **249** pp 401-408
[2] Taylor Hobson Precision 2003 *Exploring surface texture* (TH Ltd)
[3] Stawicki T, Sedlak P, Koniuszy A 2010 The testing of the influence of the roughness of the crankshaft journal upon the durability of the crankshaft bearing in engines of agricultural machines *Sci. probl. of mach. op. and maint.* **4** pp 7-17
[4] Allwood J, Ciftci H 2005 An incremental solution method for rough contact problems *Wear* **258** pp 1601-1615
[5] Allwood J 2005 Survey and performance assessment of solution methods for elastic rough contact problems *J. Tribol.* **127** pp 10-23
[6] Bakolas V 2003 Numerical generation of arbitrarily oriented non-gaussian three-dimensional rough surfaces *Wear* **254** pp 546-554
[7] Patrikar RM 2004 Modeling and simulation of surface roughness *App. Surf. Sci.* **228** pp 213-220
[8] Jamari J 2006 *Running-in of rolling contacts* (PhD Thesis Twente)
[9] Jamari J, de Rooij MB and Schipper DJ 2007 Plastic deterministic contact of rough surfaces J. *Tribol.* **129** pp 957-962
[10] Benchea M, Iovan-Dragomir A, Crețu S 2014 Misalignment effects in cylindrical roller bearings *App. Mech. and Mat.* **658** pp 277-282
[11] Crețu S, Benchea M, Iovan-Dragomir A 2016 On basic reference rating life of cylindrical roller bearings. Part II - Elastic-Plastic Analysis *J. Balk. Tribol. Assoc.* **22/1** pp 272-280
[12] Chaboche JL 2008 A review of some plasticity and viscoplasticity constitutive theories *Int. J. Plasticity* **24** pp 1642-1693
[13] Besson J, Cailletaud G, Chaboche JL, Forest S and Blétry M 2010 *Non-linear mechanics of materials* (Springer)
[14] ISO 16281:2008, Rolling bearings-methods for calculating the modified reference rating life for universally loaded bearings
[15] Linares Arregui I Alfredsson B 2010 Elastic-plastic characterization of a high strength bainitic roller bearing steel - experiments and modelling *Int. J. Mech. Sci.* **52** pp 1254-1268
[16] ISO 281:2007, Rolling bearings-Dynamic load ratings and rating life
[17] SKF Group 2013 *Rolling bearings catalogue*

IOP Conf. Series: Materials Science and Engineering **724** (2020) 012026 doi:10.1088/1757-899X/724/1/012026

The Fretting Contact of Coated Bodies. Part I – Contact Parameters

S Spinu[1,2], **D Cerlinca**[1,2]

[1] Department of Mechanics and Technologies, Stefan cel Mare University of Suceava, 13th University Street, 720229, Romania
[2] Integrated Center for Research, Development and Innovation in Advanced Materials, Nanotechnologies, and Distributed Systems for Fabrication and Control (MANSiD), Stefan cel Mare University, Suceava, Romania

E-mail: sergiu.spinu@fim.usv.ro

Abstract. In case of a mismatch between the elastic properties of the contacting bodies, the solution of the fretting contact can only be achieved numerically due to the connection between the contact pressure and the shear tractions. The numerical treatment of the contact processes employs an iterative strategy, and therefore requires the computation of the displacement response of the elastic body, assumed as an elastic half-space, subjected to general loadings. Unlike the case of homogeneous bodies with known Green's functions, a closed-form expression of the response of layered semi-infinite solids to unit point loads has only been attained in the frequency domain. The latter formulas are used in this paper to assess the displacement due to arbitrary normal and shear surface tractions, thus empowering the application of a trial-and-error approach in finding the parameters of a fretting contact: the contact area, the slip and the stick regions, and the distributions of pressure and shear tractions. The contact parameters are assessed in a nested loop strategy, involving three levels of iterations. The inner level, based on the conjugate gradient method for linear systems of equations, finds the pressure when solving the contact problem along the normal direction, disconnected from any tangential effects, or the shear tractions when considering the contact equations in the tangential direction, disconnected from the pressure influence. The middle level stabilizes the pressure with respect to the shear tractions, thus assuring that the global instantaneous contact solution is achieved. The outer level manages the reproduction of the loading history in a fretting loop, by load incrementation. The proposed strategy proves itself as a robust tool for the prediction of the fretting contact process involving coatings.

1. Introduction

Modern engineering applications, including bearings, gears, dental crowns, hip prostheses, hard disks or electronic parts, involve functionally graded materials and/or coatings that prolong the service life of critical components by preventing surface damage. In many contact scenarios, the service conditions involve oscillatory relative movement of small amplitudes, also known as fretting. The contact stresses developed in the layered body under contact load are of chief importance for predicting the service life of the contacting element. Giving the complexity of the fretting phenomenon, closed-form solutions are available for very selective cases, and numerical techniques such as the finite element analysis (FEM) or the semi-analytical methods (SAM) present themselves as attractive alternatives for calculating the fretting contact stresses. Whereas the advantage of FEM resides in its generality, SAM brings excellent computational efficiency, allowing [1] the calculation of a 3D contact scenario with the computational resources needed for a 2D FEM simulation.

Gallego, Nélias and Jacq [2] first used a fully grown SAM capable of repeated resolution of the contact problem in the normal direction, while accounting for the change in conformity due to wear. The model was later refined by Gallego and Nélias [3], who advanced a numerical model for the fretting wear either under partial or gross slip conditions, assisted by efficient three-dimensional algorithms for both the normal and the tangential contact problem. Chen and Wang [4] presented a three-dimensional numerical model for the simulation of the contacts of elastically dissimilar materials based on the Boussinesq–Cerruti integral equations. An elastic contact model for three-dimensional layered or coated materials under coupled normal and tangential loads, with consideration of partial slip effects, was further developed by these authors [5]. Gallego, Nélias and Deyber [6] advanced a general algorithm for fretting contact problems with application to fretting modes I, II and II, in which the coupling between the normal and the tangential contact problems can or not be taken into account. A numerical analysis of the partial slip contact under a tangential force and a twisting moment was also performed [7].

The load path dependence is a distinctive feature of the fretting contact, which was addressed in more detail by Spinu and Glovnea [8], and by Spinu and Frunza [9], who studied the hysteretic behaviour of partial slip elastic contacts undergoing a fretting loop.

The fretting contact of layered materials is less covered in the literature giving the additional difficulty in expressing the elastic response of the contacting bodies. In this paper, a technique for the computation of displacement and stresses in bilayered half-spaces is combined with an iterative solver for the fretting contact. The resulting numerical analysis technique is both fast and accurate, and can be used with various contact configurations. The relation between the contact processes in the normal and in the tangential directions, often neglected, are accounted for in this contact study.

2. Contact model review and solution

In the absence of friction, a point contact under normal load has a contact area given by the Hertz theory. The mutual contact pressure established between the contacting surfaces yields normal compression but also shear displacements. However, if the contact interface can sustain friction, the shape and size of the contact area are affected only if the elastic constant of the contacting materials are different. The latter case is assumed in the present work, focused on the indentation of a coated flat by a rigid sphere.

The different displacements of dissimilarly elastic materials, induced in each body by the equal (in magnitude) but opposite (in direction) shear tractions, lead to a relative peripheral velocity at the interface, i.e. local or micro-slip, if not opposed by friction. Practically, the contact area of the three-dimensional point contact under normal load consists in a central region of stick enveloped by slip zones towards the edge of the contact. On the stick region the surfaces adhere in the tangential direction, meaning that mating points on the two surfaces stick together after the deformation, and therefore have vanishing relative displacement. As opposed to the slip peripheral zone, where mating points undergo different tangential displacements and consequently slips occurs. As the contact area grows with the normal load increase, points initially laying outside the adhesion region will progressively be enveloped in the stick region.

A subsequently applied tangential force may generate a global sliding motion, if greater than the force of limiting friction (i.e. that for the case when sliding is about to occur), or a tendency to slide. In the latter case, although there is no nominal relative velocity between the bulks of the bodies, local slip occurs at specific regions of the interface. The tangential tractions induced by the tangential force affect the size and shape of the contact area, and also the pressure distribution, only if the two solids have different elastic constants. Otherwise, for similarly elastic materials, given that the normal displacements induced by the shear tractions will be equal, the contacting surfaces will warp conformingly. Consequently, the pressure distribution and the contact area will be given by the initial contact geometry, and will be independent of the tangential force.

Considering the aforementioned interdependencies, it can be concluded that the tangential tractions interact with pressure when the solids have different elastic properties. The practice of contact problems involving homogenous solids usually assumes that pressure and tangential tractions are independent of each other, as the influence of the latter on pressure is small, especially for low frictional regimes. For the contact of coated materials, the aforementioned assumption may be too strong, and therefore it is discarded in the present study. The interdependence between the normal and the tangential tractions is overcome by numerical analysis, in an iterative approach whose convergence is dictated by the stabilization of pressure with respect to the shear tractions.

One distinctive feature of the contact process is that, although the contact is assumed quasi-static and in the frame of linear elasticity, due to the irreversibility of friction, the final stress state will depend upon the loading history rather than solely on the final magnitudes of the normal and tangential loads. Consequently, the surface tractions can only be found by following the complete loading history in incremental steps.

Given the aforementioned contact process features, the model discretization has to be made in both spatial and temporal dimensions (although the latter does not imply that the time parameter is explicitly needed, as long as the loading history is reproduced). Two or three indexes, respectively, may be needed to substitute the continuous coordinates, allowing for positioning on the surface (e.g., i, j in equations (1)-(5)) or in the volume computational domain, respectively, whereas an additional index (e.g., k in equations (1)-(5)) may be needed for the sequence of the loads application. In order to speed up the computation of convolutions resulting from the superposition of effects, techniques derived from the digital signal processing are applied, which require a uniform rectangular mesh and piecewise constant distributions. This state-of-the-art approach is sustained by the analytical solutions existed in the literature for the stress or displacement effect induced in a half-space by a pressure or shear stress excitation uniformly distributed on a rectangular patch located on the half-space boundary. Given the aforementioned solutions, an additional assumption is needed, commonly used in contact mechanics, i.e. the approximation of solids of general profiles with half-spaces. This assumption is reasonable in case of concentrated contacts, when the contact stresses are confined to a limited body domain, and not influenced by the proximity of its boundary. The latter assumption warrants the calculation of displacements using superposition of fundamental solutions derived for the elastic half-space, and also forces contact pressure to act perpendicular to the interface regardless of the deformation it produces. Moreover, the strains developed in the contact process are assumed sufficiently small to fit the framework of the linear theory of elasticity.

The contact model [10,11] for the dry, frictional contact, under normal and tangential load, the latter generating a partial slip regime, is restated here for clarity and completeness. The equations are reported to a Cartezian coordinate system with the origin in the initial point of contact, and with the x_1-axis conveniently aligned with the direction of the tangential force, without losing generality. The contacting bodies are compressed by a steady normal load W, giving birth to a contact area $A(k)$, while an oscillating tangential force $\mathbf{T}(T_1, T_2 = 0)$ is subsequently applied, having an amplitude insufficient to cause the sliding of the two surfaces. The static force equilibrium equates the applied normal and tangential loads to the contact pressure p and the shear tractions $\mathbf{q}(q_1, q_2)$, respectively:

$$W(k) = \Delta \sum_{(i,j)\in A(k)} p(i,j,k); \quad T_n(k) = \Delta \sum_{(i,j)\in A(k)} q_n(i,j,k), n = 1,2, \tag{1}$$

where Δ is the elementary cell area. On the computational domain P, the displacement equation in the normal direction results from comparison of contact geometry before and after the deformation, whereas that in the tangential direction from consideration of subsequent load increments:

$$u_3(i,j,k) = h(i,j,k) - hi(i,j) + \delta_3(k), (i,j) \in P; \tag{2}$$

$$\begin{bmatrix} u_1(i,j,k)-u_1(i,j,k-1) \\ u_2(i,j,k)-u_2(i,j,k-1) \end{bmatrix} = \begin{bmatrix} s_1(i,j,k)-s_1(i,j,k-1) \\ s_2(i,j,k)-s_2(i,j,k-1) \end{bmatrix} + \begin{bmatrix} \delta_1(k)-\delta_1(k-1) \\ \delta_2(k)-\delta_2(k-1) \end{bmatrix}, (i,j)\in A(k). \quad (3)$$

In equation (2), h is the separation between the two surfaces, whereas hi denotes the sum of initial (i.e., in unloaded state) digitized body profiles, u_3 the composite (i.e., relative) displacement in the normal direction, and δ_3 the rigid-body approach, measured as the approach of body points distant to the contact zone, measured along the normal direction. Equation (3) relates the increments of the tangential displacements u_i with the those of the tangential rigid displacements δ_i and with those of slip s_i, $i=1,2$. As δ_i are position-independent, it follows from the latter equation that surface points with vanishing slip (i.e., in stick) undergo the same tangential displacement.

The contact problem solution is difficult to achieve because: (1) the division of the computational domain into the contact and non-contact zones is not known in advance, such as that of the contact area into slip and stick regions, and (2) these separations keep changing during the loading history due to the interaction between pressure and the tangential traction. In the numerical model, these divisions are expressed by the boundary complementarity conditions, and are found by trial-and-error.

The boundary conditions express the complementarity between the contact and the non-contact regions, as well as between the slips and stick zones:

$$\begin{cases} p(i,j,k)>0 \quad \text{and} \quad h(i,j,k)=0, \quad (i,j)\in A(k); \\ p(i,j,k)=0 \quad \text{and} \quad h(i,j,k)>0, \quad (i,j)\in P-A(k); \end{cases} \quad (4)$$

$$\begin{cases} \|\mathbf{q}(i,j,k)\| \le \mu p(i,j,k) \quad \text{and} \quad \|\mathbf{s}(i,j,k)-\mathbf{s}(i,j,k-1)\|=0, (i,j)\in S(k); \\ \|\mathbf{q}(i,j,k)\| = \mu p(i,j,k) \quad \text{and} \quad \|\mathbf{s}(i,j,k)-\mathbf{s}(i,j,k-1)\|>0, (i,j)\in A(k)-S(k). \end{cases} \quad (5)$$

Equation (4) implies that tensile normal tractions cannot be sustained, which is reasonable for contact process involving metallic solids with inherent microtopography that inhibit the contribution of the adhesion force in the normal direction. The relationship between pressure and the tangential tractions is assumed according to the Amontons'a law of sliding friction, as shown in equation (5). The frictional coefficient μ, determined by the materials properties and by the physical state of the interface, is assumed uniform over the whole contact region and constant during the loading history, although this is not a limitation imposed by the numerical scheme, which can handle mapped friction data with no additional algorithm modifications.

To solve the contact model, it is required to find the discrete distributions of surface tractions transmitted between the two solids at their surface of contact. To this end, the contact model (1)-(5) is first divided into two submodels for the normal and the tangential direction. Given the fact that calculation of displacement in any direction require knowledge of all contact tractions, the submodels will not be independent. However, an algorithm for the solution of each model with known, yet arbitrary output from the other, is readily available. The scheme was originally proposed for the frictionless contact or rough bodies by Polonsky and Keer [11], and was subsequently adapted [12] to the submodel in the tangential direction. An outer loop [8] is required to stabilize the inputs of the two submodels, i.e. the tangential tractions with respect to the contact pressure. The latter algorithm strategy applies in full to the contact of coated bodies under the assumption that the elastic displacements are accurately computed. An algorithm for the computation of the elastic response of coated bodies to general loading is detailed in the companion paper. From the point of view of the latter contact solver strategy, it is important to note that the method to solve each of the two submodels neither requires nor calculates the rigid-body displacements δ_i, $i=1,2$. In other words, only the relative values of displacements are needed for the contact process solution, which warrants a simplified approach in displacement calculation, as detailed in the companion paper.

3. Results and discussions

For benchmarking purposes, two simpler scenarios with well-known solutions were replicated with the newly proposed computer program. Firstly, the pressure profiles in the spherical indentation of a coated flat by a rigid indenter were obtained for the frictionless contact. The Young modulus of the coating E_1 was varied with respect to that of the substrate, E_2, which was kept constant. The Hertz contact parameters a_H and p_H, calculated with the substrate elastic parameters, were used as normalizers for the radial coordinate and the contact pressure, respectively. The coating thickness was fixed and equal to the Hertz radius. The semi-elliptical Hertz pressure profile was also added for reference. These results depicted in figure 1 match well the pressure distributions obtained by O'Sullivian and King [13] using a different method, and validate the contact scheme for the normal contact problem, as well as the procedure for the calculation of normal displacements in coated bodies.

Secondly, simulation of a fretting loop for homogenous materials allows validation of the contact solvers in the normal and in the tangential directions. The reference closed-form relations were deduced in [10], based on the solutions of the partial slip contact obtained independently by Cattaneo [14] and by Mindlin [15]. The corresponding curves are displayed with continuous grey lines in figure 2, which shows the distributions of shear tractions at specific moments from the loading history, in a fretting loop of amplitude $T_{\lim} = 0.9\mu W$, in the time interval τ between the first and the third null of the tangential force. In this case, the pressure is independent of the shear tractions and of the loading history, and obeys the Hertz theory.

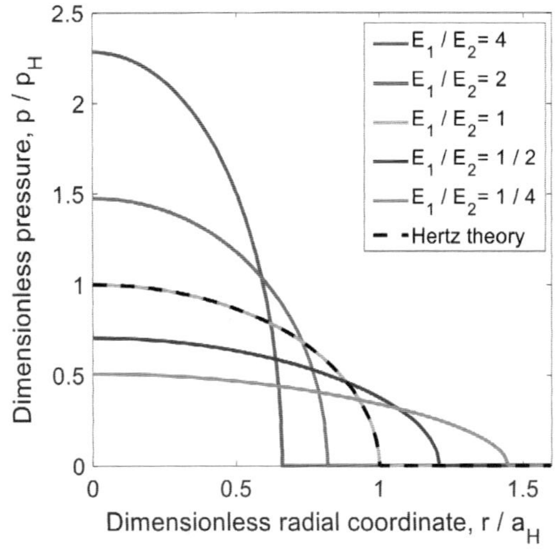

Figure 1. Influence of elastic moduli ratio on the pressure profiles.

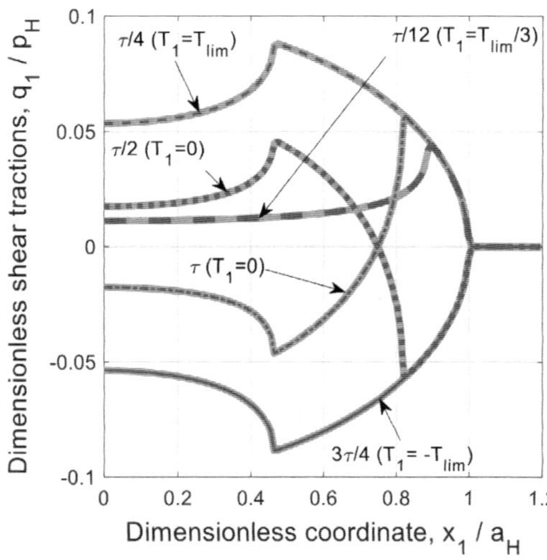

Figure 2. Shear tractions q_1/p_H in a fretting loop, $\mu = 0.1$.

The simulation of a fretting loop of amplitude $T_{\lim} = 0.75\mu W$, with $\mu = 0.1$, is performed for different elastic moduli ratios between the coating and the substrate. The loading history is reproduced with 400 loading increments, from which the first 100 are reserved for the initial normal indentation. In this stage, although no tangential force is applied, self-equilibrating shear tractions arise due to the mismatch in the elastic properties between the contacting bodies. The indenter is assumed rigid, while the ratio of the elastic moduli ratio between the coating and the substrate is varied. Whereas the contact pressure is not greatly affected, the distributions of shear tractions vary significantly from the elastically similar case presented in figure 2. More so, the maximum tangential force that the contact can withstand before going into full sliding is significantly less than the theoretical one [14,15], $T_{\lim} = \mu W$. The predicted profiles of dimensionless shear tractions are depicted in figure 3.

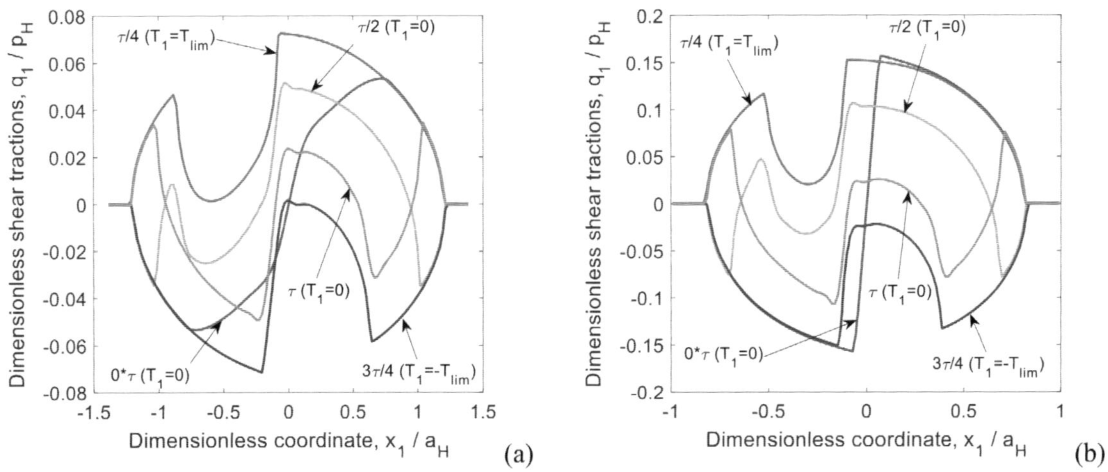

Figure 3. Shear tractions in the indentation of a coated half-space: (a) $E_1/E_2 = 1/2$; (b) $E_1/E_2 = 2$.

4. Conclusions

The lack of analytical solutions for the model of the fretting contact encourage the use of numerical methods that employ trial-and-error strategies to derive the contact area and the slip-stick boundary, as well as their evolution with the loading history. The accurate reproduction of the latter is an important prerequisite for the precise estimation of the contact normal and shear tractions.

A fretting contact solver originally designed for elastic homogenous bodies was successfully adapted to the contact of layered bodies with the assistance of a numerical procedure for the calculation of displacement in multi-layered mediums. Only the relative displacement fields are required, which avoids the evaluation of the frequency response functions in the origin of the frequency domain.

The three-level iterative strategy converged in all situations involving various elastic moduli ratios between the coating and the substrate. Well known literature results were faithfully reproduced, giving confidence in the method ability to advance the understanding of the fretting contact of coated bodies. The knowledge of contact tractions will allow the calculation of stress fields developing in the layered mediums.

References

[1] Renouf M, Massi F, Fillot N and Saulot A 2011 *Tribol. Int.* **44** 834
[2] Gallego L, Nélias D and Jacq C 2006 *ASME J. Tribol.* **128** 476
[3] Gallego L and Nélias D 2007 *ASME J. Tribol.* **129** 528
[4] Chen W W and Wang Q J 2008 *Mech. Mater.* **40** 936
[5] Wang Z J, Wang W Z, Wang H, Zhu H and Hu Y Z 2010 *ASME J. Tribol.* **132** 021403
[6] Gallego L, Nélias D and Deyber S 2010 *Wear* **268** 208
[7] Wang Z J, Meng F M, Xiao J X and Wang W Z 2011 *P. I. Mech. Eng. J.-J. Eng.* **225** 72
[8] Spinu S and Glovnea M 2012 *J. Balk. Tribol. Assoc.* **18** 195
[9] Spinu S and Frunza G 2015 *J. Phys.: Conf. Ser.* **585** 012007
[10] Johnson K L 1985 *Contact Mechanics* (Cambridge: University Press)
[11] Polonsky I A and Keer L M 1999 *Wear* **231** 206
[12] Spinu S and Frunza G 2011 *Proc. VI Int. Conf. Balttrib (Kaunas, Lithuania, 17-19 November 2011)* (Kaunas: Aleksandras Stulginskis University) p 105
[13] O'Sullivian T C and King R B 1988 *J. Tribol. – Trans. ASME* **110** 235
[14] Cattaneo C 1938 *Rend. Accad. Lincei* **27** 342-348, 434-436, 474-478
[15] Mindlin R D 1949 *ASME J. Appl. Mech.* **16** 259

The Fretting Contact of Coated Bodies. Part II – The Stress State

S Spinu[1,2], **D Cerlinca**[1,2]

[1] Department of Mechanics and Technologies, Stefan cel Mare University of Suceava, 13th University Street, 720229, Romania

[2] Integrated Center for Research, Development and Innovation in Advanced Materials, Nanotechnologies, and Distributed Systems for Fabrication and Control (MANSiD), Stefan cel Mare University, Suceava, Romania

E-mail: sergiu.spinu@fim.usv.ro

Abstract. The service life of contacting coated machine elements is ultimately determined by the distribution of stresses in the coating and in the substrate. By assuming the elastic bodies as elastic half-spaces, the contact stress computation entails the calculation of convolutions expressing the superposition of effects of unit point loads acting on the boundary. The fundamental solutions of stresses and displacements in multilayered materials have only been calculated in the frequency domain, and are known as the frequency response functions. An additional difficulty arises in the stress calculation, related to frequency response function valuation in the origin of the frequency domain, where a singularity is usually encountered. This case of un-determination is circumvented in this paper by substituting the required value with the mean value of the frequency response function over a vicinity centered in origin. The latter approach is endorsed by the fact that the frequency response function is singular, but numerically integrable in the aforementioned vicinity. The latter technique is validated by comparison with results obtained for the sliding contact, and then applied to derive the elastic stresses arising during a fretting loop in the coating and in the substrate. The stresses due to shear tractions are superimposed to those induced by contact pressure. The calculation is performed in layers of constant depth, and the algorithmic complexity is optimized by using state-of-the-art techniques for discrete convolution computation. The equivalent stress is discontinuous across interface between the two layers, and the location and intensity of the maximum von Mises stress is determined by the frictional coefficient and by the mismatch between the Young moduli of the coating and the substrate. The results obtained with the newly proposed numerical technique may extend the understanding of the fretting contact of coated materials and assist the design of improved coating configurations.

1. Introduction

The contact mechanics modelling of functionally graded materials and coatings, whose elastic properties vary in the depth direction, involves multi-layered systems consisting in perfectly bonded layers of different elastic constants, allowing for the half-space assumption valid for small contact region compared to the bulk body dimensions. The study of these materials pioneered with the stress and displacements analytical solutions given by Burminster [1] for a single layered material and axisymmetric normal loading. Chen [2], and Chen and Engel [3], extended these solutions to arbitrary normal loadings. The case of multilayered transversely isotropic half-spaces was addressed by Kuo and Keer [4] using the boundary integral method together with the Hankel transform. The frequency response functions (FRF) of bilayered materials, relating in the Fourier transform domain unit surface tractions to displacements and stresses, were derived by O'Sullivan and King [5] with the aid of the Papkovich–Neuber potentials and the double Fourier transform. Their procedure was further employed to analyse various dry [6-12] or lubricated [13] contact scenarios, in which the elastic response of the layered material is assessed based on the FRFs.

From a computational point of view, an important breakthrough in the calculation of stresses arising in multi-layered systems due to general loadings, has been achieved with the development of the spectral techniques for the calculation of convolution products in the Fourier transform domain. Nogi and Kato [6] first employed the aforementioned FRFs in the derivation of the solution for the coated contact of rough surfaces with the aid of the fast Fourier transform (FFT). The source of the periodicity error linked to the application of the FFT to non-periodic problems such as the concentrated contact was discussed [14-16], and different correcting procedures were advanced. The extension of the physical computational domain coupled with the zero-padding of the excitation are now generally applied in evaluation of the discrete cyclic convolutions expressing stresses and displacements in contact problems.

Considering the complexity of the arising mathematical models, the derivation of the FRFs is an important prerequisite in stress calculation. Liu and Wang [16] obtained the FRFs for both normal and shear tractions in a closed-form that is well adapted to numerical implementations. In the same manner, the FRFs of tri-layer materials were derived [10], as well as the recurrence relation for the FRFs of multilayered materials [17].

In this paper, the stress fields in the fretting contact are obtained following the general procedure for stress calculation established in the literature of layered materials. The numerical implementation uses the FRFs derived by Liu and Wang [16], whereas the aliasing in the frequency domain is controlled by an extension of the computational target domain. The computer program is validated by comparison with results obtained for the sliding spherical contact of bilayered materials.

2. Elastic response of coated bodies

The half-space approximation employed in the framework of contact mechanics, together with the Linear Elasticity fundamental solutions, also referred to as the Green's functions, allow for the calculation of the response of an elastic half-space to arbitrary, yet known, loadings. In particular, displacements and stresses induced by surface general loadings result as integrals superimposing the individual contributions of concentrated forces applied on the boundary. Mathematically, the latter integrals are convolution products of the relevant Green's function with the contact tractions. The Fourier transform of a Green's function is also known as the frequency response function (FRF). Explicit expressions were derived in the literature for both the Green's functions and the related FRFs of homogenous semi-infinites, but in case of layered materials, the Green's functions still lack explicit expressions. The derivation of the FRFs suggests the numerical calculation of the layered half-space response in the frequency domain, assisted by the discrete convolution theorem, which brings additional advantages regarding the algorithmic efficiency: the convolution product of series with N terms is reduced in the frequency domain to an improved $O(N\log N)$ order of operations, which is a substantial saving from the $O(N^2)$ specific to the time/space domain. This decrease stems from the conversion of the convolution product in an element-wise product in the Fourier transform domain.

The main disadvantage of model digitization in the frequency domain is the problem periodization: the discrete Fourier transform of a discrete series, calculated with the fast Fourier transform (FFT), tacitly assumes the series as periodical. The physical meaning of the periodization is that a bogus periodical surface load is assumed instead of the actual one. The fake neighbouring periods of contact tractions are likely to perturb the half-space response to the non-periodical load. Important research efforts aimed to reduce or control this periodicity error. Nogi and Kato [6] protected the displacement solution of both homogenous and bilayered bodies by a grid refinement with a factor of two along each direction, whereas Liu et al. [15] addressed the issue of the conversion of a linear convolution into its discrete cyclic counterpart, concluding in the Discrete Convolution FFT method (DCFFT) that allegedly brings no additional error beside the discretisation error in convolution computation. The aliasing and the Gibbs phenomena were identified [16] as the main sources of perturbation when the half-space response is calculated based on the FRFs directly in the Fourier transform domain. Based on their findings, the algorithm accuracy is assured in this paper by extending the target computational domain, resulting in a refinement of the mesh in the spectral domain that reduces aliasing, and by the zero-padding of the excitation (i.e., the contact tractions, pressure or shear), aiming to move away from the target domain the spurious neighbouring periods, thus limiting their contribution.

An algorithm for a more general problem is thus advanced, allowing the convolution computation when the convolution members are arbitrary, yet known functions, and one of the convolution members is only known in the frequency domain. In particular, the latter algorithm can be applied to derive numerically the elastic response (i.e., displacement and stresses) of a coated half-space to a general surface excitation in the form of pressure or shear tractions. It should be noted that, although the method implies digitization of the excitation, the employed fundamental solution (i.e., the FRF), is derived in closed-form. Therefore, similar methods are known in the literature of contact mechanics as semi-analytical [18].

For clarity and simplicity, but without losing generality, the method presentation is restricted to the line contact of infinitely long cylinders (i.e., the two-dimensional problem). Extension to the three-dimensional case is achieved by adding a new dimension, matching the second tangential direction from the contact 3D problem. The algorithm needs the following input: (a) the target domain L, (b) the digitised surface loading p_i, $i=1\cdots N$, (c) the FRF closed-form for the needed response, $\tilde{f}$, and (d) the extended domain χL, defined with the aid of an extension ratio χ. It should be noted that the algorithm output should only be considered for the target domain L, whereas results at the edge of the extended domain may be of a questionable precision. For the Fourier transform domain, the tilde (~) symbol is employed to denote continuous functions, as opposed to the hat (^) mark for discrete series. The extension of the target domain L into χL aims at the refinement of the mesh in the Fourier transform domain. This can only be achieved by preserving the original sampling interval in the spatial domain, $\Delta = L/N$, and leads to a decrease of the sampling interval in the frequency domain from $2\pi/L$ to $2\pi/(\chi L)$. Thus, the number of samples in the spatial domain needs also to be increased from N to χN. The value of the extension ratio χ thus has a negative impact on the needed computational resources, and should be chosen as a compromise. Values between 2 and 8 make good candidates.

The spectral coordinates ζ_i of the refined spectral mesh result as:

$$\zeta_i = 2\pi(i-\chi N/2)/(\chi L), \quad i=1\cdots\chi N, \tag{1}$$

and represent discrete frequencies at which the FRF $\tilde{f}$ is evaluated. A discrete spectral series $\hat{f}_i$ characterising the spectral response to a unit impulse response is achieved:

$$\hat{f}_i = \tilde{f}(\zeta_i), \quad i=1\cdots\chi N. \tag{2}$$

One additional difficulty arise, as the stresses and displacements FRFs for both homogenous and layered materials are singular at $\zeta=0$, i.e. the origin of the frequency domain. This shortcoming can be overcome depending on whether the calculated elastic response is needed in absolute or in relative terms. It was stated in the companion paper that, due to a particularity of the contact solver in both normal and tangential direction, only relative displacements are needed to achieve the contact solution. Therefore, if $\tilde{f}$ is a displacement FRF, $\tilde{f}(0)$ is not needed, as shown below, as it only introduces a constant to the calculated response field, and any value can be considered instead, e.g. $\tilde{f}(0)=0$. The situation is similar to the 2D contact problem, in which the displacement field is known except for a constant, but the contact pressure can be assessed in absolute terms. Relative calculation is not satisfactory in stresses computation, when absolute values are required to assess the propensity of plastic yielding via an equivalent stress tensor intensity such as the von Mises criterion. In the latter case, by taking advantage of the fact that the FRF is numerically integrable on a vicinity centred in origin, a mean value of the FRF can be considered as the best available indicator of the local function behaviour:

$$\tilde{f}(0) \cong \frac{\chi L}{2\pi} \int_{-\pi/(\chi L)}^{\pi/(\chi L)} \tilde{f}(\tau) d\tau . \quad (3)$$

Following the requirements of the discrete cyclic convolution theorem [15], the series $\hat{f}_i$ is subsequently rearranged in wrap-around order, with the terms corresponding to negative frequencies shifted at the end of the series. Consequently, the first position in the series $\hat{f}_i$ will be occupied by the term calculated in eq. (3). Symmetrical zero-padding of the excitation in the extended domain is also required [15], resulting in a series of vanishing terms, apart from:

$$p_{i+\frac{(\chi-1)N}{2}} \leftarrow p_i, \quad i = 1 \cdots N , \quad (4)$$

series which is subsequently transferred to the frequency domain, $\hat{p} = \mathrm{FFT}(p)$. The convolution result σ_i is first calculated as element-wise product in the frequency domain, then transferred to the space domain via inverse FFT, as shown in eq. (10). The middle N terms match the target domain and are retained as output.

$$\hat{\sigma}_i = \hat{f}_i \cdot \hat{p}_i, \quad i = 1 \cdots \chi N , \ \sigma = \mathrm{IFFT}(\hat{\sigma}) . \quad (5)$$

One can verify that, when the first term of a series is perturbed with an arbitrary constant, all the terms in the FFT or the IFFT of the series are perturbed with a constant. Considering the way the convolution result is achieved in eq. (5), it can be seen that misevaluating the FRF in the origin of the frequency domain results in a constant added to the computed stress or displacement field.

The proposed algorithm not only circumvents the computation of the Green's functions for the coated half-space, but also reduces the computational complexity due to the convolution calculation in the frequency domain in $O(N)$ operations, as opposed to $O(N^2)$ in the space domain. Considering that both FFT and IFFT are of order $O(N \log N)$, the latter is the overall order of operations for displacement or stress calculation in layers of constant depth. The decrease is of paramount importance as the displacement field needs to be calculated many times during the iterative search of the contact area and the stick region. More so, the stress tensor components evaluation implies six convolutions for each traction component and each considered depth. The gap between $O(N^2)$ and $O(N \log N)$ becomes significant for $N > 10^6$, which is considered as a minimum threshold for the treatment of rough contact problems. Practically, a three-dimensional contact simulation with a surface grid of $N = 128^2$ points, multiplied 64 times in depth for stress calculation, and with a domain extension ratio $\chi = 4$, is performed on a quad-core 3.2 GHz CPU in less than 30 minutes.

The FRFs required in eq. (2) are given in the literature [5-7,16]. The closed-form expressions were obtained by taking the double Fourier transform of the Papkovich-Neuber potentials expressing the stress and displacement fields in layered materials, and by imposing the boundary conditions and the continuity conditions at the layers interface. From a computational point of view, the formulas obtained by Liu and Wang [16] are more convenient to use in a numerical implementation, as certain derivatives are not involved. The proposed algorithm can equally handle multilayared materials provided the required FRFs are made available, such as in [17].

3. Results and discussions

The validation of the sequence for the computation of stresses is attained by benchmarking the distributions of the calculated J_2 stress tensor second invariant with the data obtained by O'Sullivian and King [5] with a different approach. In this simpler case, the contact is in a full sliding, and the tangential tractions are simply proportional to the contact pressure via the frictional coefficient μ. Two typical cases are presented in figure 1, but it should be noted that all results presented in [5] were closely reproduced with the newly advanced computer program. The Hertz parameters a_H and p_H, calculated for the indentation of a homogenous half-space having the same elastic modulus as the substrate, by a rigid sphere, are used as normalizers. The coating thickness is denoted by h_c. The location of the maximum value in each distribution is depicted with an X mark, and its dimensionless magnitude is indicated on each plot. The comparison gives confidence in the sequence for the stress field computation in bilayered materials. The discontinuity in the equivalent stress at the interface between the coating and the substrate, i.e. at $z = h_c = a_H$, suggests that, although the σ_{zx}, σ_{zy} and σ_{zz} stress tensor components are continuous across the interface, the remaining ones are not.

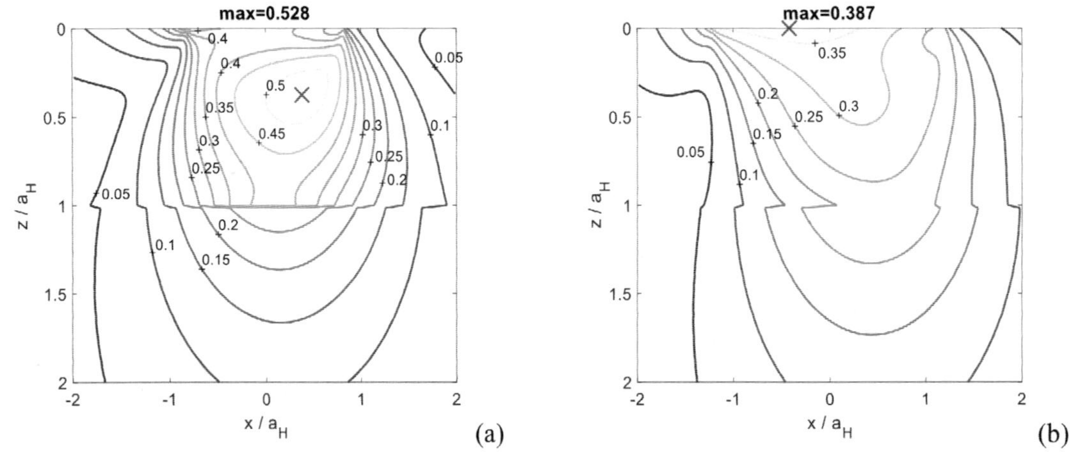

Figure 1. J_2/p_H contour plots in the plane $y=0$, obtained for the sliding contact of coated bodies: (a) $E_1/E_2=2$, $\mu=0.25$; (b) $E_1/E_2=1/2$, $\mu=0.5$.

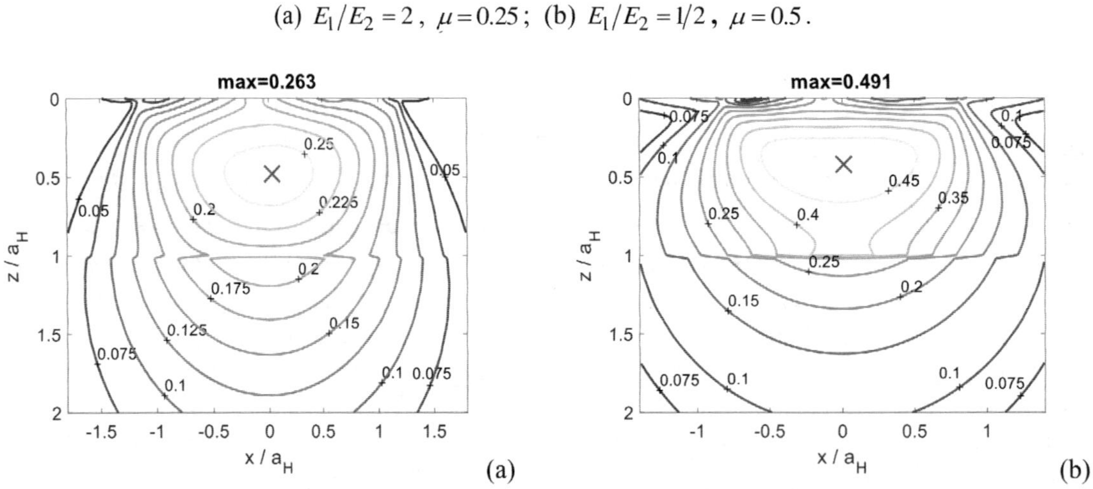

Figure 2. J_2/p_H contour plots in the plane $y=0$, obtained for the fretting contact of coated bodies: (a) $E_1/E_2=1/2$; (b) $E_1/E_2=2$.

The stress state in the fretting contact is then evaluated with the newly advanced computer program, corresponding to the contact tractions presented in the companion paper. The simulation parameters are: frictional coefficient $\mu = 0.1$, the coating thickness $h_c = a_H$, the tangential force $T_1 = 0.75\mu W$, corresponding to the second pas of the increasing tangential force through zero (or the third zero value). Iso-contours of the dimensionless J_2 are depicted in figure 2, for different elastic moduli ratios between the coating and the substrate.

The computational process can take advantage of the fact the values of the FRF for the q_2 shear stress are permutations of those calculated for q_1. The numerical integration of the FRF in the vicinity of origin in the frequency domain was performed with the Matlab function "quad2d" with the default absolute and relative precisions, and no convergence issues were encountered.

4. Conclusions

The Green's functions for the layered half-space are difficult to derive in closed-form expression, and therefore the method of influence coefficients cannot be directly applied as in the case of homogenous mediums. The stress and displacement elastic response of coated bodies is computed in this paper in the frequency domain, based on closed-form solutions for displacement and stresses induced in a coated half-space by unit point force, also known as the frequency response function. The evaluation of the latter leads to a series that is multiplied in the frequency domain with the transform of the excitation, thus substituting the convolution in the space domain. The rearrangement in wrap-around order and the zero-padding of the excitation are well-known techniques applied in discrete cyclic convolutions computations. The singularity in the origin is circumvented by replacing the missing sample with a mean value calculated around the vicinity of the singularity.

The program validation is achieved against classic results for the sliding contact of coated bodies. The simulations show the location and intensity of the maximum stress tensor second invariant, which is an indicator of the plastic flow susceptibility in the elastic body according to the von Mises criterion. The strong point of the proposed method consists in its computational efficiency, allowing for increased resolution in stress calculation. A map of the possible yield positions in the coated body is proposed for a future study, leading to the quick estimation of the probable yield inception in the coated system subjected to fretting.

References

[1] Burmister D M 1945 *J. Appl. Phys.* **16** 126
[2] Chen W T 1971 *Int. J. Eng. Sci.* **9** 775
[3] Chen W T and Engel P 1972 *Int. J. Solids Struct.* **8** 1257
[4] Kuo C H and Keer L M 1992 *J. Tribol. – Trans. ASME* **114** 253
[5] O'Sullivian T C and King R B 1988 *J. Tribol. – Trans. ASME* **110** 235
[6] Nogi T and Kato T 1997 *J. Tribol. – Trans. ASME* **119** 493
[7] Wang Z J, Wang W-Z, Wang H, Zhu D and Hu Y-Z 2010 *J. Tribol. – Trans. ASME* **132** 021403
[8] Wang, Z J, Wang W Z, Meng F M, and Wang J X 2011 *J. Tribol. – Trans. ASME* **133** 031401
[9] Zhang P Y, Diao D F and Wang Z J 2012 *J. Tribol. – Trans. ASME* **134** 021301
[10] Yu C, Wang Z, Liu G, Keer L M and Wang Q J 2016 *J. Tribol. – Trans. ASME* **138** 041402
[11] Spinu S and Cerlinca D 2018 *IOP Conf. Ser.: Mater. Sci. Eng.* **400** 042054
[12] Spinu S and Cerlinca D 2018 *IOP Conf. Ser.: Mater. Sci. Eng.* **400** 042055
[13] Wang Z, Yu C, and Wang Q 2014 *J. Tribol. – Trans. ASME* **137** 011501
[14] Polonsky I A and Keer L M 1999 *J. Tribol. – Trans. ASME* **122** 30
[15] Liu S B, Wang Q and Liu G 2000 *Wear* **243** 101
[16] Liu S B and Wang Q 2002 *J. Tribol. – Trans. ASME* **124** 36
[17] Yu C, Wang Z and Wang Q J 2014 *Mech. Mater.* **76** 102
[18] Renouf M, Massi F, Fillot N and Saulot A 2011 Tribol. Int. **44** 834

Fractal evaluation aspects in characterizing the roughness of a driving wheel from a locomotive

L. Babici[1,2], A. Tudor[1], J. Romeu Garbi[2] and M. Stoica[1]
[1]Department of Machine Elements and Tribology, Polytechnic University of Bucharest, 313Spl.Independentei, Bucharest, Romania
[2]Acoustical and Mechanical Engineering Laboratory (LEAM),Polytechnic University of Catalonia (UPC), 11 C/Colom, Terrassa (Barcelona), Spain

Abstract. For comprehending tribological phenomena such as wear, friction, contact deformation and tightness of contact joints is the essential characterisation of the multiple scale topography of roughness surfaces. Wheel roughness obtained through direct measurement and then a description of wheel roughness has been obtained using the fractal function. The statistical parameters and fractal sizes for drive wheels which ran 2000 km and another ones new wheels have been analysed.By registering the results and programmable automaton of the roughness, the principal statistical characteristics were determined to be stands out the Abbott-Firestone curve and respectively, the fractal character. The appearance of roughness on the surface of the wheel causes the production of vertical vibrations that act on the complex wheel-rail system.In the analysis of the wheel irregularity, the fractal geometry was applied, so that to obtain the fractal parameters D_w, respectively L_{tw} (opothesy length), the method of the structure function was used [1, 2]. Therefore, the experimental measurement and obtained results processed presented.

1. Introduction

Nowadays, one of the essential sources of vibration in urban areas is the railway transports. It has been observed that aspects regarding the vibrations, can be harmful to the human body, in particular in sectors like rail and construction [3].

In the railway field, vibrations caused by the passing of trains arise as one of the primary sources of potential disturbance for the environment near railway lines. Roughness plays a crucial role in the generation of vibration. The vibrations levels of surroundings and its nearby buildings induced by running trains considered in relation with speed [4] rail roughness [5, 6] and general of vibratory behaviour of the superstructure [7]. Studies have been made concerning the similarity of the state of the surface, in particular, geometry roughness, comparing of multiple scale length in fractal terms. A discrete structure-function gives a suitable description. Majority of surfaces used this structure functions, and it was characterised with the r.m.s.-value of the roughness, transition length between fractal behaviour at high wavenumbers and the fractal dimension in the fractal region, respectively[8]. The first works in which the asperities were treated with the help of the Gaussian distribution were those of Greenwood and Williamson [9] followed by those of Nayak [10], Bush et al. [11], Greenwood [12] and McCool [13]. The roughness determined by the fractals taking into account an elastic deformation was presented by Ciavarella et al. [14], Persson [15] and Archard [16].

On the other hand, Ciavarella and Demelio [17, 18] and subsequently Gao and Bower [19] extended their calculations taking into account a plastic deformation and analysed the contact between a rigid surface and an elastic surface, both surfaces with roughness with fractal character.In Romania, D. Pavelescu, together with A.Tudor [20] dealt with the micro-metric roughness of the engineering surfaces of some machine elements.The proposed paper describes, for the first time, the roughness of a driving wheel with the help of fractal theory.

2.Roughness parameters of wheel

Surface roughness is formed by fluctuations of short wavelength in the surface, characterised by hills or valleys of varying amplitude and spacing. The irregularities of rail (roughness and corrugation) and wheel (roughness and plane defect) induce the vibrations, mainly, as an effect of contact load. The corrugation could have values typically 50 μm and the wavelength of 50 mm. The amplitude of corrugation can be higher than wavelengths when these are superior to 50 mm [21]. To describe the roughness of a surface, in general, it is the possibility to choose the amplitude parameters, spacing parameters and hybrid parameters [22]. To characterise surface topography, to measure the vertical characteristics there are used the amplitude parameters. For horizontal aspects, there are spacing parameters and the combination of the first two, are hybrid parameters.

3.Fractal geometry of wheel roughness

To characterise the multiscale self-affine topography for a railway wheel, the fractal geometry was used, using the fractal parameters that are independent of the scale. Also, the fractal character of the roughness of the wheel surfaces was investigated by analysing the structure-function using the fractal parameters DwandG$_w$[1, 2, 19, 20]. The fractal approach to other approaches such as the power of spectral density has the potential to predict the behaviour of a surface phenomenon at any length scale.
This paper aims to analyse the roughness of the wheel surface through different methods: fractal parameters, bearing area curve and amplitude parameters.
Analysing the surface of the wheel, the method algorithm was similar to the one used in our paper [1] where has been developed for the rail.
Based on measurements made with m | wheel device which offered the dates of amplitude parameters, it has been developed the fractal character derived from surface topography. According to the vertical characteristics of the surface of the system wheel-rail, the vibrations occur in vertical coordinated in movement.

4.Roughness measurement and results

Roughness on the wheel surface performed with m | wheel device (Figure 1). For the roughness of the wheel, the bogie of the LE-MA locomotive from the Softronic Craiova was used. (figure 2).

Figure 1. m|wheel device for wheel.

Figure 2. Bogie mounted on the locomotive

Thewheel roughness measuring device m|wheelused to perform the tests belongs to the laboratoriesRomanian Railway Notified Body (RRNB).This equipment has been used to record the acoustically relevant roughness for the investigation of the polygonization (out-of-roundness), wheel flat and the exact wheel diameter to examine the wear behaviour.To determine the roughness of the wheel, scanning in running direction of the wheel of 1 mm was carried out (determination of the diameter with an accuracy of ± 0.15 mm.For the roughness of the wheel, the bogie of the LE-MA locomotive from the Softronic Craiova used. Measurements for half of the bogie mounted with roughness wheels and another half of the bogie attached with new wheels were carried out.Theobtainedresults were processedwith one soft developed in Mathcad program.

4.1.Hybrid parameter- Bearing Area Curve (BAC)
The roughness parameters of the wheel allow describing the topography of the surface. The analyzes based on the BAC for tribological characterization of the surface topography revealed information regarding the wheel's rolling capabilities. BAC shows us the dependence of the height of the roughness of the real contact area [1]. The BAC curves are shown in Figure 3 and 4 for the roughness of a new wheel, respectively, for used wheel.

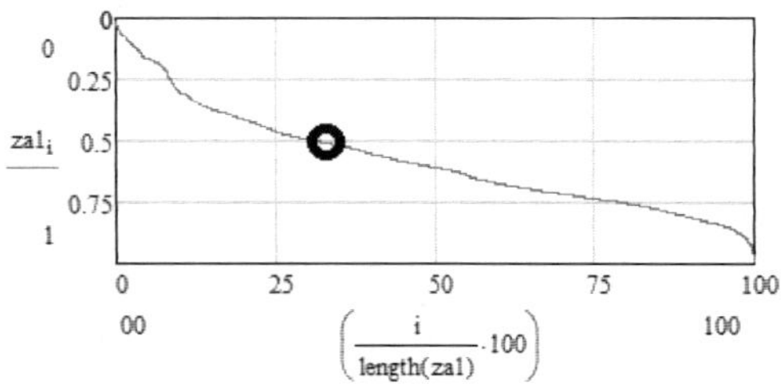

Figure 3. BAC (za1) for roughness of used wheel

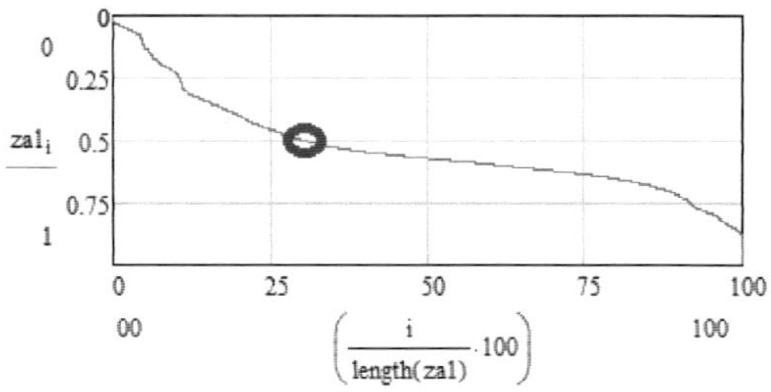

Figure 4. BAC (za1) for roughness of new wheel

4.2.Amplitude parameters
The amplitude parameters are useful for the qualitative and quantitative assessment of roughness wheel. These parameters have been analysed (ISO 4287-1:1984 Surface roughness) are listed in table 1. The results are obtained for three new wheels and three used wheels, respectively.

Table 1.Resultsforamplitudesparameters

Parameter/Wheel type	[1]nw$_1$	nw$_2$	nw$_3$	[2]uw$_1$	uw$_2$	uw$_3$
Ra [μm]	0.697	11.769	14.055	-24.181	-46.624	-127.400
Rq [μm]	45.470	19.055	20.879	124.529	133.051	178.165
Ten points ISO [μm]	1.000	0.100	4.080	54.260	27.360	-24.140
Ten pointsDIN [μm]	0.620	0.170	2.120	27.130	27.140	-12.070
Skewness (Rsk) -	0.175	2.500	2.168	-1.261	-1.499	-1.370
Kurtosis (Rku) -	2.114	4.177	8.390	3.738	4.177	2.116

[1]*new wheel (nw)*
[2]*used wheel (uw)*

5. Characterisation geometry of surface topography by fractal

The characterisation of surface topography with fractal geometry it is an option for many researchers in the study the contact surfaces more detailed[1, 12-15].
The fractal theory has been applied to characterise the surface of the wheel, and also in the algorithm from [1] where has been developed for the rail. The fractal analysis methods can replace the disadvantages of the surface analysis methods by the conventional method. Lately, various methods have been developed to characterise the size of a fractal set, such as the size of the box, the size of the compass etc. [1, 11, 12].To calculate the size of the fractal parameters for the roughness of a wheel (D_w)

and the topothesy length (L_{tw}), we used the structure function method.Theroughness wheel signal $y_w(x_i)$ is investigated as a function to delay (t).
The structure-function for the wheel roughness is presented in equation 1 and it is in continuum form:

$$S_w(t) = E[(y_w(x) - (y_w(x+t))^2] = c_w(|t|)^{(4-2D_w)} \quad (1)$$

It is known that the topothesy length of roughness wheel (L_{tw}) is defined that displacement which r.m.s. slope has a unity statistically:

$$S_w(L_{tw}) = L_{tw}^2, \quad (2)$$

The structure-function for wheelfractal roughness is presented in equation 3:

$$S_w(t) = L_{tw}^{2D_w-1}.(|t|)^{4-2D_w} \quad (3)$$

where $D_w$and $L_{tw}$are fractal parameters of wheel roughness. These parameterscan be determined by graphical form (log-log) of the structure-function, similarly to roughness of rail [1].
The structure-function of roughness wheel, in discrete form, $S_w(N,k)$, of a signal $y_w(x_i)$ is:

$$S_w(N,k) := \frac{1}{N-k}\sum_{i=1}^{N-k}\left(y_{w(i+k)} - y_{w_i}\right)^2. \quad (4)$$

For example, in Figures 5 a and b show the structure functionof the roughness profile corrected for the used and new wheel and figure 6hightliting the comparingstructure-function for bothwheel.

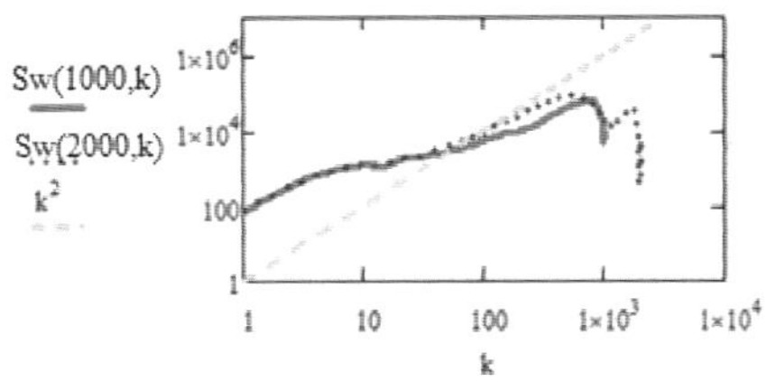

Figure 5a. Structure function for used wheel

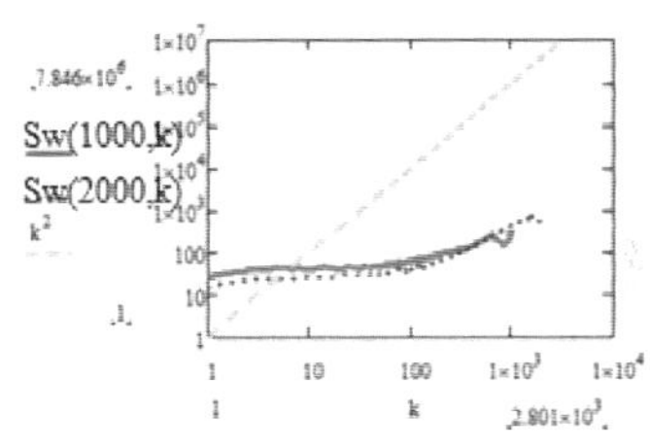

Figure 5b. Structure function for new wheel

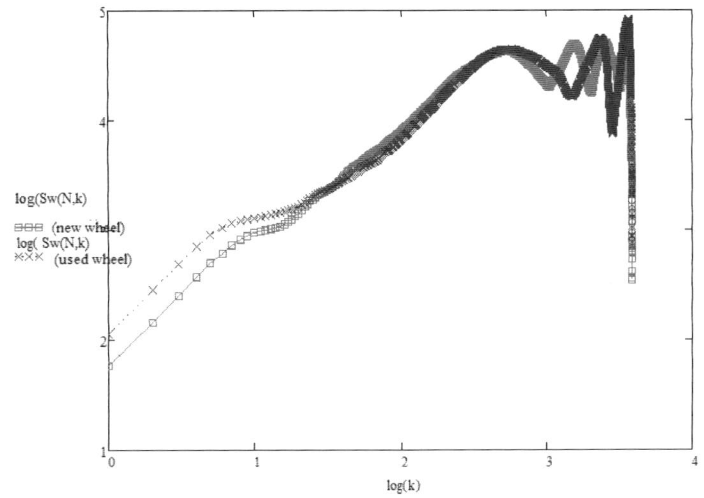

Figure 6. Comparing structure-function for new and used

In table 2 are presented the results for fractal parameters. It can be seen that the slope of the curve for the two wheels is different up to around 1.5, both increasing exponentially and then overlapping them to the value of 3.5. It is furthernoticeable for the new wheel a sharp decrease for structure-function in discrete form and an increase for the used wheel. In terms of the topological length L_{tw} and the fractal parameter, D_w has a higher value for the used wheel compared to the new wheel. It is observed that structure function in log-log coordinate $\left(log\left(S(N,k)\right) - log\left(x(k)\right)\right)$ has a linear function, having slope β. In this case, can be nominate that roughness wheel has a fractal character [1, 9-16].

Tabel2. Fractal parameters

Parameter wheel type	[1]nw1	nw2	nw3	[2]uw1	uw2	uw3
B_w	0.816	0.872	0.737	0.369	0.43	0.559
D_w	1.592	1.564	1.632	1.815	1.785	1.720

[1]*new wheel (nw)*
[2]*used wheel (uw)*

For the fractal character roughness (parameters D_w,L_{tw}, G_w (function to L_{tw})), it was associated with a continuous and undifferentiatedWeierstrass-Mandelbrot (W-M) function y_w (x) [1,9, 14, 16, 18]:

$$y_w(x, D_w, G_w, \gamma_w) = G_w^{D_w-1} \sum_{n=n_1}^{\infty} \frac{cos(2\pi \gamma_w{}^n x)}{\gamma_w{}^{(2-D_w)n}}, \quad 1<D_w<2 \text{ and } \gamma_w>1, \tag{5}$$

with frequency modes γ_w^n. This frequency is a function to wavelength (λ_{nw}) of wheel roughness and be calculated similarly to our paper [1].

Figures 7 a and bexemplify the fractal evaluation of the wheel roughness by the Weierstrass-Mandelbrot (W-M) function for small part of wheel.

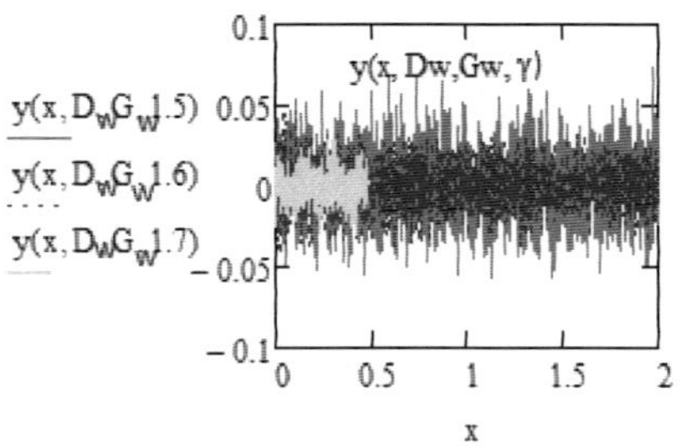

Figure 7 a. W-M (y(x,D_w,Gw,γ)) for roughness of used wheel

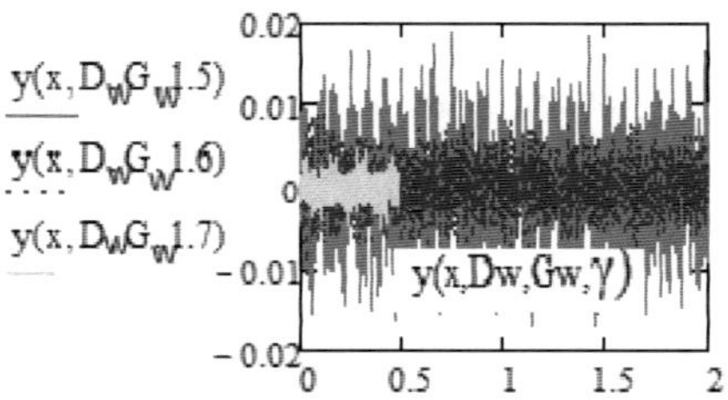

Figure 7 b.W-M (y(x,D_w,Gw,γ)) for roughness of new wheel

6. Conclusions

The roughness, corrugations and circularity of features wheel are the fundamental cause of rolling noise and vibrations.

The amplitude parameters, through its six two-dimensional statistical parameters, provided us with information regarding the average statistical properties, the shape of the height distribution histogram, and about the extreme properties (peak and valley). BAC shows how the profile bearing length ratio varies with a level in both situations for a new and used wheel and also could be observed that the area. Structure-function comes to rest on towards the end of increment size for each scan length. The WM function presents for both types of wheels, new and respectively used, profiles using three values of the gamma scaling parameter Mathcad's computational program has been allowed statistical analysis, correction of aberrant spikes and valley, deduction of the BAC, deduction of power spectral density, and fractal parameters of the microgeometry of new and used wheels.

Acknowledgements

The authors feel thankful for the support from Softronic Craiova, Romania and Romanian Railway Authority.

7. References

[1] Babici L, M, Tudor A, 2019 Some aspects regarding the roughness of the railway surface and rolling noise at locomotives, *Materials Science and Engineering* **514**01210.

[2] ZhouGY, LeuM C, Blackmore D, 1993 Fractal geometry model for wear prediction, *Wear*, **170**, p 1-14.

[3] Sica G, Peris E, Woodcock J S, Moorhouse A T, Waddington D C, 2014 Design of measurement methodology for the evaluation of human exposure to vibration in residential environments, *Sci. Total Environ* p 461–483

[4] Xia H, Zhang N, Cao Y M, 2005 Experimental study of train-induced vibrations of environments and buildings, *J. Sound Vib.* 280 p 1017–1029

[5] Zougari A, Miralles J M, 2012 Modelos numéricos para el análisis del comportamiento vibratorio de vías férreas, *Asoc. Sp. de Ing. Mech,* XIX Congreso Nacional de Ingeniería Mecánica

[6] Orta J, Cardona J, Romeu J, 2008 Influencia de la Rugosidad del Carril en la Generación de Vibraciones en Infraestructuras Ferroviarias, Metro, *Acústica,*Coimbra, Portugal

[7] Arcos R, 2011 A model for railway induced ground vibrations in the frame of preliminary assessment studies, Thesis, Universitat Politècnica de Catalunya, Spain

[8] Goerke D, Willner K, 2008 Investigations of the influence of geometrical irregularities and roughness on the normal contact stiffness of joints,*Proc.Appl.Math.Mech.***8**p 10275–10276

[9] Greenwood J A, Williamson J B P, 2006 Contact of nominally flat surfaces, *Proc. R. Soc. London. Ser. A. Math. Phys. Sci.***295**p 300–319

[10] Nayak P R, 2010 Random Process Model of Rough Surfaces, J. *Lubr. Technol.* ***93***

[11] Bush A W, Gibson R D, Thomas T R, 1975 The elastic contact of a rough surface, *Wear* ***35*** p 87–111

[12] Greenwood J A, 2006 A Unified Theory of Surface Roughness, *Proc. R. Soc. A Math. Phys. Eng. Sci.* ***393***p 2006 133–157

[13] McCool J I, 1986 Comparison of models for the contact of rough surfaces, *Wear* ***107***p 37–60.

[14] Ciavarella M, Demelio G, Barber J R, Jang Y H, 2000 Linear elastic contact of the Weierstrass profile *Proc. R. Soc. A Math. Phys. Eng. Sci* p 387–405

[15] Persson B N, Tosatti E, Fuhrmann D, Witte G, Wöll C, 1999 Low-frequency adsorbate vibrational relaxation and sliding friction, *Phys. Rev. B - Condens. Matter Mater. Phys.* ***59*** p 11777–11791

[16] Lincoln B, 1953 Elastic deformation and the laws of friction, *Nature* ***172*** p 169–170

[17] Ciavarella M, Murolo G, Demelio G and Barber J R, 2004 Elastic contact stiffness and contact resistance for the Weierstrass profile, *Elsevier Science Ltd* ***52***p 1247–1265

[18] Ciavarella M, Murolo G and Demelio G, 2006 On the elastic contact of rough surfaces: Numerical experiments and comparisons with recent theories, *Wear* ***261*** p 1102–1113

[19] . GaoY.F, BowerA.F, 2006 Elastic-plastic contact of a rough surface with Weierstrass profile, *Proc. R. Soc. A Math. Phys. Eng. Sci.* **462**p 319–348

[20] Pavelescu D, Tudor A, 2004 On the roughness fractal character, the tribological parameters and the error factors, *Proc. Rom.***5**p 1–6.

[21] Thompson DJ, 2009 Railway noise and vibration: Mechanisms, modelling and means of control, *Elsevier, Oxford*

[22] Gadelmawla E S, Koura M M, Maksoud T M A, Elewa I M and Soliman H H, 200 Roughness parameters, *Journal of Material Proccesing Technology* **123**p 133-145

IOP Conf. Series: Materials Science and Engineering **724** (2020) 012029 doi:10.1088/1757-899X/724/1/012029

Comparative study on the behaviour of a serial robot RRR and hand-arm human systems

A. F. Cristea[1]

[1] Technical University of Cluj-Napoca, Building of Machines Faculty, Mechanical Engineering Systems, Romania

Abstract. The paper creates a preamble in the comparative study regarding the loading of vibrations on a robotic system RRR or 3R and hand-arm homologous system. In this study, vibration measurements and comparative analyze are conducted for the two systems, trying to correlate similarities between them, but of the different scale.

1. Introduction

The subject of this paper has as a starting point the analysis of the mechanical vibration measurements transmitted from a machine - tool to the human operator. It has been observed that transmitting vibrations, at low frequencies, in this paper up to 25Hz, could cause it the negative effects on individual's health, respectively professional diseases (for example Vibration White Finger, joint disorders, etc.). Specialty literature [6] reminds that vibrations between certain limits <25Hz are destructive to humans, as the highest 2000 Hz (in special conditions) are harmful to robots, especially in couplings / joints. As a means of protection against vibration in humans, protective equipment or avoidance of prolonged exposure to them, but the effect of vibrations on machines and machines and robots is only felt at maintenance checks, through the wear on the rotation couplers or in case of high precision work, where their reliability should be 100% completely.

2. Notions about hand-arm system

To begin with, we will briefly present some data about the mechanical hand-arm system, thus, according to figure 1, the mechanical system takes into account anthropometric parameters such as masses, lengths, given in the tables 2, for a human operator who work under the conditions given in Table 1.

Table 1. Initial conditions.

Direction of excitation	z_h (along of the forearm)
Excitation	Displacement $z(t) = z_0 \sin \omega t$
Subject Position	(vertical position)
Frequency scale	0 – 20 Hz
Pressure Force	25 N, according to ISO 15230/2007
Place of pressure	Hand

Table 2. Anthropometrical parameters (determined in the other studies).

Anthropometrical parameters		
$m_1 = 0.45$ k	Forearm length	$J_{c3} = 0.0149$ kgm^2 (mechanical moment of inertia in the elbow joint)
$m_2 = 1.15$ k	0.98m	
$m_3 = 1.9$ k		

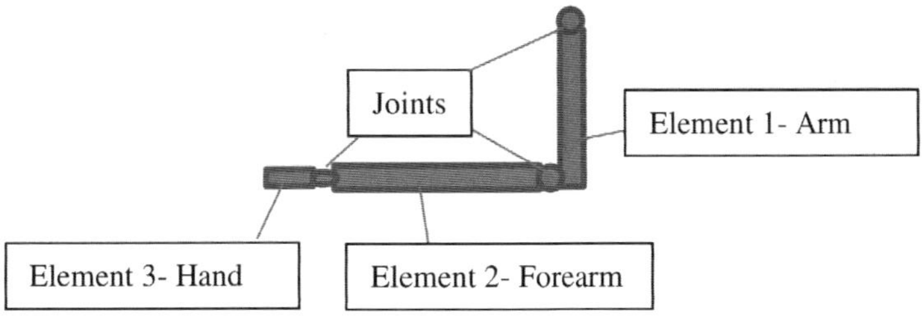

Figure 1. Schematic model of the hand-arm system.

This type of tool has been chosen, because it uses frequencies appropriate to those of small robots. Machine operation is tested by drilling through a chemically treated MDF board using a 10cm drill. The experiment measured the vibrations transmitted by the machine when operating without percussion (Fig.2-4). If should describe it the conditions of vibration measurements, they are effectuated about human operator, he is a 40-year-old male, weighing 85 kg and 1.85m tall, used and handled the drilling machine. The measurements used an inductive triaxial transducer (accelerometer type K_3), being able to simultaneously measure vibrations on three axes: Ox_h – perpendicular of the hand palm, Oy_h – direction along of the thumb, Oz_h – direction to from the hand to the shoulder. The study was done only for the vibration transmission on the Oz_h axis. This was connected to a vibration measuring device type SVAN 958.

The accelerometer was mounted on the anatomical locations: the wrist and the elbow were fastened directly under the metallic bracelet by tightening it, and directly on the arm (taped), it is trying the keep the position of the robot joints.

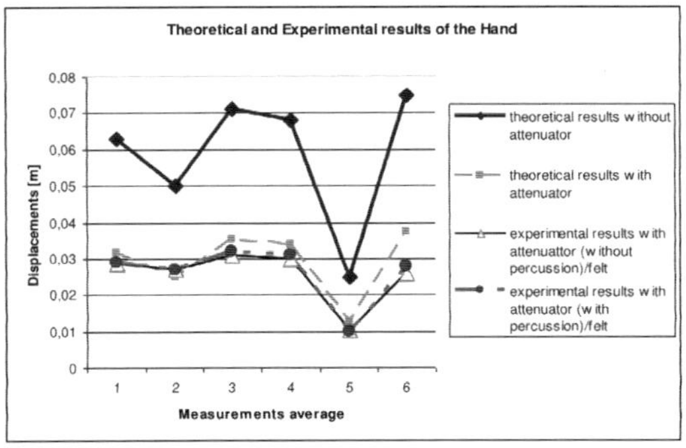

Figure. 2 Vibrations transmitted to the *hand* (theoretical and experimental) with and without attenuator with felt padding and using a drilling machine with and without percussion.

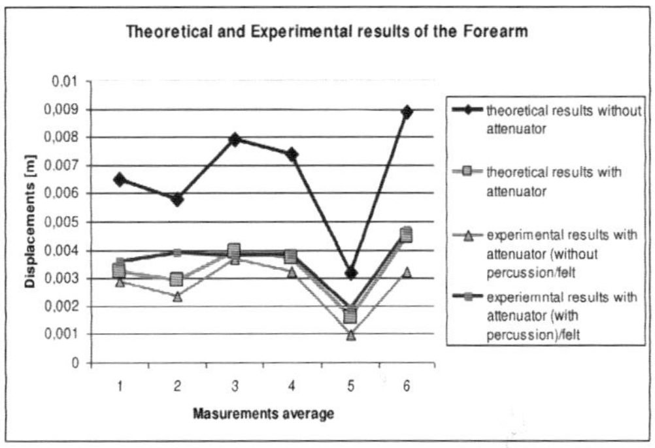

Figure. 3 Vibrations transmitted to the *forearm* (theoretical and experimental) with and without attenuator with felt padding and using a drilling machine with and without percussion.

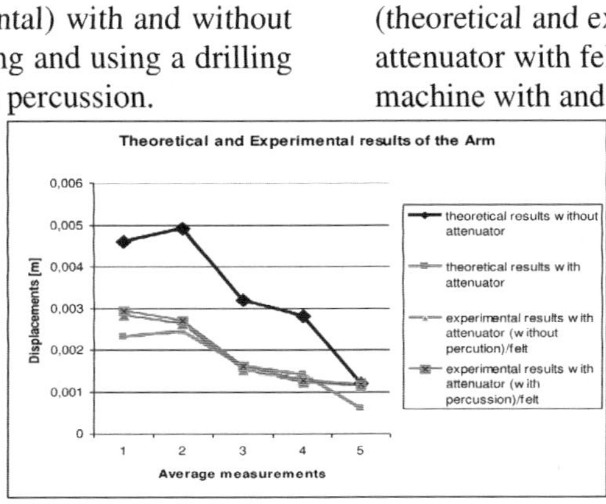

Figure. 4 Vibrations transmitted to the *arm* (theoretical and experimental) with and without attenuator with felt padding and using a drilling machine with and without percussion.

IOP Conf. Series: Materials Science and Engineering **724** (2020) 012029 doi:10.1088/1757-899X/724/1/012029

The figures 2-4 present the vibration influence of the hand-arm system, note that the bigger value is for the hand (superior curve of the graph) following of the forearm and arm, indifferently of the dumping system used it. That aspect shows us that exist owner dumping in our body (like muscular, sanguine systems).

It to keep a certain connection with the robot in terms of deacceleration, it was mounting one auxiliary vibration attenuator device (500 g) along the forearm; this could minimize the transmitted vibration, by hand to the arm (Fig.4). The vibration attenuator device (containing 2 dampers) was mounted along the forearm and fixed with a complex shank (bracelets). The interior padding of the metallic bracelets has been made to prevent skin lesions and prevent blocking the worker's forearm movements. Figures 5a and 5b present the way the dampers (element 1) are fixed in the structure. The results are analysed in figures 2-4 for all cases considered by the experiment and considering the anatomical location (the wrist, elbow and shoulder), the way in which the bracelets were fastened on the skin, and whether drilling machine without percussion. The measured results were analysed, and an average was calculated. They have been recorded directly by the machine and then downloaded on the computer using a special program.

The first analysis of the results presented in the figures 2-4 show that the mechanical vibration was transmitted along the hand-arm system to a lower extent if:

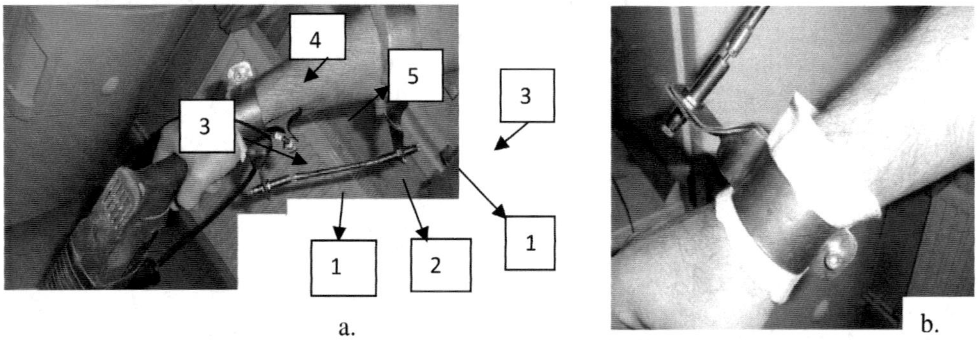

a. b.

Figure. 5 Fastening the attenuator device on the forearm. a. Technical characteristics of mini-attenuator (element 1); b. Interior padding with sponge.

A1. a vibration attenuator is mounted along the forearm (starting with 0.07 till 0.03m for the hand, 0.08-0.03 m for the forearm and 0.005 till 0.0025 m for the arm);

A2. the percussion of the drilling machine is not used;

A3. the internal padding of the bracelets that are part of the attenuator device is done with sponge and not felt.

Figures 5 *a* and *b* present the fastening methods for the attenuator device on the forearm and the interior padding of the bracelets with felt.

3. Robots 3R generalities and behavior

If we refer to the similarity of the 3R (model FANUC, small class) and the hand-arm system, we can see that the robot studied (Fig. 6) has three rigid-arm elements, which as with humans are assimilated to the arm, forearm and hand. The study will lead us to higher acceleration values transmitted to the robot arms, but these values, reported on a certain scale, should show similitudes in the vibration-like mode of transmission of the hand-arm vibration. Robot vibrations are transmitted from the engine at the base of the robot to his arms, observing that element 3, which includes the prehension, has a similar behavior to its hand-to-hand transmission, but it remains to be demonstrated, also the robots include technical characteristics such as those related to gauge, trajectory, precision, profitability, number of degrees of freedom etc. [1, 2, 3, 6].

Among the steps followed in this paper were studied and designed the components of a robot (Fig.6), in this case consisting of three rigid elements and three rotation couplers (base - element 1, element 1 - element 2, element 2 - element 3).

A1. Robot component design was performed in SolidWorks (Fig. 6) that respects the overall dimensions of a small FANUC robot.

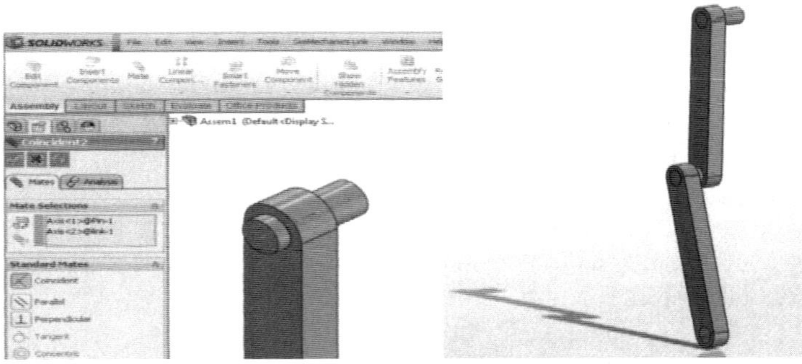

Figure 6. Designing of robot.

A2. The components of the 3R designed robot, respectively the base and the three arms called elements 1, 2 and 3, are taken over and assembled in CATIA (Fig.7 a-b).

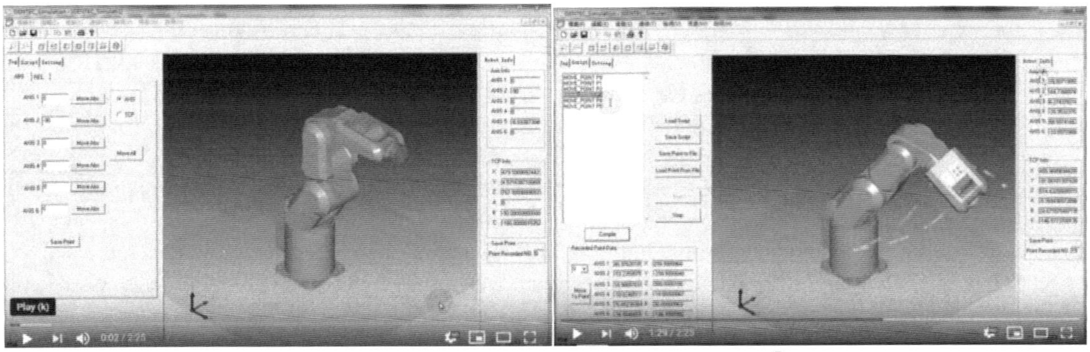

a. b.

Figure 7. Designing of robot and working his area simulating.

A2.1 The simulation of the robot behavior in terms of the resistance of the elements (arms) and their response to vibration exposure was performed with a software called FEA (Finite Element Analysis) (Fig. 8) under CATIA.

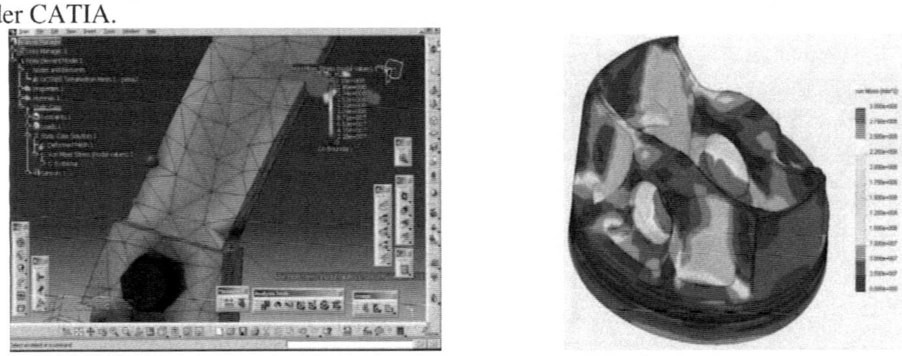

a. b.

Figure 8. Legend: blue - small vibrations (<500 Hz), yellow-green - moderate vibrations (500-2000 Hz), red - large loads (> 2000 Hz).

A simulation of the rigidity of the robot arms, especially of the element 1, will be carried out so that its resistance to vibration loads is observed. These data, estimated from the hand-arm system, on a larger scale of 10^3 times, that are recorded in special windows in the simulation environment. It is specified that the study took into account the weight and the power of the engines at the base of the small and medium robots, where the power required for the operation does not exceed the order of 3-5 kW, in which case the gauge fits in the desiderata of the form and suppleness of the mechanical structure [4], [5]. Thus, on a loaded arm and subject to a maximum load of 5 kg, the engine 1 in the base mass, 1-2 kg and supporting element 1, can carry such a load and its effort does not produce vibrations small or not to produce at all. Figures 8a. represents a simulation in the FEA of loading the robot arms to the vibrations (elements 1 near base, 2 intermediaries, 3 including prehension), as well as the beam, or fixed support thereon. It is noted that elements 1, 2 and 3, there are loaded on vibrations in the following order: bumper, element 1, element 2 and element 3. This load is due to the power of the engines that fall in power and the mass from the frame to the prehension element. Compared with robotic arms, study its resistance to the frame, namely in the joint basis, it showed that this is one of the most requested areas (Fig. 8b.) the efforts and loads, possibly due to weakening material for drilling holes fixing element 1 (area marked with point near t = 2s).

A3. In the following, a real-time measurement study will be carried out during robot arm movements. Analysis of vibration measurements. The measurements of vibrations or accelerations measured on small robots have shown that the effect of the vibrations is amplified, and bouncing at high speeds (for example, the acceleration peaks $a_{rms} = 80 m/s^2$ shown in figure 9 at the start of the movement, which then attenuates. Unlike the robot [4] the hand-arm system does not have these start-up shocks, or even if it exists, they are on a very small scale, unimaginable compared to the robot. Analog measurements of vibrations from hand-arm and the robot, the measurements were performed with a triaxial fixed with magnetic base, successively, the three elements of the robot, inquiring us the vibrations about the axis Oz, as shown in figure 5 and similitude with hand-arm system analysis. The arm 1 performs a 45-degree rotation motion, the arm 2 performs a 30- degrees, and the arm 3 at 45 degrees counter-clockwise. Figures 9a - b will show the theoretical vibrations compared to those measured on the manned system and those measured on the robot elements and are aggregated as maximum and minimum values in Table 3. To be comparable acceleration from hand-arm and measured the robot, displacement values [meter] in charts 2-4 are integrated through a data acquisition system and have emerged RMS accelerations. (root mean square) [m/s^2], who maximum and minimum are given by the table 3.

Table 3. Comparison between experimental and theoretical accelerations.

Robot		**a_{max} [m/s^2]**		**a_{min} [m/s^2]**
Element 1		10		25
Element 2		5		22
Element 3		15		25
Hand-Arm System (with attenuator)	**a_{max} [m/s^2] theoretical**	**a_{max} [m/s^2] experimental**	**a_{min} [m/s^2] theoretical**	**a_{min} [m/s^2] experimental**
Arm ~ element 1	0.015	0.018	0.010	0.015
Forearm ~ element 2	0.012	0.015	0.015	0.022
Hand ~ element 3	0.020	0.025	0.025	0.035

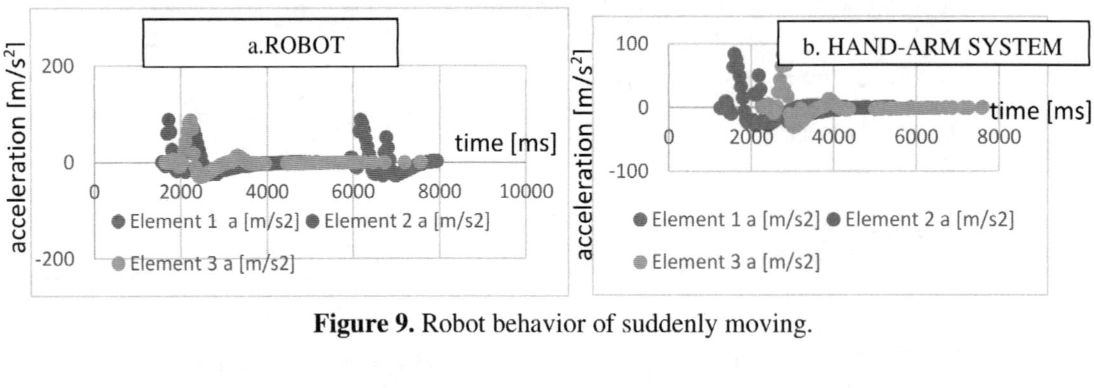

Figure 9. Robot behavior of suddenly moving.

The $80m/s^2$ acceleration leaps of the robot elements are omitted, occurring in the movement of the robot movement until the values of the movement move (Fig. 9a, b).

The comparative analysis of the vibration behavior of a robot and hand-arm system showed that:
- Vibrations are transmitted to both systems from the source of the motion generator, so at the robot the highest vibrations are at element 3 and element 1 the base articulated, where the exciting engine is. In the hand-arm system, the highest vibrations occur at the hand, where the exciting tool is and descending towards the shoulder.

Therefore, the largest robot vibration is obtained in element 3, followed by element 1, and in arm in element 3.
- Their transmission order is different in intensity (this is suspected) due to the stiffness of the component elements and the gauge, but the *new* consists in the fact that there is a scale of proportionality in transmitting the vibrations to the two systems, which seems to be orderly 10^3 times.

4. Conclusions and discussions

The work shows that there is an equivalence between the way the vibrations from the generating source to the robot and the human are transmitted. Respectively, in our case, there is a correlation between the transmitter elements 1 and 3 of the robot, where higher values were obtained at a scale 10^3 times larger than those obtained in the hand-arm system. This validates the concept that, the two systems are homologous as a construction to the extent that the robot construction materials can be adapted.

The FEA simulation study shows that resistance and vibrations in the 1, 3 and robot elements are very large (red (inside dark) area), a fact like the hand and source of excitement of the hand-arm system.

The paper wanted to highlight the fact that the two studied systems have similar structures, at different scales, so that each one meets the needs for which they were designed.

If the vibration damping / shrinking aspect is to be discussed, the literature shows that the protective means against mechanical vibration transmission at the workplace can be divided into two sub-categories.

The study can continue with concrete research on small robot types on the maximum vibration transmission a robot can withstand, so its reliability is not influenced, of course, considering industry standards.

5. References

[1] Davidoviciu A 1983 *Roboti industriali* (Bucureşti: ed Stiinta si Tehnica pentru Toti) p 107
[2] Blebea I and Ispas V 1995 *Calculul și Constructia Robotilor Industriali* (Cluj-Napoca: ed Dacia) p 322
[3] Eparu I and Badoiu D 1997 *Elemente de mecanica teoretica si de modelare a structurilor de roboti industriali* (Bucureşti: ed Tehnica) p 248
[4] Simionescu I, Ion I and Ciupitu L 2008 *Mecanismele robotilor industriali* vol.I, (Bucureşti: ed A.G.I.R) p 330
[5] McComb G 2013 Arduino Robot Bonanza *McGraw Hill Professional* p 416
[6] Cristea A F 2013 Study regarding the transmissibility vibrations about hand-arm system using vibration attenuation 27 June Sp Issue *Ad Eng Forum J* **8-9** (Switzerland: Trans Tech Publications) pp 421-36

IOP Conf. Series: Materials Science and Engineering **724** (2020) 012030 doi:10.1088/1757-899X/724/1/012030

Tooth contact analysis of spur and helical gears with crowned profile for pinion

S Cazan[1], C Dănilă[1, 2] and S Crețu[1]

[1] Department of Mechanical Engineering, Mechatronics and Robotics, Technical University of Iași, Romania
[2] Continental Automotive Romania, Iași, Romania

Abstract. The paper presents tooth contact analysis (TCA) of modified gears using semi-analytical methods (SAMs). The semi-analytical methods use Love's analytical formulation for Boussinesq equation of linear half-space theory and Hartnett approach for numerical formulation. The linear contact between two teeth has a finite length and during meshing process the edge effect appears, with a harmful influence on the gear reliability. To eliminate this, some tooth profile modifications are needed. To turn a line contact into a point contact, the crowned lead profile surface is considered. The ways used to obtain the matrices of separation are discussed for both spur and helical cylindrical gears. The large pressures' systems of equations were solved using the method of conjugate gradients and fast Fourier transform. For validation, in cases of hertzian geometries the SAM results have been compared with results obtained analytically, while for non-hertzian geometries, the SAM results have been compared with FEM results as presented in the specific literature. The developed SAM proved to be an accurate, robust and very fast tool for TCA of spur and helical cylindrical gear.

1. Introduction

The mechanics of concentrated contact is a fundamental aspect regarding gear design. Some aspects with a profound impact on gear reliability are derived from the postulates of concentrated contact, Johnson [1]. There exist various factors that affect gear reliability. Some authors, Qin and Guan [2] tried to emphasize how contact fatigue can be predicted. An unwanted influence upon gear reliability is the edge effect that manifests as a high gradient increase of contact pressures at the edge zones, Cretu [3]. In order to provide accurate values of the contact pressures. Pedrero et al. [4] proposed a combination between the Hertz equation and the non-uniform model of load distribution along the contact line. Casanova, [5], established that there are interdependences between the values of the face load factor and the gear width, the position of the gear on the shaft, the shaft's length and the ratio between the pitch radius and the shaft radius.

It has been found that the usage of SAM for tooth contact analysis is getting attention among recent researches. This happens because the SAM computing time is at least two orders of magnitude shorter than the necessary time when FEM is used, Guilbault and Najjari [6,7]. Casanova [8] presented various algorithms and computerized ways dedicated to perform tooth contact analysis. The results obtained with SAM were compared with FEA results or with analytical results, provided by AGMA [9]. In Pop et al.[10] it is exemplified a spur cylindrical gear case where contact pressures calculation was accomplished using semi-analytical methods (SAM). Proper lead modifications able to change the pure linear contact into a point contact are discussed and exemplified. In Litvin and Fuentes [11] and Feng, [12] different ways of determination of instantaneous contact ellipse for crown profiled teeth were proposed. Another study with regard TCA of spur gears using SAM or FEM is presented in Ye, Tsai [13], for different gear types, with different geometry modifications. In the current paper, the tooth flank geometry was modelled using the methodology presented in Litvin and Fuentes, [14]. The usage of SAM is also presented in Cretu et al. [15], taking into account different profile modifications for spur gear pairs. In this paper the needed modifications of the flank longitudinal profile have been done according to ISO 6336 [16].

2. Tooth contact analysis (TCA) using semi – analytical methods (TCA)

2.1. The non – Hertz method

When two bodies are in a line contact and a normal force is applied, these two bodies suffer deformations. The shape of the real contact area and the pressures distribution on that area are unknown. These two surfaces have the same tangent and the same normal in the contact point or line.

On the tangent plane of the two surfaces that are in contact, it is considered a rectangular virtual contact area chosen so it can cover the real contact area, figure 1.

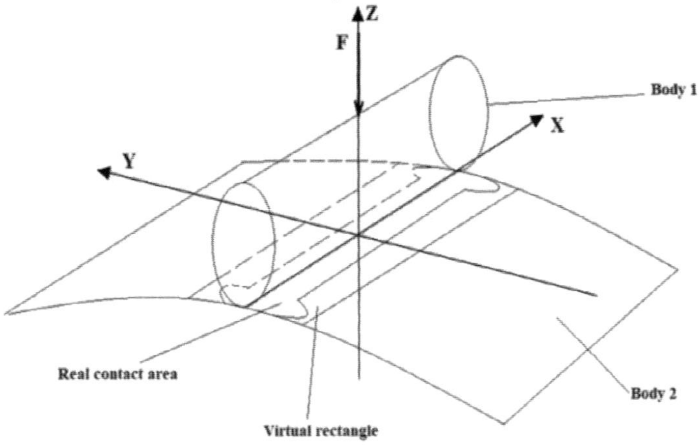

Figure 1. Two bodies in contact.

The elastic model of surface deformation under the normal load F is defined by the following equations:

- geometric equation of elastic contact:

$$g_{ij} = h_{ij} + w_{ij} - \delta_0 \tag{1}$$

- integral equation of normal displacement of elastic half – space frontier:

$$w_{ij} = \sum_{k=1}^{Nx} \sum_{l=1}^{Ny} (K_{i-k,j-l} \cdot p_{kl}) \tag{2}$$

- static equilibrium equation

$$\Delta x \Delta y \sum_{i=1}^{Nx} \sum_{j=1}^{Ny} p_{ij} = F \tag{3}$$

where for the point with coordinates (*i, j, 0*). The notations h_{ij}, w_{ij}, g_{ij} and δ_0 represent the initial separation, the elastic deformation along the normal direction, the gap resulted and the rigid displacement, respectively. The equations (1) – (3) represents an algebraic system with (n + 1) equations and unknowns: *n* pressures p_{ij} and the displacement δ_o. The hypotheses regarding the linear elasticity lead to the following restrictions:

- non - adhesion and non -penetration conditions lead to:

$$g_{ij} = 0 \xrightarrow{yields} p_{ij} > 0, \ (i,j) \in A_r \tag{4}$$

$$g_{ij} > 0 \xrightarrow{yields} p_{ij} = 0, \ (i,j) \notin A_r \tag{5}$$

- elastic-perfect plastic behaviour of the material:

$$p_{ij} > p_Y => p_{ij} = p_Y \tag{6}$$

More details can be found in [3]. The values of initial separations h_{ij} are determined considering the equations of each contacting surfaces. The high accuracy and very fast convergence speed recommended the Conjugate Gradients Method for solving the above equations [3]. The validation of the accuracy of pressures distribution subroutine was done by comparisons with:

i) pressures distributions provided by Hertz relations [3], when the contact is of hertzian type, and *ii)* FEM results obtained by neutral researchers.

In figure 2, it is exemplified the comparison between FEM results obtained by J. de Mul et al.[17] and the result provided by SAM. Very good fit between these approaches and the developed SAM have been obtained.

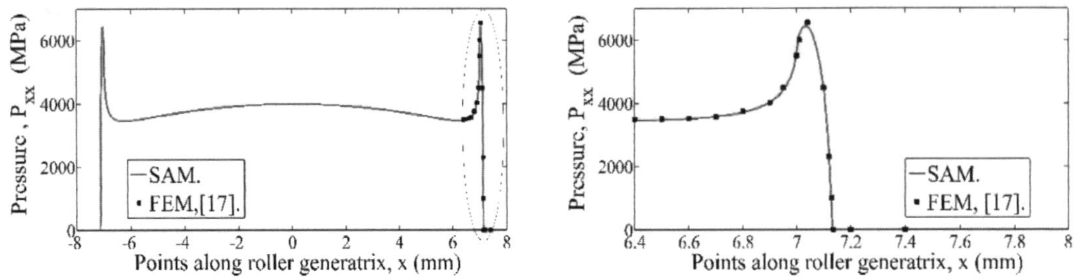

Figure 2. The synchronization between the non-Hertz algorithm results and FE, results [17].

2.2.Matrix of separations for contacts between teeth with crowned profile in helical gears

The equations that define a typical surface of a helical gear tooth flank are as follows:

$$\begin{cases} x_1 = r_{b1}\, cos(\theta_1 + \mu_1) + u_1 cos\lambda_{b1} sin\,(\theta_1 + \mu_1) \\ y_1 = r_{b1}\, sin(\theta_1 + \mu_1) - u_1 cos\lambda_{b1} cos\,(\theta_1 + \mu_1) \\ z_1 = -u_1 sin\lambda_{b1} + p_1\theta_1 \end{cases} \tag{7}$$

The meaning of each term is according to Litvin and Fuentes, [14].

To calculate the matrix of separations in such a case, the curve resulted from the intersection between the described surface and the plane $Z = 0$ (red dashed involute curve, figure 3) is rotated with respect to OY'' - axis, figure 3. In this way, a crown profiled pinion is obtained.

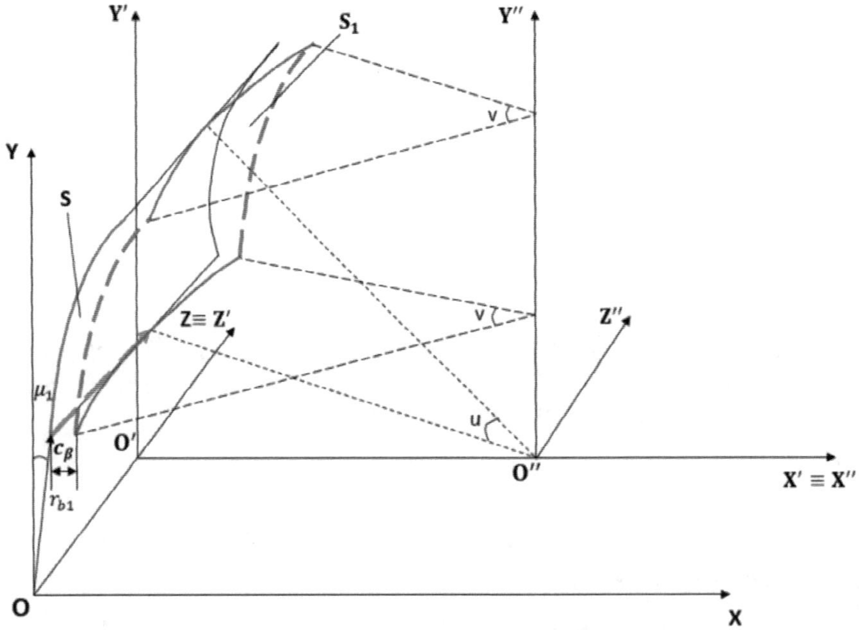

Figure 3. Schematic representation of crowned profile tooth geometry.

Surface S is the surface defined by eq. (7). The continuous red curve on the plane $Z = 0$ is the intersection between surface S and this plane. This curve is then translated along the Z – axis and OX' -

axis and its equations are written with respect to $X''O''Y''Z''$. The surface that is obtained by rotating this curve with respect to Y'' - axis has the following equations:

$$\begin{cases} x'' = \left[r_{b_1}\sin(\theta_1+\mu_1) - p_1\theta_1\cot\lambda_{b_1}\cos(\theta_1+\mu_1) - r_{b_1}\sin\mu_1 - R\right]\cdot\cos v \\ y'' = \cos(\theta_1+\mu_1) + p_1\theta_1\cot\lambda_{b_1}\sin(\theta_1+\mu_1) \\ z'' = \left[r_{b_1}\sin(\theta_1+\mu_1) - p_1\theta_1\cot\lambda_{b_1}\cos(\theta_1+\mu_1) - r_{b_1}\sin\mu_1 - R\right]\cdot\sin v \end{cases} \quad (8)$$

where v is the angle of rotation of the involute curve with respect to OY'' - axis and u is the angle between any point of surface S_1 and OX' - axis, which represents the crown profiled surface. The limits of these two angles can be easily found by calculating the angle between the straight lines that intersect each corner of the flank surface (for v) and the angle between the two straight lines that pass through the top and the bottom of surface, figure 3. The value of crowning height –figure 3 – is taken according to [16]. A tangent plane is defined so that it pass through the maximum point of the corrected surface, corresponding to the value of B/2 on the Z – axis (half pinion width). This surface is meshed in a number of points – using equations (8) – and the matrix of separations is calculated as the distances between each point and the tangent plane. In the above equations, R is the radius of the lowest circle arc that passes through the corrected surface.

3. SAM algorithms results

3.1. TCA of a spur gear pair

The initial data for the current spur gear pair are identical with those presented in [10], under the same conditions. This paper is taken as reference for comparison with the current results, figure 4. To obtain a spur gear pair, in the computerized program, the helix angle takes the value of 0 degrees.

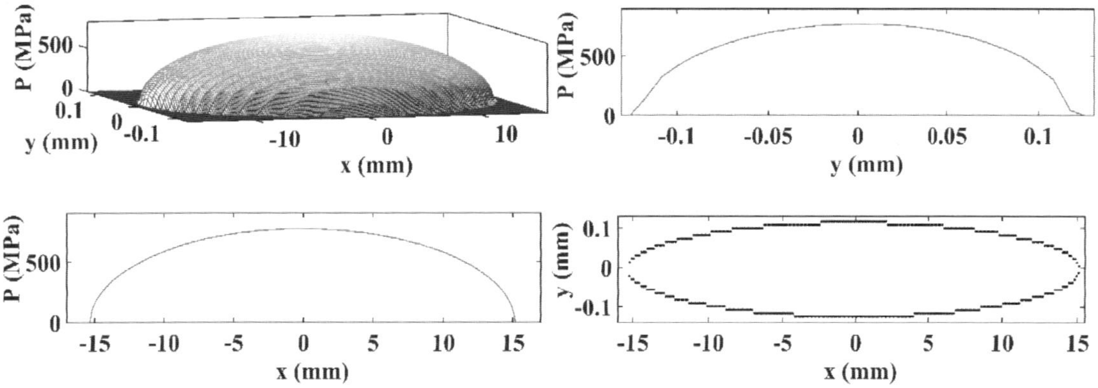

Figure 4. The 3D, transversal and longitudinal pressures distribution and contact area for crowned teeth of a spur gear pair

The maximum pressure value in this case is 772.2 MPa, while in [10] the maximum value is 748 MPa. The maximum error between these two approaches is 3.13%, which is a very good correlation. In [10], the matrix of separations has a different method of determination.

3.2. TCA of a helical gear pair

As long as the validation of the pressures determination subroutine was accomplished, according to figure 1, and the accuracy of the matrix of separations calculation procedure for crown profiled teeth was proven in the previous case (spur gear pair), the following results are provided for a right hand helical gear pair without validation. The initial data, in this case, are given in table 1. The most important aspect with respect these results are the median pressures and the fact that the edge pressures, with two zones of peak pressures, have been avoided.

Table 1. Initial data for the current helical gear pair

Data	Values
No. of teeth – z_1, z_2	$34, 57$
Helix angle – β	15°
Module – m_n	3 mm
Pressure angle – α	20°
Gear width – B	25 mm
Crowning height – C_β	5 μm
Point action radius – r_{x_1}	Meshing pole

The results are presented in figure 5.

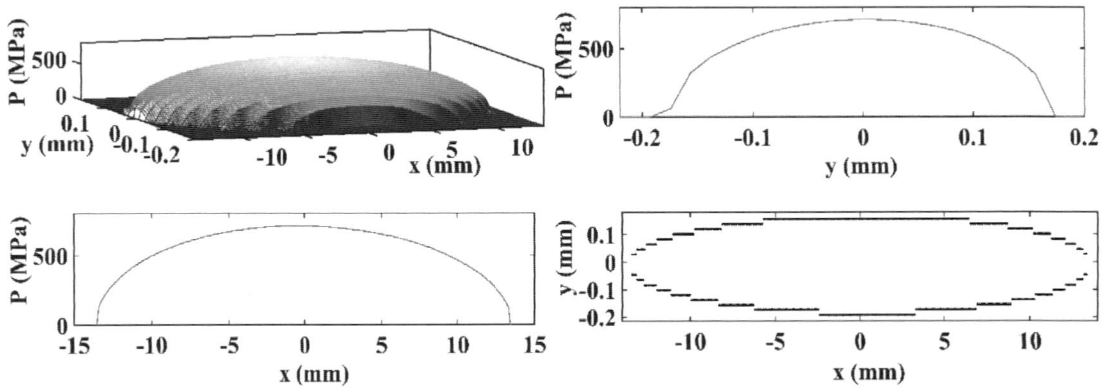

Figure 5. The 3D, transversal, and longitudinal pressures distribution and contact area for crowned teeth of a helical gear pair, helix angle $\beta = 15^0$.

The maximum pressure value is 712 MPa, in this case.

4. Conclusions

Usually, the cylindrical gears have teeth with a slightly modified longitudinal lead profile in order to transform the initial line contact into a point contact, so that the edge effect can be avoided.

To carry out case studies including teeth contact analyses, the finite element method proved to be time consuming.

A semi-analytical method has been developed to obtain the pressures distributions for concentrated contacts developed during meshing processes of spur and helical cylindrical gears.

To obtain the coefficients for matrix of separations at the contact between teeth helical gears, an original way was presented and exemplified for pressures distributions in teeth contacts of helical gears.

If the contact was of hertzian type, the developed SAM would be validated by comparisons of its results with pressures distributions provided by Hertz relations. For non-hertzian type of contacts, the FEM results obtained by neutral researchers were used for comparations and validation.

The main advantages in using semi-analytical methods consist in much lower computation time, accuracy, robustness and a good compliance. These qualities recommend SAM to perform fast TCA of spur and helical cylindrical gears.

5. References

[1] Johnson K L 1985 *Contact Mechanics* ed Cambridge University Press Cambridge pp 32 - 55

[2] Qin W J and Guan C Y 2014 An investigation of contact stresses and crack initiation in spur gears based on finite element dynamics analysis *International Journal of Mechanics Sciences* **83** 96 – 103

[3] Cretu S S 2009 *Contactul concentrat elastic – plastic, ed* Polytehnium Iasi pp 139 – 214

[4] Pedrero J I, Pleguezuelos M and Muñoz M 2011 Critical stress and load conditions for pitting calculations of involute spur and helical gear teeth *Mechanism and Machine Theory* **46** 425 – 437

[5] Roda – Casanova V, Sánchez – Marin F T, Gonzalez-Perez I, Iserte I and Fuentes A 2013 Determination of the ISO face load factor in spur gear drives by the finite element modeling of gears and shafts *Mechanism and Machine Theory* **64** 1 – 13

[6] Najjari M and Guilbault R Modeling the edge contact effort of finite contact lines on subsurface stresses *ASME Journal of Tribology (2015)*

[7] Najjari M and Guilbault R Edge contact effect on thermal elasto – hydrodynamic lubrication of finite contact lines *Tribology International* **71** 50 – 61

[8] Roda – Casanova V 2015 *Simulación del engrane y análisis del contacto en sistemas de transmission por engranajes mediante la modelización avanzada del conjunto ejes-engranajes* Thesis doctoral Universitat Jaume

[9] AGMA Standard 2001 – DO4 Fundamental Rating Factors and Calculation Methods for Involute Spur and Helical Gear Teeth American Gear Manufacturers Association Alexandria VA 2004

[10] Pop N, Cretu S, Yaldiz S and Iovan – Dragomir A 2017 A semi – analytical method for teeth contact analysis of spur gears with modified profiles *Proc. of 9'th Int. Conf. on Trib. (BalkanTRib'17)* 267 – 288

[11] Litvin F L, Gonzalez-Perez I and Fuentes A 2005 Topology of modified surfaces of involute helical gears with line contact developed for improvement of bearing contact, reduction of transmission errors and stress analysis *Mathematical and Computer Modelling* **42** 1063 – 1078

[12] Feng P H, Litvin F L, Townsend D P and Handschuh R F A 1999 Determination of Principal Curvatures and Contact Ellipse for Profile Crowned Helical Gears *Journal of Mechanical Design* **121** *(1)*

[13] Ye S and Tsai S 2016 A computerized method for loaded tooth contact analysis of high-contact-ratio spur gears with or without flank modification considering tip corner contact and shaft misalignment *Mechanism and Machine Theory* **97** 190 – 214

[14] Litvin F L and Fuentes A 2004 *Gear geometry and applied theory* (Cambridge University press second edition) pp 1 – 450

[15] Cretu S, Pop N and Cazan S 2018 Considerations regarding the pressures distributions on lead of spur gears IOP Conf. Ser.: Mater. Sci. Eng.

[16] ISO Standard 2007 – 1 Calculation of load capacity of spur and helical gears – Part 1: Basic principles, introduction and general influence factors Corrected Version 2007

[17] De Mul J M, Kalker J J and Fredriksson B 1986 The contact between arbitrarily curved bodies of finite dimensions *ASME Trans. J. Lub. Tribol* **108** 140 – 148

The influence of operating temperature and bearing steel structure in shakedown of three-dimensional elastic-plastic rolling contacts

G Popescu[1]

[1] "Gheorghe Asachi" Technical University of Iasi-Romania, Faculty of Mechanical Engineering, Blvd. Mangeron, No. 59A, 700050, Iasi, Romania

Abstract. This paper describes a model for the prediction of micro-plastic material transformation and associated residual stress field development in through-hardened and case-hardened bearing steels during over-rolling, considering the influence of operating temperature. The main challenge in analysing three-dimensional elastic-plastic rolling contacts is the accurate description of material properties in case of two-phase material, either is a constant amount of retained austenite (through-hardened AISI 52100 bearing steel) or variable amount of retained austenite along depth (case-hardened steel, i. e. 16MnCr5 bearing steel). Material parameters at room temperature, 65^0 and 100^0 C degrees were obtained by half-compressive cycles tests on an INSTRON^{TM} apparatus with a climate cell and further verified by performing experimental and numerical analyses in case of static indentation between a ceramic ball and a flat steel plate. To check the ability of the proposed models to simulate the elastic-plastic rolling contact in various loading condition, experiments were made on a special designated test rig and recorded values were compared with the numerical simulations obtained by using a semi-analytical solver. Measurements of residual cyclic profiles and residual stresses show a good agreement with theoretical values, leading to a better understanding of the fatigue process in hardened bearing steels.

1. Introduction

Through-hardened and case-hardened bearing steels were investigated. Their elastic-plastic behaviour shows important differences regarding the shakedown phenomenon. Retained austenite level, either is constant (through-hardened steel) or variable (case-hardened steel), plays a major role on the final values of plastic quantities (plastic displacements and residual stresses).

Through-hardened AISI 52100 bearing steel with a structure consisting of martensite and 13% retained austenite was modelled as combined isotropic and kinematic material, with non-linear hardening. Due to the cyclic stress-induced transformation of retained austenite (depending on the local value of the equivalent stress) the micro-yield limit increases and shakedown will be faster compared to the steels with low retained austenite, [1]. Because the austenite level is almost constant along the depth, the material's plastic parameters can be obtained by performing monotonic or half-compressive tests. Parameters describing the influence of retained austenite (the size of the yield surface and the plastic strain rate) are modelled by using the Armstrong-Frederick non-linear law, [2]. In the elastic-plastic analysis, Prandtl-Reuss equations [3] were used in association with von Mises flow rule. 16MnCr5 case-hardened steel also presents combined isotropic-kinematic non-linear behaviour. Because of a wide range of typical carburizing steels, every grade must be analysed individually. The case microstructure consists of high-carbon martensite with retained austenite. The initial yield depth profile can be associated with martensitic structure, but also the retained austenite contributes to hardness. The interdependence between initial yield limit, carbon content, and retained austenite cannot be easily quantified. The alloying elements may also play a significant role in the plastic behaviour. Böhmer et al. [4] investigated the rolling contact fatigue behaviour of heat resistant bearing steels at high operational temperatures and shows the influence of load and temperature on the characteristic deformation number for SAE 52100 at low rotational speed. The characteristic deformation number is defined as the Hertzian pressure at which the rolling contact fatigue endurance limit supersedes due to the occurrence of plastic deformation affecting the residual stresses.

Voskamp [5] made extensive research to investigate the relation between microstructure and the rolling contact fatigue properties of bearing steel at various operating temperatures, showing the alterations of the steel microstructure at various contact loads. No material models were used, but the experimental values of the residual stresses were of great importance in validating new material models and numerical programs in analysing the elastic-plastic cyclic rolling contact.

2. Theoretical model for elastic-plastic material properties as affected by temperature

In case of AISI 52100 through-hardened bearing steel the initial level of retained austenite depends mainly on the austenitizing temperature. Since the carbon distribution and initial yield limit present small variations along the depth, the material plastic behaviour can be obtained from monotonic or half-cycle compressive tests. In order to quantify the influence of retained austenite in case of cyclic loading in the elastic-plastic domain, a non-linear relation proposed by Armstrong and Frederick was used. As effect of stress-induced transformation of retained austenite, the micro-yield limit increases locally, based on the relation:

$$\sigma_{YIELD} = \sigma_E + Q_\infty \cdot (1 - \mathrm{e}^{-b \cdot \varepsilon^{pl}}) \tag{1}$$

where Q_∞ represents the size of the yield surface, σ_E is the elastic limit, b is the plastic strain rate and ε^{pl} is the total equivalent plastic strain. If Q_∞ is positive, then is isotropic hardening, while if Q_∞ is negative is soft hardening.

Performing repeated half-compression cycles on cylindrical bars 3 mm round and 7 mm length, on an INSTRON™ device equipped with a climate cell, equivalent stress-equivalent plastic strain dependence as affected by temperature is given in figure 1. The influence of temperature can be seen in small variations of Young's modulus, E, as in Gupta [6], and Poisson's coefficient, υ, but the major change was in the size of the initial yield surface (table 1). Even the initial amount of retained austenite was around 13% in all samples, increasing the operating temperature results in a drop of this parameter (i.e. temperature-induced transformation). In the same time, the value of the plastic strain rate could be considered constant, assuming that depends only of the type of loading (tensile or compressive). The obtained material parameters which describe the plastic behaviour based on non-linear Ramberg-Osgood strain-stress relation (2) are also given in table 1.

$$\varepsilon = \frac{\sigma_E}{E} + \left(\frac{\sigma_E}{B}\right)^N \tag{2}$$

Table 1. Material parameters for kinematic and isotropic hardening behaviour of 52100 bearing steel.

Parameter	Room temperature	65^0 C	100^0 C
E [MPa]	214,000	212,000	210,000
σ_E [MPa]	1,400	1,150	950
Q_∞ [MPa]	2,200	1,950	1,700
B [MPa]	4316.67	8386.26	5608.16
N [-]	12.603	10.965	9.177
b [-]	245	245	245
υ [-]	0.281	0.2816	0.2825

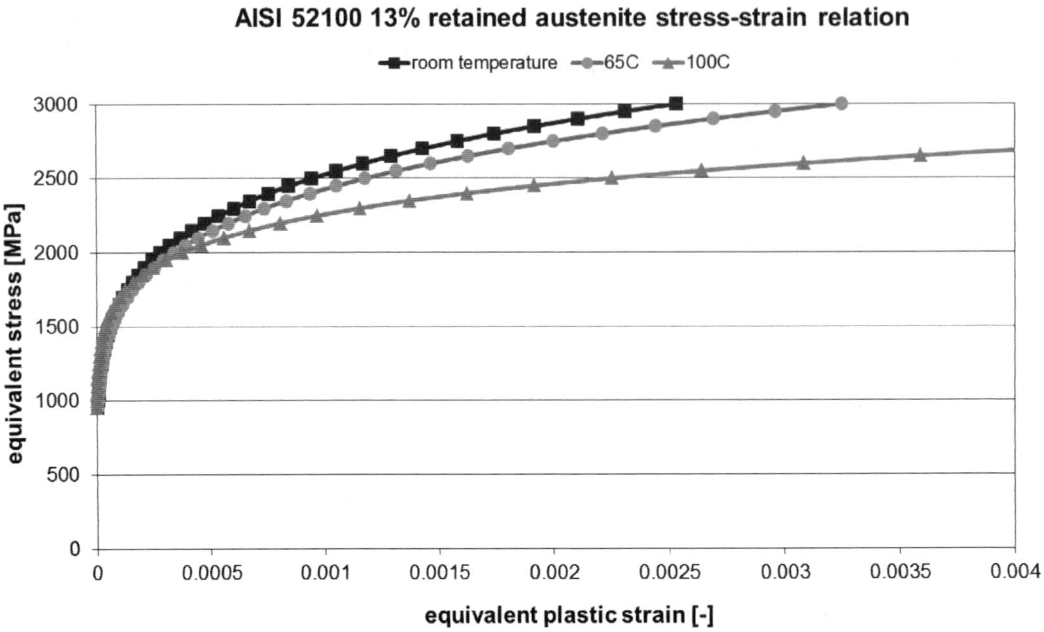

Figure 1. Ramberg-Osgood non-linear representation of the flow curve for 52100 bearing steel.

The 16MnCr5 carburized bearing steel is characterized by a large number of material parameters that depend on the carbon concentration along the depth, variable hardness of the core and the case, and the percentage of the retained austenite that also varies along depth. More, the initial yield limit presents variations due to variable carbon content along depth, [7]. Vincent [8] investigated the influence of carbon content in 16NiCrMo13 case-hardened steel and, similar as Petterson [7], showed that Young's modulus is weakly affected by the surface treatment, and can be considered constant for all carbon concentrations along depth. A material model that incorporates all the above mentioned parameters previously developed by the author [1, 9] was extended to include the influence of the operating temperature in case of elastic-plastic static and rolling contact.

3. Experimental results and correlation with theoretical results

Test roller specimens round 18 mm and 30 mm length were used for both materials. A special test rig with two ceramic rollers 19.05 mm diameter (Young modulus 315,000 MPa and Poisson coefficient 0.26) pressing on the test roller with various contact pressures: 3.2, 4, 4.5 and 5 [GPa].

The operating temperature inside the sealed chamber is obtained using a heat gun and is measured by an infrared sensor (65 and 100 Celsius degrees). Under load, the roller will suffer permanent deformation and residual profiles (along x-transverse axis) were recorded by laser profilometry.

Some results are presented in the next figures. Residual circumferential stress (y-axis, rolling direction) for various loading cases, theoretical and experimental values are presented in the next figures.

Theoretical contact model is developed based on Prandtl-Reuss equations associated with von Mises flow rule, [3]. The plastic deformation model consists of an isotropic part, as a result of cyclic austenite transformation, and a kinematic part, as a result of cyclic residual stresses. All plastic quantities (plastic strains, residual stresses and deformations) are derived analytically, based on an improved Chiu's model [10, 11].

In all cases the shakedown is attained very fast, lest than 100 rolling cycles for the AISI52100 steel and less than 20 rolling cycles for 16MnCr5 case-hardened steel. This particular aspect is related to the

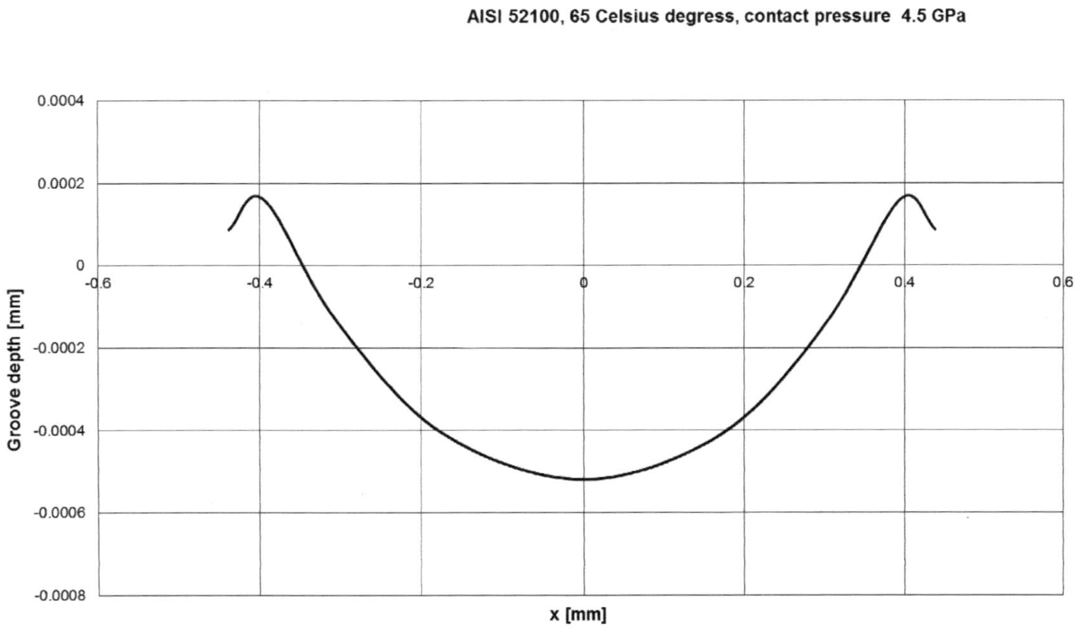

Figure 2. Theoretical residual roller profile at shakedown, AISI bearing steel loaded at 4.5 GPa and 65^0 C.

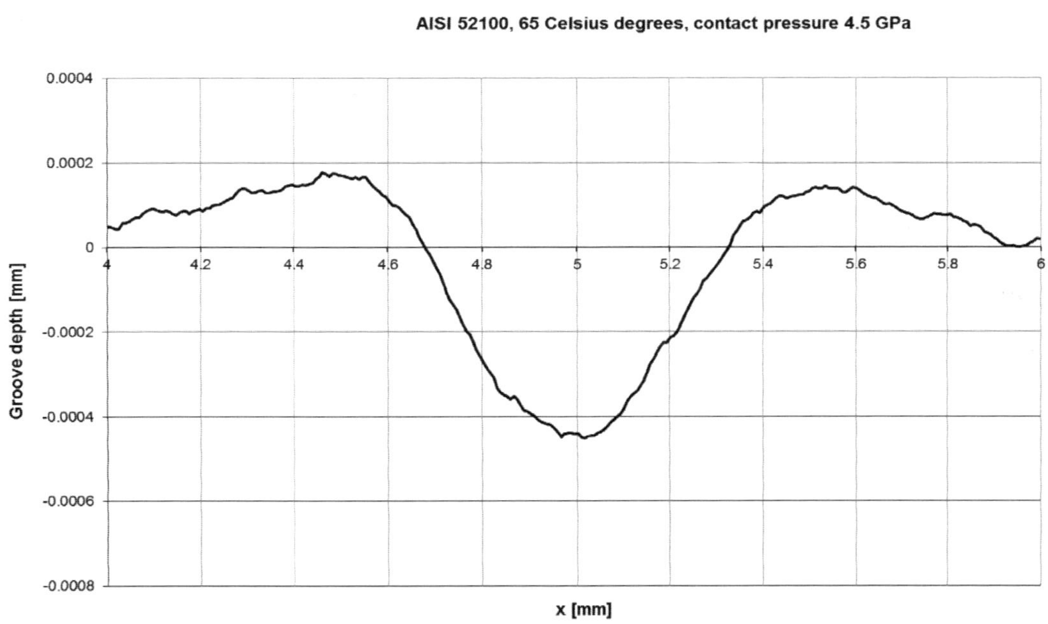

Figure 3. Measured residual roller profile at shakedown, AISI bearing steel loaded at 4.5 GPa and 65^0 C.

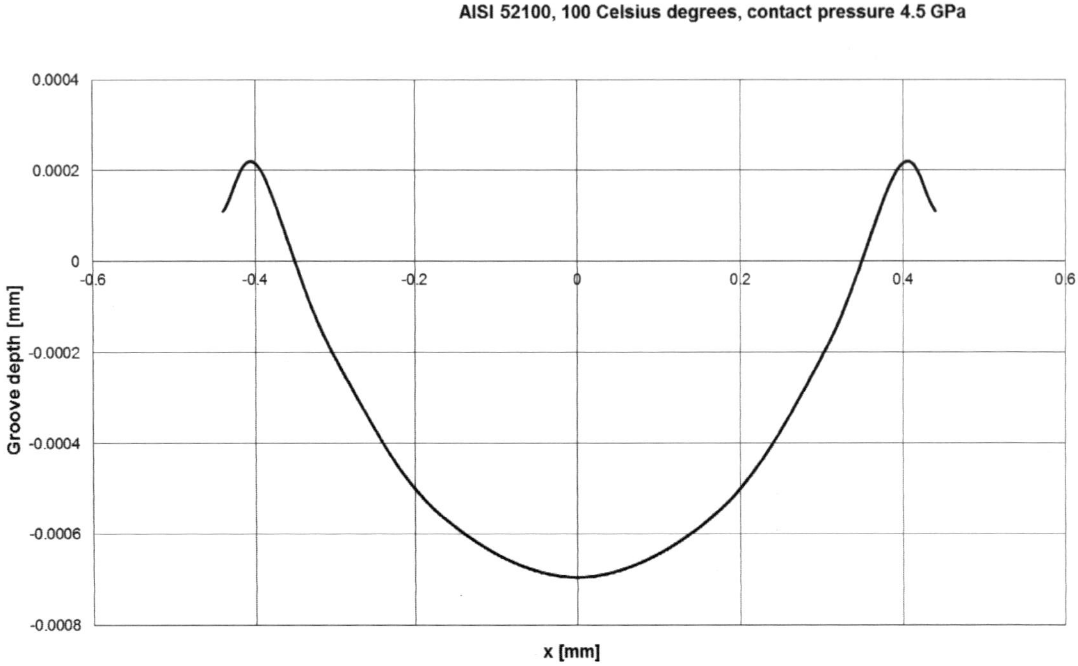

Figure 4. Theoretical residual roller profile at shakedown, AISI bearing steel loaded at 4.5 GPa and 100^0 C.

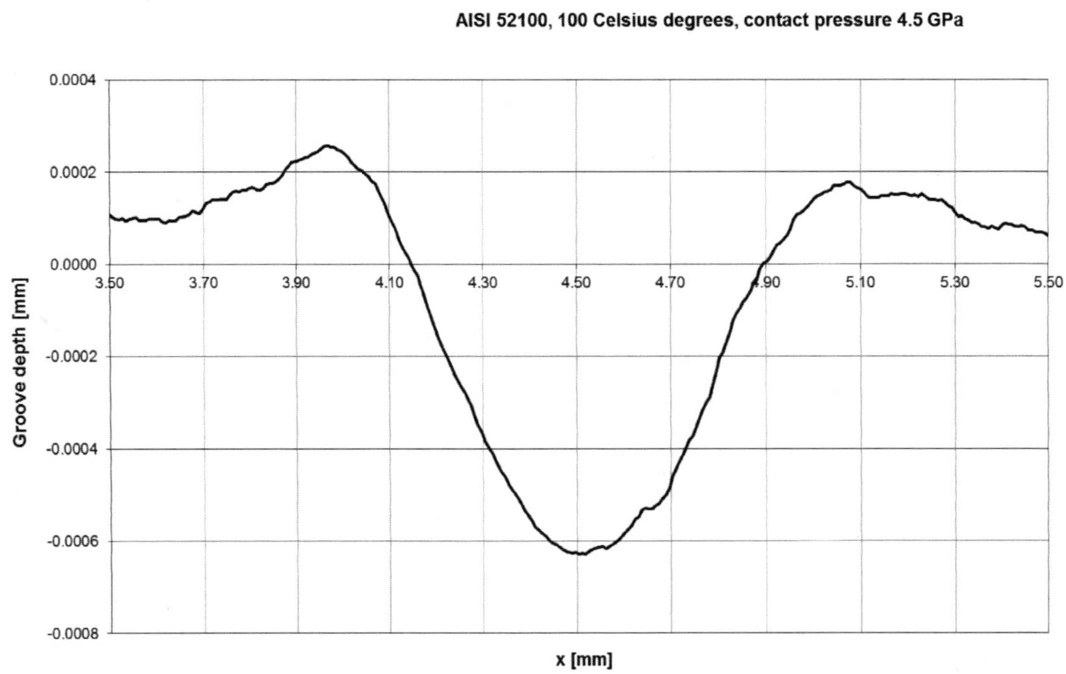

Figure 5. Measured residual roller profile at shakedown, AISI bearing steel loaded at 4.5 GPa and 100^0 C.

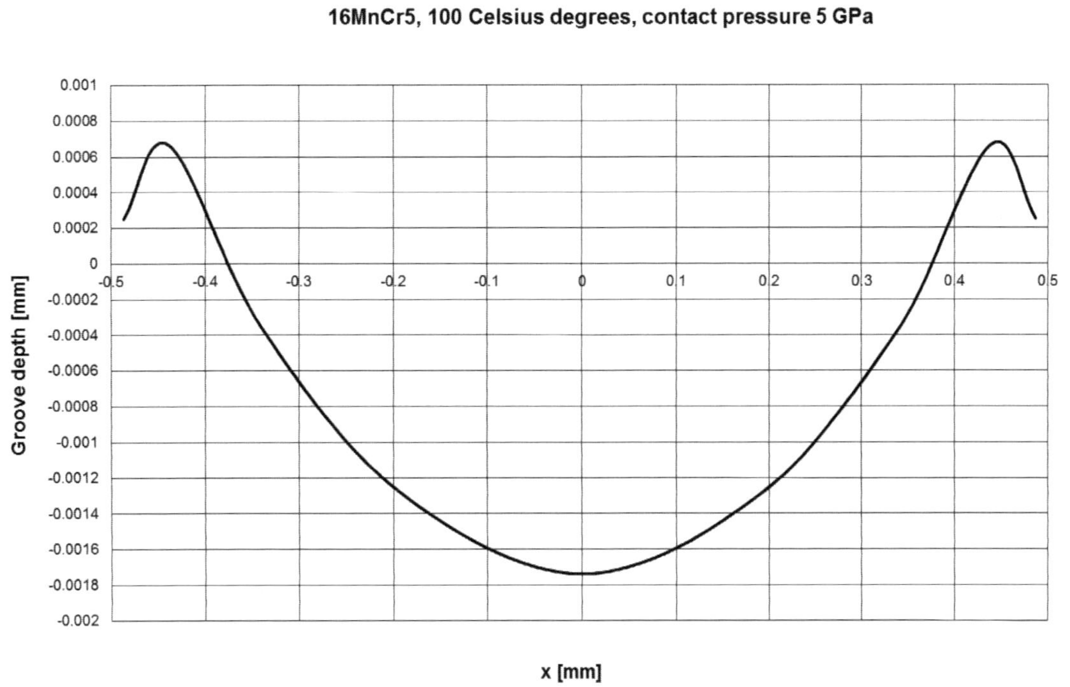

Figure 6. Theoretical residual roller profile at shakedown, 16MnCr5 bearing steel loaded at 5 GPa and 100^0 C.

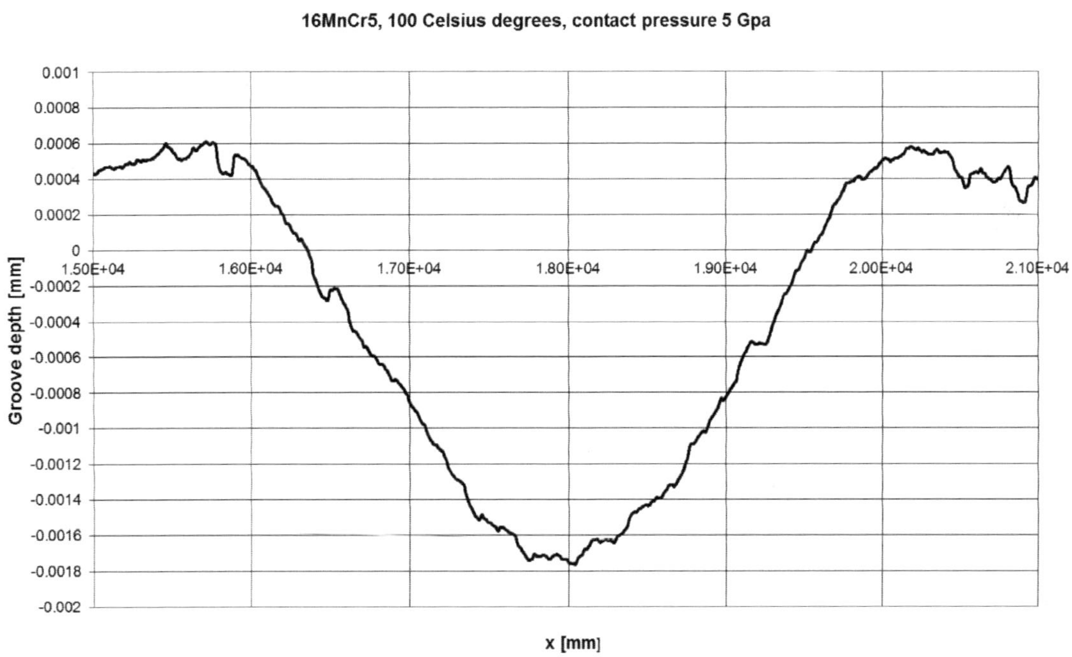

Figure 7. Measured residual roller profile at shakedown, 16MnCr5 bearing steel loaded at 5 GPa and 100^0 C.

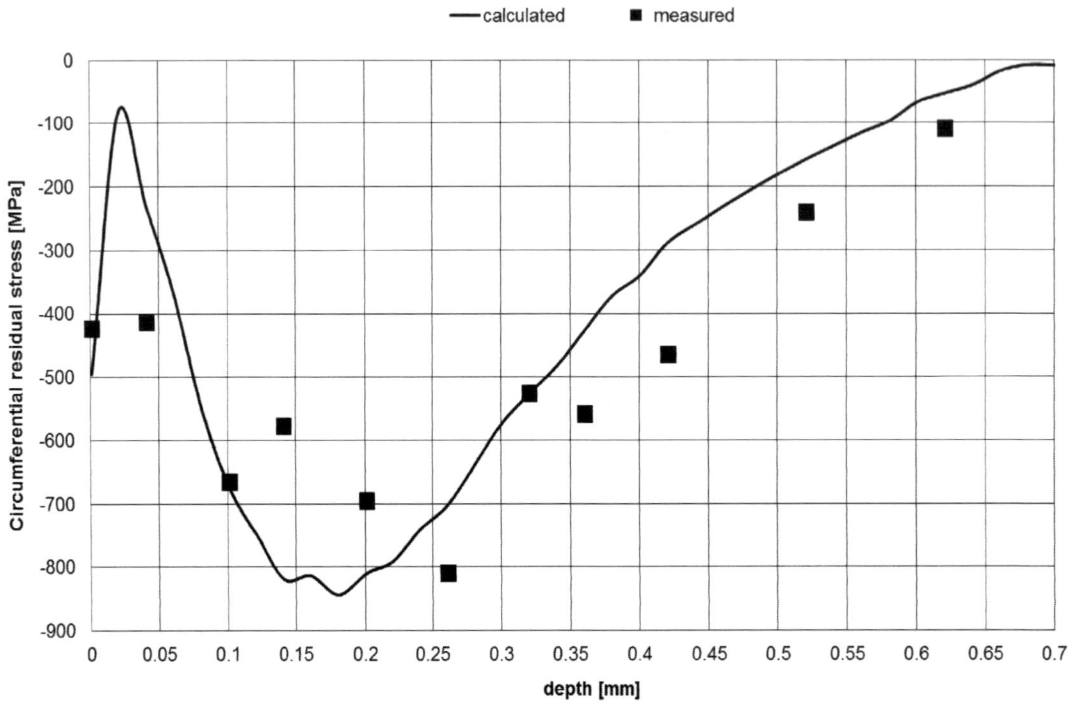

Figure 8. 16 MnCr5, residual circumferential stress at shakedown in case of 5 GPa contact pressure operating at room temperature.

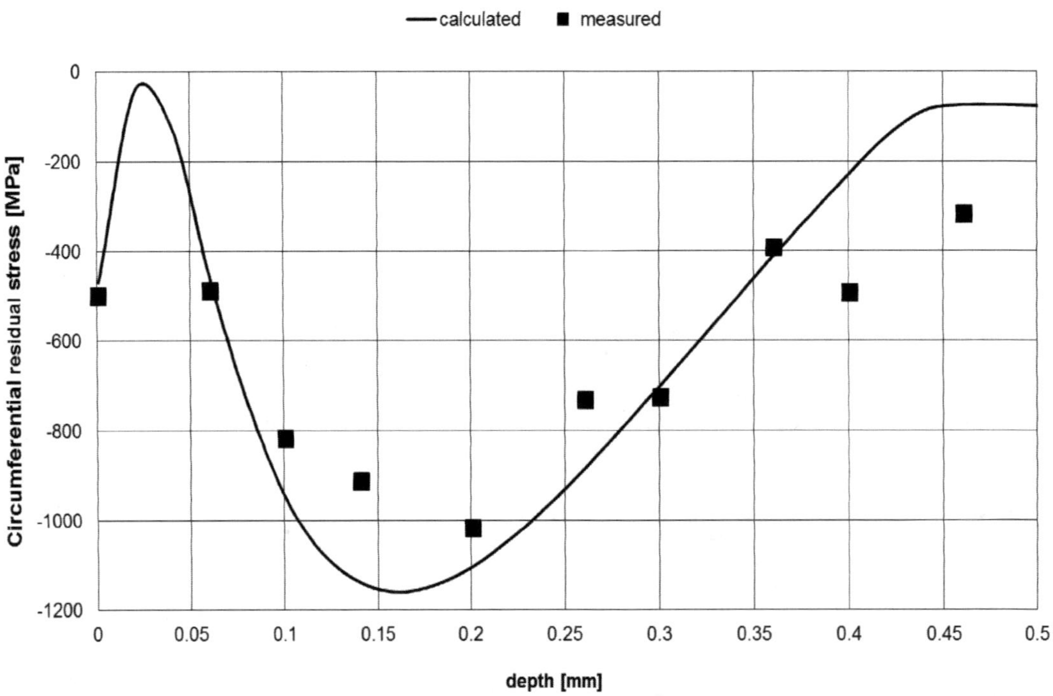

Figure 9. 16 MnCr5, residual circumferential stress at shakedown in case of 5 GPa contact pressure operating at 100^0 C temperature.

bigger initial amount of retained austenite in case-hardened steel; that transforms in martensite faster than in case of through-hardened steel and the major result is a higher increase of the local yield stress and subsequent lower plastic deformation in the following rolling cycles. Also, the groove depth is higher in case of case-hardened material and, naturally, higher residual stresses were measured, no matter the operating temperature. In case of AISI 52100, the operating change in temperature gives a permanent groove profile of about 0.5 micron depth and around 0.15 micron ridge height at 65 Celsius degrees, as compared to 0.7 micron groove depth and 0.25 micron ridge height at 100 Celsius degrees; similar theoretical and experimental results for 4.5 GPa contact pressure (figures 2 to 5). 16 MnCr5 bearing steel exhibits a more pronounced plastic deformation, the results obtained at 5 GPa contact pressure and 100 Celsius degrees operating temperature are at least double than for the AISI 52100 bearing steel (groove depth around 1.7 micron and ridge height about 0.7 micron, as in theoretical and experimental results, figures 6 and 7).

4. Conclusions

A complex model for the investigation of elastic-plastic rolling contact in case of bearing materials that exhibit combined isotropic and kinematic hardening behaviour has been developed.

Numerical simulations are in good agreement with experimentally recorded values of deformed profiles and residual stress measurements.

The model will allow further development of retained austenite transformation cycle-by-cycle, leading to a better understanding of the fatigue phenomenon in bearings. From this point of view, the current relations for bearing life calculation should be properly modified.

5. References

[1] Popescu G, Gabelli A, Morales-Espejel G E and Wemekamp B 2006 *Journal of ASTM International (JAI)* **3** (5) 1-12

[2] Armstrong P J and Frederick C O 1966 *Central Electricity Generating Board & Berkeley Nuclear Laboratories, Research & Development Department Report*

[3] Johnson K L 1985 *Contact Mechanics* (UK: Cambridge University Press)

[4] Böhmer H J, Hirsch T and Streit E 1998 Rolling contact fatigue behavior of heat resistant bearing steel at high operational temperature *Bearing steels: into the 21st century* ed J J C Hoo and W B Green (USA: ASTM International) p131

[5] Voskamp A 1997 Microstructural changes during rolling contact fatigue, *PhD Thesis Delft Technical University The Netherlands*

[6] Gupta V, Bastias P, Hahn G T and Rubin C A 1998 *Wear* **169** 251

[7] Petterson J 1998 Finite element analysis of case-hardened roller bearings *PhD Thesis Linköping Sweden*

[8] Vincent A, El Ghazal H, Lormand G, Hamel A and Girodin D 2002 *Bearing steel technology* ed J M Beswick (USA: ASTM International) p 427

[9] Popescu G 2012 *Proc. of the VARED 16* ed I Musca et al (Suceava: University of Suceava Publishing House) p 3

[10] Chiu Y P 1977 *ASME J. Appl. Mech.* **44** 587

[11] Chiu Y P 1978 *ASME J. Appl. Mech.* **45** 302

A study of the crack initiation angle in the substrate of spur gear teeth subjected to rolling contact fatigue

C O Popa[1]

[1]Department of Mechanical Systems Engineering,, Technical University of Cluj-Napoca, 103-105 Muncii Bvd, Cluj-Napoca, Romania

Abstract. Fracture failure of engineering structures is caused by cracks that extend beyond a safe size. For modeling the stable fatigue crack propagation it is necessary to determine the angle of crack initiation, i. e. the angle under the crack growth takes place. There are several criteria that try to model this aspect, such as: Sih, MTS, M, T, Ip.
The S-criterion (Sih) was used to simulate the fatigue growth of an internal crack situated in the substrate of a certain spur gear, taking into account the rolling contact fatigue. S-criterion states that the direction of crack initiation coincides with the direction of minimum strain energy density along a constant radius around the crack tip. Since the stress intensity factor K_I is negligible (because of the compression stresses in the substrate of the gear tooth), calculations were made for these conditions and it was obtained a constant and unique angle of crack initiation angle $\theta \approx 80°$, that is in concordance with the experimental observations. This study serves to determine the fatigue crack initiation angle in a compression stresses field, as well as its use in modeling the propagation of the fatigue cracks.

1. Introduction

Fracture failure of engineering structures is caused by cracks that extend beyond a safe size. It is a catastrophic event that takes place very rapidly and is preceded by crack growth which develops slowly during normal service conditions, mainly by fatigue due to cyclic loading.

The main parameters that govern the stable crack propagation are the stress intensity factors $K_{I.}$ and $K_{II.}$, Depending on their variation, the rate of the crack growth may be determined according to Paris or similar laws [1]:

$$\frac{da}{dN} = C \cdot \Delta K^n \tag{1}$$

where: a – semilenght of the crack, mm, N – number of loading cycles, n – Paris Law exponent; C – Paris Law coefficient, (mm/cycle)/[MPa·(mm)$^{1/2}$]^n;

ΔK – stress intensity factor range, MPa·(mm)$^{1/2}$:

$$\Delta K = K_{\max} - K_{\min} = \beta \cdot \Delta S \cdot \sqrt{a} \tag{2}$$

ΔS – stress range, MPa; β – constant depending on the geometry of the body.

Stress intensity factors K_I and K_{II} are related to the fundamental fracture modes: K_I corresponds to the "opening mode" and K_{II} to the "sliding mode" or "in-plane mode" (figure 1) [1].

But for modelling the stable fatigue crack propagation it is not sufficient to know only the value of the crack growth rate, it is also necessary to determine the angle of crack propagation (figure 2), i. e. the angle under the crack growth takes place [3, 5, 6].

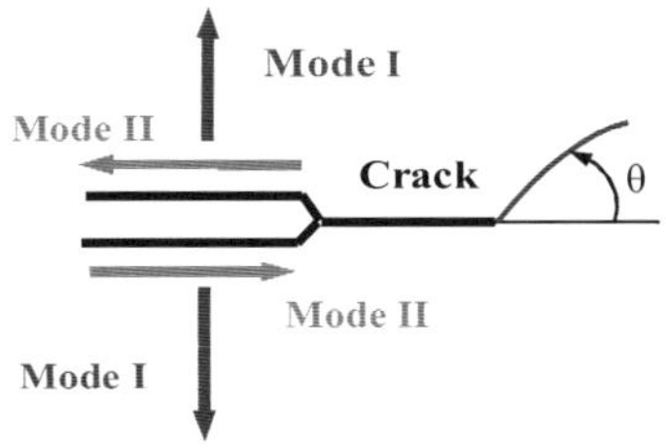

Figure 1. Fundamental fracture modes

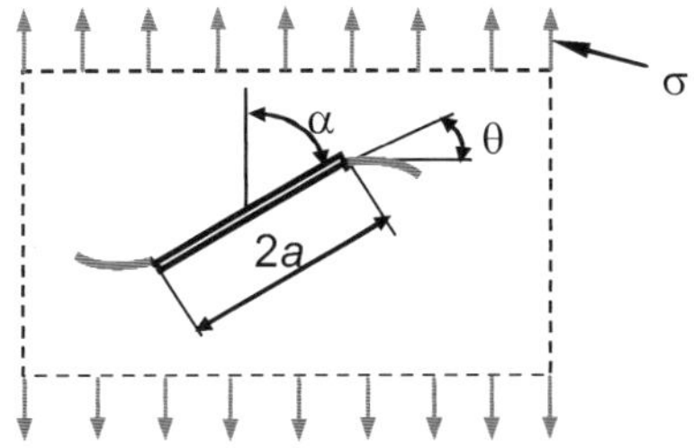

Figure 2. Fatigue crack initiation angle

On a macroscopic scale, in a field of compressive stresses, the crack will extend along a curved path that maximizes the mode II stress intensity factor (figure 3) [3].

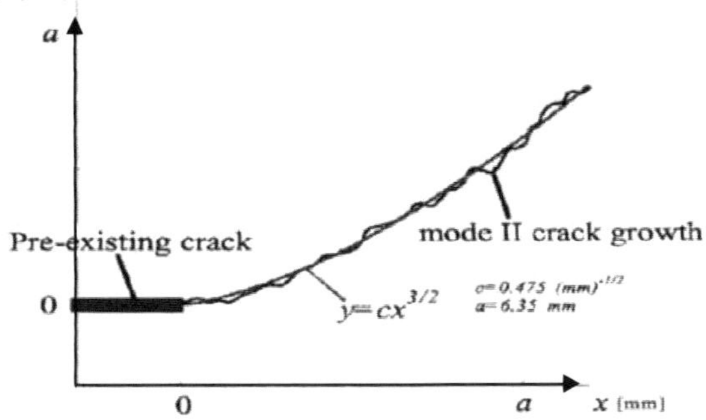

Figure 3. Experimentally observed crack growth under the compression stresses field [3]

Some criteria for determining the angle of crack growth will be presented below, as well as their application to the specific case of the substrate of a gear tooth, a zone in which the compressive stresses predominate.

2. The direction of crack initiation criteria

One cannot predict the crack propagation path without having the knowledge of the angle of crack initiation. The problem needs to be studied in mixed mode, since the state of stresses at the crack tip could be very complex and may result in mixed mode fracture.

The presented criteria can be grouped under three headings: stress-based criteria, energy based criteria, and strain-based criteria. The critical condition refers to one of the extremum of the stated parameter, i.e., stress, energy, or strain. The most commonly used criteria are the ones based on stress and energy, hence only these criteria will be presented..

2.1 MTS-criterion

The Maximum Tangential Stress criterion is the simplest of all and it states that direction of crack initiation coincides with the direction of the maximum tangential stress along a constant radius around the crack tip. It is assumed that the crack propagates perpendicularly to the direction of maximum normal stress, so that the tangential tension on the crack propagation line is zero.

In practice, this requirement gives a unique direction, regardless of the length of the crack development.

On the condition that the tangential stress is null, correlated with the expressions of the stress intensity factors, the equation defining the direction after which the crack extension is initiated [4] is (figure 2):

$$\sin\theta + (3\cos\theta - 1)\cdot \mathrm{ctg}\,\alpha = 0 \tag{3}$$

This relationship offers solutions close to the experimentally determined angles for crack initiation angle.

2.2 M-criterion

M-criterion states that the direction of crack initiation coincides with the direction of "maximum stress triaxiality ratio" along a constant radius around the crack tip.

M-criterion can be stated mathematically as [4]:

$$\frac{\partial M}{\partial \theta}=0,\ \frac{\partial^2 M}{\partial \theta^2}<0 \tag{4}$$

where M is the stress triaxiality ratio, defined as:

$$M=\sigma_H/\sigma_{eq} \tag{5}$$

where σ_H is the hydrostatic stress and σ_{eq} is the equivalent stress.

2.3 T-criterion

T-criterion (Theocaris) states that direction of crack initiation coincides with the direction of maximum dilatational strain energy density along the contour of constant distortional strain energy around the crack tip. It uses elastic plastic boundary as given by von Mises flow rule, to define the radius of the core region at the crack tip. In mathematical form, T-criterion can be stated as [4]:

$$\frac{\partial T_V}{\partial \theta}=0,\ \frac{\partial^2 T_V}{\partial \theta^2}<0 \tag{6}$$

$$T_V=\frac{1-2v}{6E}(\sigma_x+\sigma_y)^2 \tag{7}$$

3. Using the S criterion for a crack situated in the substrate of a gear tooth

S-criterion (Sih) is the only one that shows the dependence of crack initiation angle on the material elastic properties denoted by the Poisson's ratio, ν and the state of stress.

S-criterion states that the direction of crack initiation coincides with the direction of "minimum strain energy density" along a constant radius around the crack tip. In mathematical form, S criterion can be stated as [4, 5]:

$$\frac{\partial S}{\partial \theta}=0,\ \frac{\partial^2 S}{\partial \theta^2}>0 \tag{8}$$

where S is the strain energy density factor, defined as:

$$S=r_0\cdot\frac{dW}{dV} \tag{9}$$

where dW /dV is the strain energy density function per unit volume, and r_0 is a finite distance from the point of failure initiation. Using the stress field in Cartesian co-ordinates, the strain energy density function can be written as [4, 5]:

$$S=a_{11}K_I^2+2a_{12}K_IK_{II}+a_{22}K_{II}^2 \tag{10}$$

where

$$a_{11}=\frac{1}{16G}(\chi-\cos\theta)(1+\cos\theta) \tag{11}$$

$$a_{12}=\frac{1}{16G}[2\cos\theta-(\chi-1)]\cdot\sin\theta \tag{12}$$

$$a_{22}=\frac{1}{16G}[(\chi+1)(1-\cos\theta)+(1+\cos\theta)(3\cos\theta-1)] \tag{13}$$

where G is the modulus of rigidity, χ is a constant depending upon stress state and is defined as [4]:

$$\chi = \frac{3-\nu}{1+\nu} \text{ for plane stress and } \chi = (3-4\nu) \text{ for plane strain} \tag{14}$$

From equations (8), (10), (11), (12) and (13) results:

$$\begin{aligned} &2(1+\chi)\varsigma \cdot \mathrm{tg}^4\frac{\theta}{2} + \left[2\chi\left(1-\varsigma^2\right) - 2\varsigma^2 + 10\right]\cdot \mathrm{tg}^3\frac{\theta}{2} - 24\varsigma \cdot \mathrm{tg}^2\frac{\theta}{2} + \\ &+\left[2\chi\left(1-\varsigma^2\right) + 6\varsigma^2 - 14\right]\cdot \mathrm{tg}\frac{\theta}{2} + 2(3-\chi)\varsigma = 0 \end{aligned} \tag{15}$$

and the inequality become:

$$2(\chi-1)\varsigma \cdot \sin\theta - 8\varsigma \cdot \sin 2\theta + (\chi-1)\left(1-\varsigma^2\right)\cdot\cos\theta + 2\left(\varsigma^3 - 3\right)\cdot\cos 2\theta > 0 \tag{16}$$

in which ς is the ratio of the stress intensity factors, $\varsigma = K_I / K_{II}$.

It was considered that the gear tooth is subjected to a rolling contact fatigue. Since the material is a steel, knowing that the value of the Poisson coefficient is $\nu = 0{,}3$, for an easier calculation, the value of the χ constant (eq. 14) depending on the plane stress or plane strain is approximated at the value $\chi = 2$.

With this specification, the components that depend on the angle of the functions a_{ij}, defined by the relations (11), (12), (13) become:

$$a_{11} = (1+\cos\theta)\cdot(2-\cos\theta) = \frac{3}{2} + \cos\theta - \frac{1}{2}\cos 2\theta \tag{17}$$

$$a_{12} = \sin\theta \cdot (2\cos\theta - 1) = \sin 2\theta - \sin\theta \tag{18}$$

$$a_{22} = 2 - \cos\theta + 3\cos^2\theta = \frac{7}{2} - \cos\theta + \frac{3}{2}\cos 2\theta \tag{19}$$

By deriving the above relationships, it can be obtained:

- the first order derivatives:

$$a'_{11} = -\sin\theta + \sin 2\theta \tag{20}$$

$$a'_{12} = 2\cos 2\theta - \cos\theta \tag{21}$$

$$a'_{22} = \sin\theta - 3\sin 2\theta \tag{22}$$

- the second-order derivatives:

$$a''_{11} = -\cos\theta + 2\cos 2\theta \tag{23}$$

$$a''_{12} = -4\sin 2\theta + \sin\theta \tag{24}$$

$$a''_{22} = \cos\theta - 6\cos 2\theta \tag{25}$$

Thus, the equation (8), correlated with (10) whose solution represents the possible values of the crack initiation angle is:

$$K_1^2 \cdot (-\sin\theta + \sin 2\theta) + 2K_1K_2 \cdot (2\cos 2\theta - \cos\theta) + K_2^2 \cdot (\sin\theta - 3\sin 2\theta) = 0 \tag{26}$$

and inequality becomes:

$$K_1^2 \cdot (-\cos\theta + 2\cos 2\theta) + 2K_1K_2 \cdot (-4\sin 2\theta + \sin\theta) + K_2^2 \cdot (\cos\theta - 6\cos 2\theta) > 0 \tag{27}$$

Since in the substrate of the gear tooth the stress field is a compressive one, therefore $K_I = 0$, from (26) the obtained equation is:

$$\sin\theta \cdot (1 - 6\cos\theta) = 0 \tag{28}$$

The only solution that respects the inequation (27) is:

$$\theta = \arccos(1/6) \simeq 80^\circ \tag{29}$$

The above relationship shows that in the case of a crack located in the substrate of the gear tooth (made of steel), the angle at which it extends has a unique value, of about 80 ° (figure 2). It can be noted that this value falls within the range of experimentally determined values [2].

4. The simulation of crack propagation in the substrate, according to the crack initiation angle

A crack situated in the substrate of a spur gear tooth is envisaged. Its length is $2a_0$ and is inclined with an angle of $\alpha_0 \le \pi/2$ (figure 4), or $\alpha_0 > \pi/2$ (figure 5). The coordinates of the crack center are X_{crack0} and Y_{crack0} respectively and is delimited at the extremities by the point TS_0 (the tip of the crack closer to the surface of the tooth) and by the TI_0 (the tip of the crack located inwards) [6].

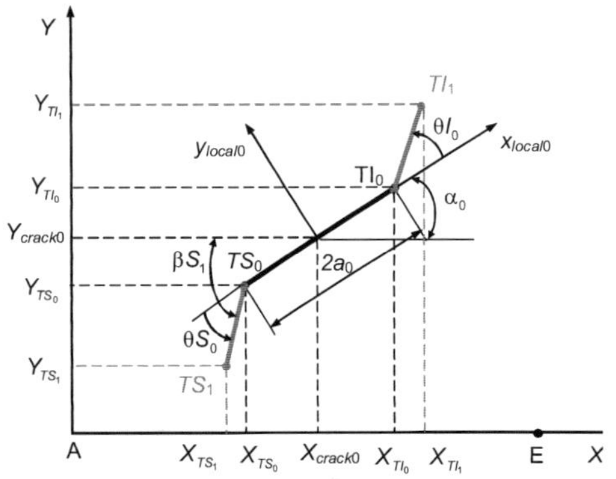

Figure 4. Crack inclined with $\alpha_0 \le \pi/2$

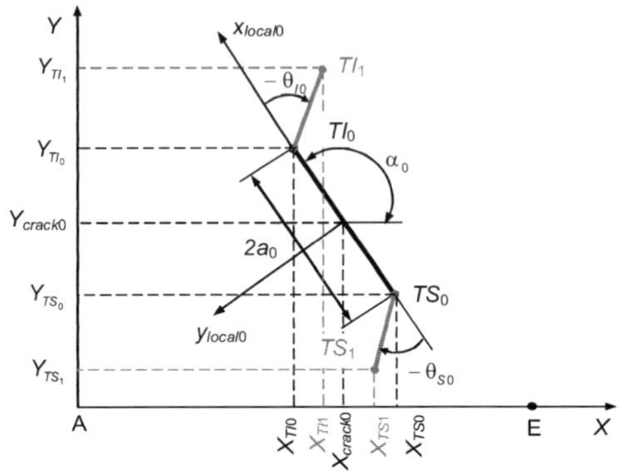

Figure 5. Crack inclined with $\alpha_0 > \pi/2$

Considering that the angle of inclination of the initial crack is $\alpha_0 \le \pi/2$ (figure 4), the coordinates of the end points of the initial crack, TS_0 and TI_0 can be deduced, depending on the coordinates of the crack centre, the angle α_0 and half the length of the initial crack a_0 [6].

$$\begin{cases} X_{TS_0} = X_{crack0} - a_0 \cdot \cos\alpha_0 \\ Y_{TS_0} = Y_{crack0} - a_0 \cdot \sin\alpha_0 \end{cases} \text{and} \begin{cases} X_{TI_0} = X_{crack0} + a_0 \cdot \cos\alpha_0 \\ Y_{TS_0} = Y_{crack0} + a_0 \cdot \sin\alpha_0 \end{cases} \tag{30}$$

If, after the first loop calculation, the initial crack has grown with sloping segments $\lambda_0 = TS_0TS_1 = TI_0TI_1$ inclined with the constant angles $\theta S_0 = \theta I_0 = \theta S_i = \theta I_i = \theta$ and if the crack has not reached the outside of the piece, the co-ordinates of the tips of the new crack configuration TS_1 and TI_1 (respectively) are:

$$\begin{cases} X_{TS_1} = X_{TS_0} - \lambda_0 \cdot \cos(\alpha_0 + \theta S) \\ Y_{TS_1} = Y_{TS_0} - \lambda_0 \cdot \sin(\alpha_0 + \theta S) \end{cases} \text{and} \begin{cases} X_{TI_1} = X_{TI_0} + \lambda_0 \cdot \cos(\alpha_0 + \theta I) \\ Y_{TI_1} = Y_{TI_0} + \lambda_0 \cdot \sin(\alpha_0 + \theta I) \end{cases} \tag{31}$$

If noted:

$$\begin{cases} \beta S_0 = \alpha_0 \\ \beta S_1 = \beta S_0 + \theta S = \alpha_0 + \theta S \\ \ldots\ldots \\ \beta S_{i+1} = \beta S_i + \theta S \end{cases} \tag{32}$$

in the case of crack propagation to the surface with the segment TS_iTS_{i+1} without it reaching the surface, the coordinates of the extremity of the crack can be written, in the general form, as:

$$\begin{cases} X_{TS_{i+1}} = X_{TS_i} - \lambda_i \cdot \cos(\beta S_i + \theta S) \\ Y_{TS_{i+1}} = Y_{TS_i} - \lambda_i \cdot \sin(\beta S_i + \theta S) \end{cases} \tag{33}$$

must be respected

$$Y_{TS_{i+1}} = Y_{TS_i} - \lambda_i \cdot \cos(\beta S_i + \theta S) \ge 0 \tag{34}$$

Similarly, in the case of crack development of the TI_i point one can write:

$$\begin{cases} X_{TI_{i+1}} = X_{TI_i} + \lambda_i \cdot \cos(\beta I_i + \theta I) \\ Y_{TI_{i+1}} = Y_{TI_i} + \lambda_i \cdot \sin(\beta I_i + \theta I) \end{cases} \tag{35}$$

If the crack has reached the surface of the piece, the coordinates of its tip will be:

$$\begin{cases} X_{TS_{i+1}} = X_{TS_i} - Y_{TS_i} \cdot \text{ctg}(\beta S_i + \theta S) \\ Y_{TS_{i+1}} = 0 \end{cases} \tag{36}$$

and the TI point will not propagate anymore.

Using such a simulation program, a propagated crack is presented in figure 6 (the variation of the depth of the crack tip Y_{TS}, according to the abscissa of the same point, X_{TS}, toward the tooth surface).

The obtained shape of the propagated crack is very close to the one presented in figure 3.

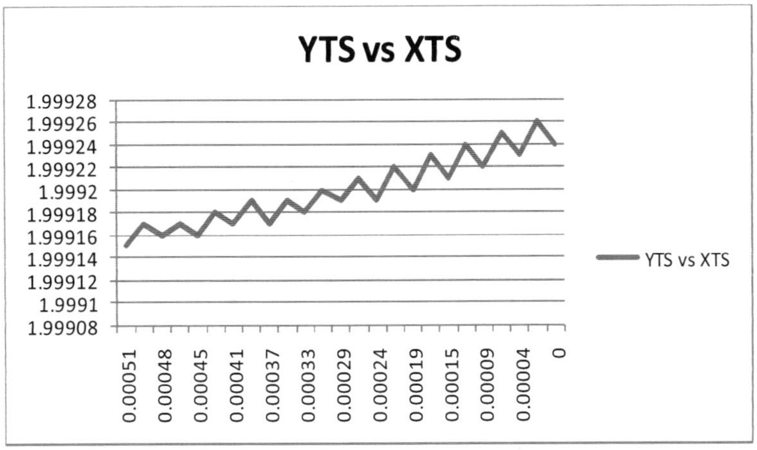

Figure 6. The simulation of the crack propagation in the substrate tooth, according to the S criterion

5. Conclusions

This article aims to highlight the main factors contributing to the propagation of fatigue cracks in the substrate of a gear tooth. From these, the crack initiation angle which leads to crack propagation was detailed. Of the many existing criteria, the S criterion has been customized.

It has been shown that in the substrate of the teeth, due to the fact that there are compression stresses, so stress intensity factor K_I is negligible, a constant value of the initiation angle of $\theta = 80^{\circ}$ was obtained. It is noted that this value falls within the range of experimentally determined values.

Based on this aspects, it was presented how the crack in the substrate propagates, determining the coordinates of the crack tips after each cycle to be requested.

6. References

[1] Anderson T L 2005 *Fracture Mechanics, Fundamentals and Applications*, 3rd Edition, (Boca Raton - CRC Press, Taylor & Francis Group, USA).

[2] Berto F, Lazzarin P, Elices M and Gómez F J 2009 *Proc. Int. Conf. on Crack Paths CP 2009* (Vicenza – University of Padua), 993.

[3] Isaksson P 2001 *On Crack Growth under Compressive Stresses*, Doctoral Thesis, (Lulea - Lulea University of Technology, Sweden).

[4] Khan S M A and Khraisheh M K 2000 Analysis of Mixed Mode Crack Initiation Angles Under Various Loading Conditions, *Engineering Fracture Mechanics*, **67**, **5**, 397-419.

[5] Sih G C 1973 Some basic problems in fracture mechanics and new concepts, *Eng. Fracture Mech.*, **5**, 365–377.

[6] Tudose L M, Popa C O and Jichişan-Matieşan D 2009 *Machine Design - 49th Anniversary of the Faculty of Technical Sciences* (Novi-Sad - Faculty of Technical Sciences), 211.

IOP Conf. Series: Materials Science and Engineering **724** (2020) 012033 doi:10.1088/1757-899X/724/1/012033

Some considerations regarding the influence of working conditions on the corrosion wear of the injection water treatment plant equipment

C Dudu[1], A C Drumeanu[2,3], R G Ripeanu[2] and A Dinita[2]

[1] Construction Department, OMV Petrom, Romania
[2] Mechanical Engineering Department, Petroleum-Gas University of Ploiesti, Romania

Abstract. One of the methods used for the secondary exploitation of oil deposits is the water injection. The injected water is obtained after the separation of the well extracted fluid that contains a mixture of crude oil, water, gases and solid particles. Prior to injection, water must meet the following requirements: a high degree of purity obtained thought the lowest possible mechanical and oil content; lower degree of aggressiveness (corrosion); the highest stability; compatibility with the fluids and minerals in the reservoir formation; low oxygen content and bacteria. The fulfilment of these requirements must be ensured by the water treatment plant whose equipment is subjected to corrosion wear. The equipment degradation are occurring mainly at pipelines and connecting elements of the flotation skid, in the form of advanced corrosion wear because the water injection contains an inconsistent composition mixture of dissolved gases, hydrocarbons, solid particles. The paper presents the regression analyses that emphasize the influence of chemical water composition and fluid speed on the corrosion rate of the different equipment parts. The corrosion rates were obtained using metallic samples according with the NACE SP0775-2013 standard that were placed in different points of the plant. The paper results can be used both at the water plant exploitation and also to next studies concerning the effectiveness of the corrosion inhibition on the corrosion rate.

1. Introduction

The exploitation of oil deposits presumes at least two stages. In the first stage, so called primary exploitation, the fluid that consists in a mixture of crude oil, water, gases and suspended solids is pushed out from the subterranean deposits to the surface under the effect of deposit internal pressure. After the subterranean pressure is exhausted, the fluid is extracted by pumping using different types of pumps. This stage is called secondary exploitation. One of the methods used for displacement the crude oil from the rock of the deposit is water injection. On both stages, the fluid mixture from the wells is directed to the collection park where the constituent phases are separated [1]. It has to be mentioned that under the exploitation conditions in Romania, the water from deposits represents on average about 60 ... 70% of the total amount of fluid extracted from the wells.

After separation from the other components of the fluid (crude oil and gas) on a plant called Separation Park, the water is sent to the treatment plant where it is treated to meet the following

[3] E-mail: drumeanu@upg-ploiesti.ro

conditions [2]: a high degree of purity obtained thought the lowest possible mechanical and oil content; lower degree of aggressiveness (corrosion); the highest chemical stability; compatibility with the fluids in the subterranean deposit; compatibility with the minerals in the subterranean oil deposit; low oxygen content and bacteria.

The water resulting from separation and treatment is sent to specially prepared tanks from where is injected into specially prepared wells (technological water) in secondary exploitation process or is eliminated by injection (waste water) in specially designed wells [2].

2. Injection water treatment plant

Factors that are taken into account when an injection water treatment system it is chosen are related to the distances to sources and injection points and also to the possibilities of location on the ground so that gravitation can be used in the process of pumping water.

The treatment and injection systems (open, closed or semi-closed systems) are selected according to the characteristics of the injection water.

The influence of working conditions on the corrosion wear was determined in different points of a water treatment plant that are considered to belong at the last generation of this kind of equipment.

A modern water treatment plant (figure 1), is composed from a chemical conditioning kit, a dissolved gas flotation unit (SPINSEP) and a nutshell filter (MONOSEP) [3].

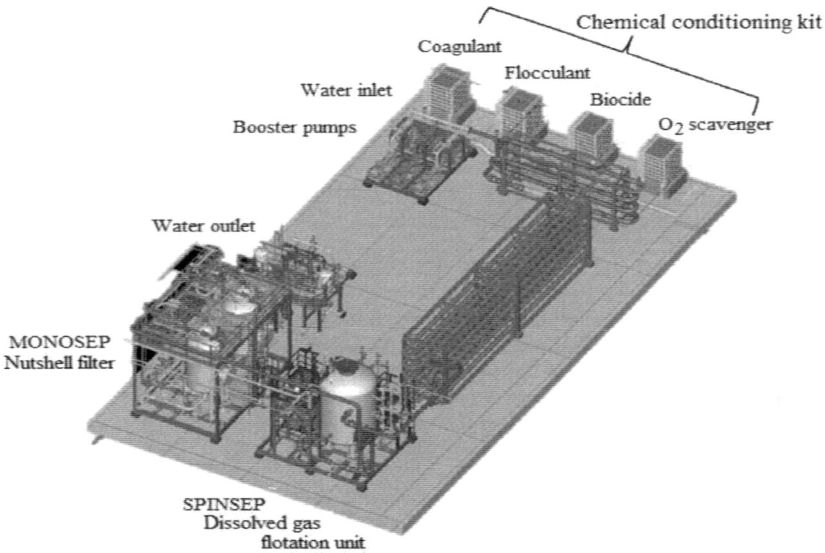

Figure 1. Water treatment plant design [3].

The water that was separated from crude oil and gas is pumped from storage tanks to the treatment plant, with three centrifugal water transfer pumps (booster pumps). In the first stage the water is treated in the chemical conditioning subsystem by injection with flocculants and coagulant substances and, in addition, with an oxygen inhibitor to remove dissolved oxygen in water and a biocide to remove bacteria. After this treatment, the water is introduced in the SPINSEP flotation system where the finest crude oil drops are separated, and finally in the nutshell filter MONOSEP where it carried out the finest ultimate filtration [3].

The main degradations occurred at the equipment in the water treatment plants were recorded mainly in the pipes and the connecting elements of the SPINSEP gas flotation skid, in the form of their advanced corrosion. Locations of the these degradations were found at bends and elbows where the direction of the pipes is changed to 90° in order to ensure a turbulent tangential flow of the fluid at the entrance to the flotation vessel, at section changes of the pipes, to the connecting elements (flanges, threads) of the pipes of the gas flotation skid and at the welding joints [4], [5], [6].

3. Experimental conditions

The corrosion rates were obtained using metallic samples (corrosion coupons) [7] according with the NACE SP0775-2013 standard [8] that were placed in different points of the water circuit between the separation parks and the outlet of the treatment plant (figure 2).

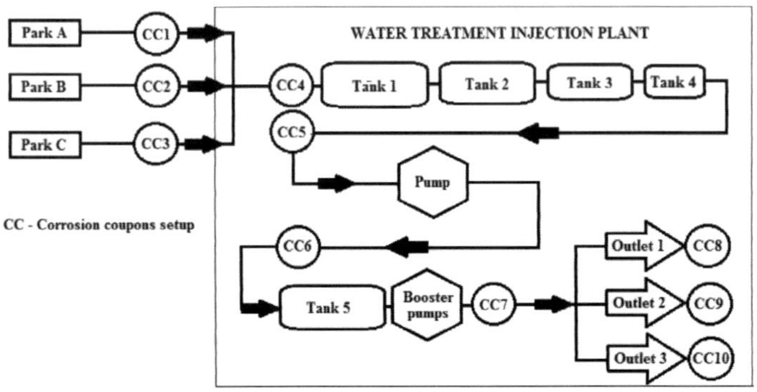

Figure 2. The setup points of corrosion coupons.

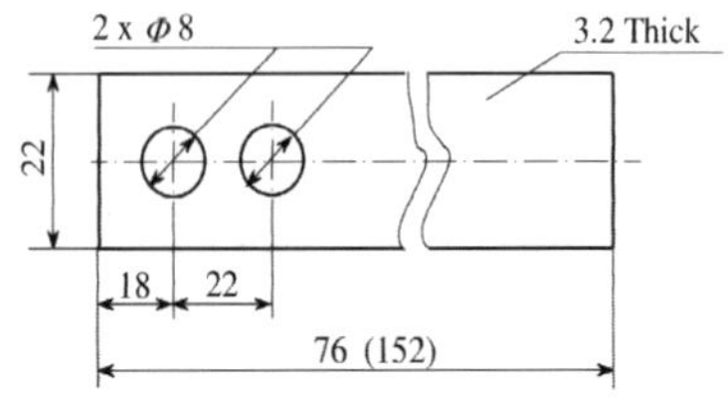

Figure 3. The sample (corrosion coupon) piece [10].

There were used strip coupons (figure 3) fixed in every setup point in pairs [8].

Coupons were made of the same steel, P285NH type, with the pipe elements used in the water treatment plant. Chemical composition of P285NH steel is presented in table 1.

Table 1. Chemical composition (wt.%) of steel P285NH (1.0477).

	C	Si	Mn	Ni	P	S	Cr	Mo	V	N
EN 10222-4	max 0.18	max 0.4	0.6 - 1.4	max 0.3	max 0.025	max 0.015	max 0.3	max 0.08	max 0.05	max 0.02
Experimental	0.189	0.23	0.908	0.006	0.007	0.003	0.02	0.001	0.003	-

Coupons were processed without affecting the structure of the material with a water jet cutting machine and subsequently by grinding with low intensive parameters using cooling fluids. Finally the surface exposed to corrosion test medium was polished with abrasive paper of 500 Mesh gradually [8].

Exposure time must be considered when interpreting corrosion coupon data. Short-term exposure (15 to 45 days) provides quick answers but may give higher corrosion rates than long-term exposures. Aggravating conditions, such as bacterial fouling, may take time to develop on the coupon. Short exposure times may be advantageous when evaluating inhibitor effectiveness. Longer exposures (60 to 90 days) are often required to detect and define pitting attack. Because exposure time affects test results, exposure periods should be as consistent as practical [9], [10]. Taking account of the above statements, the corrosion coupons used for determining corrosion rate were exposed for 90 days of water plant functioning.

For emphasizing the influence of injection water on corrosion rate were determined the chemical and electrochemical characteristics of water (table 2) in each point where the corrosion coupons were located (figure 2).

In the same time with the exposure of corrosion coupons to effect of injection water, it was determined the reducing of pipes wall thickness. Determination of wall thickness was done for different pipes diameter, especially at the elbows, where the erosion phenomenon is amplified due of the flow regime changed.

Table 2. Chemical and electrochemical characteristics of water for corrosion coupons location place.

Characteristic	Location place of the corrosion coupons									
	1	2	3	4	5	6	7	8	9	10
Density, g/cm^3	1.0407	1.0516	1.0465	1.0482	1.0495	1.0505	1.051	1.0494	1.0493	1.0495
pH	6.71	6.69	6.65	6.39	6.52	6.55	6.57	6.49	6.35	6.52
Conductivity, µs/cm	81800	96600	93000	94600	94400	95800	99100	93900	92800	92300
CO_2, mg/l	44	79	88	114	101	141	110	97	150	150
H_2S, mg/l	0.31	0.07	0.04	0.49	0.14	0.17	1.35	0.14	0.17	0.17
Chlorides, mg/l	34330	45720	39440	42600	43980	44760	44450	44120	43900	44220

4. Results and discussions

According with the NACE SP0775-2013 standard, the calculation of the average corrosion rate expressed as a uniform rate of thickness loss per unit time in millimeters per year or millimeters per annum (mm/y or mm/a), is shown in equation (1) [8]:

$$CR = \frac{W \cdot 365 \cdot 1000}{A \cdot T \cdot D} = \frac{3.65 \cdot 10^5 \cdot W}{A \cdot T \cdot D} \quad (1)$$

where: *CR* is average corrosion rate, millimeters per year (mm/y or mm/a); *W* – mass loss, grams (g); *A* – initial exposed surface area of coupon, square millimeters (mm^2); *T* – exposure time, days (d); *D* – density of coupon metal, grams per cubic centimeter (g/cm^3).

Corrosion rate was determined, using equation (1), for coupons placed in the points shown in figure 2. Regarding the influence of water characteristics (table 2) on corrosion rate it can be done the following remarks:

- the highest variations of water characteristics were registered for concentration of CO_2, H_2S and chlorides;
- the other parameters like conductivity, pH and concentrations of Ca, Mg, Ba, Na, Fe, K, Li do not registered major variations that can influenced on corrosion rate.

Taking account of above remarks, there were done regression analyses that evidencing the influence of CO_2, H_2S and chlorides concentration on corrosion rate.

Thus, the influence of CO_2 on the corrosion rate is modeled by equation (2):

$$CR = -15.514 \cdot \ln(CO_2) + 3.185 \cdot (CO_2)^{0.5} + 39.652 \quad (2)$$

where: *CR* is the corrosion rate (mm/year); CO_2 – concentration of carbon dioxide (mg/l).

The dependence expressed by equation (1) is in accordance with experimental results presented in figure 4. The determination coefficient for regression analysis expressed with equation (2) is R^2 = 0.98.

Similarly, the dependence between corrosion rate and hydrogen sulphide concentration is shown in figure 5, and is emphasized by equation (3) obtained after a regression analysis.

$$CR = 5.6 \cdot \ln(H_2S + 0.614) + 2.021 \quad (3)$$

where: *CR* is the corrosion rate (mm/year); H_2S – concentration of hydrogen sulphide (mg/l).

The determination coefficient for regression analysis equation (3) is R^2 = 0.978.

The influence of chlorides concentration on corrosion rate is shown in figure 6, and is evidenced with the regression analysis equation (4).

$$CR = -0.232 \cdot \ln(Cl - 34.327) + 0.751 \quad (4)$$

where: *CR* is the corrosion rate (mm/year); *Cl* – concentration of chlorides (g/l).

The determination coefficient for regression analysis equation (4) is R^2 = 0.968.

The reducing of pipes wall thickness as the effect of corrosion and the erosion of the water was determined for different flow speeds of the fluid (figure 7). The flow speeds were calculated for different sections values of the pipes considering the constant flow of 35 m^3/hour value.

IOP Conf. Series: Materials Science and Engineering **724** (2020) 012033 doi:10.1088/1757-899X/724/1/012033

The influence of flow speed on the reducing of pipe wall thickness can be expressed with equation (5) obtained as a result of a regression analysis.

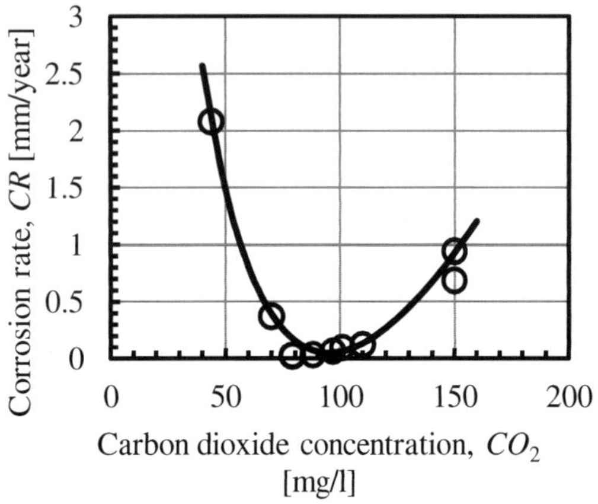

Figure 4. The dependence between corrosion rate and carbon dioxide concentration.

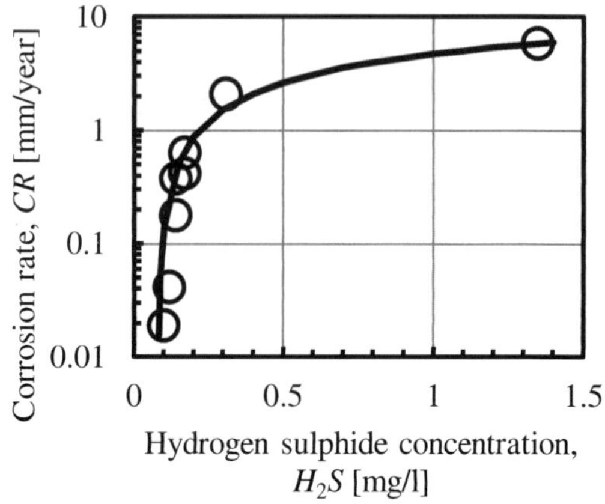

Figure 5. The dependence between corrosion rate and hydrogen sulphide concentration.

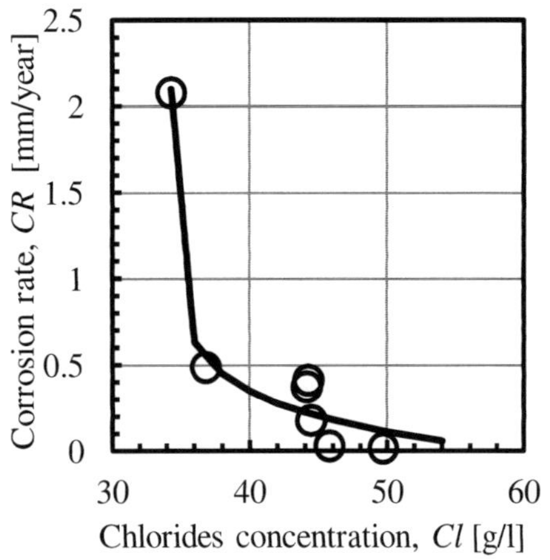

Figure 6. The dependence between corrosion rate and chlorides concentration.

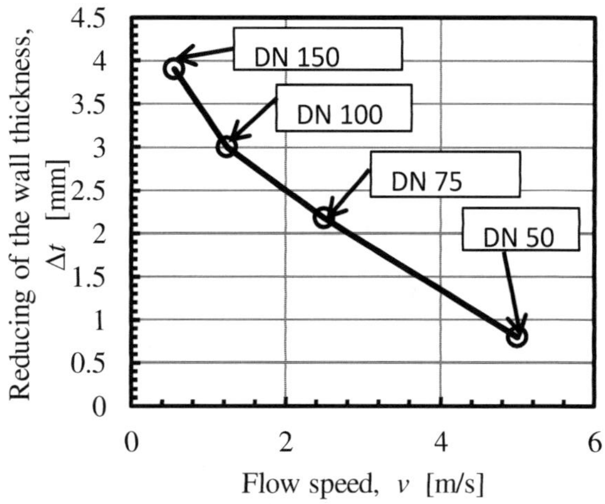

Figure 7. The influence of flow speed and nominal diameter of the pipes (mm) on reducing of the wall thickness.

$$\Delta t = 4.819 \exp(-0.353 v) \tag{5}$$

where: Δt is the reducing of pipe wall thickness (mm); v – the flow speed (m/s).

The determination coefficient for regression analysis equation (5) is $R^2 = 0.9929$.

About the above presented experimental results the following observations can be done:

- minimum corrosion rates were registered for CO_2 concentrations in the range of 80 … 110 mg/l (figure 4) that corresponding of the corrosion coupons placed in points 5, 6, 7, (figure 2 and table 2), while the maximum values correspond to minimum and maximum CO_2 concentrations;
- maximum corrosion rate value corresponding to minimum CO_2 concentration is in accordance with the flow speed (figure 7) that has minimum value for points 1, 2, 3 (figure 2);
- the increase of H_2S concentration implies the corrosion rate fast increase (figure 5) especially in the range of 0.04 … 0.3 mg/l, the highest value being registered for point 7 corresponds to a pipe diameter of 150 mm and to a low flow speed (figure 7);

- flow speed influences, also the effect of chlorides on the corrosion that is reduced for high chlorides concentrations (figure 6) corresponding to highest flow speed while for the low values of flow speed and chlorides concentration the corrosion rate has the maximum value (figure 7);
- determination coefficients of regression analyses indicate a high level of correlation between the experimental data and analytic results obtained with equations (2), (3), (4) and (5);
- flow speed value has a major influence both on reducing of the pipe wall thickness and corrosion produced by CO_2 and chlorides that registered low values for the highest flow speeds.

5. Conclusions

Effectiveness exploitation of a water treatment plant presumes not only the knowledge of the corrosion rate of equipment but also the influence of working conditions on the corrosion.

In the water treatment plant, the pipes and the connecting elements like elbows are the parts most damaged by corrosion and erosion phenomena.

The main conclusions of the paper are the following:

- in the case of CO_2 and chlorides, the high flow speeds that are associated with a turbulent regime, implies the corrosion rate decrease, while for the H_2S this is not influenced by flow regime;
- the equations obtained as the result of regression analyses can be used both for chemical conditioning of the water and establishing a proper flow regime;
- because the damage of the pipes are produced by corrosion and erosion phenomena, for separating the effects of these components, the corrosion rate has to be determinates by electrochemical techniques;
- for a better exploitation of water treatment plant it is imposing determination the influences of corrosion inhibitors, flocculants, biocides, coagulants and oxygen inhibitor on corrosion rate.

References

[1] Thomas S 2008 Enhanced Oil Recovery – An Overview *Oil Gas Sci Technol* **63** 9
[2] Reynolds R R 2003 *Produced Water and Associated Issues* (Petroleum Technology Transfer Council)
[3] Dudu C, Ripeanu R G, Drumeanu A C, Dinita A, and Lospa A M 2019 Some considerations regarding the influence of working conditions on the corrosion wear of the injection water treatment plant equipment *Xth Product design, robotics, advanced mechanical & Mechatronic Systems and Innovation Conf. PRASIC'18* IOP Conf. Series: Materials Science and Engineering (Brasov: Transilvania University of Brasov) **514** (2019) 012008
[4] Ionescu O N, Ionescu G C, Minescu M and Nae I 2013 Research on life prolongation for temperature sensors placed in highly erosive environment *J Balk Tribol Assoc* **19** 214
[5] Papavinasam S 2014 *Corrosion Control in the Oil and Gas Industry* (Gulf Professional Publishing)
[6] Heidersbach R 2011 *Metallurgy and Corrosion Control in Oil and Gas Production* (New York: Wiley)
[7] Roberge P R 2000 *Handbook of Corrosion Engineering* (New York: Mc.Graw-Hill)
[8] NACE SP0775 2013 *Standard Practice, Preparation, Installation, Analysis, and Interpretation of Corrosion Coupons in Oilfield Operations* (NACE International)
[9] Byars G H 1999 *Corrosion Control in Petroleum Production* (Houston Texas: TPC Publication 5 NACE International)
[10] Dawson J, John G and Oliver K 2010 *Management of Corrosion in the Oil and Gas Industry* (Elsevier B.V.)

IOP Conf. Series: Materials Science and Engineering **724** (2020) 012034 doi:10.1088/1757-899X/724/1/012034

Considerations regarding determination of optimal preventive maintenance periodicity for subsurface sucker rod pumps

A C Drumeanu[1,2]

[1] Mechanical Engineering Department, Petroleum-Gas University of Ploiesti, Romania

Abstract. The subsurface sucker rod pumps are important equipment used for the secondary oil recovery. In their structure, there are friction couples, which are strongly subjected at wear like barrel-plunger, traveling valve and standing valve. The operating durability of subsurface sucker rod pumps is practically determined by the reliability of these couples depending on the work conditions.
The paper presents a case study on the optimization of preventive maintenance of the subsurface sucker rod pumps. The optimal periodicity, of the preventive maintenance of the above mentioned friction couples, is determined on the basis of their operating reliability. There were analyzed two types of preventive maintenance, at fixed time, and at fixed age respectively. The paper results are embodied in practical charts which can be used for choosing optimal preventive maintenance period of the subsurface sucker rod pumps.

1. Introduction

The deep-well pumping system is used for the oil wells exploitation. In our country, this system is used at least for 75 % of the oil wells. The deep-well pumping exploitation is applied in a finally phase, when the natural energy of the stratum decreased and its value is insufficiently for surface oil ascending, or in the case when the artificial lifting can't be done). The efficiency of the deep-well pumping system is influenced by the operating life of the pumps, and in the case of subsurface sucker rod pumps, by the consumption of the high wear repair parts (barrels, plungers and valves [1, 2].

Subsurface sucker rod pumps operating life depends on: the service conditions (the depth of the hole, the aggressiveness of the pumped fluids, the pumping parameters); the subsurface sucker rod pumps construction (the type of the pump, the materials used at pumps construction, the quality of the manufacture etc.); the maintenance activities. Regarding the subsurface sucker rod pumps operating life, it can be done the remark that this is determined mainly by the high wear couples which are: barrel-plunger, standing valve and traveling valve. The failure of these couples implies the subsurface sucker rod pump replacement [3, 4, 5].

Taking into account of the above statements, it can be concluded that it is necessary to establish a subsurface sucker rod pump optimum maintenance program that can reduces at maximum the operating expenses. One of the maintenance strategies that achieve this goal is preventive maintenance based on operational reliability [6].

The paper aim is to present a general methodology to determine the optimal duration of preventive maintenance as a function of the component replacing cost. This methodology is exemplified for the

[2] E-mail: drumeanu@upg-ploiesti.ro

high wear couples above mentioned that are the most often replaced components. The paper results are embodied in practical charts which can be used for choosing optimal preventive maintenance period of the subsurface sucker rod pump.

2. Construction and operation of the sucker rod pumps

The considered subsurface sucker rod pump in this study is 25-175-RHAC 20-8-2 type and it equips the oil-wells of the Ciuresti oil field. According API Specification 11 Ax standard [7], the 25-175-RHAC 20-8-2 subsurface sucker rod pump main characteristics are the following:

- nominal tubing size 73.0 mm (2 $^7/_8$ in);
- basic bore diameter 44.5 mm (1 ¾ in);
- type of the pump: rod, stationary heavy wall barrel, top anchor, introduced in oil well with the sucker rods;
- barrel length 6.096 m (20 ft);
- plunger length 2.438 m (8 ft);
- total length of extensions 0.610 m (2 ft);
- standing and traveling valves - globe valve type;
- component parts of the valves (the valve seat and the ball) made from anticorrosive stainless steel AISI 410-440, steel type X40Cr14 (UNI 6901) respectively X38Cr13 (DIN 17440) that have the chemical composition (wt.%): C - (0.35 .. 0.44), Si - max 0.6, Mn - max 0.6, Cr - (12.0 ... 14.0) [3];
- hardness of the valve components, ball – Rc 58-65, respectively seat – Rc 52-56;
- pump barrel made from steel chrome plated, thickness of coating 0.076 mm, hardness of coating Rc 67-71, base material 10XX steel, base core hardness Rb 90-Rc 23;
- pump plunger coated with spray metal, thickness of the coating 0.254 mm, hardness of the coating Rc 58, base material 1018-1045 steel, base core hardness Rb 70-Rc 23.

The construction and working principle of the subsurface sucker rod pump is presented in figure 1. In this figure, it can be observed the high wear couples of the pump: barrel-plunger, standing valve and traveling valve.

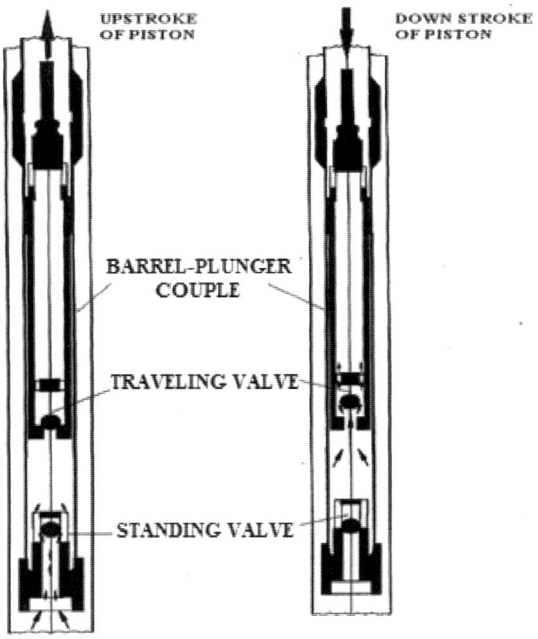

Figure 1. The construction and the working principle of the subsurface sucker rod pump.

Figure 2. Some examples concerning the working damage of valves.

The pump working is characterized by two strokes: one for the upward of the plunger and the other for the downward of the plunger. At the upstroke of the plunger, the standing valve is opened and the traveling valve is closed. Thus, the fluid enters in pump and the fluid, which is situated inside of the plunger, is pushed up. At the down stroke of the plunger, the standing valve is closed and the traveling valve is opened. Thus, the fluid enters in piston chamber because of the pressure, which appears in the pump chamber.

The subsurface sucker rod pumps work in different agents who are characterized by the fluid properties (the content of impurities, the corrosiveness etc.) [2, 3, 4]. In figure 2 there are represented some examples concerning the damage of the valves. From this figure, it can be observed that valves can be damaged by erosion and abrasion. Also, the wear of barrel-plunger couple is produced because of erosion and abrasion [3, 4]. The wear of barrel-plunger and valves has like effect the decrease of the flow and implicitly replacement of these [6].

3. Reliability modeling of barrel-plunger and valves of the subsurface sucker rod pump

The reliability modeling of the high wear couples of the subsurface sucker rod pumps was done using the data concerning the operating durability of these. So, there were taken into consideration the data, which were obtained, from 7 subsurface sucker rod pumps used in the oil field Ciuresti. These pumps are characterized by the following characteristics concerning the construction and the operating:

- type of the pumps 25-175-RHAC 20-8-2;
- anchorage depth between 1900 and 2600 m;
- flow rate in the range of 5 … 30 m^3/24 hours;
- double strokes number in the range of 8 … 10 double strokes/minute;
- operating durability between replacements of considered friction couples in the range of 36 … 213 days;
- type of the agent – hybrid (mixed).

The operating durability registered for the considered couples are presented in table 1.

Table 1. Working durability of high wear couples of subsurface sucker rod pumps.

Couple	Durability (days)										
Barrel-Plunger	36	44	48	81	86	111	123	125	190	213	
Standing valve	36	48	106	107	111	123	167	169	190		
Traveling valve	36	44	50	54	75	88	113	123	125	159	190

For modeling the survival processes which characterize subsurface sucker rod pumps couples, it was used the Weibull partition law with two parameters [8, 9]. The expression of the Weibull law indicators are:

- the probability density:

$$f(t,\eta,\beta)=\frac{\beta}{\eta}\left(\frac{t}{\eta}\right)^{\beta-1}\exp\left[-\left(\frac{t}{\eta}\right)\right]^{\beta} \tag{1}$$

- the reliability:

$$R(t)=\exp\left[-\left(\frac{t}{\eta}\right)\right]^{\beta} \tag{2}$$

- partition function

$$F(t)=1-\exp\left[-\left(\frac{t}{\eta}\right)\right]^{\beta} \tag{3}$$

- failure rate

$$z(t) = \frac{\beta}{\eta}\left(\frac{t}{\eta}\right)^{\beta-1} \tag{4}$$

- mean time between failures

$$MTBF = \eta \cdot \Gamma\left(\frac{1}{\beta}+1\right) \tag{5}$$

where $\Gamma\left(\frac{1}{\beta}+1\right)$ is first species Euler function

Analytical determination of β and η parameters was done using the method of least square fit. In table 2 there are presented the values of Weibull parameters.

Table 2. Parameter values of Weibull law for subsurface sucker rod pumps couples.

Couple	Weibull law parameter values		Determination coefficient r^2	*MTBF* (days)
	β	η		
Barrel-Plunger	1.8	123	0.968	109
Standing valve	1.75	154	0.988	137
Traveling valve	2.0	108	0.976	204

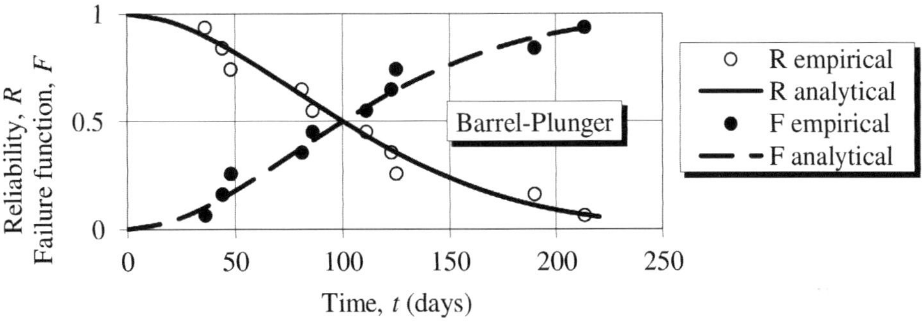

Figure 3. Empirical and analytical reliability and failure function for barrel-plunger couple.

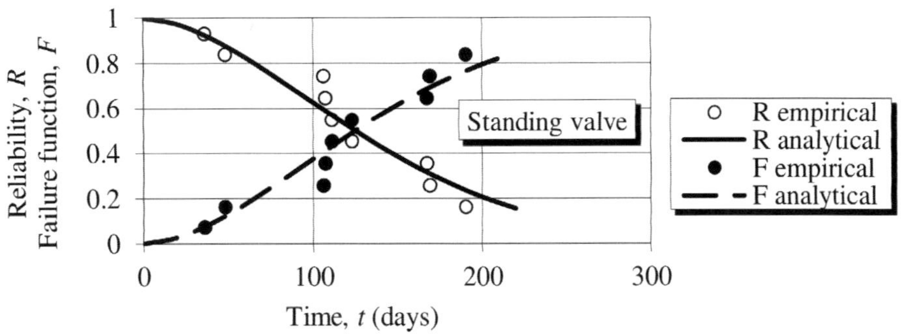

Figure 4. Empirical and analytical reliability and failure function for standing valve couple.

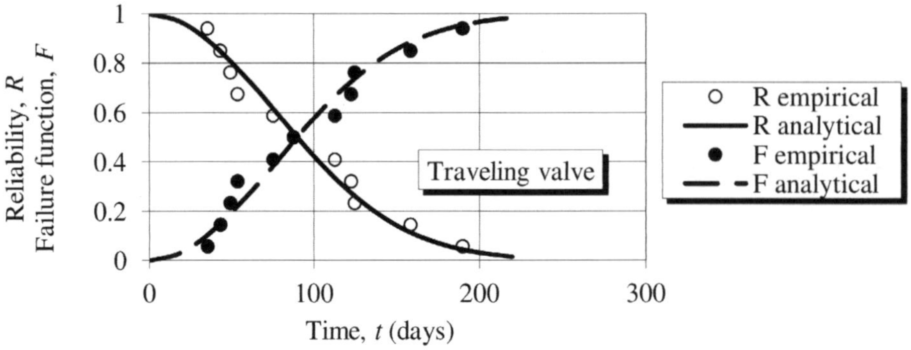

Figure 5. Empirical and analytical reliability and failure function for traveling valve couple.

Using calculated values of η and β parameters there were plotted the diagrams for reliability and failure function (figure 3, figure 4 and figure 5), failure rate (figure 6) and probability density (figure 7). The obtained results concerning the Weibull law parameters and the presented diagrams from above mentioned figures will be used for determination of preventive maintenance periodicity.

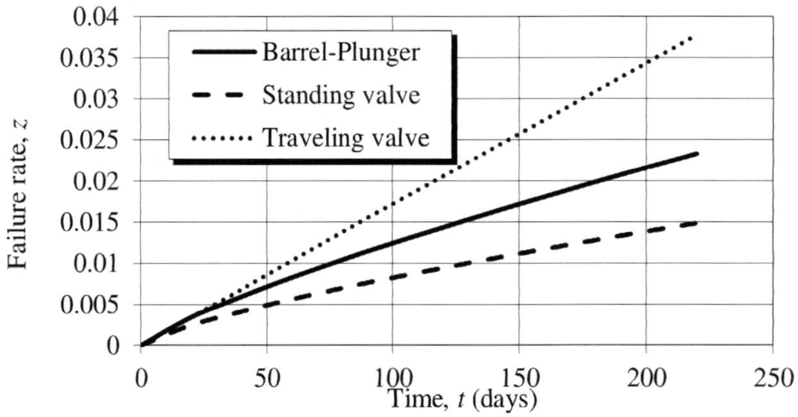

Figure 6. Weibull law failure rate for high wear couples of subsurface sucker rod pumps.

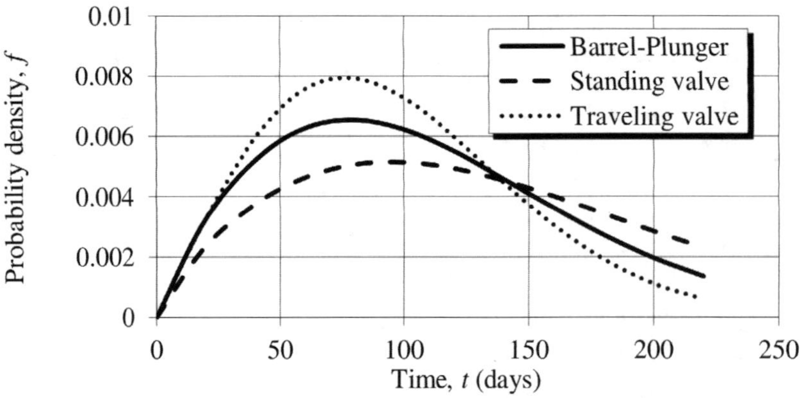

Figure 7. Weibull law probability density for high wear couples of subsurface sucker rod pumps.

4. Optimal preventive maintenance periodicity determination

One of economic problems that can be solved by knowing of operating reliability is the determination of the optimal preventive maintenance periodicity. This is justified from economic considerations when the following two conditions are met:

- the preventive and planed replacement of a device at fixed date has to cost less than the random replacement made when the device fails;
- the failure rate of the programmed preventive maintenance device has to be increasing.

Further will be used the following notations: c_1 – replacement cost for a good condition device; c_2 – cost of an unexpected failure ($c_2 > c_1$); T – the period when the device attains T age and it is replaced with a new one; $f(t)$ – the probability density; $R(t)$ – the reliability function.

There are two possible maintenance policies, which will be further presented. These policies have to be considered since the product design, and the technical documentation has to contain the maintenance indicators and their determination procedure in addition to reliability indicators [5].

4.1. Maintenance at fixed time

In this case all devices are replaced with a periodicity T, whatever their age. In other words, even if at $(T - \varepsilon)$ moment a device was individually replaced, at the fixed time will be replaced all devices again.

It presumes being replaced, then new, all devices at the 0 moment and it is calculated mathematical expectation (average) of the device replacement cost at the T moment and for a cost c_1. Then, the probability for having one failure in time T, with $c_1 + c_2$ cost is:

$$\int_0^T F(t_1)R(T-t_1)\mathrm{d}t_1 \tag{6}$$

Probability for two failures with $c_1 + 2c_2$ cost is:

$$\int_0^T F(t_1)\mathrm{d}t_1 \int_{t_1}^T F(t_2-t_1)R(T-t_2)\mathrm{d}t_2 \tag{7}$$

Continuing the reasoning, the cost replacement average for a device is:

$$M(c) = c_1 + c_2\left\{\int_0^T F(t_1)R(T-t_1)\mathrm{d}t_1 + 2\int_0^T F(t_1)\mathrm{d}t_1\int_{t_1}^T F(t_2-t_1)R(T-t_2)\mathrm{d}t_2 + \ldots\right\} \tag{8}$$

An optimistic account, which presumes that the global replacements frequency is well chosen, so that never exist more than two consecutive replacements, give a unitary cost equal with:

$$K = c_1 + \left[1 - R(T)\right] c_2 \tag{9}$$

This corresponds at a price for unit time and device:

$$c = \frac{c_1 + \left[1 - R(t)\right] c_2}{T} \tag{10}$$

Deriving relation (10) and equalizing with zero it can be determined the minimum cost which corresponds for an optimum replacement time T^*, then:

$$R\left(T^*\right) + Tf\left(T^*\right) = \frac{c_1 + c_2}{c_2} \tag{11}$$

The T^* value can be obtained using a graphic method (figure 8) which consists in searching minimum value applying relation (11).

For drawing the curves presented in figure 8 there are used the reliability expressions previous determined. The curves were plotted for a time range of 0 – 80 days, and they can be used for determining of ratio $(c_1 + c_2)/c_2$. Has to be mentioned that the maximum value of this ratio is 2 that corresponding to situation when $c_1 = c_2$, and the minimum value (theoretical) is 1.

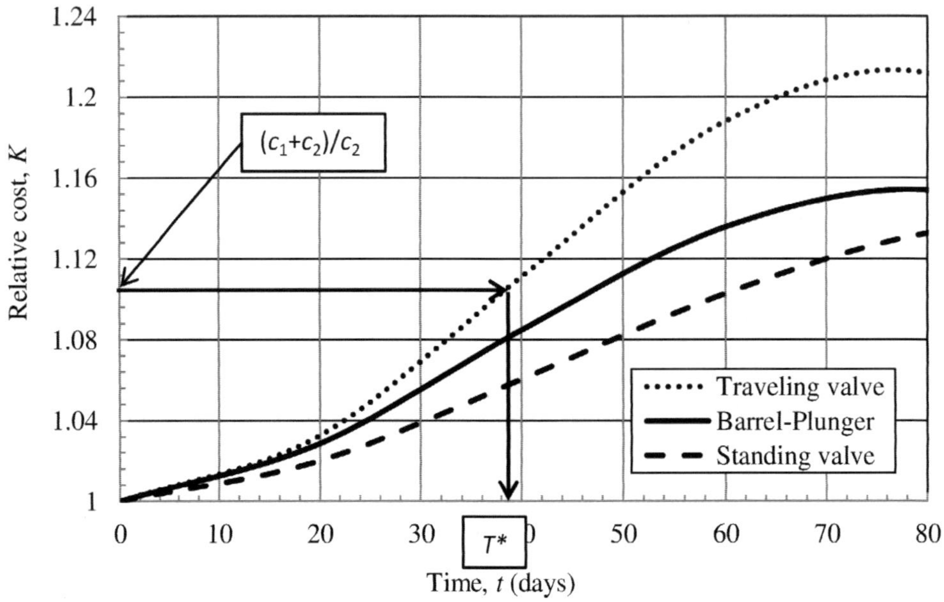

Figure 8. Maintenance at fixed time for considered couples of the subsurface sucker rod pump.

4.2. Maintenance at fixed age

The second case of maintenance is characterized by a continuous surveillance of the each element (device) age, so this is replaced when attains the age *T*. From economic reasons, in comparison with the previous case, the differences are the following [5]:

- the number of elements replaced (devices) is reducing, because if one is replaced like being defect, it will not be replaced automatically in a preventive way at a fixed time (that could be immediately after - the remedial of defect), just in the moment when attains the age *T*;
- the device's age has to be known, this implies the existence of a special monitoring;
- replacements at a fixed age are more expensive because they are done only for a single device at a time.

The calculus can be made in a manner that put in evidence the difference between the two types of maintenance. Whether:

$$\frac{1}{M(t)} = \frac{1}{\int_0^T R(t)\mathrm{d}t} = \frac{1}{MTBC} + \frac{1}{MTBP} \tag{12}$$

where:

$$MTBC = \frac{\int_0^T R(t)\mathrm{d}t}{1 - R(t)} \tag{13}$$

is the average interval between corrective maintenances and

$$MTBP = \frac{\int_0^T R(t)\mathrm{d}t}{R(t)} \tag{14}$$

is the average interval between preventive maintenances.

The cost per unit time is:

$$K_u = \frac{c_1}{MTBP} + \frac{c_2}{MTBC} \tag{15}$$

so:

$$K_u = \frac{c_1 R(T)}{\int_0^T R(t)\mathrm{d}t} + \frac{c_2 F(T)}{\int_0^T R(t)\mathrm{d}t} \tag{16}$$

with $F(T) = 1 - R(T)$.

The determination of K_u value is done, also, in a graphic manner.

In figure 9, figure 10 and figure 11 are plotted the curves corresponding on concrete operating conditions of the barrel-plunger, standing valve and traveling valve respectively. Has to be mentioned that in above mentioned figures are plotted many curves that corresponding to different values of ratio c_1/c_2, because, as been stated, generally is difficult to establish a unique value of this ratio.

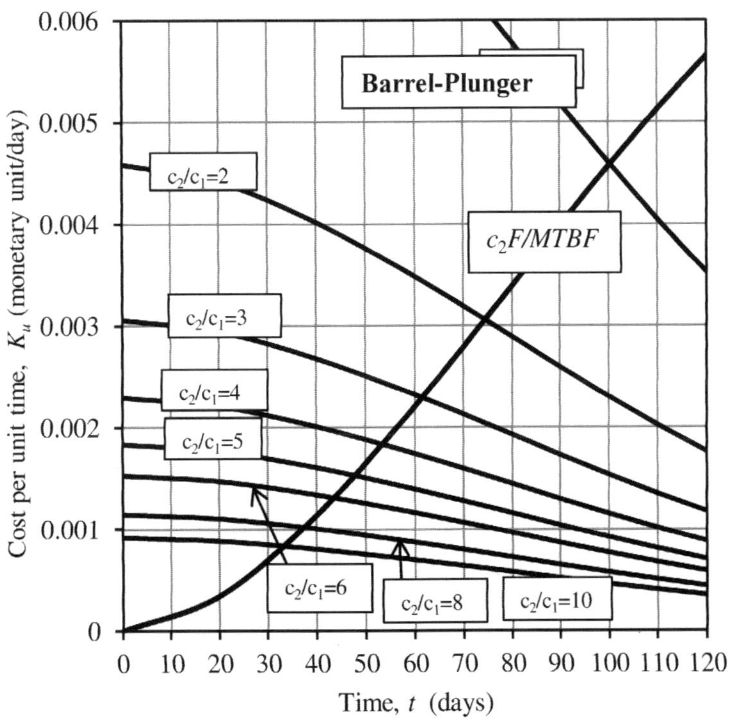

Figure 9. Graphical determination of minimum cost for maintenance at fixed age for barrel-plunger and for different values of ratio c_2/c_1.

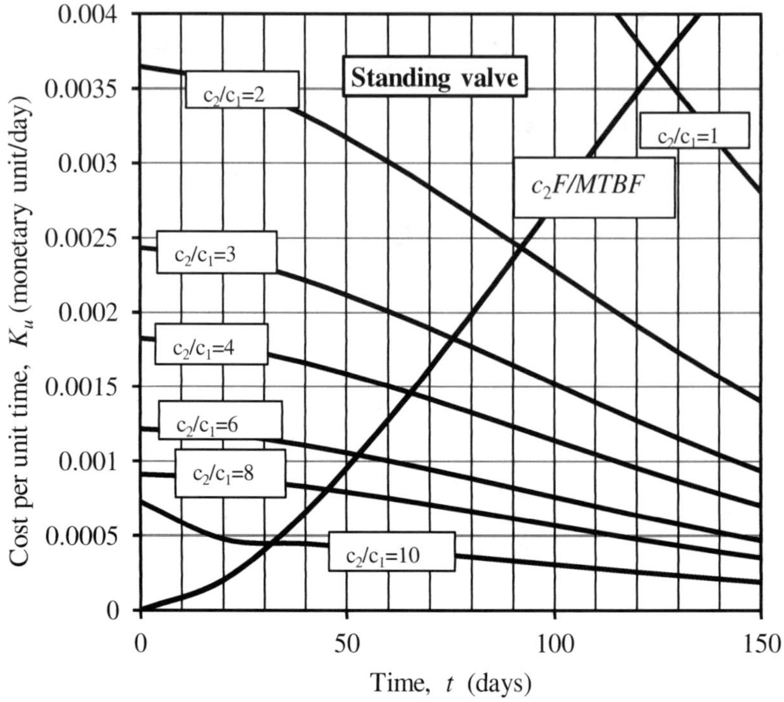

Figure 10. Graphical determination of minimum cost for maintenance at fixed age for standing valve and for different values of ratio c_2/c_1.

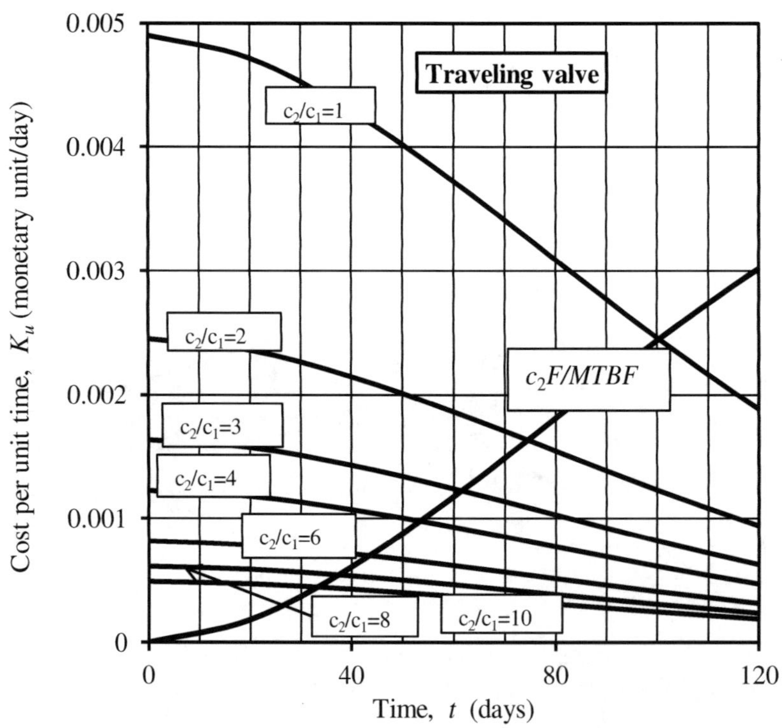

Figure 11. Graphical determination of minimum cost for maintenance at fixed age for traveling valve and for different values of ratio c_2/c_1.

5. Discussions

For comparing the two preventive maintenance variants, at fixed time and fixed age, from the point of view of costs and durations, in figure 12, figure 13 and figure 14, there were represented the dependence curves between cost ratio c_2/c_1 and maintenance time.

From above mentioned figures it can be observed that for the considered couples when the cost ratio c_2/c_1 has values higher than 10, maintenance at fixed time is recommended while for $c_2/c_1 < 10$ the maintenance at fixed age is more economical.

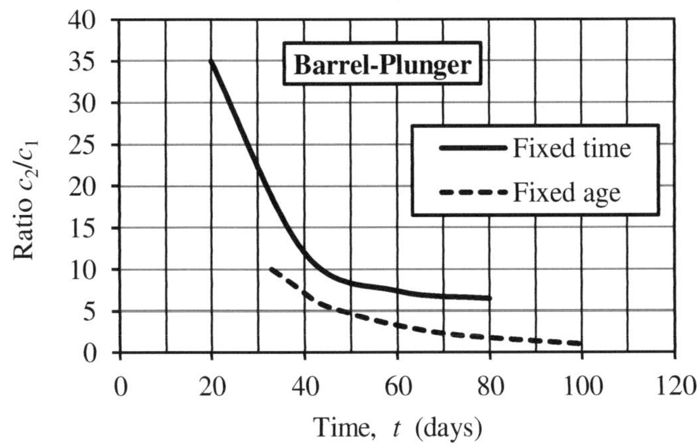

Figure 12. Comparison between maintenance at fixed time and at fixed age for barrel-plunger couple.

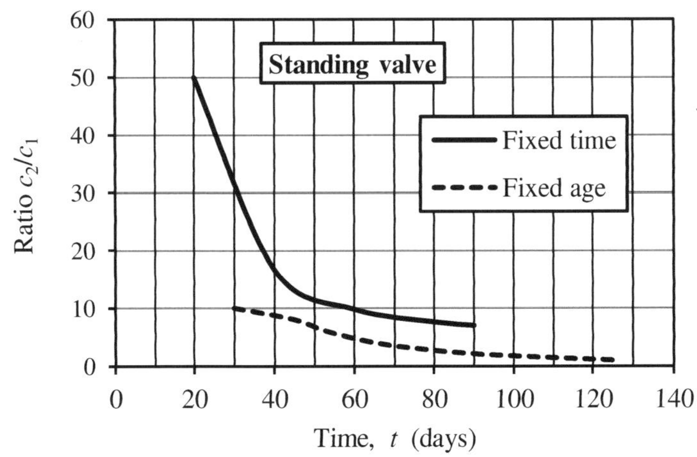

Figure 13. Comparison between maintenance at fixed time and at fixed age for standing valve couple.

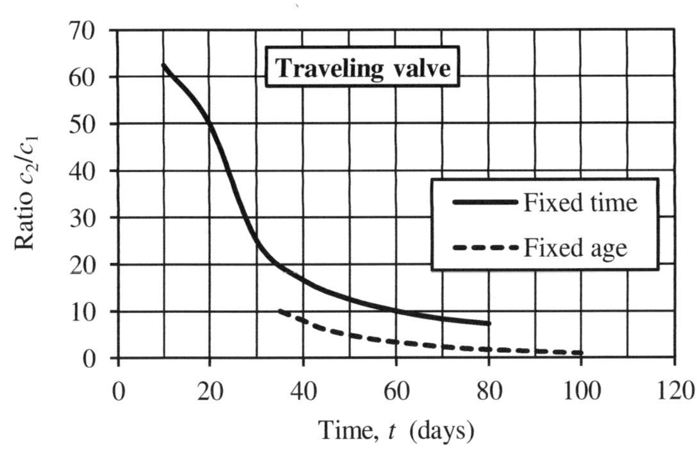

Figure 14. Comparison between maintenance at fixed time and at fixed age for traveling valve couple.

For considered couples (figure 12, figure 13 and figure 14) in the time range of 40 ... 80 days the economic value of the ratio c_2/c_1 is higher than 8 for the fixed time maintenance and smaller than 5 for the maintenance at fixed age.

In figure 12 and figure 14 it can be observed that for the same ratio c_2/c_1 the smallest replacement periods at fixed time are registered for barrel-plunger and traveling valve, while for standing valve they are the biggest (figure 13). These results are in accordance with their values of operating reliability.

Regarding the maintenance at fixed age the biggest replacement periods are registered for the same c_2/c_1 ratio in the case of standing valve (figure 13). Anyway, the minimum values of each curve indicate optimum values of ratios c_2/c_1 for each component.

6. Conclusions

The main conclusions, which can be detached from this paper, are following:

- paper presents an original approach of preventive maintenance periodicity determination, applied for subsurface sucker rod pumps used for oil secondary exploitation;
- because the preventive maintenance is a component of reliability centered maintenance process, the paper results can be used firstly in design activity of subsurface sucker rod pumps;
- paper presents a case study which is imposed to be done for every subsurface sucker rod pump type, because of the operating conditions which are different from one oil field to another;
- calculus manner of reliability which was presented in this paper can be used for each type of subsurface sucker rod pump, and generally for any element (device);
- paper results can be used for optimum programing of maintenance activity of subsurface sucker rod pumps, in order to obtain the highest operating durability and lowest costs.

References

[1] Cristea V, Grădişteanu I and Peligrad N 1985 *Installations and Equipment for Drilling Wells* (Bucharest: Technical Publishing)

[2] Shedid A S 2009 Effects of subsurface pump size and setting depth on performance of sucker-rod artificial lift: a simulation approach *SPE Production and Operations Symp.* (Oklahoma City: SPE) SPE-120681-MS

[3] Yang H B, Di Q F and Wang W C 2005 Prediction of serious abrasion position and mechanism of uneven abrasion between sucker rod string and tubing *Acta Petrolei Sinica* **26** 100

[4] Moises G V L, Andrade S F A, Bicharra A C and Ferreira Y S 2010 Sucker-Rod Pumping Failures Diagnostic System *SPE Annual Technical Conf. and Exhibition* (Florence: SPE) SPE-134975-PP

[5] Petre N and Chiţu – Militaru P 1986 *Extraction of Crude Oil Through Rod Pumping* (Bucharest: Technical Publishing)

[6] Ghareeb M, Mosavi M, Kennedy D and Tonks P 2018 Reliability and maintenance of long stroke hydraulic pumping units *2018 Sucker Rod Pumping Workshop* (Oklahoma City: Texas Tech University http://www.alrdc.com/theworkshopfolders/2018 accessed on February 2019)

[7] API Specification 11 Ax 1996 *Specification for Subsurface Sucker Rod Pumps and Fittings* (American Petroleum Institute)

[8] Militaru C 1997 *Reliability and Precision in Machine Building Technology* (Bucharest: Technical Publishing)

[9] Drumeanu A C 2016 *Maintenance of Technical Systems* (Ploiesti: Petroleum-Gas University of Ploiesti Publishing)

Lubricants condition monitoring by using a global performance passport

I Radulescu[1] and A V Radulescu[1]

[1]University POLITEHNICA Bucharest, Romania

E-mail: irina.radulescu@upb.ro

Abstract. Lubricants degradation processes are characterized by the physical and chemical indicators changes for oils composition, caused by internal and external factors and the operating conditions severity. Lubricants condition monitoring involves experimental methods to study their behaviour, for developing a lubricants life cycle assessment methodology. The paper proposes the lubricants behaviour complete characterization by using a global performance passport, to study the rheological, tribological and microscopic properties, for various wear stages. The global rheological lubricant feature is expressed by the maximum film pressure, by using the original rheological experimental module. The reference rheological lubricant feature is obtained by using the viscometer validation module, based on the variation of viscosity by the temperature. The tribological lubricant feature is given by the tribological experimental module, based on friction coefficient variation according to the normal pressing force and sliding speed. The structural microscopic lubricant feature is shown by the wear particles analysis module, analysing the solid particles type identified in the fresh and worn lubricant. The global performance lubricants passport assess the lifetime reserve, function the specific use conditions, to reduce operating costs. To do the initial investment for the preliminary experimental tests conduct it is recommended to determine the economic profitability threshold for the user investment.

1. Introduction

The lubricants degradation is a quantified process, by using physical characteristics and chemical composition oils changes; these are caused by internal and external factors and by the severity of operating conditions. The degradation of lubricants during time is due to the mechanical loads, to the operating temperature and contamination; the result is the lubrication deterioration and the worsening of other functions (providing cooling parts, their corrosion protection, etc.). The degradation process is progressive and it is more intense if the service conditions are more severe. There is a variety of external factors involved in determining of the lubricants degradation and their disuse [1-2]:

- the contamination with different impurities or chemical agents, during the usage, transport or storage;
- the mechanical of thermal overstressing;
- the possibility of increased aeration and water penetration into the oil, in bigger quantity.

The consequence of reported changes is the shortened lubricant use, which is associated, in some cases, with endangering the safety of the equipment [3]. The problem of a long use of industrial lubricants is transposed over the existence of a life cycle, which is economically satisfactory.

The lubricants choice has effects the longer use of the production equipment and life cycle costs.

By choosing a lubricant for the existing equipment of a factory it is necessary to take into account several factors, [4]:

- The lubrication frequency: the power generating equipment may require frequent re-lubrication, a special lubricants use and it is possible to have an impact over the production and the costs, during time.
- The extreme production environments: high temperatures and concentrated chemicals can cause lubricant breakdown, compromising performance and leading ultimately to an additional maintenance effort.
- The safety and the environmental interests: some lubricants contain volatile organic chemical materials or chlorinated materials, which can be hazardous to the environment and employees. The frequent re-lubrication can increase employee exposure to hazardous areas, with high temperatures.

The useful time identification for the causes which reduce the lubricants durability is possible by a careful laboratory control and by the determinations of specific parameters due to the specialized personnel. For this purpose there are necessary typical tests on laboratory stands, using standardized methods. The interests of specialists regarding the assessment of the lubricating environment efficacy led to the development of laboratory methodologies to simulate real conditions, which are specific to machinery exploitation, [5].

2. Methods and results at international and national levels

Researches from Western Europe, SUA or Japan signed that the oil changing moment is based on the oilcans written instructions, with a minimal recommended covered distance. To discover the most important factors of the lubricants degradation it is important to do the periodically physical-chemical oil analyses, in order to analyse the oxidation and the contamination oils. The damage of oil is pointed when its characteristics had touched some limit values, which are established for each type of engine.

The contamination level and the "lifetime" of used oil depend on the exploitation conditions, the oil quality, the construction and the technical state of engine. At international level the main directions are orientate to the preventive maintainability domain, for applying into any closed lubrication system.

Some examples from industrial domain are the MAN firm from Germany, which uses the basicity analysis method for Mobil class oils, for the oil lifetime determination for engines trucks, the firm Mercedes from Germany, which uses insoluble impurity in oil, with microscopic tests for class Shell oils. BOGE Compressors Ltd. From United Kingdom can yield up to 5% energy savings by using its special lubricants (S46 Screw Lube mineral oil, SYPREM 8000 S), which extend maintenance life to 40.000 hours. The importance of using the adequate lubricant and the oil changing moment at the proper mileage can improve engine reliability and has the potential to reduce nationwide waste and recycled oil by 325 million gallons annually, [6-7].

The problem was analysed also at national level, by making used oil changing moment evaluations at ROMAN firm on Lubrifin Romanian oils, by applying the oil spot method, which is based on the aspect of the spot made on a filter paper. Another method applied in ROMAN firm on Lubrifin oils is represented by the photoelectric method, which allows the establishment of used oil changing moment, according to the quantity of oil impurities.

At educational level there were made researches by the "POLITEHNICA" University from Bucharest and "GHEORGHE ASACHI" Technical University from Iaşi, for the determination of the rheological properties of the mixture of oil with polymers, that was emphasized their non-Newtonian thixotropic character. "TRANSILVANIA" University from Braşov made interesting results in management direction of solid residues from lubricants, by using the ferografic analysis of used oils.

Lubricants properties and durability were made in Romanian manufacturing bearing firms: RULMENTUL Braşov, TIMKEN Ploieşti or Koyo Alexandria. The industrial laboratories offered the opportunity to make physical-chemical standard tests and tribological ones (Timken stand, Amsler stand, etc). The information obtained by analysing these methods give important data from performing laboratories, at high technical level, but it involves special attention from the technical staff, a good practice experience and high level practitioners, [8].

3. Basic principles concerning the global performance lubricants passport

Experimental methods for determining the wear degree and lubricants durability, which are analysed in other works, led to the development of a global methodology for the assessment and quantification of lubricants life cycle and the achievement of a complex, service life, monitoring tool.

There are presented the basic steps required to implement the global methodology for life cycle lubricants monitoring, during their lifetime. The applied methodology consisted in following steps:

1. The acquisition of two types of industrial lubricants: one oil from internal combustion engines and one hydraulic power transmission oil;
2. The defining of two specific users for these lubricants types: one cross-country car and a construction equipment. These users will be monitored during all life cycle of the used lubricants;
3. The testing of the studied lubricants during the working operation, for all users, with regular removal of oil samples, at varying wear degrees;
4. The complex analysis of fresh and worn oil samples and the determining of the rheological, tribological and structural characteristics, by using the four basic modules of the complex monitoring instrument (Figure 1);

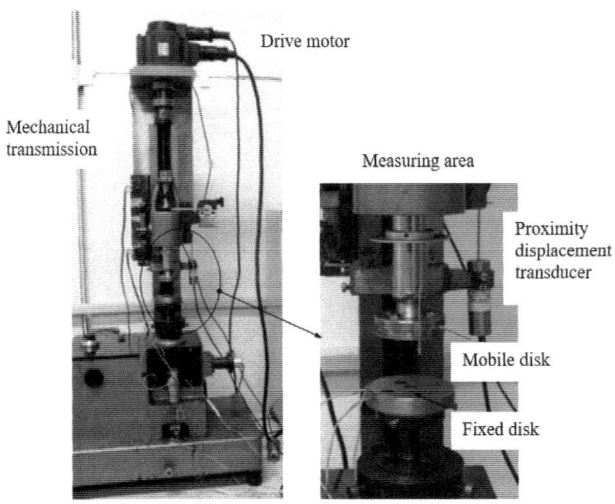

a) Original experimental rheological modulus

b) Wear particles analysing modulus

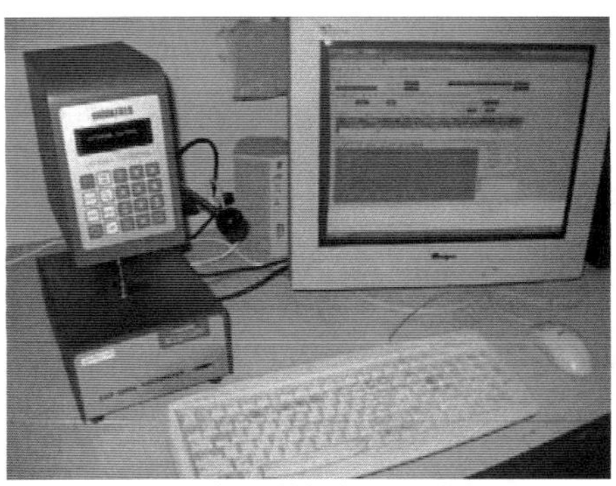

c) Viscometric validation modulus

d) Experimental tribological modulus

Figure 1. The Complex monitoring instrument

The obtaining of the performance lubricants global passport, which represents a complete characterization of the lubricants behaviour during their lifetime (Figure 2).

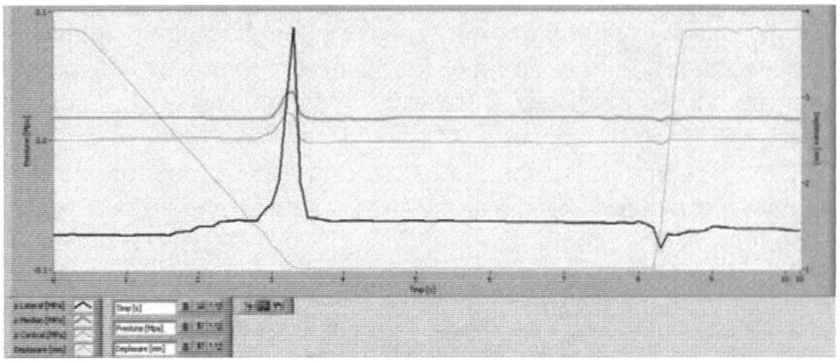

a) Characteristic squeeze film curve

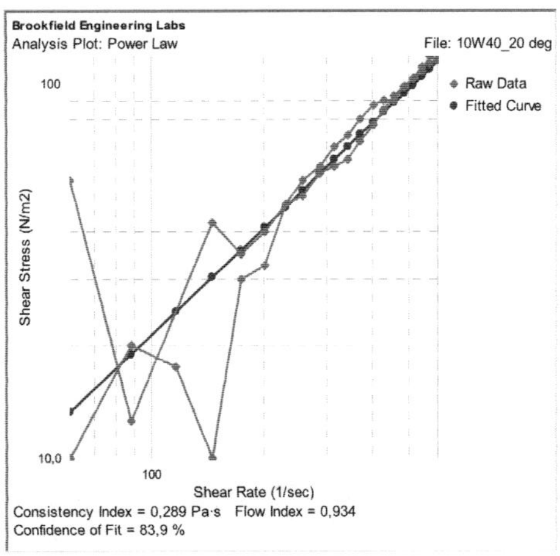

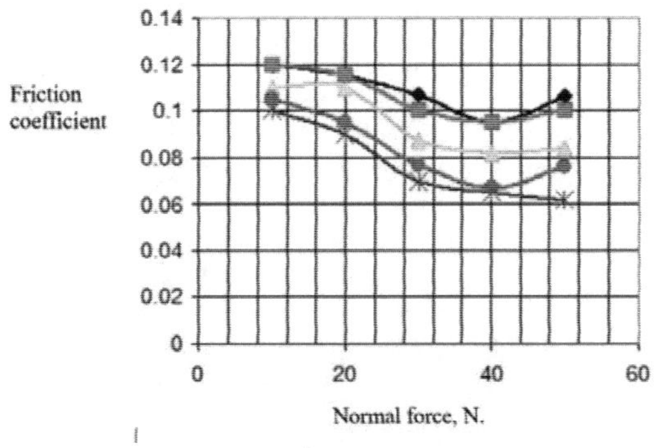

b) Oil reference rheogram c) Variation of the friction coefficient with the velocity and load

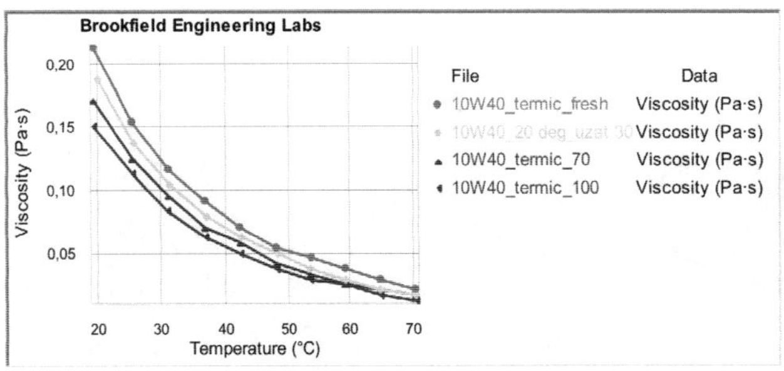

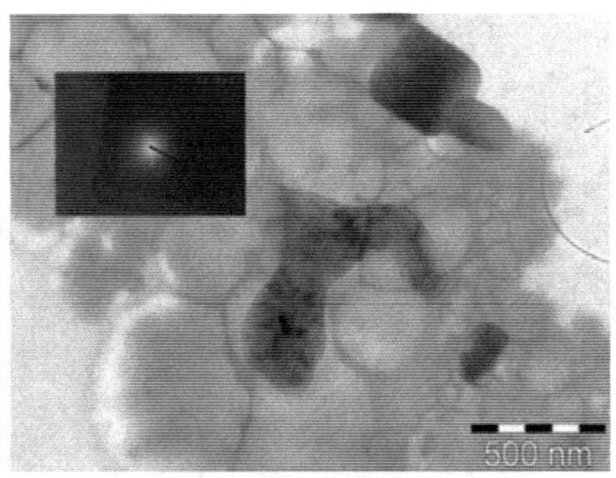

d) Variation of the viscosity by the temperature e) Wear particles identified in oil

Figure 2. Global performance passport

The core of this global methodology for the industrial lubricants monitoring is done by the complex monitoring tool, which is composed from four experimental modules: the rheological one, the viscometer one, the tribological one and the structural analysis. Briefly, the four experimental modules of the monitoring complex instrument are:

- the original rheological experimental module, which has a measuring device mounted on a Weissenberg rheogoniometer and a data acquisition and processing system. By using for the diagnose a small amount of lubricant, it is recorded the temporal variation of lubricant film thickness, by using a proximity sensor, while the oil film is expulsed with constant speed. It is recorded the squeeze-film curve, which is the "fingerprint" of the worn-out lubricant (Figure 2 a).
- the viscometer validation module, which is based on a Brookfield viscometer type. Its role is to validate the experimental results provided by the rheological experimental module and to establish the necessary correlations (Figure 2 b, d).
- the tribological experimental module consists in a friction torque pin – disk tribometer, made by Guntt company. It is possible to determine the friction coefficient variation due to contact pressure and sliding speed, for lubricants having varying wear degrees, in terms of the same friction torque use (Figure 2 c).
- the wear particles analysis module, which is based on the transmission electronic microscopy, in order to obtain information on the wear particles, which are identified in worn lubricants samples. Working element is represented by the JEOL Japan - JEM - 200 CX transmission electron microscope. By comparing two microscopically structures of the same lubricant, in various decay stages, it may be revealed the wear particles, their shape and their concentration (Figure 2 e).

Global methodology for assessing and quantifying the industrial lubricants life cycle aims, finally, the achievement of a global lubricants performance passport. This is a complete characterization of the lubricants behaviour, which are studied during their life cycle, in terms of rheological, tribological and microscopically point of view.

4. Experimental investigations

The experimental research were monitored for the lubricants, during their working time, till their complete degradation. Rheological, tribological and structural tests were conducted, over their entire service time; finally there was obtained a global performance passport, for each lubricant. This will include:

- the global rheological lubricant feature, which is expressed by the maximum pressure of fresh and worn lubricating film;
- the reference rheological lubricant feature, which is expressed by the variation of viscosity with the temperature, according to various lubricant wear stages;
- the lubricant wear feature, which is given by the viscosity variation, function the wear degree;
- The tribological lubricant feature, which is expressed by the friction coefficient variation according to the normal pressing force and the sliding speed, both for fresh and worn lubricant;
- the structural (microscopic) lubricant feature, which is done by the solid particles type which are identified in the fresh and worn lubricant.

Tables 1 and 2 show two examples of the global performance passport for two lubricants types, one for the internal combustion engine (10W40 oil – Table 1) and the other, for the hydraulic power transmission (H46 oil – Table 2).

5. Conclusions

Analysing the results presented in the previous paragraphs it is obvious the necessity of a global performance lubricants passport, which represents a complete characterization of the studied lubricants behaviour, during their life time. Each lubricant has a specific rheological, tribological and structural behaviour, which requires continuous monitoring of these properties, all the use period, from fresh stage till its complete depletion.

The developed global performance lubricant passport may be used to assess the lifetime reserve, function specific use conditions (load, speed, temperature, etc.). Its advantages are the reducing of operation costs, but it requires an initial investment for the preliminary experimental tests conduct. In conclusion, the achievement of a global performance lubricant passport should be correlated with the determination of economic profitability threshold for the investment made by a potential user.

Companies with mechanical maintenance needs might be familiar with used oil testing for their equipment, but many have not fully grasped the relevance of and the value incurred from implementing an oil testing program.

This paper provides an overview of the practice of a lubricant global performance passport and clarifies how testing and lubricant behaviour analysis imply companies benefits, with no operating equipment investments. Using passport model, it will minimize equipment downtime, permitting more efficient maintenance scheduling, protect warranty claims and increase equipment resale value.

Table 1. The global performance passport for the internal combustion engine 10W40 oil.

Global rheological lubricant feature

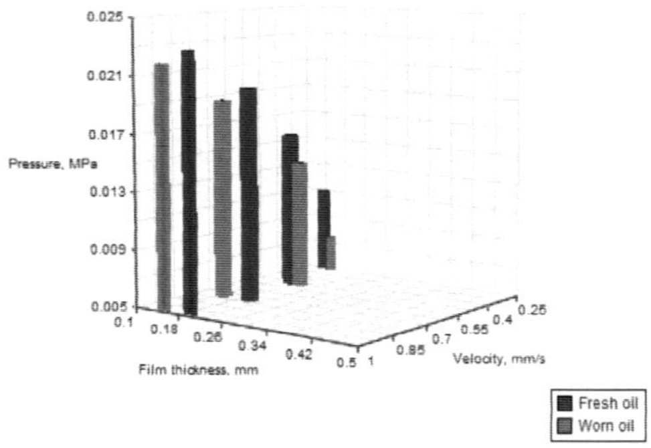

Reference rheological lubricant feature

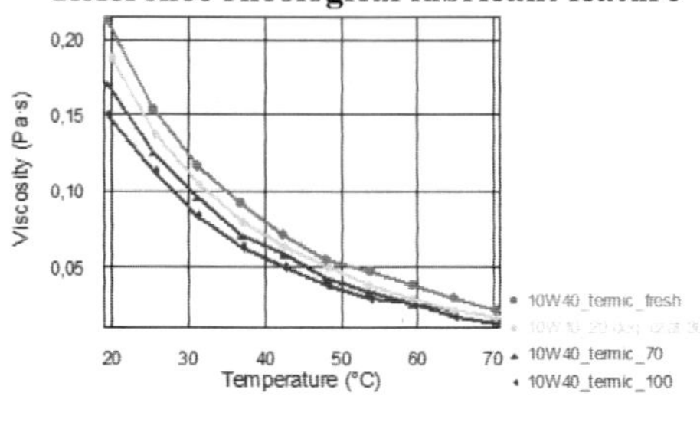

Wear lubricant feature

$\eta = \eta_0 e^{-Kd}$, where η_0 = 0,2141 Pa.s and K = 4,3243·10^{-5} km^{-1}

Tribological lubricant feature

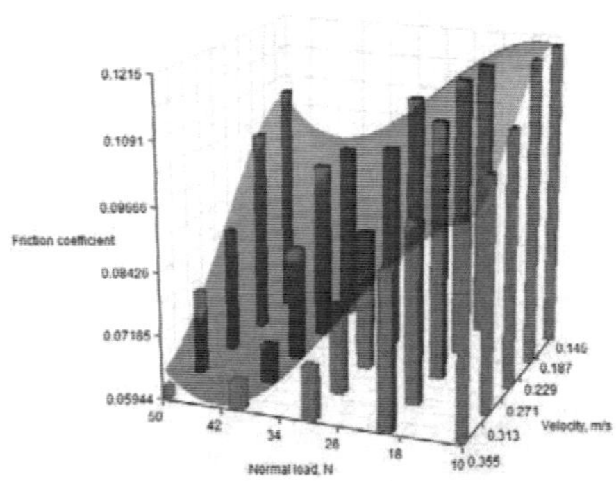

Fresh oil

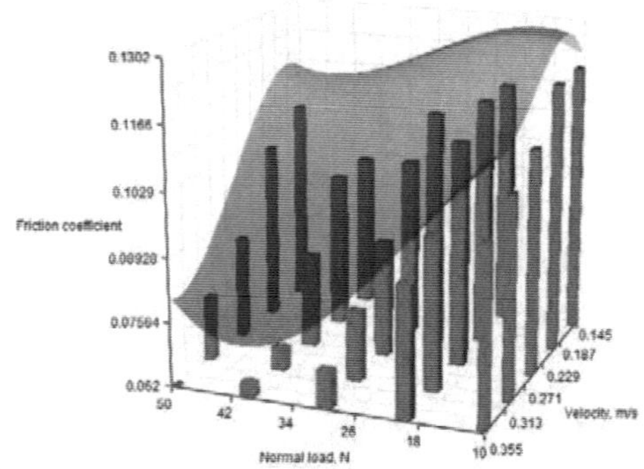

Worn oil

Structural lubricant feature

- ZnO hexagonal (ASTM 80-0075)
- $ZnSO_4$ cubical (ASTM 70-1255)
- $Ca_3(PO_4)_2$ (ASTM 02-0786)
- Amorphous phase

Fresh oil

- Cr_2O_5 unidentified cristalization type X (ASTM 28-0370)
- AlNi cubical (A) (ASTM20-0019) + V_2O_3 rhombohedral (B) (ASTM 84-0319)
- $Mn_{0,9}V_{0,1}P$ orthorombic (ASTM 78-0802)
- Amorphous phase

Worn oil

Table 2. The global performance passport for the hydraulic power transmission H46 oil.

Global rheological lubricant feature

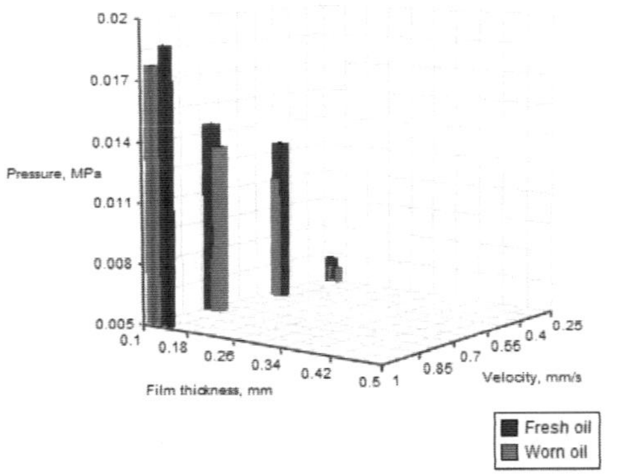

Reference rheological lubricant feature

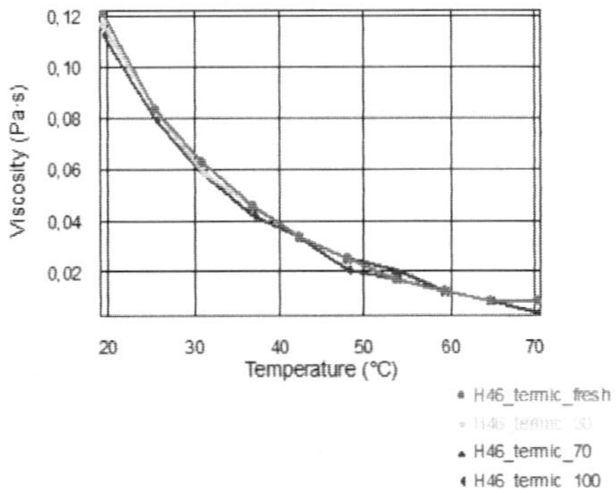

Wear lubricant feature

$\eta = \eta_0 e^{-Kd}$, where $\eta_0 = 0{,}1206$ Pa.s and $K = 1{,}4073 \cdot 10^{-5}$ hours^{-1}

Tribological lubricant feature

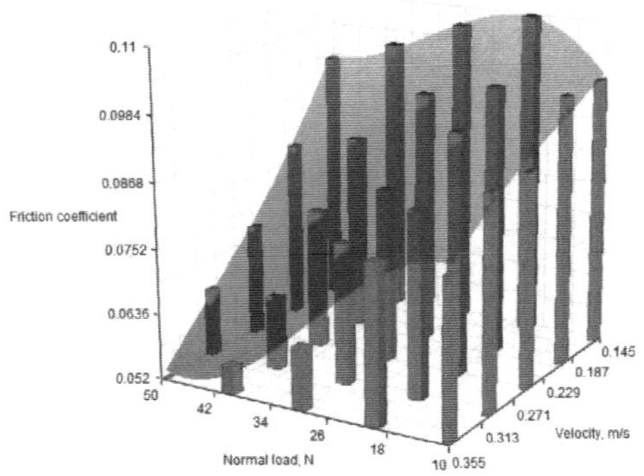

Fresh oil

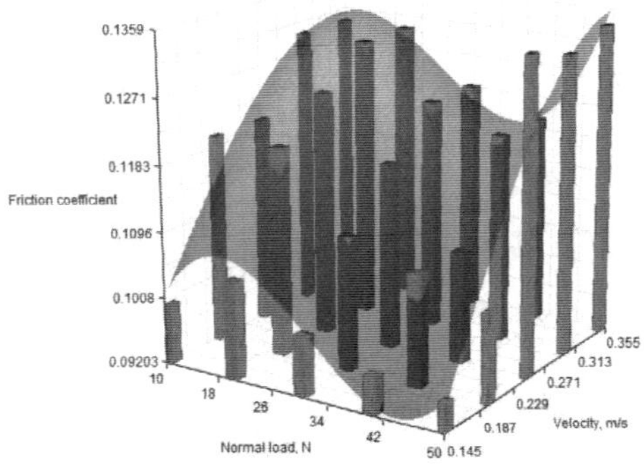

Worn oil

Structural lubricant feature

Fresh oil

- $ZnSO_4$ cubical phase (ASTM 70- 1255)
- Mo_2S_3 (X) (ASTM 12-0692)
- Unidentified cristalization phase

Worn oil

- C (diamond) cubical (ASTM 6-0675) + SiC (ASTM 02-7463)
- γ FeOOH orthorombical (Wustit) (ASTM 8-98)
- Al_4Si cubical (ASTM 24-0035)
- C cubical (ASTM 6 -0675) + $FeCO_3$ rhombohedral traces (ASTM: 29-0696)
- Amorphous phase

6. Conclusions

Analysing the results presented in the previous paragraphs it is obvious the necessity of a global performance lubricants passport, which represents a complete characterization of the studied lubricants behaviour, during their life time. Each lubricant has a specific rheological, tribological and structural behaviour, which requires continuous monitoring of these properties, all the use period, from fresh stage till its complete depletion.

The developed global performance lubricant passport may be used to assess the lifetime reserve, function specific use conditions (load, speed, temperature, etc.). Its advantages are the reducing of operation costs, but it requires an initial investment for the preliminary experimental tests conduct. In conclusion, the achievement of a global performance lubricant passport should be correlated with the determination of economic profitability threshold for the investment made by a potential user.

Companies with mechanical maintenance needs might be familiar with used oil testing for their equipment, but many have not fully grasped the relevance of and the value incurred from implementing an oil testing program.

This paper provides an overview of the practice of a lubricant global performance passport and clarifies how testing and lubricant behaviour analysis imply companies benefits, with no operating equipment investments. Using passport model, it will minimize equipment downtime, permitting more efficient maintenance scheduling, protect warranty claims and increase equipment resale value.

7. References

[1] T Mang and W Dresel 2007 *Lubricants and Lubrication* (Weinheim: Wiley-Vch) pp 7-33
[2] W J Bartz 1993 Long Life and Life – Time Lubrication – Possibilities and Limitations *Lubric. Eng.* **49** pp 518-524
[3] L Bogatu 2014 Influence of chemical structure changing on lubricants behavior in service *Rev. Chim-Bucharest* **65** pp 1230-1234
[4] H Kaleli and I Yavasliol 1997 Oil ageing – drain period in a petrol engine *Ind. Lubr. Tribol.* **49** pp 120-126
[5] H Kaleli 2004 Evaluation of additive's layer formation in engine crankcase oil using two different types of tribological test rigs *Ind. Lubr. Tribol.* **56** pp 158-170
[6] R D Youngk 2000 Automobile engine reliability, maintainability and oil maintenance *Proc. Annual Reliability and Maintainability Symposium (Los Angeles)* pp 94-99
[7] R M Gresham and G E Totten 2008 *Lubrication and Maintenance of Industrial Machinery: Best Practices and Reliability* (New York: CRC Press) pp 2-30
[8] M Lukas and D P Anderson 1998 Laboratory used oil analysis methods *Tribol. Lubr. Technol.* **54** (11) p 19

IOP Conf. Series: Materials Science and Engineering **724** (2020) 012036 doi:10.1088/1757-899X/724/1/012036

Microscopic testing and analysis of drinking water filters after the final life cycle, using an experimental stand

H G Crişan[1], O A Crişan[2], M S Pustan[1], C J Bîrleanu[1] and V V Merie[3]

[1]Mechanical Systems Engineering Department, Technical University of Cluj-Napoca, Cluj-Napoca, Romania
[2]Department of Environmental Engineering and Sustainable Development Entrepreneurship, Technical University of Cluj-Napoca, Cluj-Napoca, Romania
[3]Materials Science and Engineering Department, Technical University of Cluj-Napoca, Cluj-Napoca, Romania

E-mail: Horea.Crisan@omt.utcluj.ro

Abstract. The purpose of this paper was to develop a testing infrastructure, under laboratory conditions, for the drinking water filters used in the household. The mechanical filtration of existing nanoparticles in drinking water distributed to the large public is a very important aspect of ensuring a high level of drinking water quality, at the final consumer - the household user, regardless of the conditions of supply or its distribution. Moreover, the importance of increasing the quality of domestic drinking water is reflected by the need to encourage its consumption, to the detriment of bottled PET packaging (polyethylene terephthalate), which confers a number of disadvantages in relation to environmental protection and consumer's health. In order to achieve the proposed goal, a computer aided design (CAD) was developed and an experimental stand was designed and built to allow different categories of constructive and structural filters to be tested. The filters were chosen according to well established criteria. Their testing was aimed at identifying the potential and characteristics of mechanical filtration of existing nanoparticles in drinking water. Finally, using the atomic force microscopy technology (AFM), it was possible to microscopically investigate the nanofiltration results performed using the sample of tested filter elements and their classification according to the quality of the filtration obtained.

1. Introduction

A viable solution for increasing the quality of domestic drinking water is the purchase and use of water filters, to the detriment of the acquisition of bottled drinking water.

The necessity to purchase filters for drinking water provided by the public network has a dual role. On the one hand, it is achieved an increasing with regard to the quality of drinking water supplied from the public network, which is qualitatively affected during its supply along the distribution network. On the other hand, the population is discouraged from purchasing bottled water in Polyethylene Terephthalate (PET) packaging, which has direct and immediate negative effects on the environment [1-3].

In this sense, through the developed research it was proposed to identify the optimal filtering element, in order to improve the quality of life of the population and at the same time to support the environment in the context of sustainable development [4].

2. Materials and methods

In order to analyze the quality of water samples, two diametrically opposed test points were chosen, both having the regional operator - Someș Water Company, as the exclusive supplier. Therefore, for the research, the first test point was located in the Iris district on Muncii Boulevard from Cluj-Napoca and the second one was located in the Gilău villlage, on Braniște street. The reason for choosing these two areas consisted primarily in the fact that they are located at a considerable distance, more exactly 21 km from each other, which can lead to different water quality effects, both before and after mounting the filter elements.

On the current market, there are a variety of filters, such as ceramic, polypropylene, active carbon, textile material or washable material, etc., with particle filtering capacities ranging between 0.3 μm to 70 μm. However, their different structural properties have allowed Atomic Force Microscopy (AFM) to analyze only the ceramic and polypropylene filter elements, which are also the first two in the order of filtering capacity declared by the manufacturer [3, 5].

The CAD environment was used to design an experimental stand, which was built and utilized for testing the filters under laboratory conditions. Also, for the analysis of deposition results, the AFM technology was used, with which was performed the 3D scan of the tested samples [6].

After the experimental tests of the filters has been carried out, at the maximum life time using capacity of them, it was done a topological analysis of the filtering samples surfaces and a mechanical nanocharacterization of the polluting particles deposit. This was possible through the Atomic Force Microscopy, using the AFM Microscope XE 70 from the MINAS research laboratory (Micro and Nano Systems Laboratory), part of Technical University of Cluj-Napoca.

The tests were carried out for a relative humidity of 28 % at a room temperature of 23 °C. The topography determination and the nanocharacterization was performed using a PPP-NCHR cantilever which is defined by: thickness of 4 μm, width of 30 μm, length of 125 μm, resonance frequency of 400 kHz and force constant of 42 Nm^{-1}, also the tip height is 15 μm and radius is smaller than 10 nm.

The surface topography evaluation, in non-contact mode, supposed that a sharp tip placed on the top of the cantilever, vibrate near the surface of the sample, interacting with the sample trough the inter-atomic attractive forces principle [7, 8].

The roughness of filtering samples surfaces analyzed is directly influenced by the size and the distribution of deposited polluting particles, closely related to the filtration capacity and the filter element materials structure. The samples analyzed were taken from the same middle area of the tested filters, because due to the testing carried out on the experimental stand, the flow rate of the water was uniform and the filter section being constant [9].

3. Developing the testing infrastructure

For the testing of the filter elements, first was made a CAD model of the experimental stand, which is shown in Figure 1 (a). The CAD model has been implemented in practice, and the result is shown in Figure 1 (b).

The stand developed is composed by the following elements: 1) adjustable foot table; 2) network water inlet hose; 3) water quantity meter; 4) water pressure control valve; 5) water pressure indicator; 6) drinking water filter (consisting of a filter casing and a removable filter cartridge); 7) filtered water outlet hose; 8) water tap with double role (adjusting post filter flow and retaining post filter impurities on an auxiliary sieve).

(a) (b)

Figure 1. (a) CAD model of the experimental stand, (b) Experimental stand developed.

4. Carrying on the experimental tests and microscopic analysis

In this way, starting from the amount of minimum water required (indicated) to be consumed by a person over a day, correlated with the filter cartridge manufacturer's specification of their maximum lifetime, but also by the average number of water consumers into a family, the amount of total water consumed for each filter cartridge (intended for direct consumption or in the cooked preparations form) was determined using the calculus relation 1.

$$C_{water\_total} = C_{water\_i} \; x \; N_{days} \; x \; N_{consumers} \quad [\mathrm{m}^3] \tag{1}$$

After replacing in the calculus relation 1, it is obtained:

$$C_{water\_total} = 2 \; x \; 180 \; x \; 3 = 1080 \;\; [liters] \; \mathrm{x} \; 10^3 = 1.08 \;\; [\mathrm{m}^3]$$

where:

- C_{water_total} is the amount of total water intended for self-consumption (water or cooked water) by the average number of people in a family for a period of 6 months;
- C_{water_i} represents the daily water consumption at the individual level;
- N_{days} represents the number of days from the analyzed period;
- $N_{consumers}$ represents the average number of people in a family (consumers).

Considering that the result reflects the minimum amount of water consumed for the specified activities during the life of the filter cartridges, but also the fact that the manufacturer additionally indicates the possibility of filtering up to 3 m^3 for one filter cartridge over the lifetime of the filters, to perform the experimental procedures, it was chosen to test them for the maximum amount of water admitted.

After the testing infrastructure was completed, the next step in the filtering test phase of the analyzed cartridges was to generate samples containing elements obtained by cutting the tested filter cartridges and subjecting them to microscopic analyzes. The microscopic analyzes have the role of determining the quality of drinking water filtration (in terms of types, geometry, size and density of impurities deposited in the structure of the experimentally tested filter elements) and these objective was possible with the help of AFM. From the results obtained, it can be said that the efficiency of the filters is directly proportional to the value of the microns number held by each one, for the filtration capacity [9, 10].

The scanned area for topography of the ceramic sample is the maximum permissible of 40 µm x 40 µm, and for the topography of the polypropylene sample due to a small dispersion of the determined particles, it was enough to analyze a surface of only 16 µm X 16 µm. From Figure 2 (a) results an increased density of particle agglomeration, with approximate heights of 2000 nm, while Figure 2 (b) shows a much lower particle density agglomeration, with mean heights up to 1000 nm.

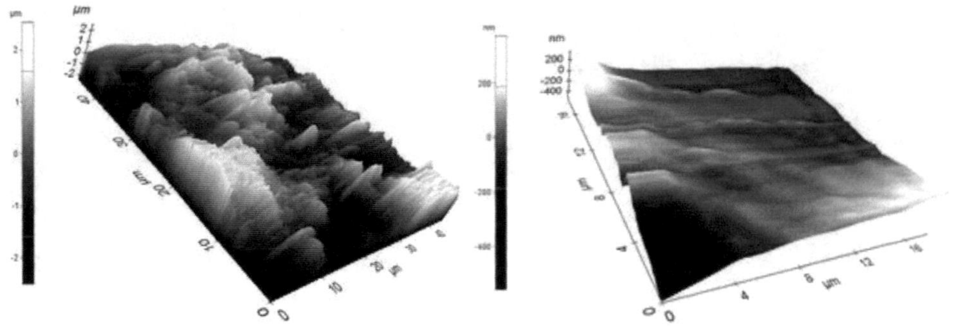

(a) (b)

Figure 2. (a) Ceramic filter topography – deposits, (b) Polypropylene filter topography.

The samples surface scanning reflects more roughness parameters, but an important one called R_a, is the arithmetic average height of roughness. R_a high values, shows the influence of the polluting particles deposit in relation to a standard filtrating surface and also the quality of the filter with regard to itself filtration capacity. In Figure 3 (a), it is observed the surface roughness of the standard sample of ceramic material, where $R_a = 0.242$ μm. In Figure 3 (b), it is observed the roughness of the ceramic surface of the tested filter element, where $R_a = 0.646$ μm. Figure 3 (c) shows the roughness of the polypropylene surface of the tested filter element, where $R_a = 0.062$ μm.

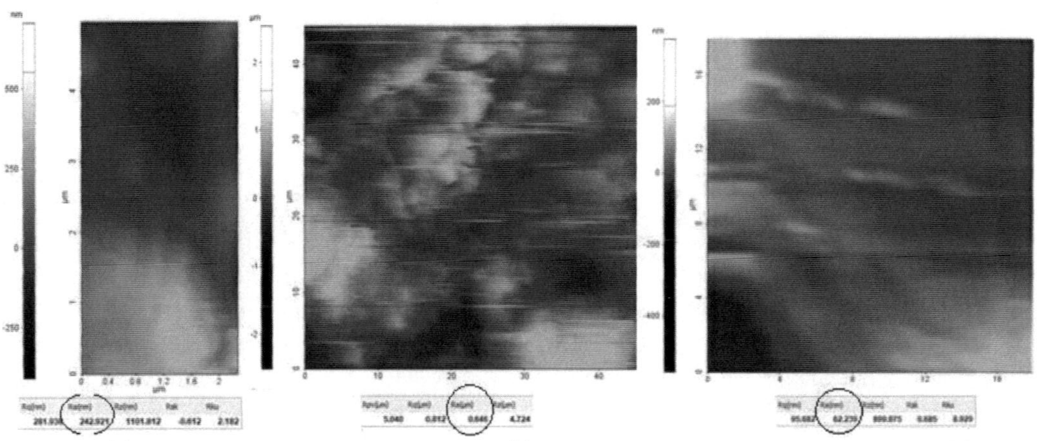

(a) (b) (c)

Figure 3. (a) Ceramic standard filter roughness, (b) Ceramic used filter roughness, (c) Polypropylene used filter roughness.

Corroborating the results of the topographic analyzes with those of the roughness of the scanned surfaces, it is observed that the highest density of particles retained by filtration is found on the ceramic sample. This density, through the particle deposition process, results in an increase of roughness R_a with 0.4 μm in the case of ceramic sample. On the opposite side, the low surface roughness of polypropylene is reflected by the lower filtration capacity of the tested filter, as can be seen also from the particle density agglomeration, highlighted in the topography from Figure 2 (b).

Farther, it was made the individual analysis of the most dimensional relevant particles, deposited on the surfaces of scanned ceramics and polypropylene samples. For this, the samples taken from the two test points were analyzed. For ceramic samples, from Figure 4 (a) results that the particle height size is 2173 nm for the first test point, while from Figure 4 (b) results a height size of 393 nm in the case of the second test point.

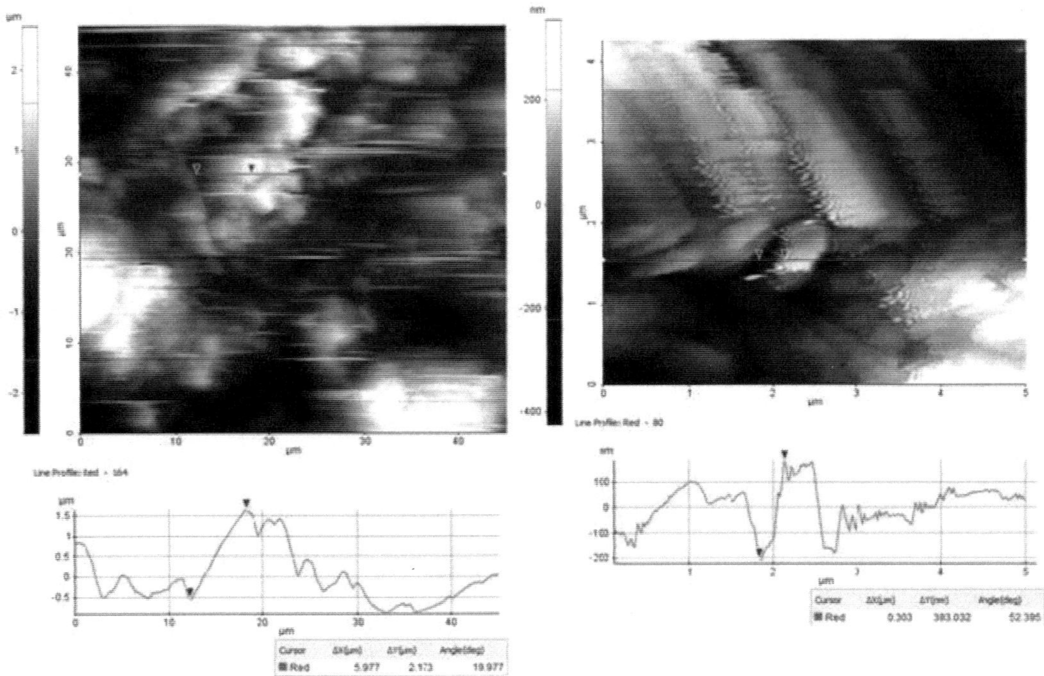

(a) (b)

Figure 4. (a) Ceramic sample Test Point 1 (Cluj-Napoca), (b) Ceramic sample Test Point 2 (Gilău).

For the polypropylene samples, from Figure 5 (a), it was observed a 120 nm particle height size for the first test point, while Figure 5 (b) shows a particle height size of 203 nm found in the case of the second test point.

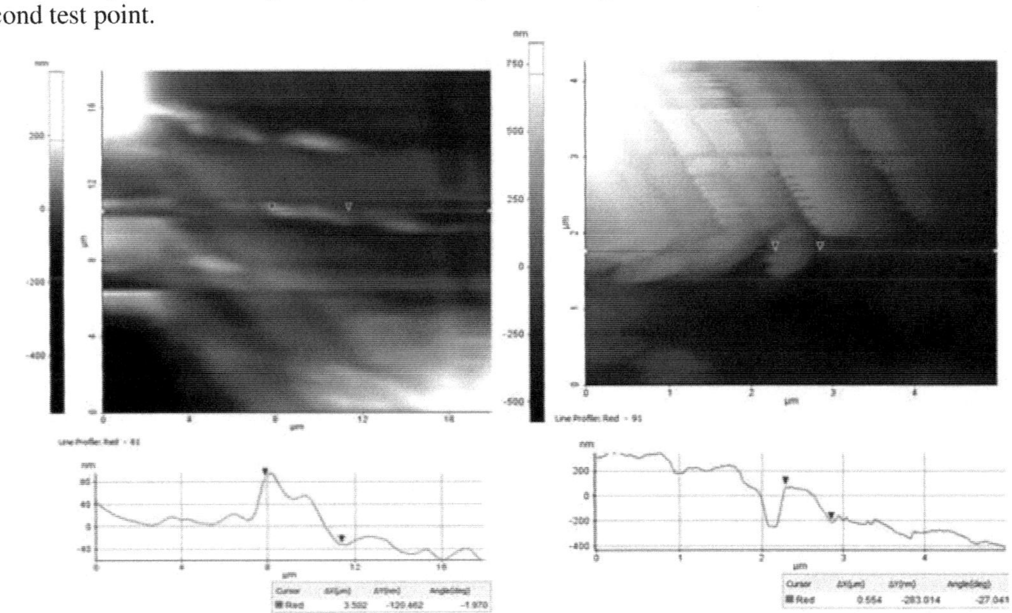

(a) (b)

Figure 5. (a) Polypropylene sample Test point 1 (Cluj-Napoca), (b) Polypropylene sample Test point 2 (Gilău).

5. Conclusions

Using the experimentally developed stand, there was made tests at the two test points, on the ceramic and polypropylene filter cartridges up to the maximum usage time of 3 m^3. Subsequently, samples of these filters were taken which were subjected to microscopic AFM analysis. The results of experimental tests and microscopic analyzes indicate an increased particles retention capacity for the ceramic filter cartridge, much higher than the polypropylene filter cartridge. This is highlighted on the one hand by topography and surface roughness analysis, resulting into an increased agglomeration density of particles deposited on the ceramic sample, leading in an increase of the surface roughness value with R_a= 0,4 μm.

On the other hand, from the dimensional analysis of the relevant particles, it is found that in the case of the ceramic sample, the structure of the material and its roughness allows to retain a much higher number of particles with dimensions up to 2173 nm, and the characteristics of the polypropylene material, although they allow the retention of particles with dimensions up to 203 nm, their number is much lower.

In conclusion, it is noted that the best filtration efficiency is owned by ceramic cartridge, which is one of the best indicators for both, producers and household consumers who want to purchase a water filter with superior characteristics.

6. References

[1] Webster K 2017 *The Circular Economy: A Wealth of Flows* - 2nd Ed. (Ellen MacArthur Foundation Publishing) 102
[2] Howe K H, Hand D W, Crittenden J C, Trussel R R and Tchobanoglous G 2012 *Water Treatment: Principles and Design* – 1nd Ed. (Wiley) 321
[3] Gray N F 2010 *Water Technology IWA Edition: An Introduction for Environmental Scientists and Engineers* – 3nd Ed. (IWA Publishing) 28
[4] Peng Z and Chen S 2011 *Phys. Rev.* E **83** (5) 051915
[5] Eftekhar B, Skini M, Shamohammadi M, Ghaffaripour J and Nilchian F 2015 *J. of Dent.* **16** (3) 278-81
[6] Lobontiu N and Garcia E 2004 *Mechanics of Microelectromechanical Systems* (Springer-Verlag, New York)
[7] Kemp R and Volpi M 2008 *J. Clean. Prod.* **16** 14-21
[8] Knauss W G, Chasiotis I and Huang Y 2003 *Mech. of Mater.* **35** (3) 217-31
[9] Bellitto V 2012 *Atomic Force Microscopy – imaging, measuring and manipulating surfaces at the atomic scale* (Naval Surface Warfare Center, United States of America)143
[10] Briggs G and Briscoe B 1977 The effect of surface topography on the adhesion of elastic solids *J. of Phys. - Appl Phys.* D **10** (18) 2453

Erosion modelling: a systematic review of available models and equations

A M Lospa[1,2]**, R G Ripeanu**[1] **and A Dinita**[1]

[1]Mechanical Engineering Department, Petroleum and Gas University of Ploiesti, Romania
[2]E-mail: adrian.lospa@gmail.com

Abstract. Across the years there was an increased attention drawn by erosion modelling, both as a general approach and also towards pipeline applications (with a concentration on pipe bends). Several authors have studied the phenomena and tried to explain the mechanisms and factors affecting the emergence and magnitude of erosion. Most of the proposed models and equations were developed for ductile materials, with a smaller part looking also at brittle materials. There are several differences observed, starting with the proposed wear mechanism and continuing with the parameters used in the equations as significant factors (for the erodent particles, target material and working conditions). The most important models and equations are analysed and presented in this article and a systematic visual representation is used to ease and simplify the understanding of the erosion modelling evolution across the years (erosion modelling in general conditions, empirical erosion modelling in pipes and bends, CFD erosion modelling in pipes and bends).

Key words: erosion, wear, pipe bend, erodent, pipeline

1. Introduction

Erosion is defined as the loss of material or loss of material integrity due to solid particle impact on the material surface [1]. Material degradation cannot be avoided, however by a proper dimensioning, material selection, use of inhibitors, by using other measures intended to reduce the erosion mechanisms or by utilizing erosion allowances, the associated effects can be minimized. Some of these measures however come with a high cost. Erosive wear can be estimated based on the speed and impact angle of the solid particles, with notable differences between ductile and brittle materials.

Salama and Venkatesh [2, 3] studied the erosion phenomena and the associated erosion mechanisms together with the parameters that are influencing the erosion rate, stating that the material erosion damage occurs as a result of the following mechanisms [2, 3]:

- Fatigue – because of cavitation (bubbles of vapours or dissolved gas collapse) or particle impingement (liquid droplet and solid particle impingement);
- Abrasive wear – as a result of repeated impingement of solid particles on ductile materials;
- Erosion – corrosion – as a result of the material protective layer deterioration by fatigue or erosive wear.

One of the most impressive studies on erosion models from 1960 to 1992 was documented by Meng and Ludema [4]. The authors checked over 5000 articles and categorized the existent equations:

- Empirical equations – developed by experimentation with varying testing conditions. This type of equations has limited validity, being more accurate than theoretical models in the testing conditions intervals. The models describe wearing under fixed sliding conditions without a proper control of testing temperature, material roughness etc.;

- Contact-mechanics-based equations – by assuming relationships between the working conditions. Many of the models are based on the assumption that one of the conventional material property (hardness, Young modulus etc.) has a big importance in the wear process;
- Equations based on material failure mechanisms.

The analysis was continued by Parsi et al. [5], taking into account also newer models (1960 – 2014). From the total of models and equations, only a part has been presented based on [4, 5]:

- Author's maturity – measured by the length of time the author has published papers and research on erosion and by how many times it was cited by other authors;
- Equation logical consistency – a reasonable and detailed explanation for the derivation of the wear equations from the initial assumptions to the final expressions;
- Historical significance and continuations of the proposed model.

In this article, the authors want to review the available erosion models that were proposed along the years, to understand the conditions in which the models were obtained, the parameters that were used as predictors in the equations and the applicability.

2. General equations and models developed for erosion phenomena analysis

Model 1 - Finnie [6] studied erosion by solid particles in case of ductile materials, for low impingement angles ($\alpha < 45°$). There is a power law between the volume of removed material and the solid particle velocity, with an exponent of 2. He proposed equations (2) and (3) to determine the erosion produced by a single erodent particle, for different impingement angles.

$$\varepsilon_{VP} = \frac{MV_p^2}{P} f(\alpha) \quad (1) \qquad \varepsilon_{VP} = \frac{mV^2}{P\Psi\kappa}\left(\sin(2\alpha) - \frac{6}{\kappa}\sin^2\alpha\right), \ \tan\alpha \le \frac{k}{6} \quad (2) \qquad \varepsilon_{VP} = \frac{mV^2}{P\Psi\kappa}\frac{\kappa\cos^2\alpha}{6}, \ \tan\alpha \ge \frac{k}{6} \quad (3)$$

Model 2 - Bitter [7] studied erosion in fluid-bed systems, assuming two wear mechanisms, deformation because of repeated solid particle collision and cutting of the free-moving particles:

$$\varepsilon_{VT} = \varepsilon_{VP} + \varepsilon_{VC} \quad (4) \qquad \varepsilon_{VC1} = \frac{2MV_p\left(V_p\sin\alpha - K\right)^2}{\left(V_p\sin\alpha\right)^{\frac{1}{2}}}\left(V_p\cos\alpha - \frac{C\left(V_p\sin\alpha - K\right)^2}{\left(V_p\sin\alpha\right)^{\frac{1}{2}}}\chi\right), \ \alpha \le \alpha_{p0} \quad (6)$$

$$\varepsilon_{VP} = \frac{1}{2}\frac{M\left(V_p\sin\alpha - K\right)^2}{\delta} \quad (5) \qquad \varepsilon_{VC2} = \frac{\frac{1}{2}M\left(V_p^2\cos^2\alpha - K_1\left(V_p\sin\alpha - K\right)^{3/2}\right)}{\chi}, \ \alpha \ge \alpha_{p0} \quad (7)$$

Model 3 - Sheldon and Finnie [8] studied the erosive cutting of a brittle material by the normal impact of a stream of solid particles (angular silicon carbide particles and spherical steel shot):

$$\varepsilon_{VP} = K_4 r^a V_p^b, \ a = \frac{3n}{n-2}, \ b = \frac{2.4n}{n-2} \quad (8) \qquad \varepsilon_{VC} = K_5 r^{a'} V_p^b, \ a' = \frac{3.6n}{n-2}, \ b = \frac{2.4n}{n-2} \quad (9) \qquad K_5 = \frac{K_4}{r^{0.6n/(n-2)}} \quad (11)$$

$$K_4 = C_2\pi\left(\frac{K_3}{K_2}\right)^{2n/(n-2)}\left(\frac{5\pi}{3}\right)^{1.2n/(n-2)}\left(\frac{\rho_p}{\gamma}\right)^{1.2n/(n-2)} \quad (10) \qquad K_3 = \frac{1-2\upsilon}{4\pi} \quad (12) \qquad K_2 = \sigma_b V_b^{1/n}\left(\frac{n-1}{4\pi(n+1)K_1}\right)^{1/n} \quad (13)$$

Model 4 - Neilson and Gilchrist [9] determined the erosive action of a particle laden gas stream on specimen materials with different physical properties, assuming that the normal component of the solid particle kinetic energy is causing deformation wear, with the parallel component causing cutting wear:

$$\varepsilon_V = \frac{1}{2}\frac{M\left(V_p^2\cos^2\alpha - V_r^2\right)}{\chi} + \frac{1}{2}\frac{M\left(V_p\sin\alpha - K\right)^2}{\delta}, \ \alpha < \alpha_{p0} \quad (14) \qquad \varepsilon_V = \frac{1}{2}\frac{MV_p^2\cos^2\alpha}{\chi} + \frac{1}{2}\frac{M\left(V_p\sin\alpha - K\right)^2}{\delta}, \ \alpha > \alpha_{p0} \quad (15)$$

Model 5 - Goodwin, Sage and Tilly [10] have studied the influence of solid particle velocity, erodent type and particle fragmenting on the erosion of ductile materials. For sand particles, the most important factors are the particle size and quartz concentration. For industrial erodent particles, the most important characteristics are particle hardness and particle sharpness.

$$\varepsilon_m = K_1 V^a \quad (16) \qquad a = 2 \text{ for } 25\mu\text{m and } a = 2.3 \text{ for } 125\mu\text{m}$$

Model 6 - Head and Harr [11] developed statistical models for erosion on ductile and brittle target materials. The main factor is represented by the energy transmitted from the erodent particles to the target material. Erosion develops only when a threshold value of the transmitted energy is reached.

$$\varepsilon_{V(DW)} = 0.000233 - 0.000160R - 0.000238\ln\left(\frac{H_p}{E_{re}}\right) - 0.000210\ln\left(\frac{H_t}{E_{re}}\right) + 0.001577\left(\frac{V^2}{E_{re}}\right) + 0.00829\sin\alpha + 0.000034\left[\ln\left(\frac{H_p}{E_{re}}\right)\right]^2 - 0.000967\left(\frac{V^2}{E_{re}}\right)^2 R + 0.000119R\ln\left(\frac{H_t}{E_{re}}\right) - 0.001380\ln\left(\frac{H_t}{E_{re}}\right)\cos\alpha + 0.0001456\left[\ln\left(\frac{H_p}{E_{re}}\right)\right]^2\cos\alpha \quad (17)$$

$$\varepsilon_{V(BS)} = \frac{V^{3.06}\alpha^{2.69}H_p^{2.08}E_{re}^{0.03}}{H_t^{2.08}} \quad (19)$$

$$\varepsilon_{V(BW)} = 0.005345 + 0.000006\frac{V^2}{E_{re}} + 0.003507R + 0.009335\sin\alpha - 0.000630\ln\left(\frac{H_p}{E_{re}}\right) - 0.004706\cos\alpha - 0.003114\sin^2\alpha - 0.010888\sin\alpha - 0.003810R\cos\alpha + 0.000495\ln\left(\frac{H_t}{E_{re}}\right)\cos\alpha \quad (18)$$

$$\varepsilon_{V(DS)} = \frac{V^{4.34}\alpha^{0.46}H_p^{0.10}E_{re}^{0.21}}{R^{2.48}H_t^{2.48}} \quad (20)$$

Model 7 – Sheldon [12] studied solid particle erosion for ductile and brittle materials and acknowledged the erosion dependence on particle diameter and velocity and target material properties.

$$W = K_1 r^a V^b \quad (21) \qquad K_1 = \frac{E^{0.8(n+1)/(n-2)}}{\sigma_0^{2n/(n-2)}} m^{1.2(n-0.67/(n-2)} \quad (22) \qquad a = \frac{3(n-0.67)}{n-2} \quad (23) \qquad b = 2.4\frac{n-0.67}{n-2} \quad (24)$$

Model 8 – Finnie [13] has resolved the motion equation for an erodent particle and proposed the following expression to determine the erosion rate:

$$\varepsilon_V = \frac{cMV_p^2}{4P\left(1+\frac{m_p r^2}{I}\right)}\left[\cos^2(\alpha) - \left(\frac{\dot{x}_t'}{V_p}\right)^2\right] \quad (25) \qquad \dot{x}_t' = V_p\cos\alpha - \frac{2V_p}{P}\sin\alpha \quad (26)$$

Model 9 – Sheldon and Kanhere [14] investigated the effect of a high dimension erodent particle on aluminium surfaces by material displacement followed by material breakage due to existing efforts.

$$\varepsilon_{VP} = \frac{d_p^3 V_p^3 \rho_p^{3/2}}{H_V^{3/2}} \quad (27)$$

Model 10 – Tilly [15] continued the research on ductile materials and acknowledged the existence of two stages: indentation and/or breakage of material because of solid particle repeated impact on the surface and the material wear because of particle fragments generated in the first stage.

$$\varepsilon_V = \hat{\varepsilon}_1\left(\frac{V_p}{V_{ref}}\right)^2\left[1-\left(\frac{d_0}{d_p}\right)^{3/2}\frac{V_0}{V_p}\right]^2 + \hat{\varepsilon}_2\left(\frac{V_p}{V_{ref}}\right)^2 F_{d,v} \quad (28) \qquad \hat{\varepsilon}_2 = \varepsilon_V - \hat{\varepsilon}_1 \quad (29) \qquad F_{d,v} = \frac{W_0 - W}{W_0} \quad (30)$$

Model 11 – Head, Lineback and Manning [16] derived the previous proposed model to take into account also higher speeds of the solid particles (3000 m/s).

$$\varepsilon_{V(DMW)}(\alpha) = \left|\frac{\varepsilon_{V(DW)}(\alpha_0) - \varepsilon_{V(DW)}(\alpha)}{F}\right| \quad (31)$$

Model 12 - Grant and Tabakoff [17] determined the erosion behaviour for Aluminium 2024.

$$\varepsilon_V = K_1 \left\{1 + C\left[K_2 \sin\left(\frac{90}{\alpha_0}\alpha\right)\right]\right\}^2 V_p^2 \cos^2\alpha\left(1 - R_T^2\right) + K_3\left(V_p \sin\alpha\right)^4 \quad (32)$$

$$C = 1,\ \alpha \le 3\alpha_0 \qquad C = 0,\ \alpha > 3\alpha_0 \quad (33)$$

Model 13 - Williams and Lau [18] studied the erosion mechanism for epoxy graphite composites.

$$\varepsilon_V = K_1 d^a \left[V_p \sin\left(\alpha + K_2\right)\right]^b \quad (34)$$

Model 14 - Jennings, Head and Manning [19] established that the erosion mechanism develops because of superficial melting of the ductile target material at the solid particle impact.

$$\varepsilon_V = \frac{K_T^{5/2}}{R} \frac{G^{1/3}}{\rho_t^{1/3} k T_m \Delta H_m} \quad (35)$$

Model 15 - Hutchings, Winter and Field [20] experimentally determined the erosion produced by spherical metallic particles at the oblique impact on mild steels, crater dimensions depending on the particle impingement angle and speed.

$$\varepsilon_m = 5.82 \times 10^{-10} V_p^{2.9} \quad (36)$$

Model 16 – Finnie [21] proposed two methods by which the cutting mechanism is finalized: when the particle tip cannot advance, the speed horizontal component becoming zero and when the particle tip leaves the material surface, while the particle still moves on horizontal trajectory.

$$\varepsilon_V = \frac{cMV_p^2}{4P\left(1 + \frac{m_p r^2}{I}\right)} \cos^2(\alpha);\ \dot{x}_t = 0,\ \alpha \ge \tan^{-1}\frac{p}{2} \quad (37)$$

$$\varepsilon_V = \frac{cMV_p^2}{4P\left(1 + \frac{m_p r^2}{I}\right)} \left[\sin 2\alpha - 2\frac{\sin^2\alpha}{p}\right];\ y_t = 0,\ \alpha \le \tan^{-1}\frac{p}{2} \quad (38)$$

Model 17 – Evans, Gulden and Rosenblatt [22] studied the wear phenomena for brittle materials based on an elastic-plastic response, the radial cracks formation determining the affected area increase and the lateral cracks formation determining the material penetration.

$$\varepsilon_{VP} \propto \frac{V_p^{19/6} r^{11/3} \rho_p^{1/4}}{K_c^{4/3} H_t^{1/4}} \quad (39)$$

Model 18 – Evans [23] proposed a new equation for brittle materials.

$$\varepsilon_{VP} \propto \frac{\left(\rho_p \rho_t \mu_p \mu_t\right)^{2/3}}{\left[\left(\rho_p \mu_p\right)^{1/2} + \left(\rho_t \mu_t\right)^{1/2}\right]^{8/3}} \rho_p^{19/12} r^{11/3} V_p^{19/6} K_c^{-4/3} H_t^{-1/4} \quad (40)$$

Model 19 - Ruff and Wiederhorn [24] proposed an equation to determine erosion for brittle materials.

$$\varepsilon_{VP} \propto \frac{\left(\rho_p \rho_t \mu_p \mu_t\right)^{2/3}}{\left[\left(\rho_p \mu_p\right)^{1/2} + \left(\rho_t \mu_t\right)^{1/2}\right]^{8/3}} \rho_p^{11/19} r^{11/3} V_p^{22/9} K_c^{-4/3} H_t^{1/9} \quad (41)$$

Model 20 - Tabakoff, Kotwal and Hamed [25] developed a semi-empirical model to determine the erosion rate for various impingement angles and speed of ash particles.

$$\varepsilon_m = K_1 f(\alpha) V_p^2 \left(\cos^2\alpha\right)\left(1 - e_t^2\right) + f\left(V_p, n\right) \quad (42) \qquad e_t = 1 - 0.0016 V_p \sin\alpha \quad (43)$$

$$f(\alpha) = \left\{1 + K_4\left[K_2 \sin\left(\frac{\pi}{2}\frac{\alpha}{\alpha_m}\right)\right]\right\}^2 \quad (44) \qquad f\left(V_p, n\right) = K_3\left(V_p \sin\alpha\right)^4 \quad (45)$$

$$K_4 = 1,\ \alpha \le 3\alpha_m \qquad K_4 = 0,\ \alpha > 3\alpha_m \quad (46)$$

Model 21 - Routbort, Scattergood and Turner [26] determined the erosion for siliconized silicon carbide SiSiC at the impact with Al_2O_3 particles, with dimensions of 23 - 270 μm, speeds of 54 – 151 m/s and impingement angles of 10 - 90°.

$$\varepsilon_m = r^{0.7-0.95} V_p^{2.0-2.5} \quad (47)$$

Model 22 - Routbort, Scattergood and Kay [27] continued the research and determined the erosion for siliconized silicon carbide SiSiC at the impact with sharp Al_2O_3 particles, with dimensions of 23 - 270 μm, speeds of 32 – 134 m/s and impingement angles of 22 - 90°.

$$\varepsilon_m = \left(V_p \sin\alpha - V_0\right)^n \left(d - d_0\right)^m \tag{48}$$

Model 23 – Hutchings [28] performed a theoretical analysis for the erosion of aluminium alloys at the normal impact of spherical steel shots.

$$\varepsilon_m = 0.033 \frac{\alpha \rho_t P^{1/2} V_p^3}{\Omega_c^2 H_t^{3/2}} \tag{49}$$

Model 24 - Sundarajan and Shewmon [29] proposed an erosion model based on a localization concept (the lip formation is the result of localization of deformation in the near-surface regions of the target and is removed either by inertial-stress-induced tensile fracture or by separation across shear bands).

$$\varepsilon_m = \frac{0.085 V_p^{2.5} \rho_p^{1/4} \rho_t^{1-b} \alpha \left(z+1\right)^{5a} \left[1 - \left(z+1\right)/\left(z+2\right)\right] F(z)}{6.06^b \left(1 - CT_c\right)^{1.5} \left[n_c C_p T_m^{0.75} \left(1 - 436/T_m\right)^{0.75}\right]^b \left(K_1 H_t\right)^{1.25-b}} \tag{50}$$

$$a = \frac{0.25 n_c^{1-a_1 S} \left(z+1\right)^S \left(K_1 H_t\right)^{a_1 S}}{\left[6.06 \rho_t C_p T_m^{0.75} \left(1 - 436/T_m\right)^{0.75}\right]^{a_1 S}} \tag{51}$$

$$b = \left(1 + 5a\right) a_1 \tag{52}$$

Model 25 – Beckmann and Gotzmann [30] tested the hypothesis by which the volume of removed material in case of abrasive wear is proportional with the shear stress generated in the affected region.

$$\varepsilon_m = \left[\tau_0 + \left(\frac{\varphi}{2}\right)^{1/2} H\right] \varphi^{5/2} \cos^2\alpha \sin^{1/2}\alpha + \left(\tau_0 + \left(\frac{2\varphi}{1-\varphi}\right)^{1/2} H\right) \varphi^2 \sin^2\alpha \tag{53}$$

$$\varphi = V\left(\frac{2\rho}{3H}\right)^{1/2} \tag{54}$$

Model 26 - Wiederhorn and Hockey [31] determined the erosion for brittle materials.

$$\varepsilon_m \propto V_p^{2.8} r^{3.9} \rho_p^{1/4} K_c^{-1.9} H_t^{0.48} \tag{55}$$

Model 27 – Ritter [32] studied the ceramic materials erosion response at the impact with Al_2O_3, SiC, Si_3N_4 and MgO particles, stating that material wear is dependent upon the solid particle kinetic energy.

$$\varepsilon_m \propto \frac{d_\gamma E U_k}{K_c^2} \tag{56}$$

Model 28 – Reddy and Sundarajan [33] determined experimentally that the main erosion mechanism consists of a lip formation at the solid particle impact on the target material followed by tearing and this generating material removal.

$$\varepsilon_{VP} \propto \frac{L^3 \Delta\Omega_m}{\Omega_c} \tag{57}$$

Model 29 – Johansson, Ericson and Schweitz [34] developed a statistical model to determine the erosion rate for single-crystal semiconductors.

$$\varepsilon_m = \left(1 - f\right) K_1 \frac{\rho_t \rho_p^{2.9} E_t^{2/3} V_p^{22/9} d_p^{2/3}}{H_t^{5/9} K_c^{4/3}} + f K_2 \frac{\rho_t \rho_p^{1/3} E_t V_p^{8/3} d_p}{H_t^{1/3} K_c^2} \tag{58}$$

Model 30 – Lhym and Wapner [35] studied erosive wear on PPS composite materials.

$$\varepsilon_m = K_1 K_c^{a-2} V_p^{5-2b} d_p^{6K_2} \rho_t^2 \rho_p \tag{59}$$

Model 31 – Sundarajan [36] proposed a new model for ductile material erosion that takes into account the whole range of impingement angles and various dimensions of the erodent particles.

$$\varepsilon_m = \left(2^{n_h} C V_p^2 \sin^2\alpha F(t) / \left(n_h C_p\right)\right) \left\{1 + \left[\left(n_h + 1\right)\left(\frac{\mu}{\mu_c}\right)\left(2 - \left(\frac{\mu}{\mu_c}\right)\right) \Big/ 4\left(1+\lambda\right) \tan^2\alpha F(t)\right] - e^2\right\} \tag{60}$$

$$e = \frac{1.9H_p^{5/8}}{E_e^{1/2}\rho_p^{1/8}V_p^{1/4}} \quad (61) \qquad F(t) \approx 0.02 - 0.05 \quad (62)$$

Model 32 – Ahlert [37] proposed an empirical correlation to determine the erosion rate for carbon steels with wet or dry surfaces.

$$\varepsilon_m = 2.17 \times 10^{-7} \times (BH)^{-0.59} F_s V_p^{2.41} F(\alpha) \quad (63) \qquad F(\alpha) = \sum_{i=1}^{5} A_i \alpha^i \quad (64)$$

Model 33 – Haugen et al. [38] studied the erosion from choke valves used in the oil and gas industry.

$$\varepsilon_m = MK_p F(\alpha) V_p^{n_p} \quad (65)$$

Model 34 – Chen et al. [39] proposed an erosion model for the normal impact of solid particles based on a residual tensile stress mechanism.

$$\varepsilon_m = 0.064 \frac{\rho_t \rho_p V_p^3}{\bar{p}^{3/2}\left[D_1 + D_2 \exp\left(D_3 \sigma^*\right)\right]^2 \left[1 + D_4 \ln \dot{\varepsilon}^*\right]^2 \left(1 + D_5 T^*\right)^2} \quad (66)$$

Model 35 – Chen et al. [40] continued the research and proposed a new general model based on the friction generated at particle impact.

$$\varepsilon_D = \frac{T_y}{m_p \delta} = \frac{\Sigma m_p V_{pn} \Delta V_{pn}}{m_p \delta} \quad (67) \qquad \varepsilon_C = \frac{T_\chi}{m_p \chi} = \frac{\Sigma m_p V_{pt} \Delta V_{pt}}{m_p \delta \chi} \quad (68) \qquad T_t = T_y + T_\chi \quad (61) \qquad \varepsilon_V = \varepsilon_C + \varepsilon_D \quad (69)$$

Model 36 – Levin et al. [41] developed an erosion model that takes into account the mechanical properties of the target material and the hardening effect at the particle impact.

$$\varepsilon_m \propto \frac{m_p V_p^2}{2} \left\{ \frac{\left[1 - \left(\frac{3.06 H_t^{5/4}}{\rho_p V_p^{1/2}}\right)\left(\frac{1-\mu_t^2}{E_t} + \frac{1-\mu_p^2}{E_p}\right)\right]}{T \cdot L_V} \right\} \quad (70)$$

Model 37 – Oka et al. [42] proposed an erosion model that considers the mechanical properties of the target material and the impact parameters. The model includes the repeated plastic deformation of the target material because of normal particle impact and the cutting process generated at low impingement angles.

$$\varepsilon(\alpha) = f(\alpha)\varepsilon_{90} \quad (71) \qquad f(\alpha) = (\sin\alpha)^{n_1}(1 + H_V(1 - \sin\alpha))^{n_2} \quad (72) \qquad \varepsilon_{90} = K_p (H_V)^{k_1} (V_p)^{k_2} (d_p \times 10^{-6})^{k_3} \quad (73)$$

Model 38 – Oka et al. [43] modified the previous proposed model for a better correlation.

$$\varepsilon(\alpha) = f(\alpha)\varepsilon_{90} \quad (74) \qquad f(\alpha) = (\sin\alpha)^{n_1}(1 + H_V(1 - \sin\alpha))^{n_2} \quad (75) \qquad \varepsilon_{90} = K_p (k_4 H_V)^{k_1 b} \left(\frac{V_p}{v}\right)^{k_2} \left(\frac{d}{d}\right)^{k_3} \quad (76)$$

Model 39 – Huang et al. [44] developed a phenomenological model to determine the erosion, assuming that particle impact generating normal and tangential forces determines material deformation.

$$\varepsilon_{VT} = \varepsilon_{VC} + \varepsilon_{VD} \quad (77) \qquad \varepsilon_{VD} = C_1 \frac{m_p \rho_p^{1/4b'} \left(V_p \sin\alpha\right)^{2+(1/2b')}}{\varepsilon_c^{1/b'} p_n^{1+(1/4b')}} \quad (78)$$

$$\varepsilon_{VC} = \frac{C_2 m_p^{1+[3(1-n_s)/4]} V_p^{2+[3(1-n_s)/2]} (\cos\alpha)^2 (\sin\alpha)^{[3(1-n_s)/2)]}}{d_p^{(1-n_s)/4)} \varepsilon_0^i p_t p_n^{3(1-n_s)/4)}} \quad (79)$$

Model 40 – Nsoesie et al. [45] studied the erosion mechanism for five Stellite alloys.

$$\varepsilon_V = \frac{C_1 d_p^3 \left(V_p \left(A \left(\sin(\alpha/2)^{1/3}\right)\right)^B\right)^3 \rho_p^{3/2}}{H_V^{3/2}} \quad (80)$$

3. Empirical equations and models developed for erosion phenomena analysis in pipe bends

Model 41 – API RP 14E [46] – as per the standard, to prevent the apparition of erosion phenomena, it is recommended to have a threshold for the flow velocity when designing the piping or pipeline system.

$$V_e = \frac{C}{\sqrt{\rho_m}} \tag{81}$$

Model 42 – Salama and Venkatesh [47] studied the limitation of the API RP 14E model, considering that the approach is conservative based on experimental testing and proposed the following equation:

$$V_e = \frac{300}{\sqrt{\rho_m}} \tag{82}$$

Model 43 – Salama and Venkatesh [48] continued the experimental testing and based on the results of Rabinowicz [49] proposed the following model for pipe bends. The model is conservative and overestimates the experimental results with a factor of 1.44.

$$ER = 1.86 \cdot 10^5 \frac{W_p V_f^2}{PD^2} \tag{83}$$

Model 44 – Bourgoyne [50] suggested two equations to determine the erosion rate, one for gas conveying systems (74), and the other for liquid flows (75).

$$ER = F_e \frac{\rho_p}{\rho_t} \frac{W_p}{A} \left(\frac{V_{SG}}{100\alpha_g} \right)^2 \tag{84}$$

$$ER = F_e \frac{\rho_p}{\rho_t} \frac{W_p}{A} \left(\frac{V_{SL}}{100 H_L} \right)^2 \tag{85}$$

Model 45 – Svedman et al. [51] studied the applicability of Bourgoyne's model for low speed flows and suggested an equation to limit the flow speed in order to limit the tolerable erosion rate at 5 MPY.

$$V_e = K_S \frac{D}{\sqrt{W_p}} \tag{86}$$

$$K_S = 1.34 \textit{ for long radius elbows} \qquad K_S = 7.04 \textit{ for tees} \tag{87}$$

Model 46 – Jordan [52] also proposed a model for erosion rate calculation.

$$ER = 10^{C_1} V_{SG}^{2.349} W_p^{0.9535} \left(1 - \left(1 + \frac{1}{2r_c} \right)^{-2} \right)^{1.885/2} \tag{88}$$

Model 47 – McLaury and Shirazi [53] from Tulsa University E/CRC – Erosion/Corossion Research Centre proposed a semi-empirical model to calculate erosion in pipe bends and tees.

$$ER = F_M F_S F_P F_{r/D} \frac{W_p V_L^n}{\left(D/D_0 \right)^2} \tag{89}$$

Model 48 – DNVGL-RP-O501 [54] - as per the standard, the following relations can be used for erosion rate calculation in straight pipes (90) and in pipe bends (91).

$$E_{L,y} = 2.5 \cdot 10^{-5} \cdot U_p^{2.6} \cdot D^{-2} \cdot m_p \tag{90}$$

$$E = \frac{K \cdot U_p^n \cdot F(\alpha)}{\rho_t \cdot A_t} \cdot G \cdot C_1 \cdot GF \cdot M_p \cdot 10^3 \tag{91}$$

4. Systematic representation of available wear models and equations

The authors came up with a visual representation of all the discussed erosion models, both general and empirical for pipe bends. Taking some ideas from the periodic table of elements, the models are represented based on number, author name, year, plus details related to the parameters used in the equations. There are three criteria used for parameter categorization, solid particle material properties, target material properties and experimental working conditions. For the cases where the authors contributed with more than one model, the same colour is used. The "Periodic table of erosion models" can be amended whenever a new model arises that ensures the conditions from the first section.

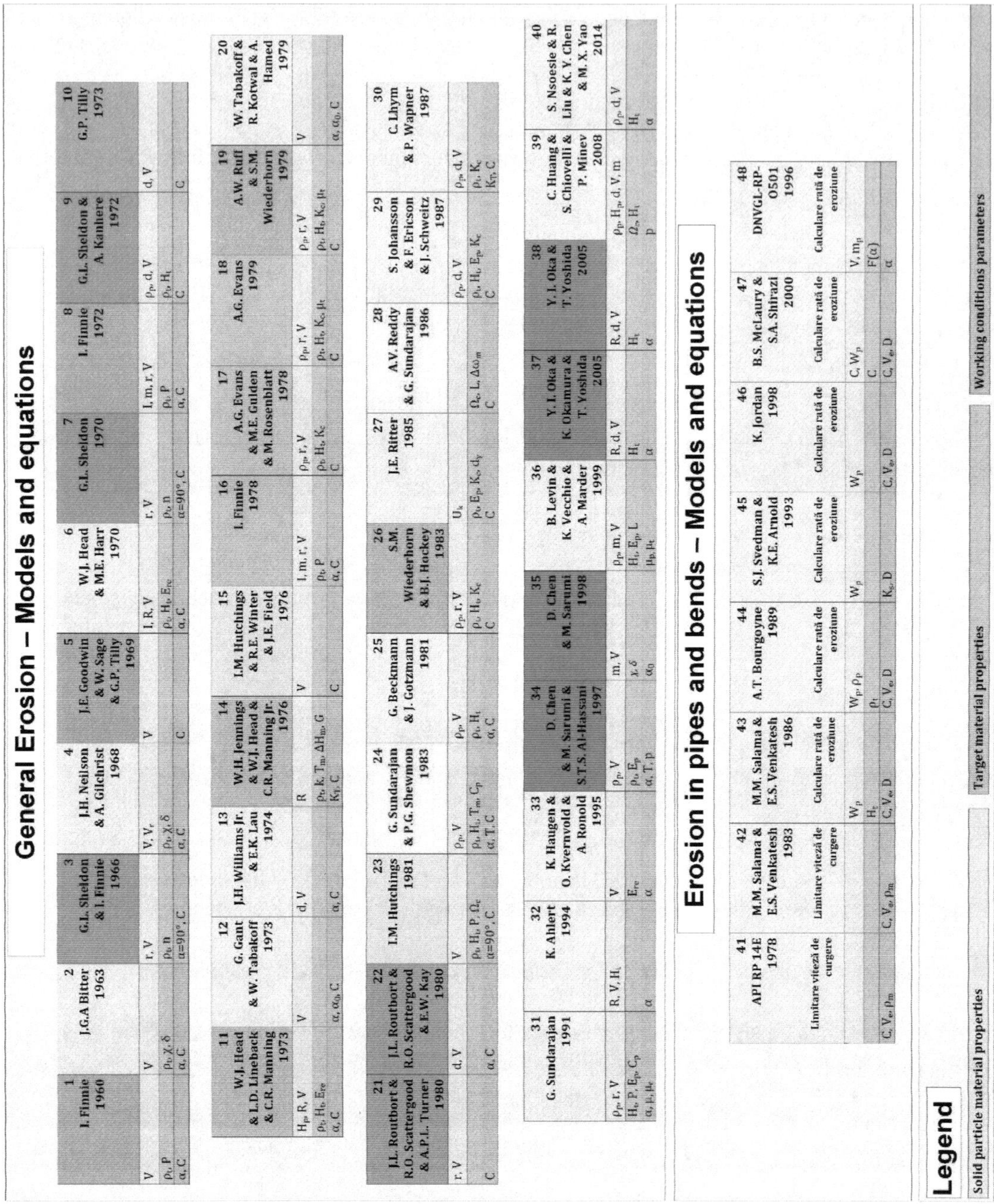

Figure 1. Periodic table of erosion models

The parameters from the "Periodic table of erosion models" have the following meaning: ρ_p – *particle density;* H_p – *particle hardness;* I – *moment of inertia;* R – *particle circularity;* m – *particle mass;* d – *particle diameter;* r – *particle radius;* V – *particle speed;* V_r - *rebound speed;* U_k - *kinetic energy;* ρ_t – *material density;* H_t – *material hardness;* P – *flow stress;* E_p - *Young's modulus;* K_c – *fracture toughness;* Ω_c – *critical strain;* L – *depth of deformation;* $\Delta\omega_m$ – *incremental strain per impact;* k – *thermal conductivity;* T_m – *melting temperature;* ΔH_m - *enthalpy of melting;* χ – *cutting energy;* δ – *deformation energy;* E_{re} – *erosion resistance;* C_p – *heat capacity;* G - *grain molecular weight;* n - *Weibull flaw* *parameter; μ_t - Lame constant; d_γ - grain diameter; μ - friction coefficient; μ_c - Critical friction coefficient; μ_p - Poisson coefficient; μ_t - critical Poisson coefficient; α - Impingement angle; α_0 - maximum wear impingement angle; K_T - Kinetic energy transfer from particle to target; T – temperature; p – pressure; C – different constants.*

5. Conclusions

The presented models have an average of 6 parameters as factors in the equations, the highest number of parameters being 11 and the lowest 2. 28 of the models take into account the working conditions by incorporating different constants and 23 out of them have included the impingement angle of the solid particles in the relations. 37 models are factoring the solid particle speed, 20 of them the particle dimension and 14 of them the particle density. 22 models included the target material density, 17 of them the target material hardness and 7 of them the fracture toughness. All the models provide acceptable results with specific materials and working conditions as per the experimental conditions, however they lack a general applicability.

For pipeline practical applications, in case of complex geometrical layouts or when the exact location of the erosion attack is important it is recommended to use Computational Fluid Analysis (CFD). There is no convergence between the models proposed by the above-mentioned authors, the parameters used as factors being quite different between equations. This shows a need for further research by looking at what parameters are really important in the erosion phenomena. The "Periodic table of erosion models" will be further amended in future work with the models used in various commercial CFD software.

6. References

[1] Det Norske Vertitas 2015 Recommended practice DNVGL-RP-O501 Managing sand production and erosion
[2] Salama M M and Venkatesh E S 1983 *15th Annual Offshore Technol. Conf. Houston Texas* pp 371 – 376
[3] Venkatesh E S 1986 *Proc. Rocky Mountain Meeting of SPE Billings MT* pp 489-497
[4] Meng H C, Ludema K C 1995 *Wear* **181–183(2)** pp 443-457
[5] Parsi M, Najmi K, Najafifard F, Hassani S, McLaury B S and Shirazi S A 2014 *J. Nat. Gas Sci. Eng.* **21** pp 850-873
[6] Finnie I 1960 *Wear* **3(2)** pp 87-103
[7] Bitter J G A 1963 *Wear* **6(1)** pp 5-21
[8] Sheldon G L and Finnie I 1966 *J. Energ. Resour. -ASME* **88(4)** pp 393-399
[9] Neilson J H and Gilchrist A 1968 *Wear* **11(2)** pp 111-122
[10] Goodwin E, Sage J and Tilly W P 1969 *Proc. Inst. Mech. Eng.* 1-196 pp 279-292
[11] Head W J, Harr M E 1970 *Wear* **15(1)** pp 1-46
[12] Sheldon G L 1970 Similarities and differences in the erosion behaviour of materials *J. Basic Eng.*
[13] Finnie I 1972 *Wear* **19(1)** pp 81-90
[14] Sheldon G L and Kanhere A 1972 *Wear* **21(1)** pp 195-209
[15] Tilly G P 1973 *Wear* **23(1)** pp 87-96
[16] Head W J, Lineback L D and Manning C R 1973 *Wear* **23(3)** pp 291-298
[17] Grant G and Tabakoff W 1973 An experimental investigation of the erosive characteristics of 2024 aluminium alloy *Dept. of Aerospace Eng. Univ. Cincinnati*
[18] Williams J H and Lau E K 1974 *Wear* **29(2)** pp 219-230
[19] Jennings W H, Head W J and Manning C R 1976 *Wear* **40(1)** pp 93-112
[20] Hutchings I M, Winter R E and Field J E 1976 *P. Roy. Soc. A-Math. Phy.* **348(1654)** pp. 379-392
[21] Finnie I, McFadden D H 1978 *Wear* **48(1)** pp 181-190
[22] Evans A G, Gulden M E and Rosenblatt M 1978 *P. Roy. Soc. A-Math. Phy.* **361(1706)** pp. 343-365
[23] Evans A G 1979 *Treatise on Mater. Sci. Technol.* **16** pp. 1-67
[24] Ruff A W and Wiederhorn S M 1979 Erosion by solid particle impact *National bureau of standards Gaithersburg md national measurement lab*
[25] Tabakoff W, Kotwal R and Hamed A 1979 *Wear* **52(1)** pp 161-173

[26] Routbort J L, Scattergood R O and Turner A P L 1980 *Wear* **59(2)** pp 363-375
[27] Routbort J L, Scattergood R O and Kay E W 1980 *J Am. Ceram. Soc.* **63** pp 635-640
[28] Hutchings I M 1981 *Wear* **70(3)** pp 269-281
[29] Sundararajan G and Shewmon P G 1983 *Wear* **84(2)** pp 237-258
[30] Beckmann G and Gotzmann J 1981 *Wear* **73(2)** pp 325-353
[31] Wiederhorn S M and Hockey B J 1983 *J. Mater. Sci.* **18(3)** pp 766-780
[32] Ritter J E 1985 *Mater. Sci. Eng.* **71** pp 195-201
[33] Reddy A V and Sundararajan G 1986 *Wear* **111(3)** pp 313-323
[34] Johansson S, Ericson F and Schweitz J 1987 *Wear* **115(1-2)** pp 107-120
[35] C. Lhymn C and Wapner P 1987 *Wear* **119(1)** pp 1-11
[36] Sundararajan G 1991 *Wear* **149(1-2)** pp 111-127
[37] Ahlert K 1994 *M. Sc. Thesis Dept. Mech. Eng. Univ. Tulsa*
[38] Haugen K, Kvernvold O, Ronold A and Sandberg R 1995 *Wear* **186–187**(1) pp 179-188
[39] Chen D, Sarumi M, Al-Hassani S T S, Gan S and Yin Z 1997 *Wear* **205(1–2)** pp 32-39
[40] Chen D, Sarumi M and Al-Hassani S T S 1998 *Wear* **214(1)** pp 64-73
[41] Levin B, Vecchio K and Marder A 1999 *Metall. Mater. Trans. A 30A* pp 1763-1774
[42] Oka Y I Okamura K and Yoshida T 2005 *Wear* **259(1–6)** pp 95-101
[43] Oka Y I and Yoshida T 2005 *Wear* **259(1–6)** pp 102-109
[44] Huang C et al. 2008 *Powder Technol.* **187(3)** pp 273-279
[45] Nsoesie S, Liu R, Chen K Y and Yao M X 2014 *Wear* **309(1–2)** pp 226-232
[46] API 14E Recommended practice for design and installation of offshore production platform piping systems 5th edition 1991 (revised 2000)
[47] Salama M M, and Venkatesh E S 1983 Evaluation of API RP 14E erosional velocity limitations for offshore gas wells *Offshore Technol. Conf*
[48] Venkatesh E S 1986 Erosion damage in oil and gas wells *Proc. Rocky Mountain Meeting of SPE Billings MT*
[49] Rabinowicz E 1979 *Intl. Conf. on Erosion by Liquid and Solid Impact 5th Cambridge England* pp. 38-1
[50] Bourgoyne A T 1989 Experimental study of erosion in diverter systems due to sand production *SPE*
[51] Svedeman S J and Arnold K E 1993 Criteria for sizing multiphase flow lines for erosive-corrosive service *Proc. of 68th SPE Annual Fall Technical Conf. Houston*
[52] Jordan K 1998 Erosion in multiphase production of oil & gas corrosion *NACE Intl. Annual Conf. San Antonio*
[53] Shirazi S A, Shadley J R, McLaury B S and Rybicki E F 1995 *J. Press. Vess. - T. ASME* **117 (1995)** pp 45-52
[54] McLaury B S and Shirazi S A 2000 *J. Energ. Resour. - ASME* **122** pp 115-122

IOP Conf. Series: Materials Science and Engineering **724** (2020) 012038 doi:10.1088/1757-899X/724/1/012038

The Effect of Working Parameters upon Oil-Lubricated EHD Contacts Subjected to Load Variation

X Zhang, R Glovnea

Department of Engineering and Design, University of Sussex, Brighton, BN1 9QT United Kingdom

E-mail: Xingnan.Zhang@sussex.ac.uk, R.P.Glovnea@sussex.ac.uk

Abstract. The technique of optical interferometry incorporated within a ball-on-disc apparatus has been widely used in the field of the experimental research in thin-film lubrication for approximately the past five decades. This paper presents the theoretical analysis of the dynamic response of an optical interferometry based experimental rig adapted for non-steady work. The dynamic response of the rig was analytically modelled as a three-degree of freedom, forced harmonic vibrating system with damping. Quantitative evaluation of damping of EHD contact was not one of the objectives of this paper, but the results obtained showed the opportunity to try a rather qualitative analysis. The aims of this experimental study are to investigate the effect of entrainment speed and overall film thickness on the oil film thickness of an EHD contact subjected to load cyclic variation. Experimental tests were carried out using different type of base oils with various viscosity. The behaviour of the EHD films was studied and analysed under forced harmonic load vibrations with frequency up to 100 Hz. The tests performed with lubricants of lower viscosity have shown that deviations from steady state film thickness become less and less pronounced as the viscosity of the lubricant decreases. When the amplitude of the oscillatory motion is significantly larger than the deformations of the surfaces under loading, it was observed that fluid entrainment is less important and film squeeze plays the dominant role in the film behaviour.

1. Introduction

Rolling bearings are the second most widely used machine components [1]. They work in what it is called elastohydrodynamic (EHD) lubrication regime. This lubrication regime is governed by the hydrodynamic action of the lubricant forced into a converging contact region, local surface deformations of similar magnitude to the lubricant film thickness and the variation of lubricant's viscosity with pressure. Under realistic working conditions, EHD contacts in rolling bearings are almost un-avoidably subjected to transient conditions such as variation of load and speed, passage of impurities, or vibrations generated by the equipment or machine they are part of. Transient conditions make the experimental measurement of EHD film thickness challenging. The influence of transient conditions of speed or load upon the behaviour of the elastohydrodynamic contacts has been investigated in the past, both experimentally and theoretically [2 – 5], however experimental studies dedicated to the effect of various working parameters on the formation of the EHD film, under load variation, are few. Glovnea and Spikes [6] and Wijnant et al [7] have shown that rapid change of speed, respective load induce fluctuations in the EHD film thickness. The amplitude of these fluctuations decreases with time, after a few cycles, very much alike the dynamic response of a mass/spring/damper system. Kilali et al [8] reported a new test rig capable of loading an EHD contact both as an impact load and as a continuously variable load. The reported results were rather limited. Ciulli and Bassani [9] also used optical interferometry to study the film thickness in a contact subjected to random vibrations. Cann and Lubricht [10] studied the effect of low – frequency loading/unloading upon the track replenishment of a rolling EHD contact. They noticed the beneficial effect of these conditions upon the lubricant film thickness.

The behaviour of EHD films subjected to vibrations normal to the plane of the contact is influenced by two mechanisms: fluid entrainment and film squeeze. These mechanisms shape two phenomena which determine the lubricant film thickness: the dynamic response of the film and its deviation from normal EHD film shape.

This paper firstly presents the analysis of dynamic response of an optical interferometry based experimental rig adapted for non-steady work in the experimental setup section. Quantitative evaluation of the damping of EHD contacts was not one of the initial objectives of this research, but the results obtained showed the opportunity to try a rather qualitative analysis. Secondly, experimental results of the effect of entrainment speed and overall film thickness on the oil film thickness in an EHD contact subjected to load variation are shown and discussed.

2. Experimental rig setup and dynamic response analysis

The experimental method for measuring film thickness in this paper is the optical interferometry technique. The principle of optical interferometry and its various ways in which it is applied to the study of elastohydrodynamic contacts has been published extensively during the past fifty years [11-15] and it will not be detailed in this paper.

Figure 1a shows a sketch of the test rig while Figure 1b shows the schematic of the vibrating masses.

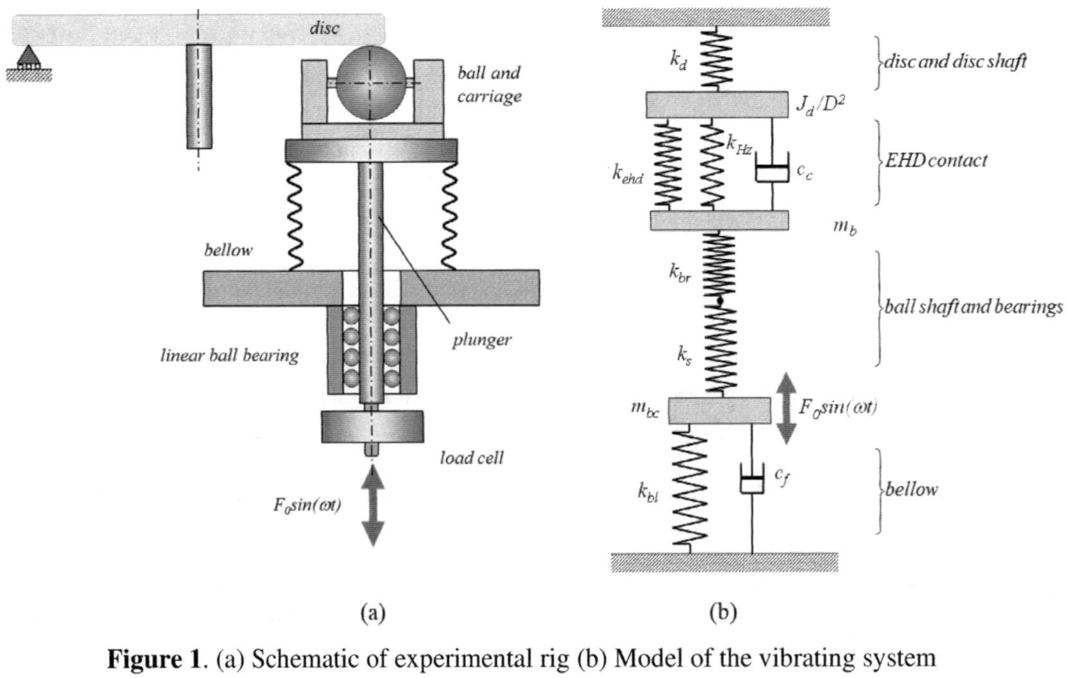

Figure 1. (a) Schematic of experimental rig (b) Model of the vibrating system

Details of film thickness calibration procedure and experimental hardware introduction are provided in [17]. In order to increase the stiffness of the disc and its shaft, the disc is supported at a location diametrically opposite the ball, by a rolling support shown schematically like a simple support in figure 1. The load cell records the variable force applied to the plunger and thus to the ball and the EHD contact. The stainless-steel bellow is intended to avoid any leakage of lubricant from the working chamber, but its construction makes it practically a linear spring. As seen in figure 1b the EHD contact acts as a spring and damper; its stiffness and damping coefficient are denoted with k_c and c_c respectively. The contact (as an elastic element of stiffness k_{Hz}) is connected to the oscillating mass of the ball m_b via the shaft supporting the ball and the bearings supporting the ball shaft. The stiffness of these elements is denoted by k_s for the shaft and k_{br} for the bearings. The contact "spring" is connected to the fixed support through the disc which is considered a spring of constant k_d. The second mass of this system is that of the ball carriage and plunger denoted by m_{bc}. These at their turn are connected to the rigid support through the bellow seen as a spring of stiffness k_b. There is also friction in the linear bearing supporting the plunger, represented here by the damping c_f. This is a three-degree of freedom, forced vibration system with damping, which can be solved for the motion of the mass of the ball and disc, which are of interest.

The stiffness of the dry Hertzian contact can be found from the relationship between the displacement of two distant points and the load, as shown below.

$$\delta = \left(\frac{9P^2}{16E^{*2}R} \right)^{1/3} \tag{1}$$

Obviously the EHD contact is a non-linear spring with the stiffness depending on the load. The stiffness of the dry Hertzian contact is obtained by inversing the ratio of the derivative of the displacement with respect to the load P, as shown below.

$$k_{Hz} = \frac{dP}{d\delta} = \left(6E^{*2}RP\right)^{1/3} \tag{2}$$

This stiffness varies with load can be seen in figure 2. After a sharp rise at lower loads the gradient of the curve becomes smaller and the curve would eventually flatten out at much higher loads.

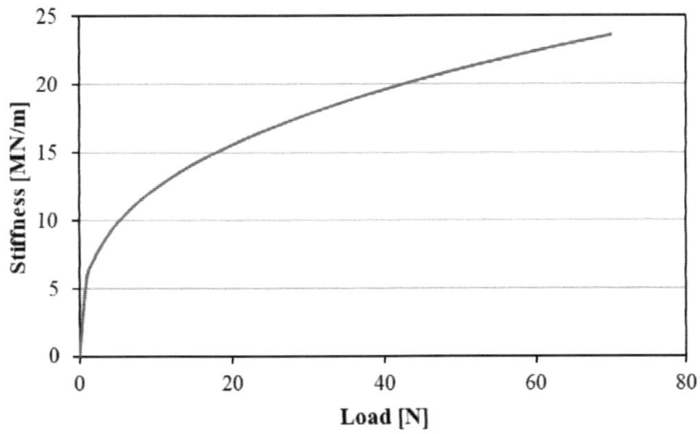

Figure 2. Stiffness of contact variation with load

This small change of the stiffness with load, at loads higher than about 4 N, prompts an approximation: the dry Hertzian contact can be considered as a linear spring, over the most part of the loading cycle.

The EHD film also depends on the load, so it too can be considered as a non-linear spring. In order to find the stiffness of this spring, the same method is applied. The central film thickness for a circular contact is given by the relationship:

$$\frac{h_c}{R_x} = 2.69\overline{U}^{0.67}\overline{G}^{0.53}\overline{P}^{-0.067} \tag{3}$$

Where $\bar{P}$ is the load factor, $\bar{G}$ is the material parameter and $\bar{U}$ is speed parameter. Grouping all factors that do not depend on the load in one constant A, and differentiating film thickness with respect to the load, the inverse ratio yields the stiffness of the EHD film:

$$k_{EHD} = \frac{\partial P}{\partial h} = -\frac{15}{A} \cdot P^{32/30} \tag{4}$$

The variation of this stiffness with the load is depicted in figure 3.

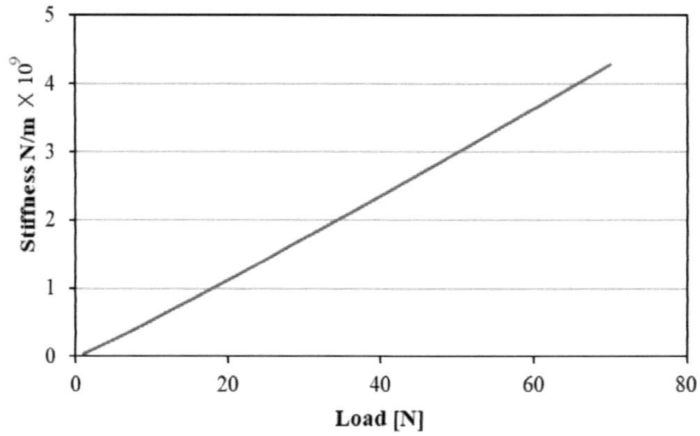

Figure 3. Stiffness of the EHD film.

The stiffness of the dry contact and that of the EHD film are connected in series thus the combined stiffness is given by:

$$\frac{1}{k_c} = \frac{1}{k_{Hz}} + \frac{1}{k_{EHD}} \tag{5}$$

By comparing the values of stiffness in figure 2 and 3, it is seen that the stiffness of the EHD film is two orders of magnitude larger than that of the dry Hertzian contact, thus from equation above it is clear that the later dominates the stiffness of the whole EHD contact. It thus appears that the stiffness of the EHD contact is equal to that of the dry Hertzian contact and it can be assumed a linear spring, according to the approximation mentioned above, which simplifies considerably the modelling of the whole system.

On the other hand, the damping constant of the EHD film (c_c) can also be assumed as non – linear and it is unknown at this stage. For the sake of simplicity of this analysis, it is further assumed that the damping is of viscous type, thus the motion – resisting force is proportional to the speed. Detailed stiffness calculations and comparisons of experimental parts showed in figure 1(b) as well as EHD friction force results are presented in [16]. With the mechanical characteristics of the system known, the equations of motion can be written as:

$$\begin{bmatrix} 1 & 0 & 0 & 0 & 0 & 0 \\ 0 & 1 & 0 & 0 & 0 & 0 \\ 0 & 0 & 1 & 0 & 0 & 0 \\ 0 & 0 & 0 & m_{bc} & 0 & 0 \\ 0 & 0 & 0 & 0 & m_b & 0 \\ 0 & 0 & 0 & 0 & 0 & J/D^2 \end{bmatrix} \frac{d}{dt} \begin{bmatrix} x_{bc} \\ x_b \\ x_d \\ \dot{x}_{bc} \\ \dot{x}_b \\ \dot{x}_d \end{bmatrix} + \begin{bmatrix} 0 & 0 & 0 & -1 & 0 & 0 \\ 0 & 0 & 0 & 0 & -1 & 0 \\ 0 & 0 & 0 & 0 & 0 & -1 \\ (k_{bl}+k_b) & -k_b & 0 & 0 & 0 & 0 \\ -k_b & (k_b+k_c) & -k_c & 0 & c_c & -c_c \\ 0 & -k_c & (k_c+k_d) & 0 & -c_c & c_c \end{bmatrix} \begin{bmatrix} x_{bc} \\ x_b \\ x_d \\ \dot{x}_{bc} \\ \dot{x}_b \\ \dot{x}_d \end{bmatrix} = \begin{bmatrix} 0 \\ 0 \\ 0 \\ 0 \\ F_0 \\ 0 \end{bmatrix} \cos\omega t \tag{6}$$

In this equation $\dot{x}_b$, $\dot{x}_{bc}$ and $\dot{x}_d$ are the velocities of the ball, ball carriage and disc respectively, while J is the moment of inertia of the disc about an axis (perpendicular to the plane of the drawing) passing through the fixed support shown in figure 1.

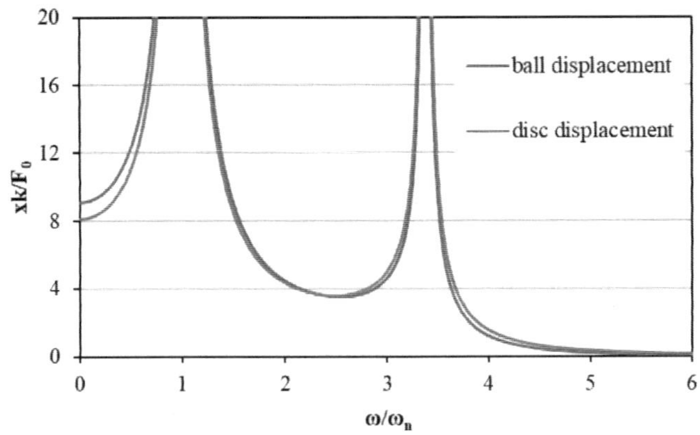

Figure 4. Displacement of ball and disc

The normalised displacements of the ball and the disc which are of interest here are shown in figure 4. Stiffness k and amplitude of the load F0, in these graphs were taken as 14,314MN/m and 20 N respectively. The first mode for the oscillations of the ball and disc is at 2140 Hz and the second is at 7270 Hz. Obviously, these are much larger than the frequency of variation of the force employed in these tests. The absolute values of the displacement of the ball and disc are of no real practical interest, but the difference between them it is. The variation of this difference is shown in figure 5, for frequencies around the first mode, which are more relevant in any applications.

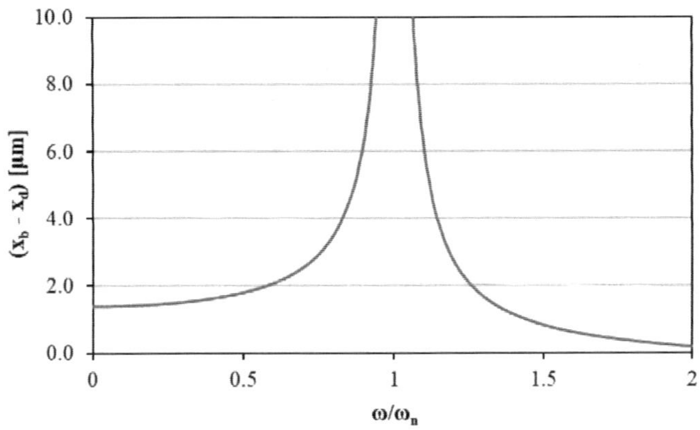

Figure 5. Displacement of the ball relative to the disc

The displacement of the ball relative to the disc is shown in real values instead of normalised ones. This gives a better idea of how realistic this simulation is. It can be seen that the value, 1.39 μm, at the largest frequency of the test that is 100 Hz ($\omega/\omega_n = 0.046$), is remarkably close to the amplitude resulted from Hertz's theory, which is 1.67 μm. The damping constant used in the above calculations varied between 0.01 Ns/m and 314 Ns/m, but the results show that at least at lower frequencies the displacement of the masses is insensitive to this parameter. There are other observations which can be withdrawn from this analysis. If the disc is made rigid, then the natural frequencies of the system shift toward larger values. For example, the first mode changes from 2140 Hz to 3600 Hz. The degree of freedom of the system reduces by one in this case. For a two–degree of freedom the system's response changes with the damping constant, but only at larger frequencies. This is shown in figure 6, where the response of a two–degree of freedom system is shown. As seen, the values of the damping constant change the response visibly, only at frequencies around the second mode.

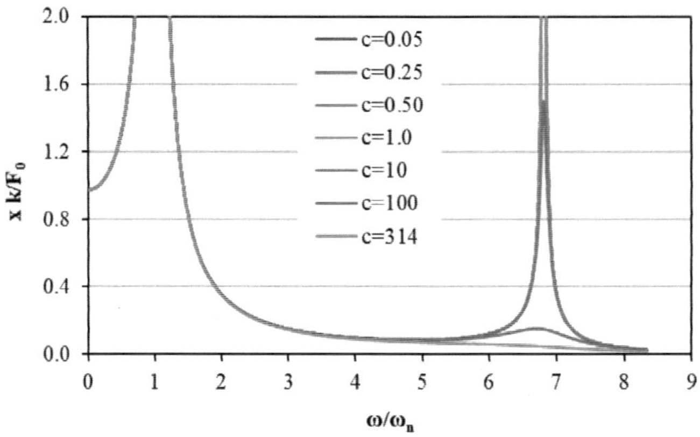

Figure 6. Response of the system

This theoretical dynamic analysis can be used to evaluate the damping constant of the EHD films. By attaching a large mass to the ball carriage, the natural frequency of the system can be lowered to values around 70 Hz – 80 Hz, which is within the range conveniently covered by the current optical system. Subsequently measuring the lubricant film thickness over a sweep of frequencies, encompassing the natural system, the damping characteristics of the EHD film can be evaluated by fitting the model to the experimental values. This is planned for future research.

3. Test condition and lubricants

The load was varied sinusoidally from zero (or close to zero) to about 40 N. For the pair of materials used which have a reduced elastic modulus $E' = 55$GPa, this means a Hertzian pressure variation between zero and around 0.64 GPa. The frequency of the oscillatory motion of the ball (19.05 mm diameter) was 100 Hz throughout these tests. In order to get a good number of images for each cycle of oscillation, even at larger frequencies, a speed of 2000 frames per second was set on the camera. Many lubricants were tested in this research, but for limitations of space only the results for lubricants of three different viscosities will be shown. The kinematic viscosities at are: *Lubricant 1* - 339.8 mm^2/s, *Lubricant 2* - 135.8 mm^2/s and *Lubricant 3* - 17 mm^2/s.

4. Result and discussion

Experimental tests regarding the effect of overall film thickness of EHD contact subjected to load variation were carried out with two types of lubricant at different experimental working conditions. The images shown in Figure 7 were recorded at 100 Hz with an entrainment speed of 0.05 m/s for *Lubricant 1*, images of figure 8 are for *Lubricant 2*, and images of figure 9 are for *Lubricant 3*. The images shown in figure 13 were also recorded at 100 Hz with a slightly increased entrainment speed of 0.1 m/s for *Lubricant 3*. Loading conditions for figure 7, figure 8 and figure 9 were identical, however the load variation for figure 13 is between 0 N and 39 N.

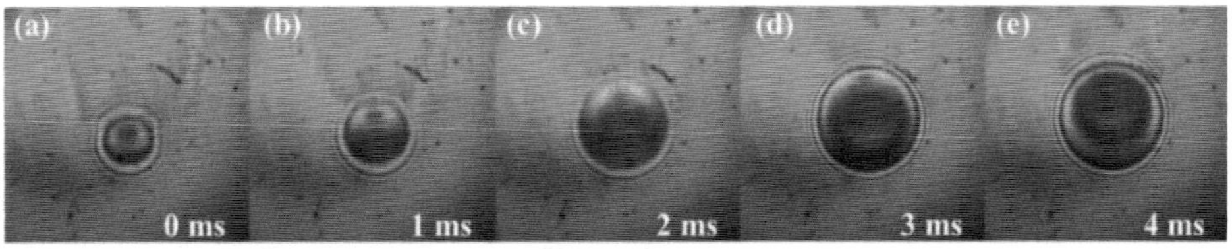

Figure 7. Selected images of the EHD contact for 100 Hz, 0.05 m/s, Lubricant 1.

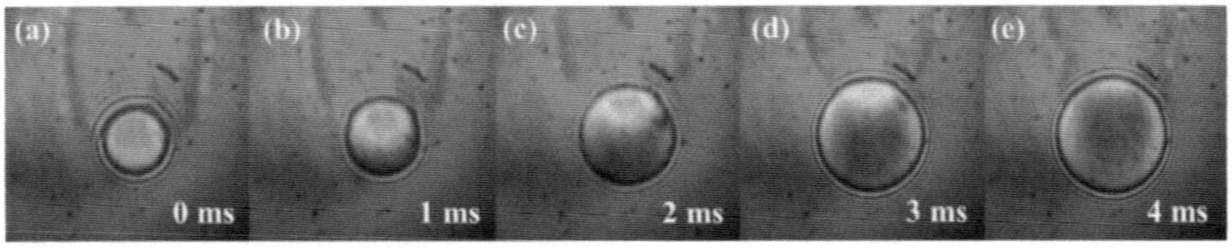

Figure 8. Selected images of the EHD contact for 100 Hz, 0.05 m/s, Lubricant 2.

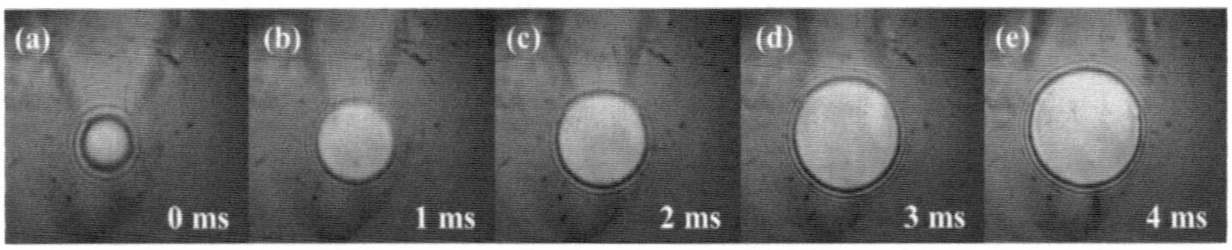

Figure 9. Selected images of the EHD contact for 100 Hz, 0.05 m/s, Lubricant 3.

It has shown that the deviations from steady state film thickness become less and less pronounced as the viscosity of the lubricant decreases. This is illustrated by the captured images shown in figures 7-9. In terms of the actual values of the film perturbation, the difference is more obvious in figures 10 – 12, which captures the film thickness profile at the same time during the loading cycle for Lubricant 1-3 respectively. As seen, film perturbations of 62 nm for Lubricant 1, around 22 nm for Lubricant 2, and around 11 nm for Lubricant 3 are formed during the load variation cycle.

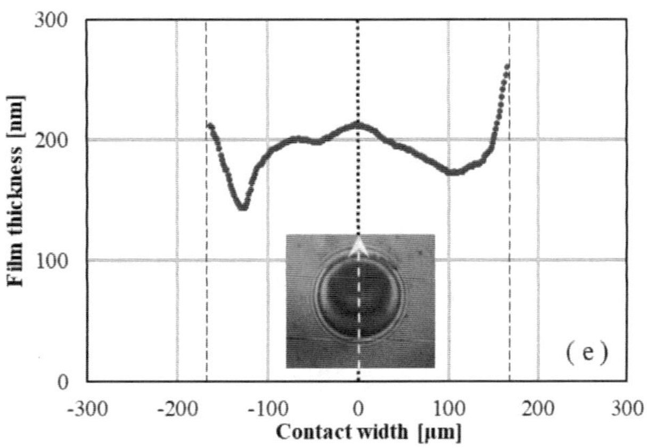

Figure 10. Film thickness profile, 100 Hz, *lubricant 1*.

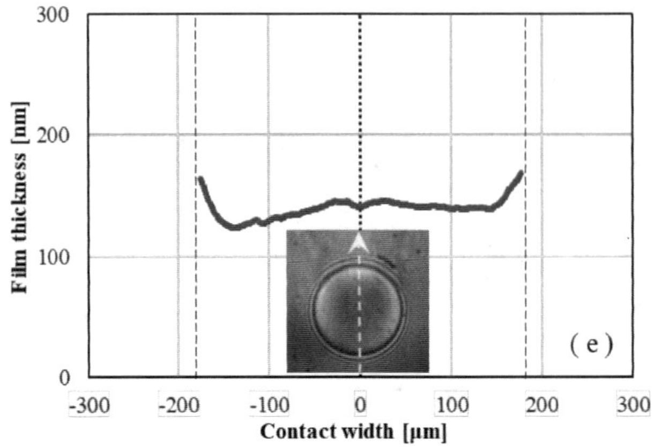

Figure 11. Film thickness profile, 100 Hz, *lubricant 2*.

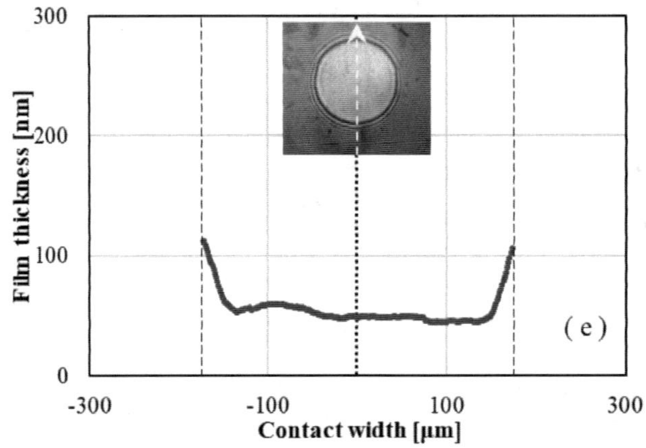

Figure 12. Film thickness profile, 100 Hz, *lubricant 3*.

From the above analysis it is clear that the lower the viscosity of the lubricant, thus the film thickness in steady-state conditions are, the smaller the film perturbation formed during transient loading conditions. This conclusion is apparently contradicted by the results shown in figures 13 and 14 which were taken for Lubricant 3 which has the smallest viscosity among the oils tested here. Even at an entrainment speed of 0.1 m/s this gives a central film thickness of about 50 nm in steady state conditions, lower than the other two lubricants. As seen in Figure 14 this lubricant forms an entrapment of 220 nm which obviously is larger than those formed by the other two lubricants.

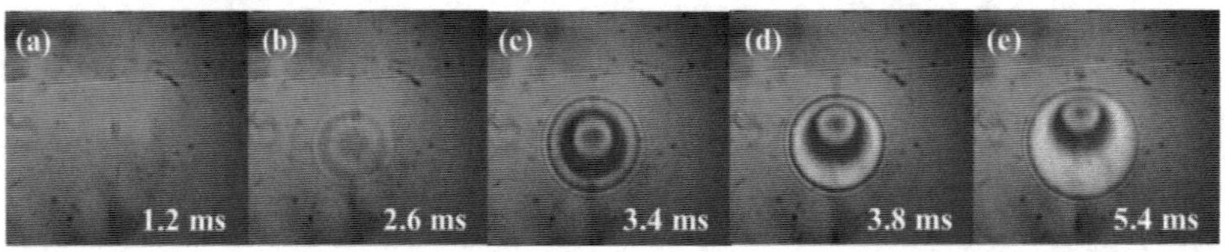

Figure 13. Selected images of the EHD contact for 100 Hz, 0.1 m/s, Lubricant 3.

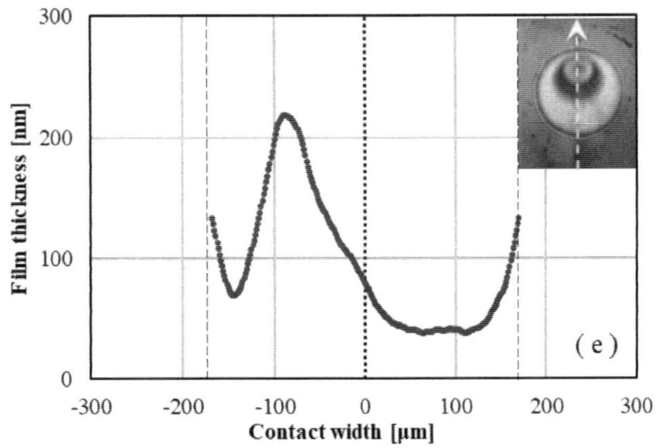

Figure 14. Film thickness profile, 100 Hz, 0.1 m/s, *lubricant 3*.

This phenomenon can only happen when the amplitude of the oscillatory motion is significantly larger than the deformations of the surfaces under loading. In this case the rebounding of the surfaces is large which in turn generate large impact forces. The fluid entrainment is less important in this situation and film squeeze plays the dominant role in the film behaviour. In a way, this is similar to pure squeeze or sudden halting of motion, where larger dimples are formed and trapped inside the contact. In this particular test the maximum height of the perturbation is nearly 4.5 times greater than the value under steady state condition. This implies that impact loading always generates perturbations to the film thickness no matter what the overall film thickness is.

5. Conclusions

This paper shows results of a larger study on the effect of cyclic load variation upon the film thickness of elastohydrodynamic contacts. It also includes a simplified analysis of the dynamic response of an optical interferometry based experimental rig. The dynamic response has been analytically modelled as a three-degree of freedom, forced harmonic vibrating system with damping. Results show that this dynamic analysis can be extended into evaluating the damping constant of EHD films by attaching a large mass to the ball carriage to lower the natural frequency of the whole system.

Experimental tests were carried out using a number of fluids with varied viscosity. The behaviour of the EHD films was studied and analyzed under forced harmonic load vibrations at a frequency of 100 Hz. The test performed with lubricants of lower viscosity have shown that deviations from steady state film thickness become less and less pronounced as the viscosity of the lubricant decreases. When the amplitude of the oscillatory motion is significantly larger than the deformations of the surfaces film perturbation still occurs because in this case film squeeze dominates the fluid entrainment as the mechanism which influences the film.

6. References

[1] Lugt, P. M., 2013, "*Grease Lubrication in Rolling Bearings*", John Wiley and Sons.
[2] Dowson, D., and Higginson, G. R., 1959, "A Numerical Solution to the Elastohydrodynamic Problem", *J. Mech. Eng. Sci.*, **1**, pp 6-15.
[3] Safa, M. M. A., and Gohar, R., 1986, "Pressure Distribution under a Ball Impacting a Thin Lubricant Layer", *Trans. ASME, J. Trib*, **108**, pp 372-376.
[4] Sugimura, J., Jones, W. R. Jr, and Spikes, H. A., 1998, "EHD Film Thickness in Non-Steady State Contacts", *Trans. ASME, J. Trib*, **120**, pp 442-452.
[5] Sakamoto, M., Nishikawa, H., and Kaneta, M., 2003, "Behavior of Point Contact EHL Films under Pulsating Loads", *Transient Processes in Tribology*, Elsevier, **43**, pp 391-399.
[6] Glovnea, R.P. and Spikes, H.A., 2001, "Elastohydrodynamic Film Collapse During Rapid Deceleration-Part I: Experimental Results", *Trans. ASME, J. Trib*, **123**, 2, pp 254-261.

[7] Wijnant, Y. H., Venner, C.H., Larsson, R. and Ericsson, P., 1999, "Effects of Structural Vibrations on the Film Thickness in an EHL Circular Contact", *J. Trib*, **121**, pp 259-264.
[8] Kilali, T. El., Perret-Liaudet, J. and Mazuyer, D., 2003, "Experimental Analysis of a High Pressure Lubricated Contact under Dynamic Normal Excitation Force", *Transient Processes in Tribology*, Elsevier, **43**, pp 409-417.
[9] Ciulli, E. and Bassani, R., 2006, "Influence of Vibrations and Noise on Experimental Results of Lubricated non-Conformal Contacts", *Proc. IMechE.*, **J220**, pp 319-331.
[10] Cann, P. M. and Lubrecht, A. A., 2003, "The Effect of Transient Loading on Contact Replenishment with Lubricating Greases", *Transient Processes in Tribology*, Elsevier, pp 745-750.
[11] Françon, M., 1966, "Optical Interferometry", *Academic Press*
[12] Cameron, A., and Gohar, R., 1966, "Theoretical and Experimental Studies of the Oil Film in Lubricated Point Contacts", *Proc. R. Soc. Lond. A*, **291**, pp 520-536.
[13] Gohar, R. and Cameron, A., 1967, "The Mapping of Elastohydrodynamic Contacts", *ASLE Trans*, **10**, pp 215-225.
[14] Foord, C. A., Hammann, W. C. and Cameron, A., 1968, "Evaluation of Lubricants Using Optical Elastohydrodynamics', *ASLE Trans*, **11**, pp 31-43.
[15] Wedeven, L. J, 1970, "Optical Measurements in Elastohydrodynamic Rolling Contact Bearings", PhD. thesis, London University, London.
[16] Zhang. X, 2017, "The Effect of Vibrations on the Behavior of Lubricated Elastohydrodynamic Contacts. PhD thesis, University of Sussex.
[17] Glovnea, R.P and X. Zhang, 2018, "Elastohydrodynamic Films under Periodic Load Variation: An Experimental and Theoretical Approach", *Trib. Lett*, **66**, pp116.

IOP Conf. Series: Materials Science and Engineering **724** (2020) 012039 doi:10.1088/1757-899X/724/1/012039

An Experimental Approach to Predict the Behaviour of EHD Contacts under Harmonic Loading

X Zhang, R Glovnea

Department of Engineering and Design, University of Sussex, Brighton, BN1 9QT United Kingdom

E-mail: Xingnan.Zhang@sussex.ac.uk, R.P.Glovnea@sussex.ac.uk

Abstract. Rolling element bearings, gears and other machine components operate in the lubrication regime known as elastohydrodynamic (EHD), where lubricant film thickness is governed by the hydrodynamic action of the converging surfaces, the elastic deformation of the non-conforming surfaces and the increase of lubricant's viscosity with pressure. Classic EHD theory indicates that under the assumption of fully flooded conditions in the contact inlet zone the central film thickness increases with the product of entrainment speed and dynamic viscosity and is very little dependent of load variations. EHD lubricating films most often work under transient conditions of load, geometry or speed. These make the behaviour of the lubricant film different from steady-state conditions. Previous studies showed that the squeeze effect generates film perturbations due to entrapment of lubricant inside the contact. This phenomenon becomes significant when both entrainment and squeeze are presented. In practical applications, this type of film perturbation will generate pressure fluctuations, which in turn may influences the fatigue life of the contact. In the present paper the results of a systematic experimental study for predicting the occurrence of film thickness perturbation phenomenon under variable load condition are shown. In the case of a mean load larger than the amplitude of the load variation, it was found that the film perturbation depends of the frequency of the load cycle, the entrainment speed of the lubricant and the overall lubricant film thickness; it also depends of the ratios between the mean load and the amplitude. Film perturbations occur in contacts for which the ratio between half the load cycle period and the average time of transit of the lubricant through the contact in the loading phase of the vibration is less than 2. In the case of mean load smaller than the amplitude, film thickness perturbation has been observed in every single test regardless of entrainment speed, frequency and lubricant viscosity.

1. Introduction

The regime of EHD lubrication under steady-state conditions has now been well understood for more than six decades. Many experimental and numerical studies have been carried out to predict EHD film formation. In reality, EHD-lubricated contacts rarely operate in steady-state conditions; in rolling bearings, it is known that the contacts between the rolling elements and raceways are cyclically loaded/unloaded as the former pass through the loaded zone of the bearing, while in gears, the geometry and operation of these elements makes them prone to cyclic load variation, shock loading and vibrations.

It is well known that the load change has small influence on steady state film thickness. Probably this is why the number of experimental studies carried out on the aspect of vibrating EHD films is much lower than the other transient effects on EHD films. However, load variation due to vibrations of EHD contacts may be more relevant to practical applications. The experimental studies carried out in the past on the effect of dynamic variation of load on the behaviour of EHD point contact can be divided into studies on, pure squeeze motion, pure squeeze motion combined with sliding motion, free vibration with pure rolling motion and forced sinusoidal vibrations.

Larson and Lundberg [1] carried out an experimental study of squeeze films resulted by the impact between a steel ball and a flat glass surface. Dimples observed at the centre of the contact were analysed with a test rig based on optical interferometry technique. A detailed discussion regarding the effect of various parameter on dimple's geometry was given. Kaneta et al [2] also applied optical interferometry technique to directly observe the effect of impact loads on point EHD contact films in two experimental working conditions, that is pure impact to an oily Hertzian contact and impact load under pure rolling conditions. For pure impact conditions, a central dimple was formed, and the maximum film thickness was found at the center of the contact region. For impact under pure rolling condition, a crescent-shaped oil entrapment formed at the inlet of the contact region, which subsequently passed through the contact at approximately the entrainment speed. Sanborn and Winer [3] carried out an experimental investigation to measure of film thickness profile of an EHD point contact during transient loading under pure sliding condition. During transient loading, Hertzian pressure of the EHD contact varied from zero to maximum of 1.03 GPa in about 45 ms. The results showed that there was a negligible effect of transient loading on the lubricant film thickness profile. Wijnant et al. [4] experimentally studied the effects of structural vibrations on the EHD film thickness in pure rolling conditions. Experimental results of film thickness were qualitatively compared with theoretical contact predictions obtained by numerical simulations. The initial load applied was 45 N and transiently increased to 165 N. Results showed that the sudden increase of the load lead to an expansion in contact region consequently inducing film thickness changes propagated along the entraining direction. Kilali et al. [5] developed an optical based EHD rig to study the dynamic response of the contact under harmonic loading. For pure rolling contact conditions, after a shock load, a disturbance of film thickness was observed travelling along the direction of rolling. Sakamoto et al. [6] studied the effect of pulsating and impact loads on EHD point contact under pure rolling condition by optical interferometry technique. More recently, the work carried out in [7] was the only one regarding the effect of sinusoidal variable loading on EHD film, however, the experimental described was rather limited. Fryza et al. [8] experimentally studied the effects of lubricant rheology and impact speed on EHD film thickness at pure squeeze action. Result showed that the entrapped film shape directly depends on loading speed and the central film thickness is mainly determined by the approaching speed.

In the present paper, a systematic experimental study for predicting the occurrence of film thickness perturbation phenomenon under variable load condition has been carried out. The range of frequencies applied in this paper is between 10 Hz and 100 Hz. A summary of occurrence of film thickness perturbation based on results of various types of lubricant is given.

2. Experimental setup and testing conditions

The experimental rig employed in this paper of measuring lubricant film thickness and observing film thickness perturbation phenomenon was built based on optical interferometry technique. The principle of optical interferometry is widely known by the current research field, which can be found from papers such as [9-11]. Spacer layer imaging technique (SLIM) [12] as showed in figure 1(a) was applied to evaluate EHD film thickness during vibration motion which takes advantage of the high precision of another method of measuring lubricant film thickness called Ultra-Thin-Film Interferometry [13]. The schematic of the employed experimental rig is showed in figure 1(b). For the detailed film thickness calibration procedure accuracy, test equipment information and dry contact dimension analysis during vibration can be found in [14]. In brief, the contact was formed between a glass disc and a 19.05 mm diameter steel ball.

Table 1 shows all the performed experiments regarding EHD film subjected to harmonic loading with different lubricants, entrainment speed between 0.05 m/s and 0.5 m/s, frequency range from 10 Hz to 100 Hz. In road vehicles, the IC engine most often runs at its peak efficiency and performance at speed between about 2700 rpm and 3000 rpm. This means that loading/unloading cycles of the shaft supporting the bearings as well as vibrations due to unbalanced masses have most often frequencies anywhere between 5 Hz and 100 Hz, which justifies the choice of the frequency range employed.

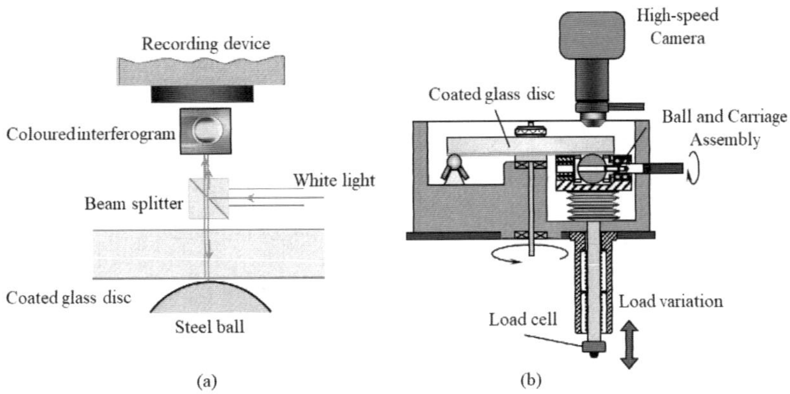

Figure 1. (a) SLIM technique (b) Schematic of experimental rig

Table 1. Performed experiments of EHD film subjected to harmonic loading.

Lubricant	Viscosity	Speed	10 Hz	25 Hz	50 Hz	100 Hz
Lubricant 1	955.7 mm^2/s	0.05 m/s	•	•	•	•
		0.1 m/s	•	•	•	•
		0.2 m/s	•	•	•	•
		0.3 m/s	•	•	•	•
		0.5 m/s	•	•	•	•
Lubricant 2	484.2 mm^2/s	0.05 m/s	•	•	•	•
		0.1 m/s	•	•	•	•
		0.2 m/s	•	•	•	•
		0.3 m/s	•	•	•	•
		0.5 m/s	•	•	•	•
Lubricant 3	339.8 mm^2/s	0.05 m/s	•	•	•	•
		0.1 m/s	•	•	•	•
		0.2 m/s	•	•	•	•
		0.3 m/s	•	•	•	•
Lubricant 4	135.8 mm^2/s	0.05 m/s	•	•	•	•
		0.1 m/s	•	•	•	•
		0.2 m/s	•	•	•	•
		0.3 m/s	•	•	•	•
Lubricant 5	77.13 mm^2/s	0.05 m/s	•	•	•	•
		0.1 m/s	•	•	•	•
		0.2 m/s	•	•	•	•
		0.3 m/s	•	•	•	•
Lubricant 6	17 mm^2/s	0.1 m/s	•	•	•	•
		0.2 m/s	•	•	•	•
		0.3 m/s	•	•	•	•
		0.5 m/s	•	•	•	•

Six different types of lubricant were selected and named as Lubricant 1 – 6. The viscosity information of those lubricants can be also found in table 1 just under lubricant's name.

According to the results of the performed tests listed in table 1, the film thickness will respond differently to the variation of load depending whether the two solid surfaces separate completely or not during harmonic loading cycles.

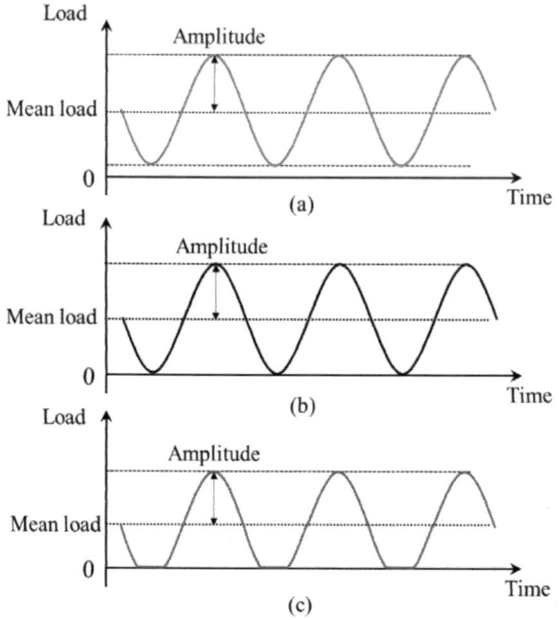

Figure 2. Variable loading conditions
(a) mean load larger than load amplitude
(b) mean load equal to load amplitude
(c) mean load smaller than load amplitude

As showed in figure 2, the experimental loading condition in this paper can be classified as: mean value of variation cycle larger than or equal to load amplitude; mean value of variation of cycle is smaller than load amplitude. In the first case, the amplitude of the displacement of the vibrating element is smaller than or equal to the cumulative approach of the solids, while in the second is larger. Evidently the second case also corresponds to impact loading of the contact. The amplitude of the load cycle was about 21 N.

3. Results and discussion

Due to limitation of space of the current paper, only selective and typical results are presented below, based on different harmonic loading type resulting on the corresponding film thickness perturbation phenomenon. The left-most coloured interferogram in figure 3 (a) shows the steady state EHD contact at an entrainment speed of 0.05 m/s, lubricated with *Lubricant 2*. The horse-shoe shape and side lobes of the classical EHD contact behaviour can be clearly viewed. The lubricant entraining direction is from the bottom to the top of the image. Selected interferograms of the EHD contact subjected to 100 Hz harmonic loading (mean load is larger than load amplitude) within a cycle are presented to the right of the interferogram of the corresponding steady state condition. Film thickness perturbation, as seen from the coloured interferogram captured at 2.5 ms of the harmonic loading cycle, is formed at the inlet of the contact region. This crescent-shaped film perturbation travels across the contact region and eventually leaves the contact when the harmonic loading completes a full cycle.

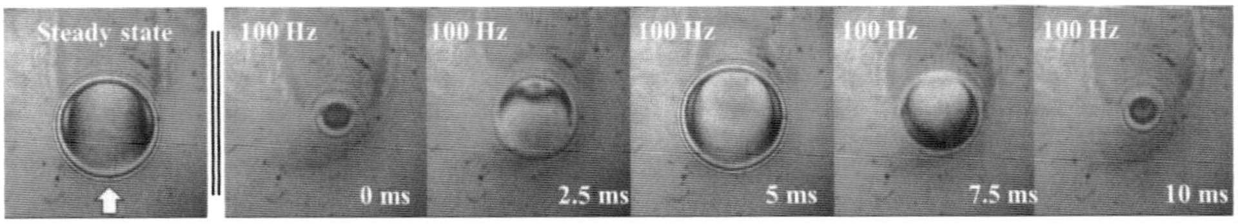

(a)

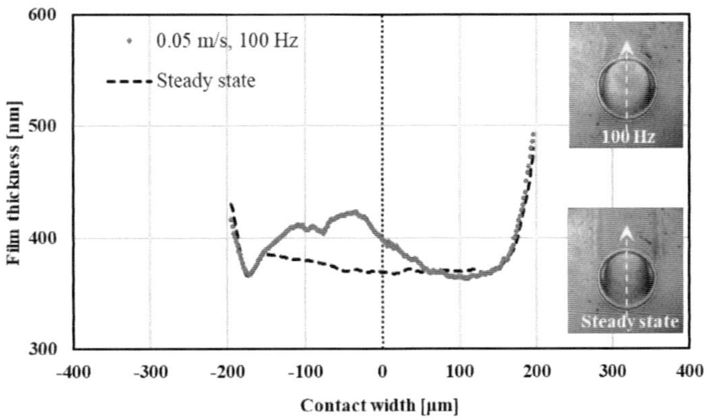

(b)

Figure 3. (a) Interferograms of EHD contact, 100 Hz-harmonic loading.
(b) Steady state and transient film thickness profiles

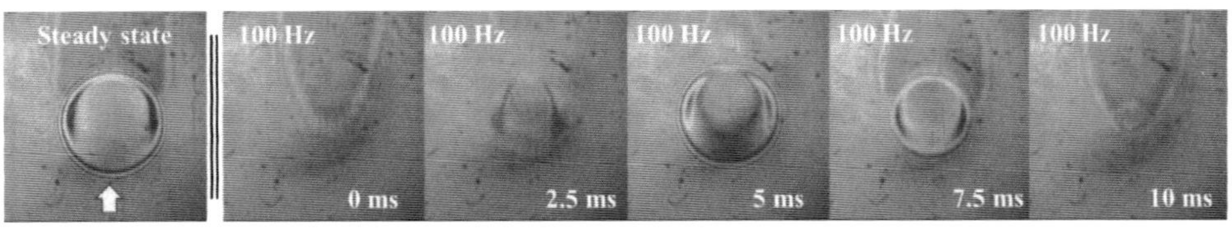

(a)

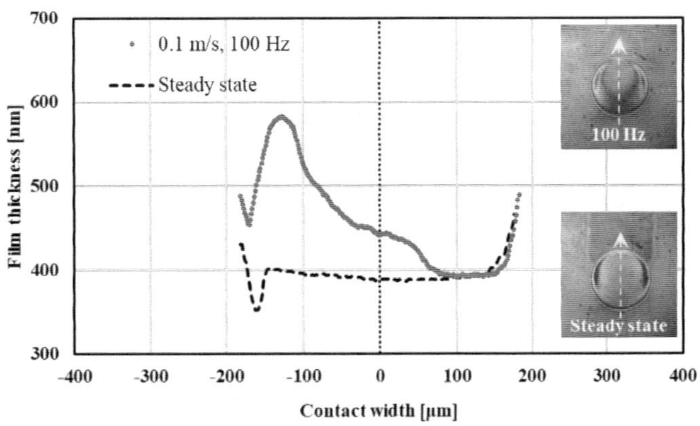

(b)

Figure 4. (a) Interferograms of EHD contact, 100 Hz-impact loading.
(b) Steady state and transient film thickness profiles

Figure 3 (b) compares the steady state film thickness profile with film thickness profile of the EHD contact within the harmonic loading cycle under peak load. The film thickness profile is calculated from along the entraining motion. The maximum generated film thickness perturbation is about 53 nm greater than film thickness under steady state condition. Figure 4 (a) shows selected coloured interferograms of the EHD contact subject to 100 Hz harmonic loading (mean load is smaller than load amplitude). This means that shock loading is applied to the contact. It can be observed that a dimple shaped film thickness perturbation has been formed in the central region of the contact and it travels across the contact along the direction of the entraining motion. In Figure 4 (b), the film perturbation has travelled towards the exit of the contact at the point when the load reaches the maximum value. It can be seen that this dimple perturbation is 193 nm greater than the steady state film thickness.

A graphical summary of film perturbation phenomenon on harmonic loading (mean load is larger than load amplitude) is showed in Figure 5.

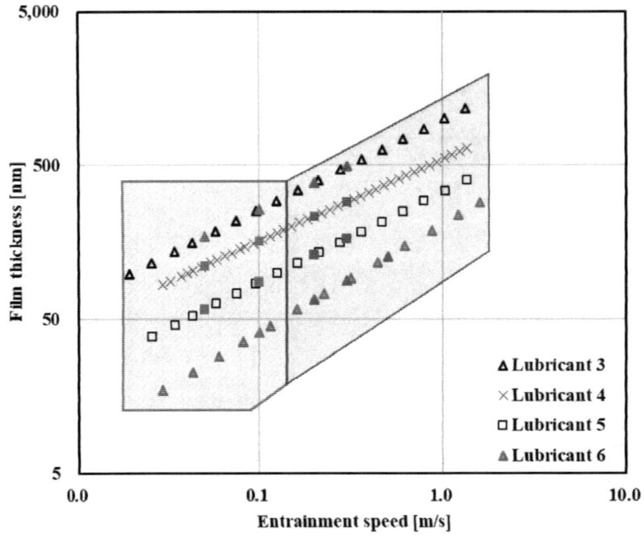

Figure 5. Film thickness perturbation map- harmonic loading

Steady state film thickness is shown for four different lubricants, which appear as four straight marker lines in the logarithmic graph. The grey markers indicate steady state values. Coloured symbols marked within the steady state values, indicating the conditions where vibration tests were performed at any of the working frequencies 10, 25, 50 or 100 Hz. The red markers show the tests when film perturbations occurred, while for the conditions corresponding to the blue markers no significant perturbations occurred. Based on previously discussed results, film thickness perturbation phenomenon is most significant at maximum frequency studied (100 Hz) in the red region, and it is less pronounced at lower test frequencies of 10 Hz and 25 Hz. In the blue region, film thickness perturbation was not observed even at the largest frequency employed in these tests. As a note, this is not a guaranty that film perturbations may not occur at frequencies larger than 100 Hz. Anyway, as seen in these experiments film thickness perturbations do not extend to entrainment speeds higher than approximately 0.1 m/s. This is marked by the blue region in the graph. It thus seems that film perturbation depends on the ratio between the average transient time of the lubricant through the contact and the period of the oscillatory cycle.

Parameter τ, defined as the ratio between the half period of oscillation $T/2$ and the average lubricant transit time t_{av} in the loading phase of vibration cycle is therefore used to provide a guideline for the occurrence of film thickness perturbation.

$$\tau = \frac{T}{2t_{av}} \tag{1}$$

Analysis of the results shown indicates that if τ is greater than 2 no film perturbation occurs, while for smaller values it does. For instance, in the 100 Hz cycle test, the load applied to the EHD contact varied from 1 N to 55 N and the entrainment speed was 0.1 m/s. The diameter of Hertzian contact, in this case, varied between 100 μm and 392 μm and the average lubricant transit time is 2.46 milliseconds. Parameter τ for this case becomes 2.03 which can be taken as 2. For the entrainment speed of 0.2 m/s, τ is obviously double to 4.06 thus results show no film thickness perturbation. Obviously, the value of 2 for parameter τ is only approximate as no tests were carried out at every possible speed in the interval between 0.1 m/s and 0.2 m/s.

More vibration tests are required to be carried out for shaping a more accurate boundary location of the red-blue regions in figure 5.

Figure 6 shows film thickness versus entrainment speed in the same steady state as in Figure 5. Results of transient experiments are again marked over the steady-state values. The transient tests were performed at speeds marked by coloured symbols at 10 Hz, 25 Hz, 50 Hz and 100 Hz for values of the amplitude of the load larger than the mean value of the cycle. As seen, there are no blue symbols, which means that film thickness perturbations occurred in every single test, no matter the speed and frequency

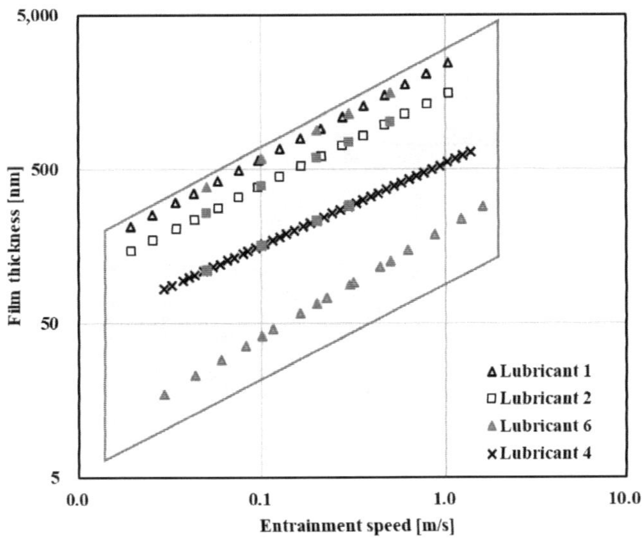

Figure 6. Film thickness perturbation map-impact loading

The red region can now be shaped as shown in figure 6. Only film thickness values at the maximum load in the cycle are shown in these graphs. It should be speculated that in case of cyclic application of load with impact, film thickness perturbation is generated in any condition of speed even for those larger than the speeds employed in these tests, however it may exit the contact before the load has reached the peak value in the vibration cycle.

4. Conclusions

In the present paper, a systematic experimental study for predicting the occurrence of film thickness perturbation phenomenon under harmonic loading condition has been carried out. The range of frequencies of the harmonic load employed in these tests ranges between 10 Hz and 100 Hz which has strong practical relevance.

In the case of a mean load larger than the amplitude of the load variation, it was found that the film perturbation depends of the frequency of the load cycle, the entrainment speed of the lubricant and the overall lubricant film thickness; it also depends of the ratios between the mean load and the amplitude. Film perturbations occur in contacts for which that ratio between half the load cycle period and the average time of transit of the lubricant through the contact in the loading phase of the vibration is less than 2. In the case of mean load smaller than the amplitude, film thickness perturbation has been observed in every single test regardless of entrainment speed, frequency and lubricant viscosity.

5. References

[1] Larsson. R and Lundberg. J, 1995, "Study of Lubricated Impact Using Optical Interferometry", *Wear*, **190**, pp 184-189.
[2] Kaneta. M, Ozaki. S, Nishikawa. H and Guo. F, 2006, "Effects of Impact Loads on Point Contact Elastohydrodynamic Lubrication Films", *Proc. I. Mech. E.*, **J221**, pp 271-278.
[3] Sanborn, D. M. and Winer., W. O., 1971, "Fluid Rheological Effects in Sliding Elastohydrodynamic Point Contacts with Transient Loading: 1- Film Thickness", *Trans. ASME, J. Lub. Tech*, **93**, pp 262-271.
[4] Wijnant., Y. H, Venner, C. H, Lassron. R and Eriksson. P, 1999, "Effects of Structural Vibrations on the Film Thickness in an EHL Circular Contact", *J. Trib*, **121**, pp 259-264.
[5] El Kilali. T, Perret-Liaudet. J and Mazuyer. D, 2003, "Experimental analysis of a high pressure lubricated contact under dynamic normal excitation force", *Transient Processes in Tribology*, Elsevier, **43**, pp 409-418.
[6] Sakamoto. M, Nishikawa. H and Kaneta. M, 2003, "Behaviour of Point Contact EHL Films Under Pulsating Loads", *Transient Processes in Tribology*, Elsevier, **43**, pp 391-399.
[7] Kalogiannis. K, 2013, "Behaviour of Elastohydrodynamic Films Subjected to Oscillatory Motion", Ph. D. Thesis, University of Sussex, United Kingdom.
[8] Fryza. J, Sperka. P, Kaneta. M, Krupka. I and Hartl. M, 2017, "Effects of Lubricant Rheology and Impact Speed on EHL Film Thickness at Pure Squeeze Action", *Trib. Int*, **106**, pp 1-9.
[9] Gohar. R and Cameron. A, 1967, "The Mapping of Elastohydrodynamic Contacts", *ASLE Trans*, **10**, pp 215-225.
[10] Foord, C. A., Hammann, W. C. and Cameron. A, 1968, "Evaluation of Lubricants Using Optical Elastohydrodynamics', *ASLE Trans*, **11**, pp 31-43.
[11] Wedeven, L. J., 1970, "Optical Measurements in Elastohydrodynamic Rolling Contact Bearings", PhD. thesis, London University, London.
[12] Cann, P. M., Spikes, H. A. and Hutchinson. J, 1996, "The development of a Spacer Layer Imaging Method (SLIM) for Mapping Elastohydrodynamic Contacts", *Trib. Trans*, **39**, 4, pp 915-921.
[13] Johnston, G. J., Wayte. R and Spikes, H. A., 1991, "The Measurement and Study of Very Thin Lubricant Films in Concentrated Contacts", *Trib. Trans*, **34**, 2, pp 187-194.
[14] Zhang. X, 2017, "The Effect of Vibrations on the Behavior of Lubricated Elastohydrodynamic Contacts. PhD thesis, University of Sussex.

Influence of lubricant on the friction in an angular contact ball bearing under low load conditions

A Popescu[1] and D N Olaru[1]

[1] "Gheorghe Asachi" Technical University of Iasi-Romania, Faculty of Mechanical Engineering, Blvd. Mangeron, No. 59A, 700050, Iasi, Romania

Abstract. The presence of the lubricant in a ball-race contact has two principal effects on friction: reduces the sliding friction, as a result of developing a film thickness between the contact surfaces and increases the rolling friction, as result of developing a hydrodynamic rolling friction. The influence of the hydrodynamic rolling friction on the torque of a modified 7205B angular contact ball bearing has been theoretically and experimentally studied. The experiments demonstrated that, for very low contact loads between balls and races, the effect of the hydrodynamic rolling force is dominant in the total friction torque. The experiments have been validated by using Biboulet's transition model for hydrodynamic rolling forces developed in ball-race contacts.

Nomenclature

d_m	Mean diameter of the angular contact ball bearing [m]	$d_m = \frac{(D+d)}{2}$
d_b	Ball diameter [m]	
dT	Friction torque generated by a single ball [N·m]	
d	Inner diameter of the angular contact ball bearing [m]	
f	Ball-race conformity [-]	
k	Radius ratio [-]	$k_{i,o} = \frac{2 \cdot f_{i,o}}{(2 \cdot f_{i,o} - 1) \cdot (1 \mp \gamma)}$
n_o	Outer race rotational speed [rpm]	
v	Tangential speed in the rolling direction [m/s]	$v = \frac{\pi \cdot n_o \cdot (1-\gamma^2)}{30} \cdot \frac{d_m}{4}$
D	Outer diameter of the angular contact ball bearing [m]	
E^*	Equivalent Young's modulus [Pa]	
MC	Curvature friction moment [N·m]	
MER	Rolling resistance moment [N·m]	
MP	Pivoting moment [N·m]	
η_o	Oil dynamic viscosity [Pa·s]	
Q	Normal load on contact ellipse [N]	$Q = \frac{F_a}{3 \cdot \sin\alpha}$
$R_{i,o}$	Radius of the inner and outer races [m]	$R_{i,o} = \frac{d_m \cdot (1 \mp \gamma)}{2}$
R_X	Equivalent radii in the rolling direction [m]	$R_{Xo,i} = \frac{d_b}{2} \cdot \left(1 \pm \frac{d_b \cdot \cos\alpha}{d_m}\right)$
R_y	Equivalent radius of the curvature perpendicular to the rolling direction [m]	
T	Friction torque of the bearing [N·m]	

α	Contact angle of the angular contact ball bearing [rad]	
α_p	Piezo-viscosity coefficient of oil [Pa^{-1}]	
γ	Dimensionless parameter [-]	$\gamma = \frac{d_b \cdot \cos(\alpha)}{d_m}$
μ	Friction coefficient [-]	
ω_0	Angular speed of the outer race [rad/s]	
ω_b	Angular speed of the balls [rad/s]	

Index

o	Outer
i	Inner
s	Sliding
p	Pivoting
x, y	Contact ellipse coordinate

1. Introduction

In a ball-race contact, the power loss is due to many factors, including both losses from sliding friction on the contact ellipse and from rolling friction. Houpert [1] and [2] described all friction forces and resistant moments in a ball-race contact from an angular contact ball bearing and proposed a general equation for the total friction torque, including both losses generated by sliding and by rolling. One of the important resistant force for a low loaded ball-race contact is the hydrodynamic rolling force. Biboulet and Houpert [3] proposed a complex model for the hydrodynamic rolling force, by including both IVR (Isoviscous Rigid) and EHD (Elasto-Hydrodynamic) effects. The new Biboulet&Houpert's model has been validated by Balan et al. [4], [5] and [6] for a modified thrust ball bearing, having only three balls, without cage and lubricated with mineral oil. Ianuş et al. [7] also validated Biboulet&Houpert's model, applied to a grease lubricated low loaded thrust ball bearing.

By using a 7205B modified angular contact ball bearing, with 3 balls and without cage, Popescu et al. [8] experimentally determined the total friction torque in dry conditions. All friction sources (sliding and rolling in the ball-race contacts) were included in the measured friction torque, except the effect of the lubricant.

In the present paper the authors determined both theoretically and experimentally the influence of the oil on the total friction torque, in a modified 7205B ball bearing with three balls and no cage. Considering two oils with different viscosities and a low axial load, the authors evidenced the important increasing of the total friction torque as result of hydrodynamic rolling forces generated in ball-race contacts. Also were verified two equations for hydrodynamic rolling forces resulted from the Biboulet&Houpert's model.

2. Theoretical model for friction torque

In an angular contact ball bearing, Houpert [1] and [2] proposed following general equation for the friction torque, generated by a single ball:

$$dT = 2 \cdot (FRo + FRi) \cdot \frac{Ro \cdot Ri}{dm} + \frac{MC_i \cdot R_o + MC_o \cdot R_i}{d_b} + \frac{MER_i \cdot R_o + MER_o \cdot R_i}{d_b} + \frac{MP_i \cdot (1+\gamma) + MP_o \cdot (1-\gamma)}{2} \cdot \sin(\alpha) \quad (1)$$

where $FR_{o,i}$ is the hydrodynamic rolling friction in ball-race contacts, $MC_{o,i}$ is the curvature friction moment developed in ball-race contacts, $MER_{o,i}$ is the rolling resistance moments in ball-race contacts and $MP_{o,i}$ is the pivoting moments. According to the ball bearing geometry, the friction coefficient on ball-race contacts and normal load, as well as the moments *MC*, *MER* and *MP* can be determined according to Houpert's relations [1]. The geometrical parameters of the angular contact ball bearing (d_m, d_b, α, $R_{o,i}$ and γ) are defined in the nomenclature.

In the absence of lubricant, the moments *MC* and *MP* can be determined by imposing a constant friction coefficient on ball-race contacts and MER depends only on the normal load and ball-race geometry [1]. In the presence of lubricant, two important parameters must be included in equation (1): the hydrodynamic rolling forces FR_o and FR_i.

As it was demonstrated in previous research [4], [5] and [6], the transition model for hydrodynamic rolling friction force, proposed by Biboulet and Houpert [3], represents the best solution in low loaded ball-race contacts. Therefore, the transition model includes both IVR and EHD regimes, which usually exist in lubricated ball bearings but in different percentages [6].

The transition model for the hydrodynamic rolling force described by Biboulet and Houpert [3] is characterized by the following general equation:

$$FR_{Trans} = \frac{FR_{IVR}-FR_{EHL}}{1+\frac{M}{6.6}} + FR_{EHL} \quad (2)$$

The two components of the hydrodynamic rolling forces FR_{IVR} and FR_{EHL} correspond to the IVR and EHD lubrication regimes, respectively. In [4] the two transition equations have been compared for rolling friction:

(i) Houpert's transition model which includes in equation (2) the Houpert's relations for FR_{IVR} and FR_{EDL} determined by the following [4]:

$$FR_{IVR\_H} = 1.213 \cdot E^* \cdot R_x^2 \cdot k^{0.358} \cdot U^{0.636} \cdot W^{0.364} \quad (3)$$

$$FR_{EHL\_H} = 2.765 \cdot E^* \cdot R_x^2 \cdot k^{0.35} \cdot U^{0.656} \cdot W^{0.466} \cdot G^{0.022} \quad (4)$$

(ii) Biboulet's transition model which includes in equation (2) the Biboulet&Houpert's relations for FR_{IVR} and FR_{EDL} determined by the following [3]:

$$FR_{IVR\_B} = 2.9766 \cdot E^* \cdot R_x^2 \cdot k^{0.3316} \cdot W^{1/3} \cdot U^{2/3} \quad (5)$$

$$FR_{EHL\_B} = 7.5826 \cdot E^* \cdot R_x^2 \cdot k^{0.4055} \cdot W^{1/3} \cdot U^{3/4} \quad (6)$$

The transition parameter M is described by Biboulet and Houpert [3] with the following equation:

$$M = 0.5549 \cdot k^{-0.6029} \cdot W \cdot U^{-0.75} \quad (7)$$

Therefore, in the proposed simulation program two models for the transition of the hydrodynamic rolling force, FR_{Trans} were considered:
Houpert's transition model:

$$FR_{Trans\_H} = \frac{FR_{IVR\_H}-FR_{EHL\_H}}{1+M/6.6} + FR_{EHL\_H} \quad (8)$$

Biboulet's transition model:

$$FR_{Trans\_B} = \frac{FR_{IVR\_B}-FR_{EHL\_B}}{1+M/6.6} + FR_{EHL\_B} \quad (9)$$

The parameters included in equations (3) - (7) are defined as follows:
U is the dimensionless speed parameter:

$$U = \frac{\eta_0 \cdot v}{E^* \cdot R_x} \quad (10)$$

W is the dimensionless load parameter:

$$W = \frac{Q}{E^* \cdot R_x^2} \quad (11)$$

G is the material parameter: $$G = \alpha_p \cdot E^* \quad (12)$$

in which η_o is the oil dynamic viscosity in Pa·s, at the operating temperature of the contact; α_p is the piezo-viscosity coefficient of oil in Pa^{-1}, $v = (v_1 + v_2)/2$ is the average entrainment speed in m/s; E^* is the equivalent Young modulus of the two elements in contact (for steel ball – race contact, $E^* \approx 2.3 \cdot 10^{11}$ Pa), k is the radii ratio R_y/R_x, with R_y defined as the equivalent radius of curvature in the y direction (perpendicular to the rolling direction) and R_x the equivalent radius in the rolling direction.

3. Simulation of the friction torque for a modified angular ball bearing

The theoretical model has been applied to a modified 7205B angular contact ball bearing, having the following geometrical parameters: d=0.025 m; D=0.052 m; d_b=0.0079375 m; d_m=0.0385 m, f_i=0.515; f_e=0.522 and a nominal contact angle α=15 degrees. Also, the roughness parameters on races and balls are: R_{qr}=0.06 μm and R_{qb}=0.03 μm. Three balls have been distributed at 120 degrees between the inner and outer race and axially loaded with a 5 N weight cylinder.

Figure 1 presents a view of a section through the modified 7205 angular contact ball bearing.

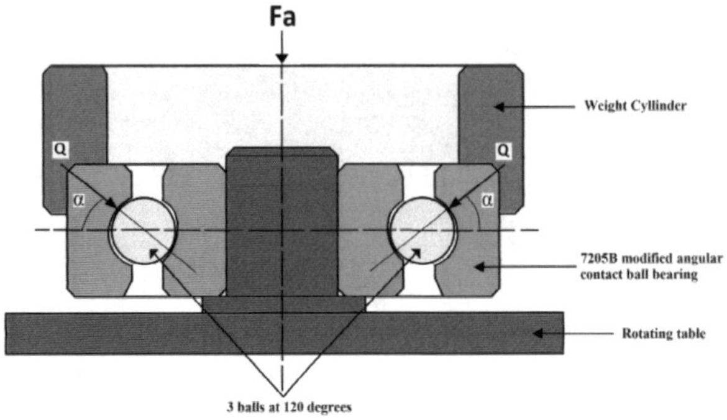

Figure 1. View of a section through the modified 7205B ball bearing assembly

The simulation of the total friction torque has been performed for the kinematic conditions generated by the spin-down method, in which the inner race is rotating with a given rotational speed, between 100 and 400 rpm. After some time, as result of the friction in all six ball races contacts, the outer race and attached cylinder rotates with a rotational speed similar to the inner race rotational speed. In that moment, the inner race is stopped and all kinetic energy of the outer race and cylinder are dissipated by friction, in all six ball-races contacts, in a deceleration process during a short or long time, depending on the friction resistance in the ball-races contacts.

The simulation of the total friction torque has been realized for two mineral oils, having the dynamic viscosity of η of 0.085 Pa·s and 0.35 Pa·s at laboratory temperature (22-24° C) and with the piezo-viscosity coefficient $\alpha_p = 2.2 \cdot 10^{-8}$ Pa^{-1}.

The friction torque generated by a single ball and given by equation (1) can be divided in two components: a component generated by hydrodynamic rolling forces dT_H and a component generated by sliding and rolling friction dT_{S+R}:

$$dT = dT_H + dT_{S+R} \tag{13}$$

where the two components are given by the following equations:

$$dT_H = 2 \cdot (FRo + FRi) \cdot \frac{Ro \cdot Ri}{dm} \tag{14}$$

$$dT_{S+R} = \frac{MC_i \cdot R_o + MC_o \cdot R_i}{d_b} + \frac{MER_i \cdot R_o + MER_o \cdot R_i}{d_b} + \frac{MP_i \cdot (1+\gamma) + MP_o \cdot (1-\gamma)}{2} \cdot sin(\alpha) \tag{15}$$

To determine the component dT_{S+R}, the friction coefficient on the contact ellipses has been calculated with equation (16) as function of the lubricant parameter λ. The component dT_{S+R} is presented in detail in reference [9], which was determined for dry condition by imposing a constant friction coefficient. The friction coefficient is determined for the inner and outer ball-races contacts by the following relation [4]:

$$\mu_s = \mu_e \cdot 0.82 \cdot {\lambda_{i,o}}^{0.28} + \mu_a \cdot \left(1 - 0.82 \cdot {\lambda_{i,o}}^{0.28}\right) \tag{16}$$

μ_a is the friction coefficient for limits conditions, usually μ_a=0,11, μ_e is the friction coefficient in EHD lubricant film, usually considered as μ_e=0.05 and the lubricant parameter $\lambda_{i,o}$ is calculated with relation:

$$\lambda_{i,o} = \frac{h_{min_{i,o}}}{\sqrt{{R_{qr}}^2 + {R_{qb}}^2}} \tag{17}$$

The minimum film thickness on inner and outer races has been evaluated with Hamrock&Dowson's equations [1] and [2]:

$$h_{min_i} = 3.63 \cdot R_{xi} \cdot {U_i}^{0.68} \cdot G^{0.49} \cdot {W_i}^{-0.073} \cdot \left(1 - 0.61 \cdot e^{-0.68 \cdot k_i}\right) \tag{18}$$

$$h_{min_o} = 3.63 \cdot R_{xo} \cdot {U_o}^{0.68} \cdot G^{0.49} \cdot {W_o}^{-0.073} \cdot \left(1 - 0.61 \cdot e^{-0.68 \cdot k_o}\right) \tag{19}$$

The variation of the minimum film thickness and lubricant parameter for inner and outer ball-races contact have been calculated for both oil viscosities between 100 and 400 rpm. The lubricant parameters obtained have values between 1 and 2.8, for the oil with the viscosity of 0.085 Pa·s. Also, the lubricant parameters obtained have values between 3 and 7, for the oil with the viscosity of 0.35 Pa·s. The correspondent friction coefficients varied between 0.06 and 0.05.

Important differences between Houpert's transition model and Biboulet's transition model can be observed for both oil viscosities. These variations can be explained by the differences between equation (3) – (4) and (5) – (6). The hydrodynamic rolling friction variation as function of rotational speed for the two oil viscosities are presented in figures 2 and 3. Equations (8) and (9) have been used for Houpert's transition model and Biboulet's transition model, respectively.

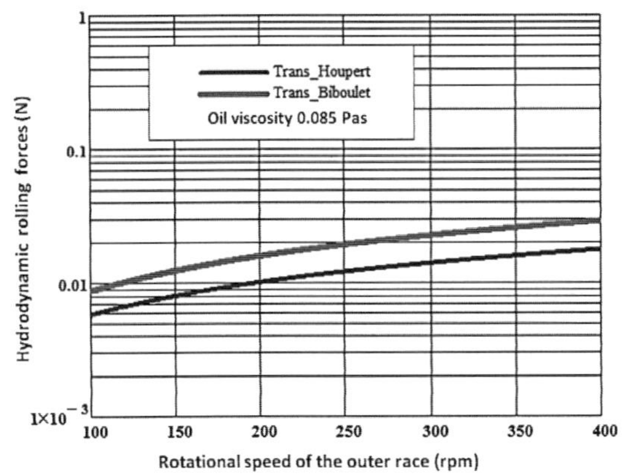

Figure 2. Variation of the hydrodynamic force FR_{Trans} as function of the rotational speed for an oil viscosity of 0.085 Pa·s.

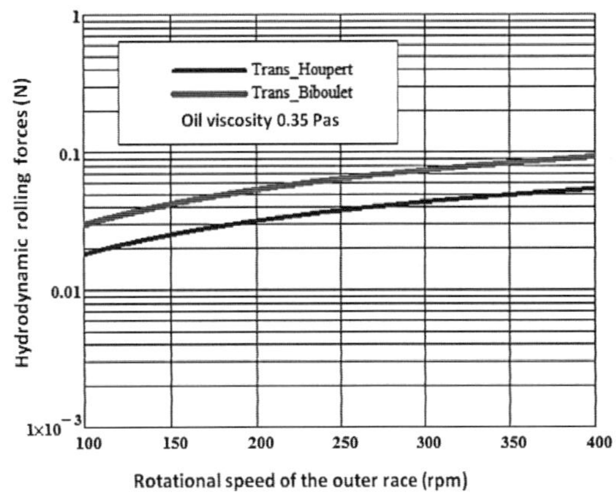

Figure 3. Variation of the hydrodynamic force FR_{Trans} as function of the rotational speed for an oil viscosity of 0.35 Pa·s.

In the absence of the cage, the total friction torque in the modified 7205B ball bearing is obtained by following equation:

$$T = 3 \cdot dT \tag{20}$$

The total friction torques have been calculated for the two oils using both simulation models for transition hydrodynamic rolling forces. These variations of total friction torque are presented in figures 4 and 5.

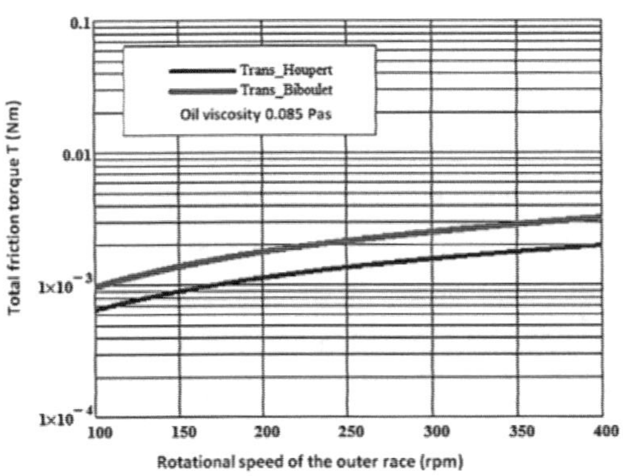

Figure 4. Variation of the total friction torque as function of the rotational speed for oil viscosity of 0.085 Pa·s.

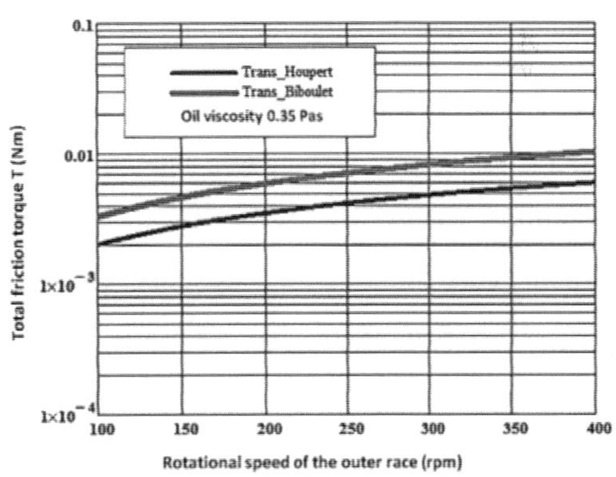

Figure 5. Variation of the total friction torque as function of the rotational speed for oil viscosity of 0.35 Pa·s.

The next step in our approach consist in validating the model by use of the experiments to adequate the transition hydrodynamic rolling force model for the modified 7205 ball bearing.

4. Experimental validation of the theoretical model

The experiments were performed in the Tribology Laboratory of the Mechanical Engineering Faculty of Iasi. The modified 7205B angular contact ball bearing presented in figure 1 has been tested. The inner ring was fixed to the rotational table of the Tribometer CETR UMT2, so that it rotates with the table (figure 7). On the outer ring a steel cylindrical support was fixed, imposing an axial load of 5.0 N, with an inertial moment J=4.354·10^{-4} kg·m^2. No cage was used in the experiments. A video camera recorded with 60 frames/second the rotation of the outer race in the deceleration process. The video captured by the camera was recorded on a computer in real time and subsequently processed with a specialized program. For every rotational speed, two experimental parameters have been determined: the total angular position cumulated by the outer race in the decelerating process, from an initial rotational speed to stop, expressed in radians and the time during the deceleration process, expressed in seconds.

In the deceleration process of the outer race, the dynamic balance of the moments acting on the rotating system has been used:

$$J \cdot \frac{d\omega_o}{dt} + T(\omega_o) = 0 \quad (21)$$

where J is the inertia moment of the outer race and cylinder and $T(\omega_o)$ is the total friction torque developed in the modified ball bearing, as a function of the angular speed of the outer race ω_o. For a low axial load, the hydrodynamic component of the torque is dominant and the total friction torque $T(\omega_o)$ can be approximated by a formal relation: $T(\omega_o) = K^* \cdot \omega_o^n$, where K^* and n was determined by integration of equation (21) and according to the measured parameters (total angular position of the outer race and deceleration time) for each rotational speed. Details of the integration of equation (21) are presented in [4]. A general view of the experimental equipment is presented in figure 6 and some details are presented in figure 7.

Figure 6. General view of the CETR UMT 2 Tribometer.

Figure 7. Details of the Tribometer and 7205B modified angular contact ball bearing [8].

Some repeated tested have been performed for rotational speeds of the outer race between 100 and 400 rpm for both viscosity of 0.085 Pa·s and 0.35 Pa·s.

The experimental friction torques obtained by using the spin-down method have been indicated in figures 8 and 9. Also, the theoretical variation of the friction torque determined by the Biboulet's transition model was superposed on the experimental results in figure 8 and 9. A very good correlation has been obtained for both oil viscosities.

Comparing these results with the results obtained by Balan et al. in [4] (figures 10, 12 and 13) on the modified 51205 thrust ball bearing with similar oil viscosities, low axial load and comparable rotational speed, the following comment can be made: the total friction torques experimentally determined in [4] have been validated with Houpert's transition model for hydrodynamic rolling forces and not with Biboulet's transition model as in our experiments.

It can be explained by following important differences for the transition parameter M. Thus, for a modified 7205B ball bearing lubricated with oil having 0.085 Pa·s, the parameter M has values between 25 and 9 when the rotational speed varied between 100 and 400 rpm, while for the modified 51205 ball bearing, for similar oil and speeds, the parameter M varied between 7 and 2. Similarly, for oil viscosity of 0.35 Pa·s the parameter M varied between 7 and 2 for the modified 7205B ball bearing and between 2 and 0.9 for the modified 51205 ball bearing.

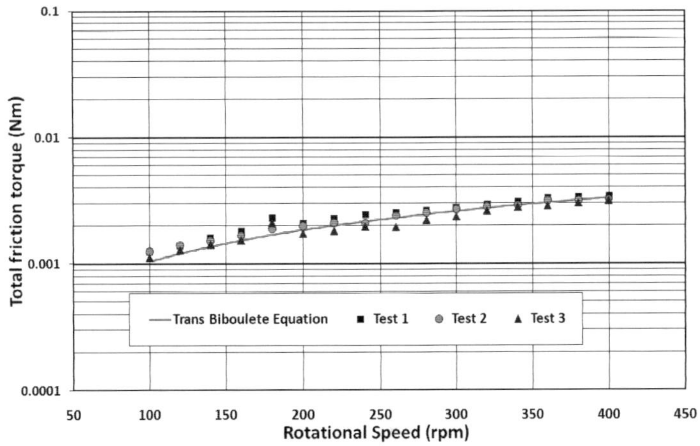

Figure 8. Variation of the theoretical and experimental friction torque for a modified 7205B ball bearing and for an oil viscosity of 0.085 Pa·s.

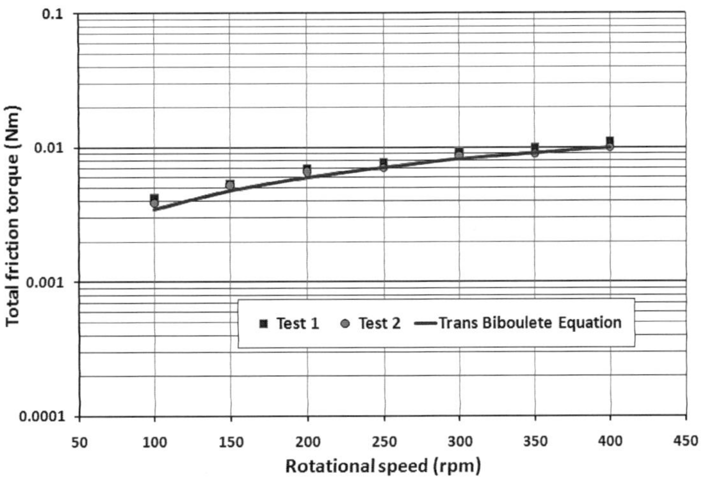

Figure 9. Variation of the theoretical and experimental friction torque for a modified 7205B ball bearing and for an oil viscosity of 0.35 Pa·s.

Supplementary the geometrical parameters which are included in the IVR and EHD rolling friction equations as Rx and k have higher values for the 7205B ball bearing than for 51205 ball bearing. Also, the normal load for each ball-race contact is 6.44N for the 7205B ball bearing and 1.42 N for the 51205 ball bearing.

All these differences lead to disparities between the total friction torque for the two modified ball bearings considered in this paper and in [4], even if the ball diameter, oil viscosity and rotational speed are similar and the applied loads are very close.

5. Conclusions

A complex model for determining the influence of the lubricant on the friction torque in a modified angular contact ball bearing has been theoretical simulated and experimentally validated.

The simulation was realised for an angular contact ball bearing, having only three balls, without cage and axially loaded with a very low load. Only the friction losses in ball-race contacts were evidenced.

Two transition models for hydrodynamic rolling force in ball-race contacts, Houpert's transition model and Biboulet's transition model obtained by combining different equations for IVR and EHD lubrication regimes have been simulated. Two different oil viscosities (0.085 Pa·s and 0.35 Pa·s) and a variation of the rotational speed between 100 and 400 rot/min have been used in the simulation.

The experiments were realised with spin-down methodology for various outer race rotational speed between 100 and 400 rpm for both oil viscosities and with an axial load of 5N.

The experimental values validated Biboulet's transition model for hydrodynamic rolling forces for both viscosities.

6. References

[1] L Houpert 1999 Numerical and Analytical Calculations in Ball Bearings *European Space Agency* ESA-SP **438** 283

[2] L Houpert 2002 Ball Bearing and Tapered Roller Bearing Torque: Analytical, Numerical and Experimental Results *STLE Trib. Trans.* **45**(3) 345-53.

[3] Biboulet N, Houpert L 2010 Hydrodynamic force and moment in pure rolling lubricated contacts: Part 2, Point contacts. *Proc. Inst. Mech. Eng, Part J, J. Eng. Tribol.* **224** 777–87

[4] Mihaela Rodica D Bălan, Vasile Ciprian Stamate, Luc Houpert and Dumitru N Olaru 2014 The influence of the lubricant viscosity on the rolling friction torque, *Tribology International*, **7** 1–12

[5] Balan Mihaela Rodica, Vasile Ciprian Stamate, Luc Houpert, Ana Tufescu and N Dumitru Olaru 2014 Influence of the Geometry on the Rolling Friction Torque in Lubricated Ball-Race Contacts *Applied Mechanics and Materials-Advanced Concepts in Mechanical Engineering I* **658** 271-76

[6] Balan M R, Houpert L, Tufescu A and Olaru D 2015 Rolling Friction Torque in Ball-Race Contacts Operating in Mixed Lubrication Conditions *Lubricants* **3**(2) 222-43

[7] G Ianuş, A C Dumitraşcu, V Cârlescu and D Olaru 2016 Friction Torque in Thrust Ball Bearings Grease Lubricated *IOP Conf. Ser.: Mater. Sci. Eng.* **147**(01) 20-26

[8] A Popescu, M I Nazare and D Olaru 2018 Friction torque in a modified angular contact ball bearing operating at low axial loads, *IOP Conf Series: Materials Science and Engineering* **444** 02

[9] A Popescu, L Houpert and D N Olaru 2019 Four approaches for calculating power losses in an angular contact ball bearing *Mechanism and Machine Theory* https://doi.org/10.1016/j.mechmachtheory.2019.103669 Article in press

A simple analytical model for an elastohydrostatic thrust bearing

T Cicone[1], A A Marinescu[1] and S Sorohan[1]

[1]University POLITEHNICA of Bucharest, Splaiul Independenței 313, Bucharest 060042, Romania

Abstract. A new design of an elastohydrostatic bearing with a structurally elastic pad is theoretically analyzed. A single recess circular thrust bearing has the upper pad provided with a large cavity, covered with an elastic metallic membrane. The latter deflects under the pressure distribution in the bearing, creating a radial convergent gap. A simplified axi-symmetric analytical model based on two assumptions is proposed: (i) parabolic approximation for the elastic component of the film thickness and (ii) the maximum elastic deformation of the membrane is calculated analytically using a simplified linear pressure variation in radial direction on the land. Thus, the coupled equations for fluid film and elastic deformation can be solved successively and the bearing performance characteristics can be easily calculated. A parametric analysis, function of a complex elastic parameter and relative recess radius was performed, showing higher load capacity and lower friction similarly to what was previously obtained using rubber-based compliant bearings. A first stage of validation of the analytical model consisted in using a structural finite element model. For the sake of simplicity, the pressure distribution was written in an APDL code using the ANSYS program. The comparison showed small differences for typical values of the recess radius.

1. Introduction

Since the dawn of hydrostatic (HS) bearing applications, the elasticity of one of the mating pads was viewed as a feasible solution for heavy loaded bearings, providing better adaptability to misalignments. After the successful employment of water-lubricated hydrostatic rubber bearings at the Mile High Stadium in Denver, Dowson and Taylor [1], and Castelli et al. [2] published experimental and theoretical studies which proved an increased load carrying capacity and reduced friction when one of the pads was compliant. However, rubber pads inflate at high pressures and cannot be used for a long period of time, due to aging. Moreover, rubber - fluid compatibility may be a major concern. The present paper focuses on a new possible design solution of a HS bearing with a structurally elastic pad. Similar design solutions have been proposed for gas bearings by Hayashi and Hirasata [3] and Hao et al. [4]. Both papers included numerical solutions of the fluid - structure interaction, which showed an improved stiffness. However, the generality of the results was limited to studying only several practical cases. The present work proposes a simplified analytical solution for the coupled fluid film - elastic deformation mechanism, based on an approximate elastic deformation of the elastic pad. The proposed solution allows a quick and comprehensive parametric analysis which reveals the improved performance characteristics of this new design of a compliant HS thrust bearing.

2. The analytical model

The model of the proposed compliant hydrostatic (HS) bearing is shown in figure 1. A classical, axi-symmetric, single recess, hydrostatic thrust bearing has a modified, deformable upper pad. A thin shell plate is placed over the upper pad, provided with a large recess of reduced depth. Thus, the plate can be deflected inside the pad under the action of the pressure generated in the bearing. The plate is bent over the extremely thin, outer ridge of the pad, and glued on the outer diameter. The lower pad is assumed rigid and concentric with the upper pad. At rest, the undeformed flexible plate is parallel with the mating face of the lower pad. The lubricant is supplied with pressure p_s into the centrally placed recess of radius r_s, cut into the lower pad; thus, axi-symmetry prevails.

For the *fluid flow model*, classical assumptions of HS lubrication theory are used:

- the lubricant is Newtonian and incompressible;
- the flow is laminar and isoviscous, with no slip;
- pressure variation across the film thickness is neglected;
- inertia, gravitational and external forces are neglected.

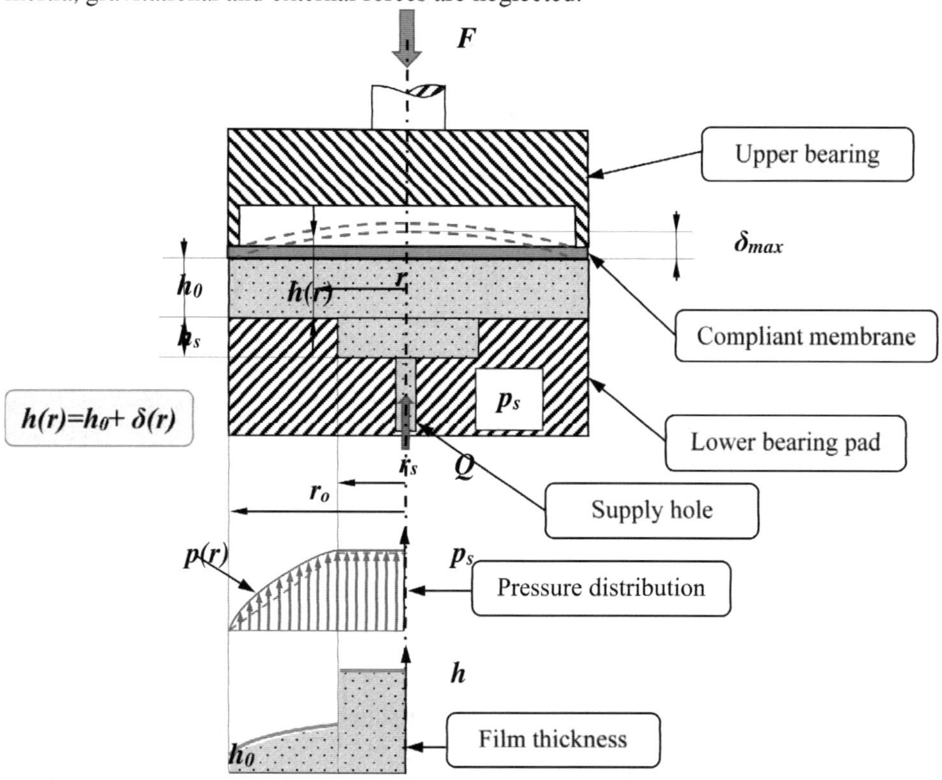

Figure 1. Schematic of the compliant bearing, including pressure distribution and film thickness variation

For the *elastic model*, classical assumptions of plane, plate theory are satisfied:

- the circular plate is perfectly flat, with a uniform thickness much smaller than the bearing radius;
- the material is homogenous and isotropic;
- deformations in the normal plane with respect to the surface of the plate are considerably smaller than the thickness of the plate.

To compute the bearing performance characteristics, the coupled equations for fluid flow and elastic deformations must be solved simultaneously.

According to the above-mentioned assumptions, the Reynolds equation takes the following simplified form for an axi-symmetric hydrostatic thrust bearing with a centrally placed recess:

$$\frac{\mathrm{d}}{\mathrm{d}r}\left(rh^3\frac{\mathrm{d}p}{\mathrm{d}r}\right) = 0 \tag{1}$$

Equation (1) can be solved in order to obtain the pressure distribution, only if the film thickness is defined. Thus, for the HS bearing, film thickness can be expressed as follows:

$$h(r) = h_0 + h_{el}(r) \tag{2}$$

where h_0 is the minimum film thickness (at the outer boundary of the bearing), and h_{el} is the elastic deformation of the upper plate of the bearing, produced by the pressure distribution. The latter can be calculated from general equations of elasticity for the pressure distribution obtained after solving the Reynolds equation.

A solution of this system of equations can be obtained only by using an iterative numerical procedure. Such a procedure does not facilitate parametric investigations; as consequence, a simplified approach is proposed, based on supplementary assumptions, which allow uncoupling the two mechanisms.

A parabolic variation of the elastic component of the film thickness is assumed along the radial direction of the bearing, with a maximum value, δ_{max}, in the centre axis of the bearing:

$$h_{el} = \delta_{max}\left[1 - \left(\frac{r}{r_o}\right)^2\right] \tag{3}$$

Further, the maximum deflection of the thin plate is calculated analytically, using a simplified pressure distribution in the bearing. Because the depth of the recess is much greater than the film thickness, the pressure distribution is assumed constant in the recess and drops down to a value corresponding to the outer atmospheric pressure (null relative pressure) following an inflated curve. A schematic of this pressure variation is shown in figure 1. The two model components will be further described in detail.

2.1 Elastic plate model

Due to the complex shape of the compliant upper plate, the deformation produced by the real pressure distribution cannot be calculated analytically. The proposed model follows a simplified approach based on the well-known plane plate theory, which considers a flat plate of uniform thickness, made of a homogeneous and isotropic material. The elastic deformation of the upper plate is analytically calculated for a simplified pressure distribution in the bearing, with linear variation on the land. Analytical solutions of the elastic deformation in plane circular plates, simply supported on the outer edge and axi-symmetrically loaded with either constant pressure or linearly variable pressure, can be found in [5]. The elastic component of the film thickness defined as a function of the maximum deflection is briefly reproduced herein, in dimensionless form.

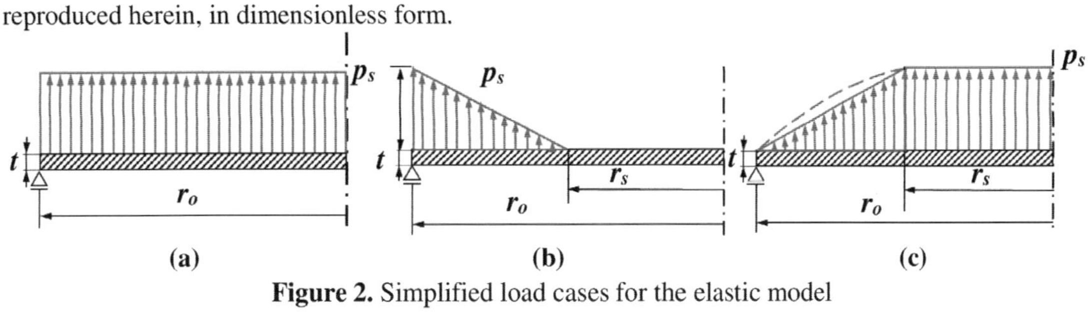

Figure 2. Simplified load cases for the elastic model

(a): constant pressure distribution on the entire plate (figure 2a)

$$\overline{\delta}_{\max\_1} = K_e \frac{3}{16}(5+\upsilon)(1-\upsilon) \tag{4}$$

(b): linear pressure distribution on an annular surface (figure 2b)

$$\overline{\delta}_{\max\_2} = 6K_e[(1-\upsilon)L_1 - 2(1-\upsilon^2)L_2] \tag{5}$$

where $\bar{\delta}_{max} = \frac{\delta_{max}}{h_0}$ and

$$L_1 = \frac{1}{720(1-R)}[(20R^3+16)(4+\upsilon) - 45R(3+\upsilon) - 9R^5(1-\upsilon) - 60R^3(1+\upsilon)\ln R] \tag{6}$$

$$L_2 = \frac{1}{14400(1-R)}[64 - 225R - 100R^3 + 261R^5 - 60R^3(3R^2+10)\ln R] \tag{7}$$

From equations (4) – (7) one can remark that the dimensionless form of the maximum deformation is expressed in terms of two important parameters: $K_e = \frac{p_s {r_o}^4}{Et^3 h_0}$ is a complex elastic parameter (which depends on the geometry and material properties of the elastic membrane), and $R = \frac{r_s}{r_0}$ is the relative recess radius.

Finally, for the simplified pressure distribution in the bearing (figure 2c), one can obtain the maximum deflection in dimensionless form, based on superposition of effects:

$$\overline{\delta}_{\max} = \overline{\delta}_{\max\_1} - \overline{\delta}_{\max\_2} = K_e\left[\frac{3}{16}(1-\upsilon)(5+\upsilon) - 6(1-\upsilon)L_1 + 12(1-\upsilon^2)L_2\right] \tag{8}$$

2.2 Fluid flow model

Using dimensionless notations, the Reynolds equation (1) becomes

$$\frac{\mathrm{d}}{\mathrm{d}\bar{r}}\left(\bar{r}\bar{h}^3\frac{\mathrm{d}\bar{p}}{\mathrm{d}\bar{r}}\right) = 0 \tag{9}$$

and film thickness can be written as:

$$h(\bar{r}) = 1 + \bar{\delta}_{max}(1-\bar{r}^2) \tag{10}$$

Considering the classical boundary conditions for pressure in single recess circular bearings ($\bar{p} = 1$ at $\bar{r} = R$ and $\bar{p} = 0$ at $\bar{r} = 1$), after integration and several algebraic manipulations, one can obtain the pressure distribution on the land:

$$\overline{p}(\overline{r}) = \frac{\left[\frac{3\Delta^2 - 2\Delta\overline{r}^2}{(\Delta-\overline{r}^2)^2} - \frac{\Delta(3\Delta-2)}{(\Delta-1)^2} + 2ln\left(\overline{r}^2\frac{\Delta-1}{\Delta-\overline{r}^2}\right)\right]}{\left[\frac{3\Delta^2-2\Delta R^2}{(\Delta-R^2)^2} - \frac{\Delta(3\Delta-2)}{(\Delta-1)^2} + 2ln\left(R^2\frac{\Delta-1}{\Delta-R^2}\right)\right]} \tag{11}$$

where $\overline{p} = \frac{p}{p_s}$, and $\Delta = 1 + \frac{1}{\bar{\delta}_{max}}$ is a dimensionless elastic parameter used to compact the equation.

Figure 3 shows the inflated pressure distribution for a bearing with the recess radius half the outer radius (R= 0.5), for different values of the elasticity parameter, K_e.

It is worth noting that for a rigid bearing, $\Delta \to \infty$ and equation (11) yields to the classical formula for rigid bearings:

$$\overline{p}(\overline{r}) = \frac{\ln\overline{r}}{\ln R} \tag{12}$$

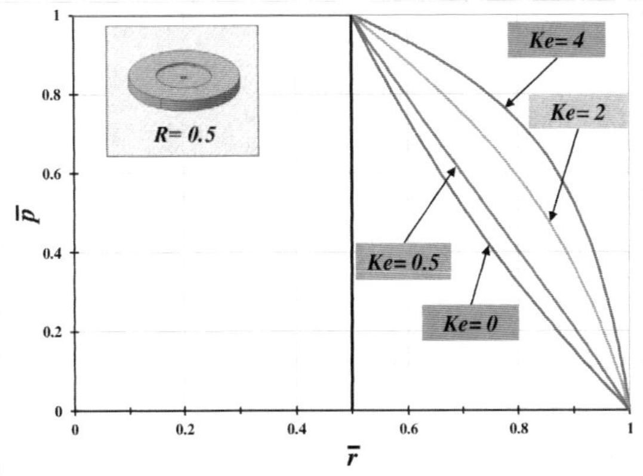

Figure 3. Dimensionless pressure distribution

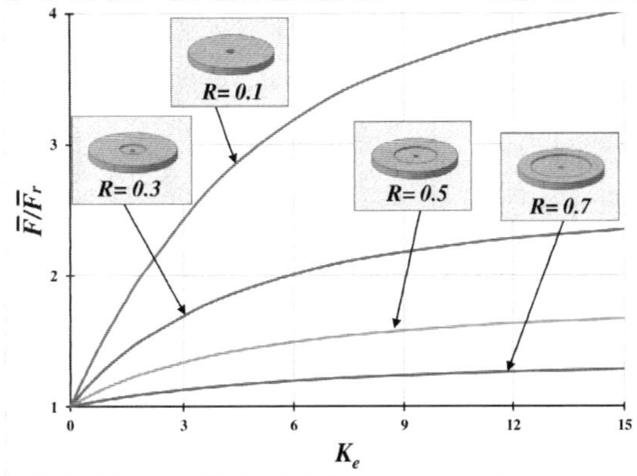

Figure 4. Relative load carrying capacity

The load carrying capacity can be obtained only by integrating numerically the pressure distribution:

$$\overline{F} = \frac{F}{\pi {r_o}^2 p_s} = R^2 + 2\int_R^1 \overline{r}\,\overline{p}\,\mathrm{d}\overline{r} \tag{13}$$

The beneficial effect of elasticity, for various sizes of the recess, is clearly depicted in figure 4 in terms of the ratio between the actual load carrying capacity and that of the rigid bearing $\overline{F}_r = \frac{1-R^2}{2\ln\frac{1}{R}}$. The greater the elasticity parameter, the greater the load capacity; however, the elasticity parameter is limited by the high stresses in the plate.

The rate of flow results after differentiating the pressure distribution:

$$\overline{Q} = Q\frac{\eta}{p_s {h_0}^3} = -\frac{2\pi}{3}\cdot\frac{\left(\frac{\Delta}{\Delta-1}\right)^3}{\frac{3\Delta^2 - 2\Delta R^2}{(\Delta - R^2)^2} - \frac{\Delta(3\Delta-2)}{(\Delta-1)^2} + 2ln\left(R^2\frac{\Delta-1}{\Delta-R^2}\right)} \tag{14}$$

The predicted values divided by the rate of flow of the rigid bearing$\left(\overline{Q}_r = \frac{\pi}{6lnR}\right)$, for different sizes of the recess, are shown in figure 5.

The friction torque for rotational speed ω and lubricant dynamic viscosity η, yields:

$$\overline{M_f} = M_f\frac{h_0}{\eta\omega {r_o}^4} = 2\pi\int_R^1 \frac{\overline{r}^3}{a - b\overline{r}^{-2}}\mathrm{d}\overline{r} = \pi(\Delta-1)\left[(R^2-1) + \Delta ln\left(\frac{\Delta-R^2}{\Delta-1}\right)\right] \tag{15}$$

Figure 6 presents the values of the friction torque divided by the friction torque of the rigid bearing $\overline{M}_{fr} = \frac{\pi}{2}(1 - R^4)$.

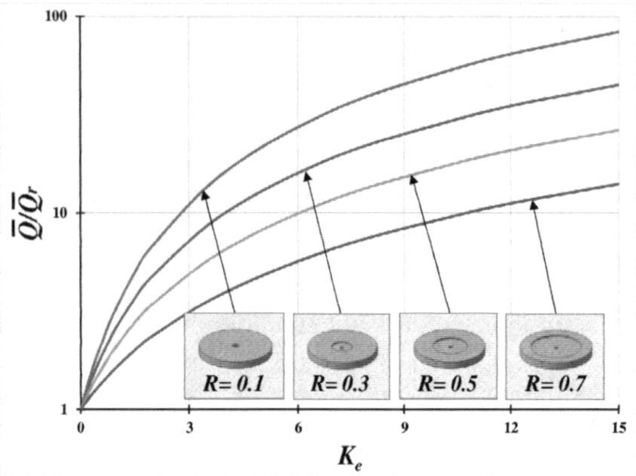

Figure 5. Relative rate of flow

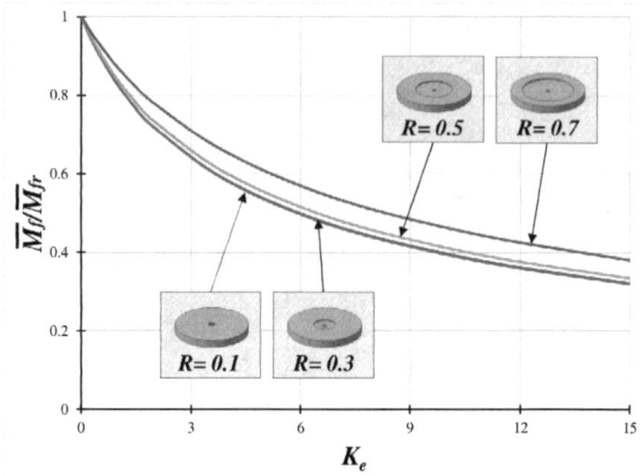

Figure 6. Relative friction torque

3. The FE model

The accuracy of the proposed analytical model of the elastic plate was evaluated using structural FE simulation. This approach uses a particular design solution of the elastic membrane (r_o=50mm, t=3mm), made of steel (E=210GPa and ν =0.3) which is meshed with regular quadrilaterals (element type - PLANE183 axi-symmetric). A total number of 5217 nodes and 1600 elements was found sufficient for yielding a reasonable accuracy, due to the fact that the model did not include an intricate geometry. The boundary conditions were identical with the ones considered for the simplified analytical model, whereas the pressure distribution applied on the bottom side of the membrane was computed using the analytical model - equation (11). Two representative sizes of the recess radius were considered, and the results were compared with the analytically predicted values, calculated with the maximum plate deformation, obtained two load cases. For load case 1 (figure 2c) the maximum deformation is calculated with equation (8) and slightly underestimates the real load distribution, whilst for load case 2 (figure 2a) the deformation is calculated with equation (4), yielding to an overestimated solution.

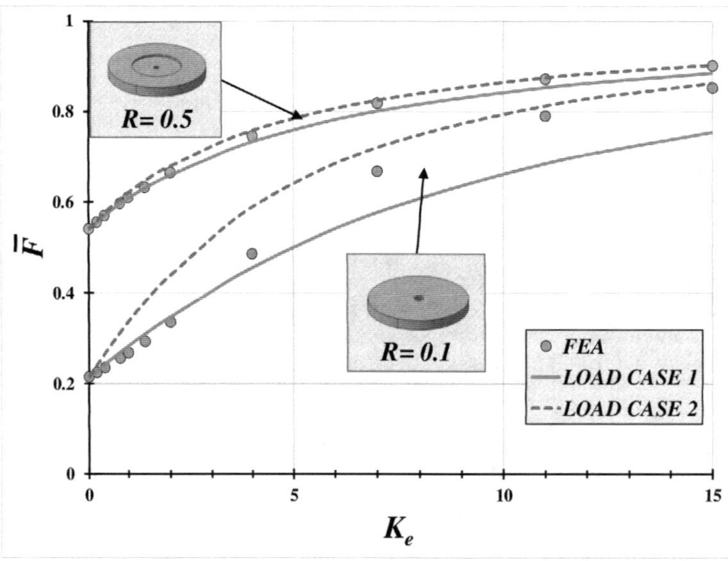

Figure 7. Analytical vs. FEA load capacity

Figure 7 represents the most valuable result of the present work. The graph depicts the load carrying capacity predicted theoretically using two models for the pressure distribution of the elastic plate (constant pressure and combined pressure, respectively), and results obtained used pseudo-numerical FE model. One can observe that, for values of $K_e < 5$, it is more accurate to use the combined load case 1, whilst for values of $K_e > 5$, a constant pressure load case 2, approximates better the elastic behaviour.

4. Conclusions

An original design solution of a compliant hydrostatic bearing is theoretically analysed. Based on simplified pressure distribution and elastic deformation, analytical formulations for bearing performance characteristics are proposed. The predicted results are shown relatively to those obtained for the rigid bearing. The superior load capacity and lower friction torque of the elastic HS bearing are put in evidence with the proposed model. The accuracy of the elastic component of the model has been evaluated by finite element analysis (FEA) and a two-range analytical formulation was proposed. The limits of elasticity due to high stresses are also defined using FEA. A coupled numerical approach for more accurate solutions of bearing performances is foreseen for the next step of the analysis.

5. References

[1] Castelli V, Rightmire G K and Fuller D D 1967 On the analytical and experimental investigation of a hydrostatic, axisymmetric compliant-surface thrust bearing *ASME J. Lubr. Technol.* **89(4)** 510-519
[2] Dowson D and Taylor C M 1967 Elastohydrostatic lubrication of circular plate thrust bearings *ASME J. Lubr. Technol.* **89(3)** 237-242
[3] Hao D, Xiao-Long Z and Juan-An Z 2015 Static characteristic analysis and experimental research of aerostatic thrust bearing with annular elastic uniform pressure plate *Advances in mechanical engineering* 1-13
[4] Hayashi K and Hirasata K 1982 Theoretical Investigation on the back-pressured elastohydrostatic thrust bearing *Journal of JSLE International Edition* **3(3)** 277-283
[5] Young W and Budynas R 2011 *Roark's Formulas for Stress and Strain* McGraw-Hill 7th Edition New York

IOP Conf. Series: Materials Science and Engineering **724** (2020) 012042 doi:10.1088/1757-899X/724/1/012042

Evaluation of inertia effects in planar squeeze flow inside soft, porous layers

G Lupu[1], P Turtoi[2], T Cicone[2]

[1] Military Equipment and Technologies Research Agency, Aeroportului 16, Clinceni, Ilfov, 077025, Romania
[2] Department of Machine Elements & Tribology, University POLITEHNICA of Bucharest, Spl. Independenței 313, Bucharest, 060042, Romania

Abstract. The resistance to flow of a fluid squeezed out from a soft and porous structure under compression, generates high load capacity. The process is dependent on permeability, which is variable during compression of the porous layer. This mechanism (named Ex-Poro-Hydrodynamic Lubrication – XPHD) plays an important role in a wide range of tribological applications (biolubrication, shock absorbers, etc.). A literature search reveals that Darcy's law is intensively used for modeling flow through porous media. However, for high porosity materials and relatively high velocity flows that may occur during impact, the inertia effects could be significant. The full squeeze solution of high velocity flow inside deformable porous materials including inertia effects and permeability variation with the level of compression could not be found in the literature. A numerical study for planar squeeze flow of a Newtonian liquid inside a soft porous layer, based on the Darcy flow model including Forchheimer correction, is presented in this paper. The porous material is compressed at high speed between two perfectly rigid, parallel and impermeable discs. Finite differences technique is used to solve the system of equations. Pressure distribution, load and Reynolds pore based number variation during compression are obtained and compared with similar values predicted by simple analytical models based on Darcy law. The results define the limits of Darcy approach in XPHD lubrication.

1. Introduction

Lubrication with fluid film is the main topic of numerous studies with various applications. A particular application based on fluids imbibed inside a highly deformable porous media is emerging and there are studies that prove higher load capacity. Load is generated by the resistance to flow inside the porous matrix with its variable (decreasing) permeability during compression. This mechanism (named Ex-Poro-Hydrodinamic Lubrication – XPHD [1]) shows great potential for squeeze dampers. For squeeze flow inside porous materials, Darcy law is widely used. However, for high velocity flow inertia effects could be important. It was already proven that Darcy law is less accurate when pore based Reynolds number Re_p becomes greater than unity [2]÷[4]. There is no well established limit or transition zone between laminar to turbulent flow regime. According to Zeng & Grigg [5] the transitional flow occurs for Re_p=1÷100, while Boomsma & Poulikakos [6] proposed a Re_p=5.

First proposal for modification of Darcy law by introducing a quadratic term was made by Dupuit [7], and Forchheimer [8] by analogy with Navier-Stokes equations. Presently the equation proposed by Ward [9] and Beavers et al. [10], based on a series of experimental results, is widely used:

$$\frac{\mathrm{d}p}{\mathrm{d}r} = -\frac{\eta}{\phi}u - \frac{C_f \cdot \rho}{\sqrt{\phi}}u^2 \tag{1}$$

where C_f is a nondimensional Forchheimer coefficient, experimentally determined, ϕ is material permeability, η dynamic viscosity, ρ density. Some studies tried to find a universal value for this coefficient. Ward [9] stated that C_f=0.55 can be used for all permeable materials. Later experimental results reported that C_f is different for various classes of materials. Beavers and Sparrow [11] found C_f=0.078÷1.32 for porous media in the form of latticework of metallic fibers and water. For particles beds, Schwartz and Probstein [12] used C_f=0.26. Other studies tried to find a connection between C_f and porosity. Ergun and Orning [13] proposed for gases that $\frac{C_f}{\sqrt{\phi}} = b''\frac{(1-\varepsilon)}{\varepsilon^3}$ where b'' is a factor of proportionality. According to Irmay [14], for any fluid, the dimensionless coefficient is defined as $C_f=\alpha\cdot(\beta\cdot\varepsilon^3)^{-1/2}$, and later, Bear [3] suggested: α=0.6 and β=180. Thus $C_f=0.045\cdot\varepsilon^{-3/2}$.

Despite a relatively great number of papers dedicated to inertia effects in porous media, no attempt could be found for the case of a highly deformable media with time variable porosity and permeability.

This study is focused on a particular application of XPHD lubrication for damping devices subjected to medium impact speed, when inertia effect can be significant. It is of major interest to evaluate the limits of Darcy flow model when squeezing speed is high. In the first stage, the analysis is performed for constant speed squeeze in a disc-on-plane configuration.

2. Problem formulation and main assumptions

A simple case of squeeze of an imbibed porous material between two circular parallel discs is considered for this study. The upper rigid disc approaches with constant speed V=dh/dt and compresses the material with initial thickness h_0. The porous material has the same radius R as the rigid discs. The model is axisymmetric with radial flow and atmospheric pressure on the outer boundaries. The geometry of the model is presented in figure 1.

The porous material is highly deformable (compression forces generated by the solid structure are neglected), homogeneous and isotropic. A reticulated structure with open pores, saturated with liquid, is considered. The fluid is Newtonian and the flow is laminar, isothermal and isoviscous.

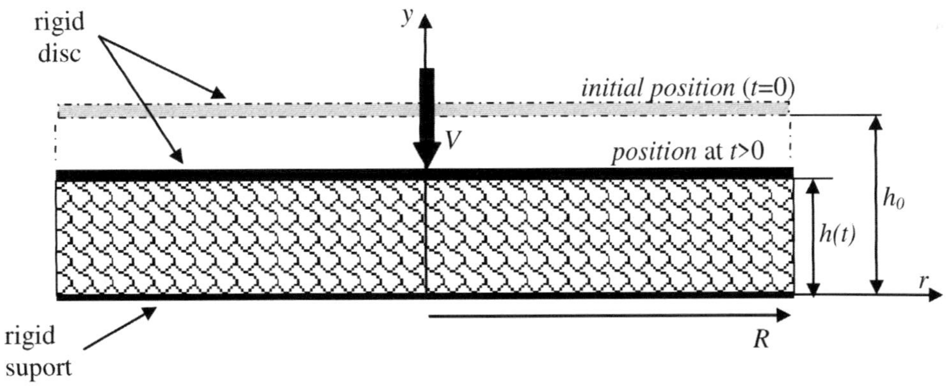

Figure 1. The geometry of the axisymmetric planar squeeze flow

The equation of mass conservation in cylindrical coordinates (r,u) is:

$$\frac{1}{r}\frac{\mathrm{d}}{\mathrm{d}r}(ru) + \frac{\mathrm{d}v}{\mathrm{d}y} = 0 \tag{2}$$

Using Kozeny-Carman equation [15] permeability-porosity correlation is:

$$\phi = \frac{D \cdot \varepsilon^3}{(1-\varepsilon)^2} \tag{3}$$

where D is a complex parameter, function of the characteristic size of the porous structure.

The material deformation is produced only in the normal direction of the porous layer and the conservation of the solid fraction is used $h \cdot (1-\varepsilon)$=ct. Thus introducing dimensionless layer thickness $\bar{h} = h / h_0$ ecuation (3) becomes:

$$\phi = \frac{D \cdot (\bar{h} - 1 + \varepsilon_0)^3}{\bar{h} \cdot (1-\varepsilon_0)^2} \tag{4}$$

The following dimensionless parameters are introduced: $\bar{y} = y / h_0$, $\bar{v} = v / V$, $\bar{u} = u / V$, $\bar{p} = p \cdot D / (\eta V R)$. Permeability ϕ is also replaced with dimensionless form (eq. (4)).

The dimensionless form Darcy-Forchheimer (D-F) equation (1) in dimensionless form is:

$$\frac{\mathrm{d}\bar{p}}{\mathrm{d}\bar{r}} = -\frac{\bar{h}(1-\varepsilon_0)^2}{(\bar{h}-1+\varepsilon_0)^3}\bar{u} - (C_f \cdot K)\frac{1}{\sqrt{\dfrac{(\bar{h}-1+\varepsilon_0)^3}{\bar{h}(1-\varepsilon_0)^2}}}\bar{u}^2 \tag{5}$$

where $\bar{y} = y / h_0$, $\bar{v} = v / V$, $\bar{u} = u / V$, $\bar{p} = p \cdot D / (\eta V R)$ and $K = \rho V \sqrt{D} / \eta$. If we integrate eq. (2) with respect to radius *r*, while considering the boundary conditions (at *y*=0, *v*=0 and at *y*=*h*, *v*=*V*) we obtain eq. (6) in a dimensionless form:

$$\frac{\mathrm{d}}{\mathrm{d}\bar{r}}(\bar{r}\bar{h}\bar{u}) = -\bar{r}\left(\frac{R}{h_0}\right) \tag{6}$$

3. Numerical solution

There is no analytical solution possible for these equations and consequently, a numerical approach based on finite differences method is proposed. Because there is no flow on $\bar{y}$ direction, the domain is split only in radial direction in small elements. Pressure and radial flow velocity is assumed to be constant across the thickness of the porous layer. An explicit first-order FD scheme with forward derivatives is used for the discretization of the eq. (5) and (6). A one-dimensional grid (along $\bar{r}$) based on *i*=1...*N*+1 equally-spaced nodes is used and leads to the following set of discrete equations:

$$\bar{p}_i = \bar{p}_{i+1} + \left(f \cdot \bar{u}_i + C_f K \frac{1}{\sqrt{1/f}} \bar{u}_i^2 \right) \Big/ \Delta\bar{r} \tag{7}$$

$$\bar{u}_i = \left(\bar{u}_{i-1} \cdot \bar{r}_{i-1} + \frac{\Delta\bar{r}}{\bar{h}_k} \cdot \frac{\bar{r}_i + \bar{r}_{i-1}}{2} \cdot \frac{R}{h_0} \right) \Big/ \bar{r}_i \tag{8}$$

where $f = \dfrac{h_k(1-\varepsilon_0)^2}{(h_k - 1 + \varepsilon_0)^3}$, $\bar{p}_i$ and $\bar{u}_i$ refer to pressure and flow velocity, respectively, on each node and *k* is the subscript that counts for time step.

Gauss-Seidel iterative method without relaxation is used. At each time step pressure distribution is integrated and load calculated. The solution convergence is checked by: the relative variation in force $\bar{F}$ between two iterations must be inferior to 10^{-9}. Grid independence solution was satisfied for $N > 1000$.

4. Results and discussion

The present analysis is focused on a particular application presented in [16]. A high porosity material (knitted fabric made of polyester yarns) and glycerin are considered for impact squeeze. First, an analysis of the values of *K* for various conditions reveals a large interval, spanning from 10^{-3} for low fluid velocity and glycerin to 10^2 for high fluid velocity and water. For a squeeze speed of 1 m/s the coefficient *K*=0.015 was found. A second less viscous fluid (e.g. synthetic oil), was considered which, for same speed, yields to a coefficient *K*=0.07. The algorithm allows determination of the pressure and flow velocity distribution with radius when the porous imbibed material is compressed and the fluid is squeezed out including the inertia influence. Dimensionless pressure distribution on the radius of the porous disc during compression for different values of $\bar{h}$ is presented in figure 2. According to continuity equation, variation of dimensionless flow velocity in radial direction is linear and reaches the maximum value when the pressure gradient is maximum at the outer boundary of the porous disk.

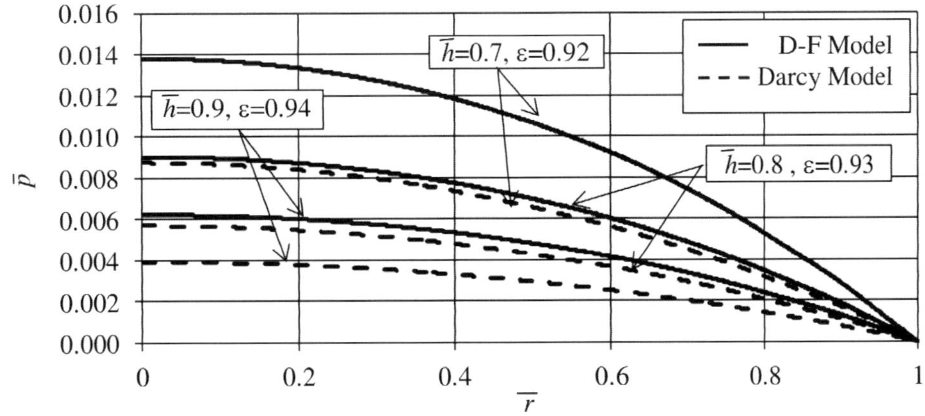

Figure 2. Variation of dimensionless pressure during compression for different values of dimensionless thickness $\bar{h}$ (ε_0=0.95, K=0.015)

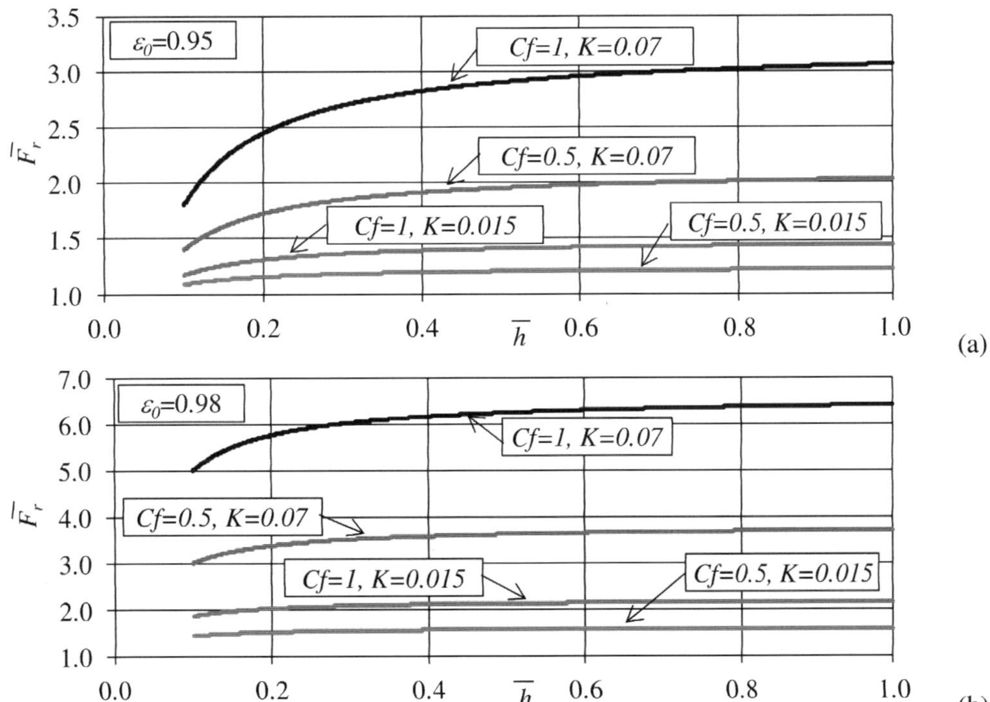

Figure 3. Variation of relative load $\bar{F}_r$ for different values of C_f , K and initial porosity a) ε_0=0.95, b) ε_0=0.98

Further, by numerical integration the load $\bar{F}$ can be determinate. To better understand the influence of inertia effects, relative load $\bar{F}_r = \bar{F} / \bar{F}(C_f = 0)$ was calculated by dividing the load obtained using the proposed model for different values of C_f and K to the load obtained for C_f=0 when no inertia effects are considered. Figure 3 shows the variation of relative load for two different initial porosities: a) ε_0=0.95, b) ε_0=0.98. One can see that for low material thickness inertia effects decrease. The drop is accentuated for low thickness due to the variation of porosity in accordance with conservation of solid phase assumption. The load also increases with increasing C_f. If a less viscous fluid when K is increasing to 0.07, the inertia effects increase and also the estimated load. Further, to validate the numerical model, the variation of load $\bar{F}$ was compared at limit, when Forchheimer coefficient C_f=0, with the analytical solution of the Darcy flow model proposed by Radu [17] for disc on plane squeeze. The difference is very small and can be neglected.

An analysis of pore based Reynolds number variation $Re_p=\rho u\phi^{1/2}/\eta$ is presented in figure 4 for the maximum value of the flow velocity at the outer boundary of the domain. For a low squeeze speed of 1 m/s and K=0.015 the Re_p exceeds the laminar flow limit. When considering the same fluid (K=0.04), if the speed increases, the Re_p increases as well. Furthermore, when a less viscous fluid is used, Re_p exceeds the limit 10. This is in accordance with the assumption that for high squeeze flow inertia must be taken into account. The variation of Rep on the radius is presented in figure 5. It can be seen that the variation is linear from centre, as no radial flow occurs, gradually increasing towards the exterior of the disc where the pressure gradient is maximum and also the flow velocity.

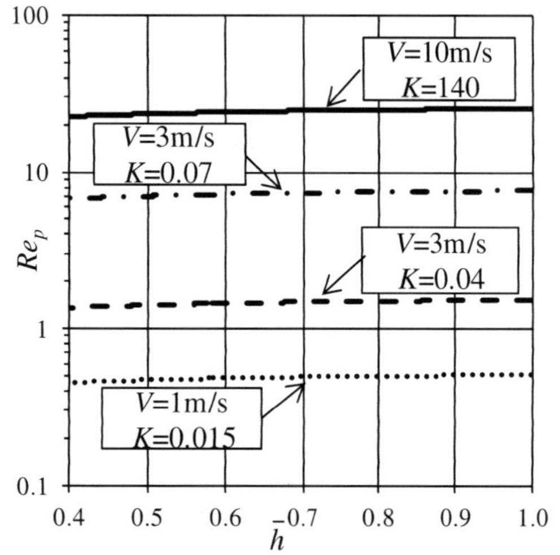

Figure 4. Variation of Re_p on compression for highest radial flow velocity u_{max}

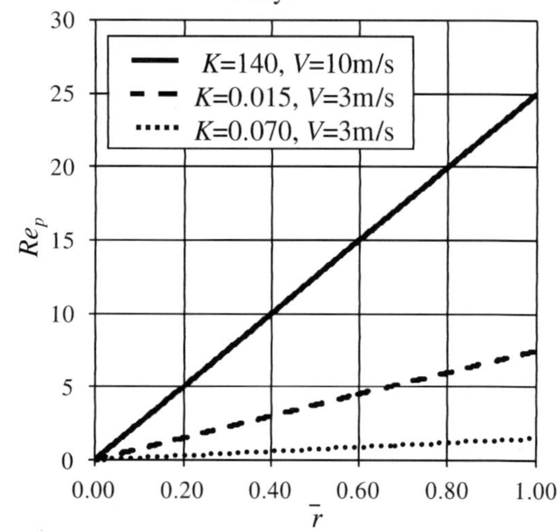

Figure 5. Variation of Re_p on the radius of the disc for $\bar{h}$ =0.8

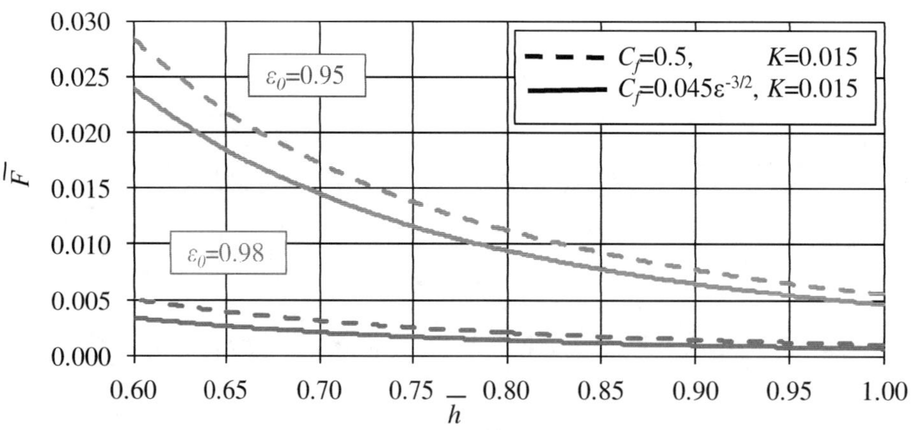

Figure 6. Load variation during squeeze

The effect of Forchheimer coefficient C_f when is variable with porosity according to definition of Irmay [13] was studied in parallel with the case of constant C_f (figure 6). One can see that when C_f is variable the predicted force is lower and the difference becomes higher when the material is more compressed. Two cases for different initial porosities ε_0=0.95 and ε_0=0.98 are used.

5. Conclusions

An evaluation of the inertia effects on force generated during high constant velocity squeezing out of a liquid imbibed in a highly porous and soft layer has been performed. The analysis has been performed numerically using finite differences method.

Pressure distribution was determined and influence of inertia effects on load was evaluated for two values of Forchheimer coefficient. The dimensionless results show that for relatively low squeeze speed inertia effects could be important especially for high porous materials and low viscosity fluids. It is also evident from these results that at high rates of compression (near the end of squeeze process) the inertia contribution vanishes.

An improved numerical solution should consider the dependence of friction factor in Forchheimmer term with Reynolds number.

Due to variable porosity specific to XPHD applications, a variable Forchheimer coefficient with porosity can be more appropriate to predict load. Further theoretical work appears necessary in order to evaluate inertia effects in the case of impact squeeze where fluid velocity decline with time due to damping effect.

6. References

[1] Pascovici MD, Cicone T, 2003, Squeeze-film of unconformal, compliant and layered contacts Tribol Int 36 791–799;
[2] Gartling DK, Hickox CE, and Givler RC, 1996 Simulation of coupled viscous and porous flow problems *Int J Comput Fluid D.* **7** 1 23-48;
[3] Bear J 1972 Dynamics of Fluids in Porous Materials *American Elsevier;*
[4] Pearson JRA, Tardy PMJ, 2002 Models for flow of non Newtonian and complex fluids through porous media *J Non-Newton Fluid.* **102** 2 447-473;
[5] Zeng Z, Grigg R, 2006 A criterion for non-Darcy flow in porous media *Transport Porous Med.* **63** 1 57–69;
[6] Boomsma K, Poulikakos D, 2002 The effects of compression and pore size variations on the liquid flow characteristics in metal foams *J Fluids Eng* **124** 263–72;
[7] Dupuit J, Estudes Theoretique et pratiques sur le movement des eaux, *Paris : Dunod ;*
[8] Forchheimer P, 1901 Wassebewegung durch Boden *Z Ver Deutsch Ing* **45** 1782-1788;
[9] Ward J, 1964 Turbulent flow in porous media *J Hydraul Div am Soc Civ Eng* **90**(HY5) 1-1;
[10] Beavers GS, Sparrow EM, Rodenz DE, 1973 Influence of bed size on the flow characteristics and porosity of randomly packed of spheres *J Appl Mech* **40** 655-660;
[11] Beavers GS, Sparrow EM, 1969 Non-Darcy flow through fibrous porous media *J Appl Mech* **36** 4 711-714
[12] Schwartz J, Probstein RF, 1969 Experimental study of slurry separators for use in desalination *Desalinisation* **6** 239-266;
[13] Ergun S, Orning AA 1949, Fluid flow through randomly packed columns and fluidized beds *Industrial and Engineering Chemistry* **41** 6 1179-1184;
[14] Irmay S, 1958 On the theoretical derivation of Darcy and Forchheimer formulas *Eos Trans AGU* **39** 702-707;
[15] Sheidegger AE, 1974 The physics of flow through porous media. *University of Toronto Press*;
[16] Cicone T, Pascovici M, Melciu C, Turtoi P, 2019 Optimal porosity for impact squeeze of soft layers imbibed with liquids *Tribol. Int.* **138** 140–149;
[17] Radu M, 2015 Modelling and simulation of the squeeze process during impact of highly compressible porous layers imbibed with fluids *PhD Thesis University POLITEHNICA of Bucharest* (*in Romanian*);

IOP Conf. Series: Materials Science and Engineering **724** (2020) 012043 doi:10.1088/1757-899X/724/1/012043

Some considerations about the effect of grooves on the viscous coupling performances

A I Stoicescu[1], **A Predescu**[1] **and V Gâtu**[1]

[1]Politehnica University of Bucharest, Department of machine elements and tribology, 313 Splaiul Independenței, Sector 6, 060042, Bucharest, Romania

Abstract. A viscous coupling transmits torque by viscous shearing of the oil film between one or multiple pairs of discs that are in relative motion one to the other. A viscous coupling uses a viscous fluid as a working medium and performs best in full film lubrication condition. A continuous film of oil must be present at all times between the two surfaces that are in relative motion. Only in this stage, when operating in full film lubrication condition, this coupling can achieve the maximum transmission of torque. This kind of study is of a great importance since this coupling is widely used all around the world in all sorts of applications. In this paper, the effect of grooves on the viscous coupling performances was carried out deeply using a Computational Fluid Dynamics Software. The flow of the fluid film and shear torque between the friction pair are complex and keep changing with the operating conditions, the geometry, the number of grooves, the properties of the working oil and the thickness of the oil film.

1. Introduction

As technology progresses, engineers are continuously seeking for new improvements in the auto industry sector in every way possible in order to increase consumer experience and comfort. Viscous couplings present a high interest when developing comfortable and performant drive trains due to their advantages. Viscous couplings are widely studied in all sorts of applications such as fan transmission in vehicles (viscous fan clutches) [1], pumps, winches (hydro-viscous winch) [2], mechanism of belt conveyors [3], wind turbines [4], tunnel boring machines (TBM with viscous clutch) [5].

In this paper the effects of grooves on the viscous coupling performances device are studied thoroughly in different settings such as different speeds, different number of grooves, the properties of the working oil and the thickness of the oil film. Below in figure 1 we can see different variants for placing a viscous coupling in a car [1].

Nomenclature

M_t (Nm) - transmission torque
η (Pas) - dynamic viscosity (oil)
ω_1 (rads^{-1}) - angular speed of the rotating grooved disc
ω_2 (rad^{-1}) - angular speed of the stationary smooth disc
r_e (mm) - external radius of the disc
r_i (mm) - internal radius of the disc
h (mm) - thickness of the oil film
z (mm) - number of oil films

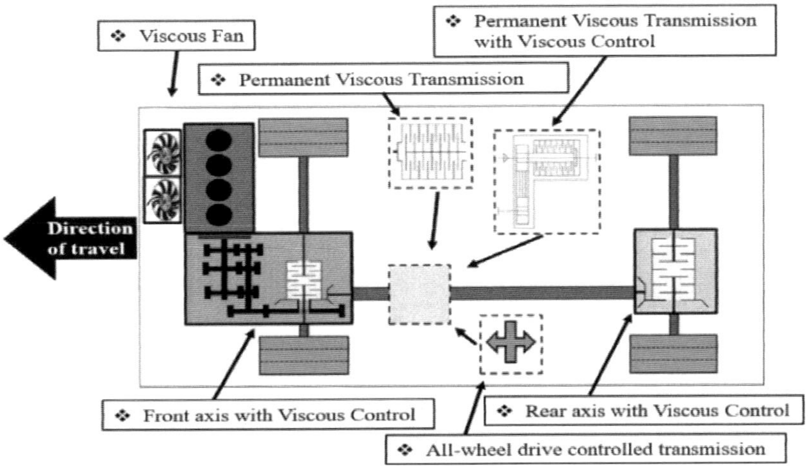

Figure 1. Different variants for placing a viscous coupling in a car

2. Mathematical modelling of a viscous coupling

2.1. Basic operating principle

The functioning of a viscous coupling is based on the Newton's law of viscosity, and works by transferring the power through the viscous shear force of the oil film. In order to transmit torque, there has to be both a thin and continuous film of oil and also an angular speed difference between the two surfaces that are in relative motion one to the other. A schematic drawing is presented in figure 2.

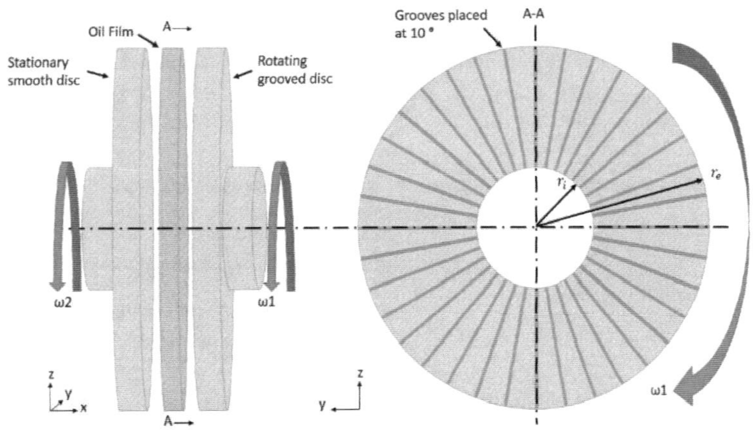

Figure 2. Schematic drawing of a viscous coupling

The torque calculation model of a viscous coupling is shown below and can be obtained by [6]:

$$M_t = z\pi\eta(\omega_1 - \omega_2)\frac{r_e^4 - r_i^4}{2h} \quad (1)$$

3. CFD model development

3.1. Designing of the 3D model

The geometrical model of the viscous coupling was built in a pre-processing software. In order to better study the phenomenon, the geometrical model was simplified and the 1/12 representing 30° of the full model was analysed using Computational Fluid Dynamics Software. Also, there were considered 5 different groove depths, 4 different oil thicknesses and there were chosen 7 different geometrical models as shown in figure 3.

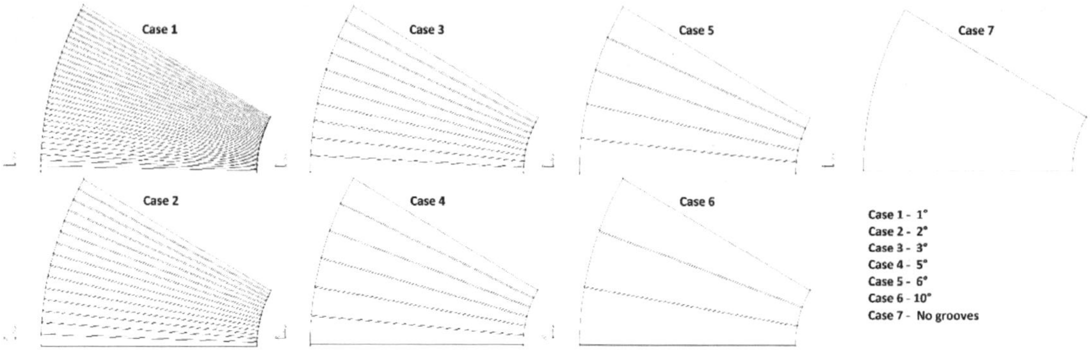

Figure 3. The geometrical models that were chosen for the CFD analysis

3.2. Boundary conditions

The boundary conditions were assigned in the pre-processing software. As shown in figure 4, stationary wall boundary condition was assigned for the external radius surface, internal radius surface and for the stationary smooth disc surface. Since the model is simplified to a 30° of the full model, periodic rotational boundary condition was assigned for both of the side surfaces and for the remaining rotating grooved disc surface, rotating wall boundary condition was selected. The internal volume of the model was considered to be of a fluid continuum type. After the assigning of the boundary conditions, the meshing of the volume was performed. Quad elements were used for the meshing of all the faces of the model and Hex elements were used for the meshing of the models volume and in both cases the type of the meshing was Map.

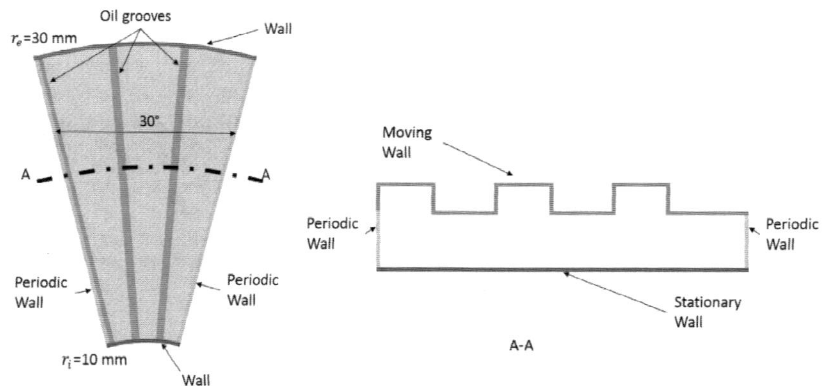

Figure 4. Assignment of the boundary conditions for the simplified model

3.3. Pre-processing

The numerical simulation was carried out in a CFD software, with 3ddp precision (three-dimensional double precision). After scaling the model to the its proper original size, the boundary conditions were reviewed and conditions were set. The entire study was carried out using speeds ranging from 10 up to 300 (rads^{-1}). An oil was selected from the material database but with the viscosity changed from constant to power-law with Three Coefficient Method using 3 reference viscosities 0.02, 0.58 and 1.06 (Pas). The reference temperature was set in all cases to be 300 (K) and. Also, in this stage the Energy Equation was selected. Furthermore, Laminar Model was chosen for the study of the simplified viscous model.

After setting all the conditions, the initialization and iteration of the model followed.

4. CFD simulation results

4.1. Analysis of transmission torque considering the different groove placements

For this analysis, there were used 6 discs with grooves placed at 1°, 2°, 3°, 5°, 6° and 10°, the textured area density is between 10% and 70%. The viscosity of the oil was chosen 0.02 (Pas) and the thickness of the oil film is 0.1 mm. The depth of the groove is 0.05 mm and the width of the grooves is 0.10 mm.

Upon analysing of the graph from figure 5 it can be seen that the torque increases with the increase of the speed. Furthermore, from the same graph it can be clearly seen that the torque increases if the number of radial grooves increases. Also, it can be seen that the torque transmitted is not linear, for example a much steeper slope is the one representing the disc with 1° grooves on it.

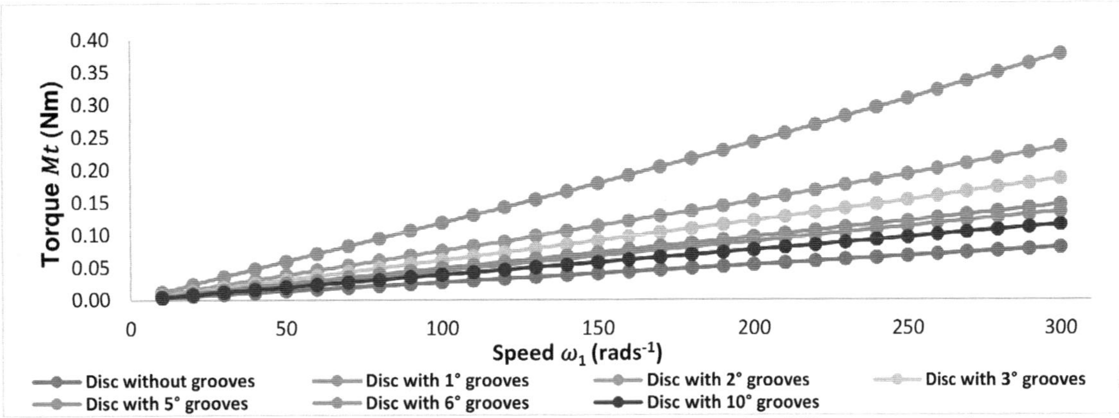

Figure 5. Torque generated at different angular speeds by discs with different groove placements

4.2. Analysis of transmission torque considering the different groove depths

For this analysis, there were used 5 discs with grooves placed at 10° and 1 disc without any grooves. The disc with grooves has depths ranging from 0.01 mm up until 0.05 mm. The viscosity of the oil is 0.02 (Pas), the thickness of the oil film is 0.1 mm and the width of the grooves is 0.10 mm.

Upon analysing of the graph from figure 6 it can be seen that the torque increases with the increase of speed and with the increase of the grooves' depths. Furthermore, from the same graph it can be clearly seen that the torque transmitted increases until it reaches about the same value at 0.04 mm ~ 0.05 mm depth of the groove.

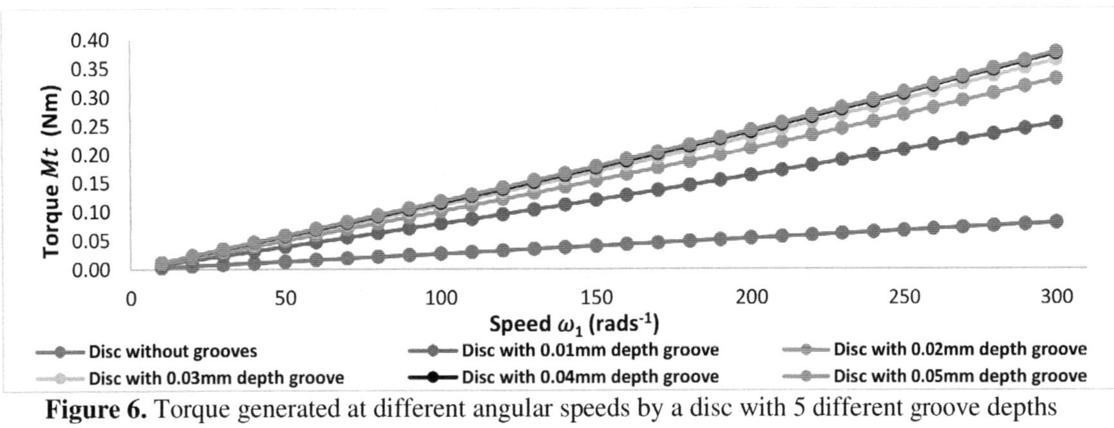

Figure 6. Torque generated at different angular speeds by a disc with 5 different groove depths

4.3. Analysis of transmission torque considering the different oil thicknesses

For this analysis, there were used 4 different oil thicknesses ranging from 0.05 mm and up to 0.20 mm. The disc used has grooves placed at 10°, the channel's depth is 0.05 mm and the width of the grooves is 0.10 mm. The viscosity of the oil was chosen 0.02 (Pas).

Upon analysing of the graph from figure 7 it can be seen that the torque increases with the increase of speed and with the decrease of the oil thickness found between the discs. Also, it can be seen that the torque transmitted is not linear, for example a much steeper slope is the one representing the oil thickness of 0.05 mm.

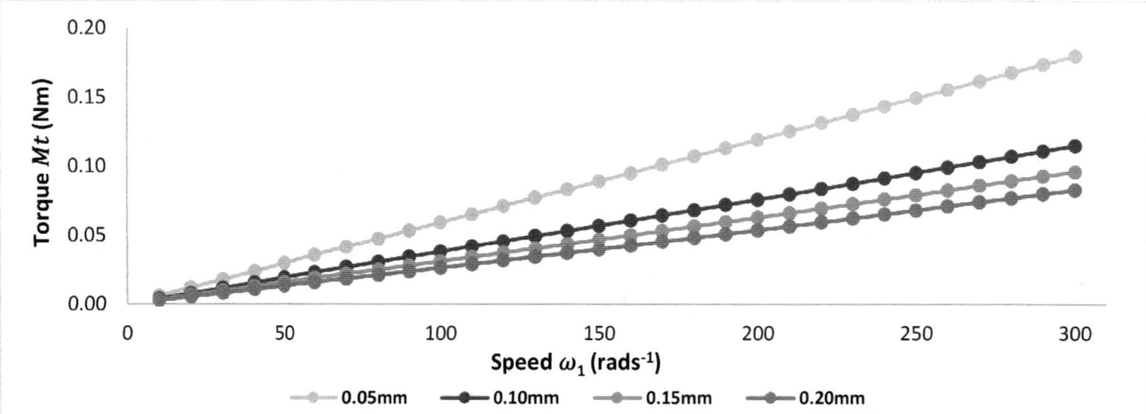

Figure 7. Torque generated at different angular speeds by a disc with 4 different oil thicknesses

4.4. Analysis of transmission torque considering the different oil viscosities

For this analysis, there were used 3 different oil viscosities: 0.02, 0.58, 1.06 (Pas). The disc used has grooves placed at 10°, the channel has depth of 0.05 mm and the width of the grooves is 0.10 mm.

Upon analysing of the graph from figure 8 it can be seen that the torque increases with the increase of speed and with the increase of the oil's viscosity found between the discs. Also from the graph it can be observed that the oil that has a viscosity of 0.02 (Pas) can barely transmit a torque of no more than 1 (Nm) at an angular speed of 300 (rads^{-1}).

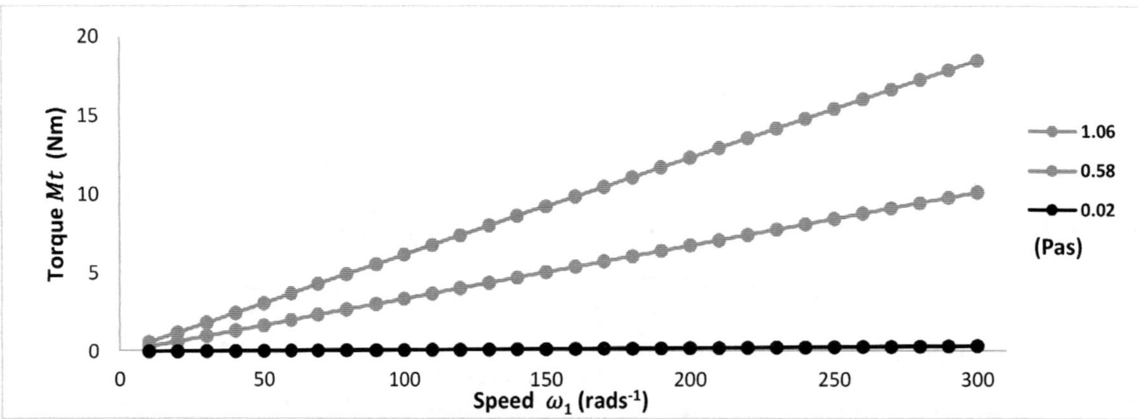

Figure 8. Torque generated at different angular speeds by a disc with 3 different oil viscosities

Due to the short operating time, the regime can be considered stationary with constant viscosity.

5. Conclusions and future work

5.1. Conclusions

Based on the present investigation that was carried out, the results are summarized as follows:

1. In all cases regardless of the number of grooves that are present on the disc or groove depths or viscosity values or oil thicknesses, the torque increases with the angular speed.
2. The torque increases if the number of radial grooves increases.
3. The torque increases with the grooves' depths until it reaches about the same values at about 0.04 mm ~ 0.05 mm depth.
4. The torque decreases with the increase of the thickness of oil film.
5. The torque values are directly proportional with the oil's viscosity.

5.2. Future work

Future work will focus on the validation of the mathematical model with an experimental test rig. Also future work will focus on the geometry of the grooves. In this paper only straight radial grooves are studied but for future work a much complex geometry and shapes will be taken into consideration. Different widths of the grooves will be also studied. In this paper only one value of the groove width 0.05 mm, is studied.

References

[1] André B, Jan S, Alexander S, Gerold S, Wolfram K and Uwe H, 2015, Hydrodynamics analysis in micro-channels of a viscous coupling using gamma-ray computed tomography, *Flow*, Pages 288-297

[2] Hou Y. F. et al., 2011, Dynamic Analysis of a New Type of Hydro-Viscous Drive Winch, *Advanced Materials Research*, Pages 328-330

[3] Qing-rui M, You-fu H, 2008, Mechanism of hydro-viscous soft start of belt conveyor, *Journal of China University of Mining and Technology*, **Volume 18**, Issue 3, Pages 459-465

[4] Xiu-xing Y, Yong-gang L, Wei L, Hong-wei L and Ya-jing G, 2014, Output power control for hydro-viscous transmission based continuously variable speed wind turbine, *Renewable Energy*, **Volume 72**, Pages 395-405

[5] Haibo X, Huasheng G, Liang H and Huayong Y, 2016, Improving the Extricating Performance of TBM Cutter-head Driving System with Hydro-viscous Clutch, *12th IEEE/ASME International Conference on Mechatronic and Embedded Systems and Applications (MESA)*, Paper 133

[6] Predescu A, 2011, *Cuplaje cu frecare vîscoasă*, Printech, Pages 315

Experimental research regarding the tribological properties of a hydraulic oil

S Haragâş[1], D Pop[1], F Cristea[1], O Buiga[1,2]

[1]Technical University of Cluj-Napoca, Romania

Abstract. This paper presents the results of the interesting experimental research regarding the influence of the additives over a hydraulic oil i.e. HFE46 (produced by ChemTrend [10]). In this vein, were used two types of additives (resulting two variants of oils denoted as sample C and D respectively) and was carried out a gravimetric determination (using the Timken test) analysing the hydraulic oils lubricating proprieties. Also, measurements regarding the vibrations in the tribotechnical system - TTS were conducted.

1. Introduction

Hydraulic oils are used in today's world in a wide range of domains, such as - *food and agriculture* - in various farming machinery; in *manufacturing engineering* - for all types of mechanical machines, presses etc.; in *constructions* - for all sorts of cranes and other pieces of equipment including here buckets, crawlers, etc. In all these applications the pressurized hydraulic oil - is carried out through the associated hydraulic circuit from the pump (driven by a power source, i.e. an electric motor) - pipes - valves - and finally to the end element - an actuator to convert the energy of the fluid into mechanical force or torque, in order to perform various tasks. The pump and the control valves create flow and pressure which are transmitted throughout the pipes to the entire system resulting vibrations and noises with harmful effects on the system's components. For example, in aircraft industry vibrations conduct to almost half of the failure of the hydraulic system [1]. Therefore, the performance of the hydraulic system is highly influenced by the hydraulic oils.

There are a series of studies regarding the influence of the hydraulic oils upon the hydraulic systems. Chai L. et al. in [1] developed a new versatile and compact hydraulic pulsation attenuator for hydraulic piston - pump systems. Chenxiao and Zhang [2] studied the vibration and noise sources into the hydraulic hoist systems. Cristea and Haragâş in [3] and [4] presented two studies regarding the oil the viscosity influence over the vibrations and noises in heavy industry machinery. Changbin and Zongxia in [5] describes an optimal design for a 3 DOF Helmholtz resonator used for a certain type of aviation piston - pump system in order to reduce the flow pulsation. Ye-qing et al. in [6] carried out a vibration analysis using the 3D model of a linear-motor-driven water piston pump. Harrison and Edge in [7] developed a novel mechanism in order to reduce the flow ripple in the case of an axial-piston-pump system. Bettig and Han in [8] described a hydro-generator rotor dynamic model for determining imbalances and shaft misalignments by running vibrations measurements.

Next the authors introduce the gravimetric determination of wear (section 2) for the case of a tribotechnical system - TTS [9] in presence of a hydraulic oil i.e. **HFE46** (produced by ChemTrend [10]) with two different types of additives (hereinafter refer as sample **C** and **D** respectively), after which the measurements carried out with a **SVAN 974 vibrometer** (section 3) are described. The fourth section contains discussions regarding the lubrication efficiency, about the variation of the friction coefficient with the speed (Stribeck curve). Finally, the authors conclude this discussion with some reflections and suggestions regarding the possible extensions of the present study.

2. Gravimetric analysis

The gravimetric analysis is the technique through which the amount of wear corresponding to a specific sample is appreciated by comparing the masses of the test specimen between and after the procedure. Obviously, the accuracy of this method depends on the precision of the measuring device (in here was used a digital analytical balance with 0.1 *mg* readability).

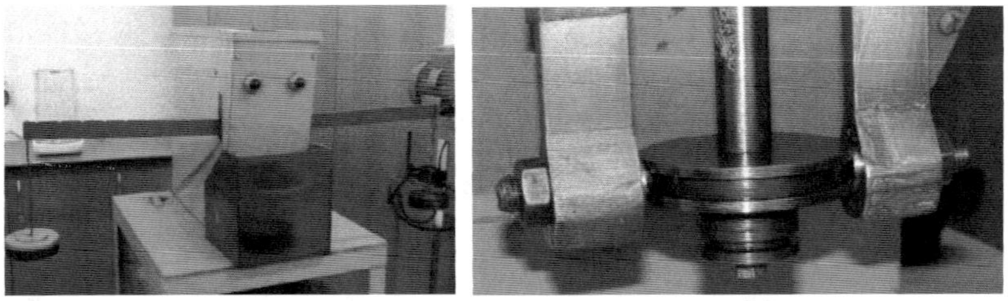

(a) **(b)**

Figure 1. The testing machines

(a) - the tribometer; **(b)** - the base body and counterbody analysed

In this paper the gravimetric wear analysis was carried out on a Timken universal wear and friction testing machine (figure 1,a).

Table 1. Experimental results.

Lubricant type	Initial mass, m_0 (g)	Mass after the test period, m_1 (g)	Wear mass loss, $m_0 - m_1$, (g)
Sample C	10.3023	9.9071	0.3952
Sample D	10.6226	10.5858	0.0368

The interfacial medium [9] i.e. the third element of the TTS as we already pointed out is represented by the **HFE46 hydraulic oil** in his two variants (i.e. sample C and D).

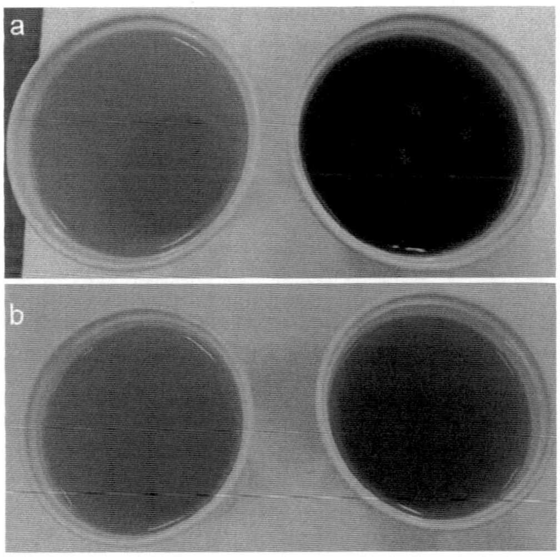

Figure 2. Initial and final colour of the samples. On the left the colour before and on the right – after the wear process

(a) - sample ***C***; **(b)** - sample ***D***

IOP Conf. Series: Materials Science and Engineering **724** (2020) 012044 doi:10.1088/1757-899X/724/1/012044

The other components of the TTS are the two cylindrical test specimens (with 16 *mm* diameter) which are in contact with a disc mounted on an electrical motor shaft having an input speed of ≈ 980 *rpm*. The rotary disc is made of 1C45 hardened carbon steel with 48 HRC. The tests are carried out for a total load of 227 *N* during 60 *min*. The experimental results are shown in Table 1.

As it can be observed from table 1 the quantity of material lost due to wear process in the case of sample **D (0,0368 g) is ten times lower than in the case of sample C (0,3952 g)**. This means that the wear rate is ten times higher for the sample C. It is obviously from this analysis that the hydraulic oil denoted as sample D has superior proprieties over the sample C.

Another important aspect represents the oxidation stability (the aging test [9]). During these experimental tests the oil temperature reached about 80^{0}C (in general the hydraulic oils are used up to 50 - 56^{0}C). As a result of this, a pronounced oxidation occurred. In figure 2 are illustrated the initial and final colours for both samples of oils. From here it can be observed that the sample D has the final colour close to the original one which indicate a lower oxidation rate.

3. Vibration determination

In this section is described the method used for measuring the vibration of the TTS. In this vein a **digital vibrometer** i.e. **SVAN 974** (figure 3,b) was used. It has a magnetic base which has been positioned on the tribometer according to figure 3,a (this position was used for both samples and was considered to be the most relevant in terms of transmission of vibrations to the tribometer's housing).

(a) **(b)**

Figure 3. Vibration determination

(a) - the vibrations measurement position; **(b)** - the SVAN 974 vibrometer

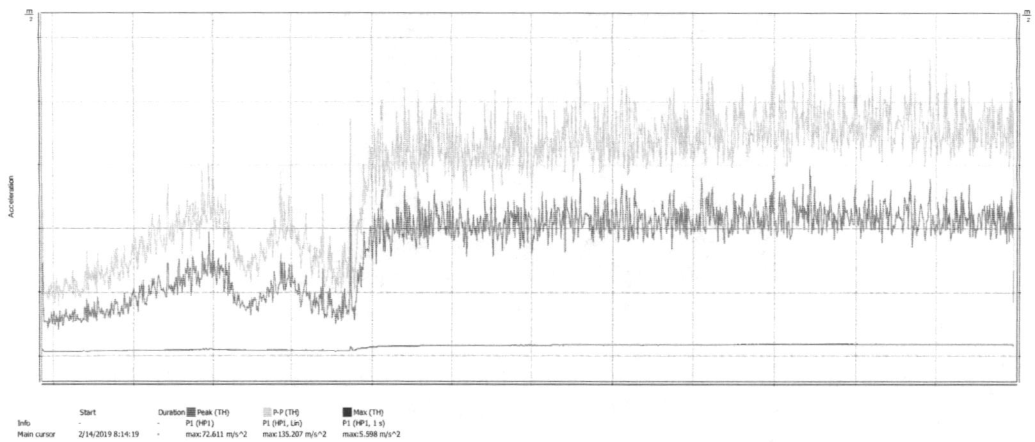

Figure 4. Peak-to-Peak (P-P) and Peak (P) acceleration for **sample *C***

In here were measured Peak-to-Peak (P-P) (which represents the distance from a negative peak to a positive one) and Peak (P) accelerations. The graphs resulted after the measurements for both samples are presented in figures 4 and 5.

It is observed from the above graphs that the P-P accelerations corresponding to ***sample C*** varies from **150-250** ***m/s***^{2} while for the ***sample D*** the P-P accelerations are in the range of **100-150** ***m/s***^{2}. Regarding the P accelerations, as it can be seen in figures 4 and 5, they vary between **100-150** ***m/s***^{2} for ***sample C*** and from **50** to **100** ***m/s***^{2} for ***sample D*** respectively.

Also, from these figures it can be seen that during 20 min of running the experiment the accelerations increase for both samples. After this interval of time for ***sample D*** the accelerations are dropping around the initial values while in the case of ***sample C*** the accelerations are continuously increasing (as a result of oil oxidation).

Regarding the vibration analysis it is clearly that in the case of the ***sample D*** they decrease considerably which means that this sample has superior lubricating proprieties over the other sample.

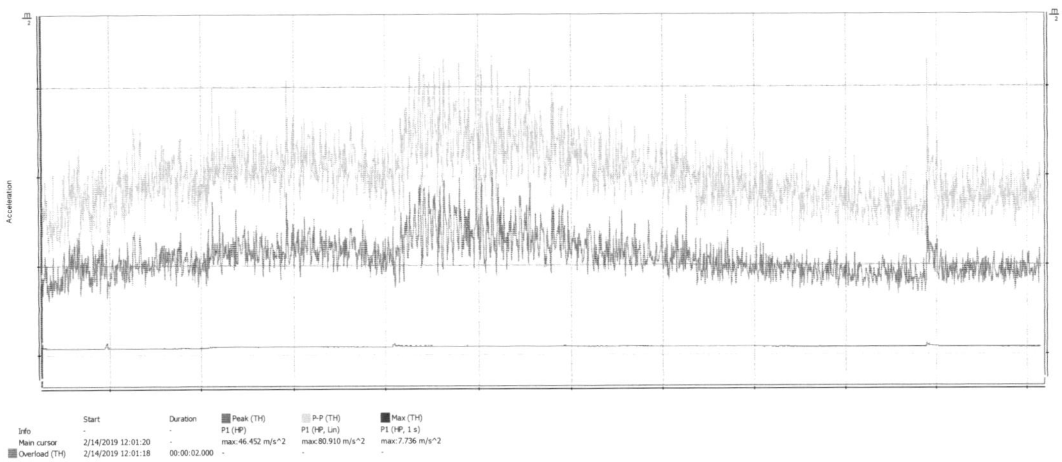

Figure 5. Peak-to-Peak (P-P) and Peak (P) acceleration for **sample *D***

4. Lubrications state simulation

In this section were carried out a set of experiments on an MCR 102 tribometer (figure 6) in order to study for both the hydraulic oil samples (C and D) the lubrication efficiency. Also, in here was analysed the variation of the friction coefficient with sliding velocity (Stribeck curve).

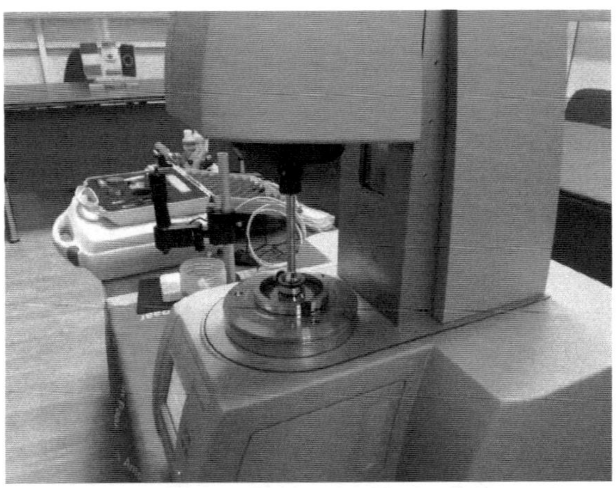

Figure 6. MCR 102 tribometer

In figure 7 the Stibeck curve is plotted. This curves in widely used [11] to describe the lubrication regimes, i.e. the boundary lubrication - *BL*, mixed lubrication - *ML* and hydrodynamic lubrication - *HL*.

It can be seen from these curves that initially; in the case of boundary and mixed lubrication the coefficient of friction is higher for the sample D (possible due to a higher kinematic viscosity of this oil). Upon passing to fluid lubrication regime, the coefficient of friction is lower for sample D than sample C.

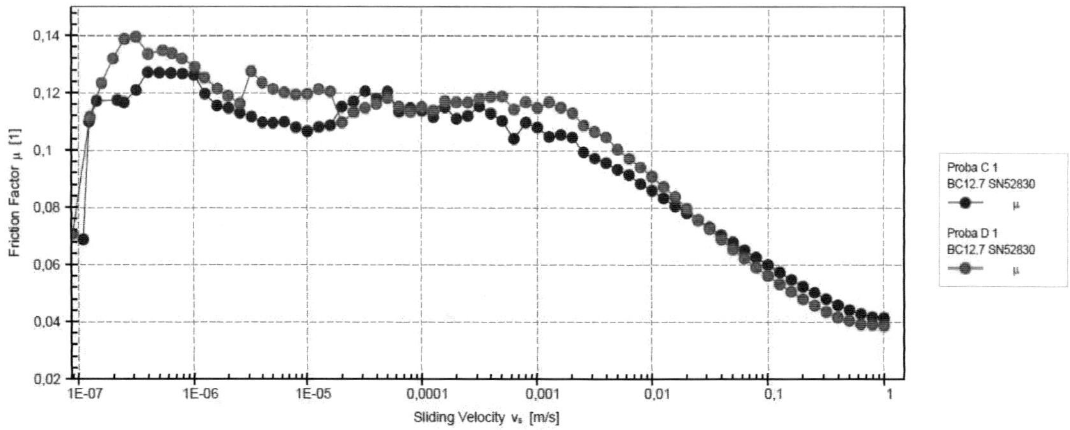

Figure 7. Stibeck curves for both samples

5. Conclusions

Experimental research carried out in this study for two hydraulic oil samples highlights the influence of additives used for hydraulic oils over theirs tribological properties. In this vein, in the case of gravimetric determination (using the Timken test) it was concluded that after the experiment the sample C was weighted ***9.9071 g*** (the initial value was ***10.3023 g***) while the mass of the sample D was ***10.5838 g*** (the initial mass was ***10.6226 g***). As compared with their initial masses it was observed that the wear rate in the case of sample C are ten time higher than for the sample D.

Regarding the vibration analysis, it is observed that the P-P accelerations corresponding to ***sample C*** varies from **150-250** ***m/s²*** while for the ***sample D*** the P-P accelerations are in the range of **100-150** ***m/s²***. The P accelerations, vary between **100-150** ***m/s²*** for ***sample C*** and from **50** to **100** ***m/s²*** for ***sample D*** respectively.

Also, after an initial interval of functioning i.e. 20 *min* in the case of ***sample D*** the accelerations are dropping around the initial values while in the case of ***sample C*** the accelerations are continuously increasing (as a result of oil oxidation). From here we can conclude that the sample D has superior lubricating proprieties over the sample C.

From the Stribeck curve analyse it can be seen that in fluid lubrication regime the friction coefficient had lower values for the sample D over the sample C.

6. References

[1] Chai L, Jiao Z, Xu Y, Zheng H 2016 A compact design of pulsation attenuator for hydraulic pumps, *IEEE International Conference on Aircraft Utility Systems Beijing*, 1111-16

[2] Chenxiao N, Zhang X 2012 Study on vibration and noise for the hydraulic system of hydraulic hoist, *Proceedings of 2012 International Conference on Mechanical Engineering and Material Science* (MEMS 2012), 126-128

[3] Cristea F, Haragâş S 2018 The lubricants influence in transmission of vibration and noises from aluminum injection machines, PROC ROM ACAD SER A, **18**, 1, 77-84

[4] Haragâş S, Cristea F 2017 Experimental research concerning the vibration of the hydraulic pump machine that injected aluminum, depending on the type of hydraulic oil used, *Acta Technica Napocensis, series: Applied Mathematics, Mechanics and Engineering*, **60**, 171-74

[5] Changbin G, Zongxia J 2012 Modelling and Optimal Design of 3 Degrees of Freedom Helmholtz Resonator in Hydraulic System, Chinese Journal of Aeronautics **25**, 776-83

[6] Ye-qing H, Song-lin N, Hui J, Shuang N 2016 Vibration analysis and experimental research of the linear-motor-driven water piston pump used in the naval ship, *Shock and Vibration*

[7] Harrison AK, Edge KA 2000 Reduction of axial piston pump pressure ripple, PROC INST MECH ENG I-J SYST C, **214**, 53-64

[8] Bettig BP, Han RPS 1999 Modelling the lateral vibration of hydraulic turbine-generator rotors, *Journal of Vibration and Acoustics - Transactions of the ASME*, **121**, 3, 322-27

[9] Grote KH, Antonsson EK 2009 Springer handbook of mechanical engineering, Berlin: Springer-Verlag

[10] https://www.chemtrend.com/

[11] Lin L, Chao Y, Yuanyuan S, 2018 Wear model based on real-time surface roughness and its effect on lubrication regimes, TRIBOL INT, **126**, 16-20, ISSN 0301-679X

IOP Conf. Series: Materials Science and Engineering **724** (2020) 012045 doi:10.1088/1757-899X/724/1/012045

Influence of ZnO concentration in rapeseed oil on tribological behavior

T F Ionescu[1], D Guglea[1], C Georgescu[1], D Dima[2] and L Deleanu[1]

[1] Mechanical Engineering Department, Faculty of Engineering, "Dunarea de Jos" University of Galati, Galati, Romania

[2] Department of Chemistry, Physics and Environment, Faculty of Science and Environment, "Dunarea de Jos" University of Galati, Galati, Romania

Abstract. This paper presents results of testing the coarse rapeseed oil additivated with different concentrations of ZnO (0.25%wt, 0.50%wt and 1%wt, with 14±5 nm in size), on a four-ball machine. The test parameters were load on main shaft of the machine: 100 N, 200 N and 300 N and the rotational speed 1000 rpm, 1400 rpm and 1800 rpm. For the tested range of ZnO concentration, the value of 1%wt does not improve the friction coefficient, but the wear rate of wear scar diameter was lower than that obtained with the neat rapeseed oil and the values are less sensitive with load and sliding speed for the more severe regimes. The additivation of rapeseed oil with ZnO is still efficient for the tested range of load and speed as compared to neat rapeseed oil. The authors formulated lubricants using as dispersant (2-methoxyphenol) in the ratio 1:1 to the additive. The technology is at scale laboratory and implies mechanical stirring and sonication. The tribological behavior was analysed based on obtained results, by mapping the friction coefficient, the wear rate of the wear scar diameter and the final temperature in the oil bath. For the additivated lubricants the following conclusions could be drawn. Friction coefficient tends to reach 0.1 only for low speed and high load (v = 0.38 m/s and F = 300 N), wear rate of wear scar diameter decreases with the increase of load, for each tested speed, probably because of a full fluid film generation. For F = 250...300 N, the map reveals a poor dependence on the additive concentration. The final temperature in the lubricant bath depends on speed, but for the same speed, it depends less on load, having a maximum value for 0.75% nano additive. These conclusions underlined the necessity of testing the new formulated lubricants because additive concentration and regime parameters could have synergic influences on the tribological parameters.

1. Introduction

The rapeseed oil is a plant oil in the focus of tribologists as it has a set of properties that recommends it for applications that do not exceed temperature limit and seizure limit. Additivation of this oil is still at the beginning as, due to their complex composition, the additive response could be unexpected. Metallic oxides are treated in literature [1], [2] [3], [4], [5] as a particular group, including ZnO [6], [7], [8], CuO, Al_2O_3, ZrO_2, TiO_2, even a combination $ZnAl_2O_4$ [9] etc. Shahnazar et al. [1] mentioned vegetal oils as base-stock for formulating additivated lubricants used in food, pharmacological industry, such as bakery ovens or kilns, where the risk of contamination needs to be minimized.

Lubricants having vegetal oils as base oil and ZnO and/or CuO as additives are biodegradable and have better performance in boundary lubrication [5].

Nano-ZnO has good characteristics for being introduced in lubricants: large surface area, high surface energy, strong adsorption (but by mechanical fixing), high diffusion, easy sintering, and a low melting point [10]. Due to the low solubility of ZnO in oil, its dispersion in the base oil could prove to be a challenge [11]. In 2011, Qian et al. [12] using lauryl sodium sulfate as the surfactant, studied the tribological properties of nano-ZnO as a lubricant additive, the average size of particles reaching 125 nm. The addition of 1.0%, 2.0%, 3.0%, and 4.0% mass fraction of ZnO resulted in oil decreases the wear by creating deposits on the sliding surfaces and forming a lubricating layer on moving surfaces.

Hernandez Battez et al. [13] reported the antiwear behaviour a polyalphaolefin oil (PAO6) with CuO, ZnO and ZrO_2 nanoparticles, respectively, in concentration of 0.5%, 1.0% and 2.0%wt, after testing on block-on-ring tribometer. Tests were done under a load of 165 N, sliding speed of 2 m/s and a total distance of 3.066 m. All formulated lubricants exhibited reductions in friction and wear as compared to the base oil; the suspensions with 0.5% of ZnO and ZrO_2 had the best general tribological behaviour, but with high friction and wear reduction values even at low deposition levels on the wear surface; CuO suspensions showed the highest friction coefficient and the lowest wear at the content of 2%. The antiwear mechanism of nano additive was produced by tribo-sintering. A previous study of Hernandez Battez et al. [14] presented the tribological behaviour of ZnO nanoparticles as an additive in polyalphaolefin (PAO6) and on the influence of dispersing agents (OL100 and OL300), based on tests run on a four-ball machine. OL300 has better dispersant properties than OL100; formation of aggregates was found for 1% and 1.5% of ZnO in PAO6 + 3% OL300, and these non-coated ZnO nanoparticles had an abrasive behaviour in the contact area, despite their lower hardness as compared to steel surfaces; the tribological behaviour improvement of PAO6 + 1% OL300 + ZnO mixtures was related to the increase of ZnO concentration and their surface deposition on the wear scar surfaces.

Sepyani et al. [15] evaluated the addition of nano ZnO in SAE 50 oil and its influence on viscosity. At different shear rates, all lubricants had Newtonian behavior. The results showed that the maximum increase in viscosity was 12% that occurs for more concentrated samples at low temperature. Thus, it is expected that the addition of ZnO in the rapeseed oil to have a minor influence of the dynamic viscosity of the formulated lubricant. In conclusion, the tribological behaviour of this metallic oxide is based on the action as friction and wear modifier and not as viscosity improver of this metallic oxide.

Bhaumik et al. [16] investigated the tribological properties of ZnO nano friction modifiers in castor oil, in different concentrations between steel surfaces, tests being done using a four-ball test rig. Anti-wear properties of the castor oil samples were improved by adding ZnO as nano friction modifier, up to a certain concentration of zinc oxide (0.1%), beyond which not much improvement in the coefficient of friction was noticed, but wear rate increased. The formation of tribo film due to the adsorption and the adherence of this nano friction modifier on the metallic surfaces were responsible in reducing wear. The results show that castor oil can be a good choice as a green lubricant for the replacement of commercial mineral oils, if optimally additivated.

From the available documentation, ZnO was rarely used in rapeseed oil for improving wear and friction characteristics. This research aims to point out the influence of the additive concentration in rapeseed oil, on tribological parameters determined on four-ball tester.

2. Lubricant formulation and test method

The additive was supplied by PlasmaChem [17] and has the following characteristics (see figure 1): average particle size ca. 14 nm, specific surface area 30 ± 5 m^2/g, purity > 99%.

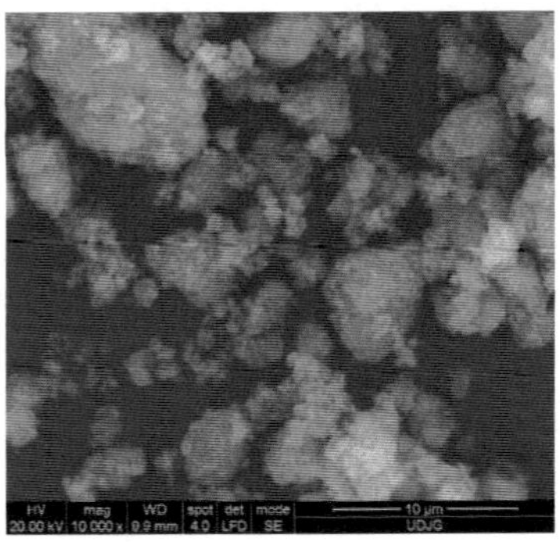

Figure 1. The nanoparticles of ZnO.

This study presents the results for the neat rapeseed oil (see Table 1) and the same vegetal oil additivated with 1% wt ZnO.

The ratio of the root-mean-square roughness of surfaces in contact to the nanoparticles size is essential because the particles have to remain in the lubricated contact in order to protect them against wear, during the system functioning. If their size is too large as compared to surface profiles, the nanoparticles will not deposit on the contact zone, they will be dragged and squeezed in contact, resulting an increase of both friction and wear [5]. The issue related to this type of nano lubricants is that, initially, the nanoparticles remain dispersed, but after an operating time, they are pressed into the solids' texture and agglomerated from nano to macro size. The dispersant and the lubricant have to prevent that, but larger particles are noticed on contact surface and even around it (see figure 2).

Table 1. Typical composition in fatty acids of the rapeseed oil (from Expur Bucharest).

Fat acid	Symbol	Composition, %wt
Myristic acid	C14:0	0.06
Palmitic acid	C16:0	4.60
Palmitoleic acid	C16:1	0.21
Heptadecanoic acid	C17:0	0.07
Heptadecenoic acid	C17:1	0.18
Stearic acid	C18:0	1.49
Oleic acid	C18:1	60.85
Linoleic acid	C18:2	19.90
Linolenic acid	C18:3	7.64
Arachidic acid	C20:0	0.49
Eicosenoic acid	C20:1	1.14
others		3.37

The formulated lubricant was obtained in small amounts of 200 g, each. The steps followed in this laboratory technology were similar to those presented by Cristea [18]:

- mechanical mixing of the additive and an equal mass of dispersing agent (guaiacol, from Fluka Chemica, with the chemical formula $C_6H_4(OH)OCH_3$ (2-methoxyphenol)), for 20 minutes;
- gradually adding rapeseed oil, mixing with a magnetic homogenizer for 1 hour;
- ultrasonication + cooling of formulated lubricant in step of 10 minutes; the fluid is heating to about 70°C during sonication; the cooling time was 1 hour; this technological step is repeated 5 times to have a total sonication time of 60 minutes. The parameters of ultrasonic regime are power 100 W, frequency 20 kHz ± 500 Hz, continuous mode.

The test balls are lime polished, made of chrome alloyed steel balls, having 12.7 ± 0.0005 mm in diameter, with 64-66 HRC hardness, as delivered by SKF. The oil volume required for each test was 8 ml ± 1 ml. The test method for investigating the lubricating capacity was that from SR EN ISO 20623:2018 [19].

The test parameters for each test were:

- sliding speeds of 0.38 m/s, 0.53 m/s and 0.69 m/s, corresponding to the spindle speeds of the four-ball machine of 1000 rpm, 1400 rpm and 1800 rpm (± 6 rpm), respectively;
- loading force on the machine spindle - 100 N, 200 N and 300 N (± 5%);
- test time - 60 minutes (± 1%).

The graphs of the wear scar diameters (WSD) as a function of speed could not reflect in a relevant way the influence of testing regimes, because all tests has 1 h (with different sliding distances for each speed), and, thus, the authors studied the influence of additive concentration with the help of wear rate of the scar diameter, noted by w(WSD). The w(WSD) is calculated with the help of the following relationship:

$$w(WSD) = \frac{WSD}{F \cdot L}\left[mm/(N \cdot mm)\right] \quad (1)$$

where WSD is the average value of six measurements of the wear scar diameter, two on each fixed ball (one along the sliding direction and the other perpendicular to it), F is the load applied on the main shaft of the tribotester (carrying the rotating ball) and L is the sliding distance. Thus, the wear rate of WSD reflects the dimensional modification of WSD for the unit of mechanical work, done by the tester system.

3. Results

From SEM investigations, the authors noticed that the nanoparticles of ZnO are spread over the contact scar of the balls. Images in figure 2 were obtained at a scanning electron microscope, after drying the balls as they were taken from their cup to point out the presence of the additive on the contact zone. The additive does not form a continuous layer on the contact and the nanoparticles agglomerate in very different size and shape, especially under low load (F = 100 N). Under higher load, the particles agglomerate less and have a spheroidal shape (figure 2a) also presents how the two scar diameters were measured. Mariani [20], Gulzar et al. [21], Chan et al. [22] and Zulkifli et al. [23] stated that metallic oxide (CuO, ZnO etc.) are able to coalesce in asperity valleys, creating a thin, smooth and solid lamellar film between contacting surfaces. Authors' SEM investigation (figure 2) revealed that this "film" is not continuous, the additive being spread in contact as small rollers or spheroidal particles that prevent wear of the triboelements and reducing friction when acting like micro rolling bearings.

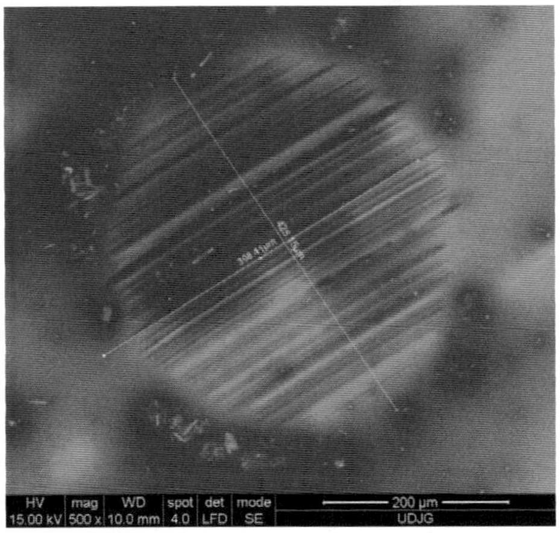

a) 1% Zn, F = 100 N, v = 0.38 m/s

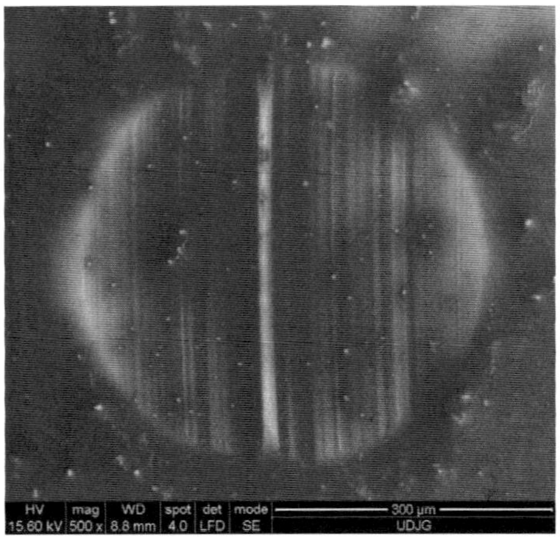

b) 1% Zn, F = 100 N, v = 0.69 m/s

Figure 2. Worn scars, dried in calm air with the lubricant on.

The obtained experimental data were included in maps that could emphasis the synergic effect of two variables, here, the load and the additive concentration, each map being drawn for the same sliding speed. The maps were done with the help of Matlab R2016a, using a spline interpolation and the surfaces are "forced" to include the experimental values. A point on these maps represents the average of two tests, with the same set of parameters (F [N], c [%wt.], v [m/s]), where F is the load on the spindle of the four ball tester and c is the mass concentration of the nano additive (0%, 0.25%, 0.5%, 1.0%), and v is the sliding speed.

Figure 3 presents the maps for the friction coefficient (COF). The qualitative modification could be allocated to change in the lubrication regime. For instance, at v = 0.38 m/s, COF has the lowest values (blue colour) for the neat oil till 200 N and for 0.75...1% additive concentration, for F = 100...150 N. The values that characterize a mixt or boundary regime (around 0.1) were obtained for concentration of 0.75...1% and for loads F = 200...300 N. Under high load the neat oil has for this tribological parameter a better (lower) value. For the additivated oil, COF decreases when the additive is towards

1% due to its rolling effect in contact, but under higher loads, the particles do not roll, being dragged in contact. They protect the surface but increase the friction. Increasing the sliding speed at v = 0.53 m/s, COF has a band of low values (blue colour and light green) for concentrations 0.6...1%. And the parameter seems to be almost insensitive to load, meaning a friction almost constant even if the contact has the load variable in the tasted range. At lower concentration (0.15...0.35%), COF increased toward 0.1, meaning that this concentration interval makes COF to increase. Probably, the presence of ZnO particles disturb the generation of the EHL film. At highest speed (v = 0.69 m/s), COF has only a small region (F,c) with values around 0.1, meaning the speed increase is beneficial for not losing power by friction. As it is given in Dowson et al. [24], speed has a more powerful influence on generating a fluid film than load and, here, additive concentration. The lowest values of COF were also obtained for the neat vegetal oil.

Similar tests done by Bhaumik et al. [16], with speed 1200 rpm (0.48 m/s), load 392 N, 1 h, at 75°C and room temperature, using balls with close characteristics and dimension, give the lowest value of COF (0.06) for castor oil + 0.1% ZnO and between 0.062...0.07 for the other additive concentrations (0.5...2%).

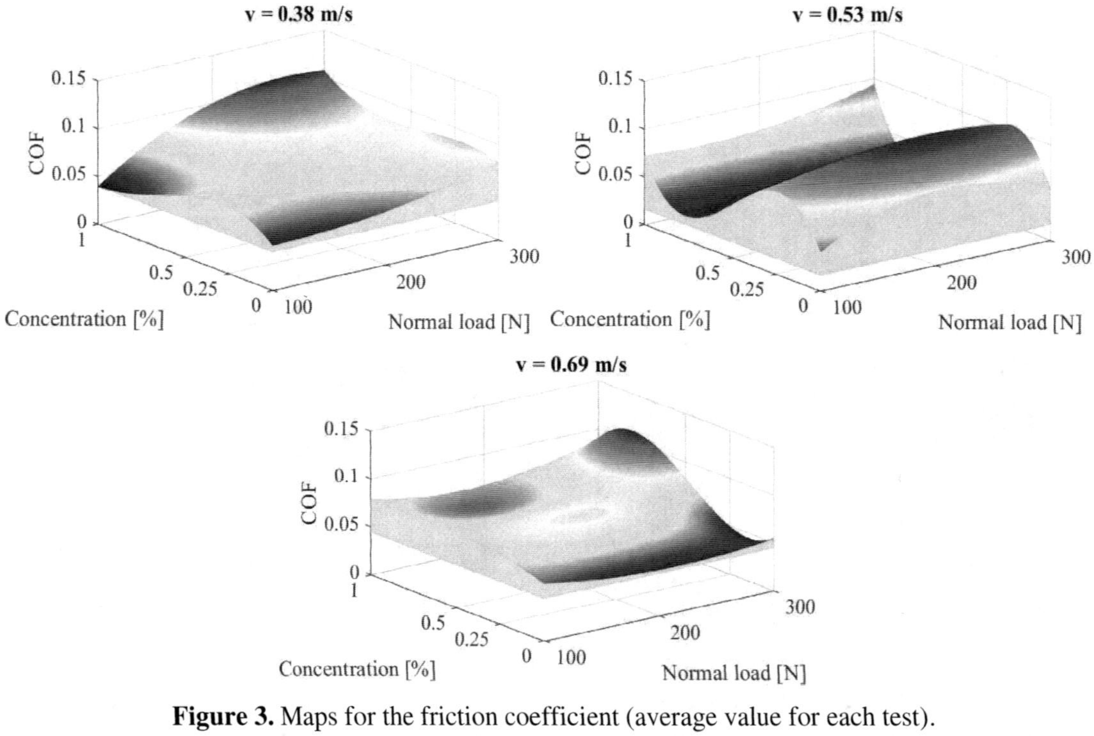

Figure 3. Maps for the friction coefficient (average value for each test).

Bhaumik et al. [16] reported a decrease of the WSD for 0.1% ZnO in castor oil, at 0.42 mm, as compared to 1 mm for castor oil + 0.5% ZnO, for intermediate test parameters as compared to authors' tests (0.48 m/s and F = 392 N). For the same concentration, the authors obtained 0.56...0.72 mm for F = 300 N, for v = 0.38 m/s and v = 0.53 m/s, respectively (Table 2).

Table 2. WSD, calculated as average of six measurements of diameters of the fixed ball

Load (N)	v = 0.38 m/s ZnO concentration (wt%)				v = 0.53 m/s ZnO concentration (wt%)				v = 0.69 m/s ZnO concentration (wt%)			
	0	**0.25**	**0.5**	**1**	**0**	**0.25**	**0.5**	**1**	**0**	**0.25**	**0.5**	**1**
100	0.37	0.386	0.438	0.379	0.351	0.3922	0.399	0.351	0.438	0.399	0.438	0.438
200	0.476	0.528	0.603	0.476	0.45	0.487	0.606	0.454	0.603	0.606	0.603	0.603
300	0.560	0.656	0.724	0.560	0.527	0.620	0.722	0.527	0.724	0.722	0.724	0.724

When analysing the wear parameter (Figure 4), two distinct zones are visible on the maps, one of high values (red and yellow colours), characteristic for low load and the other one, with low values (light to dark blue) for F = 200...300 N, less dependent on additive concentration. Speed makes the wear rate of wear scar diameter to be under 1×10^{-6} mm/N×m, a good value as comparing to other results obtained on four-ball tribotester. Higher speed produces a fluid film that is in the favour of reducing wear by keeping the separation of the solid bodies in contact.

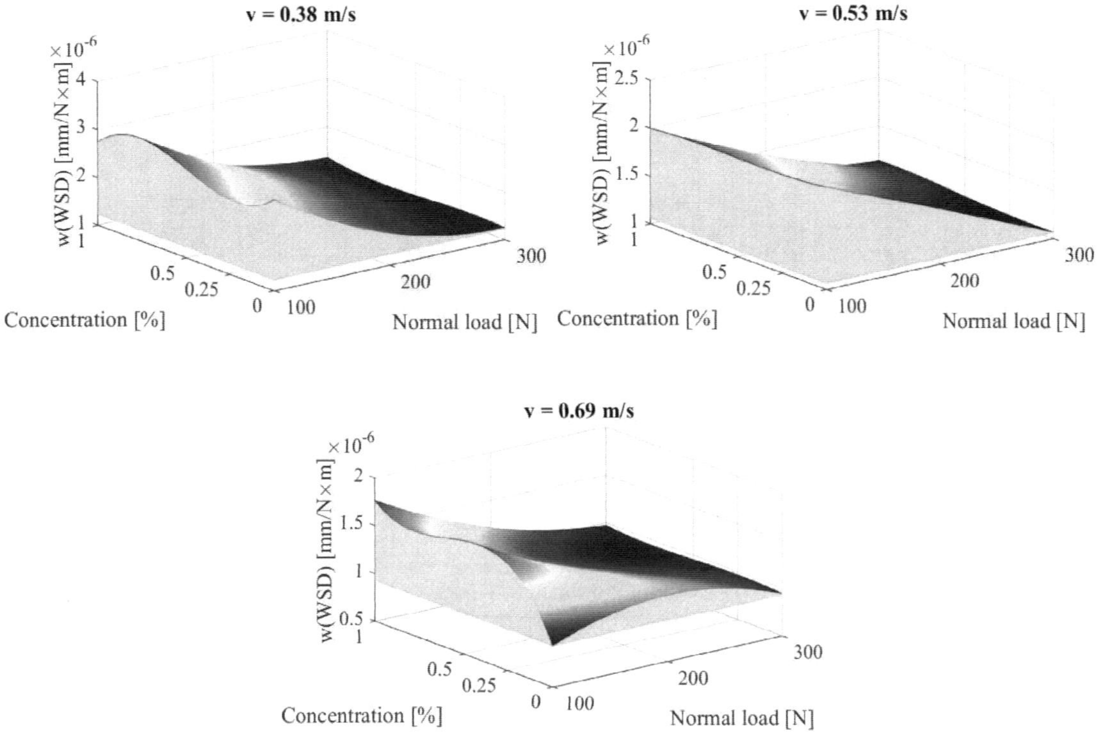

Figure 4. Wear rate of WSD.

Qian et al. [12] obtained smaller WSD for 3% ZnO in base oil, tested on four-ball tester, at 1450 rpm (0.54 m/s) at room temperature, for 30 min. For instance, the WSD for F = 300 N is 0.4 mm, but for 1%, WSD was 0.6 mm. The authors obtained 0.722 mm but for test of 1 h, meaning that continuous running is favourable to keep wear at low level.

The discussion on temperature recorded at the end of the test is important as a vegetal oil has a narrower temperature range of functioning, higher temperature inducing oxidation and other chemical changes in the components of the oil. The addition of nanoparticles increases this temperature value, especially at higher loads for v = 0.53...0.69 m/s. Observing the evolution of end test temperature for the neat rapeseed oil, the lowest values were obtained and the increase of final temperature with load is almost linear (see figure 5). For F = 300 N and v = 0.38...0.53 m/s, the temperature has just a little increase of several degrees, but for v = 0.69 m/s, this increase is of 10...14°C. For the additivated lubricant, the maximum final temperature was obtained for the additive concentration of 0.5...0.8%. A slight decrease of final temperature was obtained for 1% ZnO, meaning that this concentration (the highest tested in this study) helps to evacuate the heat generated in contact.

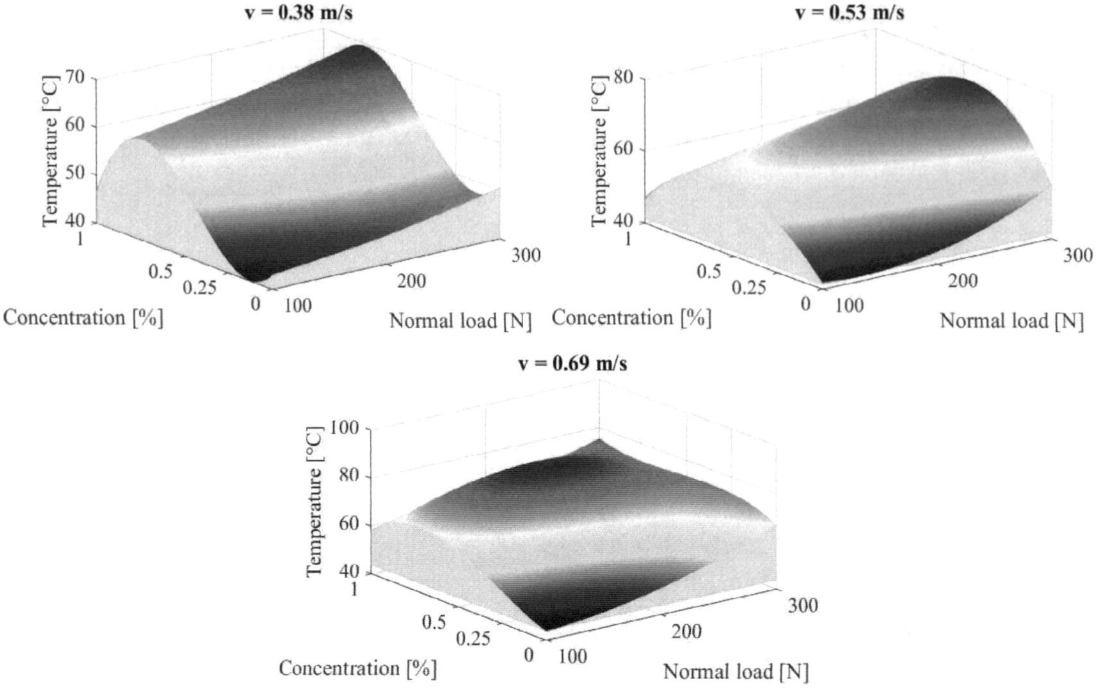

Figure 5. Final temperature in the lubricant bath, recorded at the end of 1 h test.

4. Conclusions

The authors of this research study formulated lubricants based on rapeseed oil, with addition of nanoparticles of ZnO, in different mass concentration (0.25%, 0.5% and 1%, respectively), using as dispersant 2-methoxyphenol in the ratio 1:1 to the additive. The technology is at scale laboratory and implies mechanical stirring and sonication.

The tribological behavior was analysed based on results obtained on a four-ball machine, by mapping the friction coefficient, the wear ate of the wear scar diameter and the final temperature in the lubricant bath.

For the additivated lubricants the following conclusions could be drawn only for the tested regimes:
- friction coefficient tends to reach 0.1 only for low speed and high load (v = 0.38 m/s and F = 300 N),
- wear rate of wear scar diameter decreases with the increase of load, for each tested speed, probably because of a fluid film generation; for F = 250...300 N, the map of this parameter reveals a poor dependence on the additive concentration,
- the final temperature in the lubricant bath depends on speed, but for the same speed, it depends less on load, having maximum values for 0.75% nano additive.

These conclusions underlined the necessity of testing the new formulated lubricants because additive concentration and regime parameters do not have similar influences on the tribological parameters. Thus, the designer and the operator have to make a compromise and select the lubricant and the functioning regime that less affect the entirely system exploitation and adding also some constraints concerning the environment protection and security regulations.

Acknowledgments

This work has been supported by the European Social Fund through the Sectoral Operational Programme Human Capital 2014-2020, through the Financial Agreement with the title „Burse pentru educaţia antreprenorială în rândul doctoranzilor şi cercetătorilor postdoctorat (Be Antreprenor!)" "Scholarships for entrepreneurial education among doctoral students and postdoctoral researchers (Be Entrepreneur!)", Contract no. 51680/09.07.2019 - SMIS code: 124539.

References

[1] Shahnazar S, Bagheri S and Abd Hamid S B 2016 *Int. J. Hydrogen Energ.* **41** 3153
[2] Trajano M F, Moura E I F, Ribeiro K S B and Alves S M 2014 *Materials Research* **17** 1124
[3] Gara L and Zou Q 2012 *Tribol. T.* **55** 345
[4] Asrul T M, Zulkifli N W M, Masjuki H H and Kalam M A 2013 *Procedia Engineer* **68** 320
[5] Uflyand I E, Zhinzhilo V A and Burlakova V E 2019 *Friction* **7** 93
[6] Ran X, Yu X and Zou Q 2017 *Tribol. T.* **60** 154
[7] Kumar V P S, Manikandan N, Subakaran C and Jeso Y G S 2018 *International Research Journal of Engineering Technology (IRJET)* **5** 1069
[8] Essa F A, Zhang Q, Huang X, Ibrahim A M M, Ali M K A, Abdelkareem M A A and Elagouz A 2017 *J. Mater. Eng. Perform.* **26** 4855
[9] Song X, Zheng S, Zhang J, Li W, Chen Q and Cao B 2012 *Mater. Res. Bull.* **47** 4305
[10] Soares J W, Whitten J E, Oblas D W and Steeves D M 2008 *Langmuir* **24** 371
[11] Tang Z and Li S 2014 *Curr. Opin. Solid St. M.* **18** 119
[12] Qian J, Zhang Y, Wang L and Xing J 2011 *Lubrication Research* **13** 69
[13] Hernandez Battez A, Gonzalez R, Viesca J L, Fernandez J E, Diaz Fernandez J M, Machado A, Choud R and Riba J 2008 *Wear* **265** 422
[14] Hernandez Battez A, Fernandez Rico J E, Navas Arias A, Viesca Rodriguez J L, Chou Rodriguez R and Diaz Fernandez J M 2006 *Wear* **261** 256
[15] Sepyani K, Afrand M and Esfe M H 2017 *J. Mol. Liq.* **236** 198
[16] Bhaumik S, Maggirwar R, Datta S and Pathak S D 2018 *Appl. Surf. Sci.* **449** 277
[17] *** Nanomaterials and related products, http://www.plasmachem.com/download/PlasmaChem-General_Catalogue_Nanomaterials.pdf, accessed 20.02.2016
[18] Cristea G C 2017 *Tribological characterization of soybean oil additivated with nano materials based on carbon (black carbon, graphite and graphene)* (PhD thesis "Dunarea de Jos" University of Galati)
[19] *** EN ISO 20623:2003 Petroleum and related products - Determination of the extreme-pressure and anti-wear properties of fluids - four ball method
[20] Mariani G 2009 *Lubricant Additives Chemistry and Applications Second Edition* ed Rudnick L R (New York: CRC Press) chapter 6 pp 173–194
[21] Gulzar M, Masjuki H H, Varman M, Kalam M A, Mufti R A, Zulkifli N W M, Yunus R and Zahid R 2015 *Tribol. Int.* **88** 271
[22] Chan C H, Tang S W, Mohd N K, Lim W H, Yeong S K and Idris Z 2018 *Renew. Sust. Energ. Rev.* **93** 145
[23] Zulkifli N W M, Kalam M A, Masjuki H H and Yunus R 2013 *Procedia Engineering* **68** 152
[24] Dowson D and Higginson G R 1977 *Elasto-hydrodynamic lubrication* (Oxford: Pergamon Press)

Tribological behavior of rapeseed oil additivated with boron nitride

D Guglea[1], T F Ionescu[1], D Dima[2], C Georgescu[1] and L Deleanu[1]

[1] Mechanical Engineering Department, Faculty of Engineering, "Dunarea de Jos" University of Galati, 111 Domneasca, Galati, Romania
[2] Department of Chemistry, Physics and Environment, Faculty of Science and Environment, "Dunarea de Jos" University of Galati, 111 Domneasca, Galati, Romania

Abstract. This paper presents results of testing the coarse rapeseed oil additivated with 1%wt hexagonal boron nitride (hBN). Tests are done on a four-ball machine from the laboratory LubriTest, at "Dunarea de Jos" University of Galati. Tests were done in mild regime (recommended for applications) and in severe regime. The mild test parameters were: load on the main shaft of the machine (100 N, 200 N and 300 N) and the sliding speed of 0.38 m/s, 0.53 m/s and 0.69 m/s, test duration 1 h. Severe test were run in step of 50 N till 900 N, each test being run for 1 minute at 0.53 m/s. The concentration of 1%wt hBN does not improve the friction coefficient, but it makes it stable for a larger domain of test parameters. The wear rate of wear scar diameter was lower than that obtained with the base oil, especially for high loads. Also, the paper presents the influence of the additive on the temperature of oil bath and the wear load curve of the neat oil and the additivated one. The additivation of rapeseed oil with hNB is still efficient for the tested ranges of load as compared to the rapeseed oil. The load-wear curve for the severe regime points out a sudden increase of wear scar diameter (WSD) for the rapeseed oil +1% hNB at 700 N, but the neat vegetal oil starts the increase of WSD at 550 N, followed by a greater slope, meaning the additive protect the rubbing surfaces under these high loads.

1. Introduction

Vegetal oils consist of triglycerides that influence their performance limits, as poor thermal, hydrolytic and oxidation stability, but also low viscosity. Most natural vegetal oils cannot withstand reservoir temperatures greater than 80ºC [1].

Boron nitride is a compound with heat and chemical resistance, good thermal conductivity, making it an excellent choice for lubricant applications that require rapid heat evacuation and it exists in several crystalline structures. The hexagonal structure of this boron nitride (hBN) similar to graphite is the most stable and soft and is recommended as solid lubricant or as additive in liquid lubricants [2].

Bondarev et al. [3] utilized hBN nanoparticles (NPs) with different morphologies (hollow NPs with smooth surface (H-BNNPs), solid NPs with "pompon"-like or petalled structure (P-BNNPs), and globular NPs formed by numerous thin hBN nanosheets (N-BNNPs)) as additives to PAO6 oil. Triboelements were made of steel 100Cr6, the same steel grade as in this study, were tested in lubricants with PAO6 + BNNP, with a concentration 0.1% and 0.01% of BNNPs, respectively. The positive effect of BNNP additives increased along the row P-BNNPs → H-BNNPs → N-BNNPs. Utilization of N-BNNPs allowed for decreasing the friction coefficient from 0.1 to 0.06 and for reducing significantly the wear rate.

Mariano [4] considered it an additive for niche applications when design requirements make graphite or molybdenum disulphide unacceptable. Only the hexagonal structure has lubricating ability.

Cubic boron nitride is a very hard substance used as an abrasive and cutting tool component and has no lubrication value. The bond with high strength between boron and nitrogen within hexagonal rings supplies a good load-carrying capability that is necessary to maintain the separation of rubbing surfaces. Particle size influences the adhesion to substrate and particle orientation within an already formed tribolayer. Impurities, such as boric oxide, will influence the ability of the powder to reduce the coefficient of friction.

Boron nitride was added in liquids with low viscosity for improving wear resistance of bodies in contact as full film generation to separate them is difficult to maintain in actual regimes. Cho et al. [5] added hexagonal boron nitride (h-BN, 300 nm wide and >30nm thick flakes) nano-sheets in water, in concentrations of 1%, 0.05% or 0.01wt%, without any surfactant or organic functionalization, using sonication for a good dispersion. The friction and wear were evaluated using a SiC ball sliding on a disk arrangement. Even small amounts of hBN nano-sheets improve wear resistance and reduce friction coefficient. Tribofilms (but not continuous) were observed to have formed on worn surfaces due to repeated exfoliation and deposition of hBN nano-sheets during sliding. This process is responsible for tribological improvements. hBN could be a promising "green" additive for water.

This study aims to emphasis the influence of 1% BN in rapeseed oil, both in typical regimes and severe ones, using a four-ball tester

2. Lubricant formulation and test method

The laboratory methodology of obtaining the lubricant with 1% BN is given in detail in Cristea [6] and the authors used the same regime of sonication and cooling and also the same dispersant, guaiacol, supplied by Fluka Chemica, with the chemical formula $C_6H_4(OH)OCH_3$ (2-methoxyphenol), in the ratio 1:1 to the additive.

The additive nano particles (figure 1), supplied by PlasmaChem [7], have the following characteristics: size full range 100-1000 nm, average particle size 500 ± 100 nm, specific surface 23 ± 3 m^2/g, purity > 98.5%, Nitrogen content > 55%.

Figure 1. Boron nitride, before being added in rapeseed oil.

Bondarev et al. [3] also prepared lubricating suspensions, each type of BNNPs in the amount of 0.01% and 0.1% was added into 30 ml of PAO6 oil (Neste) and then twice sonicated for 10 min with an interval of 5 min, but it is difficult to compare those tests in reciprocating mode using a 3 mm-sized 100Cr6 ball on plate with the results of these tests. The normal load and sliding speed were 1 N and 0.05 m/s. Under these conditions each sample passed 10,000 cycles.

Table 1 presents the composition in fatty acids of the rapeseed oil used in this research study.

The parameters for mild regime were: loading force on the machine spindle 100 N, 200 N and 300 N (± 5%), sliding speeds of 0.38 m/s, 0.53 m/s and 0.69 m/s, corresponding to the spindle rotational speeds of the four-ball machine of 1000 rpm, 1400 rpm and 1800 rpm (± 6 rpm), respectively, test time - 60 minutes (± 1%).

For the severe regime, tests were done in step of 50 N, from 500 N till 900 N, for 1 minute, at 0.53 m/s and there were measured the wear scar diameter (WSD) and the final temperature in the bath oil.

The test balls are lime polished, made of chrome alloyed steel, with 12.7±0.0005 mm in diameter, with 64-66 HRC hardness, as supplied by SKF. The oil volume required for each test was 8 ml ±1 ml.

The testing method for investigating tribological characteristics is following the recommendations from SR EN ISO 20623:2018 [8].

Table 1. Typical composition in fatty acids of the rapeseed oil (from Expur Bucharest).

Fat acid	Symbol	Composition (%wt)
Myristic acid	C14:0	0.06
Palmitic acid	C16:0	4.60
Palmitoleic acid	C16:1	0.21
Heptadecanoic acid	C17:0	0.07
Heptadecenoic acid	C17:1	0.18
Stearic acid	C18:0	1.49
Oleic acid	C18:1	60.85
Linoleic acid	C18:2	19.90
Linolenic acid	C18:3	7.64
Arachidic acid	C20:0	0.49
Eicosenoic acid	C20:1	1.14
others		3.37

3. Results

3.1. Mild test regime

Figure 2 presents the average value of friction coefficient (COF), for both tested lubricants. It is obvious that the neat oil has, generally, lower values, especially for low regimes, but the additivated oil has this parameter less sensitive to sliding speed for all loads. This is recommended for machines that frequently change their working regime.

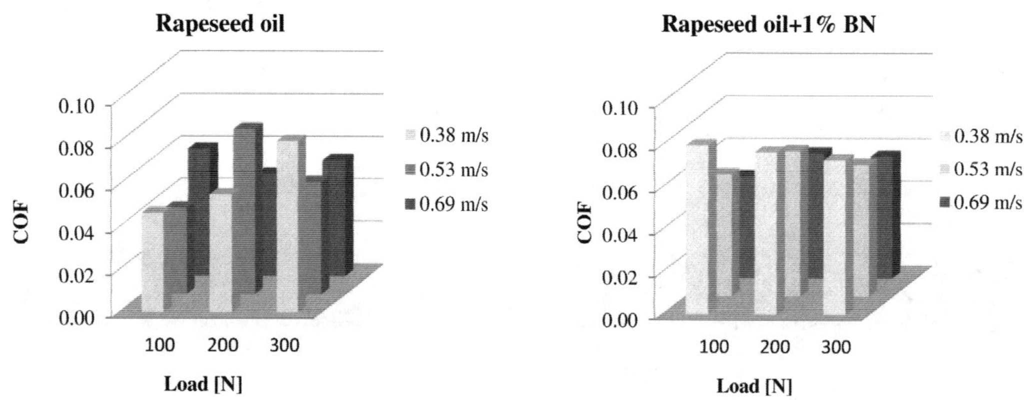

Figure 2. The friction coefficient (average value for each test of 1 h).

Diameters of wear scars were measured with an optical microscope, in accordance with the procedure given in SR EN ISO 20623:2018 [8] (for each scar, a diameter along the sliding direction and the other perpendicular to it). Each test supplies three wear scars, these being located on the three fixed balls. Wear scar diameter (WSD) is the average of the six diameter measurements and this value is given in graphs in figure 3. This value represents the diameter of the wear scar, reported for each of the tests performed.

Comparing values of WSD for the same sliding speed, for v = 0.38 m/s, WSD is almost twice lower for the additivated lubricant, but for higher loads and speed, the influence of additive is low.

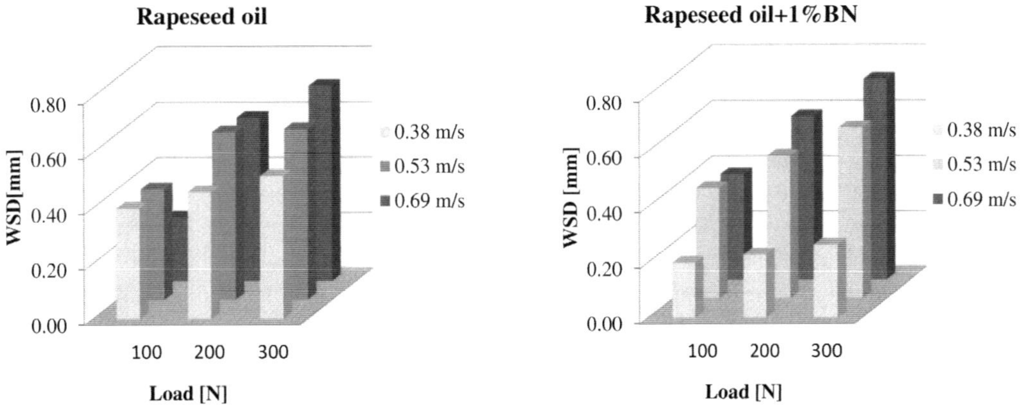

Figure 3. WSD for the rapeseed oil (left) and the rapeseed oil + 1% hBN (right).

The graphs of the wear scar diameters (WSD), as a function of speed, are difficult to be compared as they were done for all tests during 1h, meaning different sliding distances for each speed. The authors studied the influence of additive, using a tribological parameter that takes into account the sliding distance, a parameter also recommended by Holmberg [9], the wear rate of the wear scar diameter, noted by w(WSD) and calculated with the help of the following relationship:

$$w(WSD) = \frac{WSD}{F \cdot L}\left[mm/(N \cdot mm)\right] \quad (1)$$

where WSD is the average value of six measurements of the wear scar diameter, two on each fixed ball, F is the load applied on the main shaft of the tribotester and L is the sliding distance. The product F×L is the mechanical work done by the system. Thus w(WSD) is the modification of WSD for the unit of mechanical work.

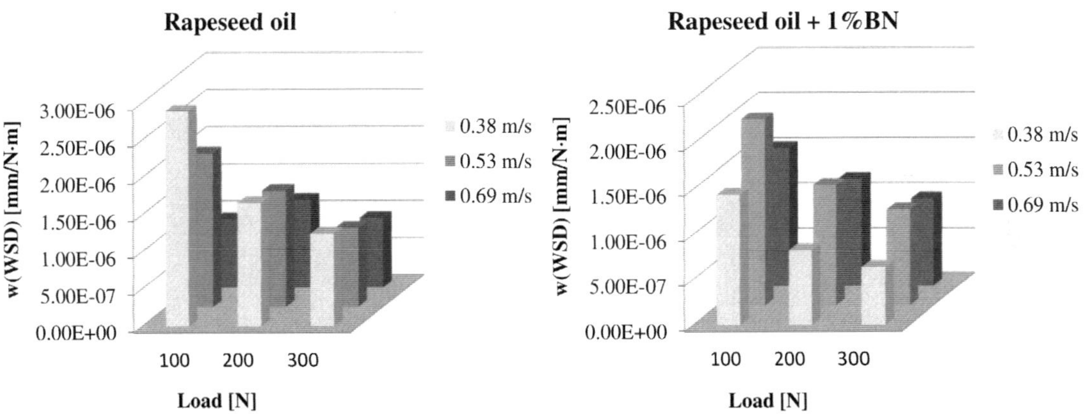

Figure 4. The wear rate of WSD for the rapeseed oil (left) and the rapeseed oil +1% hBN (right).

When calculating this parameter for these tests (figure 4), the additivated lubricant produces lower values for v = 0.38 m/s, meaning that surfaces of solid bodies are protected by hBN powder, reducing their wear by particular mechanisms of nano-rolling, smoothing/levelingor mending process by filling the valleys of the profiles, coating the higher asperities with nano sheets that reduce friction and local load, with a minor component of smoothing/leveling by abrasion. This comment is based on discussions done by Wu et al. [10], Chinas-Castillo and Spikes [11], Liu et al. [12]. The leveling by abrasion is supposed to increased under higher loads, as both graphs of WSD increase with load (figure 3), but photos in figure 8 show, for F = 500 N, qualitatively, less abrasion, for the additivated lubricant.

Figure 5 presents the evolution of temperature in the lubricant bath. At the same speed, the temperature lines are ordered from low to higher load. This similarity in load and speed dependence points out that the lubricating regime is not changing (very probably a EHL regime, with locally temporary contact among rare high asperities or due to additive agglomerations dragged through contact. It is worthy to mention that the temperature for the additivated oil is only few dregrees higher for the combinations (low load - low speed) but for (high speed - high load), this difference diminishes, probably due to the influence of additive in evacuating the heat generated by friction.

When plotting a graph with the final measured temperature in the lubricant bath (figure 6), the differences are small, several degrees Celsius higher for the additivated lubricant. For the additivated lubricant, this could be explained by two balacing processes: the friction increased by the movement of the nano sheets or their agglomerations through contact, that will increase the temperature and the other process of heat evacuation due to the good thermal properties of the additive.

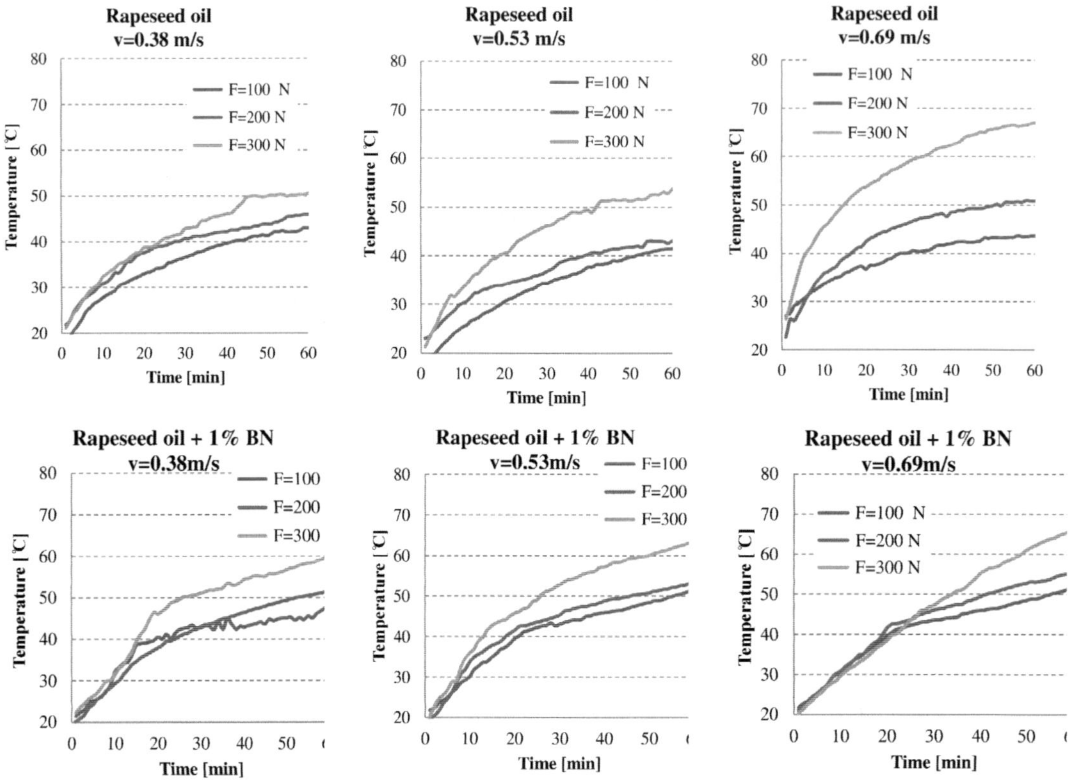

Figure 5. Temperature evolution for 1 h test (sampling at 1 minute).

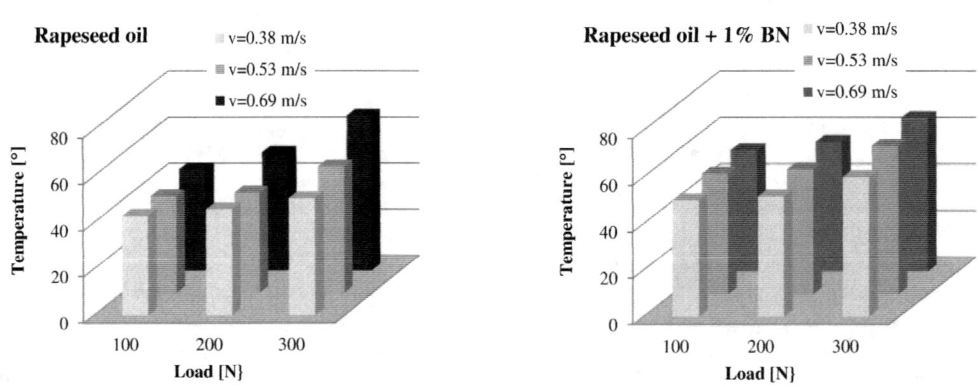

Figure 6. Temperature at the end of a test of 1 h.

3.2. Severe regime (seizure)

Bondarev et al. [3] studied the structure of wear debris and the worn surfaces of triboelements and concluded that the N-BNNPs split into thinner nanosheets due to weak Van der Waals forces, which, under applied load,

IOP Conf. Series: Materials Science and Engineering **724** (2020) 012046 doi:10.1088/1757-899X/724/1/012046

could easily be reoriented parallel to the friction direction, thus minimizing the resistance to shear. BN nanosheets formed aggregates (typically 200–800 nm in size, also retaining oil), uniformly distributed over the area of tribological contact, contributing to separate the surfaces in contact (even partially).

In 2014, Abdullah et al. [13] obtained results in the favor of adding hBN in engine oil (SAE 15W-40) The trend of the obtained curves underlined that diesel engine oil started the incipient seizure on contact surfaces at 618 N, but the nano-lubricant (0.5% vol. of 70 nm hBN in oil) had this point at 785 N, indicating that with this additive, the lubricant sustain a higher load before seizure takes place.

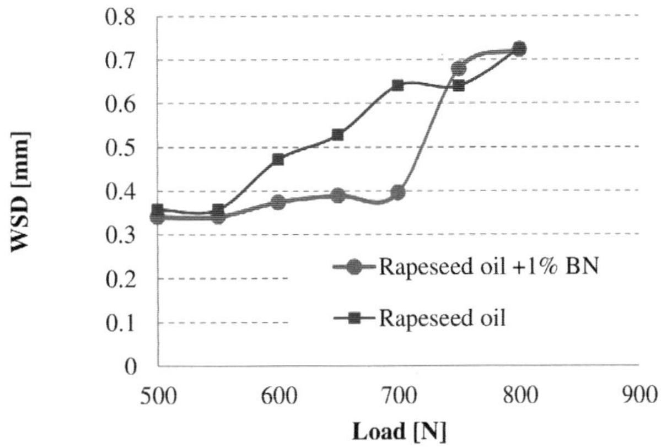

Figure 7. Wear load curve for the tested lubricants.

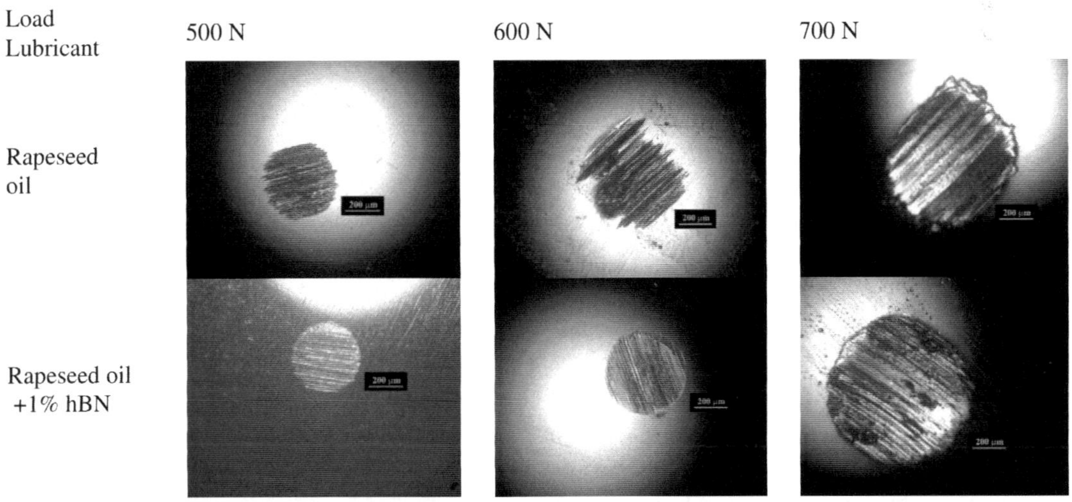

Figure 8. Images of wear scar in the severe regime.

Analysing the wear load curve in figure 7, the additivated lubricant keeps WSD stil lower at 550 N. This is a strong point in a SWOT analysis for selecting a lubricant, especially when the designer is aware of the exploitation regime. For instance, when this regime could have load fluctuations and even small shocks (from actual reasons as applied technology), this lubricant behaves better than the oil without additive. This is obvious from images of the wear scars, as presented in figure 8. One may notice that at 500 N the abrasive tracks on the wear scar resulted after testing the rapeseed oil seems rougher and the wear scar has bigger area as compared to that obtained with rapeseed oil + 1% hBN.

As compared to results in Abdullah et al. [13], the additivated rapeseed oil start seizure after 700 N, this being remarcable in comparison to the mineral oil tested by Abdullah et al.

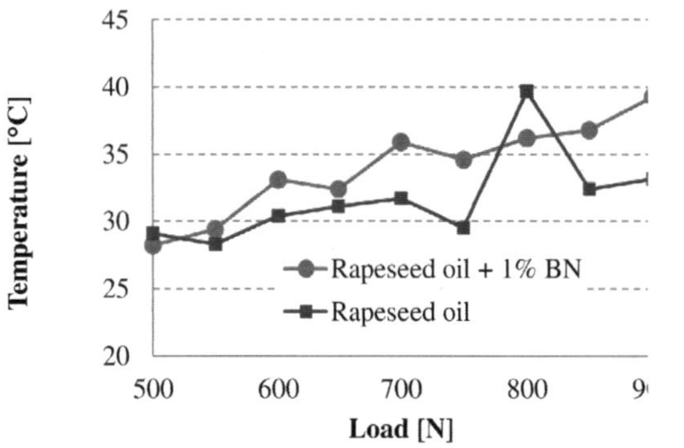

Figure 9. Bath oil temperature at the end of the test.

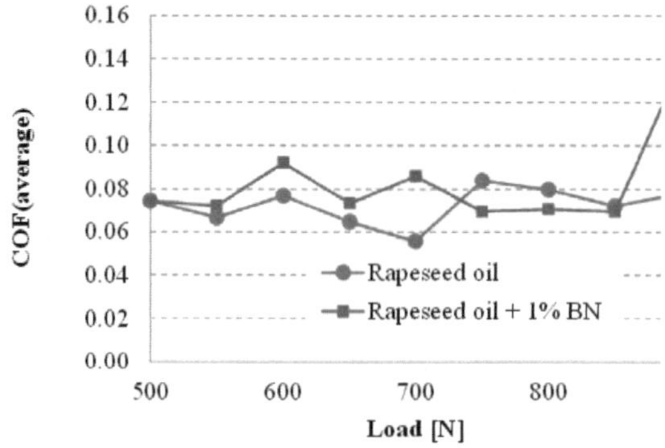

Figure 10. The average value of COF for 1 minute of severe test

Because of the short test time, the temperature in the bath oil is less important when evaluating the lubricating capability of a lubricant, figure 9 presents the evolution of the recorded temperature just at the end of a test. The additivated oil has higher values but in a trend with smaller slope. The rapeseed oil has values that evoluate in a larger band, due to a more aggressive abrasive wear, caused by the direct contact of the steel balls. The nano hBN protected the surfaces in contact and the image for F = 700 N has a different aspect as compared to the that obtained with neat oil, meaning the aditive could remain on the rubbing surfaces (see figure 8). Further studies on SEM and with EDX are necessary for proving that.

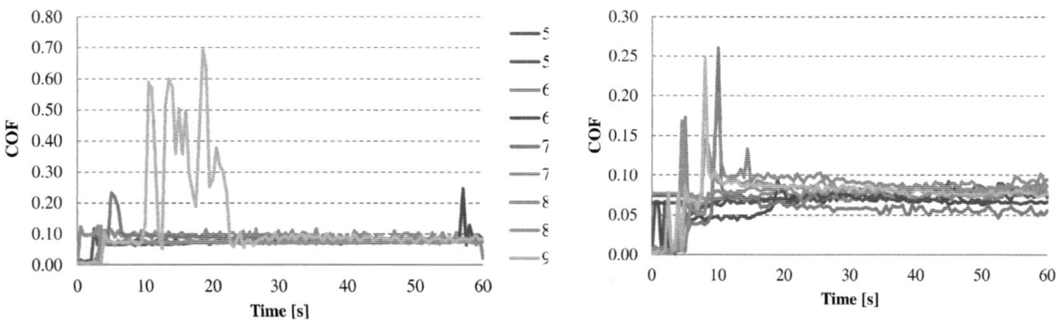

a) rapeseed oil b) rapeseed oil + 1% BN

Figure 11. Friction coefficient.

Figure 11 presents the evolution in time of friction coeffficient (COF) for loads greater than 500 N (1 minute test). High oscillating values were recorded in the first 20 s, meaning that aggressive wear could be generated at test start. The average value of COF (figure 10) is similar for both lubricants (additivated and not additivated) only under load lower than 850 N. Further, at F = 900 N, this value has a sharp increase for the neat vegetal oil.

4. Conclusions

This research presented the tribological behavior of rapeseed oil additivated with 1% hBN in comparison to the neat vegetal oil. Tested were done for normal (useful) regimes (F = 100...300 N and v = 0.38...0.69 m/s), but also for severe regime in order to determine the influence of additivation on these regimes.

For normal regimes, the value of 1% wt does not improve the friction coefficient, but it makes it stable for a larger range of test parameters. The wear rate of WSD was lower than that obtained with the base oil.

This study presents the seizure curve of the neat oil and the additivated one. The additivation of rapeseed oil with hNB is still efficient for the tested ranges of load as compared to the rapeseed oil as WSD start seizure at higher load and with a lower slope.

For severe regimes, it was obvious that the lower load wear curve recommends the additivated lubricants in systems with possible higher loads around the nominal ones as triboelements are the surfaces protected by additive deposition in the surface textures.

Acknowledgments

This work has been supported by the European Social Fund through the Sectoral Operational Programme Human Capital 2014-2020, through the Financial Agreement with the title „Burse pentru educaţia antreprenorială în rândul doctoranzilor şi cercetătorilor postdoctorat (Be Antreprenor!)" "Scholarships for entrepreneurial education among doctoral students and postdoctoral researchers (Be Entrepreneur!)", Contract no. 51680/09.07.2019 - SMIS code: 124539.

References

[1] Leugner L 2003, Biodegradable Oils - How to Apply and Maintain, Machinery Lubrication, 7/2003, https://www.machinerylubrication.com/Read/511/biodegradable-oils
[2] Tang Z and Li S 2014, *Current Opinion in Solid State and Materials Science*, **18**, pp 119–139
[3] Bondarev A V, Kovalskii A M, Firestein K L, Loginova P A, Sidorenko D A, Shvindina N V, Sukhorukova I V and Shtansky D V, *Ceram Int* **44** (2018) pp 6801–6809
[4] Mariano G 2009 *Lubricant Additives Chemistry and Applications Second Edition* ed Rudnick L R (New York: CRC Press) chapter 6 pp 173–194
[5] Cho D-H, Kim J-S, Kwon S-H, Leenn C and Lee Y-Z 2013 *Wear* **302** 981–986
[6] Cristea G C 2017 *Tribological characterization of soybean oil additivated with nano materials based on carbon (black carbon, graphite and graphene)* (PhD thesis "Dunarea de Jos" University of Galati, in Romanian)
[7] *** Nanomaterials and related products, http://www.plasmachem.com/download/PlasmaChem-General_Catalogue_Nanomaterials.pdf, accessed 20.02.2016
[8] *** EN ISO 20623:2018 Petroleum and related products. Determination of the extreme-pressure and anti-wear properties of fluids - four ball method
[9] Holmberg K 2005 Quality of reporting empirical results in tribology, IRG-OECD: New Wear Unit / Documents
[10] Wu Y Y, Tsui W C and Liu T C 2007, *Wear*, **262**, pp. 819-825
[11] Chinas-Castillo F and Spikes H 2003, *J of Tribol*, **125(3),** pp 125-552
[12] Liu G, Li X, Qin B, Xing D, Guo Y and Fan R 2004, *Tribol Lett*, **17**, p 961
[13] Abdullah M I H C, Abdollah M F B, Tamaldin N, Amiruddin H and Nuri N R M, 2016, *Ind Lub and Tribol*, **68(4)**, pp. 441-445

Preliminary results on nano-diamond and nano-graphite testing as additive for an engine lubrication oil

D Moldovanu[1], A Molea[1] and I Barabás[1]

[1]Technical University of Cluj-Napoca, Automotive Engineering and Transports Department, Bd. Muncii 103-105, Cluj-Napoca, Romania
Corresponding author: andreia.molea@auto.utcluj.ro

Abstract. An important amount of the fuel energy burned in an internal combustion engine is used to overcome the friction losses inside the engine. Developing advanced lubricants is one of the ways to improve the fuel efficiency and reducing the pollution caused by internal combustion engines. This paper presents tests done with 1% v/v nano-diamond and nano-graphite additives in engine oils, with bench tests on an air-cooled spark ignition engine (Honda GX 160) with a displacement of 118 cubic centimeters. Using the raw lubrication oil laboratory, tests were done to determine the viscosity, density and viscosity index using an Anton Paar SVM 3000 viscometer in comparison to the oil with nano-diamond and nano-graphite additives. The tests also revealed an average reduction in fuel consumption of 21% for the single cylinder engine in case of the oil with additives, for the entire speed range of the engine, showing a decrease of the fuel efficiency as the speed increases. The temperature of the exhaust manifold, outer side of the cylinder and the temperature of the oil pan were measured. By adding nano-diamond and nano-graphite additives to the lubricant oil, tests have revealed a reduction of the temperature of the exhaust manifold (with 11%), temperature on the outer side of the cylinder was reduced with 21% showing a clear friction reduction within the cylinder and piston-rings assembly. The temperature of the oil pan was reduced with 20%, showing a general reduction of the lubricating oil temperature which means a substantial reduction of the friction losses.

1. Introduction

An important amount of 33% of the fuel energy is used to overcome the friction losses in the car from which 35% is used to overcome the friction in the engine system. From this 12% of fuel energy, 45% is consumed in the piston assembly, 30% is consumed in bearings, seals, etc. (hydrodynamic lubrication), 15% is consumed in the valve train (mixed lubrication), and 10% is consumed by pumping and hydraulic viscous losses [1]. The less friction in the internal combustion engine the better fuel efficiency and less pollutant emission. The friction can be reduced using high performance engine oils.

Modern oils for internal combustion engines consist of a base oil (synthetic or semisynthetic) and an additive pack which could have a concentration up to 30-35% vol and it contains surface protective additives (anti-wear, corrosion and rust inhibitor, detergent, dispersant, friction modifier agents), oil performance improver additives (pour point depressant, seal swell agent, viscosity improver agents) and lubricant protective additives (anti-foaming, anti-oxidant, metal deactivator agents) [2].

During the past few years, various nano-materials have been studied as a new additive for lubricant oils, including polymers, metals, organic and inorganic materials [3].

In [4] the authors have studied the anti-scuffing property of lubricating oil and diamond nano-particles additive. They found out that 3% vol concentration of the diamond nano-particles had the most favourable effect on reducing the wear loss and the friction coefficient. Raina and Anand [5] have investigated the nano-diamond in combination with copper oxide and hexagonal boron nitride particles as an additive for poly-alpha-olefin. They observed a significant improvement in the lubrication properties of the base oil, the decrease of the friction coefficient was 15-25%.

Su et al. [6] have investigated the tribological properties of multi-walled carbon nanotubes with different sizes and volume fractions as an oil additive using a pin-on-disk friction and wear tester. They found that at 0.05% vol fraction the thin and short multi-walled carbon nanotubes could improve both friction-reducing and anti-wear properties of the oil more effectively than the thick and long multi-walled carbon nanotubes. Martorana et al. [7] have investigated the use of the dispersed fine graphite flake and carbon nanofiber in ethanol as a potential replacer of low pressure conventional hydraulic fluids in gear pump-driven hydraulic circuits. They found that both graphite and carbon nanofiber dispersions in ethanol within a concentration range of 195−1500 ppm can sustain hydraulic circuits with increases in pump efficiency.

Liu et al. [8] have conducted an investigation regarding the effects of nano-diamond additives in lubricating oil not only in a friction tester but also in a diesel engine. The authors ascertained a significant reduction of the friction coefficient of 30% and a slight increase on engine performances.

In our best knowledge the effect of the mixture of nano-diamond and nano-graphite as an additive for engine oil on the engine performances have not been investigated yet. The main goal of this article is to investigate the effect of this mixture on an internal combustion engine fuel consumption and friction.

2. Methodology

The test bench is presented in figure 1 and it consists of the internal combustion engine (I.C.E.) fitted with an external fuel tank, placed on an electronic scale (Gibertini Euromatic) to measure the fuel consumption. The speed of the engine was measured using an infrared tachometer (Lutron-DT 1236L), and the temperatures were measured with a digital infrared thermometer (ANENG AN550). For the Lutron-DT 1236L tachometer, the measurement range is 10 to 99999 rpm [9].

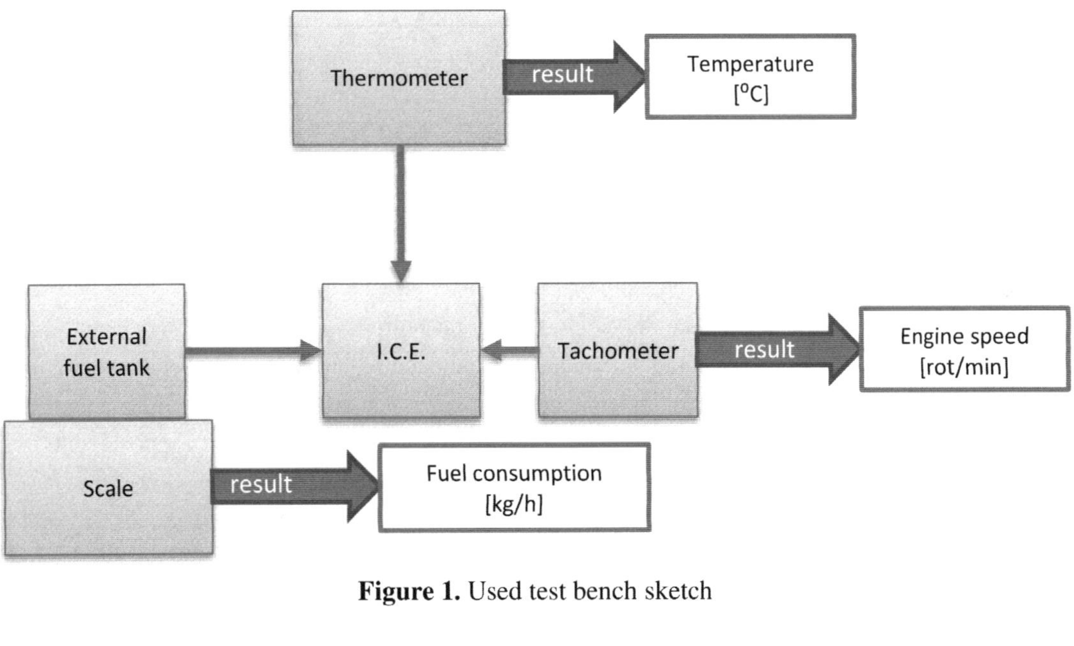

Figure 1. Used test bench sketch

The engine used for the tests was a Honda GX 160 single cylinder engine with the specifications presented in table 1.

Table 1. Honda GX 160 engine specifications.

Type	Single cylinder, 4 stroke
Displacement	163 cm^3
Bore x Stroke	68 x 45 mm
Max Power	3.6 kW @ 3600 rot/min
Max Torque	10.3 Nm @ 2500 rot/min
Cooling	Air cooled
Oil capacity	0.6 L
Fuel consumption	1.4 L/h @ 3600 rot/min

The used lubrication oil is Motul 5100 4T 10W-30, because it fulfils the API SL/SM standards. The standard properties of the lubrication oil are presented in table 2.

Table 2. Properties of Motul 5100 4T 10W-30.

Viscosity grade	SAE J 300	10W-30
Density at 20°C	ASTM D1298	0.871 g/cm^3
Viscosity at 40°C	ASTM D445	74.6 mm^2/s
Viscosity at 100°C	ASTM D445	11.5 mm^2/s
Viscosity index	ASTM D2270	147
Pour point limit	ASTM D97	-36°C
Flash point	ASTM D92	226°C
Total Base Number (TBN)	ASTM D2896	7.5 mg KOH/g

A commercially available ADDO nano-diamonds and nano-graphite particles, purchase from Plasma Chem, was used as additive. The additive contains 90% v/v motor oil and 10% v/v carbon nanoparticles, with an average of nano-particles size of 4-6 nm and this fulfils DIN EN ISO 6245 and DIN EN ISO 2160 standards. In order to prepare the oil with nano-diamonds and nano-graphite particles, a volume of 60 mL nano-additive was dispersed into 540 mL lubricant oil and the final concentration of the lubricant oil with additive was 1% v/v. The viscosity at 40°C and 100°C, viscosity index and density recorded with Stabinger viscometer SVM 3000 for the oil with nano-additve and the oil with nano-additives after used are presented in table 3.

Table 3. Properties of oil lubricant with nano-additives before and after use in engine.

Physicochemical properties	Oil with nano-additives before use	Oil with nano-additives after use
Density at 15°C	0.868 g/cm^3	0.867 g/cm^3
Viscosity at 40°C	62.175 mm^2/s	69.159 mm^2/s
Viscosity at 100°C	10.389 mm^2/s	11.178 mm^2/s
Viscosity index	156	154

The first test was done using as lubricant, only the Motul 5100 4T 10W-30 oil, with the engine speed variation from 1000 to 2500 rot/min in 100 rot/min steps, at zero load. The temperatures of the exhaust pipe (T_1 was chosen to monitor the temperature of the exhaust and therefore the fuel burn temperature loss), outer side of the cylinder (T_2 was chosen to evaluate the friction between the cylinder and the piston-rings assembly), exterior of the oil pan (T_3 was chosen to estimate the temperature of the lubricant itself) and the starter (T_4 was chosen as a reference temperature for the start of measurements) were measured, as seen in figure 2.

The second test was made using the same Motul 5100 4T 10W-30 oil, but with added nanomaterial additive (5 mL, the equivalent of 1%).

Figure 2. Measurements points for all temperatures

3. Results

First results that were extracted are the fuel consumptions for both tests, determined using the external fuel tank and scale. According with measurements, the variation of fuel consumption with speed presents a second order polynomial regression. The fuel consumption average (Fig. 3) when the oil lubricant with nano-additives was used (5.03 [kg/h]), decrease by 21% in comparrison with the average of fuel consumption when the standard oil lubricant was used (7.11 [kg/h]). This reduction is extremely high, but it is logical due to the fact that the engine was tested at zero load, where the engine only has to overcome the internal frictions, and no outer load is applied. This leaves more room for further investigations using the same additive but with an engine that can be loaded accordingly.

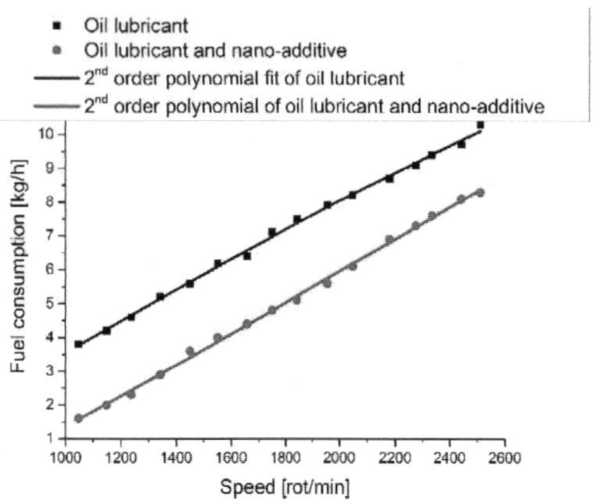

Figure 3. Fuel consumption versus speed

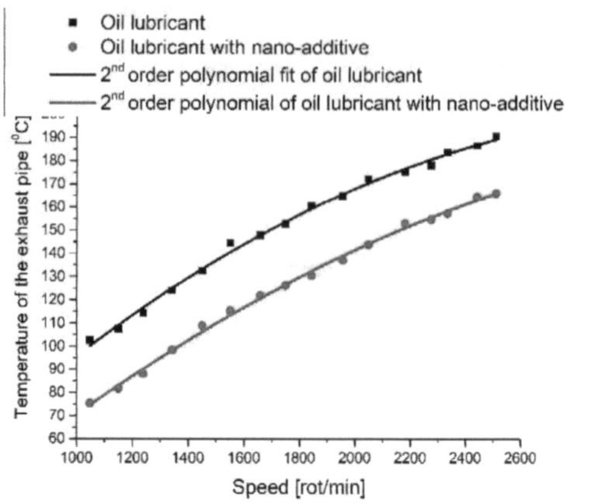

Figure 4. Temperature of the exhaust pipe versus speed

On the other hand, a reduction when nano-additives were used is explained thru the viscosity index. The results showed that the viscosity index of the oil lubricant increase with 6% when nano-additives are used (Tab. 3), which influence the fluid friction due to the formation of a better viscous film in the surface, leading to a reduction of the frictional losses. This reduction of power losses further enhances the automotive engine efficiency and our results are in agreement with [10] and [11].

The results shown in figure 4 indicate the variation of the temperature of the exhaust pipe (T_1) with speed. An increase of the temperature was observed for the oil lubricant with nano-additives. For the temperature of the exhaust pipe, an average value of 150°C was recorded for the standard lubricant oil while using the nano-additives, the average value of the temperature decreases to only 124°C, which indicates a better evolution of the temperature for the same testing time, i.e. better efficiency over time due to a lower friction coefficient.

The temperature of the outer side of the cylinder (T_2) also has second order polynomial variation, but the start temperature is around 45°C for the standard lubrication oil, and only around 37°C for the oil with nano-additive (Fig. 5). Due to the reduction of the frictions, the temperature of the outer side of the cylinder is reduced with 11% on the entire speed spectrum, when using nano-additives.

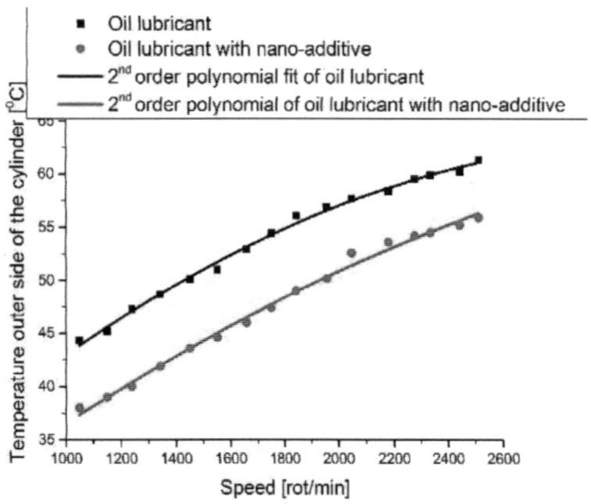

Figure 5. Temperature of the outer side of the cylinder versus speed

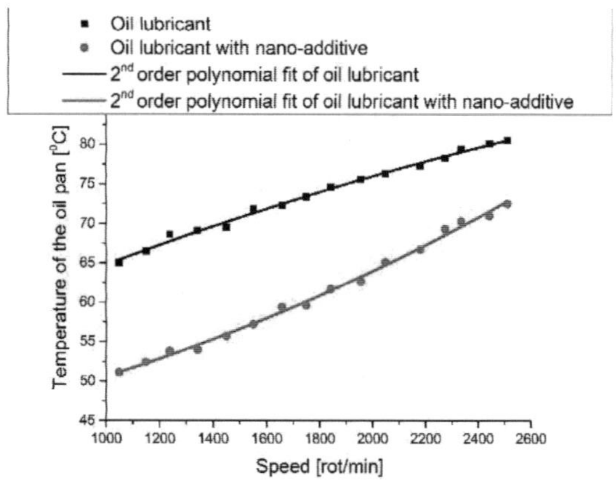

Figure 6. Temperature of the oil pan versus speed

Outside the oil pan (T_3), the temperature was monitored to understand the variation of the lubrication oil for both cases. By using the nano-additives instead of reaching an average value of 71°C, a value of 59°C was reached (Fig. 6) underlining the friction reduction with 20%.

4. Conclusions

The measurements were done with the increase of the speed at precise time intervals, with the same starting temperature of 25°C for the starter (neutral reference temperature zone).

Based on the present experimental investigations, the following conclusions can be drawn:

- ✓ by using nano-additives in the lubrication oil, an improvement of the physicochemical properties was observed, especially because of the decreasing of the viscosity and increasing of the viscosity index which lead to the formation of a better viscous film capable to reduce the frictions between the lubricated components;
- ✓ due to the reduced frictions, the fuel consumption was reduced by 21% at zero load;
- ✓ all the measured temperatures were reduced for the same exploitations of the engine.

Further investigations will be focused on loading the engine using a test bench, and the stabilization of the working temperature of the engine.

References

[1] Holmberg K Andersson P and Erdemir A 2012 *Tribol. Int.* **47** 221
[2] ***. Lubricant Additives - A Practical Guide. Available from: https://www.machinerylubrication.com/Read/31107/oil-lubricant-additives.
[3] Jatti VS and Singh TP *J. Mater. Sci. Technol.* 2015 29(2) 793
[4] Chu HY Hsu WC and Lin JF 2010 *Wear* **268(7)** 960
[5] Raina A. and Anand A *Appl. Nanosci.* 2017 **7(7)** 371
[6] Su Y, Tang Z, Wang G and Wan R 2018 *Adv. Mech. Eng.* **10(5)** 1.
[7] Martorana P, Bayer IS, Steele A and Loth E 2010 *Ind. Eng. Chem. Res.* **49(22)** 11363
[8] Liu H., Bai M and Qu Y 2012 *In: SAE-China, FISITA (eds) Proc. of the FISITA 2012 World Automotive Congress. Lecture Notes in Electrical Engineering*, Springer, Berlin, Heidelberg **190** 1313
[9] *** http://www.amt.cz/info/3/DT1236L.pdf
[10] Souza de Carvalho M.J, Seidl P, Pereira Belchior CR and Sodre JR 2010 *Tribol. Int.* **43** 2298
[11] Ali MKA Xianjun H, Mai L, Bicheng C, Turkson RF and Qingping C *Wear* 2016 **364-365** 270

IOP Conf. Series: Materials Science and Engineering **724** (2020) 012048 doi:10.1088/1757-899X/724/1/012048

Factors Influencing the Triboelectric Charging of Granular Plastics in a Rotating-cylinder-type Tribocharger

I E Achouri[1,2], T Zeghloul[1], K Medles[1], G Richard[2] and L Dascalescu[1]

[1]PPRIME Institute, UPR 3346 CNRS – University of Poitiers, ENSMA, IUT, 16021Angoulême, France
[2]CITF, Dorgeville, 16170 Saint Cybardeaux, France

Abstract. The electric charging of granular plastics in view of their electrostatic separation is one of the major industrial applications of the triboelectric phenomenon. The granules get charged by collisions and frictions with each-others and with the walls of the tribocharging device. In electrostatic separation applications, the granules of a mixture must be sufficiently charged to be attracted by electrodes of opposite polarity for their separation by type of material. The objective of the present work is to study the effects of three factors that influence the triboelectric charging of polycarbonate granules (size: 2 mm to 5 mm) originating from waste electric and electronic equipment in a rotating-cylinder-type device: particle residence time, cylinder inclination and its rotation speed. The performance of the triboelectric charger was evaluated by measuring the charge per mass ratio of the granules collected at the outlet of the device.

1. Introduction

Waste electrical and electronic equipment is composed of a wide variety of conducting and insulating materials [1]. The electrostatic separation of the latter [2] requires their charging by triboelectric effect [3]. Triboelectric charging occurs when two materials of different nature collide [4]: an electric charge is transferred from one to the other; this phenomenon is also called contact charging [5]. The higher the electric charge acquired by the particles, the more effective is the electrostatic separation.

Besides electrostatic separation [6, 7], triboelectricity is the basis for many other important technologies like electro-photography [8, 9], electrostatic precipitation [10, 11], triboelectric energy harvesting [12, 13] There are many different tribocharging devices [14], including rotary cylinders [15], fluidized beds [16], vibrating plates [17], fan-or propeller-type units and tribo-cyclones [18].

The objective of this work was to optimize the operation of a new triboelectric charging device designed by the authors and manufactured by CITF Company, in view of its utilization for the electrostatic separation granular plastics mixtures originating from waste electric and electronic equipment. The polycarbonate (PC) granules used in this study had sizes between 2 mm and 5 mm, which are typical for the considered application.

Modelling and optimization of triboelectric charging process were carried out using the design of experiments methodology [19] with respect to three factors, namely: the rotation speed of the outer cylinder, as well as its inclination angle and the residence time of the particles in the device.

2. Experimental setup

In this device (Figure 1), the particles to be tribocharged are introduced into a funnel (1). A valve (2) is used to regulate the particle flow entering into the hollow cylinder (3) which allows their triboelectric charging. This cylinder is entrained by an electrical motor (4). The rotational speed of the hollow cylinder is variable up to 120 rpm. The hollow cylinder is made of PVC, its length is adjustable (200 mm, 400 mm or 600 mm) and three or six blades of 10 mm or 15 mm height can be attached to its inner wall. Its inclination angle can also be changed between 0° (horizontal) and 45° (5). The flow rate of material entering the tube is controlled by a trapdoor. When it is closed, the particles remain inside the device. This allows the increase of the residence time of the particles in the device, which might be beneficial for the triboelectrical charging process. An exit hopper (6) prevents the projection of the particles and allows their recovery. Moreover, a control cabinet (7) is used for adjusting the speed the cylinder. At the outlet of this device, the particles can be recovered in a Faraday cage if it is desired to measure their charge or directly introduced in an electrostatic separator.

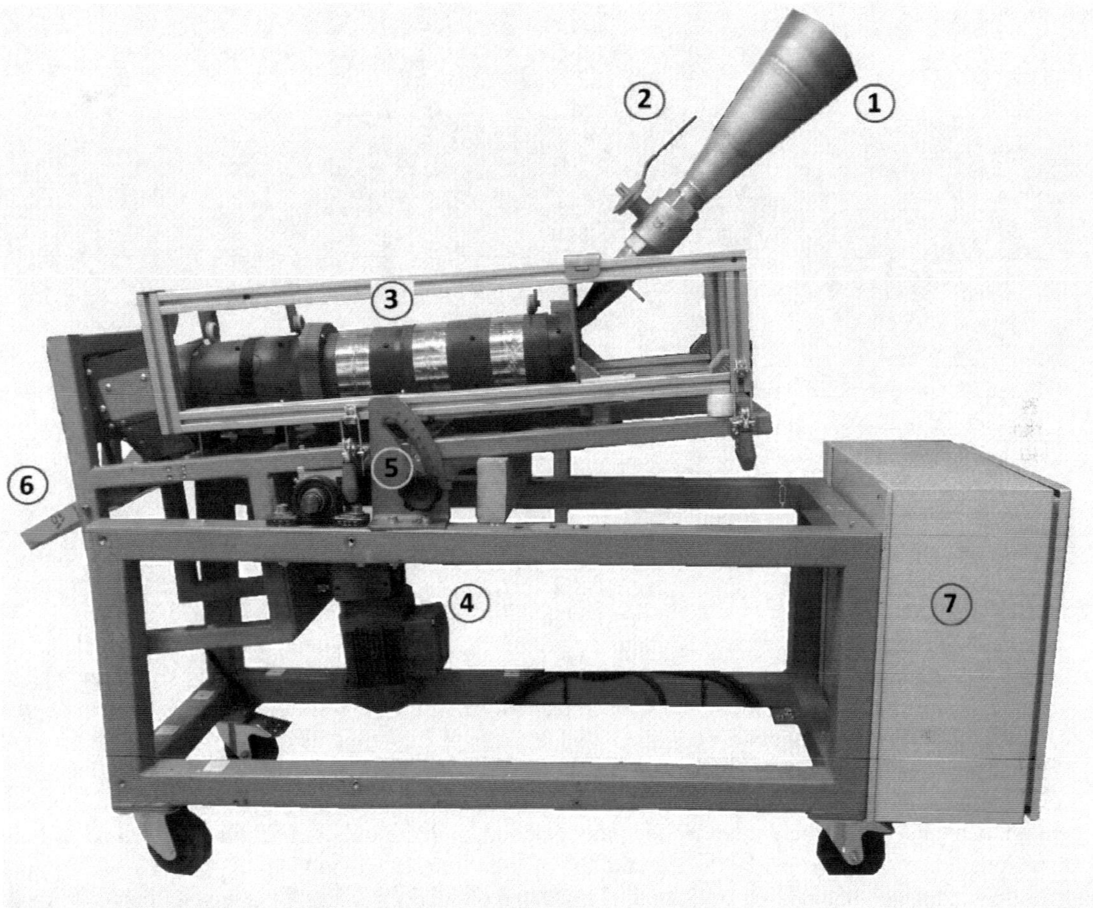

Figure 1. Charging device with rotating cylinders. (1) Funnel; (2) Valve; (3) Hollow cylinder; (4) Motor; (5) Inclination adjustment system; (6) Output hopper; (7) Electrical control cabinet.

3. Materials and method

The granular plastic waste used in the present study was white PC (Figure 2). The size of the particle was typically between 2 mm and 5 mm. The experiments were all duplicated and performed by the same operators, using similar samples and under almost constant climatic conditions: room temperature 18.3°C to 19.0°C and relative humidity of the air 40.2% to 55.7%.

In each experiment, 100 g of particles were used.

Figure 2. Aspect and size of PC particle.

Note that, according to the positioning in the triboelectric series, PC should get positively charged in contact with PVC.

For each test, an experimental procedure was elaborated and scrupulously respected:

- close the trapdoor of the cylinder;
- deposit the particle mixture in the funnel;
- adjust the angle of inclination;
- adjust the cylinder rotation speed;
- start the cylinder rotation motion;
- open the valve and at the same time measure the particles residence time in the cylinder;
- stop the rotation;
- open the trapdoor;
- recover the product in a Faraday cage connected to an electrometer to measure its electrical charge and placed on a precision electronic scale (KERN, resolution: 0,01 g) to measure its mass.

The experimental design method is used to identify the optimum operating point of the tribocharging device. This method determines the number of experiments to be performed in order to achieve a predetermined goal and predicts the behavior of the process in the domain of definition of several factors that may vary simultaneously.

In addition, it enables the evaluation of the significance of the effects of these factors and of their interactions. In the industrial field, this method is of great interest because it allows the evaluation of the influencing factors and the optimization of a process while performing a minimum number of tests.

The response y of the process can be expressed a function of several factors (the input variables u_i). Most often, the answer is expressed as a polynomial of the first or second degree:

$$y = c_0 + \sum c_{ij} u_i u_j + \sum c_{ij} u_i^2 \tag{1}$$

In the case of what is called a factorial composite design, the total number of tests N, for k factors, at two levels (LOW and HIGH) is written as follows:

$$N = 2^k + 2 * k + 3 \tag{2}$$

4. Results and discussion

4.1. Determination and modeling of the experimental domain

The domain of variation of three control variables of the process was established as follows, after preliminary tests:

- Inclination angle α [°] = [3, 7];

- Rotation speed n [rpm] = [50, 90];
- Residence time of the particles in cylinder t [min] = [1, 3].

The analysis of the experimental results given in Table 1was carried out with MODDE 5.0 program [19], which calculates the coefficients a_{ij} of the mathematical model, draws the response contours, and identifies the best adjustments of the parameters for optimizing the process.

Table 1. Charge/mass ratio (*Q/m*) of PC tribocharged particles, for the 17 runs of the composite factorial experimental design.

N	Inclination angle α [°]	Rotation speed *n* [rpm]	Residence time *t* [min]	Charge to mass ratio *Q/m*[nC/g]
1	3	50	1	5.83
2	7	50	1	3.47
3	3	90	1	6.14
4	7	90	1	4..31
5	3	50	3	7.12
6	7	50	3	3.26
7	3	90	3	7.06
8	7	90	3	4.00
9	3	70	2	7.08
10	7	70	2	4.20
11	5	50	2	5.46
12	5	90	2	5.81
13	5	70	1	5.59
14	5	70	3	6.11
15	5	70	2	6.32
16	5	70	2	6.26
17	5	70	2	6.25

4.2. Descriptive quality of the model

The evaluation of the effects of the control factors reveals the inclination angle of tribocharger as the most important. The mathematical model of the response (i.e., the charge to mass ratio) is given by:

$$Q/_m = 6.14 - 1.4 * \alpha + 0.22 * n + 0.22 * t - 0.41 * \alpha^2 - 0.41 * n^2 - 0.20 * t^2 + 0.16 * \alpha * n - 0.34 * \alpha * t - 0.06 * n * t \quad (3)$$

The statistical indicators obtained for this experiment design R^2= 0.9945 and Q^2= 0.9694, which characterize the quality of the regression and prediction, respectively, are close to 1.0, indicating that the obtained model can be confidently used for optimizing the triboelectric charging process.

4.3. Influence of selected factors on responses

From the mathematical model (1), the MODDE 5.0 software plotted the variation of the response according to each of the chosen factors(Figure 3).The results show a decrease in the charge of the entrained particles with the increase of the inclination angle of the rotating cylinder as shown in figure 3.a.When the inclination is small, the particles are more spread out along the cylinder length and less in contact with each other, which increase the number of impacts and enhance the friction between particles and wall, facilitating the transfer of electric charges.

At first, the charge increases with the rotational speed of the cylinder, then saturates for values higher than 70 rpm (Figure 3.b). This can be explained by the peculiarities of the relative sliding motion between the particles and the wall of the cylinder.

The charge is quasi-proportional to the residence time as illustrated in Figure 3.c. The increase of charge is due to the higher number of impacts and the longer time of contact between the particles and the wall.

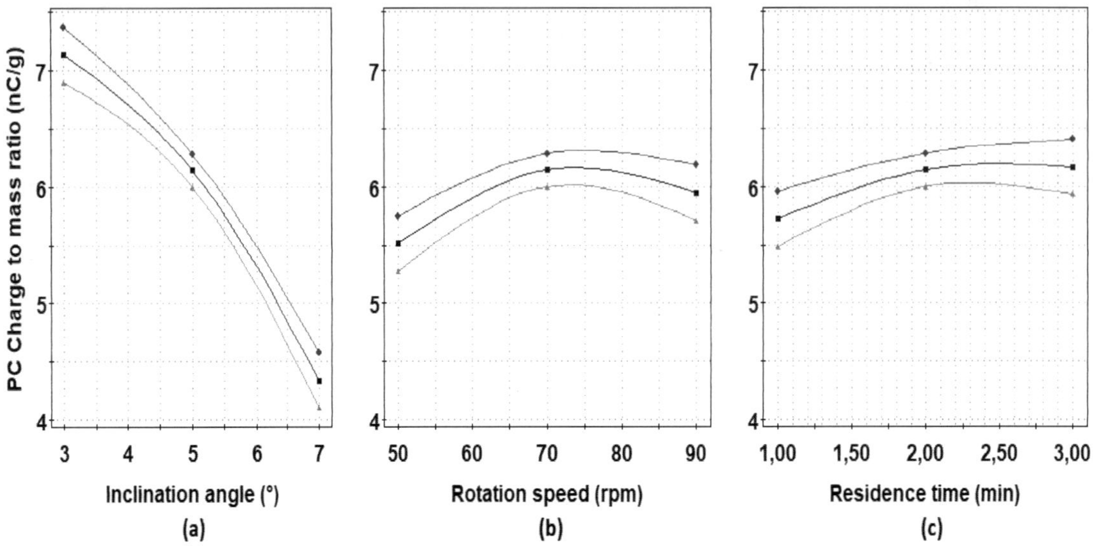

Figure 3. Charge to mass ratio predicted by MODDE 5.0, as function of: (a) inclination angle α, at n= 70 rpm, t= 2 min; (b) rotation speed n, at α=5°,t= 2 min; (c) residence time t, at α= 5°, n= 70 rpm.

4.4. Determination of optimal operating point

MODDE 5.0 software proposes graphic representations of the predicted response Q/m according to the studied factors (Figure 4).

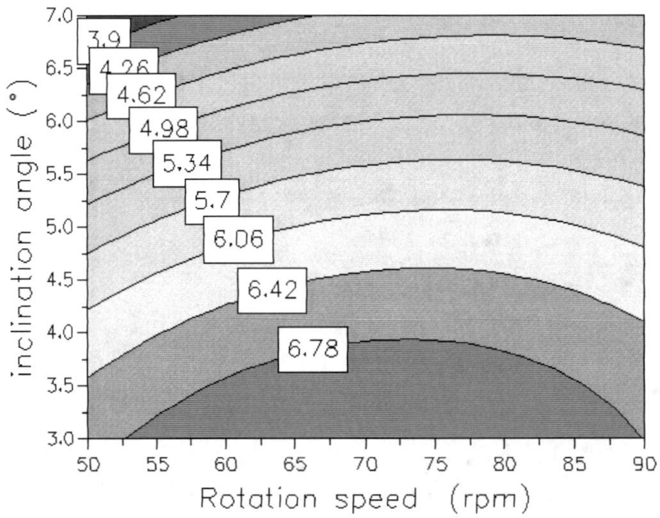

Figure 4. MODDE-predicted charge to mass ratio contours for tribocharged granular PC, depending on the control variables (rotation speed, inclination angle), for a residence time t = 2 min.

According to the predictions made by MODDE 5.0 software, the best tribocharging is obtained when the control factors are adjusted at the values given in Table 2. Two additional experiments were done for a minimum and a maximum residence time of particles. They were in good agreement with the theoretical predictions.

Table 2. Comparison between MODDE 5.0-predicted and experimentally obtained results.

Results	Inclination angle α [°]	Rotation speed *n* [rpm]	Residence time *t* [min]	Charge to Mass ratio[nC/g]
Predicted	5.5	50	1	4.68
Experimental				4.82
Predicted	4.3	90	3	6.36
Experimental				5.88

5. Conclusion

The results of the experimental study performed on mm-size PC granules authorize the following conclusions:

(1) Tribocharging device presented in this paper proved its efficiency in processing granular polymers;
(2) The experimental design methodology enables the modelling and optimization of the triboelectric charging process. The obtained mathematical models can then be used for the determination of the optimal values of the different control factors of the triboelectric charging process.
(3) The outcome of the triboelectric charging process depends on the inclination angle and the rotation speed of the cylinder, as well as on the residence time of the particles inside the tribocharger.
(4) In 1 min, PC granules can acquire a maximum charge/mass ratio Q/m= 4.68 nC/g with the rotating cylinder inclined at an angle α=5.5° and rotating at a speed n=50 rpm. Higher charge/mass ratios, close to 6.5 nC/g, can be obtained for a longer residence time (3 min).

6. References

[1] Kang H and Schoenung J M 2005 *Resour. Conserv. Recycl.* **45** 368-400
[2] Tilmatine A, Benabboun A, Brahmi Y, Bendaoud A, Miloudi M and Dascalescu L 2014 *IEEE Trans. Ind. Appl.* **50** 4245-4250
[3] Iuga A, Samuila A, Morar R, Bilici M and Dascalescu L 2015 *Part. Sci. Technol.* **34** 45-54
[4] Castle G S P 1997 *J. Electrostat.* **41** 13-20
[5] Lowell J and Rose-Innes A C 1980 *Adv.Phys.* **29** 947-1023
[6] Chagnes A, Cote G, Ekberg E, Nilsson M and Retegan T 2016 *WEEE Recycling Research, Development and Policies* (Amsterdam: Elsevier), 75-102
[7] Benabboun A, Tilmatine A, Brahami Y, Medles K, Bendimerad S and Miloudi M 2014 *Sep. Sci. Technol.* **49** 464-468
[8] Schein L B 1992 *Electrophotography and Development Physics* (Berlin: Springer)
[9] Schein L B 1999 *J. Electrostat.* **46** 29-36
[10] Achouri I E, Hamou N and Achouri F 2017 *Proc. 2016 8th Int. Conf. Model. Identif. Control. ICMIC 2016* (Algiers) 133-137
[11] Nouri H, Achouri I E, Grimes A, Said H A, Aissou M and Zebboudj Y 2015 *Int. J. Electr. Comput. Eng.* **9** 1451-1456
[12] Zhu G, Lin ZH, Jing Q, Bai P, Pan C, Yang Y, Zhou Y and Wang Z L 2013 *NANO LETTRES* **13** 847-853
[13] Wang S, Xie Y, Niu S, Lin L, Liu C, Zhou Y. S and Wang Z L 2014*Adv. Mater.* **26** 6720-6728
[14] Achouri I E, Zeghloul T, Richard G, Medles K, Nouri H and Dascalescu L 2019 *IEEE Trans. Ind. Appl.* **55** 802-811
[15] Zelmat M E, Rizouga M, Tilmatine A, Medles K, Miloudi M and Dascalescu L 2013 *IEEE Trans. Ind. Appl.* **49** 1113-1118
[16] Calin L, Mihalcioiu A, Iuga A and Dascalescu L 2007 *Part. Sci. Technol* **25** 205-211
[17] Higashiyama Y, Ujiie Y and Asano K1997*J. Electrostat.* **42** 63-68
[18] Miloudi M, Medles K, Tilmatine A, Brahami M and Dascalescu L 2011*J.Electrostat.* **69** 631-637

Standardization in testing ballistic protection systems

G G Ojoc[1], C Pirvu[1], S Sandu[3] and L Deleanu[1]

[1] Faculty of Engineering, "Dunarea de Jos" University, 111 Domneasca, Galati, Romania
[2] National Institute for Aerospace Research "Elie Carafoli" INCAS, 220 Iuliu Maniu, Bucharest, Romania
[3] Scientific Research Center for CBRN Defense and Ecology, 225 Oltenitei, Bucharest, Romania

Abstract. Even more than in other field of activities, standardization in ballistic protection is in trends of CEN/CENELEC declaration that "standards build trust", promoting innovation and introduction of new, performant solutions based on test procedures and protocols that are able to point out quality and safety for such systems. Ballistic testing standards have particular specifications for ballistic materials, differentiation of tests being made especially on the purpose of the ballistic protection product. In the last decades, test methods have been refined and measurement devices have become more complex. The main set of standards is developed by military organizations (as NATO), highly industrialized countries, where the military and research in the field are priorities (USA, European Community, Asian countries, South Africa). Some states refer to American standards by using them as guides. Depending on the destination of the ballistic product, specifications in standards are different. The authors give examples of how standards requirements are implemented in testing ballistic protection systems, in order to point out the particularities of introducing and verifying new and classical products in this domain. Generally, the results of the ballistic tests are evaluated by acceptance or rejecting the tested product, based on specific criteria, included in standards.

1. Introduction

Even more than in other field of activities, standardization in ballistic protection is in the trends of CEN/CENELEC declaration that "standards build trust" [1] and promote innovation and the introduction of new, performant solutions, based on test procedures and protocols that are able to point out quality and safety for such systems. Ballistic testing standards have particular specifications for materials, differentiation of tests being done especially based on the purpose of the ballistic protection product. In the last decades, test methods have been refined and the measurement devices have become more complex [2]. They also establish a clear terminology, useful for the quality assessment of ballistic protection systems.

Standards and specification evaluations are used for:

- assessing the quality and the performances by testing,
- evaluation of new materials and design solutions under the same conditions as for "classical" ones,
- investigating and understanding failure mechanisms of protection systems,
- a logical approach of designing solutions for more lethal ballistic threats,

- understanding and avoiding the consequences of a hit on humans and technical systems,
- protecting with keeping the performances of personnel and systems at an efficient level.

Standards supply test procedures, requirements for measuring equipment, conditions and terminology for determining the ballistic protection of different solutions and purpose (metallic, non- metallic, composites, ceramics or, more often, hybrids of the above mentioned materials).

Tests for ballistic protection always take place under particular and imposed conditions. Actually, in situ conditions may differ. If a protection glass is fixed in a frame, once rigidly and once flexibly, it may stop the projectile on one occasion, but it is penetrated on the other occasion. Factors like humidity, dust, light and chemical agents could diminish the protection performance. Specialists have to know the sensitivity of each product to the test conditions. A test condition is evaluated as sensitive if a minor modification is sufficient to produce a considerably different test result [3].

A regulation may be elaborated according to two different basic principles, taking into account either the attack potential (attack-orientated regulation) or the protection potential (protection-orientated regulation). Regulations can be divided into two groups:

– standards included in law and nationally recognized regulations; they are elaborated by the national or nationally recognized standards institutes, and become mandatory,

– guidelines not included in law; they are test instructions, established by public or private institutions in order to achieve uniform testing of certain products.

Taking into account standards and guidelines [4], [5], [6], [7], [8], [9], [10], [11] for body armors and light protection of vehicles, Table 1 presents the information included in a ballistic test report. Results have to be compared, qualitatively and quantitatively, to standard requirements and supplementary demands, if required by any interested part.

Table 1. Information to be found in a ballistic report

• Contractor information	• Personnel conducting the test and any witness
• Measuring and recording devices	• Weapon used
• Contact information	• Information of used projectiles: code, mass, design, materials, dimension, supplier
• Lot number and quantities	• Information of propellant: type, weight of propellant for each shot
• Item specification number	• Results include: impact velocity used in computing V_{50}, with the highest partial penetration, the lowest complete penetration, spread intervals for test parameters and results and velocities of each rounds and shot.
• Tested product specifications	• Witness plate characteristics, partial or complete
• Armor material description	• Remarks to the conduct of the test, or material behavior
• Material identification for each sample	• Photos, films
• Temperature and humidity at test facilities	• (If required) failure mechanisms
• Date of the test	• Standard number or regulation

Test methods have two variants, usually applied both for each product:

– testing a defined minimal mean penetration velocity (v_{50}) that provides information on a ballistic protection since it also reveals the production quality and the manufacturing variation of protection system,

– testing at a defined impact velocity at which none, for an imposed number of rounds, may penetrate the protection system.

A non-encompassing list of standards and requirements for individual armor is also given in [12]. This includes threat ammunitions and their characteristic velocities to be applied for this testing protocol. A still discussed issue is the value for which back face signature (BFS) is lethal. Interested governmental institutions have adopted a BFS standard not exceeding 44 mm. The army, for hard armor, has used a BFS standard less than 43 mm without penalty. Many specialists consider this value is too high for human trauma acceptance. Figure 1 presents several issues related to standard and non-standards tests for ballistic protection. In standards, threats are classified in levels and, consequently, the system provides protection against a certain threat level. There is no permission of supposing that a higher level recommends a lower level without testing for the required levels.

The main families of standards are developed by military organizations (NATO), highly industrialized states where military research are priorities (USA, European Community, Asian countries, South Africa). Typical examples of standards are regulations of DIN (German Institute for Standardization), Ö-Norm (Austrian Standards Institute) or SNV (Swiss Standards Association).

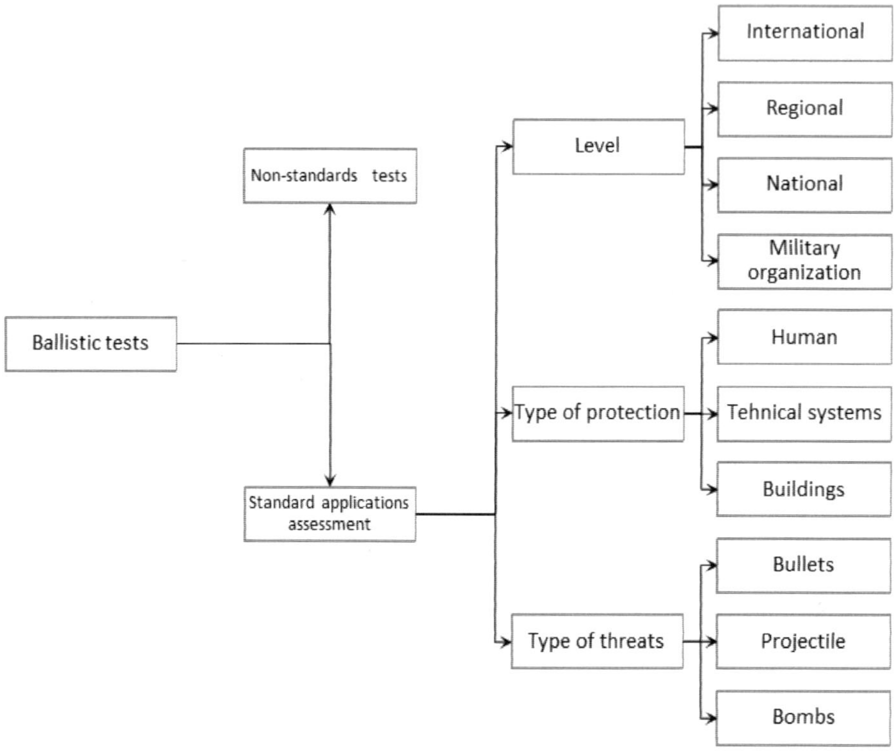

Figure 1. Standard tests for ballistic protection.

Guidelines mostly emerge as prior references for the procurement of equipment for the army or police, as NATO STANAG guidelines or guidelines of the PFA (police records academy) in Muenster (Westphalia). Some states refer to American standards by using them as reference guides.

Depending on the destination of the ballistic product, specifications in standards are different. For example, in standard NIJ 0101.04 and NIJ 0101.08 for vests impact resistance, there are specific issues related to limit the trauma the human body is subjected to. For armored vehicles, the specifications are tailored to vehicle type (truck, helicopters etc.) and the threat they face, and ballistic products may be subject to additional fire resistance, weathering etc.

Hard body armors have to comply with regulations for testing in order to decide the more suitable product. From 2007, the US Army conducted ballistic testing against hard body armor, completed analyses of test results and developed a statistically-based test procedure providing a high level of confidence for resistance to penetration and backface signature (BFS) [12]. Specialists and end-users (army and low enforcement staff) require standard test references, procedures and analyses for every type of ballistic protection, including hard body armor testing, vehicle armor or building protection.

New hybrid composites gain this market due to their improvements over the classical materials in terms of strength to weight ratio, better life expectancy and enhanced thermal properties [14].

At CEN, Technical Committee 33 is responsible for elaborating and up-dating the standards related to ballistic protection, the representative standards being: EN 1063:1999 [4], EN 1522:1998 [5] EN 1523:1998 [6]. ISO16935:2007 (reconfirmed in 2017) [10] and test procedures for evaluating resistance of security glazing materials and products (for both interior and exterior use) against ballistic impact with classification by weapon and ammunition. It is assumed the glazing is adequately fixed, but it does not apply to glazing system or surrounding materials and structure [10]. EN 1063 [5] is a security glazing standard for measuring the protective strength of bullet-resistant glass. It is commonly used with EN 1522 [5] to form a ballistic classification system, by which armored vehicles and structures are tested and rated. A similar classification system used in the United States is NIJ Standard 0108, which includes glass and armor plate.

NATO introduced STANAG 2920 [15] for quantifying materials ability to stop fragments and shrapnel. The measuring technique was initially developed for body armor, but now it could be used in events that generate fragments with high risk of impact, including for armored vehicles. Standard tests are conducted by shooting FSPs (fragment simulating projectiles) onto the target sample with different velocities, while measuring the velocity of each FSP. By varying the velocities, after a number of shots, an estimate of the ballistic limit can be obtained, which is the velocity to which the protection system defeats the fragment. Combat troops rarely suffer injury or fatality from bullets, but are at high risk from primary (direct) and secondary (environmental) fragmentation. The V_{50} test is the internationally recognized standard for assessing the fragmentation resistance of personal protection. The higher the ballistic velocity, the higher the rating of the protecting armor, The V_{50} is the average of the velocities recorded for six fair impacts consisting of the three lowest velocities for complete penetration and the three highest velocities for partial penetration, provided the spread is not greater than 40 m/s [15].

NATO AEP-55 STANAG 4569 [8] includes standard techniques and reproducible test procedures for evaluating the ballistic resistance of vehicle armor components (integral, add-on, opaque and transparent), as well as the required vehicle vulnerable area assessment.

Figure 2 presents a diagram resulted from a NIJ standard for testing protection systems and the main factors influencing the requirements, procedures and results analysis.

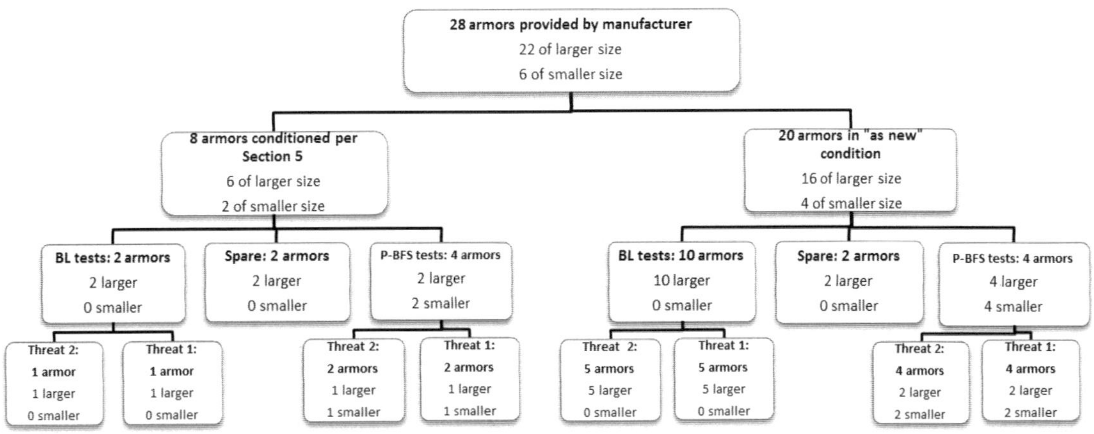

Figure 2. Sample quantity and utilization for armor types IIA, II, and IIIA [16].

The advantages of non-standard samples are:
- available for laboratory fabrication (square samples of 100 mm to 250 mm a side),
- easy to fix on the laboratory frames,
- easy to use supplementary investigation equipment like fast camera and thermal camera,
- easier to use non-destructive investigations like SEM, laser profilometer etc.

The disadvantages of using standard non-complying samples could include:
- difficulty in extrapolating the panel response to larger one,
- small dimensions of the samples could emphasis some failure processes and could hide others,
- the fixing type could induce different response as compared to that of the actual panel that could be mounted in a more sophisticated frame.

Samples for standard tests are larger. For instance, for panels of individual armors, the dimensions are 500 mm x 500 mm, as also for vehicle panels [4]. For buildings and non-mobile systems, the standards recommend larger panels. These samples, even if they react to impact more likely to panels in actual applications, are too large for small-scale investigations (SEM, FTIR, EDX) and they require to be cut with a method that does not influence (too much) the impacted zone. Panels made of aramid fabrics are difficult to be cut and it is very probably to disturb fiber positions and their matrix when cutting. Panels made of glass fiber fabrics are prone to supplementary delaminate when cutting small samples. In a review on textiles and fibre-reinforced composite impact responses, Abtew et al. [17] underlines by photos the influence of sample size, clamping and projectile design, shape, size and velocity, concluding that the more reliable tests are those "imitating" closely the field situations.

Pach et al. [18] tested ballistic resistance of hybrid shields: three types of two-layer composites. The first shield layer has plates of tungsten carbide, ceramic balls Al_2O_3 or steel balls (100Cr6), in a polymer matrix. The second layer was a laminate of 10 layers of aramid fabric in a matrix of styrene-butadiene-styrene. Samples were overshot along the ballistic track from the Beryl rifle. The 5.56x45 mm ammunition with the bullet of SS109 MESCO type was used. Non-standard methods were used, panels being small (100 mm x 100 mm), shooting distance 10 m from the target, shooting angle 90°, the projectile impact velocity 870...890 m/s.

Kilic [19] determined ballistic performance of an armor, being helped by finite element methods for approximating ballistic limit thickness of armor steels. The tests were done to validate the model and the ballistic resistance of high hardness armor steels against 7.62 mm armor piercing ammunition. Ballistic shot tests on 20 mm thickness target were performed according to standard Stanag 4569 [8].

2. Test campaign for ballistics

After reviewing the open literature, one may notice that there are differences between standards tests and laboratory tests for ballistic protection. The authors experienced tests for a ballistic panel (see figure 3), designed for vehicle protection and pointed out the differences between laboratory tests and those required by specific standards [13].

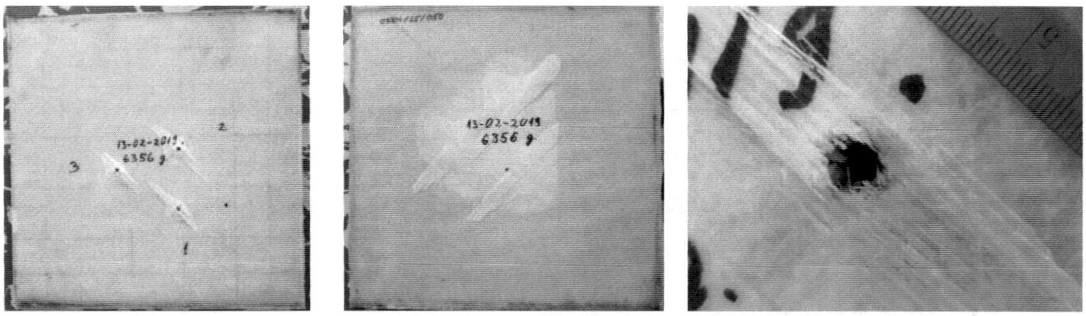

a) the panel face b) the back face of the panel c) detail

Figure 3. Tests for a protection panel for level IIIA in agreement with SR EN ISO 1523 2004 [13], at standard distance for proving or not fragments detached from the panel.

There are standards for bulletproof vests, but few are used extensively [20]. Among these, there are stab and ballistic test norms for vests, like NIJ Standard-0101.06 [16]. "Since NIJ started its testing, more than 3,000 officers' lives have been saved" [21]. Also, the standards of the United Kingdom's Home Office Scientific Development Branch are widely used in several countries. Countries will use these standards and incorporate the basics and modify them to fit the threats of interest. This ballistic standard rates vests on a scale against penetration and trauma protection.

The test campaign, regardless the standard, has the following components: method, test and in-situ investigation equipment, samples, including one to remain as whiteness, the ballistic test report. Photographs of all samples are not expressly required but, photographs at different scale (micro - obtained with scanning electron microscope and macro - obtained with high resolution camera) could reveal deficiencies, failures or unusual results, this being useful information to both the armor manufacturer, the conformity assessment body and the end-user.

For ballistic protection, test standards are elaborated in order not to be limited at conventional or already known threats, but also to be applicable for new entries of threats and materials. For instance, the ballistic-resistant protective materials covered by NIJ 0101.06 [16] are classified into the following levels of ballistic performance: type I (22 I.R; 38 Special), type IIA (9 mm FMJ, 357 Magnum), type II (9 mm FMJ, 357 Magnum), Type IIIA (9 mm FMJ, 44 Magnum), type III (M80 ball), type IV (30-06 AP), type special, this being added to face progress in threats performances.

3. Comments concerning standard test procedures

Taking into account the discussion on tests for assessing the ballistic protection, the research included in project [13], [21] did non-standard and finally standard tests.

Body armors have to be tested for two reasons: penetration resistance against bullets and the back-face signature (BFS) that is associated with the personnel's risk to be wounded. The impact energy is found by shooting the armor mounted in front of a backing material (clay), which is used at a precise temperature and confirmed for impact flow. After the test, the obtained indentation in clay is measured [20]. Comparing the BFS may be tricky if the tests are done following requirements of different standards. Even using the same type of bullets, but from different suppliers, and different clay grades (even considered equivalent) may give hidden "errors", In European, German and British standards, there is an allowable value of 20 to 25 mm for the back face signature. The American NIJ standard admits 44 mm, which could produce internal injuries and it has been controversial since its introduction. New non-destructive investigation equipment's allow for characterizing more complex the BFS. Wetting conditions degrade a vest. Room temperature water will not affect para-aramid. But either basic or acid liquids decrease strength of aramid fiber fabrics and standards demand tests on wet armors and even after exposing them in heated, airtight enclosures [22].

Gillepsie et al. [23] tested panels of dimensions 30.5 cm x 30.5 cm were impacted using a fragment simulating projectile (FSP), following a NIJ Standard. For the unstitched (baseline) and stitched panels without tile, a 12.7 mm fragment simulating projectile and a velocity of 465 m/s was used, corresponding to a kinetic energy of 5.8 kJ, while the tiled panels were impacted at a much larger velocity, i.e., 810 m/s, by a larger, 20 mm projectile (17.6 kJ impact energy). Damage size was much smaller and the residual compression strength after ballistic impact was much larger for the processed composites than for the hand laid-up glass–polyester composite. This seems to suggest that the new proposed manufacturing process has higher quality. Processed glass–vinyl ester and glass–epoxy panels with a ceramic tile bonded to the front surface were impacted by a much larger energy projectile. Despite the high impact energy, the tile prevented penetration of the projectile, and the damage pattern was quite uniform through the panel thickness due to the changed stress wave propagation through the panel. Stitching reduced the damage size and increased the residual compression strength of the tiled panels, but not very much.

Nunes et al. [24] pointed out that when the impact velocity approaches the ballistic limit, the impact process becomes very sensitive to initial conditions (e.g. material defects, impact speed or angle), and even small variations may lead to significant differences in the outcome of the test (perforation or not, of the target). Therefore, the results for this condition must be examined with care. Due to their low specific gravity, and high stiffness, hardness, strength and thermal stability, the ceramic-based systems have shown potential for improving upon current standards for ballistic performance, which includes multi-hit capability. During experimental methods, the effect of different target configuration and target mechanical properties on ballistic performance of the material can be observed when struck by various standard bullet or FSP at the different velocities [25]. The experimental approach while ballistic impact test could also help to determine the energy absorbed by the target and transmitted beyond the target based on different parameters while using in body armours [26]. After the test, the impacted projectile could be either perforated the material or trapped inside the target. In the non-perforation situation, the projectile might be either stopped inside the material or bounce back by leaving the trauma indentation at the backing material. This ability of the target which prevents the happening of internal injuries is known as trauma resistance and ballistic tests are mainly carried out according to different standards [27], [28], [29]. The two most common standards are the National Institute of Justice NIJ-standard 0101.06 [16] and Home Office Scientific Development Branch (HOSDB) ballistic body armor standards for UK Police.

A study tried to investigate the effect of thickness and to determine the minimum value to avoid perforation while ballistic impact of aramid (Kevlar®) fabric laminate, subjected to 7.62 mm ammunition, at a speed of 800 m/s, with standard procedure [30]. Based on the result, 96 layers (~50 mm) of aramid fabric thickness was required to capture the bullet, which was much higher than the necessary thickness for stopping 7.62 mm bullet at the specified speeds.

The performance of samples designed by [31], to have penetration resistance, was assessed based on NIJ standard, at test level III. Each sample was placed on a steel frame, 15 m in line with the gun barrel, and subjected by an impact with 7.62 x 51 mm bullet, at a velocity of 847±9.1 m/s, at an angle of 90° to specimen, for one shot, at the center. Based on results, S glass fiber reinforced PBA could be effectively used as strike panel to substitute the conventional ceramic tile and E glass fiber composite. An integration of S glass composite backed by aramid fiber reinforced composite was potentially used as hard ballistic armor providing full protection, based on NIJ standard level III in terms of penetration resistance, lower weight and less bulkiness as compared to E-glass composite.

Many researchers start a project with a preliminary test campaign that include several standard requirements, especially threat, distance to the target, but not necessary all the requirements of the standard. This initial campaign is important in order to establish technology, thickness and shape of the panel. After doing such an investigation and establishing the failure mechanisms, a probability to failure, the engineer could organize the campaign that will meet all the standard demand. The results of such a study for body armors made of fabrics SB1plus (from Teijin) is given in Figure 4 presenting details of the panel (face and back of each layer) for a panel made of 12 layers that bear complete penetration with 9 mm FMJ, at impact velocity of 400±15 m/s. At 8 layers there were some panels with complete penetration and some that arrested the bullet and for a panel with 12 layers all BFS were bellow the admissible one of 22...24 mm. Standards are the subject of continuing research, development and testing, review and modification, as appropriate [16].

Generally, the results of the ballistic tests are evaluated by acceptance or rejecting the tested product, based on specific criteria, included in standards. For instance, for body protection, a vest is accepted if it fulfils the following conditions:

- the projectile is arrested within the panel (partial penetration);
- the backface signature does not overpass a maximum value, given in standard,
- the test is valid only for a certain ammunition and a particular set of fire conditions, this is why fire are done with actual guns or laboratory stands that make the projectile to hit the target under the same characteristics (for instance, same impact velocity, energy, shape and components of the projectile.

For vehicle armors, the conditions may be different, depending on vehicle type, threat, particular requirements for their use etc.

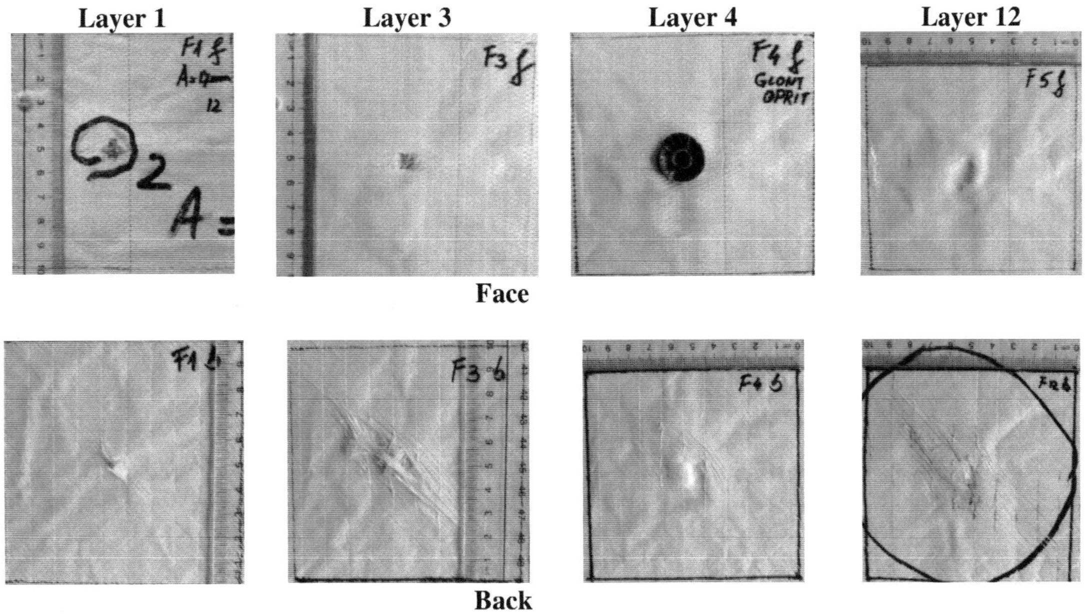

Figure 4. Preliminary tests for panel made of 12 layers of LFT SB1 plus (Teijin) [32].

4. Conclusions

Standards test procedures are usually applied after preliminary tests at laboratory scale, those fulfilling a part of the standard requirements, especially those related to threats (projectile type and its mass and impact velocity). Usually, the technology for manufacturing the samples depend on sample size and thus, the response of the non-standard test could differ when it is compared to that complying with the standard requirements. Also, the actual solution for implementing the ballistic protection could differ, including fixing design, the presence of other components of the system that will be protected (as, for instance, edges and corner for fixing the ballistic protection panels.

Acknowledgments

This paper was supported by a grant of the Romanian Ministry of Research and Innovation, CCCDI-UEFISCDI, project no. PN-III-P1-1.2-PCCCDI-2017-0404/31PCCDI/2018, within PNCDI III and by Doctoral School for Mechanical and Industrial Engineering of "Dunarea de Jos" University of Galati.

References

[1] CEN CENELEC Declaration, Standards build trust, https://www.cencenelec.eu/news/publications/Publications/CENCENELEC-DECLARATION-Final.pdf

[2] Bhatnagar A 2006 *Standards and specifications for lightweight ballistic materials*, in (Woodhead Publishing Limited, CRC Press, Boca Raton), pp 127-167

[3] Kneubuehl B P 2003 Ballistic protection, http://www.vpam.eu/fileadmin/Wissenswertes/Ballistic-protection.pdf

[4] EN 1063:1999 Glass in building. Security glazing. Testing and classification of resistance against bullet attack (under revision)

[5] EN 1522:1998 Windows, doors, shutters and blinds. Bullet resistance. Requirements and classification

[6] EN 1523:1998 Windows, doors, shutters and blinds - Bullet resistance - Test method

[7] National Institute of Justice, NIJ 0108 ± ballistic resistant protective materials

[8] STANAG 4569 2012 Protection levels for occupants of armoured vehicles, (2nd ed) NATO Standardization Agency

[9] NIJ Standard 0101.03, 1987 Ballistic Resistance of Police Body Armor, U.S. Department of Justice, National Institute of Justice
[10] ISO16935:2007 Glass in building -- Bullet-resistant security glazing -- Test and classification
[11] Testing of Body Armor Materials: Phase III, 2012, ISBN 978-0-309-25599-8
[12] Office of the Secretary of Defense, Standardization of Hard Body Armor Testing, 27.04.2010
[13] Deleanu L, contract 725/2017, "Dunarea de Jos" University, beneficiary MPS Urziceni, Romania
[14] Potluri R, Supriya K and Vittal G V G, 2018, *Mater Today-Proc* **5**, 20257
[15] STANAG 2920 Classification of personal armour. 2015
[16] Selection and Application Guide to Ballistic-Resistant Body Armor for Law Enforcement, Corrections and Public Safety NIJ Selection and Application Guide-0101.06, https://www.ncjrs.gov/pdffiles1/nij/247281.pdf
[17] Abtew M A, Boussu F, Bruniaux P, Loghin C and Cristian I, 2019, *Compos Struct* **223**, 110966
[18] Pach J, Pyka D, Jamroziak K and Mayer P, 2017, *Compos Part B-Eng* **113**, 24
[19] Kılıç N, Bedir S, Erdik A, Ekici B, Tasdemirci A and Güden M 2014 *Materials & Design*, **63,** 427
[20] The Most Widely Used Ballistic Standards For Bulletproof Vests, https://www.bodyarmornews.com/ballistic-standards/
[21] Muhlhausen D B 2018 Director's Corner: Improved Officer Safety in 2017 — and Beyond, https://nij.ojp.gov/speech/directors-corner-improved-officer-safety-2017-and-beyond
[22] Opportunities in Protection Materials Science and Technology for Future Army Applications, 2011, National Academy of Sciences, ISBN 978-0-309-21285-4
[23] Gillespie Jr J W and. Monib A M, 2003 *J of Composite Materials*, **37**(**23**), 2131
[24] Nunes G S, Scazzosi R, Manes A, Campos Amico S, W. Ferreira de Amorim Jr and Giglio M 2019, *Int J of Impact Eng*, doi:https://doi.org/10.1016/j.ijimpeng.2019.05.021
[25] Mudric T, Giacomuzzo C, Francesconi A and Galvanetto U, 2016 *J of Aerospace Eng*, **29** (**6**), 04016047
[26] Karahan M, Kuş A and Eren R 2008 *Int J Impact Eng*, **35(6**), 499
[27] Tan V B C, Lim C T and Cheong C H, 2003, *Int J Impact Eng*, **28** (2), 207
[28] Jantharat P, Mccuiston R, Gamonpilas C and S. Kochawattana, 2014, *Key Eng Mater*, **608**, 253
[29] Purushothaman P, Coimbatore G and Ramkumar S S, 2013 *J Eng Fiber Fabr*, **8(2)**, 97
[30] de Oliveira B F, Lima Jr. É P, de Sousa L E and Monteiro S N 2017 *Material Research* 2**0**, 676
[31] Okhawilai M, Hiziroglu S and Rimdusit S, 2018 *Measurement* **130**, 198–210
[32] Pirvu C, 2015 PhD thesis, "Dunarea de Jos" University, Galati, Romania

IOP Conf. Series: Materials Science and Engineering **724** (2020) 012050 doi:10.1088/1757-899X/724/1/012050

Study of the effect of duration of non-thermal plasma treatment on the surface properties of polymers

O Tilmatine, T Zeghloul, A Fatu and L Dascalescu

PPRIME Institute, UPR 3346 CNRS – University of Poitiers – ENSMA, IUT, 4 avenue de Varsovie, 16021 Angoulême Cedex, France
E-mails: thami.zeghloul@univ-poitiers.fr

Abstract. One of the factors affecting the electric charge acquired through friction between the polymers is their surface roughness. The dielectric barrier discharge (DBD) treatment is known to modify the surface characteristics of polymers. The aim of this study is to quantify the effect of atmospheric DBD exposure, in a well-defined configuration on the triboelectric behavior of polymers. Thus, Polytetrafluoroethylene (PTFE) plate samples characterized by different surface roughness were rubbed against bronze plate samples in a triboelectric test bench and then the electric potential at their surface was measured. The results were compared with those obtained for surfaces previously exposed to DBD treatment duration of which did not bring any change in their roughness. It was found that the roughness of the surface of the treated polymers hardly changed for rather short DBD exposure time (a few minutes). Significant changes were noticed for longer treatment duration (30 minutes and more). The samples that had undergone a longer treatment period modification acquired higher levels of electric charge by triboelectric effect.

1. Introduction

For several years now, whether in the field of industry or the medical field, the use of polymers is more and more widespread and all that is thanks to their mechanical, electrical and thermal properties. Therefore these new materials have become increasingly the focus of scientific research. Despite the extent of such research, knowledge about the mechanical and electrical properties of polymers as well as about the electric charge generated by friction between them are not perfectly mastered [1-3].

Polymer surface properties like morphology and topography can be modified by means of different chemical and physical treatment techniques. These techniques are used in many fields: from automotive and aerospace industries, to the manufacturing of electronic and bio-medical devices.

Among various surface modification techniques, plasma treatment is probably one of the most used for polymer surface treatment, as it is very easy to implement in the industry, requiring simple and flexible configurations with regard to electrode geometry. Moreover, it has high efficiency and processing speed, very low electrical energy consumption and excellent operating costs [4-7].

Previous studies have clearly shown the unpredictable effect of plasma surface modification on tribological behavior of polymers [8-10]. For example, Bekkara et al [8] and Natsuta et al [9] have shown that a DBD treatment modifies the wettability of certain polymers (Polypropylene (PP), Polyethylene (PE), polyethylene terephthalate (PET) and polyvinyl chloride (PVC)) and that a longer plasma exposure increases the roughness of the material. On the other hand, Al.Maliki et al [10] showed that the roughness of the materials (Polyetheretherketone and Polycaprolactame) may decrease and then increase with the duration of the treatment.

The aim of this paper is to study the effect of the duration of the DBD on the roughness of the Polytetrafluoroethylene (PTFE) known as Teflon. It also investigates the effect of variation in roughness on the triboelectric properties of PTFE in an attempt to find a relationship between the amount of charge that will be produced and the roughness of the friction surfaces.

2. Materials and experimental setup

2.1 Materials and preparation

The samples chosen during this study are a PTFE / metal sample pair. This choice is justified because the PTFE is one of the most used polymers in industry field and especially in medical applications, such as orthopedic implants and vascular grafts. [6]. It has excellent mechanical properties, high chemical resistance, a very low refractive index and the lowest known dielectric constant of all polymers.

For the metal used in this study, a bronze sample (consisting essentially of copper and tin) was selected. The dimensions of the bronze sample (A) were: 100 mm x 15 mm x 5 mm, and those of the PTFE samples (B) were: 50 mm x 180 mm x 6.5 mm. The PTFE samples (B) were first taken for a roughness measurement to ensure that the surface condition is uniform. In order to reduce their surface roughness, some of these samples have been polished with alumina abrasive discs with a P400 grain content, and continuous lubrication with water. The polishing was manual and each sample was held against the abrasive disk for a period of 5 minutes. The PTFE sample was then taken into the DBD cell for treatment with atmospheric plasma. They were then transferred to the triboelectric test bench and finally the distribution of the electric charge at their surfaces after tribocharging was scanned using an electrostatic probe. The measured values of the electric potential have been directly recorded on a PC, using LabView software.

2.2 DBD cell

The surface treatment of the polymer samples was performed under ambient temperature conditions (temperature: 20 °C – 24 °C, relative humidity: 50% - 55%).

Figure 1.a represents the configuration of the DBD cell composed essentially of two horizontally disposed parallel electrodes: an aluminum discs with 120 mm of diameter and 6 mm of thickness and the active electrode that is made mobile by a sliding system and can be adjusted up and down to change the discharge gap. The grounded electrode is covered by square glass plate as dielectric barrier of 180 mm x 260 mm with the thickness of 4 mm. The upper electrode was connected to a high-voltage supply (22.5 kV, 800 Hz), while the bottom one is grounded. The air-gap was maintained constant at 3.5 mm and the duration of plasma exposure was modified for each experiment. Three samples were treated for each of 4 different durations of treatment: 5, 10, 20, and 30 minutes.

2.3 Triboelectric test bench

The triboelectric test bench (Figure 1.b) has the role of charging the metal/polymer samples by friction with each other. Designed to ensure a plane/plane contact between the pair of samples, the bench is composed of two sample holders: the top one is for the bronze sample (A) and the bottom one is for the PTFE sample (B). These two holders will ensure the uniformity of contact between the two samples.

The bench also has a rail guide system that ensures the horizontal displacement of the lower holder in back and forth motion. The speed of this movement can be varied and controlled using an electric actuator. The bench also contains a system of vertical guide bars that maintain the normal force perpendicular to the surface of the lower sample. This force can be adjusted using the mechanical force control system and measured by a force sensor. In addition to this sensor, two other sensors for tangential force and displacement are included in the system.

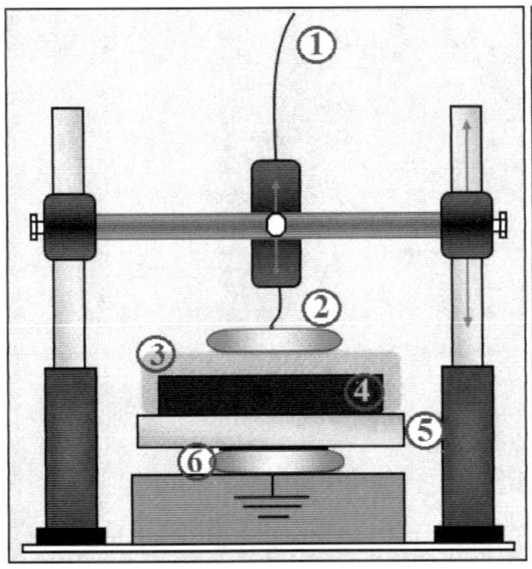

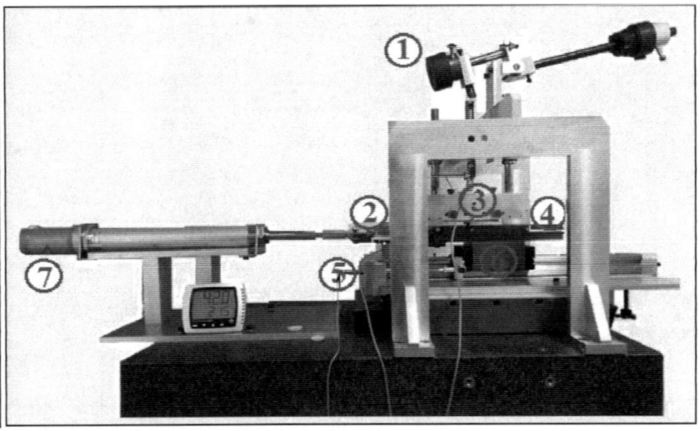

(a) (b)

Figure 1. (a); DBD cell configuration; (1) AC high voltage; (2) Active electrode; (3) Plasma; (4) Polymer sample; (5) Dielectric barrier; (6) Grounded electrode, (b); Tribolectric test bench; (1) Vertical force control system; (2) Force sensor; (3) Top sample holder; (4) Bottom sample holder; (5) Displacement sensor; (6) Horizontal guide system; (7) Electric motor.

The different parameters of the bench remained unchanged throughout the experiments: the normal force is 10 N, the translational speed is 10 mm/s, the stroke of the bottom sample holder is 55 mm and the number of a tribocharging (rubbing) cycles is 10 [1].

The electric potential at the surface of the bottom sample is measured by the auto-compensated induction probe (Trek, model 6300-7) of an electrostatic voltmeter (Trek, model P0865). 20 points of measurement were made on each sample and the highest values were recorded.

Before each experiment, any residual charges of the samples are eliminated with the charge neutralizer (ELCOWA, model ECA88-BS), equipped with a 5 kV high-voltage generator (ELCOWA model SC04B).

3. Results and discussion

In the first series of experiments, a comparative study on the effect of PTFE surface texture on the amount of charge generated after tribocharging was carried out. This results in 4 type of PTFE surface finishing. For each surface type, the same experiment was performed on 5 different samples.

The first surface condition was the "non-treated" one: a PTFE sheet was cut into 5 plates which, after neutralization, were directly taken to the tribocharger. After that, the charge generated on the surface of this sample was measured in the next minute. The measurements were recorded and the maximum value of the potential of the surface was determined for each charged sample.

The second surface condition was "polished": before taking the 5 plates of PTFE to the tribocharger they were mechanically polished with abrasive paper. Their surface state has clearly changed and their roughness has significantly decreased compared to their initial state. The same procedure as for the "non-treated" samples was then used to determine the maximum value of the surface potential generated after tribocharging.

The third and fourth surfaces finishes were raw PTFE plates and polished PTFE plates treated with DBD for 30 minutes. After the DBD, the experimental procedure was similar to that described above. Figure 2 shows the obtained result, which is the maximum potential value recorded on the tribocharged surface. The highest potential was found on the surface of PTFE which was only polished, with a maximum value of -7600 V. A DBD treatment lasting 30 minutes with a polished surface reduced the potential generated by tribocharging to -6300 V. The lowest value measured of the potential is that of a surface that has not been polished or treated with cold plasma. The maximum potential of this surface has only reached a value of -2200 V, but unlike the previous case, a DBD treatment of 30 minutes of the non-polished surface increased the potential value up to -4300 V.

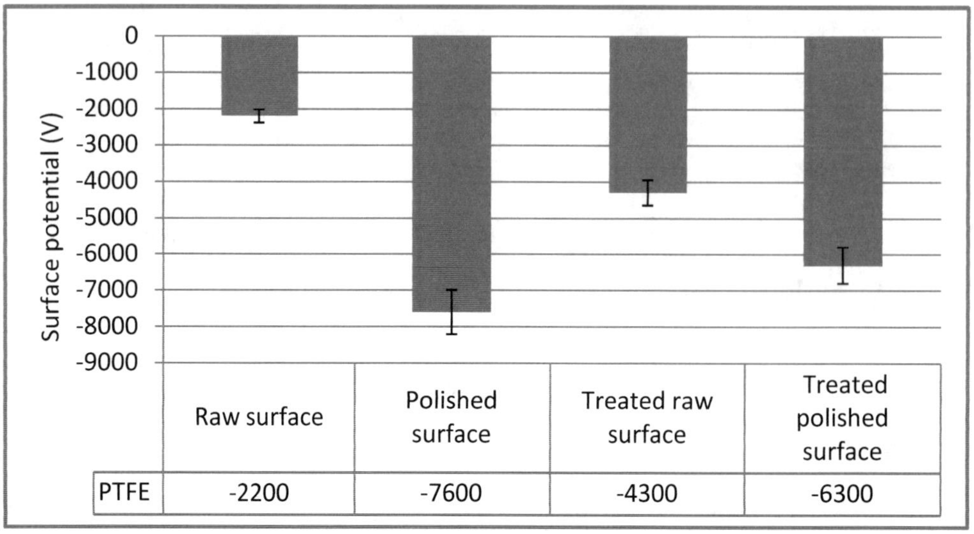

Figure 2. The maximum surface potential on different surfaces of PTFE rubbed with bronze.

The second series of experiments involved only raw (unpolished) PTFE plates, all of which underwent DBD treatment but with different treatment times. The treatment duration was 0 (untreated), 5, 10, 20 and 30 minutes. The procedure was the same as in the previous series of experiments: the same parameters of the DBD cell and of the tribocharging facility. Only the duration of the DBD treatment was varied.

Figure 3 shows that the treatment with DBD has an influence on the maximum value of the surface potential generated by tribocharging. A surface treated with DBD acquired a much greater potential than an untreated surface. The maximum value of the surface potential obtained on an untreated sample was -2200 V, while that of a sample treated for 5 minutes was -6400 V. However, a longer treatment time does not mean a higher generation of surface potential by tribocharging. The maximum recorded value of surface potential was obtained for 10 minutes treatment period. For a longer treatment time (30 minutes), the maximum value of the surface potential decreased.

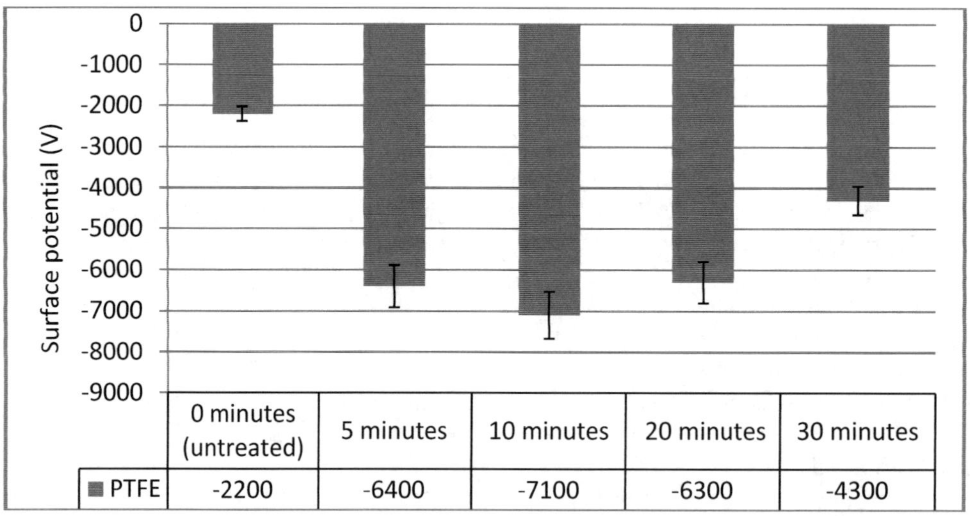

Figure 3. The maximum surface potential for different DBD processing times of PTFE rubbed with bronze.

IOP Conf. Series: Materials Science and Engineering **724** (2020) 012050 doi:10.1088/1757-899X/724/1/012050

4. Conclusions

The paper was focused on the study of the influence of the surface state of a polymer on its triboelectric properties after tribocharging, as well as the effect of the cold plasma and its treatment time duration on the triboelectric charge generation. The main conclusions that could be drawn from this study are:

- The surface state of the polymer may have a direct influence on its tribological properties: a smoother surface state allows generating a higher surface potential after tribocharging.
- Cold plasma (DBD) treatment also has a direct influence on the polymer triboelectric properties. Exposure of the polymer to the DBD allows it to gain a higher surface potential after tribocharging than that of an untreated one.
- A longer cold plasma treatment time does not necessarily mean a higher surface potential after tribobcharging. The generated surface potential increases with the duration of treatment time to a certain value and then begins to diminish despite of the fact that the treatment time is longer.

5. References

[1] Neagoe B, Prawatya Y, Zeghloul T, Souchet D and Dascalescu L 2015 *Journal of Physics: Conference Series* **646** 012058.
[2] Zeghloul T, Neagoe M B, Prawatya Y E and Dascalescu L 2017 IOP Conference Series: *Materials Science and Engineering* **174** 012002.
[3] Glor M 2005 *Journal of Electrostatics* **63** 447.
[4] Yang Y, Zhang G J, Zhao W B and Zheng N 2005 *Int. Symp. on Electrical Insulating Materials* Kitakyushu Japan.
[5] Al-Maliki H, Károly Z, Klébert S Z and Kalácska G 2018 *Int. Review of Applied Sciences and Engineering* **9** 1.
[6] Dumitrascu N, Topala I and Popa G 2005 *IEEE Transactions on Plasma Science* **33** 1710.
[7] Rymuza Z 2007 *Archives of Civil and Mechanical Engineering* **7** 177.
[8] Bekkara M F, Neagoe B, Prawatya Y, Benmimoun Y, Zeghloul T and Zouzou N 2016 *10th Conf. of the French Society of Electrostatics* Poitiers France.
[9] Nastuta A V, Rusu G B, Topala I, Chiper A S and Popa G 2008 *Journal of Optoelectronics and* Advanced *Materials* **10** 2038.
[10] Al-Maliki H, Zsidai L, Samyn P, Szakál Z, Keresztes R and Kalácska G 2018 *Polymer Engineering & Science* **58** 93.

6. Acknowledgments

This work was partially funded by the French Government program "Investissements d'Avenir" (LABEX INTERACTIFS, reference ANR-11-LABX-0017-01) and by a FEDER-CPER program "Electinnov", financed by the European Union and by the Nouvelle Aquitaine Regional Council.

A survey of Mechanical failure and design for Reliability of MEMS

A Somà

Mechanical and Aerospace Department, Politecnico di Torino, Italy

Abstract. In this paper, several experimental mechanical investigation techniques are presented to evaluate the reliability of micro-electro-mechanical systems (MEMS). Microsystems in recent years have spread in many everyday devices. We find micro-scale sensors and actuators in automotive, biomedical and aerospace applications where are demanded very strict performance requirements. Electromechanical non-linear coupling is often a crucial problem both in design and also for the reliability of the system. Mechanism of failure and failure modes has to be taken into account in order to evaluate the reliability of the final system. Focusing on device failure, it emerges that mechanical damage is the most significant source. In this paper a survey of recent advance in mechanical testing of MEMS is presented including: mechanical fatigue, mechanical strength and plasticity, surface and contact failure and creep. Different design of testing specimens is discussed to identify the material properties and failure modes behavior in order to obtain design rules and strategies.

1. Introduction

MEMS technology allows the advantage, first of all, to considerably reduce the size and weight of the devices and consequently allow an actuation using a lower electric power. In literature are mainly described aspects of micro-fabrication and technological realization of MEMS. Recently, with the significant diffusion in different application fields of MEMS sensors and actuators, several researchers concentrated to the attention to the problems of reliability and durability [1,2].

MEMS reliability is challenging but a survey of the literature pointed to a lack of experience in systematic approach. Several papers pointed out on a specific failure mode or concentrated the attention on a particular device behavior [3-5].

Wide are the applications of MEMS. Main reasons for choosing MEMS device, instead of a more mature technology, are the reduction in dimensions and weight. These are then accompanied by a more advantageous integration in an electronic system that also includes in the final device a microchip of signal processing. For example, in the telecommunications and satellite communications sector oscillators or MEMS switches have been introduced for radio frequency applications, in automotive and aerospace vehicle applications accelerometer sensors and MEMS gyroscopes have been introduced, in biomedical applications have been developed MEMS micro chambers for microfluidic analysis techniques (LAB-on a-chip).

However, MEMS has to be categorized in order to derive effective design rules. MEMS devices can be categorized based ability to interact with their environment and on their actuation sources into four main groups: electrostatic, electromagnetic, piezoelectric, and electro thermal actuators [6-9].

Unlike macro structures in micro devices and in particular in MEMS structures the study of reliability becomes more complicated due to the difficulty in generating and replicating fault events.

The following list can help to address their possible failure depending on the dynamic behavior, schematically devices with:

- deformable parts of the micro-structure that can move but without impacting on the substrate such as for example in the case of gyroscopes, accelerometers and RF oscillators;
- deformable parts with impacting surfaces including micro-mirror arrays and RF switches.

About structural reliability, it is necessary to evaluate the effects of the process parameters and load working conditions. Among main loads the electromechanical non-linear coupling is crucial for design and represents a potential source that can lead to device collapse. Mechanical damage represents the most significant source of failure usually generated by localized material deterioration. The technology realization of a micro-device is strongly influenced by the process and therefore also the final properties of the material will be due to the process. With regard to the most frequent type of localized damage in the MEMS we find the wear of moving parts due to the electromechanical adhesion of the sliding contacts. Over time the contact surfaces can change their shape due to the wear of the surfaces and therefore determine strong variations in the electromechanical coupling force with consequent failure.

Some of the most common techniques used for MEMS have been firstly developed for integrated circuits and derived from microelectronic process industry. In the present paper a survey of recent advance in mechanical testing of MEMS is presented including: mechanical fatigue, mechanical strength and plasticity, surface and contact failure and creep. Specimen design and failure modes have been listed related to a specific fabrication process in order to derive design rules strategies focusing on reliability.

2. Failure modes

A failure occurs when "*a device or a system no longer performs the required functions under the stated conditions within the stated period of time*" [10]. Due to the nature of coupling, the performance of a MEMS device depends on the mechanical and electrical characteristics of the entire system. Two main types of faults can be highlighted:

- catastrophic (or irreversible) failures, which leads the device to be completely inoperable
- degradation failure, that affects a local part of the device changing functions and parameters outside then normal range of operation.

It is possible to define failure modes and failure mechanism describing observable effects (broken structure, cracked surface, plasticity mechanism, etc.) or directly measurable degradation exceeding prescribed limits. The study of mechanical reliability strictly depends on the understanding the failure modes and mechanisms in different operational conditions. These study output could be design rules that including fabrication processes parameters allow reliable operation of the device during its operating life. In this paper main failure modes and mechanisms that occur in MEMS devices have been identified. The following sections provide a description of the primary failure modes, together with a discussion of the techniques proposed in literature to evaluate and determine them.

2.1. Material and residual stress

The failure of a device and its reliability are strongly influenced by the type of materials used and their structural compatibility. In the technological MEMS process a large number of different materials are used. Metal materials or often silicon are used to make the moving parts of the devices. For electrical and thermal isolation both silicon dioxide and silicon nitride have traditionally been used. For electrical isolation also other materials are used as aluminum oxide and polyimide.

In the realization of MEMS devices technological processes derived from micro-electronics are used. Different materials are integrated, some with structural functions and others with electrical functions, which lead to having failure mechanisms that must be interpreted before they can be traced back to the most known failure mechanisms in the traditional field of electronics or mechanics. RF MEMS technology widely uses gold membrane and beam with different layers that behave like a composite material. In recent years the studies available in literature conducted on the stability of MEMS devices in the long term and on the maintenance of properties over time are still few in number.

The mechanical properties of micro-size material used in MEMS technology could be strictly influenced by the process parameters and process recipes [11-13]. For example the presence of the residual stress generated in the materials during fabrication process could lead to unwanted behavior of membrane, beam, or cantilever structures realized with metal or dielectric thin films within the material [14-16].

In different applications and for different micro-fabrication process experimental testing of MEMS mechanical failure mechanism have been presented in literature such as: buckling [17]; fatigue [18-22]; thermal effect and creep [23-27]; wear and tribology [28,29].

3. Testing procedures, specimen and results

In the last two decades extensive modeling and experimental activities of MEMS have been done by the research group of Politecnico di Torino - Mechanical Department. The deformable elements in MEMS are usually relatively simple and analytical techniques have often been used in the design procedures. The complexity of mathematical calculations can grow due to the non-linearity of the electrostatic coupling computer models and finite element analyses are necessary to fully understand the device characteristics. Finite element can be used, before the expensive and time-consuming fabrication and testing, to evaluate the static, dynamic and thermal behavior of the system. In the case of MEMS it is preferable to use multi-physics finite element models to take into account the couplings of the different physical domains.

The coupling mechanical electrostatic behavior has been studied taking into account the non-linear effect of material [30] and large displacement in different device configuration [31]. Both for MEMS realized in poly-silicon material and in gold material for RF MEMS switch application [32]. Non-linear and linearized simulation method is developed for the static analysis, pull-in and frequency shift dynamic simulation [33,34]. The coupling problem of fluid-structure interaction has been widely studied, simulated and defined in order to evaluate the fluidic damping parameter respect the dimension structural parameters [35,36]. Modeling simulation approach has been also derived to study the termo-structural coupling behavior of MEMS [27]. Main results of these studies were to develop simulator to enable a better understanding of MEMS behaviors, suitable to assist first the design and then the fabrication process. A key feature of this method was the iterative approach involving simulation, sensitivity analysis that leads to optimal MEMS structures [37].

In order to efficiently realized a design procedures firstly modeling approach can be performed decoupling different failure modes effect.

3.1. Elastic-plastic tests of MEMS

Unlike the macro-scale domain, in micro-scale testing it generally is not possible to decouple the specimen of characterization of the material from the machine that solicits the specimen. The test micro-structures (Figure 1) are designed in such a way as to allow an actuation of the specimen in the range of behavior expected of the material. Using electrostatic actuation of the lateral plates the stress–strain behaviour and plastic strength of the microstructures is experimentally studied by measuring the static deflection of the central specimen. By means of non-linear finite element method (FEM) simulations the tensile stress behaviour in the test specimen is evaluated [38, 39].

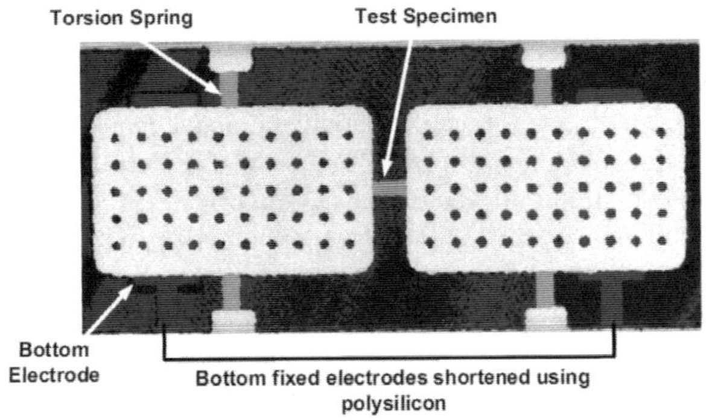

Figure 1. Test microstructure for elastic-plastic tests with the central specimen

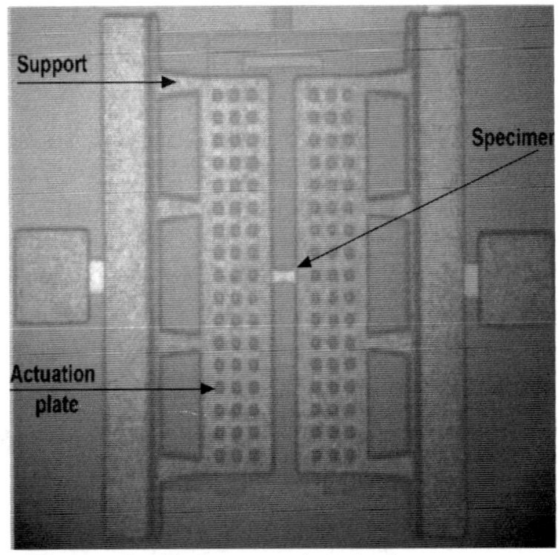

Figure 2. Fatigue testing device: actuator (perforated plates) and specimen (suspended beam).

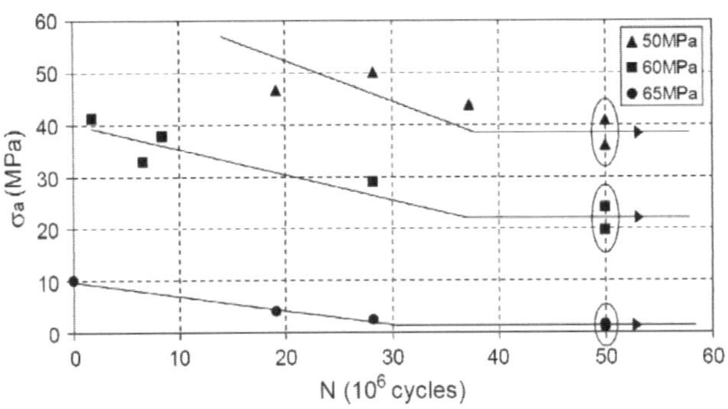

Figure 3. Experimental results of fatigue tests in the Wohler diagram for three values of mean stress. Specimens marked with an arrow did not fail.

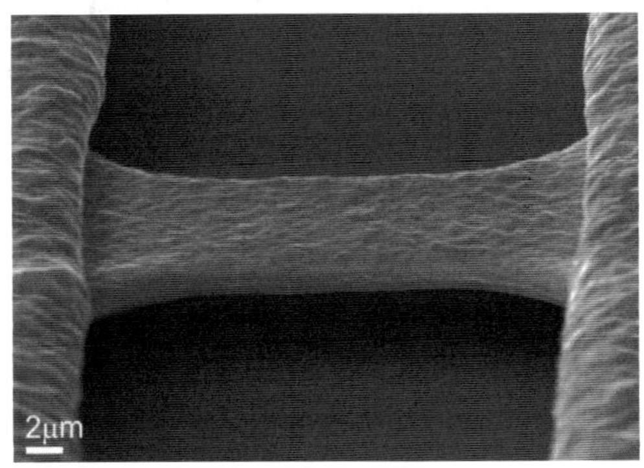

Figure 4. SEM detail of the specimen fabricated using gold electro-deposition technique.

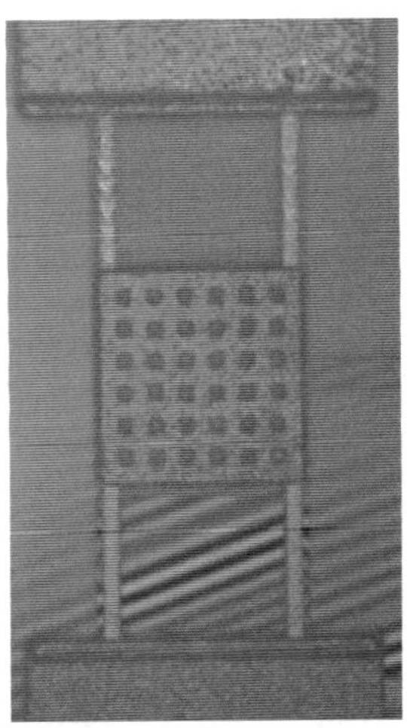

Figure 5. Electroplated gold sample for creep test

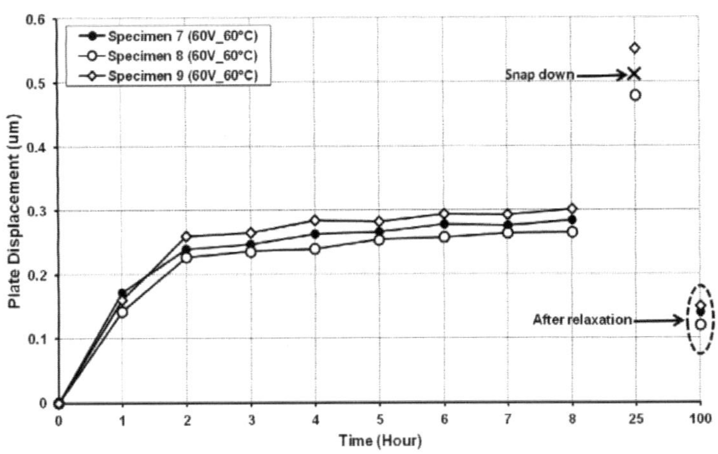

Figure 6. Experimental measurement of specimen displacement at 60 V and 60 °C and residual displacement after 100 hours

3.2. Fatigue tests
Specific specimen (schematically described in Figure 4) and extensive experimental tests have been conducted to study the fatigue behavior of MEMS [40, 41]. Dedicated fatigue test devices (Figure 2) are designed and fabricated with the aim of producing an alternate load on the embedded specimen. In the device is possible to fabricate an embedded beam specimen (Figure 3). Using electrostatic actuation as a driving force the central specimen could be loaded in tensile mode with different levels of mean and alternate stresses. By using FEM models, previously identified, is possible to evaluate the stress level produced in the specimen during the test phases. The Wohler diagrams of the experimental fatigue tests are presented in Figure 3.

3.3. Creep tests
Specific specimen (schematically described in Figure 5) and experimental test has been realized to study the thermo-mechanical behavior of MEMS in order to evaluate the material coefficient thermal expansion (CTE) and evaluate creep mechanism [42]. The evaluation of the amount of creep damage could be done, if reach a measurable magnitude, firstly through the detection of permanent deformation of specimens. In [42] was proposed also a novel indirect method associating the amount of creep damage to the variation of mechanical stiffness. This variation could be measured using electro-mechanical parameters such as resonance frequency shift, pull-in voltage. In Figure 6 results demonstrated the presence of a final plastic deformation after relaxation due to creep. Moreover the continuous curve displays reversible strain due to the viscoelastic behavior of the material.

4. Conclusion
The diffusion of MEMS devices in different advanced application depends largely on their reliability. MEMS technology is derived from micro-electronic process, but the presence of micro-movable parts determine that device level reliability is strictly influenced by mechanical behavior. In this paper, main mechanical failure modes of MEMS devices have been listed and described separately. Dedicated mechanical tests can be performed firstly to understand the associate failure mechanisms at the micro-scale and in presence of multi-physics coupled domain such as electro-mechanical coupling. It is evidenced that in MEMS specific fatigue, creep and plasticity tests are also useful to qualify the process technology and the process recipes. Final result of the combination of the test results could lead design rules suitable for a systematic modeling approach in MEMS focusing on reliability.

References
[1] Tabata O and T Tsuchiya 2006 *Reliability of MEMS : testing of materials and devices* (Wiley)
[2] Hartzell A L, da Silva M G and Shea H. 2011 *MEMS Reliability* (Springer)
[3] Tanner D M, T B Parson, A D Corwin, J A Walraven, J W Wittwer, B L Boyce and S R Winzer 2007 Science-based MEMS reliability methodology *Microelectron. Reliab.* **47** pp 1806–1811
[4] Bromley S C, L L Howell, B D Jensen, 1999 Determination of maximum allowable strain for polysilicon micro- devices *Eng. Fail. Anal.* **6** (1) pp 27-41
[5] van Spengen M W 2003 MEMS reliability from a failure mechanisms perspective *Microelectron. Reliab.* **43** pp 1049–1060
[6] P Gao, S M Swei and Z Yuan 1999 A new piezodriven precision micropositioning stage utilizing flexure hinges *Nanotechnology* **10** (4) pp 394-398
[7] L Sun, J Wang, W Rong, X Li and H. Bao 2008 A silicon integrated micro nano-positioning XY-stage for nano-manipulation, *J. Micromech. Microeng.* **18** (12) pp 125004
[8] Somà A, Iamoni S, Voicu R, Müller R, Al-Zandi and Wang C 2018 Design and experimental testing of an electro-thermal microgripper for cell manipulation, *Microsyst. Technol.* **24** pp 1053–1060
[9] Hongliang S, Hai-Jun S and Dagalakis N. 2014 A stiffness model for control and analysis of a MEMS hexapod nanopositioner *Mech. Mach. Theory* **80** pp 246-264
[10] Collins J A 1993 *Failure of materials in mechanical design. Analysis, Prediction, Prevention* (Wiley)
[11] Romig AD, Dugger MT and McWhorter PJ 2003 Materials issues in microelectromechanical devices: science, engineering, manufacturability and reliability *Acta Mater.* **12** pp 37– 66
[12] Allameh S M 2003 An introduction to mechanical properties related issues in MEMS structures *J. Mater. Sci.* **38** pp 4115-4123
[13] Espinosa H D and Prorok B C 2003 Size effects on the mechanical behavior of gold thin films *J. Mater.*

Sci., **38** pp 4125-4128

[14] Somà A, De Pasquale G, Brusa E and Ballestra A 2010 Effect of residual stress on the mechanical behaviour of microswicthes at pull-in threshold *Strain* **46** 358- 373

[15] Somà A and Ballestra A 2009 Residual stress measurement method in MEMS microbeams using frequency shift data *J. Micromech. Microeng.* **19** (9)

[16] Margesin B, Bagolini A, Guamieri I, Giacomozzi F and Faes A 2003 Stress characterization of electroplated gold layers for low temperature surface micromachining *Proceedings of DTIP* pp. 402-405

[17] Elata D, S. Abu-Salih, 2006 Analysis of the electromechanical buckling of a pre-stressed micro beam that is bonded to an elastic foundation *J. Mech. Mater. Struct.* **1** (5) pp.911-923

[18] Ando T, Mitsuhiro S. and Sato K. 2001 Tensile-mode fatigue testing of silicon films as structural materials for MEMS *Sensor Actuat. A-Phys.* **93** pp 70-75

[19] Sharpe W N and Bagdahn 2004 J Fatigue testing of polysilicon – a review *Mech. Mater.* **36** pp. 3-11

[20] Soboyejo A B, Bhalerao K D and Soboyejo W 2003 Reliability assessment of polysilicon MEMS structures under mechanical fatigue loading *J. Mater. Sci.* **38** pp 4163-4167

[21] Muhlstein C L, Brown S Band Ritchie RO 2001 High cycle fatigue and durability of polycrystalline silicon thin films in ambient air *Sensor Actuat. A-Phys.* **94** pp 177-188

[22] Saghaeian F, Lederer M, Hofer A, Todt J, Keckes J and Khatibi G 2019 Investigation of high cyclic fatigue behaviour of thin copper films using MEMS structure *Int. J. Fatigue* **129**

[23] Pustan M, Birleanu C and Dudescu C 2013 Simulation and experimental analysis of thermo-mechanical behavior of microresonators under dynamic loading, *Microsyst. Technol.* **19** pp 915–922

[24] Pustan M, Rochus V and Golinval J C 2012 Mechanical and tribological characterization of a thermally actuated MEMS cantilever *Microsyst. Technol.* **18** (3) pp 247-256

[25] Modlinski R, and al 2004 Creep as a reliability problem In MEMS *Microelectron. Reliab.* **44** (9) pp 1733-1738

[26] Shamshirsaz M and Asgari M B 2008 Polysilicon micro-beams buckling with temperature-dependent properties *Microsyst. Technol.* **14** pp 957-961

[27] Somà A and Saleem M M 2015 Modeling and experimental verification of thermally induced residual stress in RF-MEMS *J. Micromech. Microeng.* **25** (5)

[28] Ku I S Y, Reddyhoff tT, Holmes A S, and Spikes H A 2011 Wear of silicon surfaces in MEMS *Wear* **271** 1050-1058

[29] Maboudian R and Carraro C 2012 Surface engineering for reliable operation of MEMS devices *J. Adhes. Sci. Technol.* **17** pp 583-591

[30] Brusa E, De Pasquale G, Munteanu MG, Somà A 2010 FEM modelling and experimental characterization of microbeams in presence of residual stress*Analog Integr. Circ. S.* **63** (3), pp. 477-488

[31] Somà A and De Pasquale 2016 Preshaping Command Functions to Control the Dynamic Impacts in MEMS *J. Vib. Acoust.* **138** (1)

[32] Collenz A, De Bona F, Gugliotta A, Somà A 2004 A Large deflections of micro beams under electrostatic loads *J. Micromech. Microeng.* **14** pp 365-373

[33] De Pasquale G and Somà A 2010 Dynamic identification of electrostatically actuated MEMS in the frequency domain *Mech. Syst. S. Pr.* **24** pp 13 pp.1621-1633

[34] Somà A, Van der Poel Filho C J, Gugliotta A and Pavanello R 2005 Dynamic identification of MEMS by eigensensitivity and Newmark simulation *Analog Integr. Circ. S.***44** (2) pp 155– 162

[35] Somà A and De Pasquale G 2008 Numerical and experimental comparison of MEMS suspended plates dynamic behavior under squeeze film damping effect" *Analog Integr. Circ. S.* **57** pp 213–224.

[36] De Pasquale G, Veijola T, Somà A 2010 Modelling and validation of air damping in perforated gold and silicon MEMS plates " *J. Micromech. Microeng.* **20** 015010

[37] Saleem M M and Somà A, 2015 Design optimization of RF-MEMS switch considering thermally induced residual stress and process uncertainties *Microelectron. Reliab.* **55** (11) pp 2284-2298

[38] Somà, A, Saleem M. M., B. Margesin and Armando M. 2019 Pull-in tests of MEMS specimens for characterization of elastic–plastic behaviour *Microsyst. Technol.* **22** pp 2525-2533

[39] Somà A and Saleem M M 2019 Elasto-Plastic Characterization of Microstructures through Pull-in 4 points Bending Test *J. Micromech. Microeng.* **29** pp 025004

[40] Somà A and De Pasquale G. 2009 MEMS Mechanical Fatigue: Experimental Results on Gold Microbeams *J. Microelectromech. S* **18** (4) pp.828-835

[41] De Pasquale G and Somà A 2011 MEMS mechanical fatigue: effect of mean stress on gold microbeams *J. Microelectromech. S* **20** (4) pp 1054-1063

[42] Somà A, Saleem M M and De Pasquale G 2016 Effect of creep in RF MEMS static and dynamic behaviour *Microsyst. Technol.* **22** (5) pp 1067-1078

IOP Conf. Series: Materials Science and Engineering **724** (2020) 012052 doi:10.1088/1757-899X/724/1/012052

Effect of deposition oxygen pressure on the properties of cuprous oxide thin films

M Pustan[1], C Birleanu[1], V Merie[2], L Zarbo[3], S Garabagiu[3] and D Marconi[3]

[1] Department of Mechanical Systems Engineering, Technical University of Cluj-Napoca, Cluj-Napoca, 103-105 Muncii Street, 400641, Romania, Marius.Pustan@omt.utcluj.ro

[2] Department of Materials Science and Engineering, Technical University of Cluj-Napoca, Cluj-Napoca, 103-105 Muncii Street, 400641, Romania

[3] National Institute for Research and Development of Isotopic and Molecular Technologies, 67-103 Donath Street, 400293, Cluj-Napoca, Romania

Abstract. This study presents the effect of the oxygen pressure during deposition on the properties of cuprous oxide (Cu_2O) thin films. The Cu_2O is a p-type semiconductor material with a cubic crystallinity structure that has potential applications in solar cells, photo catalysis, gas sensing, superconductor, hydrogen production and thermoelectric generators. Different thicknesses of Cu_2O films and different surface parameters are obtained if the oxygen pressure is modified during the depositions process. The samples were characterized by atomic force microscopy technique (AFM) and the results clearly demonstrate that Cu_2O thin films properties are improved if the oxygen pressure during deposition decreases. The thickness of films and the surface parameters were measured using the AFM non-contact mode. The thickness and the surfaces roughness increase if the oxygen pressure during deposition decreases. This effect is based on obtaining higher particle sizes at low pressures. The modulus of elasticity and the hardness dependence on the deposition oxygen pressure is monitored by nanoindentation using a diamond Berkovich tip. The materials mechanical properties increase significantly if the oxygen pressure during deposition decreases, respectively. The results are useful to micro and nano- systems designers to fabricate reliable structures based on Cu_2O thin films.

1. Introduction

In this study, copper (Cu) metal was used to deposit cuprous oxide (Cu_2O) thin films for thermoelectrical applications. Cu_2O thin films were studied as low-cost power generation because they are nontoxic materials and their deposition process is harmless [1-3]. Cu_2O is typical of the oldest p-type semiconducting material and it is capable of absorbing thermal energy and transform in electricity useful in thermoelectrical generators. The raw materials for making Cu_2O thin films are readily available and the oldest semiconducting materials known to solid-state physicists and using them to fabricate thermoelectrical cells will be low-cost.

The temperature and oxygen flow rate during deposition are the main parameters with influence on the crystalline phase of copper oxide thin films with effect on the material properties [1]. The microstructural and nanomechanical properties of Cu_2O thin films also are modified if the growth temperature is changed during deposition process. The hardness of Cu_2O decreases with increasing the growth temperature [2]. The other important parameter with influence on the properties of Cu_2O thin films is the deposition time [4]. One way to control the thicknesses of thin films is by changing the oxygen pressure during deposition as presented in this paper. The O_2 flow ratio affects the phase of deposited copper oxide thin films and also their surface morphologies [1].

The Cu–O system has two stable oxides as cupric oxide (CuO) and cuprous oxide (Cu_2O). Each of them has different crystal structures and properties. However cuprous oxide has much better crystallinity, bigger grains and direct band gap structure and has being frequently studied as absorber for low cost solar cell [5-8].

First, this paper presents the deposition process of investigated Cu_2O thin films on MgO/Si substrate and the optical images of samples. Secondly, the experimental characterization methodology is described and the obtained results of films thickness, surfaces characteristics and the materials mechanical properties. The nanoindentation technique was used to determine the dependence of mechanical properties of Cu_2O thin films (hardness and modulus of elasticity) on the deposition oxygen pressure. The results and their interpretations are included at the end of paper as well as the conclusions.

2. Samples description

The electrodeposition of Cu_2O on MgO/Si substrate was carried out at different oxygen pressure namely 0.01mbar, 0.05mbar, 0.1mbar and 0.2mbar. The Cu_2O on MgO/Si thin films were deposited by Pulsed Laser Deposition (PLD) technique. The PLD system is equipped with a laser excimer (KrF, λ=248nm, pulse duration of 20ns). The target to substrate distance is set to 55mm. The Cu_2O films were deposited on an MgO buffer layer on Si (100). The MgO buffer layer was deposited from an MgO target in two-step deposition procedure. Laser fluency was set to 2.3J/cm^2 with a frequency of 1Hz. The first step was carried out in vacuum (10^{-6} mbar), using 180 pulses and the substrate temperature was set to 550 °C. In the second step oxygen was introduced into the deposition chamber, at a pressure of 0.1mbar, and deposition was made with 1800 pulses at a substrate temperature of 650 °C. On top of this layer Cu_2O was deposited at a temperature of 700 °C, and various oxygen pressures in the deposition chamber. The laser parameters used for the deposition were: 2.3J/ cm^2 fluency, frequency of 2Hz, and the deposition time were set for 1 hour. The oxygen depositions pressures were 0.01 mbar; 0.05 mbar; 0.1 mbar and 0.2 mbar. The optical images of obtained Cu_2O thin films are presented in figure 1.

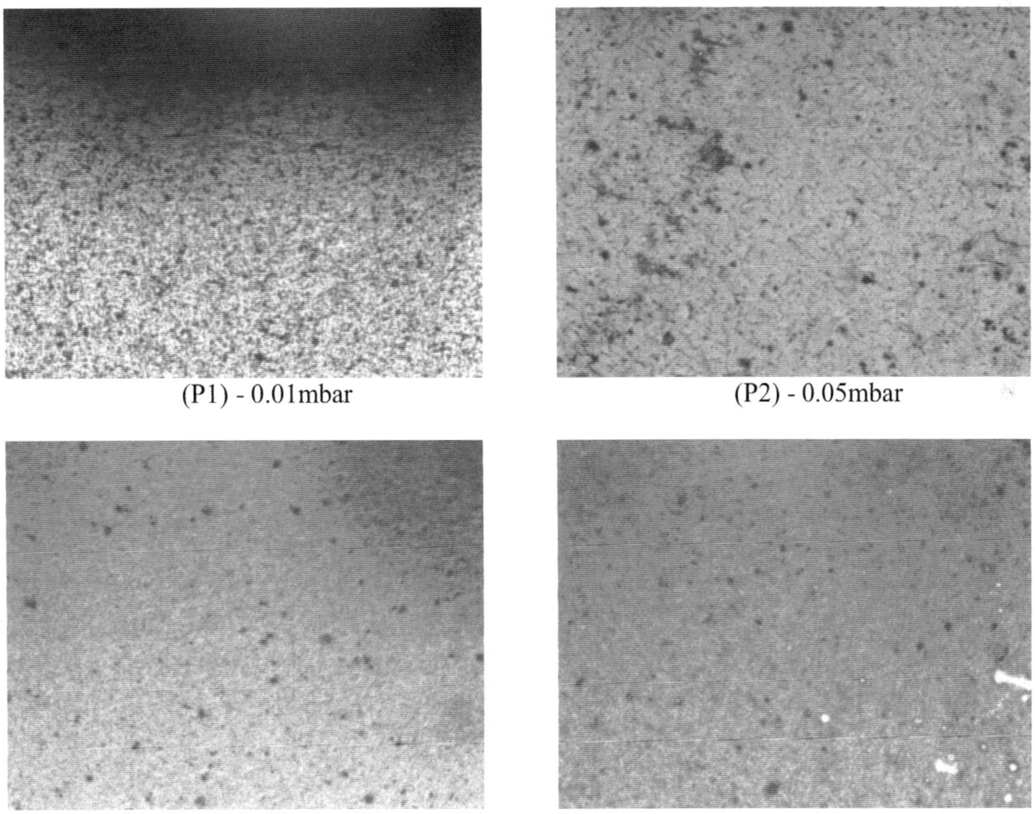

Figure 1. Optical images of deposited Cu_2O thin films.

3. Experimental procedure

The scope of experimental tests is to determine the effect of deposition O_2 pressure on properties of investigated Cu_2O thin films. The tests are performed using an atomic force microscope XE70 with a nanoindentation module. The testing temperature in cleanroom was kept constant at 20°C and the relative humidity was 40%RH. The tapping mode of AFM was used to scan the Cu_2O thin films. After scanning the variation of the roughness parameters as a function of the deposition pressure were determined as well as the films thickness. In order to estimate the film's thickness the AFM tip is positioned at the edge of the deposited thin films. The scanning area includes the investigated thin film and the substrate (figure 2a). The selected scanning area is 5μm × 45μm (figure 2b). The cross-section of the scanning area provides information about the film thickness ΔY(nm), as the difference between the substrate and the bottom surface of films (difference between cursors from figure 3). The measurements were performed in different location of the edges of thin films and the results confirm the film uniformity during the deposition process.

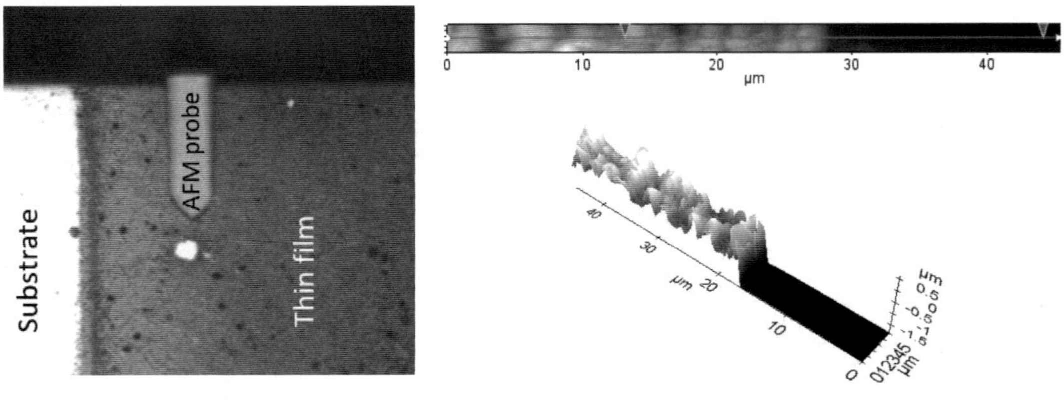

Figure 2. Thickness measurement of Cu_2O thin films: (a) scanning of the border area between the substrate and the deposited film; (b) the selected scanning area (5μm × 45μm) and 3D image.

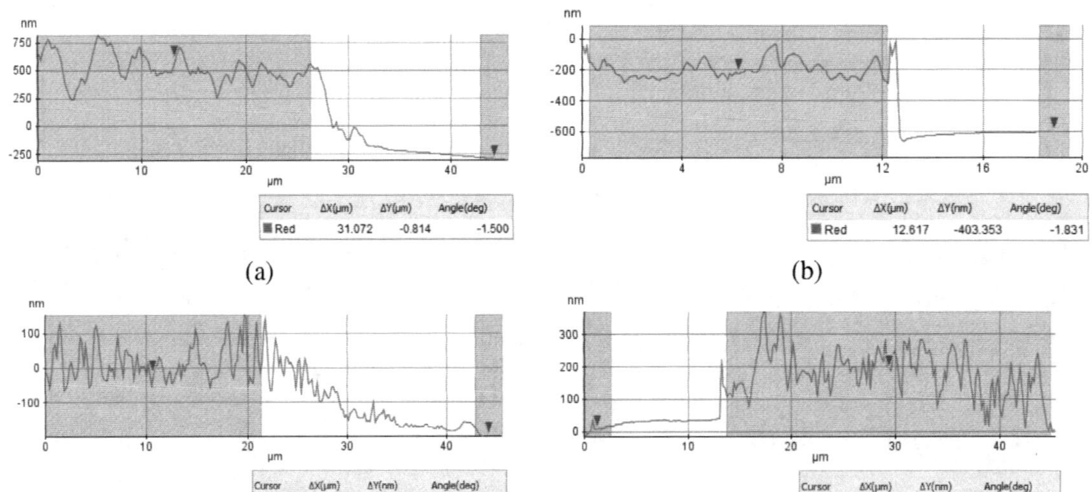

Figure 3. Thickness measurements of $Cu_2O/MgO/Si$ for different oxygen pressure during deposition process: (a) 0.01mbar, (b) 0.05mbar, (c) 0.1mbar, (d) 0.2mbar.

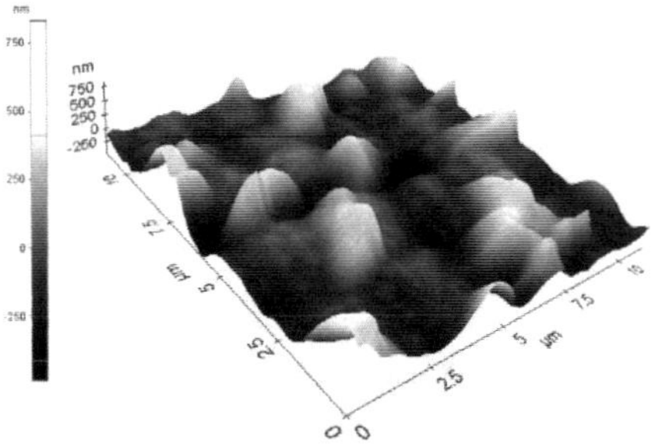

Statistics

Region	Min(nm)	Max(nm)	Mid(nm)	Mean(nm)	Rpv(nm)	Rq(nm)	Ra(nm)	Rz(nm)	Rsk	Rku
Red	-485.221	831.042	172.910	0.000	1316.263	211.402	161.420	1269.857	-1.176	4.386

(a)

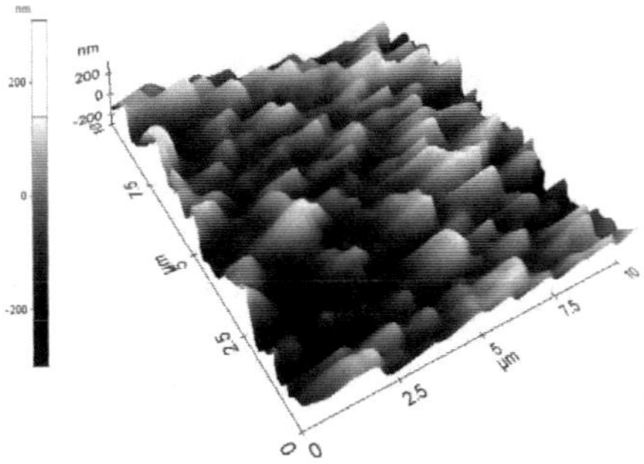

Statistics

Region	Min(nm)	Max(nm)	Mid(nm)	Mean(nm)	Rpv(nm)	Rq(nm)	Ra(nm)	Rz(nm)	Rsk	Rku
Red	-299.290	310.037	5.373	-39.495	609.328	91.430	73.603	557.340	-0.214	2.934

(b)

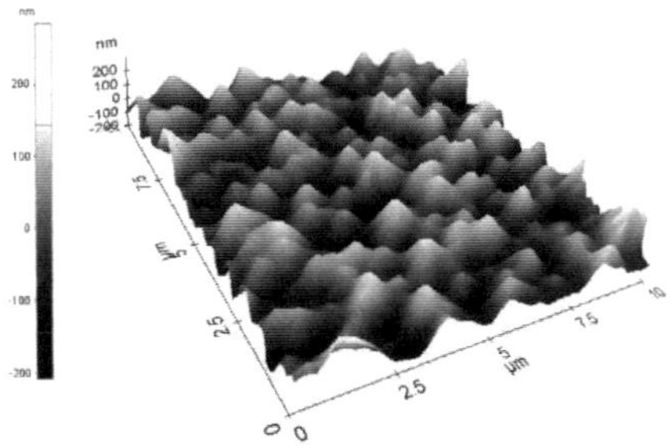

Statistics

Region	Min(nm)	Max(nm)	Mid(nm)	Mean(nm)	Rpv(nm)	Rq(nm)	Ra(nm)	Rz(nm)	Rsk	Rku
Red	-207.324	283.617	38.146	0.000	490.941	73.338	59.650	426.271	-0.310	2.801

(c)

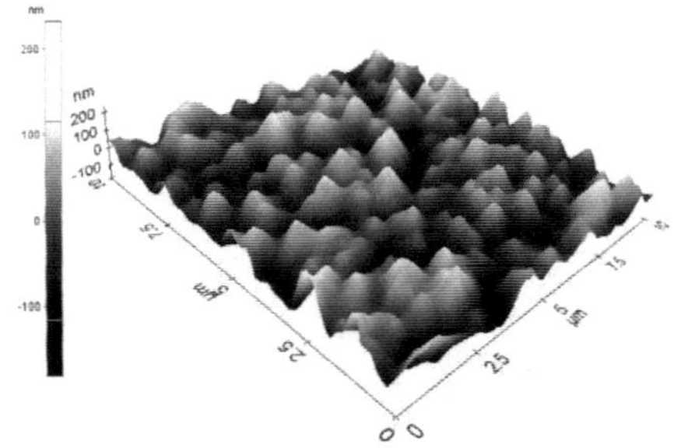

Statistics

Region	Min(nm)	Max(nm)	Mid(nm)	Mean(nm)	Rpv(nm)	Rq(nm)	Ra(nm)	Rz(nm)	Rsk	Rku
Red	-180.785	232.549	25.882	0.000	413.334	59.022	47.576	360.149	-0.357	2.978

(d)

Figure 4. Roughness parameters of investigated thin films deposited at different O_2 pressure: (a) 0.01mbar, (b) 0.05mbar, (c) 0.1mbar, (d) 0.2mbar.

After selection of the scanning surfaces the thickness values of thin films are indicated by the XEI interpretation software. The difference between cursors from figure 3 gives information about the Cu_2O thickness. As the O_2 pressure increases during the deposition process, keeping constants the other deposition parameters, the thickness of thin films decreases, respectively. Using the same tapping mode of the AFM the surfaces parameters of investigated thin films deposited at different O_2 pressure were measured. The 3D images of Cu_2O thin films and the roughness parameters are presented in figure 4.

For instance, the precise determination of modulus of elasticity for samples represents a crucial point for calculating shear stress. To approach the modulus of elasticity the load-displacement curves obtained by nanoindentation have been interpreted using the Oliver & Pharr contact model. The Poisson's ratios of the films in the interpretations on the experiment data are taken as 0.3, since literature has shown that the Poisson's ratio has a minor effect on the indentation results. Using the nanoindentation module of AFM and a Berkovich diamond tip TD23838 with a stiffness of 272N/m, the measurements of hardness and modulus of elasticity were performed for investigated thin films deposited at different oxygen pressure. The set point was 5μN and the loading force 40μN. Figure 5 presents the results of the indentation depth, hardness and modulus of elasticity under applied force in the case of the Cu_2O deposited at 0.2mbar (figure 5a) and in the case of Cu_2O obtained at 0.01mbar (figure 5b).

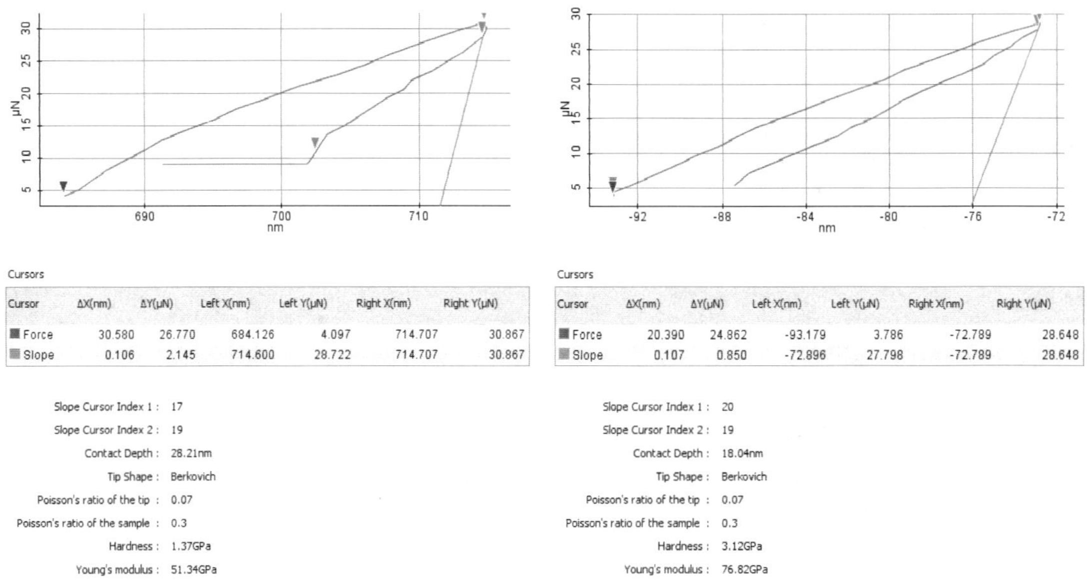

(a) (b)

Figure 5. Nanoindentation of the Cu_2O thin films deposited at different O_2 pressure: (a) 0.2mbar and (b) 0.01mbar.

The indentation was performed in 4 different points for each sample and the average results were considered. Conforming to figure 5, it can be observed that the hardness and modulus of elasticity increase if the oxygen pressure decreases during the deposition process.

4. Results and discussions

By modifying the oxygen pressure during the deposition process, keeping constant the other parameters, the characteristics of Cu_2O thin films can be changed. The experimental tests developed in this paper were directed toward:

- Determination of the effect of the deposition O_2 pressure on the thickness of Cu_2O thin films,
- Investigation of the deposition O_2 pressure on the surface parameters,
- Measurement of the Cu_2O thin films properties (hardness and modulus of elasticity).

The tests were done using the AFM technique. The tapping mode was applied to scan the thin films in order to measure the thickness and the roughness parameters. After, the nanoindentation mode of AFM was used to determine the materials mechanical properties. Table 1 presents the experimental results obtained for the investigated thin films. It can be observed that, as the oxygen pressure decreases toward vacuum, the thickness of thin films increases as well as the R_a roughness. Moreover, we can affirm that the modulus of elasticity and hardness are improved if the oxygen pressure decreases during the deposition process.

Table 1. Experimental results of Cu_2O thin films as a function of the deposition oxygen pressure.

Deposition pressure [mbar]	Investigated parameters*			
	h [nm]	R_a [nm]	E [GPa]	H [GPa]
0.2	171.1	47.57	51	1.47
0.1	211.7	59.65	59.6	1.84
0.05	403.3	73.60	64.3	2.45
0.01	814	161.42	76	3.59

* h - thickness, R_a - average roughness, E - modulus of elasticity, H - hardness.

5. Conclusions

The effect of oxygen pressure during deposition on the material properties and surface characteristics of Cu_2O thin films was investigated and presented in this paper. This material can be implemented in thermoelectrical application. The scope of research work was to identify the influence of the deposition parameter on the characteristics of thin films in order to obtain reliable systems. After deposition process the samples were experimentally investigated using the atomic force microscopy technique.

The modification of oxygen pressure during deposition process change the properties of Cu_2O thin films. Moreover, as the deposition pressure decreases, the thickness and the roughness increase. This effect is based on obtaining higher particle sizes at low pressures. The hardness and modulus of elasticity increase as the deposition pressure decreases, respectively.

Acknowledgments
This work was supported by a grant founded by the Romanian Space Agency, STAR Program C3, project no. 193/2017.

References

[1] Pan J, Yang C and Gao Y 2016 *Sens. Mater.* **28** 817
[2] Jian S R, Chen G J and Hsu W M 2013 *Mater.* **6** 4505
[3] Hartung D, Gather F, Hering P, Kandzia C, Reppin D, Polity A, Meyer B K and Klar P J 2015 *Appl. Phys. Lett.* **106** 1
[4] Hanif A S M, Azmal S A, Ahmad M K and Mohamad F 2015 *Appl. Mech. Mater.* **773-774** 677-681
[5] Mingxing J, Peibo C and Kegao L 2017 Research Status and Development of Cu_2O Thin Film Solar Cells *Proc. Asia-Pacific Eng. Tech. Conf. (Malaysia)* (DEStech Publications, Inc) 2447
[6] Zhiming B and Yinghua Z 2017 *J. Alloy Compd.* **698** 133
[7] Paracchino A, Brauer J C and Moser J E 2012 *J. Phys. Chem. C* **116** 7341
[8] Olsen L C, Addis F W and Miller W 1982 *Sol. Cells* **7** 247

A study for an aluminum electro-thermally actuated U-shaped microtweezer

R-C Voicu[1], M Pustan[2], C Birleanu[2], R Gavrila[1], C Romanitan[1], O Tutunaru[1], A Cismaru[1]

[1] National Institute for Research and Development in Microtechnologies-IMT Bucharest, Erou Iancu Nicolae str., 126A, 077190, Romania
[2] Technical University of Cluj-Napoca, Faculty of Building Machine, 103-105, Bd. Muncii, 400641, Cluj-Napoca, Romania
rodica.voicu@imt.ro

Abstract. In this paper we present the investigations of two aluminium layers with different thicknesses in order to be used for manufacturing of MEMS and the design, simulation, fabrication and characterizations of a microtweezer designated for micromanipulation applications. Electro-thermo-mechanical finite-element simulations were performed in order to describe the behaviour of MEMS devices. The microtweezers have been fabricated in aluminium, as structural material, by surface micromachining processes. Different characterizations of the structural material by Atomic Force Microscope (AFM), X-ray Diffraction System (XRD) and Scanning Electron Microscopy (SEM) characterization tools were presented.

1. Introduction

In the last decades the micro-electro-mechanical systems (MEMS) production increased a lot. Silicon, its derivatives and metals have been extensively used as structural materials with high mechanical properties [1-3]. High thermal conductive metals with low costs such as aluminium and copper have been used widely to manufacture micro heaters or MEMS devices [3-4]. Aluminum films and Al alloys are widely used for electronic device applications and include the majority of the interconnections used in the semiconductor chips [5].

Microtweezers, as MEMS devices, provide advantages in terms of compact size and low cost having an extensive role in microassembly and micromanipulation fields for manipulating micromechanical or bio components [6-8]. Different actuation principles, such as shape memory alloys, electrostatic, electro-thermal, electro-magnetic, piezoelectric, pneumatic, were proposed to drive MEMS microtweezers. Electro-thermal microactuators are usual components in microsystems and can be power-driven electrically through Joule heating or optically with a laser.

MEMS devices offer great prospective for vibrational or micromanipulation applications due to their remarkable features like small size, low power consumption, large frequency-quality factor product, low cost production, etc [9]. Investigations and controlling of different uncertainties, occurring in the fabrication process, the geometric and environmental/processing parameters effects on MEMS devices is of significant importance for increasing the production yields and reducing the fabrication costs of different micromachined sensors and actuators [10-12].

In this work we investigated two aluminium layer with different thicknesses in order to be used as structural materials for the fabrication of a microtweezer based on electro-thermal actuators. A design, simulation and the fabrication process of the microtweezer were presented. Different characterizations of the structural materials and of the fabricated microtweezer were realized.

2. Materials and Characterizations

In this paper two single layers of aluminium were deposited on (100) oriented 4" silicon wafers using E-beam high vacuum thin film deposition system. The layers were deposited with a thickness of 1 µm and with 2 µm, respectively. The physical and tribological properties of the Al films usually depend intensely on their microstructure, which can be characterized using different techniques.

Aluminium layers obtained with different thickness on silicon substrates were analysed by SEM, atomic force microscopy and the X-ray techniques.

The deposited layers were characterized using SEM by measuring the thickness for each layer and analysing the top surface of the aluminium (Figure 1). The top surface view show some defects of the aluminium layer surface (Figure 1b) due to the deposition process or after processing the structures.

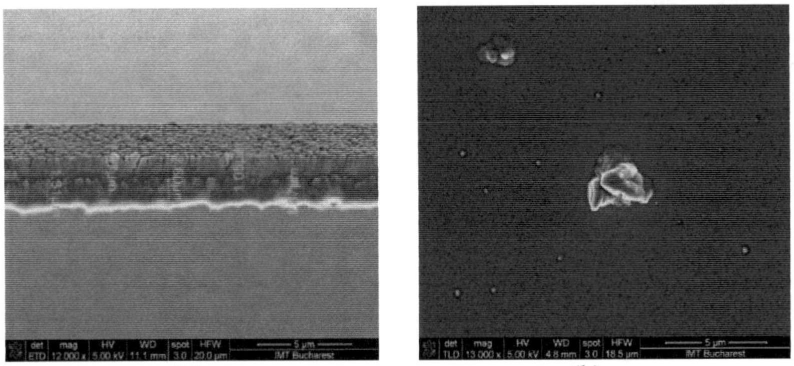

(a) (b)

Figure 1. SEM images of the Al layers after processing: (a) measurements of the thickness for the 2 µm layer; (b) top view of the Al surface with1 µm thickness.

Corresponding characterization of the surface topography were performed using AFM. The surface morphology of the aluminum layers was characterized by an Ntegra Aura Scanning Probe Microscope (NT_MDT Spectrum Instruments) operated in intermittent-contact mode. AFM probes with 9 N/m spring constant and 10 nm nominal radius were used in the measurements. The scanned areas were 7 µm x 7 µm for different location of the samples (Figure 2).

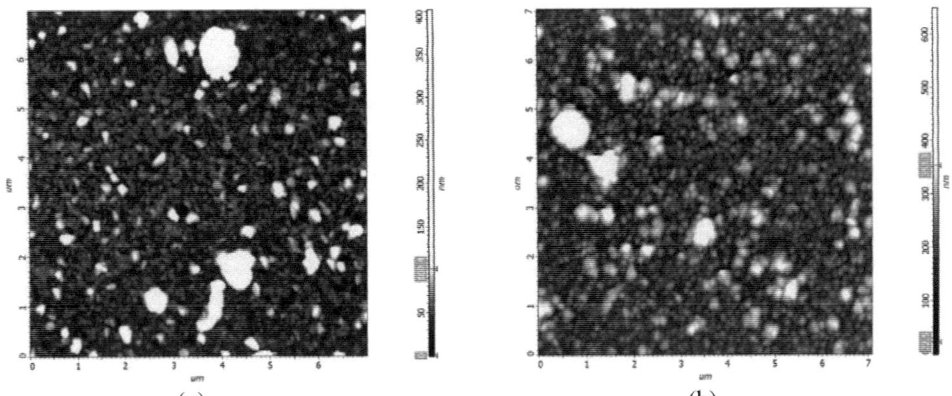

(a) (b)

Figure 2. AFM 2D images for aluminum layers with a thickness of: (a) 1 µm; (b) 2 µm (scanned size 7 µm x 7 µm).

We analized the root mean square (Sq) for each sample. The root mean square increase while the thickness of the layer increase. The root mean square for the aluminum layer with 1 μm thickness is Sq = 25.6 nm, while for the aluminum layer with 2 μm thickness Sq = 57.9 nm.

X-ray diffraction measurements were carried out using 9kW Rigaku SmartLab diffractometer, equipped with CuKα1 source, that provides monochromatic X-rays with wavelength 0.1506 nm. In order to obtain information from layer, it was employed GI-XRD (Grazing Incidence X-ray diffraction) measurements. In this configuration, the source was kept at 0.5°, while the detector position moves from 20 to 95°. The obtained pattern indicates multiple diffraction peaks, such as (111), (200), (220), (331) and (222). According to ICDD database, card number 230-0250, the obtained peaks correspond to aluminium. To gain a deeper understanding of the aluminium microstructure, the experimental X-ray patterns (black curve) were refined using Rietveld procedure (red curves). After fitting, we conclude, on the one hand, that the lattice strain is very small in both cases (e.g. 0.06% for 1 μm and 0.09% for 2 μm, respectively) and on the other hand, the mean crystallite size decreases from 26.1 to 24.7 nm. A decreasing of the mean crystallite size is ascribed with a higher dislocation density and a worse crystallinity. In this context, a different layer thickness is related to different both lattice strain and mean crystallite size.

As the film thickness increases, the (111) peak intensity increases in comparison with the (200) and (220) peaks (Figure 3).

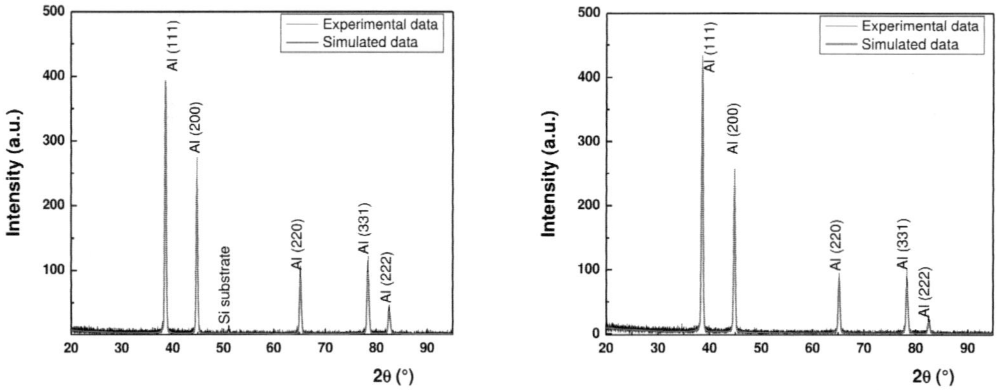

(a) (b)

Figure 3. (a) Difraction spectra XRD measurements for 1 μm aluminium layer; (b) Difraction spectra XRD measurements for 2 μm aluminium layer.

3. Design and Simulation

As a test structure, a microtweezer was designed based on the U-shape electro-thermal actuators in order to be fabricated using aluminium as structural material with a 2 μm thickness and to study his behaviour in air after actuation. The U-shaped actuator is formed by a traditional pair of a hot arm and a cold arm. The cold arm were designed with holes in order to be easily release over the silicon substrate (Figure 4). The hot arms have a width of 5 μm and the total length of the structure is 580 μm. The initial opening of the free arms is 5 μm. During the actuation the tips of the microtweezer will open in order to grip a micro-object.

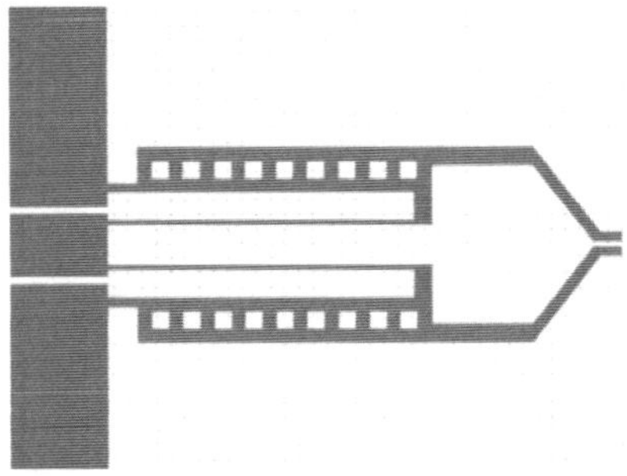

Figure 4. The Al microtweezer design (Coventorware modelling).

The electro-thermo-mechanical behaviors of the microtweezer in air have been 4hermos4 using the FEM simulations of the software tool CoventorWare. Electro-thermomechanical coupled simulations were performed in order to analyse the temperature and displacement values during the actuation. The boundary conditions of the simulations used for experimental thermal actuator were the following: the initial temperature of the whole structure and the temperature of the environment were considered to be T_0=20°C; the air convection coefficient was set to 3000 W/m^{2}K. For the material properties settings of the Aluminum a Young's modulus of 77 Gpa, a Poisson's ratio of 0.3, a thermal coefficient of expansion of 2.3×10-5 K^{-1} and a thermal conductivity depends on the temperature values were set [13]. The MemMech solver which is dedicated for mechanical and 4hermos-mechanical analyses was used for simulations. The simulated 3D microtweezer model has been meshed using hexahedral elements (Extruded bricks), with 27 nodes/element, which generate accurate results for all solvers of Coventorware program.

Due to the Joule effect the hot less wide arms will be heated and the material will expand then an in plane displacement occurred for free tips. The simulations results show that the structure can operate for an electrical current up to 250 mA and the temperatures vary from 20 to 550°C (Figures 5-6). Considering that the melting point of aluminum is 660°C, the microtweezer can operate with success for a higher electrical current in order to obtain the desired opening. The simulation results show that an opening of 55 μm can be obtained for 197 mA and a maximum temperature in the hot arm of 525 °C.

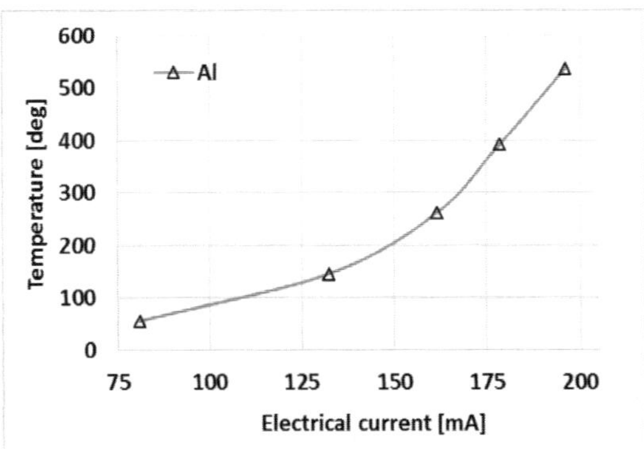

(a)

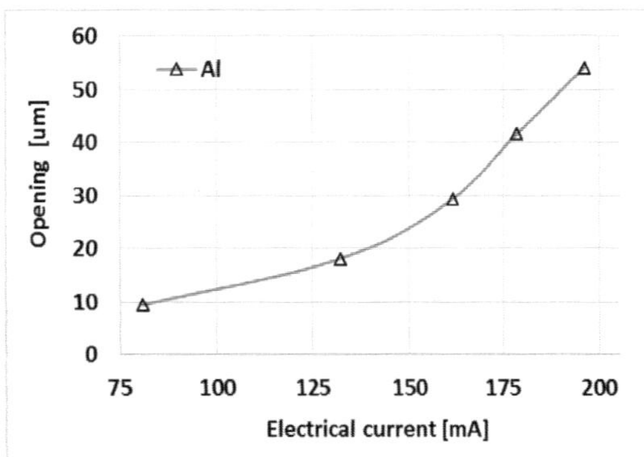

(b)

Figure 5. Simulation results: (a) The maximum temperatures in the hot arms; (b) The opening of the microtweezer tips (Coventorware simulation).

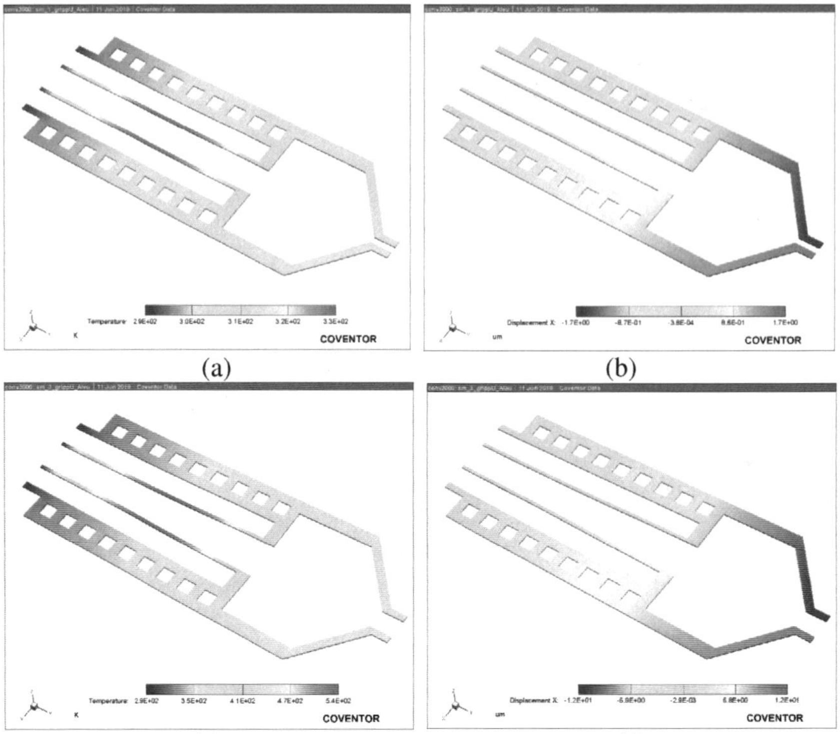

(a) (b) (c) (d)

Figure 6. Simulation results: (a) Temperature distribution for 80 mA; (b) In-plane displacement for 80 mA; (c) Temperature distribution for 161 mA; (d) In-plane displacement for 161 mA (Coventorware simulation).

4. Fabrication and characterization

Aluminum was chosen for manufacturing also due to its high electric conductivity and mechanical properties. The 2 μm aluminum layer has been deposited directly on the silicon substrate. The microtweezer configuration was obtained using a photolithographic mask. A wet isotropic etching of the aluminum was realized in order to obtain the chosen layout. The release of the arms has been done using dry etching (RIE) of the silicone substrate. Isotropic etching of silicon was performed on a RIE system (Sentech Instruments GmbH, Germany), using SF6 gas. A SEM characterization shows that the lateral etch rate was less than the vertical etch rate (Figure 7).

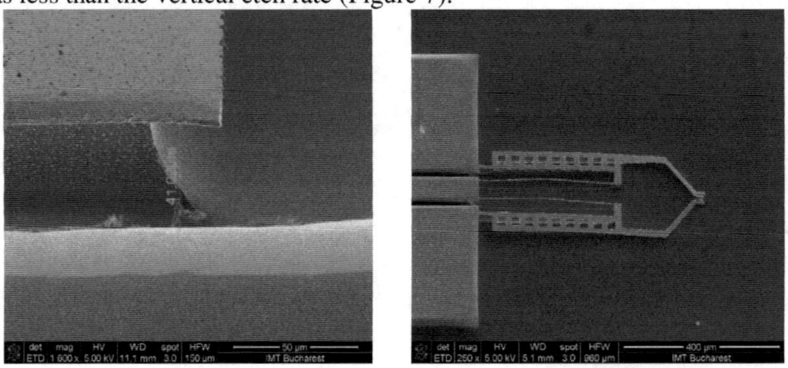

(a) (b)

Figure 7. SEM images of the released microtweezer: (a) detail with the Al pad; (b) the Al released structure.

The gap obtained after releasing was around 41 μm (Figure 7a) and a stress in the whole structure were observed (Figure 7b).

5. Conclusions

Atomic Force Microscope (AFM), X-ray Diffraction System (XRD) and Scanning Electron Microscopy (SEM) characterizations of the aluminium structural materials with 1 μm and 2 μm thicknesses were presented.

A microtweezer design based U-shaped actuators was proposed as test structure. The simulations, fabrication and characterization of the structure were realised, using 2 μm aluminum thickness. The simulation results show an opening of the microtweezer tips of 55 μm for 197 mA and a maximum temperature in the hot arm of 525 °C.

In order to improve the device performance, a further thermal treatment will be performed for the aluminum structural layer in order to reduce the stress and a careful settings of the material deposition parameters will be analysed and improved to reduce the defects of the Al surface and to reduce the stress also.

Acknowledgments

This research was supported by the National Core – Program MICRO-NANO-SIS PLUS, Contract No 14N/2019 financed by the Romanian Ministry of Research and Innovation.

References

[1] Lockwood A J et al 2012 Deformation behaviour of polysilicon components for MEMS *J. Micromech. Microeng.* **22** 105016
[2] Lofroth M and Avci E 2019 Development of a Novel Modular Compliant Gripper for Manipulation of Micro Objects *Micromachines* **10** (5):313 DOI: 10.3390/mi10050313
[3] Xiao G Q et al 2010 Fabrication of MEMS components using ultra fine grained aluminium alloys *J. Micromech. Microeng.* **20** 045029 (9pp)
[4] Baracu A, Voicu R, Müller R, Avram A, Pustan M, Chiorean R, Birleanu C and Dudescu C 2015 Design and fabrication of a MEMS chevron-type thermal actuator *AIP Conference Proceedings* 1646, 25 doi: 10.1063/1.4908578
[5] Quintana P, Oliva A I, Ceh O and Corona J E 1999 Thickness effects on aluminum thin films *Superficies y Vacío* 9, 280-282
[6] Belfiore N P (Ed.) 2018 Micromanipulation, *Actuators 7 12, 27, 38* ISBN 978-3-03897-503-8 (Pbk); ISBN 978-3-03897-504-5 (PDF), https://doi.org/10.3390/books978-3-03897-504-5
[7] Yukun J and Qingsong X 2013 MEMS Microgripper Actuators and Sensors: The State-of-the-Art Survey *Recent Patents on Mechanical Engineering* **6** 2 DOI: 10.2174/2212797611306020005
[8] Wang Z et al 2015 Design, modelling, and characterization of a MEMS electrothermal microgripper, *Microsystem Technologies* **21** 11, 2307-14
[9] Basu J and Bhattacharyya T K 2013 Microelectromechanical system cantilever-based frequency doublers *J. Intell. Mater. Syst. Struct.* **24**, 240-246
[10] Pustan M, Müller R and Golinval J C 2012 Nanomechanical and nanotribological characterization of microelectromechanical system, *J. Optoelectron. Adv. Mater.* **14**, No. 3-4, 401 – 412
[11] Voicu R, Rymuza Z, Michalowski M, Obreja C, Gavrila R and Müller R 2013 Investigations of surface properties of SiO_2 and Si_3N_4 thin layers, used for MEMS vibrating structures applications, *IEEE Proc. CAS 2013* Vol. 1 14-6 October 2013 Sinaia, Romania p 111
[12] Voicu R-C, Gavrila R, Obreja A C, Baracu A, Dinescu A and Muller R 2015 Design, microfabrication and analysis of polysilicon thin layers for MEMS vibrating structures, *Analog Integr. Circuits Process.* **82**, 611–620
[13] CoventorWare 2014 Documentation and Material Database

IOP Conf. Series: Materials Science and Engineering **724** (2020) 012054 doi:10.1088/1757-899X/724/1/012054

Nanoscale friction and wearing of octadecyltrichlorosilane covered silicon surface

S Teodoroff-Onesim, A Besleaga, L Sirghi[1]

Plasma Advanced Research Center (IPARC), Faculty of Physics, Alexandru Ioan Cuza University of Iasi, Blvd. Carol I nr. 11, Iasi 700506, Romania
[1]E-mail: lsirghi@uaic.ro

Abstract. Silicon wafers were silanized by plasma hydroxylation and chemical vapour deposition of octadecyltrichlorosilane [OTS, $CH_3(CH_2)_{17}SiCl_3$]. Then, the OTS covered silicon samples were used in atomic force microscopy (AFM) measurements of wearing and friction at nanoscale. Wearing experiment performed by 1000 cycles of forwards and backwards movements of sharp AFM tip revealed formation of a wearing track with a depth that was much smaller than the thickness of OTS coating, which confirmed a very good wearing property of OTS coating. Friction versus normal loading force experiments performed with the same AFM tip revealed a nonlinear dependence of friction force on the normal loading force in agreement with prediction of Bowden and Tabor's theory of friction and an increase of friction with the moving speed. Dependence of friction force on the sliding speed is attributed to a transfer of kinetic energy from the moving AFM tip to individual molecules of OTS coating.

1. Introduction

In their seminal work published in 1954, Bowden and Tabor [1] pointed towards the microscopic foundations of friction. They treated friction as result of elastic deformation of interlocking spherical asperities with the result that the macroscopic friction force is proportional to the total contact area of individual asperities on surfaces of the macroscopic bodies. Increase of the normal loading force determines increase of the real contact area of surface asperities and, therefore, this theory predicts a nonlinear dependence of friction force on normal loading force, in contradiction with Amontons' first law of friction. The experimental investigation of microscopic nature of friction has been facilitated by the inventions of surface force apparatus [2] and atomic force microscope (AFM) [3] and such studies laid foundations of nanotribology [4]. The development in the early 1990's of the AFM technique of friction force microscopy (FFM) allowed many studies of friction at level of single asperity contacts, since the nanoscale sharp tip of the AFM probe is considered as physical model of a single asperity. However, besides asperity interlock origin, friction may result from a mixture of complex processes as plastic deformations, material plough, local welding, molecular interactions between surfaces in relative motion, or by viscous movement of surface adsorbed molecules [5].

In this paper wearing and friction at level of single-asperity contact between octadecyltrichlorosilane (OTS) covered silicon wafers and a sharp silicon AFM tip is studied by atomic force microscopy techniques. Deposition of octadecyltrichlorosilane (OTS) molecular layers is commonly used in electronic industry and in various applications that require modification of silica surfaces as in biosensors, micro electro mechanical systems (MEMS) and actuators [6]. Most MEMS components are manufactured from silicon at nanometer or micrometer sizes [7]. The smallness of these components and gaps formed by their surfaces determines undesired high surface forces [8]. Use of common lubricants in MEMS is impossible because of capillary forces. Therefore, use of self assembled monomolecular coatings (SAM) of silane molecules is an important method for reducing capillary and friction forces in MEMS. In this context, covering with OTS SAMs confers to silica surfaces a hydrophobic character with an important reduction of adhesion and friction forces between moving parts of MEMS [9]. Macroscopic tribology measurements in ball-plane geometry on OTS SAM revealed relatively low friction coefficient at small normal loading force, increase of friction with the sliding velocity and poor wearing properties at relatively large normal loading forces [10]. These measurements may be affected by the poor quality of the OTS SAM, the experiments showing transfer of OTS material from plane substrate to the ball probe of the trybometer. However, the trybologic properties at nanoscale of OTS coatings have not been investigated so far. In the present study we show that OTS coated silicon have good wearing property at nanoscale, the wearing track formed after one thousands wearing cycles having a depth much smaller than the thickness of OTS coating. Also, it was notice a decrease of friction force during first cycles of wearing. The friction versus normal loading force measurements revealed a nonlinear dependence that confirms proportionality relationship between friction force and real contact surface area. Also, the molecular origin of friction force is proved by dependence of friction on the sliding speed, a noticeable increase of friction with the sliding speed being observed.

2. Materials and methods

2.1. Materials

Octadecyltrichlorosilane ($CH_3(CH_2)_{17}SiCl_3$, OTS, > 90%) and ethanol (> 99.8%) were purchased from Sigma-Aldrich and used as received. Distillate water was freshly purified and deionised using an Ultrapure Academic Milli-Q system. Single side polished N-type silicon <100> wafers (0.5 mm in thickness) were procured from Sigma-Aldrich. Silicon AFM probes HQ:NSC35/No Al were purchased from Micromasch.

2.2. Deposition of OTS molecular layers

Before OTS depositions, the silicon substrates (1 cm × 1 cm) were cleaned by sonication in pure ethanol and distillate water to remove large dust particles from surfaces. Then, the substrates were loaded on the chamber of plasma and chemical vapour deposition reactor. Depositions were made in two steps: 1) cleaning and chemical activation of silicon surfaces by hydroxylation under the action of discharge plasma and 2) exposure of activated surface to OTS vapour. In first step, the silicon substrates were loaded on small stainless steel cathode and the chamber was vacuumed down to 10^{-2} Torr for about 30 minutes. Activation of silicon surface was done by exposing the substrate for 2 minutes to the negative glow plasma of a d.c. luminescent discharge (voltage around 380V and current intensity around 5 mA) in air and water vapour mixture at low pressure (0.3 Torr). In the second step, OTS coatings were fabricated by chemical vapour deposition (CVD) at room temperature in the same reactor chamber by exposing the plasma-activated silica surfaces to OTS vapour for 24 hours. Then, the OTS coated silica substrates were baked in atmospheric air at 100°C for another 24 hours. Then, the probes were used for AFM friction and wearing measurements. Figure 1 shows an AFM topography image of OTS covered silicon surface and the probability distribution of surface height values. The topography image shows a very smooth OTS coating, which can be attributed to a good packing of OTS molecules in a SAM. The probability distribution has a Gaussian shape with a full width at half maximum (FWHM) of 0.19 nm. This computes for the OTS coating a root mean square (RMS) roughness value of 0.16 nm if it is considered that RMS value of noise in the AFM height data of 0.1 nm

2.3. Atomic Force Microscopy wearing and friction measurements

AFM force wearing and friction measurements were performed by a commercial AFM apparatus (XE 100 from Park Systems, South Korea) with silicon AFM probes (HQ:NSC35/No Al from Micromasch). The force constant of the AFM probes were determined by thermal noise method and the curvature radius of the AFM tip was determined by analysis of high-resolution images of the sharp edges of a standard grating probe (TGG1 from NT-MDT, Russia). The topography images of the OTS modified surfaces were obtained by AFM scans in tapping mode using the same silicon AFM probe. The friction force was computed by analysis of lateral force loops recorded during lateral forwards and backwards movements of the AFM tip (see Fig. 2 for an example). The friction force versus normal load curves were obtained by scanning an area of 2 μm × 2 μm of OTS surface with different speed and normal force values.

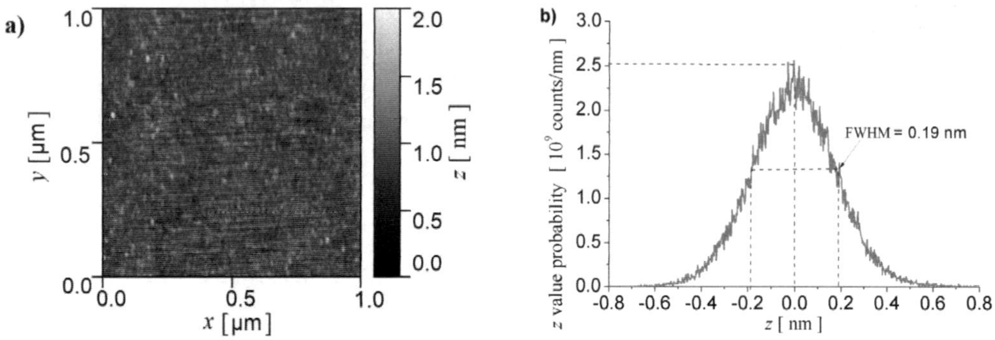

Figure 1. a) AFM topography image (1 μm×1μm) of OTS coated silicon and b) probability of surface height (z) values in the topography image showing a FWHM value of 0.19 nm.

Figure 2. Example of a friction loop, i. e. lateral force versus linear displacement of the AFM tip during forward and backward movements along a path with length of 3 μm. Due to friction, the lateral force is displaced with, F_f, above the null lateral force level during forward movement and with $-F_f$ below the null lateral force level during backwards movement. The area enclosed by friction loop determines the mean value of friction force.

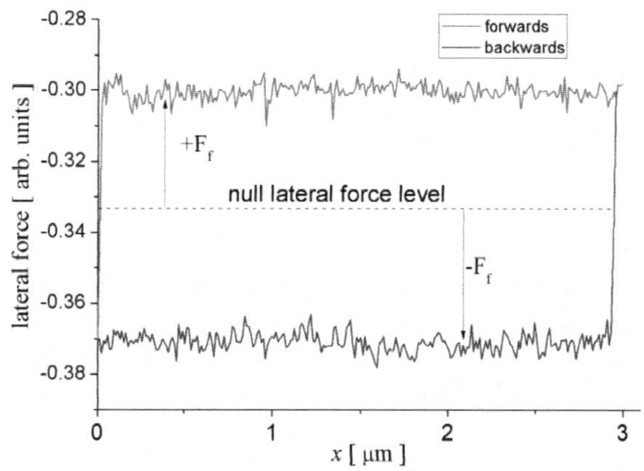

3. Results and discussion

3.1. Wearing experiment

Figure 3 shows the wearing track formed on the surface of OTS coating after 1000 wearing cycles. This experiment was performed by moving a sharp silicon AFM tip (curvature radius of 17 nm) under a normal load of 50 nN forwards and backwards in 1000 cycles along a path with length of one micron. The large width of wearing track (estimated to about 100 nm from the height profile of the wearing track in Figure 3 b)) is due partly to a drift in *y* position of the AFM tip during wearing measurements and partly to tip-sample convolution effects. Estimation of the tip-sample contact area to about 100 nm^2 indicates a compressive stress applied to OTS surface during wearing of 0.5 GPa. The depth value of 0.8 nm of the wearing track is much smaller than the thickness of OTS coating (estimated to about 2.4 nm for OTS SAM). This means that the wearing track is formed by alignment and bending of OTS molecules along the wearing track without their removal from the substrate. This is also indicated by variation of friction force during wearing cycles depicted in Figure 4, which shows decrease of the friction force during first 200 cycles and almost a constant value afterwards. In spite of formation of the wearing track and corresponding increase of tip-sample contact area, the friction force decreased noticeably during first wearing cycles because the loss of energy in interactions between OTS molecules and AFM decreased during formation of the wearing track.

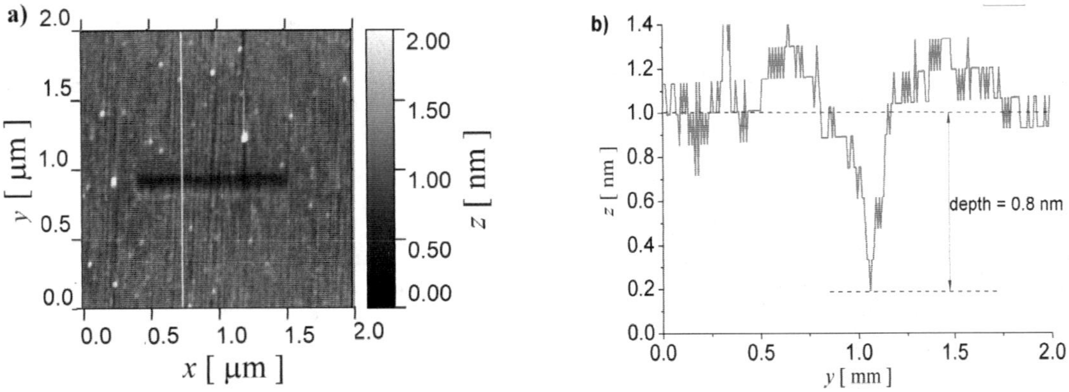

Figure 3. a) AFM topography image of wearing track formed during 1000 cycles of forwards and backwards sliding movements of an AFM tip under a normal loading force of 50 nN along a path with length of 1 μm. b) Height profile of the wearing track along the scanning line showed in the AFM topography image.

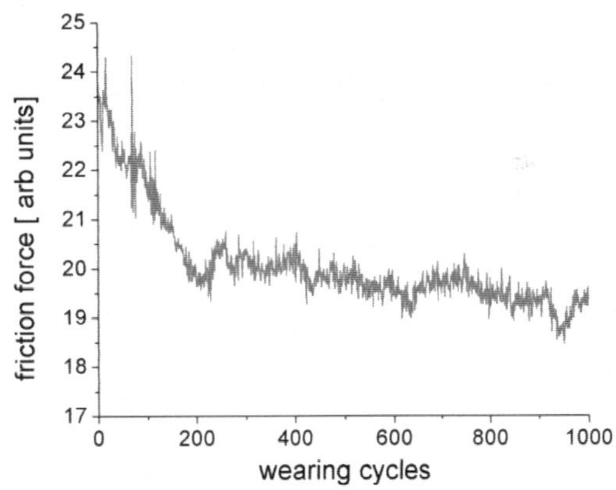

Figure 4. Variation of friction force during 1000 cycles of forwards and backwards sliding movements of an AFM tip under normal loading force of 50 nN along a path with length of 1 μm.

3.2. Friction experiment

Figure 5 shows friction versus normal force curves recorded at different sliding speed values. The curves shows a nonlinear dependence of friction force on normal loading force, the friction force decreasing to zero while the normal force is negative (pulling) with absolute value close to tip-sample adhesive force, F_{ad}, which in this experiment was around 17 nN. When the normal pulling force is equal to F_{ad}, the AFM tip detaches from the surface and friction and normal forces jumped to zero. The continuous black lines on the graph are fits of the experimental data with prediction of Bowden and Tabor's theory of friction, which for elastic deformations of surfaces foresee a dependence of friction force to 2/3 power of normal force ($F_f \alpha N^{2/3}$). The relative good agreement of the experimental data with theory predictions confirms that the deformation of OTS coating is elastic. It should be mentioned that the normal force is sum of externally applied force (through the AFM cantilever) and tip-sample adhesion force, F_{ad}. Dependence of friction on sliding speed reveals the molecular origin of friction in this particular case. In the process of friction, the AFM compresses the molecular layer of OTS molecules, each molecule being deformed elastically and released by sliding movement of the tip. In this process the tip transfers also kinetic energy to molecules by collisions and this kinetic energy increases with the tip sliding speed.

4. Conclusion

In summary, high quality OTS coatings were obtained on silicon substrates by plasma activation (hydroxylation) of substrate surfaces and chemical vapour deposition. The OTS covered surfaces were used in AFM wearing and friction measurements at level of a single-asperity contact.

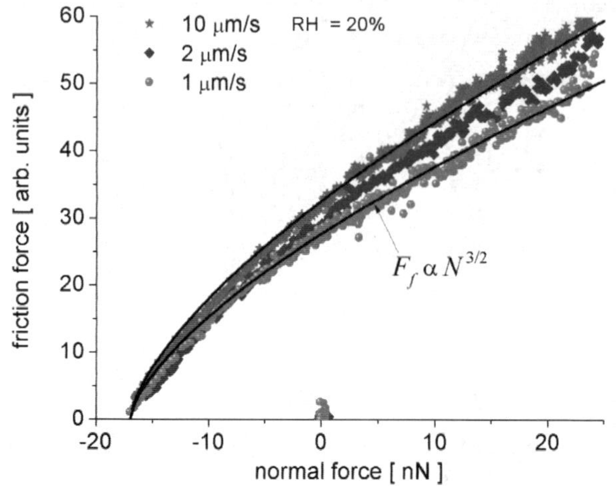

Figure 5. Dependence of friction force on normal loading force at different values of sliding speed

Wearing experiment performed by 1000 cycles of forwards and backwards movements of sharp AFM tip (curvature radius of 17 nm) under a normal applied force of 50 nN revealed formation of a wearing track with a depth that was much smaller than the thickness of OTS coating. This indicated a very good wearing property of OTS coating. The AFM friction investigation revealed a good agreement of experimental data with prediction of Bowden and Tabor's theory of friction, which confirms that friction rise from elastic deformation of the layer of OTS molecules. Dependence of friction force on the sliding speed revealed also a kinetic mechanism of friction, each molecule being collided by the moving tip with a transfer of kinetic energy.

5. References

[1] Bowden F P and Tabor D 1954 *The friction and lubrication of solids* (Oxford, Clarendon, UK).
[2] Tabor D, Winterton R H S 1969 *Proceedings of the Royal Society A: Mathematical, Physical and Engineering Sciences* **312** (1511) 435–450.
[3] Binnig G, Quate C F, Gerber C 1986 *Physical Review Letters* **56** (9) 930–933.
[4] Mate C M *2008Tribology on the Small Scale: A Bottom Up Approach to Friction, Lubrication, and Wear* (Oxford: Oxford University Press).
[5] Wang H, Zhang T, Hu Y 2004 *Science & China Ser. O Physics, Mechanics & Astronomy* **47** Supp. 8—14.
[6] Azadi M, Nguyen A V, Yakubov G E 2015 *Langmuir* **31**(6) 1941-1949.
[7] Kompovopoulos K 1996 *Wear* **200** 305–327.
[8] Maboudian R and Hove R T 1997 *J. Vac. Sci. Technol.* B15 1–20.
[9] Grigoriev A Ya, Kovaleva I N and Myshkin N K 2008 *Journal of Friction and Wear* **29** 434–440.
[10] Ren S, Yang Sh, Zhao Y, Zhou J, Xu T and Liu W 2002 *Tribology Letters* **13** 233.

Research Concerning the End-effectors for SiMFlex Microgripper

S Noveanu[1,2], C Rusu[1,2], C Rancea[1,3] and D Lateş[4]

[1] Technical University of Cluj-Napoca, 28, Memorandumului Street, 400114, Cluj-Napoca, Romania
[2] Department of Mechatronics and Machine Dynamics, 103-105, Muncii Avenue, 400641, Cluj-Napoca, Romania
[3] Department for Research, Development and Innovation Management, 15, Daicoviciu Street, 400020, Cluj-Napoca, Romania
[4] University of Medicine, Pharmacy, Science and Technology of Târgu Mureş, 38, Nicolae Iorga Street, 540139, Târgu Mureş, Romania

Abstract. In this paper the SiMFlex microgripper are designed and modelled. The result for FEM analysis for SiMFlex was presented. The prototype for SiMFlex microgripper is realized with non-conventional technologies and the design included flexure hinges and piezoelectric actuators. A study for different surface qualities for end-effectors is presented and used two materials for active parts of jaws manufactured. Simulations and experimental part were generally performed on the system and special attention was given to jaws that were illustrated and clarified by these studies. Based on the results obtained the optimization solutions for end-effectors for SiMFlex microgripper will be realize.

1. Introduction

The compliant mechanisms are a distinct field in research, and they use elastic deformation for the transmission of the forces and movements. The compliant mechanisms are built by flexure hinges and connecting elements, so the characteristics of manufacturing material are a very important influence for transmission the displacements in the structure. All the displacements are done from flexure hinge and the deformation will respect de Hook law [1]. Also, the flexure hinge can be designed in different shapes in this way, various displacements for same overall dimensions will be done [2]. There are many advantages of using the compliant mechanisms: they can be built in monobloc structures, they have reduced dimensions, they don't need lubrication, they reduce the noise and frictions, etc.

The trend during the last years is the miniaturization of the product and of the tools necessary in industrial process, including the micro-assembly. For micro-assembly the best device are microgrippers built such as the compliant mechanisms [3]. Research in microgrippers had already started, mainly lead in Europe, SUA and Japan teams and usually, at microgrippers it used the micro-actuation with specially proprieties so-called smart materials (shape memory alloys, piezoelectric, magnetoelectric, ferrofluidic, etc) [4], [5]. The manipulation of micro-objects, are perturbed by so-called surface forces such as capillary forces, due to scaling laws, when they are to assemble the small components. These forces are usually neglected in usual macroscale assembly, which is ruled by gravity [6 - 12].

In this paper, a new design for a micromanipulation system is proposed with compliant microgripper and piezoelectric actuator. The compliant microgripper has attached the changeable jaws. The active parts of the jaws were manufactured from different materials, in order to use them for a large field of applications.

2. Analysis of SiMFlex

In this part a system for micromanipulation with monolithic and symmetric compliant mechanical structure is designed and analysed. The proposed whole system is shown in figure 1, where, there are three main parts: piezoelectric actuator, compliant microgripper and microgripper's jaws.

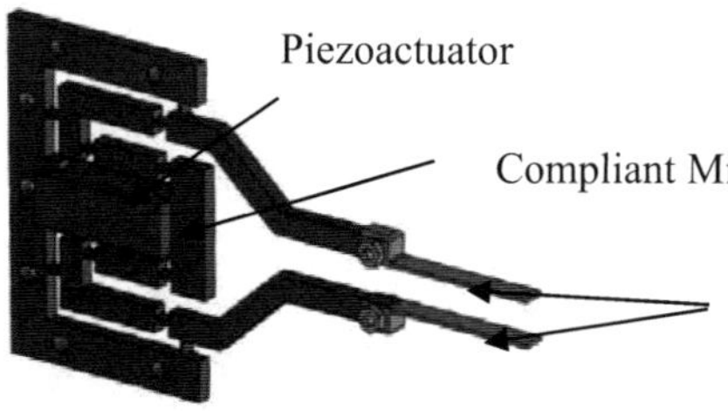

Figure 1. 3D Model for SiMFlex Microgripper.

The piezoelectric actuator has an axial deformation which represents the input for the compliant microgripper. The compliant microgripper was built with ten flexure hinges in order to eliminate the assembly errors. In the end, the changeable jaws are attached to the microgripper end-effectors, for being used for a large variety of micro-objects with different surfaces and materials for micromanipulation.

2.1. Piezoelectric actuator analysis

At first, the piezoelectric actuator was analysed with its axial deformation capability and then everything is incorporated into the model of the compliant frame. The analysis of the stack piezoelectric actuator with following specifications was realized: maximum voltage 150 V; maximum stroke 18 μm; capacitance 1350 nF; blocking force: 1000 N; resonant frequency: 65 KHz; Young's modulus: 4.4×10^{10} N/m^2 and overall dimensions are 6.5mm x 6.5mm x18 mm [13].
The analytical equations for displacements piezoelectric actuator are:

$$\Delta L = \mathrm{E} \cdot \mathrm{d_{ij}} \cdot L_0 + \frac{F}{c_T} \quad (1)$$

where: E is electric field; d_{ij} - piezoelectric coupling coefficient; L_0 – actuator length; F – axial force; c_T - rigidity

$$\Delta L = \Delta L_0 + \frac{F}{c_T} \quad (2)$$

$$\Delta L = (q \cdot U_a - F)/c_T \quad (3)$$

where: q is the factor for force and U_a is voltage actuator.

$$\Delta L_0 = \frac{F_0}{c_T} \quad (4)$$

$$F_0 = qU_{max} \quad (5)$$

The displacement for piezoelectric actuator was calculated for the input voltage U (0 ÷150) in a ramp signal, which resulted as input displacement for microgripper.

2.2. Displacement Amplification

Monolithic structure as compliant microgripper SiMFlex is manufactured with ten flexure hinges by Polyetheretherketone (PEEK) and overall dimensions 54mm x 41mm x 3mm. The jaws are fixed on compliant microgripper and have overall dimensions 20mm x 3mm x1mm.

Finite element simulation has been performed using the whole mechanical structure that is obtained from the microgripper and jaws, as it shown in figure 2. There are holes fixed and one guided member where the object will be positioned. In this case of the simulations considered, the output point is free, a condition for which the displacement amplification has been evaluated by means of the resulting input and output displacements.

The result for numerical analysis for maximum input displacements can be shown in figure 2 and the displacement histories of the output versus the input displacement is shown in figure 3.

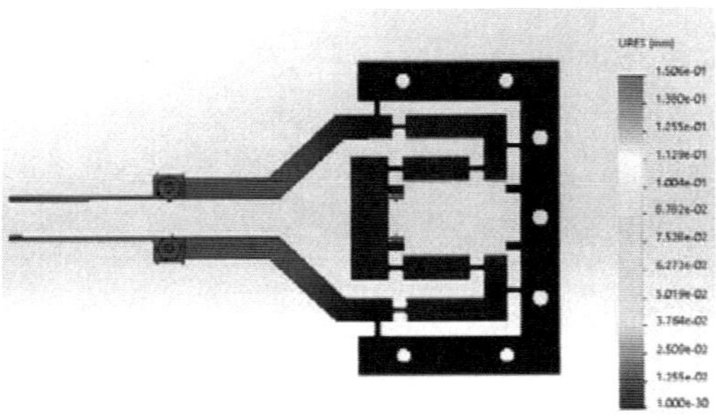

Figure 2. The maximum displacement of SiMFlex microgripper.

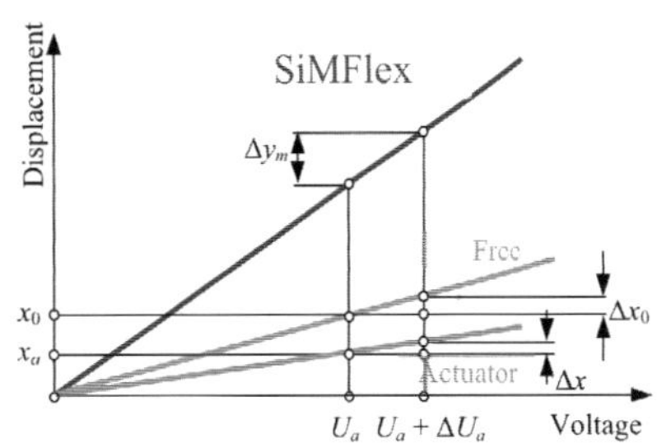

Figure 3. SiMFlex microgripper and piezoelectric actuator characteristics Displacements -Voltage.

One functioning characteristic of the compliant microgripper is the so-called geometrical advantage (GA), which is the output displacement to input displacement amplification ratio. Taking any two measurement points defined by voltages U_a and $U_a + \Delta U_a$, that are applied to the actuator, displacement Δx_a is obtained. The displacement Δx_a is input for the compliant microgripper. The output displacement Δy_m results from the two linear characteristics that define the output and input displacements of the system.

The geometrical advantage [14] is therefore calculated as:

$$GA = \frac{\Delta y_m}{\Delta x_a} \tag{6}$$

When the geometrical advantage was calculated only for the compliant microgripper without jaws, its value was: $GA = 3.47$.

2.3. Surface quality for jaws

When a manipulation of micro objects is performed, there is often a disturbance given by the adhesion between the micro-object and the jaws of the microgripper. Electrostatic forces are among the phenomena responsible for this adhesive effect.

For the increasement of the variety of the manipulated micro-objects the SiMFlex active parts of the jaws are changeable. The active parts of the jaws are manufactured out of two different materials: cooper and aluminium.

For a better control of picking and releasing micro-objects, in the next chapter, the surface roughness for each active part of jaws prototype will studied.

3. Experimental Results

The structure of the test bench for measurements contains a Laser displacement sensor with resolution of 0,01 μm and the response time of 100 μs. For collecting data of piezoelectric actuator and compliant microgripper, the command in Simulink and the application ControlDesk were realized.

IOP Conf. Series: Materials Science and Engineering **724** (2020) 012055 doi:10.1088/1757-899X/724/1/012055

3.1. Characterization for the actuator and compliant microgripper

The experimental tests were started with the stack piezoelectric actuator which was powered by the input voltage U = 0 ÷ 150V on the ramp signal which was recorded together with the output actuator's displacement. The measured results for piezoelectric actuator can be seen in figure 4, and the comparison between calculated and measured displacements is presented in figure 5.

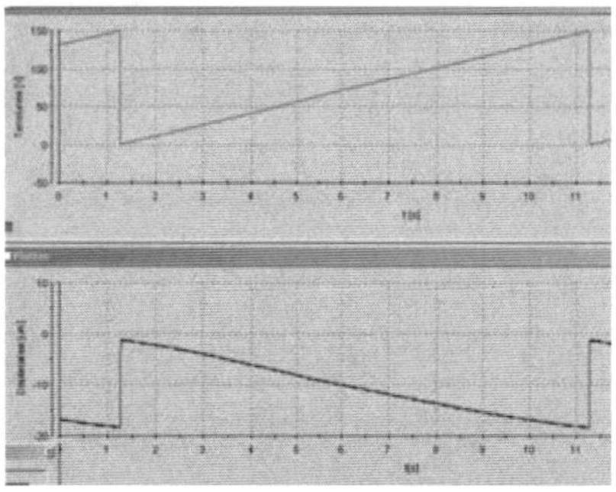

Figure 4. The experimental results of the ramp signal for stack piezoactuator.

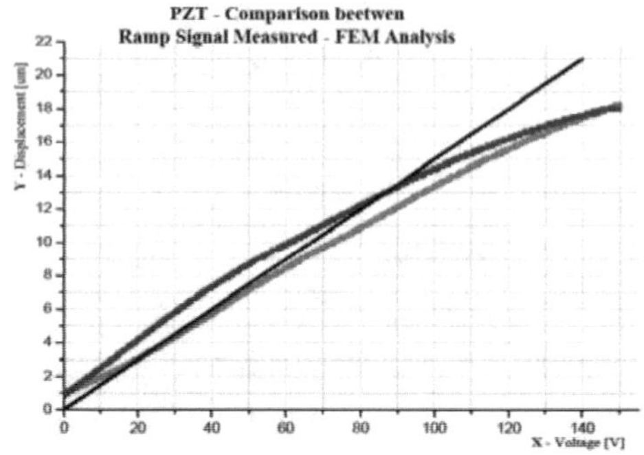

Figure 5. FEM analysis versus experimental results displacements piezoactuator.

In the second part, the out displacements for SiMFlex microgripper was measured. For this study, the same testbench with laser sensor and control desk application were used. The results obtained after the measurements were performed for SiMFlex microgripper without jaws.

This can be seen in figure 6.

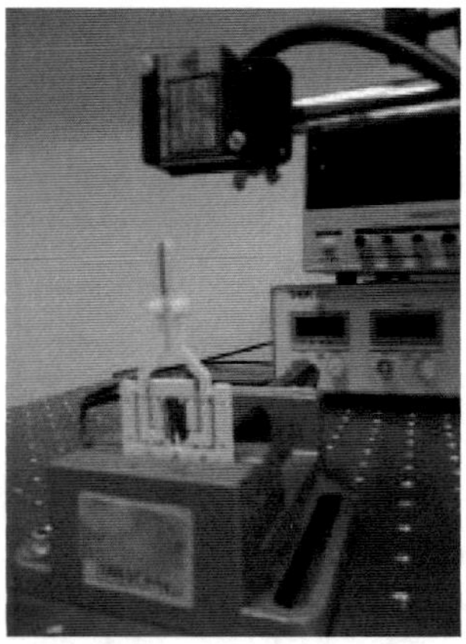

Figure 6. The gripper prototypes.

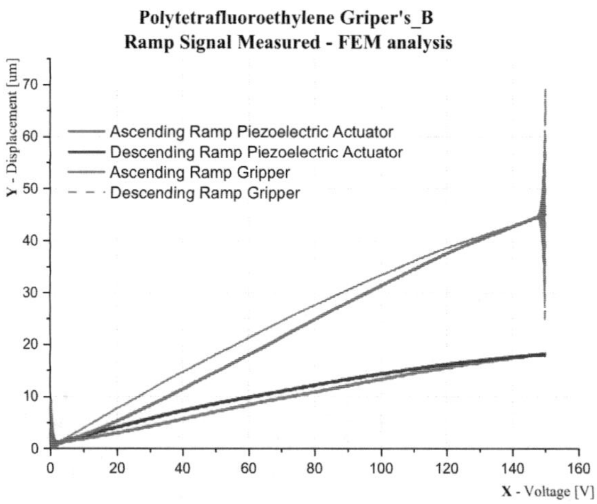

Figure 7. The experimental results of the ramp signal for stack piezoactuator versus SiMFlex microgripper without jaws.

The experimental results show the input and output displacements (figure 7) as such, the geometrical advantage can be estimated. Its value was: $GA = 3{,}11$.

The finite element simulations are performed to verify the accuracy of the model. The results are in a good agreement with those produced by the experimental approach.

3.2. Characterisation of the surface quality for jaws

An objective for SiMFlex microgripper is to use in more fields of activities, and the active parts of jaws were manufactured from two materials with different characteristics: bras and aluminium.

The AFM analysis was performed using a Veeco Dimension 3100 atomic force microscope, operated in contact mode. The vertical resolution of the microscope is < 1 nm. The cantilevers used for imaging were silicon nitride, Bruker SNL-10, having a nominal tip radius of 2 nm.

In figures 8 and 9 it can show the quality for each surface.

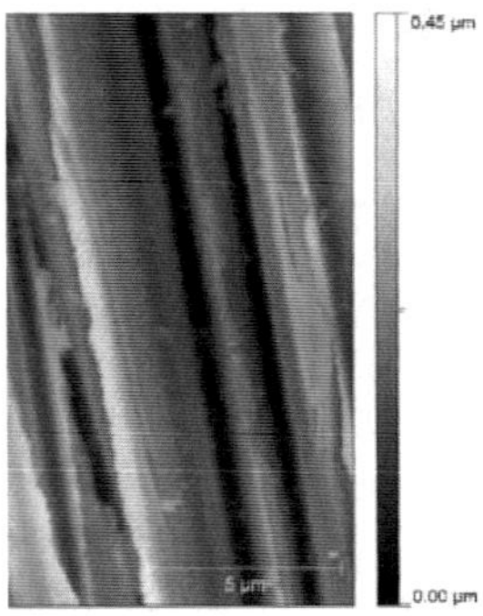

Figure 8. The surface analysis of bras jaws SiMFlex microgripper.

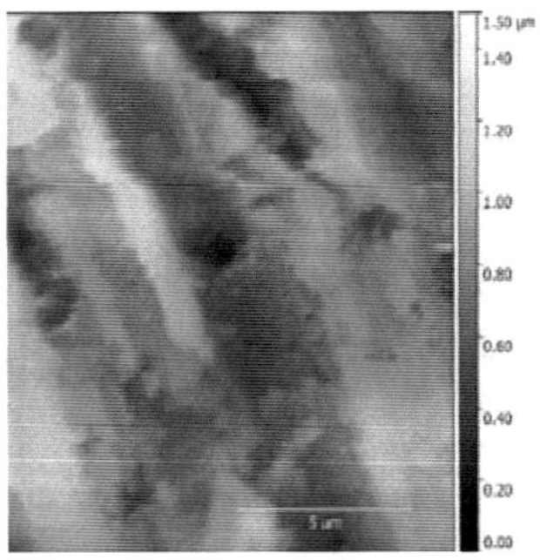

Figure 9. The surface analysis of aluminium jaws SiMFlex microgripper.

In the end, the quality of jaws surfaces is analysed. The result for brass, had a rugosity: RMS: 69.2 nm for peak-to-valley distance: 400.5 nm. On the other hand, for aluminium jaws, the surface had a rugosity RMS: 194 nm for Peak-to-valley distance: 1.406 µm.

In the future, the application for each active part of jaws will be studied.

4. Conclusion

This paper presents a compliant microgripper which has a monobloc geometry that minimizes construction complexity and increases robustness, while preserving the advantages of flexure joints.

This is a study based on small-displacement response of SiMFlex compliant microgripper with ten flexure hinges. The microgripper's geometrical advantage was emphasized. The finite element analysis predictions are confirmed by experimental testing of the SiMFlex microgripper prototype.

The structure and function of the experimental test bench for a compliant minigripper are presented. The measurements validate the FEM analysis for the Polyetheretherketone material. In the end, the surfaces used for jaws manufactured, made of two different materials, were studied. The jaws will be used for many types for microbject manipulation.

Acknowledgments

The results presented in this paper were obtained in the framework of the GNaC 2018 ARUT grant "Sistem de Micromanipulare cu Actuatori Piezoelectrici si Flexibilitate Funcţională – SiMFlex", research Contract no. 168/2019, with the financial support of the Technical University of Cluj-Napoca".

5. References

[1] Lobonţiu N 2002 *Compliant Mechanisms: Design of flexure hinges* (CRC Press) p 1-30

[2] Lobontiu N, Cullin M, Petersen T and Noveanu S 2014 Planar compliances of symmetric notch flexure hinges: the right circularly corner-filleted parabolic design, *IEEE Trans. Autom. Sci. Eng.* **11** p 169-176

[3] Noveanu S, Lobonţiu N, Lazaro J and Mândru D 2015 Substructure compliance matrix model of planar branched flexure hinge mechanisms: Design, testing and characterization of a gripper *Mech. Mach. Theory* **91** p 1-20

[4] Rakotondrabe M, Ivan I A, Khadraoui S, Clevy C, Lutz P and Chaillet N 2010 Dynamic displacement self-sensing and robust control of cantilever piezoelectric actuators dedicated for microassembly *IEEE/ASME Int. Conf. Adv. Intell. Mechatron.* (Montreal, ON, Canada) p 557-562

[5] Ivan I A, Rakotondrabe M, Lutz P and Chaillet N 2009 Current integration force and displacement self-sensing method for cantilevered piezoelectric actuators *Rev. Sci. Instrum.* **80**, 2126103

[6] Agnus J, Chaillet N, Clévy C, Dembélé S, Gauthier M and al 2013 Robotic microassembly and micromanipulation at FEMTO-ST, *J. Microbio. Robot.* **8** p 91-106

[7] Rakotondrabe M, Ivan I A, Stihi V, Noveanu S and Minca E 2011 Design and modelling of a piezoelectrically actuated microvalve *Romanian J. Phys.* **56** p 141-149

[8] Lates D and Casvean M 2019 Experimental study regarding the influence of compliant hinges geometry in mechanical structures, *Int. J. Eng. Manag. Sci. (Debr.)* **4** p 125-131

[9] Lin C, Dolovich A T, Chen A and Zhang W 2018 Topology optimization of efficient and strong hybrid compliant mechanisms using a mixed mesh of beams and flexure hinges with strength control *Mech. Mach. Theory* **121** p 213-227

[10] Selcuk E, Dogan S, Ulus S 2015 Effects of joints clearance on the dynamics of a partly compliant mechanism: numerical and experimental studies *Mech. Mach. Theory* **88** p 125-140

[11] Dearden J, Grames C, Brian D. and al 2017 Inverted L-arm gripper compliant mechanism, *J. Med. Devices* **11** 034502

[12] Lin S, Xu J, Wang C 2017 Synthesis and experiment of a novel parallel compliant mechanism with one translation and two rotations, *J. Mech. Eng. Sci.*, **231** p 2284-2290

[13] https://www.thorlabs.com/

[14] Lobontiu N and Garcia E 2004 *Mechanics of microelectromechanical systems* Springer Science & Business Media p 89

IOP Conf. Series: Materials Science and Engineering **724** (2020) 012056 doi:10.1088/1757-899X/724/1/012056

Validation of professional tooth brushing test device, test methodology and analysis

B M Achimescu, S Ciortan and M Buciumeanu

"Dunărea de Jos" University of Galati, Faculty of Engineering, Domneasca 47, 800008 Galati, Romania

E-mail: mihaela.buciumeanu@ugal.ro

Abstract. The present work is focused on developing a test apparatus to simulate the contact situation of the toothbrush used in professional toothbrushing (cleaning and polishing) and a dental biomaterial. Cylindrical samples (8×5 mm) were manufactured from a resin composite (frequently used in practice) in similar manner with the preparation of dental restorations in clinical conditions. The tests were carried by using two different toothpastes: a professional toothpaste and a home used toothpaste. All the experiments were carried out for 15 min, under 280 g normal load applied on the brush and the operating speed was around 5000 rev/min. The preliminary results on the dental restorative material selected for this study indicated that the proposed system is consistent and useful for the investigation of the microabrasion processes that arise at the surface of dental biomaterials.

1. Introduction

Some of the most common procedures in clinical practice in dentistry are direct restorations, to replace lost or damaged dental tissue. At present, a wide range of direct restoration products are available on the market, with multiple indications and adapting to the requirements of both dentist and patient [1 - 3].

It is well known that the oral cavity is a complex system, as there it is the action of several factors, such as saliva (acts as lubricant), temperature (ranging from - 10 to 50 °C), pH variation (ranging from 1 to 9), contact load (ranging from 3 to 1000N), etc. [4]. Besides, there is the influence of chemical components coming from food, drinks, and also from toothpaste, mouthwash, prophylactic gels etc. [5]. Thus the restorative materials suffer degradation due to synergetic effect between mechanical and chemical interaction.

If the regular toothbrushing used minimum twice a day for 2-3 minutes is manly for cleaning, thus preventing diseases, the professional toothbrushing is fundamental for both cleaning and/or polishing results [6]. As specified by providers those products have a low abrasion of enamel and dentine in a one-step procedure.

During the professional / regular toothbrushing the dental restorations are exposed to the action of abrasive particles / compounds from the toothpastes, such as: hydrated silica, calcium carbonate, sodium bicarbonate, etc. In some cases, these may increase the surface roughness of the restorative biomaterials. An increase of roughness with time was also reported for all dental restorations [1]. Cavalcante et al. [7] reported in their study about the effect of fillers' size on surface properties after toothbrushing, that the larger filler sizes caused a higher surface roughness after toothbrushing. This

IOP Conf. Series: Materials Science and Engineering **724** (2020) 012056 doi:10.1088/1757-899X/724/1/012056

increase on roughness was also confirmed by Jain and Wadkar [8] in their study about the effect of fillers on surface roughness of a nanofilled composite and a nanohybrid composite. A rougher surface, affects the aesthetic quality of a restoration, leads to perfect place for bacteria colonization site and also increases of wear loss of the antagonist, as the asperities acts like abrasive particles [9].

The objective of this study was to design a very simple and versatile device that is able to simulate the processes that arise at the contact situation of the tooth / restorative biomaterial surface and brush during the professional cleaning or polishing. This may be also important in the case of orthodontic appliances, as the oral hygiene in this case should be more carefully carried out, as around brackets it is easy to form a bacterial colonization site.

2. Materials and preparation of the samples

2.1. Materials

In this study commercially available resin composites for dental restorations was selected, namely Filtek™Ultimate Universal dental restorative composite (3M-ESPE, St. Paul, MN, USA). As specified by the manufacturer, this type of dental restorative composite has the lowest wear rate of all leading composites. It has been mentioned that this is due to the fact that the nanoparticles used are not obtained by grinding process and some nanoparticles are fused into nanoclusters. More details regarding the compositions and also, size and volume of the fillers of this material are provided by the manufacturer. As reported by Ferracane [1] there are many applications in dentistry where resin composites are used and are subjected the polishing or brushing. For examples cavity liners, inlays, onlays, crowns, provisional restorations etc.

2.2. Production of the samples

Cylindrical samples (8 mm in diameter and 5 mm in height) were produced from resin composite by following all the instructions provided by the manufacturer. To be consistent, the same person prepared all the samples. The sample were formed in a yellow polycarbonate mould. The resin composite was placed into the mould in layers of approximately 2 mm and each layer was light-cured for 40 s. The dental curing light device used was DENTMATE WL-070 (DENTMATE TECHNOLOGY CO., LTD., Taiwan). According to the manufactured, the output wave length range of the curing light varies from 440 and 480 nm, and the intensity up to 1200 mW/cm^2.

After polymerization, the samples were removed from the mould and were wet ground with SiC abrasive papers (down to 4000 mesh) and then polished with diamond paste. Samples were subsequently ultrasonically cleaned in an alcohol bath for 15 min and then in distilled water for 15 min. Before the toothbrushing tests, the resin composite samples were stored in modified Fusayama artificial saliva solution for 7 days. The composition of the modified Fusayama artificial saliva solution is given elsewhere [10].

3. Results and discussions

Is well known that toothbrushing is used for preventing first the dental diseases. Besides other ingredients, the content and size of the abrasive compounds added to professional toothpastes may not affect the enamel, as they are selected to have least damaging effect, but in many situations has been reported that it affects the surface of dental restorations [8].

3.1. Presentation of the new designed toothbrush device

Figure 1 (a) shows a global view of the tooth brushing test device, while figure 1 (b – d) presents some detail views of the device during the tests. These images allow to see all the components of the device. It can be seen from figure 1 (b) that the sample (9) is mounted horizontally in the samples block holder (1). The toothbrush (3) is fixed and vertically loaded against the sample by a calibrated weight (5). The speed of the electrical motor (4) is controlled by a speed controller (7).

The device was conceived to work at ambient temperature, in laboratory environment. It is a simple and versatile device. All types of dental materials can be tested, from ceramics, polymers and also metals. In the case of metallic biomaterials used for restorations (for example for dental crowns - metallic, metallic - ceramic, metallic - acrylic, metallic – composite, but also for orthopedic applications), it is also possible to evaluate their electrocorrosion behavior during the brushing test. In this case, the brushing device should work together with a potentiostatic assembly. This is an important aspect, as there are electrochemical reactions that occur at the surface of the metallic restorative material in the presence of physiological fluid and it may lead to degradation by corrosion and as a consequence metallic ions are realized into the body [10, 11].

Any type tool used in common practice for polishing or cleaning, such as dental polishing brushes (toothbrushes with different bristle stiffness) or rubber cups can be used with this new device. These tools for polishing / cleaning are used together with prophylactic paste which contains different amount abrasives (with various size, shape, and hardness).

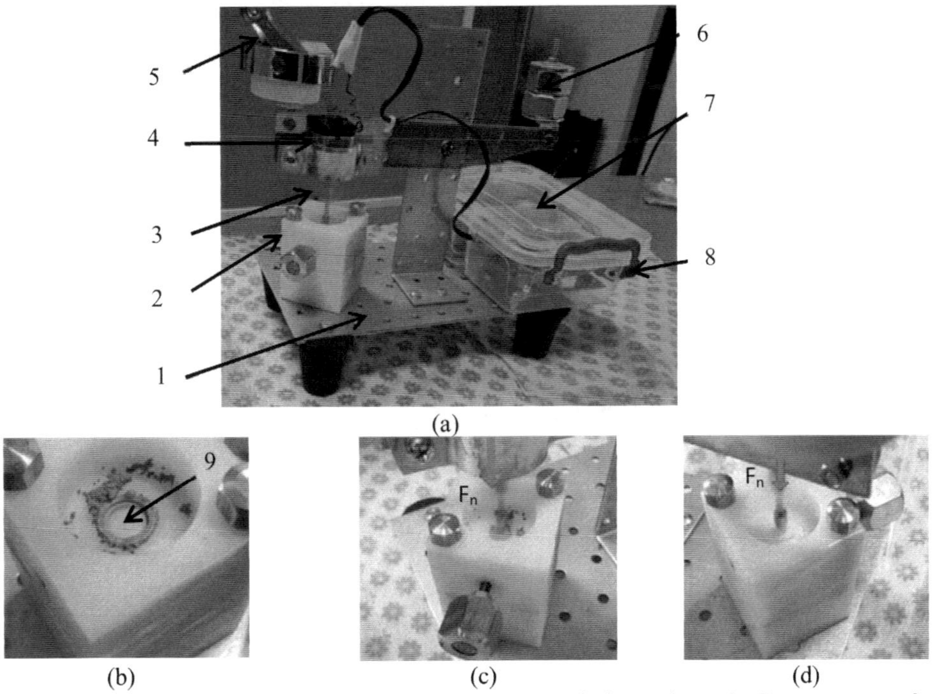

Figure 1. Images of the tooth brushing test device: (a) General view, where: 1 - Base structure; 2 - Sample block holder; 3 – Toothbrush; 4 – Electrical motor; 5 - Calibrated weight to apply normal load; 6 – Counterweight 7 – Speed controller 8 - Rotary potentiometer; 9. Sample; Fn – Normal load; (b) Detail view of the sample block holder, and (c) and (d) Detail views during the brushing tests.

3.2. Details of the toothbushing tests

Conditions for tests were selected to be as close as it is possible to the clinical conditions used by a dentist during professional cleaning and/or polishing. All the experiments were carried out with a normal load applied on the brush approximately 280 g and the operating speed was around 5000 rev / min. According to the manufacturer the time for polishing / cleaning the teeth / restorative material with Kerr Cleanic Prophylaxis Paste is around 15 min, thus the duration of each test was 15 min.

In order to see if this new device is able to simulate the professional toothbrusing, the tests were carried out by using two different toothpastes: a professional toothpaste with abrasive particles (Kerr Cleanic Prophylaxis Paste Single Dose) and a home used toothpaste (Sensodyne, GlaxoSmithKline Consumer Healthcare S.R.L). As mention by the manufacturer the Kerr Cleanic Prophylaxis Paste during the initial seconds of use has a very high cleaning power, but after that it changes into a gentle polishing action as the Perlite particles are fragmented.

As toothbrush was selected a professional perfect filament shape and for each test was used a new toothbrush.

For comparison purposes some test were also carried out in artificial saliva solution, in order to understand the influence of the toothbrush.

Five tests were carried out for each studied condition.

3.3. Morphological analysis of the samples

Figure 2 shows the SEM images of resin composite samples before the brushing tests with two different magnifications.

From the SEM images with lower magnification (figure 2 (a)) can be seen that the surface is relatively smooth, while from the SEM image with higher magnification (figure 2 (b)) it can be observed some channels from the polishing. The intention was not to have a perfectly smooth surface, as in clinical practice after the insertion of a resin composite into a cavity, the surface will not be perfectly smooth. On the other hand, as reported by Ferracane [3] generally these composite resins are quite difficult to polish and most of the time impossible to maintain surface smoothness due to the fillers content, distribution and size.

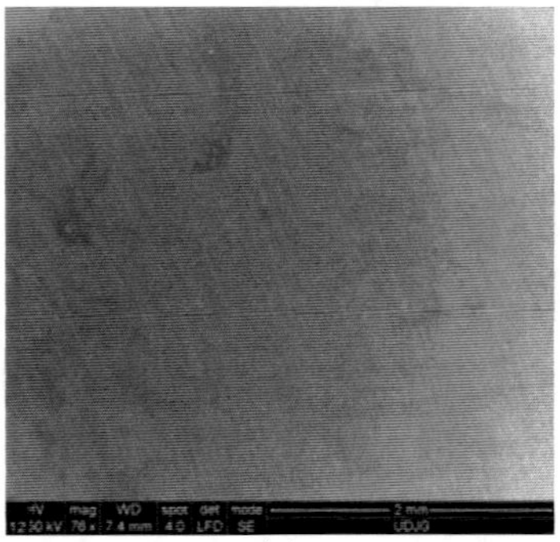
(a)

(b)

Figure 2. SEM images of as-processed resin composite representing general views at two different magnifications: (a) low magnification (76 x) and (b) higher magnification (500 x).

From the SEM image with higher magnification can be also observed some porosities (figure 2 (b)). These tiny pores are induced during the insertion of the composite resin into the cavity / mould, when air bubbles are incorporated into the material [12]. These may also increase the surface roughness of the composite.

Figure 3 shows the SEM images with two different magnifications 67 x and 500 x of the representative wear marks produced on the resin composite samples after the brushing tests for all three cases studied: professional toothpaste, Sensodyne and artificial saliva.

From the SEM images with lower magnification (figure 3 - top line), it can be seen that all the channels from the polishing were removed by the synergetic action of the brush and paste used. Only in the case of artificial saliva it still can be observed the channels from polishing (figure 3 (c)). This means that the influence of the brush is less pronounced (figure 3 (c) – bottom line) as compared to the action of the toothpastes (figure 3 (a) and (b)). After the brushing test when was used artificial saliva it was not observed any deterioration of the toothbrush, this was as stiff as in the beginning of the test.

From SEM images presented in Figure 3 can be observed grooves that are perfectly aligned to the movement of the toothbrush in all studied cases. The grooves are attributed to the abrasive wear mechanism (three-body abrasion – due to the presence of the abrasive particles from the toothpaste).

Based on qualitative analysis, the worn surface of the resin composite samples when was used the professional toothpaste reveals (figure 3 (a)) a rougher appearance than that in the case of home used toothpaste. Thus, it indicates a higher level of surface degradation when the professional toothpaste is used.

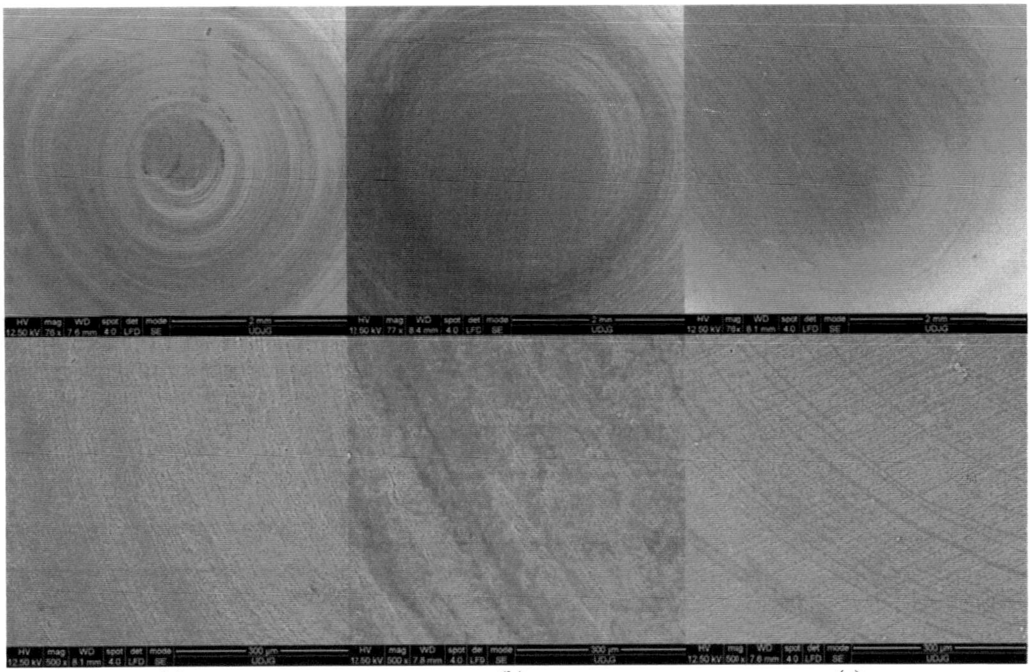

(a) (b) (c)

Figure 3. SEM images of the centre area of the wear track at two different magnifications (76 x – top line and 500 x - bottom line) formed after brushing tests on resin composite samples in the presence of different solutions: (a) professional toothpaste, (b) a home used toothpaste (Sensodyne) and (c) artificial saliva.

Even if locally in the case of Sensodyne toothpaste it can be observed a deeper groove in the middle of the wear track, in the case of the professional paste can be observed more grooves distributed on the wear track. The deeper groove observed in the case of Sensodyne toothpaste is due to the fact that it has a more sticky consistency then the professional toothpaste. The consistency of the Sensodyne toothpaste does not allow to the bristles of the toothbrush to spread out, the bristles remain stick and they slide in the same place. In the case of professional toothpaste, as mentioned before, it has a very high cleaning power in the beginning of the test, but after that it changes into a gentle polishing action due to the fragmentation of Perlite particle, thus changing the consistency of the paste. Thus, after few minute the bristles of the toothbrushes are more flexible and their action is spread out over the samples surface.

4. Conclusions

The design and the utilities of a newly developed test apparatus to simulate the processes that arise at the contact situation of the tooth / restorative biomaterial surface and brush in professional toothbrushing have been described.

- It is a very versatile apparatus and suitable for the study of any dental material. It can be adapted also in the case of metallic biomaterials to measure the electrochemical parameters.

- Significant differences were observed between the three studied cases. Within the limitations of the study it can be concluded that this device allows a better understanding of the influence of the tools (brushes, cups) used by dentists for cleaning, polishing etc.; the influence of toothpastes used in dentistry for cleaning the natural tooth, and also for polishing of any biomaterials used to replace a tooth (inlays, onlays, crowns, provisional restorations etc.).
- It seems that influence of the toothbrush alone without any paste is very small, as on the samples surface were observed only some slight scratches, and no damage of the toothbrush.
- Based on a qualitative analysis, the most sever abrasive wear was observed in the case materials tested with professional toothpaste, even though in the specification of the provider of the toothpaste it is mentioned that this has a low abrasion of enamel, which is the hardest substance in the human body. Maybe this degradation is due to a high amount and/or size of abrasive particles added to the paste.

Further tests and calibrations are needed in order to properly quantify the variables.

Acknowledgments
This work was supported by the project "Excellence, performance and competitiveness in the R&D&I activities at "Dunarea de Jos" University of Galati", acronym "EXPERT", financed by the Romanian Ministry of Research and Innovation, Programme 1 – Development of the national research and development system, Sub-programme 1.2 – Institutional Performance – Projects for financing excellence in R&D&I, Contract no. 14PFE/17.10.2018.

5. References

[1] Ferracane J L 2011 *Dent. Mater.* **27** 29–38
[2] Souza J C M, Bentes A C, Reis K, Gavinha S, Buciumeanu M, Henriques B, Silva F S and Gomes J R 2016 *Tribol. Int.* **102** 154–60
[3] Rodrigues D S, Buciumeanu M, Martinelli A E, Nascimento R M, Henriques B, Silva FS and Souza J C M 2015 *J. Bio- Tribo-Corrosion* **1** 24
[4] Antunes P V, Ramalho A 2003 *Wear* **255** 990–8
[5] Xavier A M, Sunny S M, Rai K and Hegde A M 2016 *J. Clin. Exp. Dent.* **8** e312–7
[6] Sawai M A, Bhardwaj A, Jafri Z, Sultan N and Daing A 2015 *J. Indian Soc. Periodontol* **19** 375–80
[7] Cavalcante L, Masouras K, Watts D, Pimenta L and Silikas N 2009 *Am. J. Dent.* **22** 66–4
[8] Jain N and Wadkar 2015 *Int. J. Dent. Oral Heal.* **1.1** 1–5
[9] Buciumeanu M, Queiroz J R C, Martinelli A E, Silva F S and Henriques B 2017 *Tribol. Int.* **116** 192–8
[10] Buciumeanu M, Bagheri A, Souza J C M, Silva F S and Henriques B 2016 *Tribol. Int.* **97** 423–30
[11] Reclaru L, Unger R E, Kirkpatrick C J, Susz C, Eschler P Y, Zuercher M H, Antoniac I and Lüthy H 2012 *Mater. Sci. Eng. C* **32** 1452–60
[12] Fleming G J P, Farooq A A and Barralet J E 2003 *Biomaterials* **24** 4173-9

Fretting manifestations in total hip prosthesis

L Capitanu[1], L L Badita[2*], V Florescu[3], C Tiganesteanu[1] and L F Isvoranu[1]

[1]Tribology Department, Institute of Solid Mechanics of the Romanian Academy, 15 Constantin Mille Street, Bucharest 010141, Romania
[2]National Institute of Research and Development in Mechatronics & Measurement Technique, 6-8 Pantelimon Road, Bucharest 021263, Romania
[3]Mechanical Department, University of Civil Engineering, 59 Plevnei Way, Bucharest 050153, Romania

[*]E-mail: badita_l@yahoo.com

Abstract. This paper analyses the fretting and fretting wear phenomena to the classical and modular hip prostheses. Theoretical and experimental manifestations are studied, at the interface of the femoral stem with the bone cement mantle, at the taper junction between the femoral stem and the femoral head of the prosthesis, as well as at the tapering of the femoral neck adapter's connections with the stem and the femoral head. The aim of the paper is to highlight the modularization implications of the femoral component of the Total Hip Prosthesis, especially its tribological implications (fretting wear of modular junctions, fatigue and corrosion by fretting) which can contribute to wear and increasing ion concentrations of the metal in the blood, and the possibility of the femoral shaft fracture at the femoral head junction. Theoretical approach described, allows the "Archard" law implementation to predict the fretting wear. Experimental laboratory studies focused on the analysis of the tribological phenomena occurring at the taper junctions between the femoral head and neck, as well as the neck and femoral stem, highlighting the relevant qualitative tribological aspects. For this purpose, experimental junctions were fabricated and tested on a universal MTS mechanical machine, equipped with a fatigue testing device.

1. Introduction

Modularization of total hip prostheses (THP) has started from the need to simplify the implants revision and during the years it has proven useful and reliable. This simplification was possible through the option of maintaining the femoral stem and making only a femoral head change. If the stem is to be kept under review, exposure can be improved by removing and replacing the head with a new one. Starting from the new process, this paper analyses the phenomenon of fretting and fretting wear to the classical and modular hip prostheses.

Based on the modularization reliability, many researches were made in the last years on positive effects of THP modularizations [1], [2]. It has been shown that an increase in the use of modular interfaces can have negative effects. This because, it can produce an increase of the fretting corrosion and corrosion cracking at the conical junction. In this way, corrosion products of taper junctions may contribute to the joint wear with the third body and, finally, to the joint's rejection.

Fretting corrosion can be confirmed only by revision surgery. Over the years, clinically relevant issues reported were examined but the failure mechanism could not be identified. [3] This is important to determine whether femoral stems with modular neck will be used in the future and how patients who already have implants should be monitored.

In other researches [4], [5] it has been shown that the primary micro-movements initiated the fretting within the modular connection of taper neck. There was a continuous process of abrasion and repassivation with a subsequent cold welding to the modular titanium alloy interface. Surface cracks caused by fretting or fretting corrosion eventually lead to fatigue fracture of titanium alloy modular neck adapters. Cobalt-chromium alloy neck adapters have reduced three times the micro-movements compared to the titanium neck, especially in the case of contaminated taper connection. The incidence of fretting corrosion was also substantially lower in the case of cobalt-chromium neck. In this way, it has been showed that THP corrosion reduction seems to be related to materials combinations.

Other experimental studies [6] showed that corrosion and fretting of modular taper surfaces may be influenced by different factors, like materials combination, bending stiffness, head and neck moment arm, neck length and implantation time. For example, in vivo corrosion of hip modular taper interfaces is attributed to a mechanically assisted corrosion process. The higher diameter necks will increase their rigidity and reduce the fretting and subsequent corrosion of the taper interface, regardless of the alloy used.

Micro-movements at the stem – neck interface were investigated on a Metha (Aesculap AG) and a H-Max M prostheses [7]. Metha prostheses demonstrated a substantial number of in vivo fractures for Ti-Ti couplings, but no fractures for Ti-CoCr couplings. H-Max M prostheses with Ti-Ti couplings showed only a clinical failure. Metha's prosthesis presented a tendency towards higher micro-movements compared to H-Max M (6.5 ± 1.6 μm vs. 3.6 ± 1.5 μm). Independent to design, prostheses with Ti neck have caused significantly higher micro-movements at the interface than those with a CoCr adapter (5.1 ± 2.1 μm vs. 0.8 ± 1.6 μm). Micro-movements between the Metha prosthesis with CoCr neck and H-Max M with Ti neck were identical (2.6 ± 2.0 μm).

Models of femoral stems structured using taper junctions have shown an increase in implant rupture in the recent past. Capitanu et al. showed that the fretting in a THP results from the destruction of the passive oxide layer from the metal [8]. This leads to increased corrosion and to the residues generation, such as polymer and/ or metal oxides particles, resulting in serious dysfunctions of the hip joint.

2. Fretting between femoral stem and bone cement mantle

Fretting tests were performed on an Instron® device for fatigue testing of the cemented fixture of the THP femoral stem, installed on a MTS Bionix multiaxial dynamic servo-hydraulic testing machine. It simulates fatigue loading of a hip cemented stem during a walking cycle. The machine and the device, as well as their mode of use, are shown elsewhere [8]. Figure 1 shows photographic images of the Ti6Al4V femoral stem contact and the PMMA mantle after the fretting experiment.

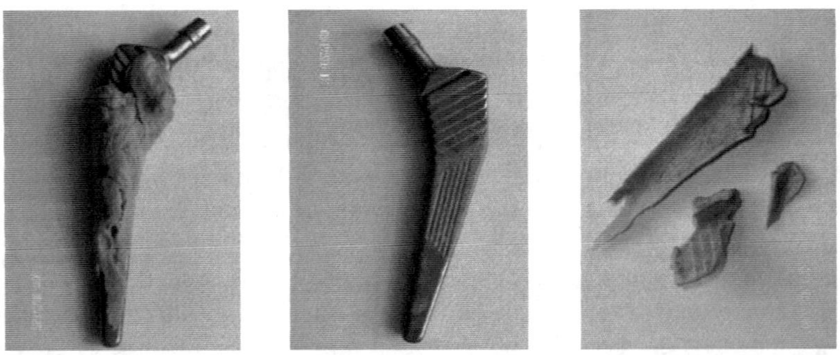

(a) (b) (c)

Figure 1. The cemented prosthesis stem (a) together with the contact sides of the cemented stem (b) and parts of the cement mantle (c) removed after 2450000 testing cycles, from the Instron® device for the fatigue test.

The presence of a visible corrosion of the femoral stem surface beneath the cement mantle and red deposits of Fe_2O_3 on the inner surface of the femoral stem (figure 1(b)), and through transfer on the inside of the PMMA mantle (figure 1(c)) were observed. These are due to fretting corrosion. For a more detailed investigation, the recovered stem and mantle were inspected further by optical microscopy.

3. Micro-movement and mechanical loading, causes of fretting in THP

An axial sinusoidal load between 230 N and 2300 N (ISO 7206-8) was applied to 1Hz for a total of 10000 cycles using the servo-hydraulic machine on which the fretting testing device was mounted, now shown in view (figure 2).

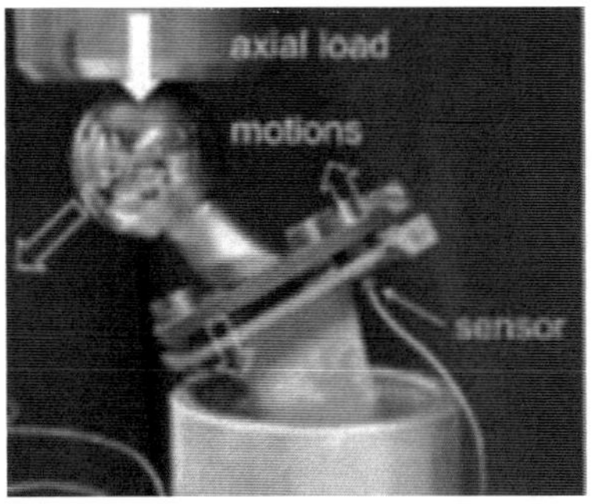

(a)

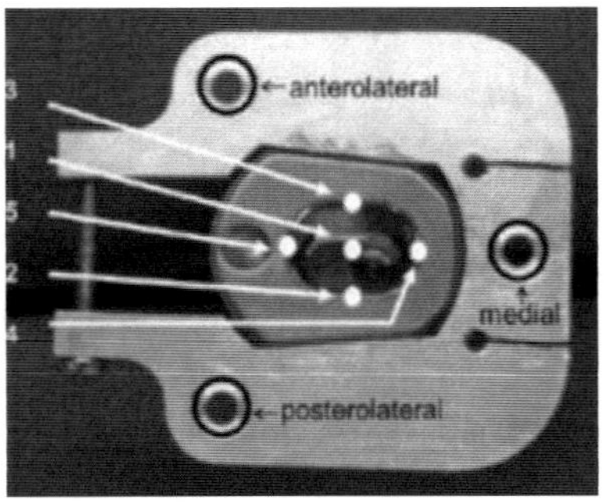

(b)

Figure 2. (a) A prosthesis mounted in the material testing machine. The servomotor was attached to an *xy* mass. The prosthesis system incorporates three eddy current sensors. Directions of motion are indicated by the light colour arrows. (b) Installation of the three eddy current sensors, with the holes of the grip screws on the sensor support and the five reference points from the mounting hole.

The micro-movements from the stem – neck interface were determined on the basis of linear displacements measured by three eddy current sensors with a 500 μm measuring range and a resolution of 500 μm – figure 2(a). This type of sensors were used due to their size and weight and to the non-contact measurement principle. The displacement in five reference points from the assembly gap was calculated from the displacements recorded by coordinate transformation – figure 2(b). For the cleanly assembled taper, the micro-movement accounted for only 27.1% / 24.9% of the displacement, while the larger part was due to elastic deformation (Co-Cr29-Mo / Ti6Al4V neck adapter). For the contaminated taper, the contribution of micro-movement was much higher (Co-Cr29-Mo: 48.1% / Ti6Al4V: 47.0%). Clear taper showed less micro-movements than the contaminated one (clean: 3.4 μm ± 2.5 μm, contaminated: 8.7 μm ± 7.41 μm, $p < 0.001$), and Co-Cr29-Mo adapters caused lower micro-movements than those of Ti6Al4V (7.8 μm ± 4.9 μm; Co-Cr-26-Mo: 7.28 μm ± 4.3 μm; $p = 0.005$).

4. Taper junctions fretting of the modular hip prosthesis

In the review published by Krishnan et al. [3] in 2013, it has been shown that any modular junction introduces the risk of micro-movement and fretting. Also, the tribological implications (fretting wear of modular junctions, fatigue and corrosion by fretting) of the THPs modularization can contribute to wear and increasing ion concentrations of the metal in blood, and to possible fracture of femoral shaft at the femoral head junction.

4.1. Theoretical approach

In the literature, there is no theoretical approach to fretting wear. This is the reason why the Archard's classic approach is applied to quantify the wear rates. It reports the amount of wear as the product between the sliding distance and the normal load. A wear coefficient is then extrapolated and it is assumed that it determines the wear resistance of the studied material. This approach does not work when the friction coefficient is not constant. The methodology described allows the implementation of the "Archard" or "dissipated energy" law to predict the fretting wear. However, the wear energy approach is presented here as a unified prediction of a single wear energy coefficient in a wider range of strokes (from 50 μm to 1.3 mm) than Archard's law and as such has no a wide range of applications [9].

Wear energy law – equation (1) – is the basis for the calculation of the volumetric wear. In this equation the mechanical interfacial shearing is the predominant parameter for determining the wear [10]. Total volumetric wear W_v is obtained from the total accumulated local energy E, that is dissipated and a worn energy coefficient α

$$W_v = \alpha \cdot E \tag{1}$$

where

$$E = Q \cdot s \tag{2}$$

and Q is the shear traction, and s is the relative displacement of the surfaces in contact, giving

$$W_v = \alpha \cdot Q \cdot s \tag{3}$$

Dividing both members of equation (3) to the contact area, the depth of linear wear W_d can be calculated using equation (4), where τ is the shear stress of the contact surface

$$W_d = \alpha \cdot \tau \cdot s \tag{4}$$

A loading cycle over several time intervals n is necessary to be digitized to accurately model the effect of the variable load distribution on the wear over time, during the loading cycle (as it appears during driving). As such, the depth of wear for a single cycle of loading (the depth of cyclic wear W_c) can be calculated using equation (5), in which both the shear stress and the relative displacement, calculated at the end of a certain time interval, i,

$$W_c = \sum_{i=1}^{n} \alpha \cdot \tau_i \cdot s_i \tag{5}$$

The total depth of W_d wear generated in a specified total number of loading cycles N can be determined from equation (6), where j is the specific "step of analysis" which reflects the evolution of wear, β is a wear coefficient.

$$W_c = \sum_{j=1}^{(N/\beta)} \beta \sum_{i=1}^{n} \alpha \cdot \tau_i \cdot s_i \tag{6}$$

4.2. Experimental approach

In order to study the fretting of THP's modular junctions, taper and trapezoidal junctions were realized to fit the stems with which current modular prostheses were equipped. They were subjected to dynamic fatigue tests, up to 2500000 cycles, on the Instron® device [8]. The neck adapters, made of CoCr and Ti6Al4V, were used in combination with femoral heads of CoCr, Ti6Al4V and ceramics. Three of these combinations and the fretting wear images on the surfaces of these junctions, are shown in figure 3.

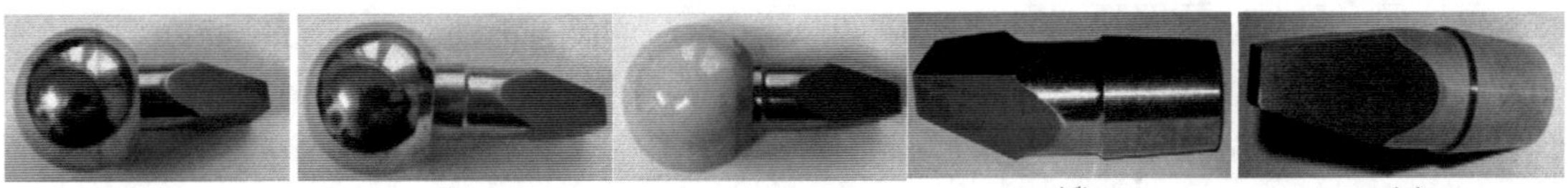

(a) (b) (c) (d) (e)

Figure 3. The femoral head – femoral neck combinations that have been tested made of CoCr (a) Ti6Al4V covered by TiN (b), ceramics (c), and taper and trapezoidal junctions of a CoCr (d) and Ti6Al4V (e) femoral neck adapter after the fretting test.

As shown in Căpitanu et al. [8], for cemented stem prostheses, the relatively large difference between the maximum and minimum recorded friction torque and the visual observation of the prosthesis functioning through the transparent fixation support, seemed to be a sign of the fretting wear manifestation. When this difference occurred, at 2450000 cycles, testing was stopped, and the prosthesis was removed from the Instron® device and qualitatively analysed. After visual and microscopic inspection visible traces of fretting were revealed both on the taper and trapezoidal junction of the femoral neck adapters as well as on the conical holes of the femoral heads. In figure 4, there are presented images of the inner taper of the CoCr, Ti6Al4V, Ti6Al4V coated with TiN, and ceramics femoral heads, after 2450000 fretting cycles.

(a) (b) (c) (d)

Figure 4. The fretting and corrosion of the inner taper of the femoral heads made of Co-Cr (a), Ti6Al4V (b), Ti6Al4V coated with TiN (c) and ceramics (d), after 2450000 fretting cycles.

5. Results and discussion

Theoretically, the results of this work can be justified by the mechanistic model of tribocorrosion [11] which is known to take into account all possible mechanical and chemical interactions as follows:

$$W_T = W_m + \Delta W_{cw} + W_c + \Delta W_{wc} \quad (7)$$

in which W_T is total tribocorrosion wear, W_m and W_c are material losses due to pure mechanical wear and pure corrosive wear. The terms ΔW_{cw} and ΔW_{wc} represent the wear enhanced by corrosion and corrosion enhanced by wear, respectively.

5.1. Mechanical wear model

Mechanical wear model uses a local form of Archard's equation to calculate the wear depth. In a ball on plane configuration, the local wear depth of each point on the surface is given by:

$$\Delta h(x, y) = \frac{K}{H} \cdot P(x, y) \cdot \Delta t \cdot v \quad (8)$$

in which H, K, P, v and Δt are the hardness of the material, the Archard's dimensionless wear coefficient, the local contact pressure, the sliding speed and the time step [8] respectively.

5.2. Corrosive wear model

Corrosive wear model is based on Faraday's law and the calculation of the metal ions volume transferred to the surface and used to form the oxide [12]. The formulation used for corrosive wear is as follows:

$$V_c = \frac{QM}{nF\rho} \quad (9)$$

where V_c is the volume of metal removed by the anodic reactions. Q is the total electrical charge passed and is calculated by integrating the current over time when an excessive potential is being applied. M is the atomic mass of the metal, n is the charge for the oxidation reaction, ρ is the passive metal density, and F is Faraday's constant.

IOP Conf. Series: Materials Science and Engineering **724** (2020) 012057 doi:10.1088/1757-899X/724/1/012057

6. Conclusion

Based on the literature, it was considered that the micro-movements at the stem – neck interface were responsible for failures of modular prosthetic neck adapters observed for a number of different models. Titanium neck adapters have shown significantly larger micro-movements than CoCr neck adapters. Interfaces contaminated with fats or debris also showed significantly higher micro-movements. Because excessive micro-movements at the stem – neck interface may be involved in the implant failure process, special care must be taken to clean the interface before assembly, and generally titanium neck adapters and titanium stems should be used with caution.

Fretting and corrosion have occurred on all modular neck – stem recoveries, regardless of model. However, mixed metal couples exhibited more corrosion than homogeneous couples. This is due to the lower modulus of elasticity of the titanium alloy used for the stem, which allows for greater metal transfer and surface damage when loaded on a modular neck from CoCr alloy.

This paper shows that when the friction coefficient is not constant, it seems more relevant to consider the dissipated energy law to predict the wear fretting. By identifying the wear energy coefficients, the quantification of wear can be rationalized, and the wear resistance of the studied tribosystems can be classified. This also seems to be a convenient approach to interpret various wear mechanisms. Micro-movements associated with the modular components can also lead to fretting corrosion and consequently to the release of debris that can cause negative local tissue reactions in the human body.

7. References

[1] Cooper H J, Della Valle C J, Berger R A, Tetreault M, Paprosky W G, Sporer S M and Jacobs J J 2012 Corrosion at the head-neck taper as a cause for adverse local tissue reactions after total hip arthroplasty *J. Bone Joint Surg. Am.* **94** 1655-61

[2] Srinivasan A, Jung E and Levine B R 2012 Modularity of the femoral component in total hip arthroplasty *J. Am. Acad. Orthop. Surg.* **20** 214-22

[3] Krishnan H, Krishnan S P, Blunn G, Skinner J A and Hart A J 2013 Instructional review: Hip. Modular neck femoral stems. *Bone Joint J.* **95-B** 1011-21

[4] Hussenbocus S, Kosuge D, Solomon L B, Howie D W and Oskouei R H 2015 Head-neck taper corrosion in hip arthroplasty *BioMed Res. Int.* Article ID 758123, 9 pages

[5] Grupp T M, Weik T, Bloemer W and Knaebel H P 2010 Modular titanium alloy neck adapter failures in hip replacement – failure mode analysis and influence of implant material. *BMC Musculoskelet Disord* **11** 3

[6] Goldberg J R, Gilbert J L, Jacobs J J, Bauer T W, Paprosky W and Leurgans S 2002 A multicenter retrieval study of the taper interfaces of modular hip prostheses. *Clin. Orthop. Relat. Res.* **401** 149–161

[7] Jauch S Y, Huber G, Hasche H, Sellenschloh K and Morlock M M 2014 Design parameters and the material coupling are decisive for the micromotion magnitude at the stem–neck interface of bi-modular hip implants *Med. Eng. Phys.* **36** 300–307

[8] Capitanu L, Badita L L and Florescu V 2017 Stability loss of the cemented stem of hip prosthesis due to fretting corrosion fatigue *Tribol. Ind.* **39** 536-46

[9] Liskiewicz T and Fouvry S. 2005 Development of a friction energy capacity approach to predict the surface coating endurance under complex oscillating sliding conditions *Tribol. Int.* **38** 69–79

[10] English R, Ashkanfar A, Rothwell G 2015 A computational approach to fretting wear prediction at the head–stem taper junction of total hip replacements *Wear* **338** 210-20

[11] Ghanbarzadeh A, Wilson M, Morina A, Dowson D and Neville A 2016 Development of a new mechano-chemical model in boundary lubrication *Tribol. Int.* **93** B 573-82

[12] Ghanbarzadeh A., Salehi F M, Bryant M, Neville A 2018 A new asperity-scale mechanistic model of tribocorrosive wear: synergistic effects of mechanical wear and corrosion *J. Trib.* **141** 021601-021601-12

Dynamic and tribological contact study for human fingertips at very low speeds

A Călin[1,3], A Tudor[1], M Stoica[1], N Stoica[1], and K Subhi[2]
[1] Department of Machine Elements and Tribology, Politechnica University Bucharest, 313 Splaiul Independentei, Bucharest Romania
[2] Al-Furat Al-Awsat Technical University, Kufa, Iraq
[3] Author to whom any correspondence should be addressed

andrei.calin@upb.ro

Abstract. Nonlinear elastic models have been widely used to describe the mechanical behaviour of complex materials such as skin. The subject of skin modelling is fairly new and of great importance to fields such as robotics and the development of skin-like materials to orthopaedics and prosthetics. The present article serves to add to the findings of previous authors and present experimental values gathered using the "Stick-Slip Experimental Stand" within the Machine Elements and Tribology Department on the tribological aspects regarding friction at very low speeds. The scope of this article is to evaluate the static and kinetic friction coefficients of the human skin in contact with different materials. A total of 5 different materials were tested to find the static and kinetic friction coefficient. The tested materials were steel, bronze, and 3 types of materials found on the inside of working gloves in direct contact with human skin: leather (bovine origin) and 2 types of textiles (cotton and nylon). All tests were performed at 1.8 mm/min, with an initial load of 7.5 Newton and a test time of 1000 seconds. The total distance evaluated was 9.25 mm.

1. Introduction

The evolution process and natural selection have provided human beings with an excellent tool for manipulation that can be mimicked artificially. Skin is the only part of the human body that interacts with the environment directly and has adapted to perform difficult tasks in the most challenging situations. Its behaviour was therefore analysed in contact with 5 materials and the coefficient of static friction, as well as for kinetic friction were determined. The results gathered can be used to compare to the performance of artificial grabbing tools designed to be replacements for loss of limb or even for robotic manipulation of objects. As such this report aims to make strides forward for a reliable model of the skin behaviour in contact with different artificial materials to be produced. Several authors studied in vivo testing of skin [1-8], utilising different measuring methods and on different parts of the body.

2. Experimental Setup

2.1. Experimental Stand Description

All experiments were performed on the "Stick-Slip experimental stand" in the Tribology Laboratory within the Machine Elements and Tribology Department at Politehnica University, Bucharest. The "Stick-Slip experimental stand" is capable of generating linear movement in the range of 0-50 mm/min, a range that is ideal for studying contact phenomenon at very low speeds. The experimental stand is specially designed for studying interactions between surfaces at very low speeds. The slider is actuated through a reduction drive by a motor that is driven by a variable DC transformer and is isolated from parasitic vibrations by the use of roller bearings.

A schematic of the experimental stand is presented in Figures 1, 2 and 3 below along with a Table 1, describing the components.

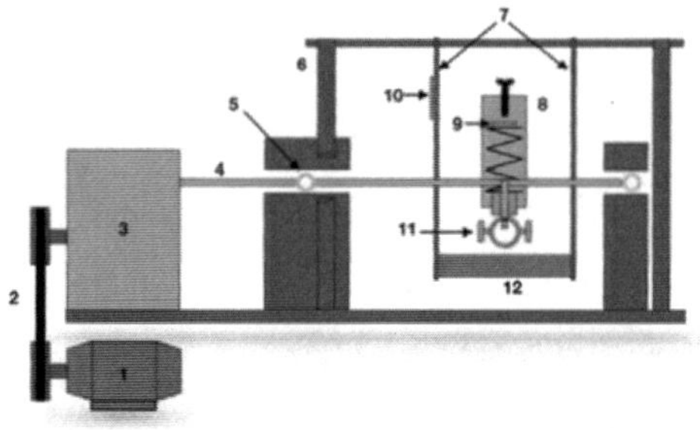

Figure 1. Schematic representation of the experimental stand.

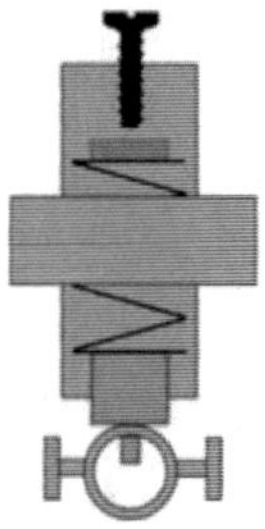

Figure 2. Holding device for finger.

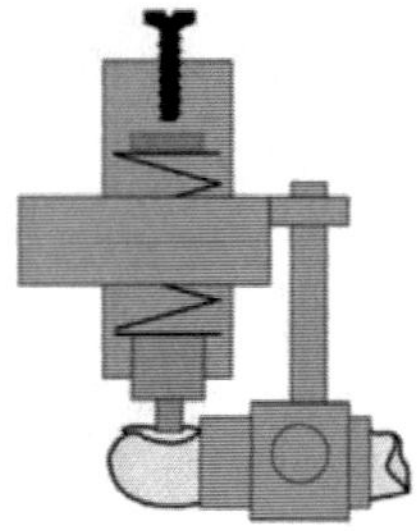

Figure 3. Position of the finger in the holding device.

Table 1. Component Table

No.	Component Name	Role
1	Electric motor	Generates rotational movement
2	Belt drive	Transfers the rotational movement from the electric motor to the reduction gearbox
3	Linear motion gearbox	Converts rotational motion to linear motion
4	Push/Pull rod	Supports and drives the normal force generating device and the finger
5	Guides and roller bearings	Guide, isolate and reduce friction
6	Struts	Framework
7	Rods with known rigidity	Support the test piece and measure the tangential contact force
8	Normal force generator and sensor device	Applies a measurable normal force on the finger
9	Normal force sensor	
10	Tensometric sensors	
11	Finger fixing device	
12	Sample holder	

2.2. Measuring Method

The measuring method is relatively simple, the finger was clamped to the moving part of the machine as seen in the figure above, and the material sample was held stationary. The normal force was measured using a force sensor and the tangential force that corresponds to the friction force generated by the contact was measured by a set of tensometric sensors positioned on the rods supporting the sample material.

The finger was immobilised and secured by the use of a plaster casing inside an aluminium tube, letting out only the distal phalange of the finger for measurements. The aluminium tube and finger were then fixed to the moving part of the machine by the use of a device that allows for optimal height and depth adjustment.

The finger was then placed in contact with the sample material and adjusted for height and depth, clamped into position and a normal force was applied by the use of a screw and was continuously measured by the force sensor.

All tests were performed using the same index finger of one young adult male individual. The finger did not present any callouses, bruises, cuts or any other skin lesions.

All tests had the same initial conditions, with a velocity of 1.8 mm/min, an initial load of 5 Newtons and a test time of 1000 seconds. The total driven distance was 9.25 mm. Before every test, the test surfaces were cleaned and degreased using alcohol.

2.3. Material samples

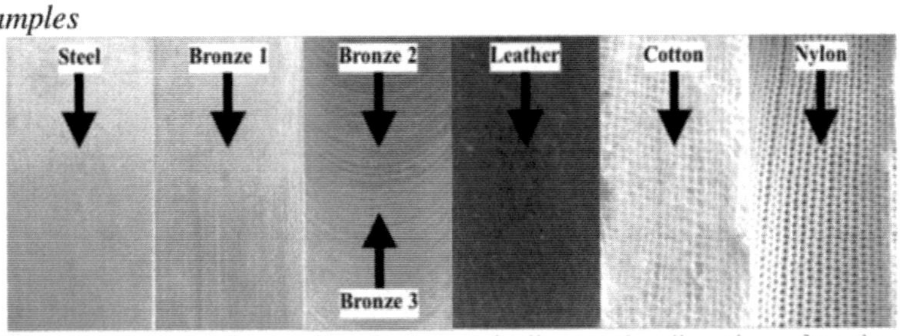

Figure 4. Material samples, the arrow indicates the direction of testing.

3. Experimental Results

3.1. Evaluation of the static and kinetic coefficients of friction

The static coefficient of friction was calculated as the ratio between the recorded tangential force (tensometric sensors) and the recorded normal force (normal force sensor) at the moment of the beginning slip. The kinetic coefficient of friction was evaluated in the same manner at the beginning of the stick period. Table 2 below presents the static and kinetic friction coefficients calculated by taking the average of at least 10 relevant data points.

Table 2. Static and Kinetic Friction Coefficients

Material	Static friction coefficient	Kinetic friction coefficient	Amplitude of stick-slip	Frequency of stick-slip (Hz)	Surface roughness Ra (µm)	Surface roughness Rz (µm)
Steel	1.1196	0.5809	0.5386	0.03	0.89	5.63
Bronze 1	0.8472	0.8245	0.0227	0.05	1.03	5.26
Bronze 2	1.14612	0.8734	0.27272	0.05	3.22	15.6
Bronze 3	1.27967	1.13098	0.14869	0.1	3.6	17.5
Leather	0.87486	0.7055	0.16936	0.11	12.6	57.8
Cotton	1.26212	0.53486	0.72726	0.025	13.5	57.1
Nylon	0.60075	0.4211	0.17965	0.01	12.2	50.8

Surface roughness was measured using the Surtronic 25 Surface Roughness Tester manufactured by Taylor Hobson Ltd, which provides a resolution of Ra 0.01 μm - Rz, 0.1 μm and an accuracy of 2%. The stylus is diamond tipped with a 5 μm radius and the digital filter is Gaussian (ISO 11562).

A sample of 200 seconds was used for each material to determine and extract the data presented in Table 2, above. The time snippet was selected after the contact was stabilised.

3.2. Steel

As it can be observed from Table 2 above, the amplitude of the "stick-slip" motion was the highest in this case. The difference between the mean static friction coefficient and the mean kinetic friction coefficient is almost 50% meaning that steel is prone to a "stick-slip" response in contact with skin.

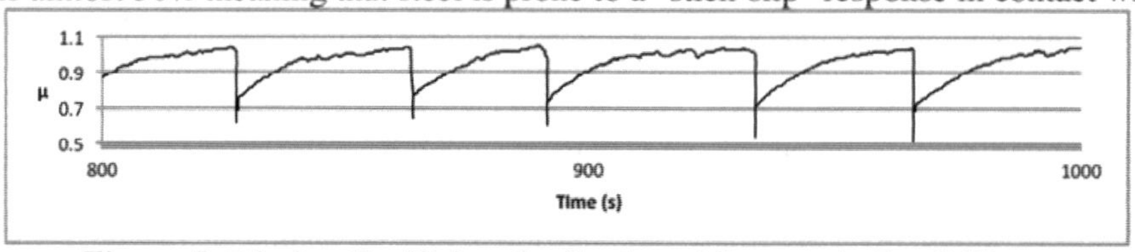

Figure 5. Friction Coefficient for Steel, v=1.8 mm/min Ra=0.89 μm.

3.3. Bronze

A more interesting approach was for bronze. The first test was performed on a sample that had a surface roughness measured at Ra=1.03 μm, as seen in Figure 6 below, and had no detectable imperfections to the touch and the results showed very little "stick-slip" behaviour. The variation for the coefficient of friction was, within reason, almost non-existent. Consequently, a decision was made for two additional tests using the same material but with altering the surface finish. Milling produced a surface that could be tested in two directions and the end-milled surface can be observed in Figure 4 below. This alteration to the surface produced a significant change in the way in which the friction coefficient varies with time. Figures 7 and 8 below depict a more prominent "stick-slip" behaviour and the direction of testing had a profound impact on the amplitude and frequency of the oscillatory response of friction.

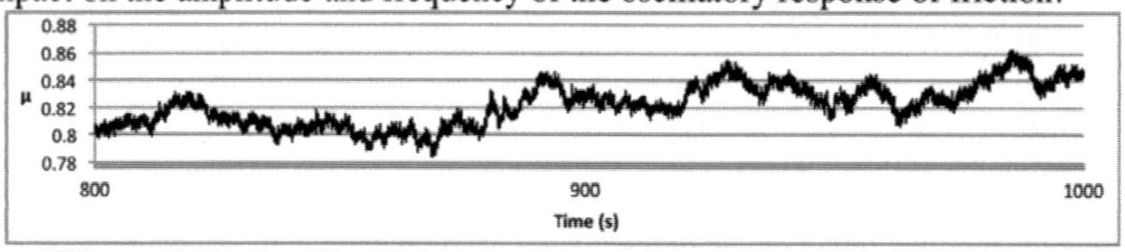

Figure 6. Friction Coefficient for Bronze 1, v=1.8 mm/min Ra=1.03 μm.

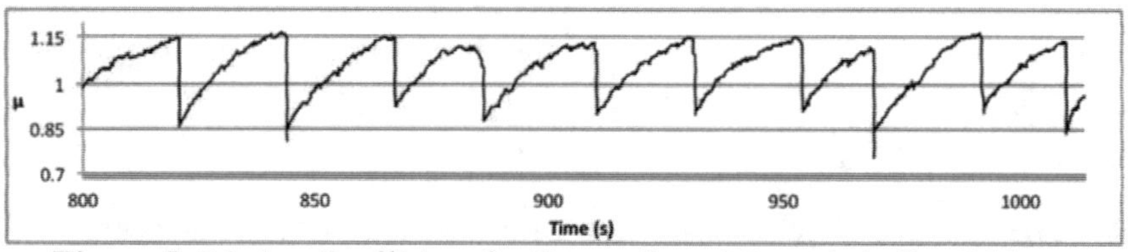

Figure 7. Friction Coefficient for Bronze 2, v=1.8 mm/min Ra=3.22 μm.

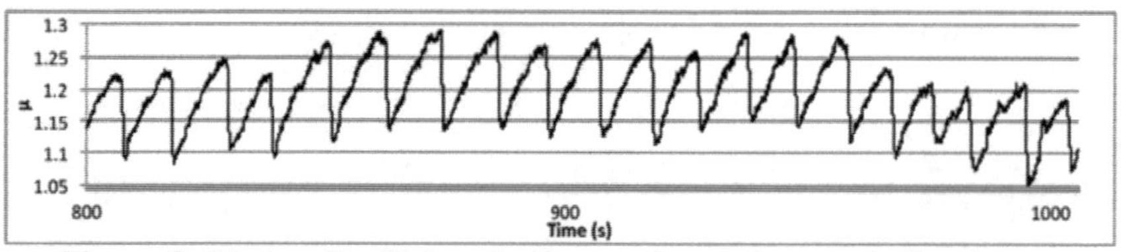

Figure 8. Friction Coefficient for Bronze 3, v=1.8 mm/min Ra=3.6 μm.

3.4. *Leather*

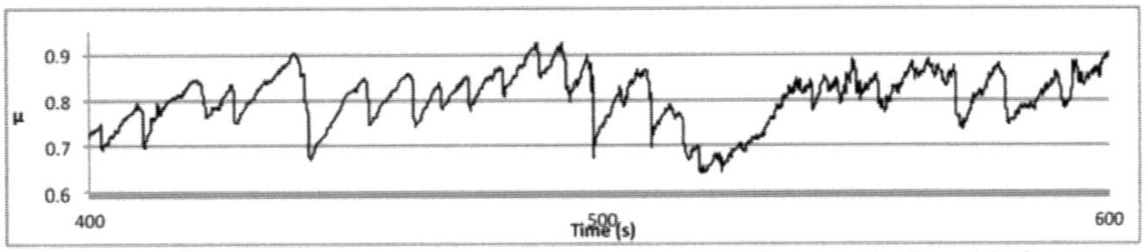

Figure 9. Friction Coefficient for Leather, v=1.8 mm/min Ra=12.6 μm.

3.5. *Cotton*

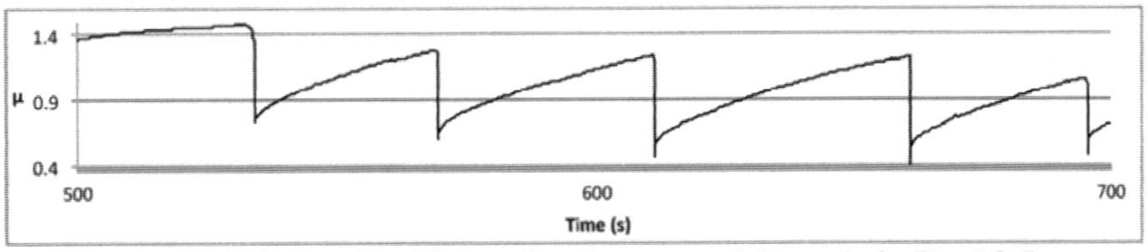

Figure 10. Friction Coefficient for Cotton, v=1.8 mm/min Ra=13.5 μm.

3.6. *Nylon*

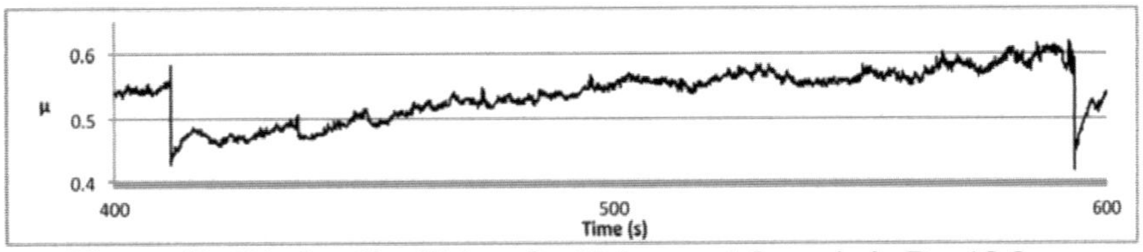

Figure 11. Friction Coefficient for Nylon, v=1.8 mm/min Ra=12.2 μm.

4. Theoretical model

4.1. *Experimental Stand Description*

The proposed dynamic model for the sample material in relative motion to the fingertip is [9-11]:

$$m_s\ddot{x} + F_f(\dot{x},\gamma_s) + \frac{h}{\omega}\dot{x} + kx = kv_o t + \frac{hv_o}{\omega_n} \quad (1)$$

where: m_s is the mass of the specimen, x - the displacement in the direction of slip, t - time, $\dot{x} = dx/dt$ - the instantaneous linear velocity and $\ddot{x} = d^2x/dt^2$ acceleration, $F_f(x;\gamma)$ is the friction force that acts to resist the motion of the mass and depends on the sliding velocity ($v_o - \dot{x}$) and the state of contact area ("age" of contact or saturation contact area) (γ_s), k is the axial stiffness, h is hysteresis damping in sample material, ω is the frequency of oscillation if the system is excited from the outside or frequency of natural oscillation and ω_n if the system is a free vibratory system.

Static friction reaches its maximum value when the contact between the finger and the sample material is saturated. This translates to the real contact area not being able to become larger over time under a specific set of initial conditions.

For the variable of state γ_s, we assume the following simple kinetic equation [9-11]

$$\frac{d\gamma_s}{dt} = \frac{1-\gamma_s}{t_{cr}} - \frac{\dot{x}}{D} \quad 0 \le \gamma_s \le 1 \quad (2)$$

The solutions of this differential equation for the two extreme cases are:

1) For the period of "stick", $\frac{dx}{dt} = 0$

$$\gamma_s\,(t) = 1 - e^{\frac{-t}{t_{cr}}} \quad (3a)$$

2) For $\frac{dx}{dt} = v_o$(steady-state) γ_s increases exponentially with time.

$$\gamma_s(t) = (1 - \frac{v_o t_{cr}}{D})(1 - e^{\frac{-t}{t_{cr}}}) \quad 3b)$$

Approximating the saturation parameter (real contact area γ_s) by developing the exponential function (3) in series and considering the first two terms.

1) For the period of "stick":

$$\gamma_s \approx \frac{t}{t_{cr}} \text{ if } \frac{t}{t_{cr}} \leq 1 \text{ and } \gamma_s \approx 1 \text{ if } \frac{t}{t_{cr}} > 1 \quad (4a)$$

2) For steady-state:

$$\gamma_s \approx 1 - \frac{v_o}{D} t \quad (4b)$$

If the body is put into motion, then the variables of state decrease faster at higher velocities. The physical meaning of t_{cr} in (3a) is the characteristic relaxation time of the parameter γ_s when the system is at rest, while D is the characteristic "relaxation length" at the initiation of motion. In a physical point of view, the contact between two rough surfaces, t_{cr} can be understood as the characteristic time of the creep process and D, as the average real contact diameter. The creep velocity can be considered as $v_{cr} = \frac{D}{t_{cr}}$.

5. Conclusions

The present results show that the proposed technique to measure both the static and kinetic friction coefficients on the fingertip skin is of consistently good quality and performance.

Surface finish plays an important role in the behaviour of friction, as seen from the bronze samples.

It can be stated that an ideal material that allows for a stable manipulation has to have a high coefficient of friction (preferably >1) and, ideally, without "stick-slip".

The microscopic region of contact between the skin and sample materials, the surface roughness of the sample material, the direction of contact and the mechanical characteristics of the skin layers are of great importance and require further examination in order to enhance the knowledge of the complex mechanics of the human hand grasping.

References

[1] Bobjer O, Johansson S E, Piguet S. 1993 Friction between hand and handle. *Applied Ergonomics 24: 190-202*

[2] Asserin J, Zahouani H, Humbert P, Couturaud V, Mougin D. 2000 Measurement of the friction coefficient of the human skin în vivo *Colloids and Surfaces B: Biointerface 19: 1–12*

[3] Comaish S, Bottoms E. 1971 The skin and friction: Deviations from Amonton's laws. *British Journal of Dermatology 84: 37–43*

[4] Cua, A.B., Wilhelm, K.P., Maibach, H.I. 1990 Frictional properties of human skin: relation to age, sex and anatomical region, stratum corneum hydration and transepidermal water loss. *British Journal of Dermatology 123(4), 473–479)*

[5] Cua A B, Wilhelm K P, Maibach H I. 1995 Skin surface lipid and skin friction: Relation to age, sex and anatomical region. *Skin Pharmacology and Physiology Journal of Pharmacological and Biophysical Research 8: 246–251*

[6] Gee M G, Tomlins P, Calver A, Darling R H, Rides M. 2005 A new friction measurement system for the frictional component of touch. *Wear 259: 1437–1442*

[7] Ramalho A, Silva CL, Pais AACC, Sousa JJS. 2007 In vivo friction study of human skin: Influence of moisturizers on different anatomical sites. *Wear 263: 1044–1049*

[8] Derler S, Preiswerk M, Rotaru GM, Kaiser JP, Rossi RM, 2015 Friction mechanisms and abrasion of the human finger pad în contact with rough surfaces, 2015 Tribology International

[9] Popov, V.,"Contact mechanics and friction. Physical Principles and Application", 2010 *Springer, 2010.*

[10] Wensrich, C., 2006 „Slip-stik motion in harmonic oscillation chain subject to Coulomb friction", *Triboogy International 39, 490-495.*

[11] A Tudor, K Subhi, A Calin, H.S. Wahad and N. Stoica, 2017 The friction effects in the stick slip phenomena of the human skin, *World Tribology Congress 2017*

IOP Conf. Series: Materials Science and Engineering **724** (2020) 012059 doi:10.1088/1757-899X/724/1/012059

Friction forces on human finger skin

C M Oprisan [1], B Chiriac[1], V Carlescu [1] and D N Olaru [*]

[1]"Gheorghe Asachi" Technical University of Iasi-Romania, Faculty of Mechanical Engineering, Blvd. Mangeron, No. 59A, 700050, Iasi, Romania
*dolaru@mail.tuiasi.ro

Abstract. The prehension of the small objects with the human fingers is depending on the friction and adherence between finger's skin and object's surface. The design of finger pads for robotic and prosthetic hands must simulate both the human finger friction and adherence and elastic deformation, especially for small objects. In this paper, the authors have experimentally investigated the friction forces and friction coefficients in a dry sliding contact tribosystem using Microtribometer CETR UMT-2. A steel cylinder was sliding on the human finger with speeds between 0.5 to 10 mm/s at normal loads up to 10 N. The results showed that the friction coefficient increases with the sliding speed and decreases with the normal load. Also, it was evidenced the presence of the adherence and sliding zones during the laterally cyclic movement of steel cylinder as function of the normal loads and sliding speeds. A general equation for the friction coefficient as function of sliding speed and normal load has been obtained. Therefore, the experimental results offer important information to select the adequate artificial skin materials for robotic finger pads, often overlooked, with relatively simple shapes and mechanical properties typically used.

1.Introduction

The friction and adherence between human fingers and various surfaces has been studied using several methods.

The friction and stick-slip behavior between the index finger and a plane wet glass probe have been studied by [1]. They determined the mean friction coefficient variation as function of normal finger force and sliding speed. Values of friction coefficient in range of 1.4 to 0.2 for the normal force increasing from 0.1 N to 10 N have been obtained. Also, the friction coefficient values between 1.4 to 0.2 have been obtained by the sliding speed increasing from zero to 1.5 m/s [2]. It was studied in vivo the friction and deformation of the human forearm skin by using a pin with a smooth ball in the top. The authors evidenced the two components of the total friction force as a sum of an adhesion force and a skin deformed force in front of the ball during the sliding process. The skin deformation and relaxation forces for a reciprocating-sliding of the ball in a cycle with positive and negative speed has been obtained by the authors for various ball diameters between 2 and 5 mm and normal force between 0.19 to 0.5 N. The friction coefficient resulted from the adhesion force varied between 0.7 and 1.5. Using a portable handheld device Veijgen *et al.* [3] studied the friction coefficient between some human skin surfaces and a cylinder realized from various materials. Depending of the cylinder material (PTFE, Aluminium, Polyamide), person gender, hydration of the skin and various skin surfaces on forearm and fingers, the friction coefficient have an important variation between 0.02 to 1.5. Another study [4] simulated the friction between the textiles materials and foot skin for the sports equipments. The authors experimentally studied the friction between artificial leather materials with a friction behavior similar to human skin and textile materials. The obtained friction coefficient varied between 0.28 to 0.38 as function of apparent contact pressure. In his PhD thesis Thomlinson [5] investigated the friction between human fingers and flat surfaces and obtained the friction coefficients for a lot of materials (glass, steel, plastics, rubber). The author evidenced that the total friction force in a finger skin-plate surfaces tribological system is a sum of adhesive, skin deformation, capillary and deformation of finger ridges forces. To avoid the skin deformation in the front of a ball Barnea *et al.* [6] experimentally studied the friction between various small cylinders and finger skin. The method consists, in a first step, to introduce an initial deformation on the finger tissue with the cylinder using small normal loads and, in the second step, it was realized the cyclic sliding of the cylinder on the finger, maintaining the normal load initially imposed. The minimum friction coefficient values have been obtained for PTFE cylinder, from 0.2 to 0.45 for normal load of 4 N and 1 N, respectively.

In the present paper the authors developed the methodology and a lot of experimental results for the friction and adherence between a small steel cylinder and the human finger in a cyclic sliding motion under normal loads between 1 to 10 N and sliding speeds between 0.5 to 10 mm/s.

2. Experimental methodology and equipments

The experiments were realized in the Tribology Laboratory from the Mechanical Engineering Faculty of Iasi using the facilities of the microtribometer CETR UMT-2. In figure 1 (a) is presented a general view of the microtribometer with the directions of the normal and tangential forces and in figure 1 (b) is presented a detail with the cylinder and finger contact. The forces were detected by a special sensor and all data including normal and tangential forces, friction coefficient, sliding speed and time are captured in the computer by an adequate program.

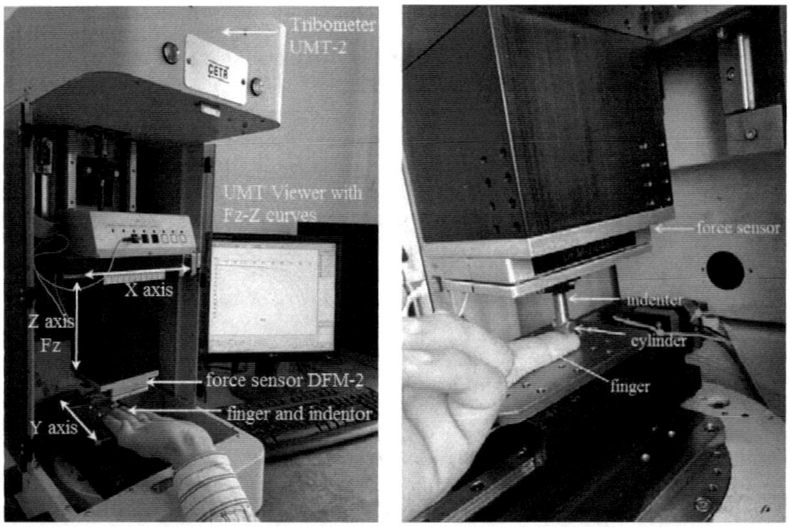

(a) (b)

Figure 1. Experimental equipments.

The steel cylinder has the diameter of 6.96 mm and the length of 38.6 mm with the roughness Ra = 0.06 μm. The direction of the cylinder is perpendicular on the finger and the cylinder sliding on the finger surfaces have imposed constant sliding speeds of 0.5, 1, 5 and 10 mm/s. The cylinder is pressed on the finger with three normal forces Fz: 1 N, 5 N and 10 N. Depending on the sliding speeds, the number of the sliding cycles on the laterally distance from -10 mm to 10 mm varies between 2.5 to 50 cycles. The tangential forces Fx are measured in both directions, positive and negative values have been obtained, according to sliding direction. The friction coefficient, indicated by COF in the program has only positive values.

3. Experimental results and comments

To evidence the influence of the normal load and sliding speed on the friction forces and friction coefficient between cylinder and dry finger skin experimental diagrams have been obtained. Figures 2 and 3 describe the registered variation of the tangential forces Fx as function of sliding time for two normal loads: Fz = 1 N and Fz = 10 N, respectively. So, it is evidenced the influence of the normal load on the variation of the tangential force Fx for positive and negative sliding cycles. The sliding speed was 1mm/s for both two tests.

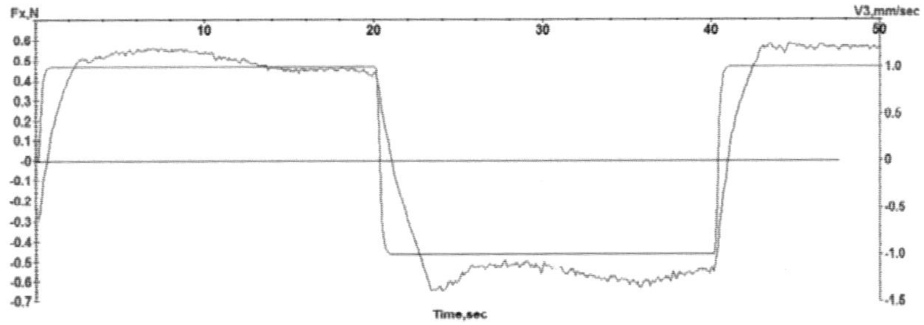

Figure 2. Friction force and sliding speed variations with the time for normal load Fz =1 N and for a sliding speed of 1 mm/s.

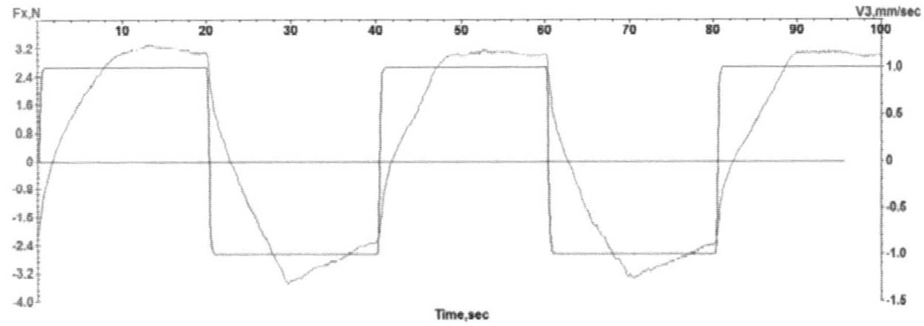

Figure 3. Friction force and sliding speed variations with the time for normal load Fz = 10 N and for a sliding speed of 1 mm/s.

The diagrams evidenced that, in a sliding cycle, the following mechanical phenomena have been produced: an adherence phase with deformation of the finger tissue, a sliding phase with a relative constant friction force Fx, a relaxation phase of the finger tissue by changing the direction of the sliding speed, a new adherence phase with increasing of the force by finger tissue deformation in opposite direction, a new sliding phase with a relative constant friction force Fx and the mechanical processes continue. For the normal force Fz = 1 N the deformation and relaxation phases are realized in 3.5 – 4 seconds and the sliding phase is dominant for complete cycle. For a normal force Fz = 10 N, the deformation and relaxation phases are realized in 9 – 10 seconds and the sliding phase is realised in 10-11 seconds. So, increasing of the normal load leads to the time increasing for deformation and relaxation phases and the time for sliding phase is decreasing. Also, increasing of the normal load Fz to 10 N, an asymmetric variation of the positive and negative friction forces has been observed. The asymmetric evolution of the friction forces obtained through the increasing of the normal force from 1 N to 10 N can be observed in figure 4 where are superposed the force-time diagrams for Fz = 1 N, 5 N and 10 N.

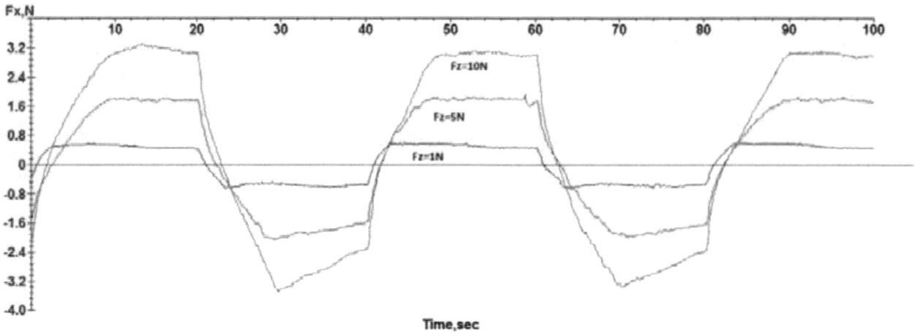

Figure 4. Variations of the friction force as function of sliding time, for normal loads Fz = 1 N, 5 N and 10 N and for a sliding speed of 1 mm/s.

In our opinion, the increasing of the normal force Fz determines the increases of the finger tissue deformation with a quantitative difference between the two sliding directions as result of no uniform finger internal structure.

The friction coefficient defined as the ratio between absolute value of tangential force Fx and normal load Fz has been considered only in the sliding phase and it is determined only by adherence between the finger skin and the cylinder.

Two diagrams for variation of the friction coefficient COF as function normal loads for two sliding speeds (2mm/s and 10 mm/s) have been presented in the figures 5 and 6.

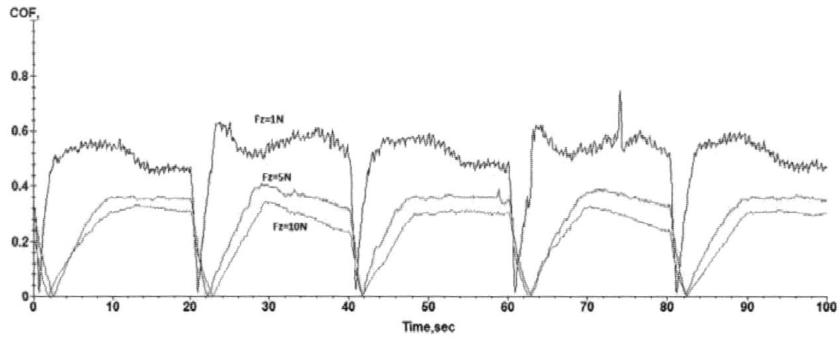

Figure 5. Variations of the friction coefficient as function of sliding time, for normal loads Fz = 1 N, 5 N and 10 N and for a sliding speed of 1 mm/s.

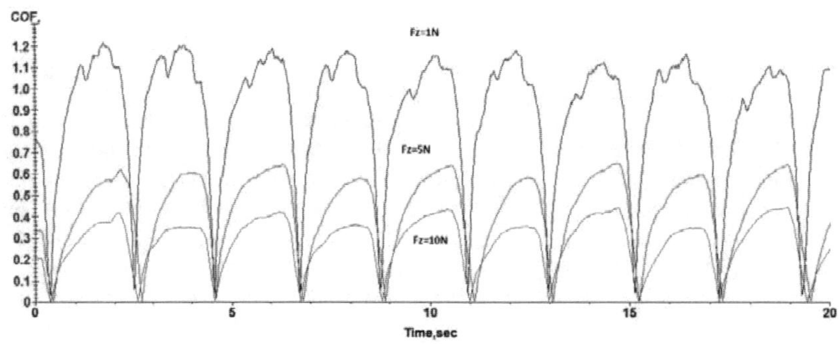

Figure 6. Variations of the friction coefficient as function of sliding time, for normal loads Fz = 1 N, 5 N and 10 N and for a sliding speed of 10 mm/s.

It can be observed important variations of the friction coefficient as function of normal load, Fz, and sliding speed, v. Average values of the friction coefficient experimentally determined for five sliding speeds and three normal loads have been presented in the table 1.

Table 1. Average values of the friction coefficient, COF

v,mm/s	Fz=1N	Fz=5N	Fz=10N
0.5	0.46	0.32	0.27
1	0.498	0.36	0.31
5	0.83	0.48	0.32
10	1.15	0.62	0.35

The variations of the friction coefficient as function of sliding speed and normal load have been presented in figure 7.

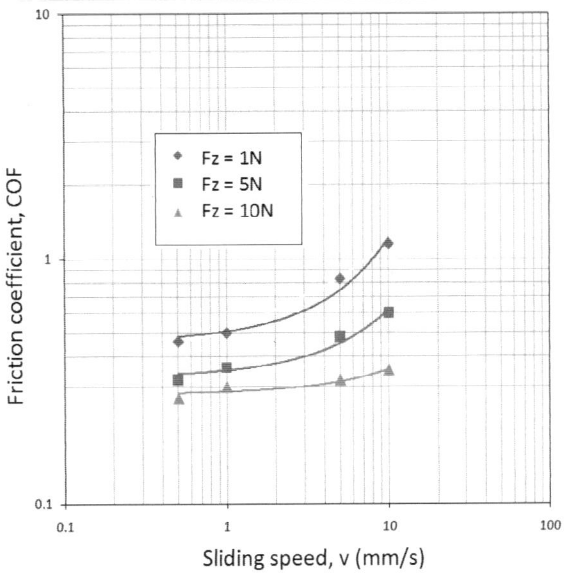

Figure 7. Variations of the friction coefficient as function of sliding speed and normal load.

By curve fitting the experimental values for friction coefficient COF, the following general equation for COF as function of sliding speed and normal load has been developed by authors:

$$COF(v, Fz) = F_1(Fz) \cdot \exp[v \cdot F_2(Fz)] \quad (1)$$

In equation (1) the functions $F_1(Fz)$ and $F_2(Fz)$ are depending only of the normal load and have the following equations:

$$F_1(Fz) = 0.459 \cdot Fz^{-0.212} \quad (2)$$

$$F_2(Fz) = 0.1049 - 0.0082 \cdot Fz \quad (3)$$

The errors between experimental values of COF presented in table 1 and similar values obtained by equations (1) - (3) are not exceeding 10%.

Based on the experimental results presented in table 1 and figure 7 following observations can be made:

(i) The sliding speed increasing leads to the friction coefficient increasing. The increasing rate is dependent of the normal load.

(ii) Increasing the normal force Fz determines the friction coefficient decreasing.

(iii) The dependence between friction coefficient and normal load obtained by authors are in good agreement with the results obtained by Derler and Rotaru [1] for sliding of index finger on wet flat glass in the similar normal load range 0.1 to 10 N.

4. Conclusions

Experimental investigations to determine the friction and adherence behavior of the human finger skin in contact with a steel cylinder have been realized by the authors.

The experimental method developed by the authors allows the friction coefficient to be determined by maintaining the contact pressure constant.

The experiments were realized for normal loads between 1N and 10N and for sliding speeds between 0.5mm/s and 10mm/s.

Both normal load and sliding speed influence the friction and adherence behavior of the finger skin.

When the normal load increases, the friction coefficient decreases. The sliding speed increasing leads to the friction coefficient increasing.

Friction coefficients between 0.27 and 1.15 have been obtained for the normal forces and sliding speed studied interval.

A general equation for the friction coefficient as function of sliding speed and normal load has been obtained.

5. References

[1] Derler S and Rotaru G M 2013 *Wear* **301** 324–329

[2] Kwiatkowska M, Franklin S E, Hendriks C P and Kwiatkowski K 2009 *Wear* **267** 1264–1273

[3] Veijgen N K, Masen M A and van der Heide E 2013 *Tribol Lett* **49** 251–262

[4] Baussan E, Bueno M A, Rossi R M and Derler S 2010 *Wear* **268** 1103–1110

[5] Tomlinson S E 2009 Understanding the friction between human fingers and contacting surfaces, *PhD. Thesis*, The University of Sheffield

[6] Barnea A, Oprisan C and Olaru D 2014 Prehension of the Small Cylindrical Objects by the Human Fingers. Friction and Adherence Processes, *Applied Mechanics and Materials-Advanced Concepts in Mechanical Engineering* I 658 721-727

Temperature effect on the mechanical behaviour of niobium nitride thin films

V V Merie[1], M S Pustan[2], G Negrea[1], C J Bîrleanu[2] and H G Crişan[2]

[1]Materials Science and Engineering Department, Technical University of Cluj-Napoca, Cluj-Napoca, Romania
[2]Mechanical Systems Engineering Department, Technical University of Cluj-Napoca, Cluj-Napoca, Romania

E-mail: violeta.merie@stm.utcluj.ro

Abstract. Niobium nitride is used in a wide range of application due to its mechanical, physical, chemical, electrical and optical properties. The main aim of this paper is to deposit niobium nitride thin films by direct current magnetron sputtering and to characterize them at nanoscale. The films were deposited on silicon Si (100) substrates. Three types of films were deposited by modifying the deposition temperature. In this regard, some of the samples were deposited at room temperature (25 °C), some were deposited when the substrate was preheated at 200 °C, and the rest at substrate temperature of 400 °C. Regarding the topography of the samples, an increase in the average roughness was determined with the increase in deposition temperature. The nanoindentation tests were carried out at temperatures between 20 and 100 °C so that to emphasize the change in hardness and modulus of elasticity in terms of testing temperature. The results pointed out an important influence of both the deposition temperature and testing temperature on the topography, adhesion and mechanical properties of the niobium nitride thin films.

1. Introduction

Transition metal nitrides present, in general, some mechanical, electrical, chemical, physical properties that recommend them for a wide range of application [1-4]. Titanium nitride (TiN), chromium nitride (CrN) and niobium nitride (NbN) are some of the most studied transition metal nitrides. Niobium nitride is characterized by high melting point, high hardness, good superconducting properties, good wear resistance, good stability at high temperatures and so on [5-7].

Niobium nitride is used in application such as wear resistance coatings, diffusion barriers against copper migration, single-photon detectors, superconducting microelectronics applications, Josephson junctions etc. [8-9]. It can be deposited by different techniques such as pulsed laser deposition, reactive magnetron sputtering, cathodic arc evaporation, ion beam assisted deposition, molecular beam epitaxy and all that [10-11].

The deposition conditions such as nitrogen flow rate, deposition temperature, deposition time and so on have a strong influence on the properties of the deposited thin films. This paper is a study concerning the influence of testing temperature on the mechanical properties (hardness and modulus of elasticity) of niobium nitride thin films deposited by direct current (DC) magnetron sputtering at different deposition temperatures.

2. Materials and experimental procedure

Niobium nitride thin films were deposited by direct current (DC) magnetron sputtering on silicon substrates to achieve the purpose of this paper.

2.1. Materials

A high purity niobium target was employed for deposition the niobium nitride thin films. The purity of the target was 99.95 %. The thin films were deposited on Si (100) substrates.

2.2. Experimental procedure

The deposition process took place in a reactive sputtering facility from the Materials Science and Engineering Department, Technical University of Cluj-Napoca. A base pressure of 10^{-7} torr was kept constant inside the deposition chamber. First the silicon substrates were cleaned to remove any possible impurities in an ultrasonic bath with acetone and isopropyl alcohol and blown with compressed air. A mixture of argon and nitrogen was introduced inside the deposition chamber. The pressure and the discharge current were kept constant at 2 mtorr and 350 mA. The argon flow rate and the nitrogen flow rate were 40 $cm^3 \cdot min^{-1}$ and 1.5 $cm^3 \cdot min^{-1}$ respectively. 60 mm were maintained between the niobium target and the silicon substrates. The deposition process ran for 20 minutes. Three types of films were deposited. Some of the samples were deposited at room temperature (25 °C), some were deposited when the substrate was preheated at 200 °C, and the rest were deposited at substrate temperature of 400 °C. The thickness of the deposited niobium nitride thin films was determined using a JEOL 5600LV electron microscope. Its mean value was 0.35 μm.

Once the niobium nitride thin films were deposited, they were characterized at nano scale using a XE 70 atomic force microscope (AFM) from the Micro and Nano Systems Laboratory, Technical University of Cluj-Napoca. All the tests were carried out for a relative humidity of 20 %. The topography of the films was investigated in the non-contact of the AFM using a PPP-NCHR cantilever. The dimensions of this cantilever are: thickness of 4 μm, width of 30 μm, length of 125 μm, resonance frequency of 400 kHz and force constant of 42 $N \cdot m^{-1}$. The tip radius is smaller than 10 nm while its height is 15 μm.

The determination of the mechanical properties was done by nanoindentation tests. These tests were performed using a TD23838 nanoindentor. The stiffness of the nanoindentor is 272 $N \cdot m^{-1}$, its frequency is 50 kHz. The height, thickness and length of the cantilever are 90 μm, 41 μm and 1050 μm respectively. The tip radius is smaller than 25 nm. The tests were carried out at temperatures between 20 and 100 °C so that to establish the temperature influence on the mechanical characteristics.

3. Results and discussion

First, we investigated the topography of the deposited thin films. This characteristic is strongly influenced by the deposition conditions and the nature of the deposited material. On this line the thin films were scanned in the non-contact mode. The scans were performed on square areas having the dimensions of 2.75x2.75 μm. The tests were done in several areas of each sample to obtain accurate values. The 3D images were obtained using the XEI Image Processing Tools for SPM Data. Figure 1 presents the 3D images of the investigated materials. It can be noticed that the niobium nitride thin films deposited at room temperature have a smoother surface than the thin films deposited on preheated substrates. Instead, the niobium nitride thin films deposited on substrates preheated at 400 °C are characterized by the rugged surface.

The roughness parameters were determined to support what we noticed on the 3D images. The average roughness, R_a, the root mean square, R_z, the skewness, R_{sk}, and the kurtosis, R_{ku}, are given for the three type of niobium nitride thin films in table 1. If the skewness is an asymmetry indicator, the kurtosis is an indicator of the shape of the distribution's tail. The average roughness increases with the increase in deposition temperature. A small increase of this roughness parameter was determined when the substrate temperature increased from 25 to 200 °C. Instead, the increase in substrate temperature from 200 to 400 °C led to the increase in average roughness of about 2.5 times.

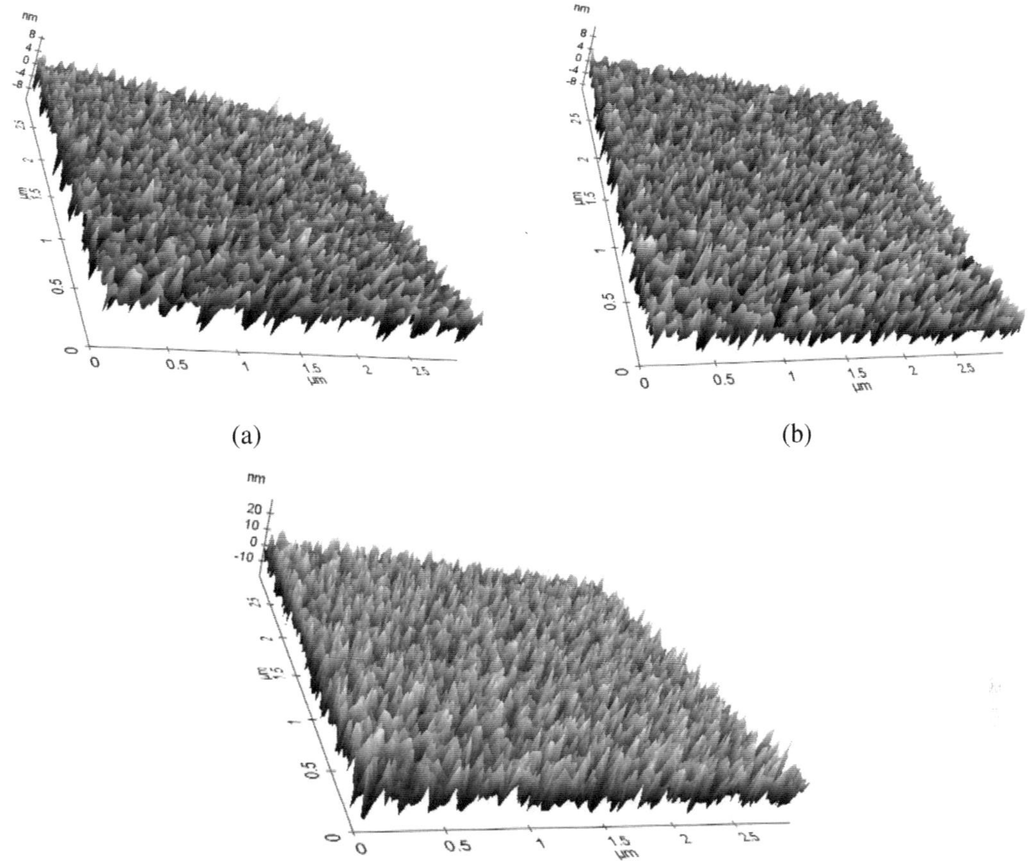

(a) (b)

(c)

Figure 1. 3D images of the niobium nitride thin films deposited on silicon substrates at (a) 25 °C, (b) 200 °C and (c) 400 °C.

Table 1. Roughness parameters of the deposited niobium nitride thin films.

Deposition temperature (°C)	Roughness parameters			
	R_a (nm)	R_z (nm)	R_{sk} (-)	R_{ku} (-)
25	1.9	18.9	0.7	3.9
200	2.1	18.2	0.2	3.0
400	5.2	42.3	-0.1	2.4

Z scan vs. force curves were achieved after each indentation (Figure 2). These curves were further interpreted using the XEI Image Processing Tools for SPM Data. The interpretation was realized using both the Oliver and Pharr method and the Hertzian method. If the first model allows us to determine the hardness, the second one allows us to determine the modulus of elasticity. The main difference between these two models is given by the presence/absence of the plastic deformation. On this regard, the Herztian method assumes that there is only elastic deformation, while the Oliver and Pharr method assumes that the material suffers both elastic and plastic deformation. The interpretation takes into account for both methods the Poisson's ratio specific to the tested material – niobium nitride, in our case. If the Oliver and Pharr does not require any other characteristics, the tip shape and the face angle have to be specified in the Hertzian model.

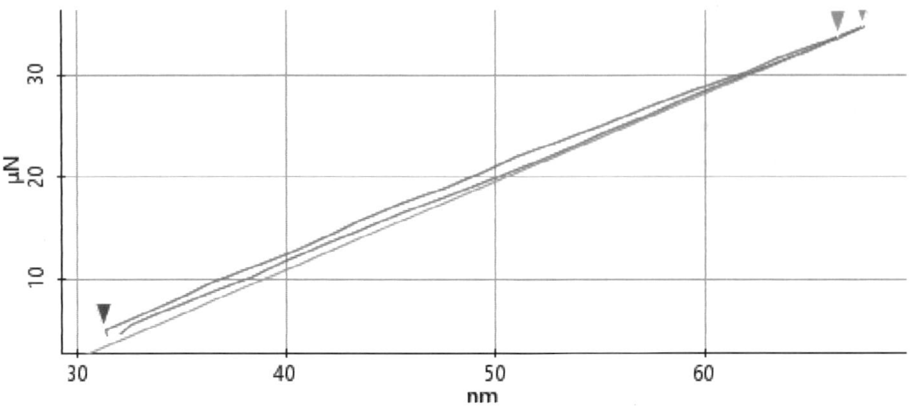

Figure 2. Z scan vs. force curve for a niobium nitride thin film deposited on silicon substrate at room temperature and tested at 20 °C.

The fluctuation of the modulus of elasticity for the niobium nitride thin films deposited at 25, 200 and 400 °C and tested at temperatures between 20 and 100 °C is graphically given in Figure 3. We can affirm that the niobium nitride thin films deposited on silicon substrates at room temperature are characterized by the highest value of the modulus of elasticity regardless of the testing temperature. When varying the deposition temperature, the decrease in this mechanical property ranged between 10 % (when the films were tested at 100 °C) and 15 % (when the films were tested at 20 °C). If the modulus of elasticity of niobium nitride deposited at 25 and 200 °C decreased with about 22 % when the testing temperature increased from 20 to 100 °C, the modulus of elasticity of niobium nitride deposited at 400 °C decreased less (about 17 %). When the tests were performed at 20 °C, the modulus of elasticity ranged between 117.6 GPa (for the films deposited on silicon substrates at 400 °C) and 138.8 GPa (for the films deposited on silicon substrates at room temperature).

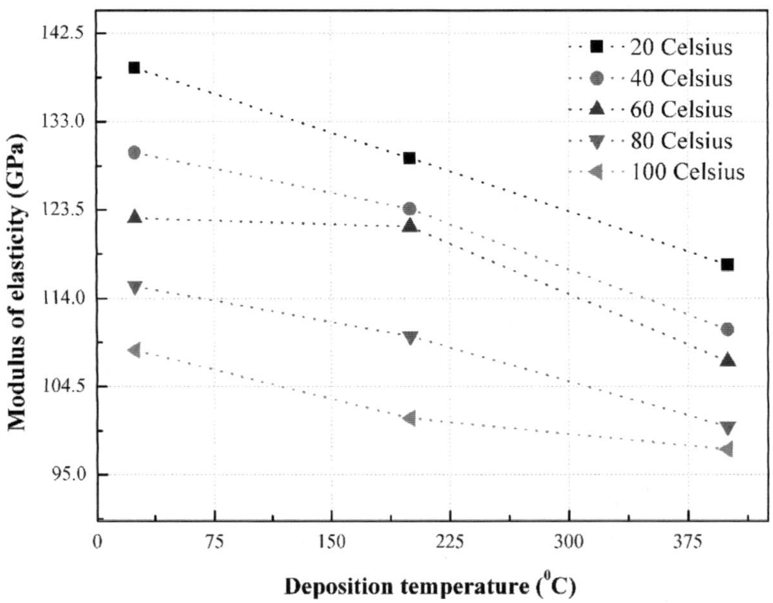

Figure 3. The fluctuation of the modulus of elasticity for niobium nitride thin films deposited at different substrate temperatures and tested at temperature between 20 and 100 °C.

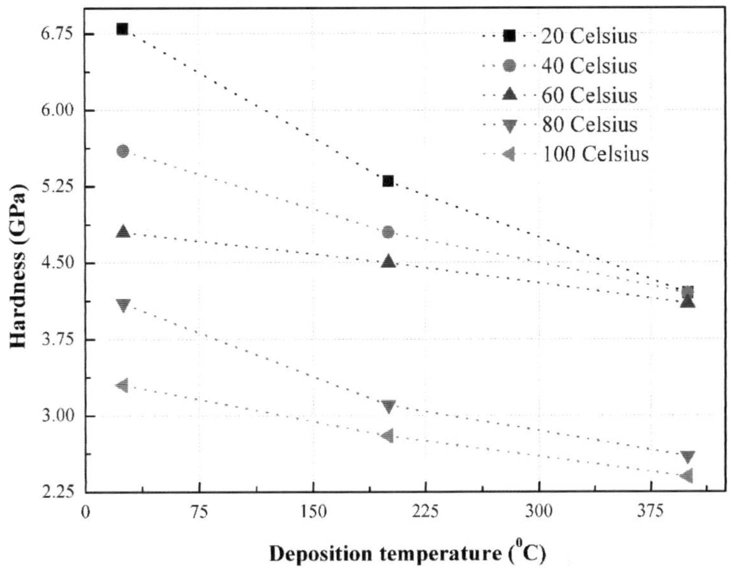

Figure 4. Hardness for niobium nitride thin films tested at temperatures between 20 and 100 °C.

The hardness of the thin films is another mechanical property of interest. The hardness fluctuation for the deposited niobium nitride thin films investigated at temperatures between 20 and 100 °C is graphically given in Figure 4. In general, the same trend as for the modulus of elasticity can be observed. We must bring out into relief that the decrease of hardness with the increase in testing temperature up to 100 °C is much more pronounced. This decrease is about 51 % in the case of the films deposited at room temperature and 43 % in the case of the films deposited on substrates preheated at 400 °C. When discussing about the same testing temperature, the hardness is still decreasing but the decrease is less accentuated, varying between 15 % (for testing temperature of 60 °C) and 38 % (for testing temperature of 20 °C). When tested at 20 °C, the hardness ranged between 4.2 GPa for the films deposited at 400 °C and 6.8 GPa for the films deposited at room temperature.

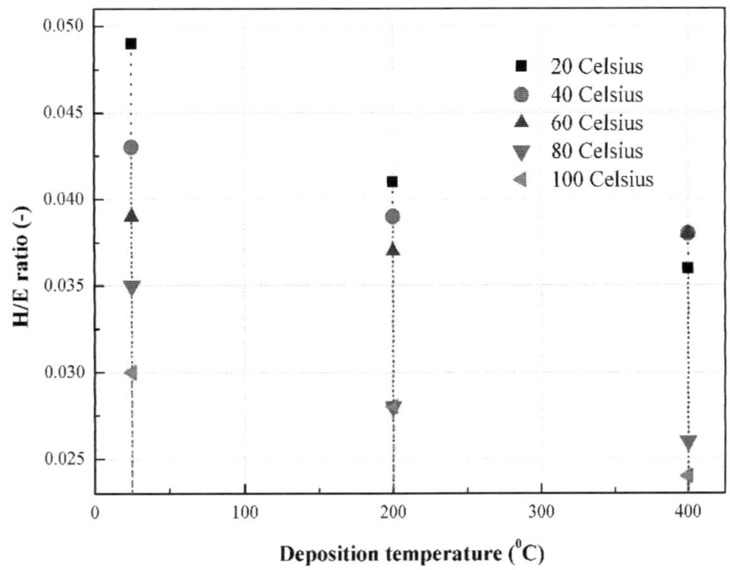

Figure 5. Fluctuation of the hardness/modulus of elasticity ratio for the investigated niobium nitride thin films tested at temperatures between 20 and 100 °C.

The hardness/modulus of elasticity ratio was calculated for each sample to determine the material characterized by the best mechanical behaviour. The fluctuation of this ratio is given in Figure 5. We notice that the value of this ratio decreases, in general, with the increase in testing temperature from 20 °C to 100 °C regardless of the deposition temperature. Its values ranged between 0.024 and 0.049. The niobium nitride thin films deposited on silicon substrate at room temperature and tested at 20 °C present the highest value of the ratio and, implicitly, are characterized by the best mechanical characteristics.

The change in niobium nitride thin films properties is probably due to their crystal structure that may contain different phases such as cubic δ-NbN, hexagonal β-Nb_2N, tetragonal γ-Nb_4N_3 and so on [7, 12, 13]. Moreover, the preheating of the substrate at 200 and 400 °C induces some thermal stresses which also affect the structure of the deposited materials.

4. Conclusions

Niobium nitride thin films were deposited by DC magnetron sputtering on silicon substrates at temperature between 25 and 400 °C. The films were further investigated by atomic force microscopy analyses. We determined that the increase in deposition temperature led to the obtaining of rough surfaces. The highest mechanical characteristics are specific to the niobium nitride thin films deposited at room temperature. The modulus of elasticity and the hardness decreased with the increase in both deposition temperature and testing temperature. However, the testing temperature had a stronger downward effect than the deposition temperature, leading to a decrease of up to 50 % of the hardness. We assume that the crystal structure of the deposited films and the thermal stresses are causing the change in these properties.

5. References

[1] Gotoh Y, Nagao M, Ura T, Tsuji H and Ishikawa J 1999 *Nucl. Instrum. Methods Phys. Res. B* **148** 925

[2] Escobar C A, Caicedo J C and Aperador W 2014 *J. Phys. Chem. Solids* **75** 23

[3] Xi Y, Gao K, Pang X, Yang H, Xiong X, Li H and Volinsky A A 2017 *Ceram. Int.* **43** 11992

[4] Singh A, Kuppusami P, Khan S, Sudha C, Thirumurugesan R, Ramaseshan R, Divakar R, Mhandas E and Dash S 2013 *Appl. Surf. Sci.* **280** 117

[5] Singh K, Krishnamurthy N and Suri A K 2012 *Tribol. Int.* **50** 16

[6] Farha A H, Er A O, Ufuktepe Y and Elsayed-Ali H E 2012 *Mater. Chem. Phys.* **132** 667

[7] Al Mamun M A, Farha A H, Ufuktepe Y, Elsayed-Ali HE and Elmustafa AA 2015 *Appl. Surf. Sci.* **330** 48

[8] Bekermann D, Barreca D, Gasparotto A, Becker H W, Fischer R A and Devi A 2009 *Surf. Coat. Technol.* **204** 404

[9] Bendavid A, Martin P J, Kinder T J and Preston E W 2003 *Surf. Coat. Technol.* **163-164** 347

[10] Cansever N 2007 *Thin Solid Films* **515** 3670

[11] Krishnan R, David C, Ajikumar P K, Dash S, Tyagi A K, Jayaram V and Raj B 2011 *Surf. Coat. Technol.* **206** 1196

[12] Farha A H, Er A O, Ufuktepe Y, Myneni G and Elsayed-Ali H E 2011 *Surf. Coat. Technol.* **206** 1168

[13] Cansever N, Danişman M and Kazmanli K 2008 *Surf. Coat. Technol.* **202** 5919

IOP Conf. Series: Materials Science and Engineering **724** (2020) 012061 doi:10.1088/1757-899X/724/1/012061

Effect of microstructure on wear behaviour of aluminium 2014 (Al2014)

S Bhaumik[1], V Paleu[2*], D Chowdhury[1], C Cîrlan Paleu[2] and S Datta[1]
1 Tribology and Surface Interaction Research Laboratory, Department of Mechanical Engineering, SRM Institute of Science and Technology, Kattankulathur, IN
[2] Gheorghe Asachi Technical University of Iaşi, Mechanical Engineering Faculty, Mechanical Engineering, Mechatronics and Robotics Department, 63 D. Mangeron Blvd., 700050, Iaşi, RO, E-mail: vpaleu@tuiasi.ro

Abstract. The paper investigates the effect of ageing on the hardness and wearing of Al2014. Precipitation hardening and ageing were carried out on the samples. The samples were solution treated at 550°C and were naturally aged. Quenching was carried out using ice-brine solution. Vickers Hardness were taken at regular intervals till 900 hours. The peak hardness value of the natural aged sample was found to be 86HV (93 hours) as compared to the untreated sample (32.4HV) The wear tests were conducted at 10N, 500rpm for 30 minutes. After the tribo test it was observed that the coefficient of friction increased by 7.06% in the naturally aged sample, but the wear rate in case of as-received samples was 46% higher than that of the naturally aged sample. The change in the microstructure of Al2014 is the primary reason for the difference in wear rate of the samples. The present work will help to understand the influence of heat treatment of aluminium alloy on wear properties.

1. Introduction

Aluminium alloys are strengthened by heat treatment method subject to alloy should have such alloy element which can form coherent precipitate. Aluminium 2014 responds heat treatment. It contains copper, magnesium and silicon as principle alloying element. The strength of the alloy is improved by grain refinement and precipitation hardening techniques significantly [1,2].

According to International Alloy Designation System (IADS) aluminium alloys of 2XXX, 6XXX and 7XXX respectively are heat- treatable. Tempering is an effective form of heat treatment for strengthening of the above aluminium alloys. According to IADS tempering designation T4 indicates alloy has been solution treated, quenched and naturally aged [3].

A similar trend in hardness can also be observed when the heat- treatable aluminium alloys are made into composites. Pavithouran et al studied the hardness properties of Al6061 reinforced with silicon carbide (SiC) and graphite particles.

Hardness increased with increase in SiC but decreased when graphite content increased to a certain level, which is due to the fact that binding energy and also hardness of graphite is less [4]. Chacko et al [5] studied the ageing behaviour of Al2014- 15 vol% SiC composite and found same variation of hardness in T4 and T6 conditions. Jha et al [6] showed that by adding strontium to Al2014- SiC composite hardness can be increased by artificial ageing.

2. Experimental details

2.1. Material

The aluminium alloy Al2014 was bought as extruded cylindrical rods of diameter 30 mm from a local vendor at Chennai. Its chemical composition, as shown in Table 1, was obtained by using ASTM E1251- 2011 standard.

Table 1. Chemical composition of Al2014 as- received sample.

Elements	Al	Cu	Si	Mn	Mg	Fe
Weight %	Remainder	4.639	0.799	0.638	0.594	0.187

2.2. Heat Treatment Process

Samples of 10 mm diameter and 30mm length were cut from the as- received sample. They were then kept in an alumina crucible and solution treated to 550°C for 1 hour in a furnace. They were then quenched in ice- brine solution. Six of them were put in an alumina crucible and kept in the furnace at 550°C for 1 hour in the furnace.

2.3. Measurement of Hardness

The sample heated to 550 °C (T4 samples) was polished in emery papers of grades 220, 400 and 600 and its hardness was measured using a Vickers Hardness Testing Machine (Make: FIE). A load of 50N was applied. Time of contact of the sample and the indenter was kept as 4 seconds. This process was carried out at intervals for a period of 845 hours.

2.4. Measurement of Roughness

The surface roughness of the samples was measured using a SurfCom contact type of profilometer. The roughness values were kept between 0.2- 0.3 μm.

2.5. Investigation of Microstructures of the Samples

The T4 and as- received samples were again polished on the circular cross- section using emery papers of with grit size 220, 400, 600, 800 and 1000. The samples were then polished on a disc polishing machine, where diamond lubricant spray (Hiffin fluid) was used to moist the velvet cloth and diamond paste was applied on the metal surface which acted as an abrasive material. The polished samples were then washed with running water and then dried. It was then etched with Keller's Reagent with an etching time of 10 seconds. It was again washed with running water and then dried. The microstructures of the samples were studied using an Olympus optical microscope.

2.6. Investigation of Wear Characteristics

The wear characteristics of the samples were tested on a pin-on-disc tribometer manufactured by Magnum in dry condition. The disc was of EN 31 with hardness 58-50 Rc. A load of 10 N was applied on the pin (Al 2014) during the 30 minutes test. The disc was rotated at 500 rpm.

2.7. Surface characterization

In order to understand the severity of the wear, the wear tracks on Al 2014 pin were analysed using a using a FEI Quanta 200 Field Emission Scanning Electron Microscopy.

3. Results and Discussions

3.1. Hardness

Figure 1 shows the effect of natural ageing on the sample, which was solution treated to a temperature of 550°C. The hardness first increased till 42 hours and then maintained a constant value. This was due to the gain of activation energy at room temperature of the super- saturated metal which led to the formation of Guinier-Preston (GP) zones. GP zones are fine-scaled near the order of 3–10 nm in size of solute enriched regions of the material, which offer physical obstructions to the motion of dislocations. Hardness of the extruded sample was measured as 35.6 HV. Hence the value of hardness increased on T4 treatment. Table 2 shows the values of the hardness measured.

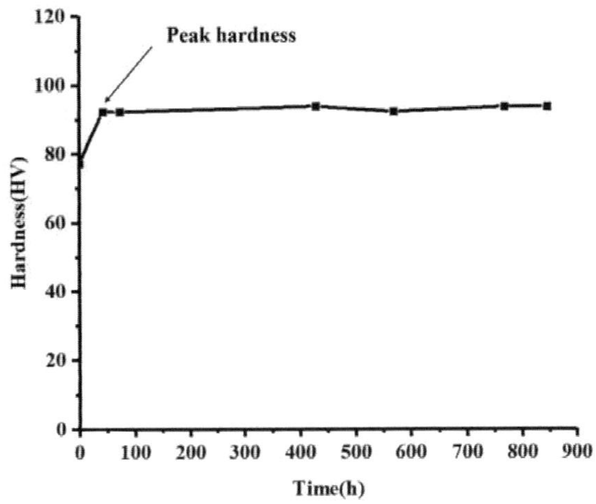

Figure 1. T4 hardness curve of the sample.

Table 2. Measurement of Vickers Hardness of T4 Sample.

Time (hour)	Hardness (HV)
0	77.3
42	92.27
73	92.23
429	93.87
569	92.23
767	93.67
845	93.60

The extruded sample hardness was measured as 32.6 HV. The peak hardness of T4 samples show that it is 93.87 HV, which occurred due to GP Zone formation.

3.2. Microstructure

Figure 2 shows the microstructures of the samples. Figure 2(a) shows the microstructure of the alloy in as received condition (extruded), where the grains are heavily elongated and the θ phase is seen to be elongated at the grain boundary. In Figure 2(b), the second phase is seen to be in lesser amount due to solution treatment. Though, the grain boundaries are not revealed properly, but the shape of the grains seems to be equiaxed. T4 heat treatment have resulted in formation of GP zones, which is not visible in the optical microstructure.

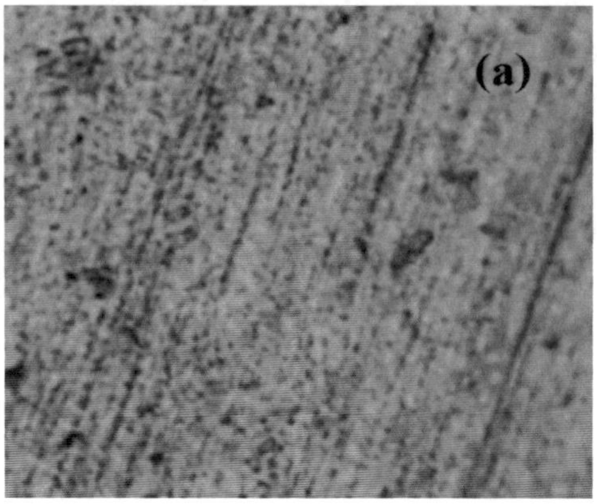

Figure 2 (a). Microstructure of extruded sample at 100X.

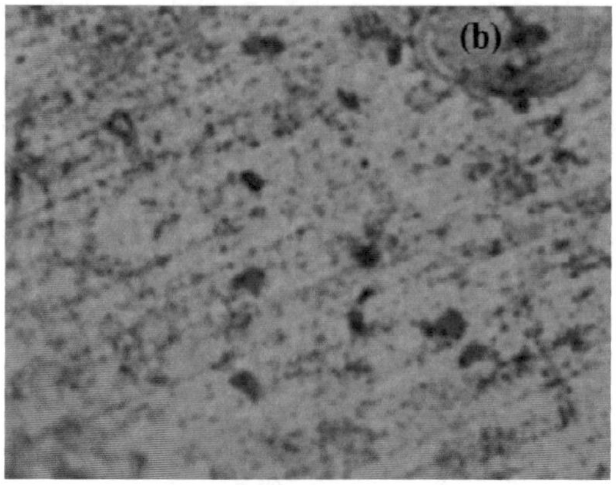

Figure 2 (b). Microstructure of T4 sample at 100X.

3.3. Wear test

Figure 3 shows that due to ageing treatment (T4) the coefficient of friction is increased, though not significantly. But the wear rate has decreased appreciably. During T4 treatment fine GP zones are formed, which act as harder dispersions in the aluminium matrix. This hard phase reduces the wear rate, but could not influence the coefficient of friction. Due to the heat treatment and formation of GP zones, the hardness of Al2014 has also increased. This increase in hardness could reduce the wear rate to a significant extent.

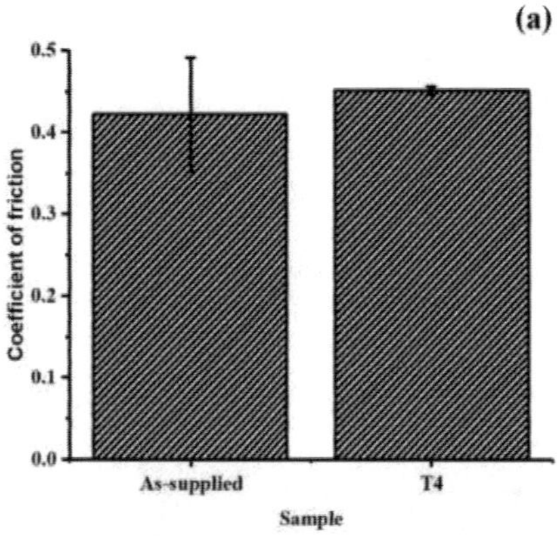

Figure 3 (a). Variation of coefficient of friction with the heat treatment.

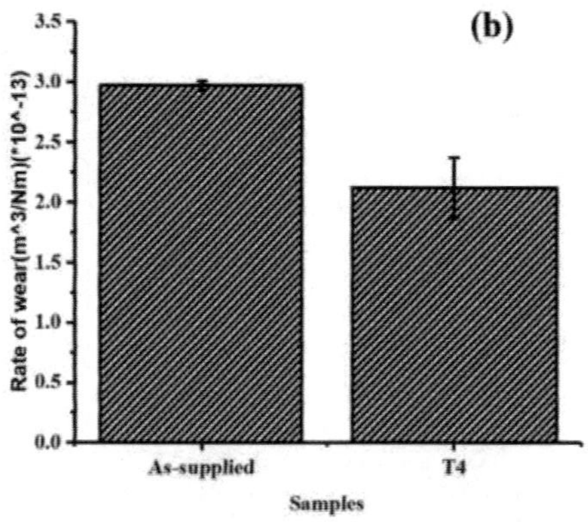

Figure 3 (b). Variation of wear with heat treatment.

3.4. *Surface characterization*

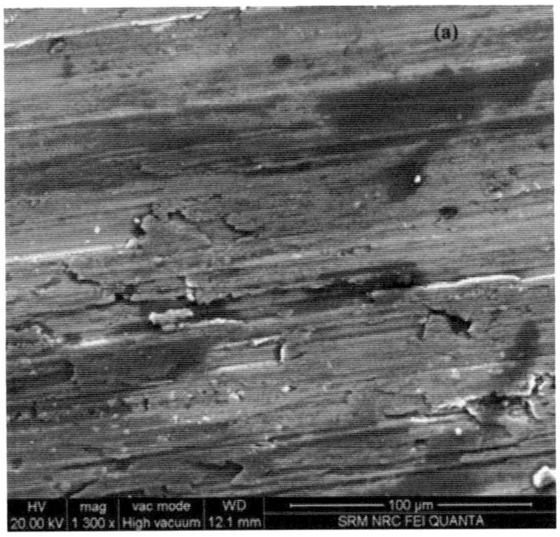

Figure 4 (a). FESEM image of extruded sample.

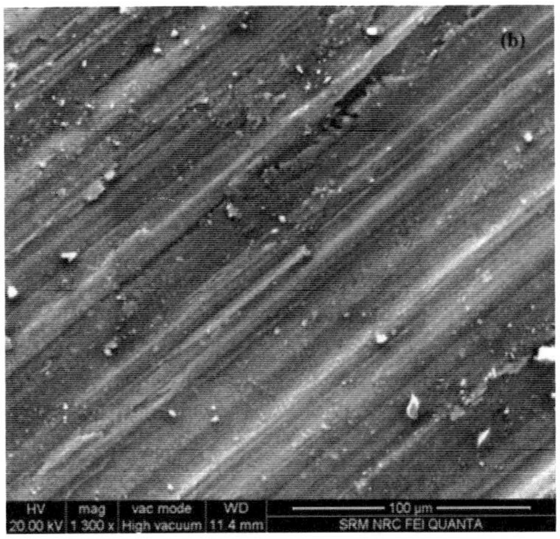

Figure 4 (b). FESEM image of T4 sample.

Figure 4 (a) and figure 4(b) show the surface degradation in the two samples. Figure 4(a) shows higher amount of crack as there is more surface-to-surface contact, as compared to T4 sample (figure 4b). This is due to the low hardness of the extruded sample.

4. Conclusions

The hardness of T4 sample increased due to the solution treatment of the extruded sample. The results of tribo-tests indicated that the coefficient of friction increased by 7.06% for the hardened sample, but the wear rate in case of as received samples was 46% more than natural aged sample. T4 treatment can be used to increase hardness of an Al2XXX sample and also improve wear properties. The present work will help to understand the influence of heat treatment of aluminium alloy on wear properties. Future test must focus on the wear rate of Al2XXX in long-lasting tests, both for dry and lubricated conditions.

5. References

[1] Gadpale V, Pragya B, and Manoj M 2018 *IOP Conf. Series: Mater. Sci. & Eng.* **338**.
[2] Chand S, Madhusudhan D, and Uma A 2016 *Int. J. Adv. in Mech. and Civil Eng.* ISSN: 2394-2827 **3** (3).
[3] Polmear I J 2006 *Light Alloys-4th Ed.* Butterworth-Heinemann eBook ISBN: 9780080496108.
[4] Pavithouran B. and Swathanandan J 2015 *Int. J. Technol. Enhance. & Emerg. Eng. Research* **3** (04).
[5] Chacko M and Nayak P 2014, *Int. J. of Mater. and Metal. Eng.* **8** (3).
[6] Jae - Ho J and Dae - Geun H 2013 *Trans. Nonferrous Met. Soc. China* **23** 631–635.

Temperature influence on mechanical properties of chromium nitride thin films

F M Şerdean[1*], V V Merie[2], G Negrea[2], C Bîrleanu[1] and I Mărieş[1]

[1]Technical University of Cluj-Napoca, Department of Mechanical Systems Engineering, 103-105 Muncii Avenue, 400641, Cluj-Napoca, Romania
[2]Technical University of Cluj-Napoca, Department of Materials Science and Engineering, 103-105 Muncii Avenue, 400641, Cluj-Napoca, Romania

E-mail: florina.rusu@omt.utcluj.ro

Abstract. Thin films present interest for researchers due to their applicability in areas such as developing microelecromechanical devices, hard coatings, cutting tools and diffusion barriers. This paper is focused on studying the influence of the testing temperature on the mechanical properties of chromium nitride (CrN) thin films. For this purpose, CrN thin films were deposited by direct current magnetron sputtering on silicon substrate using a chromium target of high purity. All the films were obtained by keeping almost all deposition parameters constant, while the nitrogen flow rate was varied. Hence, three types of samples of CrN thin films were obtained using different nitrogen flow rates namely 2 cm^3/min, 4 cm^3/min and 6 cm^3/min. The mechanical properties of these thin films such as the hardness and Young's modulus were investigated by performing nanoindentation using an atomic force microscope. All experimental investigations were conducted by varying the testing temperature between 20 °C and 100 °C. The purpose was to find the dependency between the testing temperature and the mechanical properties of CrN films deposited at different nitrogen flow rates. The obtained experimental results indicate a significant influence of the testing temperature on the properties of the studied CrN thin films.

1. Introduction

Thin films industry has presented interest both to researchers and companies due to the many real-world applications that require thin films. In particular, CrN thin films can be used in different applications due to their good corrosion and oxidation features [1]. Several studies have been conducted on CrN thin films with respect to the deposition procedure which can vary from direct current (DC) or radio-frequency (RF) magnetron sputtering [2] to vacuum arc plasma deposition [3] or ion beam assisted deposition [4]. The main purpose of such studies was to characterize the thin films using different techniques ([1, 5-8]) or to determine the influence of the deposition parameters on the film structure [9] or the adhesion force [10]. Nevertheless, there aren't studies available in the open literature that target the influence of the deposition parameters on the mechanical properties of CrN thin films determined using the atomic force microscopy (AFM).

The AFM has been used as a complementary method to other techniques with the purpose of ensuring a more complete characterization of the CrN thin film samples. Relevant examples encompass the experiments conducted by X-ray diffraction and scanning electron microscopy [11], by X-ray photoelectron spectroscopy [12] or by transmission electron microscopy and glow-discharge optical emission spectroscopy [13].

The purpose of this work is to deposit the CrN thin films by DC reactive magnetron sputtering using different nitrogen flow rates and to characterize them using the nanoindentation module of the AFM at different testing temperatures. The obtained experimental results will allow to identify the temperature effect on the deposited samples, as well as to determine the thin films characterized by the best mechanical behavior.

2. Materials and experimental procedure

The samples of CrN thin films were deposited on silicon Si (100) substrates. The deposition process employed a chromium target with a purity of 99.95 % and was conducted by DC magnetron sputtering. The deposition chamber atmosphere contained a mixture of argon and nitrogen. The nitrogen flow rate varied for each sample as it can be seen in table 1, while the argon flow rate was kept constant at 40 $cm^3 \cdot min^{-1}$. All other deposition paraments were also kept constant.

The experimental investigations on the deposited CrN thin films were conducted using a XE 70 AFM at the Micro and Nano Systems Laboratory, from the Technical University of Cluj-Napoca. First all samples were scanned in the non-contact mode of the microscope using a PPP-NCHR cantilever in order to determine the topography of each sample. Then, using the nanoindentation module of the AFM the mechanical properties of the samples were experimentally determined at different temperatures namely 20, 40, 60, 80 and 100 °C respectively. The tests were performed at a relative humidity of 20 %.

Table 1. Deposition conditions for investigated samples.

Sample	T (°C)	Time (min)	P (mtorr)	Id (mA)	Q_{Ar} (cm^3/min)	Q_{N2} (cm^3/min)
CrN_2						2
CrN_4	200	20	2.2	300	40	4
CrN_6						6

3. Results and discussions

3.1 Topography

After the deposition of the samples of CrN thin films, they were characterized using the AFM. First, the topography was investigated. Each sample was scanned in non-contact mode and, using the XEI Image Processing Tools for SPM data, 3D images such as the one in figure 1 were obtained.

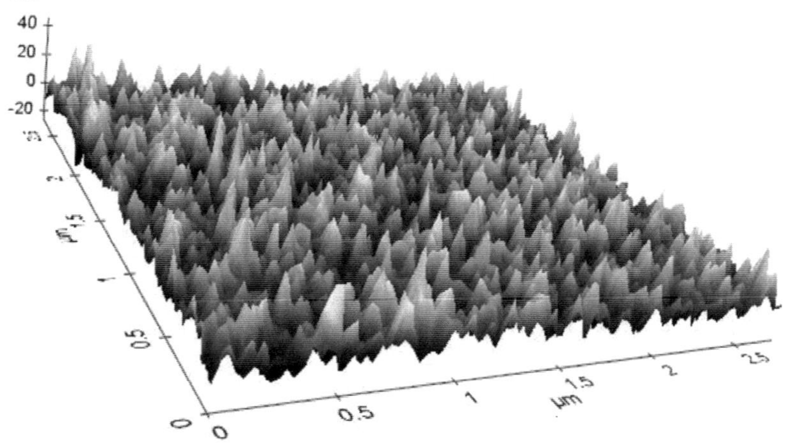

Figure 1. The 3D image of the sample CrN_6.

Also, the software provided statistical roughness parameters for each thin film such as the ones presented in figure 2. The values of the average roughness, R_a, the average maximum height of the profile, R_z, the skewness of the roughness, R_{sk}, and the kurtosis of the roughness, R_{ku}, determined at a testing temperature of 20 °C for all samples are encompassed in table 2. As it can be observed, the sample deposited at the smallest nitrogen flow rate has the smallest average roughness. The increase in the nitrogen flow rate up to 4 cm^3/min determines an increase in this roughness parameter up to 2.8 nm. However, a more significant increase in the average roughness is recorded when the nitrogen flow rate was further increased to 6 cm^3/min.

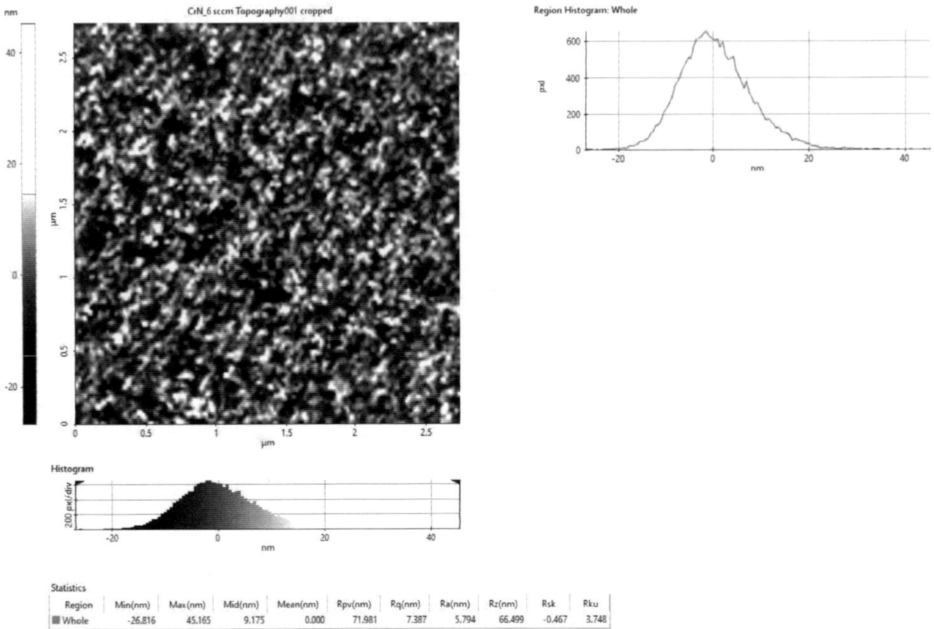

Figure 2. The surface of the sample CrN_6 and the statistical parameters provided for its roughness by the AFM software.

Table 2. Statistical roughness parameters for all samples.

Parameter \ Sample	CrN_2	CrN_4	CrN_6
R_a (nm)	1.1	2.8	5.8
R_z (nm)	15.0	30.5	66.5
R_{sk} (-)	0.4	-1.2	-0.5
R_{ku} (-)	4.7	5.3	3.7

3.2 Nanohardness and Young's modulus

In order to characterize the samples from the mechanical point of view, nanoindentation tests were performed. After these tests, force vs. Z scan curves similar to the one presented in figure 3 were obtained. These curves were interpreted using the Oliver and Pharr method in order to determine the nanohardness of the thin films and the Hertzian model in order to determine Young's modulus.

Figure 4 presents the variation of the nanohardness for all samples with respect to the testing temperature. It can be seen that this mechanical characteristic increased with the increase in nitrogen flow rate when tested at 20 °C, 80 °C and 100 °C. For the other two testing temperatures the largest values for the nanohardness have been obtained for the CrN_4 sample, while for all temperatures the lowest values for the nanohardness were obtained for the CrN_2 sample. When analyzing the results for each sample with respect to the variation of the testing temperature, it can be observed that the nanohardness values decrease for all samples. First, the decreasing trends are steep, then they become mild and in the end the decrease in the values of the nanohardness becomes strong again for all samples. The largest total decrease of almost 76% was registered for the CrN_2 sample, while the lowest of almost 66% was registered for the CrN_6 sample.

Model : Oliver and Pharr Model

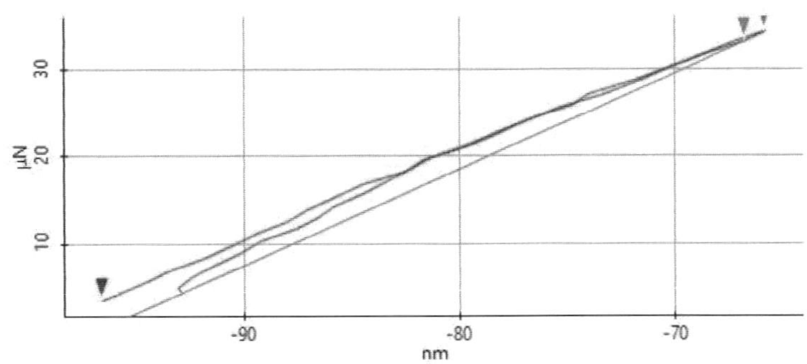

Cursors

Cursor	ΔX(nm)	ΔY(μN)	Left X(nm)	Left Y(μN)	Right X(nm)	Right Y(μN)
Force	30.880	30.922	-96.585	3.329	-65.705	34.251
Slope	0.984	1.089	-66.689	33.162	-65.705	34.251

Slope Cursor Index 1 : 23
Slope Cursor Index 2 : 22
Contact Depth : 9.93nm
Tip Shape : Berkovich
Poisson's ratio of the tip : 0.07
Poisson's ratio of the sample : 0.2
Hardness : 12.8GPa

Figure 3. Image of the XEI Image Processing Tools for SPM Data for determining the hardness of the sample CrN_6 and tested at 20 °C.

Decreasing trends with respect to the increase in the testing temperature were obtained for all samples when analyzing the experimental results for the modulus of elasticity. There is an almost linear decreasing trend for the CrN_2 sample, with a step of almost 9% decrease with every increase of 20 °C in the testing temperature. For the CrN_4 sample the largest decrease is around 15% and it is observed when increasing the temperature from 80 to 100 °C. For the CrN_6 sample the largest decrease in the values of Young's modulus is around 17% and it is observed when increasing the temperature from 20 to 40 °C. The total decrease in the values of the modulus of elasticity was lower for each sample when compared to the decrease registered for the values of the nanohardness. The largest total decrease of around 37% was registered for the CrN_6 sample, while the lowest of around 30% was registered for the CrN_4 sample. The change that occurs in the values of the mechanical properties is due to the growth after different preferential directions. Hence, the values of the studied mechanical properties may fluctuate depending on the main growth direction.

To identify the thin film with the best mechanical behavior, the nanohardness/modulus of elasticity ratio was computed. For each of the three types of thin films the ratio was computed for all testing temperatures and the results are presented in table 3. As it can be observed the ratio decreases for the first two samples with the increase in testing temperature. For the third sample the exception from the decreasing the trend is due to the ratio of the mechanical properties obtained when the tests were performed at 40 °C. For the same testing temperature, the increase in nitrogen flow rate determines an increase in the studied ratio for the tests performed at 80 and 100 °C. This increase is rather significant at 80 °C when the ratio doubles its value. For the tests performed at 20, 40 and 60 °C the largest values of the ratio are obtained for the CrN_4 sample, while the lowest for the CrN_2 sample. The chromium nitride thin film deposited at a nitrogen flow rate of 4 $cm^3 \cdot min^{-1}$ and tested at 20 °C has the highest value of the ratio between the nanohardness and the modulus of elasticity and therefore, the best mechanical behavior for applications such as hard coatings.

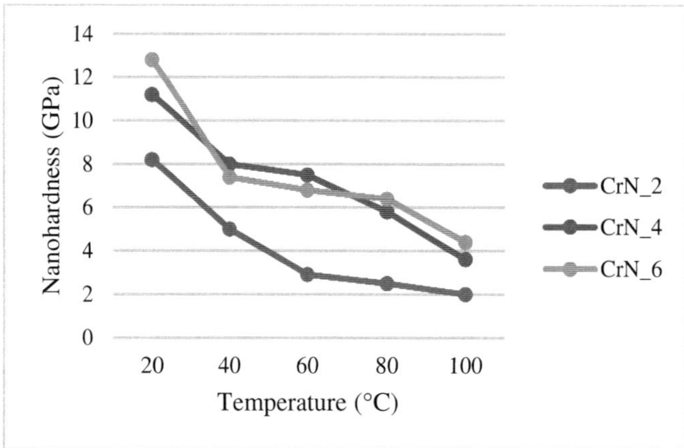

Figure 4. Nanohardness variation with respect to testing temperature.

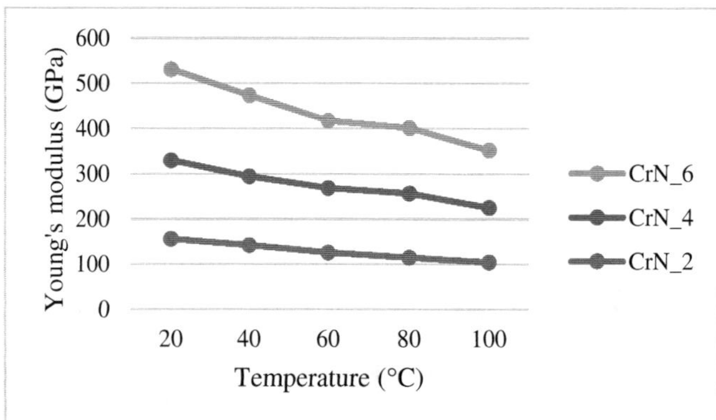

Figure 5. Young's modulus variation with respect to testing temperature.

Table 3. Nanohardness/Young's modulus of the investigated samples.

Sample	Nanohardness/Young's modulus, H/E (-)				
	20 °C	40 °C	60 °C	80 °C	100 °C
CrN_2	0.053	0.035	0.023	0.022	0.019
CrN_4	0.065	0.053	0.053	0.041	0.030
CrN_6	0.063	0.041	0.046	0.044	0.035

4. Conclusions

This paper presents the deposition by DC magnetron sputtering of three sets of samples of CrN thin films using different nitrogen flow rates. The deposited thin films were characterized from the topographical and mechanical point of view using the AFM. An increasing trend of the roughness was observed with respect to the increase of the nitrogen flow rate. Moreover, an increasing trend was observed for the modulus of elasticity values with respect to the increase of the nitrogen flow rate regardless of the testing temperature and an increasing trend was identified for the values of the nanohardness only when the tests were performed at 20, 80 and 100 °C.

Also, the work conducted for this paper encompassed the study of the influence of the testing temperature on the mechanical properties of the CrN thin films. Moreover, the ratio between the nanohardness and modulus of elasticity was determined for all samples at all testing temperatures in order to emphasize the sample with the best mechanical behavior, which proved to be the CrN_4 sample.

5. References

[1] Mammeri F Z, Chekour L and Rouag N 2013 Characterization of nitride thin films using SEM and EDX *Acta Phys. Pol. A* **123** pp 294-5

[2] Tan S, Zhang X, Wu X, Fang F and Jiang J 2011 Comparison of chromium nitride coatings deposited by DC and RF magnetron sputtering *Thin Solid Films* **519** pp 2116-20

[3] Ovcharenko V D, Kuprin A S, Tolmachova G N, Kolodiy I V, Gilewicz A, Lupicka O, Rochowicz J and Warcholinski B 2015 Deposition of chromium nitride coatings using vacuum arc plasma in increased negative substrate bias voltage *Vacuum* **117** pp 27-34

[4] Ensinger W, Flege S, Kiuchi M and Honjo K 2012 Chromium nitride films formed by ion beam assisted deposition at low nitrogen ion energies in comparison to high energies *Nucl. Instrum. Methods Phys. Res. B* **272** pp 437-40

[5] Lippitz A and Hübert Th 2005 XPS investigations of chromium nitride thin films *Surf. Coat. Tech.* **200** pp 250-3

[6] Hones P, Sanjines R and Levy F 1997 Characterization of sputter-deposited chromium nitride thin films for hard coatings *Surf. Coat. Tech.* **94–95** pp 398-402

[7] Arif M, Sanger A and Singh A 2018 Sputter deposited chromium nitride thin electrodes for supercapacitor applications *Mater. Lett.* **220** pp 213-7

[8] Fryska S, Słowik J and Baranowska J 2019 Structure and mechanical properties of chromium nitride/S-phase composite coatings deposited on 304 stainless steel *Thin Solid Films* **676** pp 144-50

[9] Khan S, Mahmood A, Shah A, Raza Q, Rasheed M A and Ahmad I 2015 Structural and optical analysis of Cr2N thin films prepared by DC magnetron sputtering *Int. J. Min. Met. Mater.* **22** pp 197-202

[10] Rusu F M, Merie V V, Pintea I M and Molea A 2016 Adhesion analysis for chromium nitride thin films deposited by reactive magnetron sputtering *IOP Conf. Ser.: Mater. Sci. Eng.* **147** 012023

[11] Lin T-N, Han S, Weng K-W and Lee C-T 2013 Investigation on the structural and mechanical properties of anti-sticking sputtered tungsten chromium nitride films *Thin Solid Films* **529** pp 333-7

[12] Lazauskas A, Grigaliūnas V, Guobienė A, Andrulevičius M and Baltrusaitis J 2012 Atomic force microscopy and X-ray photoelectron spectroscopy evaluation of adhesion and nanostructure of thin Cr films *Thin Solid Films* **520** pp 6328-33

[13] Qi J and Thompson G E 2016 Comparative studies of thin film growth on aluminium by AFM, TEM and GDOES characterization *Appl. Surf. Sci.* **377** pp 109-20

Technological and structural analysis of Al_2O_3 $40TiO_2$ coating deposited on a shaft sleeve of hydraulic pump

C Cîrlan Paleu[1], **B Istrate**[1], **V Paleu**[1] **and C Munteanu**[1]
[1] Gheorghe Asachi Technical University of Iaşi, Mechanical Engineering Faculty, Mechanical Engineering, Mechatronics and Robotics Department, 63 D. Mangeron Blvd., 700050, Iaşi, RO, E-mail: cornelmun@gmail.com

Abstract. In the last decades, mechanical engineering applications have extensive use in many subdomains. Many researchers have used Al_2O_3 $40TiO_2$ (AMDRY6250) type coatings that have a wide use of plasma jets, significantly improving their wear, corrosion resistance and heat shock properties. It is especially used for hydraulic components such as pumps, impellers, and blowers. The performance of Al_2O_3 $40TiO_2$ layer is conditioned by a number of technological parameters of deposition, powder granulation and deposition incidence angle. In the present study, the authors will make sequential depositions of Al_2O_3 $40TiO_2$ on the surface of a material coming from a hydraulic system. Microstructural analysis of optical microscopy, scanning electron microscopy (SEM), and energy dispersive spectroscopy (EDAX) are performed both in the surface of the deposited material and in the section of the coated specimens, identifying the microstructure and the component phases. The authors highlighted the emergence of the "splat" type of formations that are specific to the ceramic coatings in the deposited layers and a good adhesion to the base material authorizing the use of this powder in the reconditioning of the hydraulic systems.

1. Introduction

Pumping station are vital key points of modern urban life, their continuously running in good state being a necessity for the people. Paleu et al. [1] analysed the main failures modes of pumping stations components. A weak spot of pumping stations are the pump themselves, especially their impellers, seals, bearings, casings, shafts, bushes, and sleeves. The latest of them (bushes and sleeves) can be easily reconditioned by proper choose of wear resistant coating materials and deposition method. In case of impellers and casings maintenance a special attention must be payed if coatings are used, as the thickness of these coatings can change the small interstitial space at the entrance of impellers, accentuating the cavitation phenomena [2-3].

Between coatings, Al_2O_3 $40TiO_2$ powder, also known as AMDRY6250, is recommended by producers for atmospheric plasma spray (APS) deposition on components of pumps. Recent research [4-11] indicated that there are few aspects known regarding the influence of aluminium titanate (Al2TiO5) in Al_2O_3-TiO_2 thermal spray coatings. Three Al_2O_3-$13TiO_2$ and three Al_2O_3- $40TiO_2$ from feedstock powder were analysed. The conclusion is the aluminium titanate ($Al_2TiO_5/Al_6Ti_2O_{13}$) can improve the hardness of Al_2O_3-$13TiO_2$ coating if found in the composition of feedstock powder, but it is detrimental to mechanical properties of Al_2O_3- $40TiO_2$. It was also underlined that it is possible to obtain Al2TiO5 during spraying, as long as it is provided a sufficient intermixing of the single oxides.

The behaviour of Al_2O_3–$40TiO_2$ and Cr_3C_2–20NiCr coatings deposited by HVOF spray technique on low cost carbon steel petroleum oil piping was studied by Zavareh et al. [5]. Results of corrosion and tribological tests show that Al_2O_3–$40TiO_2$ is superior to Cr_3C_2–20NiCr from corrosion resistance viewpoint, both coatings proving good resistance to wear. Vasudev et al. [6] deposited WC-Co-Cr and Al_2O_3 + $40TiO_2$ coatings by detonation gun spray process on cast iron. Tribological tests revealed better wear behaviour of WC-Co-Cr. In XRD analysis of Al_2O_3 + $40TiO_2$ coating can be observed the presence of aluminium titanate [6]. If results of [4] and [6] are correlated, the worse wear behaviour of Al_2O_3 + $40TiO_2$ reported in [6] is explainable.

Znamirowski et al. [7] studied the field electron emission of Al_2O_3+$13TiO_2$ and Al_2O_3+$40TiO_2$. The study was intended for parabolic antenna application. Better emissivity of Al_2O_3+$13TiO_2$ is explained by thinner structure of the coating, composed by small spheres.

Positive results were related on hot corrosion resistance in aggressive environment of Al_2O_3+$40TiO_2$ coating on Superni 718 and AE 435 superalloys deposited by detonation gun method [8-9]. The application was envisioned for boilers and gas turbines components.

The microstructural transformations of TiO2, Al_2O_3+$13TiO_2$ and Al_2O_3+$40TiO_2$ at plasma spraying and laser engraving were presented in [10], Al_2O_3+$40TiO_2$ being the best candidate with low porosity (1.9%) and acceptable hardness ($HV_{0.3}$=1083).

Szala and Hejwowski [11] studied cavitation erosion resistance of flame-sprayed Al_2O_3-40%TiO_2/NiMoAl cermet coatings (low-velocity oxy-fuel (LVOF)), with respect to cavitation erosion mechanism versus feedstock powder ratio (Al_2O_3-TiO_2/NiMoAl). It seems that the structure of the obtained coating is very important on cavitation resistance, but a content increase up to 80% of Al_2O_3-TiO_2 in the feedstock powder was positive.

Modern pumps have their shafts and bushes made of AISI 303 (EN 1.4305) and AISI 316 (EN 1.4401), and their impellers, casings and sleeves of AISI 304 (EN 1.4301) and AISI 316 L (EN 1.4404). An economical method to obtain acceptable shaft sleeves is electrochemically deposition of chrome coatings, an element resistant to abrasive wear and corrosion. This paper studies the microstructure of the Al_2O_3+$40TiO_2$ coating deposited by APS technique in different number of layers (5, 7, and 9 layers) on steel samples from a reconditioned pump shaft sleeve. Microstructural investigations are welcome in selecting the best ceramic coating for pump components maintenance.

2. Experimental section

2.1. Materials

The steel substrate belongs to a shaft sleeve of a vertical pump with electrochemically deposited chrome layer of thickness 100 μm. An examination of the pump shaft sleeve taken from years of exploitation proved that its surface had abrasive wear traces and even rusted parts (figure 1 and figure 2). These pumps are usually lubricated by processed water.

Figure 1. Worn shaft sleeve outer surface.

Figure 2. Cut worn shaft sleeve inner surface.

Parallelepipeds samples of 100 x 10 x 5 mm were cut from the shaft sleeve and chromed surface of the outer cylinder was completely removed by milling, its surface becoming flat. Samples preparation technology included sandblasting and polishing. Coated pads are made by atmospheric plasma spray (APS) deposition of Al_2O_3 40TiO_2 (AMDRY6250) on steel substrate.
Three coated flat samples were realised with different number of Al_2O_3 40TiO_2 deposited layers: 5, 7, and 9. These samples will be denoted as P1, P2, and P3, respectively.

2.2. Atmospheric Plasma Spray (APS) equipment and optimum deposition parameters
Table 1 presents the technological parameters used for the APS coating process.

Table 1. Technological parameters used for coating process.

Powder	Gun	Ar		H_2		Electric		Powder feeder 9MP			Spray distance (inch)
		Pressure (psig)	Gas flow (SCFH)	Pressure (psig)	Gas flow (SCFH)	DC (A)	DC (V)	Carrier gas flow (SCFH)	Air Preasure (psig)	Rate (lb/h)	
Al2O3 40TiO2 coating	9MB	75	110	50	10	500	60-70	13.5	20	5.6	4.5

2.3. Methods of analysis
The deposited coating surfaces for the three different samples (P1, P2 and P3) were analyzed by optical microscopy, scanning electron microscopy (SEM), and Energy Dispersive Spectroscopy (EDAX). SEM microscopy was used also for transversal images acquisition of different AMDRY6250 coatings.

3. Results

3.1. Results of optical microscopy
Figure 3 highlights the optical microstructure of the base material at different magnification powers. A recrystallized biphasic structure consisting of ferrite alpha and perlite grains is highlighted. The grain distribution is relatively uniform, with slight parallel orientations of alpha ferrite grains.

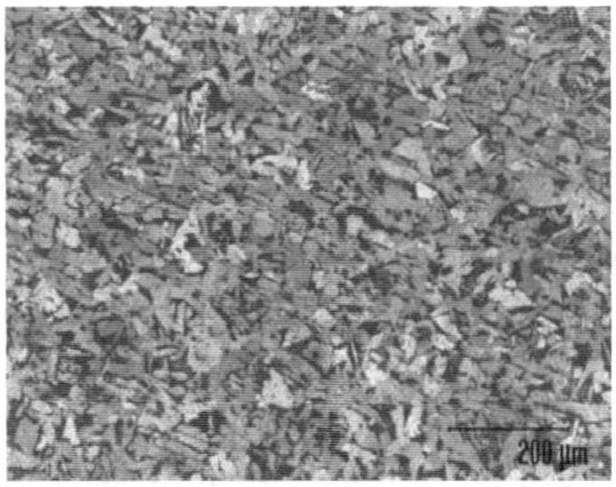

a)

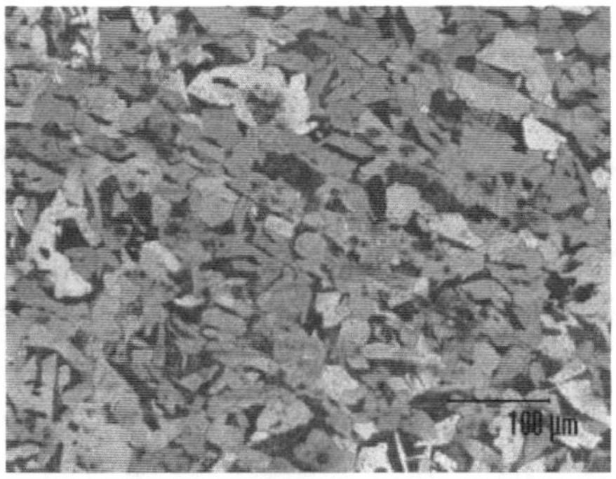

b)

c)

Figure 3. Optical images of based material microstructure: a) 100X; b) 200X; c) 500X.

3.2. SEM analysis of coated surfaces

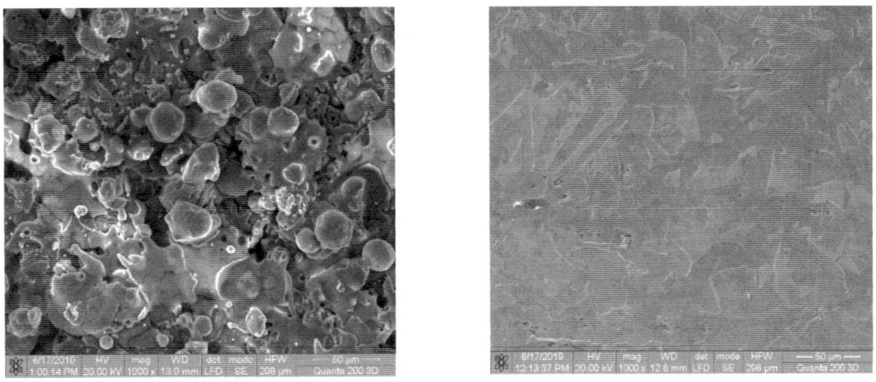

a) **b)**

Figure 4. SEM images of surface of the experimental samples: a) coated surface; b) uncoated surface.

In figure 4, SEM images indicate the morphology of the coated surface compared to the base material. These are similar with the optical micrographs and the surface of the deposited coating indicates the presence of a partially homogeneous layer with highlighting of some microcracks and unmelted particles.

The presence of microcracks may also be due to the different thermal contraction coefficient between the base material and the Al2O3-TiO2-coated layers during the cooling process.

3.3. Cross-section SEM analysis of coatings

The cross-section scanning electron images in figures 5, 6 and 7 reveal the morphology appearance of the coated samples. The thickness variation is between 55 microns (5 passes) and 116 microns (9 passes) - Table 2. It results that a single pass of the spraying process produces a layer thickness of about 12 microns. In all three cases the layers are adhered to the base material. The structure is specific to ceramic coatings with the presence of oxides in the form "splats" type due to successive passes during the deposition process.

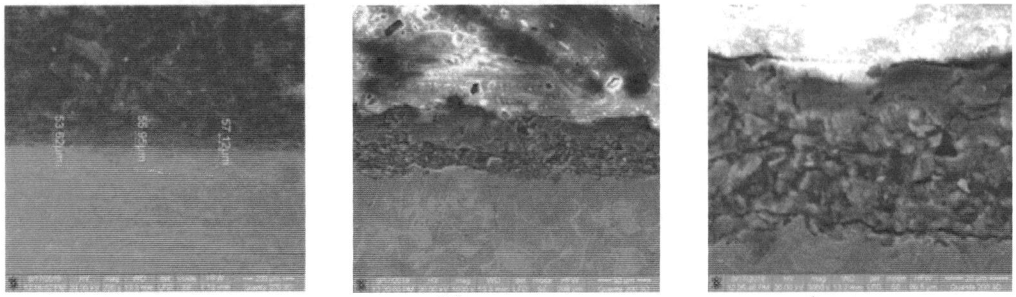

a) b) c)

Figure 5. Cross-section SEM images of 5 passes coated sample: a) 250X; b) 1000X; c) 3000X.

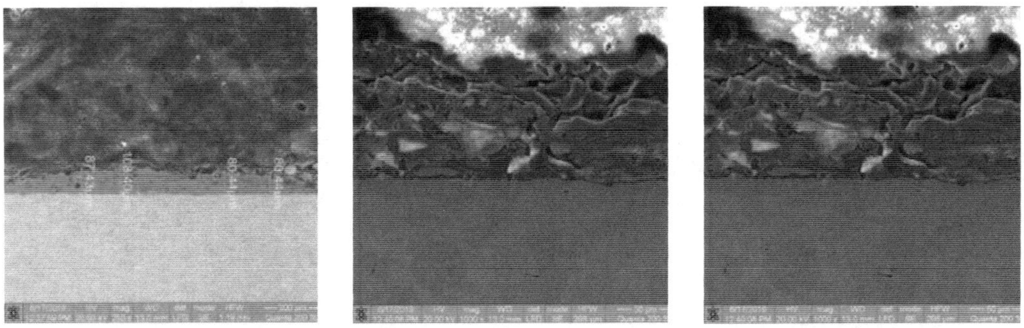

a) b) c)

Figure 6. Cross-section SEM images of 7 passes coated sample: a) 250X; b) 1000X; c) 3000X.

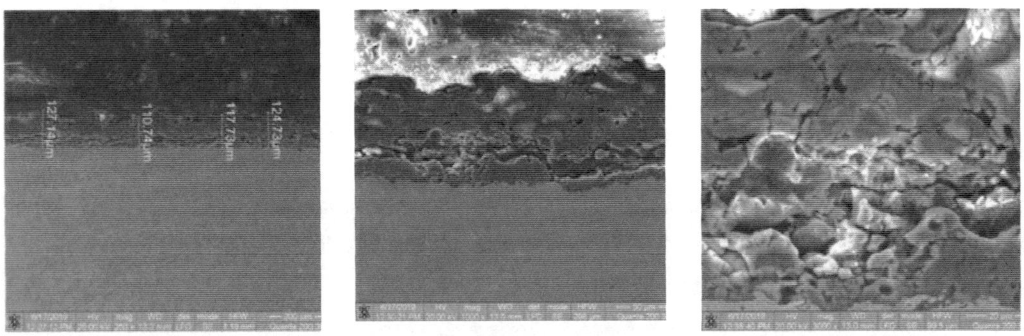

a) b) c)

Figure 7. Cross-section SEM images of 9 passes coated sample: a) 250X; b) 1000X; c) 3000X.

Table 2. Measurements of coating thicknesses.

Coatings	5 passes	7 passes	9 passes
Average thickness (μm)	55.56	87.89	116.58

3.4. EDAX analysis

Table 3 presents the chemical compositions and the distribution of the elements in the coatings section. Al, O, and Fe are evenly distributed in the base matrix, whereas titanium is found in isolated areas of the ceramic layer.

Table 3. Chemical compositions of the coated samples.

Chemical Elements (%wt.)	%O	%Al	%Ti	%Fe
5 passes	27.11	44.92	05.60	22.38
7 passes	29.80	46.06	07.73	16.41
9 passes	30.09	45.13	08.63	16.14

4. Conclusions

Different thickness layers of Al_2O_3 $40TiO_2$ (AMDRY 6250) were deposited by APS method on flat steel samples obtained from reconditioned shaft sleeve of a vertical pump. The quality of 5, 7 and 9 passes of coatings was analysed by optical microscopy, scanning electron microscopy (SEM), and Energy Dispersive Spectroscopy (EDAX).

The results show that the coatings are relatively homogenous and adherent to substrate. The thickness varies between 55 microns and 116 microns. The coatings produces by atmospheric plasma spraying (APS) are composed by individual splats of fast solidified spray powder particles. The presence of microcracks may also be due to the different thermal contraction coefficient between the base material and the $Al_2O_340TiO_2$-coated layers during the cooling process.

5. References

[1] Paleu Cîrlan C, Paleu V, Istrate B, Cimpoesu N and Munteanu C 2019 *IOP Conf. Series: Mater. Sci. Eng.* **591:012007** DOI: 10.1088/1757-899X/591/1/012007
[2] Bloch H P and Budris A R 2014 *Pump user's handbook life extension* - 4th Ed. (Lilburn: The Fairmont Press / Taylor & Francis) p 556
[3] Budris A R 2013 *WaterWorld* **28 – 4** 18
[4] Richter A, Berger L-M, Conze S, Sohn Y J and Vaßen R 2019 *IOP Conf. Series: Mater. Sci. Eng.* **480:012007**
[5] Zavareh M A et al. 2016 *Ceramics Int.* **42** (4) 5203-5210
[6] Vasudev H, Thakur L, Singh H and Bansal A 2018 *Kovove Mater.* **56** 55–63
[7] Znamirowski Z, Pawlowski L, Cichya T and Czarczynski W. 2004 *Surf. & Coat. Technol.* **187** 37–46
[8] Mishra N K, Kumar Naveen and Mishra S B 2014 *Indian J. Mater. Sci.*, **Article ID 453607**, 5 p., http://dx.doi.org/10.1155/2014/453607
[9] Mishra N K and Mishra S B 2015 *Bull. Mater. Sci.* **38** 7 1679–1685
[10] Tomaszek R, Pawlowski L, Zdanowski J, Grimblot J and Laureyns 2004 *Surf. & Coat. Technol.* **185** 137–149
[11] Szala M and Hejwowski T 2018 *Coatings* **8** 254; doi:10.3390/coatings8070254

Friction and wear resistance of Al_2O_3 $40TiO_2$ (AMDRY 6250) coating of a pump shaft sleeve bearing

V Paleu[1], C Cîrlan Paleu[1*], B Istrate[1], S Bhaumik[2] and C Munteanu[1]
[1] Gheorghe Asachi Technical University of Iaşi, Mechanical Engineering Faculty, Mechanical Engineering, Mechatronics and Robotics Department, 63 D. Mangeron Blvd., 700050, Iaşi, RO
[2] Tribology and Surface Interaction Research Laboratory, Department of Mechanical Engineering, SRM Institute of Science and Technology, Kattankulathur, IN

Abstract. The equipment of wastewater treatment plants is continuously subjected to different kinds of wear, as corrosion, abrasion, cavitation, erosion etc. The most affected components are those of slurry pumps, the slurry producing intense abrasive wear. The service life of pumps (wear rings, sealing, bearings, impellers, and casings) can considerably be extended by proper coating. Before the implementation of a new coating in equipment, its friction and wear behaviour must be determined beside corrosion resistance. This paper presents the results of friction tests carried out on AMSLER machine. The tribological pair was composed by Al_2O_3 $40TiO_2$ (AMDRY 6250) coating deposited on substrate of a pump shaft sleeve running against a rolling bearing steel disk in dry lubrication conditions. The experimental results are compared against those obtained for uncoated steel.

1. Introduction

Wastewater treatment stations are key points in modern life of urban society, assuring the wellbeing of the people and also the requirements for a healthy and ecological life. The flaws of such systems are pipelines and pumping stations, but especially the machine parts of pumps themselves: impellers, rolling and journal bearings, seals, bushes and sleeves. Slurry pumps and irrigation pumps are usually subjected to various kind of wear as abrasion, corrosion, erosion, and cavitation. The vertical irrigation pumps are usually lubricated by the circulating agent (water). Regarding the latest issue, Paleu et al. [1] investigated wear behaviour of aluminium oxide coated steel for both dry and water lubrication. Water lubrication of Al_2O_3 coated steel reduced the friction coefficient about 45%. The aluminium oxide is known for its corrosion resistance [2], being mostly used as coating for pipes of sewage systems. The latest of them (bushes and sleeves) can be easily reconditioned by proper choose of wear resistant coating material and deposition method. The deposited coatings can change impeller and casing interstitial gap, so the thickness of the coating must be carefully chosen to avoid cavitation phenomena [3-4].

Al_2O_3-$40TiO_2$ powder is also known as AMDRY6250 being recommended by manufacturer for atmospheric plasma spray (APS) deposition on pump components. Recent research [5-12] investigated aspects related to the influence of aluminium titanate (Al_2TiO_5) in Al_2O_3-TiO_2 thermal spray coatings, the structure of Al_2O_3–$40TiO_2$ for different deposition methods, corrosion and erosion resistance, tribological performances, and emissivity. The microstructure of TiO2, Al_2O_3+$40TiO_2$ and Al_2O_3+$13TiO_2$ at plasma spraying and laser engraving were investigated in [11]. According to results, Al_2O_3+$40TiO_2$ is the best of three, possessing low porosity (1.9%) and good hardness (HV0.3=1083).

Moreover Al_2O_3–40TiO_2 has corrosion resistance over Cr_3C_2–20NiCr, both coatings demonstrating good resistance to wear.

The shaft and bushes of modern vertical pumps are made of AISI 303 (EN 1.4305) and AISI 316 (EN 1.4401), and their impellers, casings and sleeves of AISI 304 (EN 1.4301) and AISI 316 L (EN 1.4404), but these materials are expensive and usually supposed to failure due to abrasion, erosion, corrosion and cavitation wear.

This paper investigates dry friction behaviour of steel samples coated by multiple layers of Al_2O_3+40TiO_2. An AMSLER machine is used as tribometer and scanning electron microscopy (SEM) and EDAX analysis are employed to inspect the worn surfaces of tested samples.

2. Experimental procedure

2.1. AMSLER machine

AMSLER machine is a tribometer used to perform tests in dry and lubricated conditions for different configurations of the contacts: disk on disk, roller on disk orthogonally disposed, brake block on disk, and fixed pad on rotating disk. A general view of AMSLER machine and testing pad-on-disk arrangement are presented in figures 1 and 2, respectively. The testing arrangement is presented in figure 2.

Figure 1. General view of testing machine.

Figure 2. Testing arrangement.

The AMSLER machine was upgraded with data acquisition system based on half-bridge strain gauges disposed for bending beam arrangement, a Vishay P3 compact module data acquisition system with 4 channels, and LabVIEW interface allowing the display of friction torque, friction coefficient, and statistical parameters of data acquisition (arithmetic mean, standard deviation, signal-to-noise ratio SNR, Kurtosis and Skewness etc.). Also, the mean values of friction torque T_f and coefficient of friction μ are automatically computed for all acquired samples and displayed. An image of data acquisition and signal processing software interface is given in figure 3.

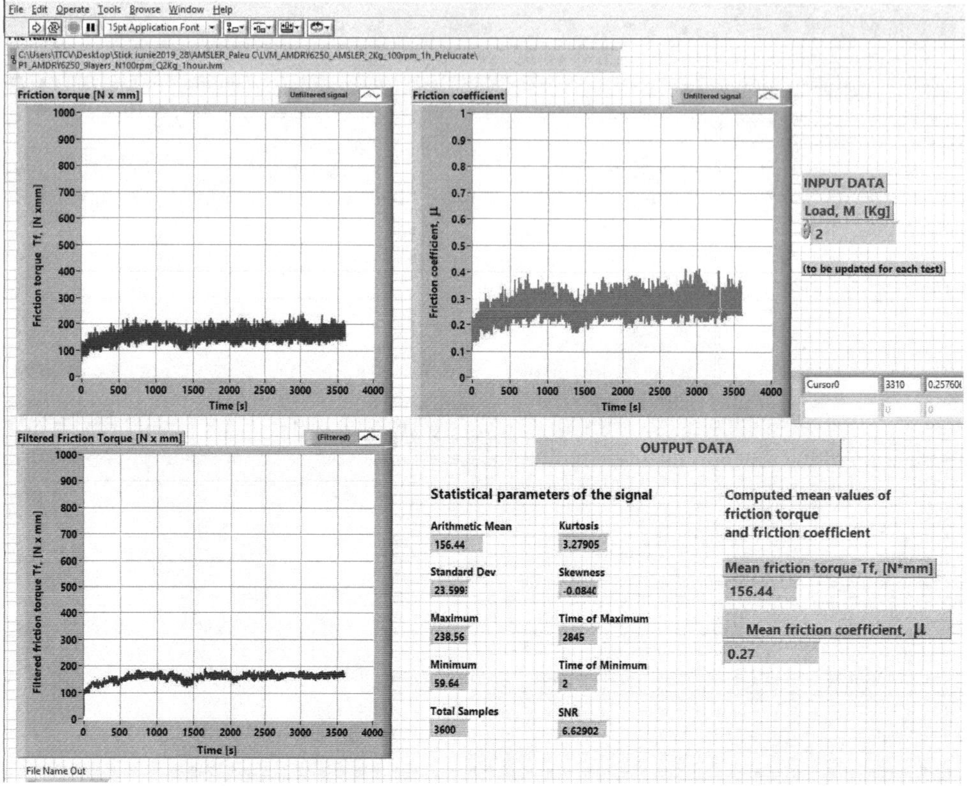

Figure 3. LabVIEW data acquisition and signal processing interface (test P1-9L).

As can be observed from statistical results, signal-to-noise ratio SNR=6.6 is high, skewness is almost nil and kurtosis factor is near to 3, indicating good quality of data acquisition and Gaussian form of diminished noise. The presence of Gaussian noise means that all the factors influencing data acquisition process were aleatory, probably generated by small vibrations noise presented due to vibrations of metallic leaf with strain gauges brazed on it, produced by variable dry friction forces between the tested rough samples and by kinematics chain of gears, no systematic errors being present.

2.2. Testing conditions

Each coated sample was tested at 2 KgF (20 N) normal load. The lower disk turned at constant speed of N=100 rpm, the span of each test being of 3600 s. Before tests the samples were cleaned with acetone and dried, all the tests being carried out in dry conditions (no lubricant).

2.3. Materials

Three sample pads of steel were coated by atmospheric plasma spay APS method (on equipment made Sulzer Metco 9MB) with 5 layers, 7 layers, and 9 layers of of Al_2O_3 – 40 TiO_2 coating, denoted as P1 – 5L, P1-7L, and P1-9L, respectively. Supplementary information on parameters of APS process are given in [13]. Samples P00 and P01 were obtained by removing the chrome layer of original sleeve taken from a damaged pump. P00 was finished by grinding and P01 was machined by milling, having different roughness (see table 1). The turning disk used in tribological tests was made of AISI52100 steel, hardness 62-64 HRC.

2.4. Profilometry

Taylor-Hobson profilometer was used to measure the roughness of the tested samples. Table 1 indicates the roughness values of tested samples on transversal (axial) and longitudinal (rolling) directions.

Table 1. Roughness of the tested samples.

Sample	P1-5L	P1-7L	P1-9L	P00	P01	Disc AISI 52100
Longitudinal roughness, Ra [μm]	5.401	6.722	7.464	0.085	1.231	0.9 – 1.0
Transversal roughness, Ra [μm]	3.382	7.729	7.285	0.065	0.334	1.1 - 1.3

It can be observed that the roughness of coating on P1-5L sample is better than the others (P1-7L and P1-9L). After each test the rotating disc of AISI 52100 steel was polished by grinding, the obtained mean roughness being also indicated in table 1.

3. Results

3.1. Friction tests

Results of friction tests consist in mean friction torque and mean friction coefficients for each test. As can be observed from figure 4 and figure 5, the least friction coefficient and friction torque are obtained for P1-7L (7 layers of AMDRY6250 coating). These are preliminary results; future tests must concentrate also on oil and water lubrication of coated samples.

One can remark that even the roughness of uncoated samples P00 and P01 was 5 times better than of coated samples, mean dry friction torque and coefficient are lower for P1-7L probe. Friction results must be correlated with SEM and EDAX analysis, which are presented in the following section.

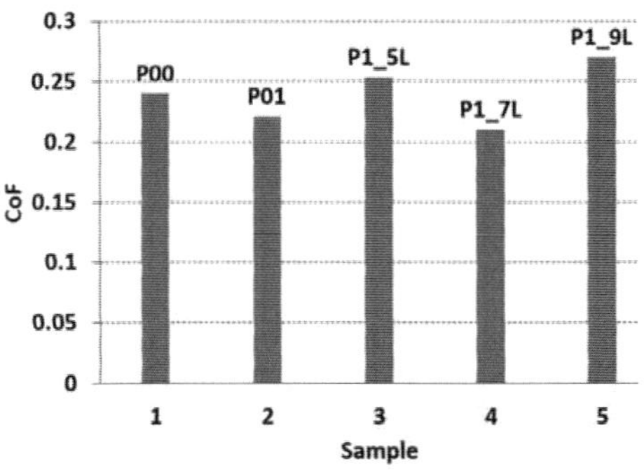

Figure 4. Mean friction coefficient for different samples.

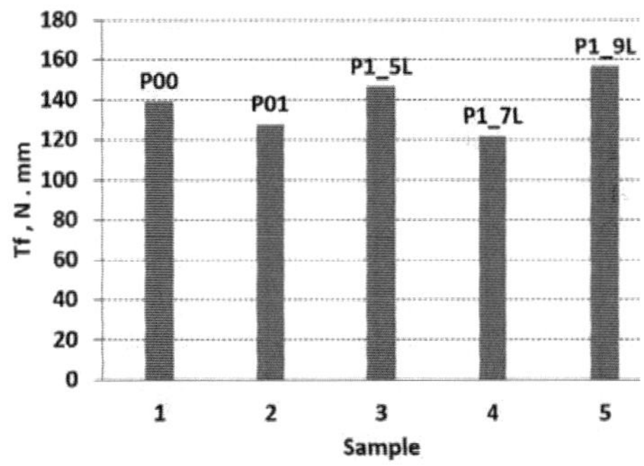

Figure 5. Values of mean friction torque for various test.

3.2. SEM analysis

SEM analysis revealed the wear modes of tested samples (figures 6-11). It seems that in some cases the coated samples present scratches (P1-5L). At magnification of 200x, the wear product seems transferred on the coated sample, the results of EDAX analysis confirming this assumption. A protective layer of aluminium oxide seems beneficial for friction decrease, but in certain limits (P1-7L

IOP Conf. Series: Materials Science and Engineering **724** (2020) 012064 doi:10.1088/1757-899X/724/1/012064

and P1-9L samples, see Table 2 – EDAX analysis). In figure 11 the glassy nature of aluminium oxide formed on sample P17-L was highlighted at higher magnification (200 x).

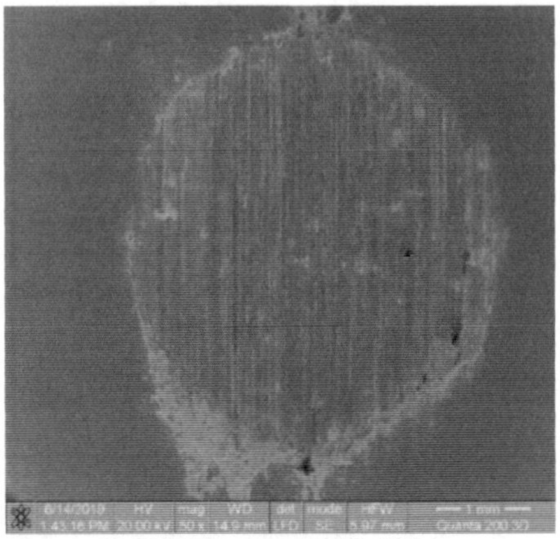

Figure 6. SEM image of P01 (50x).

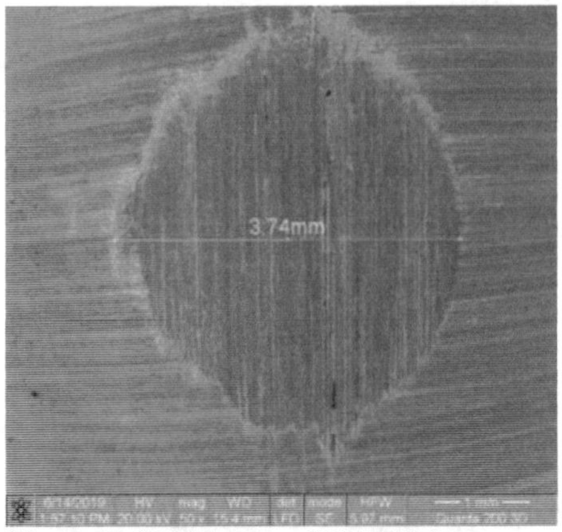

Figure 7. SEM image of P02 (50x).

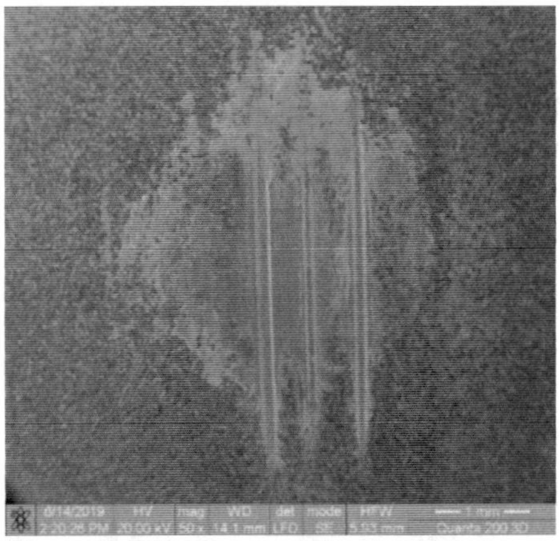

Figure 8. SEM image of P1-5L (50x).

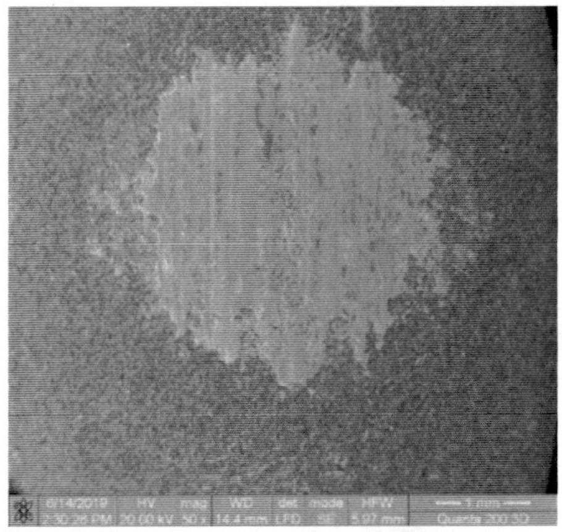

Figure 9. SEM image of P1-9L (50x).

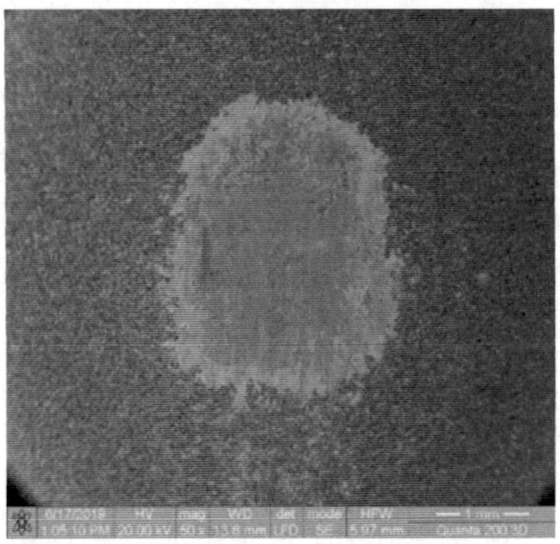

Figure 10. SEM image of P1-7L (50x).

Figure 11. SEM image of P1-7L (200 x).

3.3. *EDAX analysis*

EDAX analysis results from the wear spots are presented in table 2.

Table 2. EDAX results of tested samples.

Element,	P00		P01		P1-5L		P1-7L		P1-9L	
Wt %	new	worn	new	worn	new	worn	new	worn	new	worn
OK	0.00	5.96	0.43	5.91	35.06	13.64	35.11	26.41	36.08	28.01
CrK	0.65	0.26	0.43	0.43	-	0.61	0.34	1.41	-	1.12
FeK	99.35	93.78	99.14	93.65	8.55	80.32	6.33	62.95	6.49	48.87
TiK	-	-	-	-	11.99	0.68	13.24	1.63	12.10	3.60
AlK	-	-	-	-	44.4-	4.75	44.97	7.6	45.33	18.4

According to EDAX results, higher friction of 5 layers sample is due to coating removal, proved by the abundant iron found on worn surface. In comparison with P1-7L, the 9 layers coated sample had higher friction because of abundant aluminium which was identified in augmented proportion on the wear scar at the end of the test.

4. Conclusions

Al_2O_3-40TiO_2 powder is also known as AMDRY6250, being recommended by manufacturer for atmospheric plasma spray (APS) deposition. Various layers of coating (5, 7 and 9) were deposited on steel samples. Preliminary dry tests were carried out on AMSLER machine at constant load and constant speed. The lowest values of friction torque and friction coefficient were obtained for the sample coated with 7 layers of AMDRY6250. SEM and EDAX analysis revealed that higher friction of 5 layers sample is due to coating removal from abrasive wear, proved by the abundant iron found on worn surface. In comparison with P1-7L, the 9 layers coated sample had higher friction because of abundant aluminium, which was identified in augmented proportion on the wear scar at the end of the test. Future tests must concentrate on oil and water lubricated coatings.

5. References

[1] Paleu Cîrlan C, Paleu V, Istrate B, Cimpoesu N and Munteanu C 2019 *IOP Conf. Series: Mater. Sci. Eng.* **591:012007** DOI: 10.1088/1757-899X/591/1/012007
[2] Bejinariu C, Munteanu C, Florea C D, Istrate B, Cimpoesu N, Alexandru A and Sandu A V 2018 *Revista de Chimie,* **69** (12) 3586-3589
[3] Bloch H P and Budris A R 2014 *Pump user's handbook life extension - 4th Ed.* (Lilburn: The Fairmont Press / Taylor & Francis) p 556
[4] Budris A R 2013 *WaterWorld* **28 – 4** 18
[5] Richter A, Berger L-M, Conze S, Sohn Y J and Vaßen R 2019 *IOP Conf. Series: Mater. Sci. Eng.* **480** (2019) 012007
[6] Zavareh M A et al. 2016 *Ceramics Int.* **42** (4) 5203-5210
[7] Vasudev H, Thakur L, Singh H and Bansal A 2018 *Kovove Mater.* **56** 55–63
[8] Znamirowski Z, Pawlowski L, Cichya T and Czarczynski W. 2004 *Surf. & Coat. Technol.* **187** 37–46
[9] Mishra N K, Kumar Naveen and Mishra S B 2014 *Indian J. Mater. Sci.*, **Article ID 453607**, 5 p., http://dx.doi.org/10.1155/2014/453607
[10] Mishra N K and Mishra S B 2015 *Bull. Mater. Sci.* **38** 7 1679–1685
[11] Tomaszeka R, Pawlowski L, Zdanowski J, Grimblot J and Laureyns 2004 *J Surf. & Coat. Technol.* **185** 137–149
[12] Szala M and Hejwowski T 2018 *Coatings* **8** 254; doi:10.3390/coatings8070254
[13] Paleu Cîrlan C, Istrate B, Paleu V and Munteanu C 2019 Paper **ROTRIB 2019-69,** *Cluj*

AUTHOR INDEX